Geometric Modeling with Splines: An Introduction

Geometric Modeling with Splines: An Introduction

Elaine Cohen
University of Utah
School of Computing

Richard F. Riesenfeld
University of Utah
School of Computing

Gershon Elber
Technion
Israel Institute of Technology

CRC Press
Taylor & Francis Group
Boca Raton London New York

CRC Press is an imprint of the
Taylor & Francis Group, an **informa** business
AN A K PETERS BOOK

First published 2001 by A K Peters, Ltd.

Published 2018 by CRC Press
Taylor & Francis Group
6000 Broken Sound Parkway NW, Suite 300
Boca Raton, FL 33487-2742

First issued in paperback 2019

ISBN 13: 978-0-367-44724-3 (pbk)
ISBN 13: 978-1-56881-137-6 (hbk)

Library of Congress Cataloging-in-Publication Data

Cohen, Elaine, 1946–
Geometric modeling with splines : an introduction / Elaine Cohen, Richard F. Riesenfeld, Gershon Elber.
p. cm.
Includes bibliographical references and index.
ISBN 1-56881-137-3
1. Curves on surfaces–Mathematical models. 2. Surfaces–Mathematical models. 3. Spline theory. 4. Computer-aided design. I. Riesenfeld, Richard F. II. Elber, Gershon, 1960– III. Title.

QA565 .C656 2001
516.3′52–dc21 2001032921

Dedicated

to Pierre Bézier, Steven A. Coons, I. J. Schoenberg, and Ivan E. Sutherland, the giants on whose shoulders we stand. Their diverse, fundamental works and powerful visions provided the critical elements that have been forged into this increasingly important, multi-disciplinary field.

Foreword

Geometric modelling with splines has been an important and exciting field for many years covering numerous applications. The subject draws on various topics from mathematical approximation theory, numerical analysis, classical and discrete geometry, engineering, and computer science.

The field grew out of pioneering work in the 1960s on modelling of complex objects like ship hulls and car bodies. At that time, the topic of spline functions developed into an active area of research in approximation theory due to the fundamental work of Schoenberg. His work on B-splines opened new perspectives, and it gradually became clear that these functions were well suited for geometric modelling of physical objects. The seminal work of one of the authors played a crucial role in this development.

Geometric modelling with splines has been a significant area of research for almost 40 years with applications ranging from animated films to simulated surgery. Although many of the fundamental mathematical results have been established, the field is still burgeoning due to a continuous need for new techniques. This book is a welcome text and is written by well-known experts in the field. The authors, and their many accomplished students, have significantly influenced the advancement of this subject. Combining mathematical rigor and the science of modelling in a fruitful way, the book is well-suited as a textbook for a course on this topic, per se, or one that draws on aspects of this material.

The text contains a comprehensive treatment of curves and surfaces with emphasis on B-spline techniques. In addition the book contains a wealth of material ranging from classical techniques to a broad coverage of more specialized topics including new techniques like subdivision of surfaces, all of which add to its value as a reference for professionals and researchers working in the field.

The authors are to be congratulated for writing such a comprehensive text.

Tom Lyche
University of Oslo

Preface

This book has evolved out of lecture notes created to introduce students to various aspects of geometric modeling with splines. The shape-mimicking properties of the NURBS control polygon allows the user to create a smooth curve by manipulating a simple polyline.

While B-spline and NURBS mathematics can seem unnecessarily abstract to students, the mathematical formalisms effectively camouflage some rather formidable machinery that allows geometric shapes to be expressed and manipulated through what appears as a rather straightforward and intuitive geometric design scheme. The constructive nature of many computational B-spline algorithms can enhance a student's intuitive sense for the shape properties of the curve under design. Many algorithms for representation, computation, and querying of B-spline models can be implemented as intuitive, efficient algorithms executing at interactive rates. In light of these characteristics, we have stressed the mathematical soundness of spline methods. We present algorithms, and in some cases, pseudo-code. Since many of the surface methods, i.e., representations and algorithms, rely on curve algorithms and properties, the earlier chapters of the book stress rigorously establishing the properties that will be used throughout. Generally, later chapters address material based on schemes described in the earlier chapters.

The goal of the book is to act both as a text and a reference book. We believe the breadth and depth of the included material are sufficient to give the reader a background suitable for implementing splines and for designing with splines. In addition, the reader who attains a solid understanding of the underlying mathematical approaches, concepts, and logic, as well as a practical understanding, will have a sufficient background to conduct geometric modeling research using splines.

The book is structured as follows:

- A background review of mathematics used in this book is presented in Chapter 1. Chapter 2 presents an overview of the most common types of representations and their characteristics.

- It is in Chapter 3 that we begin the discussion of curve forms for geometric modeling with the oldest representation: conic sections. We show the equivalences of various traditional definitions, the ones most likely to have been seen by readers through the middle of the undergraduate college years. Then, a constructive geometric approach is presented. As the sections progress, we develop this approach into the typical *blending* formulation, and then derive a parametric *blending* representation, which enables writing a curve as a convex combination of geometric points. The last formulation allows us to develop curve properties which will recur in the Bézier formulation (Chapter 5) and the B-spline representation (Chapter 6). Subdivision algorithms are first presented with respect to conic sections in Chapter 3.

- Elements of differential geometry for curves, a branch of mathematics concerned with characterizing the behavior (and shape) of a parametric curve by its differential properties, are presented in Chapter 4. Concepts that recur throughout geometric modeling, including *regularity*, curvature, torsion, and Frenet equations are covered. In *bottom up* design, it is sometimes necessary to piece together pre-designed curves. We discuss methods for determining when a compound curve exhibits various types of parametric smoothness.

- The constructive approach first introduced with conics is generalized to the constructive approximation curves in Chapter 5. Then it is shown that these curves are Bézier curves, and we treat many characteristics for Bézier curve. We also introduce the idea of subdivision of Bézier curves by developing a special case algorithm for subdivision at a curve's midpoint. The Bernstein blending functions are the blending functions used in the Bézier method. Having developed formulations for the blending functions, we discuss using them for approximation and interpolation, and relate approximating a pre-existing parametric function by a Bézier curve to the Bernstein approximation method from classical approximation theory. Finally we apply results from Chapter 4 to discuss smoothly piecing together Bézier curves.

- In Chapter 6 the constructive approximation curves, or Bézier curves, are further generalized to piecewise smooth constructive approximation curves, which are then shown to be equivalent to B-spline curves. Basic properties of B-spline curves are shown. Chapters 7 and 8 reveal more characteristics and properties of spline spaces. Proving the representational power of splines and showing the linear independence of B-splines, Chapter 7 develops the more abstract properties. This is accomplished using inductive proofs like those used in Chapter 3. The idea of refinement is introduced in its simplest form, that of knot insertion of a single knot. This result is further developed for quadratic and cubic subdivision curves, which in Chapter 20 are shown to be the basis for Doo-Sabin and Catmull-Clark surfaces, respectively. In Chapter 8, frequently occurring knot vector configurations for spline curves are presented and their effects on curve shape are discussed. Rational spline curves are finally taken on, as are methods for their computation.

- Various forms of interpolation and approximation with splines are presented in Chapter 9. Two widely used interpolation methods, nodal and complete spline interpolation, are covered. In addition to basic discrete and continuous least squares approximation, the Schoenberg variation diminishing spline, the abstract quasi-interpolation method, and a more widely employed multiresolution constrained decomposition method are treated.

- Chapter 10 is devoted to various types of interpolation using classical polynomial bases.

- In Chapter 11 we present other derivations of B-splines to give a flavor of the origins of B-spline methods. One derivation manifests splines as shadows of higher-dimensional simplices, while another uses generalizations of divided differences to higher dimensions. These can be shown to be equivalent. A third view develops B-splines in terms of signal processing and filtering. The last approach, the original one presented by Schoenberg many decades ago, is once again becoming topical.

- The section on surfaces starts in Chapter 12 with differential geometry for surfaces. Using the results of Chapter 4 for curves, this chapter develops formulations for first and second fundamental forms, normal and geodesic curvation, principal curvatures, and principal directions. Surface shape analysis for design depends on such variables.

- Chapter 13 defines a tensor product surface and presents methods for evaluating position and derivatives using properties of curve forms. Emphasis is placed on B-spline and Bézier tensor product surfaces. Matrix methods for transforming between various bases are described. Based on their tensor product structure, surface forms are generalized from quadratic and cubic subdivision curves schemes.

- In Chapter 14, the ideas presented in Chapter 9 for fitting curves to data are generalized to surfaces. The classis Coons surface is detailed. Finally, an *operator* approach for transforming methods for fitting curves into methods for fitting surfaces is described.

- In Chapter 15, practical aspects of actually creating representations for specific surfaces are confronted. Methods to create representations for ruled surfaces and various surfaces of revolution are given.

- Starting with Chapter 16, more advanced techniques are taken on. These approaches and methods are necessary to create, manipulate, render, query, and fabricate B-spline representations for the complex shapes needed in applications ranging from animation to solid modeling.

- In Chapter 16 general algorithms for subdivision and refinement for B-splines are developed as well as specialized algorithms for subdivision of Bézier curves and surfaces. Pseudo-code clarifies the algorithms. Chapter 17 presents algorithms and pseudo-code of methods, based on the refinement approach, for rendering, computing intersections, and adding degrees of freedom to support hierarchical top-down design.

- It is rare that a single surface can be used to model a complicated object. In earlier chapters we presented methods for piecing together curves and surfaces. Chapter 18 includes issues that arise when attempting to combine arbitrary pieces of tensor product surfaces to define complex models. The important topic of finding curve and surface intersections is introduced. We give methods for finding such intersections, as well as criteria for determining convergence. The idea of a well-formed 3-D model is developed. Detailed algorithms for constructing a model by applying Boolean operations on existing models are described for planar polygonal models. We then show how algorithms for polyhedral models are hierarchical and rely on the results from the planar cases.

- Out of the discussion of creating actual models comes an important realization that we need data structures to traverse models. In Chapter 19 we present the winged-edge data structure, a well-known and widely used topological representation, and then discuss a data structure suitable for models bounded by trimmed surfaces, the kind that result from Boolean operations on models bounded by sculptured surfaces.

- In Chapters 20 and 21 we describe generalizations of B-splines that are just starting to be more widely taught and adopted. Chapter 20 is focused on subdivision surfaces. We show that the *template* algorithms for generating the refined meshes are generalizations of spline refinement algorithms for Catmull-Clark subdivision surfaces (a generalization of bi-cubic uniform floating spline surfaces), Doo-Sabin subdivision surfaces (a generalization of bi-quadratic uniform floating spline surfaces), and Loop subdivision surfaces (a generalization of box spline surfaces and refinement algorithms). Chapter 21 is focused on algorithms and uses for trivariate volumetric splines.

Versions of this manuscript have been used in teaching quarter, semester, and year long course sequences. Advanced undergraduate and beginning graduate students with good backgrounds in mathematics, i.e., advanced calculus and matrix algebra, and programming experience have been typical participants in these classes. The material can be taught with emphasis on the behaviors, algorithms, and implementations of splines in geometric modeling schemes, or with an emphasis on the proofs and proper mathematical development of the subject matter, depending on the goals of the course at hand.

If Chapters 1 and 2 are considered as background material, Chapters 3, 4, 5, 6, 7 (without Section 7.4), 8, 12, and 13 (without Sections 13.5 and 13.6) could form the basis for a one semester introductory class. A two semester class would include Chapters 9, 14, 15, 16, 17, as well as a selected subset of Chapters 10, 18, 19, 20, and 21, depending on the interests of the class.

We have tried to make this book broad enough to be appealing for many readers and many class situations. While some topics rely on earlier material, other chapters can be read directly. Many variations are possible in choosing a rewarding path though this book. Although considerable effort has been devoted to "debug" the text, errors will inevitably turn up. We will try to provide corrections on a web site as we become aware of any inaccuracies, so check our personal web pages for the most current information. (`www.cs.utah.edu/~cohen` or `www.cs.utah.edu/~rfr`)

This field has brought us into contact with many challenging and important problems, and provided a rich area of research. We are hopeful that the perspective and understanding that we have gained over many years contributes to making this a good book for others to learn the fundamentals more quickly, or to refer to for specific information while engaged in the subject.

Acknowledgments

This book has grown out of lecture notes from an advanced undergraduate and graduate course taught repeatedly both at the University of Utah and the Technion. Over the years, numerous students, staff, and colleagues at Utah and the Technion have stimulated and challenged our understanding of fundamental issues, particularly the members of Utah's Alpha_1 Research Group. Interacting with them has provided a daily education leading to many advancements of understanding reflected in this volume. Thanks are due to all those students at Utah and the Technion who provided critical responses to various earlier stages of the book. Their comments have led to significant improvements in organization and detail. We are particularly grateful to Bill Martin for his enormous effort in providing detailed proofreading of the material, and to Matt Kaplan for making extensive contributions to the figures.

This volume would not be going to press on schedule without extraordinary efforts of our wonderful publisher Alice Peters, who worked long days, weekends, and holidays in order to meet critical publication deadlines. We have felt fortunate to be working with her.

Mentors and teachers, Steve Coons, Bill Gordon, Robin Forrest, Pierre Bzier, Charles Lang, and Ivan Sutherland, have inspired us directly and indirectly. Finally, we wish to thank our colleague Tom Lyche at the University of Oslo, with whom two of us have collaborated periodically over the last 20 years, for all we have learned from him about the art of splines. He has regularly encouraged this book's completion, and helped with reading of an early draft.

Cohen and Riesenfeld express appreciation to the National Science Foundation for The Science and Technology Center for Computer Graphics and Visualization, which has provided a rich intellectual environment.

Elaine Cohen
Richard F. Riesenfeld
University of Utah

Gershon Elber
Technion
Israel Institute of Technology

June 2001

Contents

I

Introduction

1

Review of Basic Concepts

In the sections that follow we provide a short presentation of some of the basic material which will be needed in various other chapters of the book. If the reader is familiar with the contents of some sections, he may prefer to skip those sections.

1.1 Vector Analysis

We shall provide a brief review of some necessary concepts and manipulation techniques. Many of these concepts are general and not dependent on any particular vector space. The concept of cross product, however is defined only in $\boldsymbol{R}^3$.

Definition 1.1. *A* vector space $\boldsymbol{V}$ *is defined over a set of elements, the vectors, that have two operations, addition* $+ : \boldsymbol{V} \times \boldsymbol{V} \to \boldsymbol{V}$*, and scalar multiplication* $\cdot : \boldsymbol{R}^1 \times \boldsymbol{V} \to \boldsymbol{V}$ *which satisfy the following rules:*

If $\boldsymbol{u}$, $\boldsymbol{v}$, $\boldsymbol{w} \in \boldsymbol{V}$ *and if* r, $s \in \boldsymbol{R}^1$, *then*

1. $r\boldsymbol{u} + s\boldsymbol{v} \in \boldsymbol{V}$;
2. *There is an element* $\boldsymbol{0} \in \boldsymbol{V}$ *such that for all* $\boldsymbol{v} \in \boldsymbol{V}$, $\boldsymbol{0}+\boldsymbol{v} = \boldsymbol{v}+\boldsymbol{0} = \boldsymbol{v}$;
3. $r(\boldsymbol{u} + \boldsymbol{v}) = r\boldsymbol{u} + r\boldsymbol{v}$;
4. $(r + s)\boldsymbol{u} = r\boldsymbol{u} + s\boldsymbol{u}$.

Definition 1.2. *A finite subset of vectors* $\boldsymbol{C}$ *in* $\boldsymbol{V}$ *is called* independent *if for every choice of* n *less than or equal to the number of elements of* $\boldsymbol{C}$*, and for all arbitrary choices of* $r_1, \ldots, r_n \in \boldsymbol{R}^1$ *and* $\boldsymbol{v}_1, \ldots, \boldsymbol{v}_n \in \boldsymbol{C}$ *then*

$$r_1\boldsymbol{v}_1 + \cdots + r_n\boldsymbol{v}_n = \boldsymbol{0}$$

implies that

$$r_1 = \cdots = r_n = 0.$$

This states that no element of the set $\boldsymbol{I}$ can be written as a finite linear combination of other elements of the set, i.e., it cannot depend on a finite number of the other elements.

Definition 1.3. *Let* $\boldsymbol{S}$ *be a subset of vectors of* $\boldsymbol{V}$*. The* span *of* $\boldsymbol{S}$*, written span* $\boldsymbol{S}$*, is the set of all finite linear combinations of elements of* $\boldsymbol{S}$*. That is, for an arbitrary integer* $n > 0$*, select* n *arbitrary vectors* $\boldsymbol{v}_1, \ldots, \boldsymbol{v}_n \in \boldsymbol{S}$ *and* n *arbitrary values* $r_1, \ldots, r_n \in \boldsymbol{R}^1$ *then* $r_1\boldsymbol{v}_1 + \cdots + r_n\boldsymbol{v}_n \in$ *span* $\boldsymbol{S}$.

Example 1.4. For vector space $\boldsymbol{R}^2$, and $\boldsymbol{I}_1 = \{(1,0),(1,1)\}$, $\boldsymbol{I}_1$ is independent, but $\boldsymbol{I}_2 = \{(1,0),(3,0)\}$ is not since $(3,0) = 3(1,0)$. The span $\boldsymbol{I}_1 = \boldsymbol{R}^2$, but span $\boldsymbol{I}_2 = \{\, r(1,0) : r \in \boldsymbol{R}^1 \,\}$. □

Definition 1.5. *A* basis $\boldsymbol{B}$ *for a vector space* $\boldsymbol{V}$ *is a set of vectors that is independent and such that span* $\boldsymbol{B} = \boldsymbol{V}$.

It can be shown that all bases of the same vector space have the same number of elements. For infinite dimensional vector spaces, one must use techniques which show equivalence of the *size* of the infinity. It is left as an exercise for the reader to show that this is true for finite dimensional vector spaces.

Definition 1.6. *If* $\boldsymbol{B}$ *is a basis for* $\boldsymbol{V}$ *and has a finite number of elements, we say that* $\boldsymbol{V}$ *is a* finite dimensional vector space *with dimension equal to the number of elements of* $\boldsymbol{B}$.

Example 1.7. Examples of Vector Spaces:

1. $\boldsymbol{R}^1$, $\boldsymbol{R}^2$, $\boldsymbol{R}^3$.

2. Function space: polynomials of degree 1, 2, …, n.

3. $C^{(0)}[a, b]$ the set of continuous functions on the interval $[a, b]$. Addition is standard function addition, scalar multiplication is standard multiplication of the function value at the value of x. It can easily be shown that the rest of the properties follow.

4. 2×2 matrices. □

Definition 1.8. *A (real)* inner product space *is a vector space with a second vector operation, $<,> : \boldsymbol{V} \times \boldsymbol{V} \to \mathbf{R}^1$, defined such that the following holds true for $\boldsymbol{u}, \boldsymbol{v} \in \boldsymbol{V}$ and $r, s \in \mathbf{R}^1$:*

1. $\langle r\boldsymbol{u}, s\boldsymbol{v} \rangle = rs\langle \boldsymbol{u}, \boldsymbol{v} \rangle$;

2. $\langle \boldsymbol{u}, \boldsymbol{v} \rangle = \langle \boldsymbol{v}, \boldsymbol{u} \rangle$;

3. $\langle \boldsymbol{u} + \boldsymbol{v}, \boldsymbol{w} \rangle = \langle \boldsymbol{u}, \boldsymbol{w} \rangle + \langle \boldsymbol{v}, \boldsymbol{w} \rangle$;

4. $\langle \boldsymbol{v}, \boldsymbol{v} \rangle \geq 0$, *and* $= 0$ *if and only if* $\boldsymbol{v} = 0$.

The idea of length or magnitude of a vector can be introduced into an inner product space and a *norm* can be defined.

Definition 1.9. *The* length *or* magnitude *of a vector $\boldsymbol{v} \in \boldsymbol{V}$ is defined as $\|\boldsymbol{v}\| = \sqrt{\langle \boldsymbol{v}, \boldsymbol{v} \rangle}$. The* distance *between two vectors $\boldsymbol{u}$, $\boldsymbol{v}$ is defined as the magnitude of the difference vector, that is, $\|\boldsymbol{u} - \boldsymbol{v}\|$.*

Definition 1.10. *Some further definitions which are based on the inner product and have a geometric interpretation when applied to $\mathbf{R}^2$ and $\mathbf{R}^3$ are:*

1. *Vectors $\boldsymbol{u}$ and $\boldsymbol{v}$ are said to be* orthogonal *if for $\boldsymbol{u} \neq \boldsymbol{v}$, $\langle \boldsymbol{u}, \boldsymbol{v} \rangle = 0$;*

2. *A vector $\boldsymbol{u}$ is a* unit vector *if $\langle \boldsymbol{u}, \boldsymbol{u} \rangle = 1$;*

3. *A collection of unit vectors $\mathcal{V}$ is* orthonormal *if $\langle \boldsymbol{v}, \boldsymbol{w} \rangle = 0$ for all $\boldsymbol{v}, \boldsymbol{w} \in \mathcal{V}$.*

Example 1.11. Let the vector space under consideration be $C^{(0)}[a, b]$, the space of continuous functions on the interval $[a, b]$. Define

$$\begin{aligned} \langle f, g \rangle &= \int_a^b f(t)g(t)dt \\ &= \text{net area under curve } fg. \end{aligned}$$

To check that this is an inner product requires verifying that C is a vector space and exhibits the other properties of inner product. However,

1. For $f \not\equiv 0$, $\langle f, f\rangle > 0$;
2. Additivity, scalar multiplication properties, commutativity, and the distributive property all follow from properties of the integral;
3. The uniqueness of the zero element follows from properties of the integral. □

Lemma 1.12. *If $\boldsymbol{W}$ is a finite collection of orthonormal vectors in a space V, then the vectors in $\boldsymbol{W}$ are* linearly independent.

Proof: Suppose $\boldsymbol{W} = \{\boldsymbol{e}_1, \ldots, \boldsymbol{e}_n\}$ and

$$0 = r_1\boldsymbol{e_1} + r_2\boldsymbol{e_2} + \cdots + r_n\boldsymbol{e_n}$$

then

$$0 = \langle 0, \boldsymbol{e}_j\rangle = r_1\langle \boldsymbol{e}_1, \boldsymbol{e}_j\rangle + r_2\langle \boldsymbol{e}_2, \boldsymbol{e}_j\rangle + \cdots + r_i\langle \boldsymbol{e}_i, \boldsymbol{e}_j\rangle + \cdots + r_n\langle \boldsymbol{e}_n, \boldsymbol{e}_j\rangle.$$

Since $\langle \boldsymbol{e_i}, \boldsymbol{e_j}\rangle = \delta_{i,j}$, $0 = r_j$. Letting $j = 1, \ldots, n$, gives the result that if $\boldsymbol{0}$ is a linear combination of the elements of $\boldsymbol{W}$, then all coefficients must be zero. ■

Lemma 1.13. *If $\boldsymbol{W} = \{\boldsymbol{e}_1, \ldots, \boldsymbol{e}_n\}$ is a collection of orthonormal vectors with*

$$\boldsymbol{v} = r_{v,1}\boldsymbol{e_1} + \cdots + r_{v,n}\boldsymbol{e_n}$$

and

$$\boldsymbol{w} = r_{w,1}\boldsymbol{e_1} + \cdots + r_{w,n}\boldsymbol{e_n},$$

then

$$\langle \boldsymbol{v}, \boldsymbol{w}\rangle = \sum_{i=1}^{n} r_{v,i}r_{w,i}.$$

Proof: The proof uses straightforward properties of inner product and is left as Exercise 7 for the reader. ■

Corollary 1.14. *If $\boldsymbol{W}$ and $\boldsymbol{w}$ are defined as above in Lemma 1.13 then*

$$\|\boldsymbol{w}\| = \sqrt{\sum_{i=1}^{n} r_{w,i}^2}.$$

1.1.1 $\mathbb{R}^2$ and $\mathbb{R}^3$ as Vector Spaces

In the special cases where the vector space is $\mathbb{R}^2$ or $\mathbb{R}^3$, we note an especially simple representation. In physics and mathematics one says that vectors are uniquely defined by direction and magnitude. Two vectors with the same direction and magnitude are the same—no matter where they are located. If a vector is located with its tail at the origin and its head (arrow) at the position (x, y, z), then the head position uniquely defines the direction and the magnitude of that vector. Thus, by convention, every point in the plane or in 3-space has a one-to-one and onto correspondence with the vector space $\mathbb{R}^2$ or $\mathbb{R}^3$, respectively. This correspondence matches the point (x, y, z) with the vector whose tail is at the origin and head at the point (x, y, z). Sometimes the two notations are used interchangeably which can lead to great confusion on the part of the novice. We shall use the term *free* vectors to mean those whose position is not bound, and *fixed* vectors to mean those with a bound position.

Let $\boldsymbol{e}_1 = (1,0,0)$, $\boldsymbol{e}_2 = (0,1,0)$, and $\boldsymbol{e}_3 = (0,0,1)$. Then the set $\boldsymbol{E} = \{\boldsymbol{e}_1, \boldsymbol{e}_2, \boldsymbol{e}_3\} \in \mathbb{R}^3$ is orthonormal. For all real numbers x, y, and z, $x\boldsymbol{e}_1 + y\boldsymbol{e}_2 + z\boldsymbol{e}_3$ represents a unique vector in the span of $\boldsymbol{E}$ which is contained in $\mathbb{R}^3$, since the set $\boldsymbol{E}$ is linearly independent. Further, to every point in $\mathbb{R}^3 = \mathbb{R}^1 \times \mathbb{R}^1 \times \mathbb{R}^1$ there corresponds a vector in span $\boldsymbol{E}$. Hence, $\boldsymbol{E}$ is a basis for $\mathbb{R}^3$. Thus, every set of three orthonormal vectors forms a basis for $\mathbb{R}^3$.

Corollary 1.15. *In general, $\boldsymbol{E}$ is a set of three orthonormal vectors in $\mathbb{R}^3$, so for every pair of vectors $\boldsymbol{v}$, $\boldsymbol{w} \in \mathbb{R}^3$ with*

$$\boldsymbol{v} = r_{v,1}\boldsymbol{e}_1 + r_{v,2}\boldsymbol{e}_2 + r_{v,3}\boldsymbol{e}_3 \quad \text{and} \quad \boldsymbol{w} = r_{w,1}\boldsymbol{e}_1 + r_{w,2}\boldsymbol{e}_2 + r_{w,3}\boldsymbol{e}_3,$$

we have

$$\langle \boldsymbol{v}, \boldsymbol{w} \rangle = \sum_{i=1}^{3} r_{v,i} r_{w,i} \quad \text{and} \quad \|\boldsymbol{v}\| = \sqrt{\sum_{i=1}^{3} r_{v,i}^2}.$$

Whenever the particular choice of $\boldsymbol{E}$ is understood, one can write (r_1, r_2, r_3) to mean $r_1\boldsymbol{e}_1 + r_2\boldsymbol{e}_2 + r_3\boldsymbol{e}_3$. Frequently the $\boldsymbol{e}_1$ direction is denoted as the "x" direction, the $\boldsymbol{e}_2$ direction is denoted the "y" direction, and $\boldsymbol{e}_3$ direction is called the "z" direction. The triple (x, y, z) is then used to mean the vector $x\boldsymbol{e}_1 + y\boldsymbol{e}_2 + z\boldsymbol{e}_3$.

Define scalar multiplication as $c\boldsymbol{v} = (cx, cy, cz)$. In Exercise 3 the reader must show that this definition makes the vector c times longer, but does not change the direction.

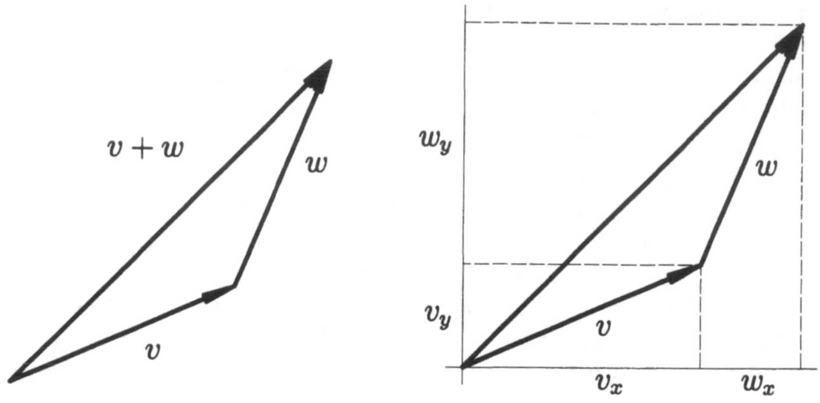

(a) (b)

Figure 1.1. The geometric meaning of adding $\boldsymbol{v}$ and $\boldsymbol{w}$ in (a), and vector addition in (b).

Geometrically, vector addition can be interpreted as positioning $\boldsymbol{v}$ arbitrarily in space and then placing $\boldsymbol{w}$ so its tail is at the same position as the head of $\boldsymbol{v}$. Then the vector with its tail in the same position as the tail of $\boldsymbol{v}$ and its head in the same position as the head of $\boldsymbol{w}$ is the sum of the vectors $\boldsymbol{v}$ and $\boldsymbol{w}$, see Figure 1.1 (a).

To derive a quantitative formula, consider fixed formulations of the vectors $\boldsymbol{v} = (x_v, y_v, z_v)$ and $\boldsymbol{w} = (x_w, y_w, z_w)$. We wish to determine a method of finding the coordinate representation for $\boldsymbol{s} = \boldsymbol{v} + \boldsymbol{w}$, that is (x_s, y_s, z_s).

We know that $\boldsymbol{s} - \boldsymbol{v} = \boldsymbol{w}$. Given the positions, we know that the change of position in the x-direction must be x_w. Similarly for y and z. But the tail of $\boldsymbol{w}$ in this position is at (x_v, y_v, z_v). Thus its head must be x_w units over, or have an x coordinate of $x_v + x_w$. Similarly, the y and z coordinates of the head must be at $y_v + y_w$ and $z_v + z_w$, respectively. But the head of $\boldsymbol{w}$ in this position is at the same place as the head of $\boldsymbol{s}$ when $\boldsymbol{s}$ starts at the origin. Thus, in general $\boldsymbol{v} + \boldsymbol{w} = (x_v + x_w, y_v + y_w, z_v + z_w)$, as shown in Figure 1.1 (b).

A similar result may be derived for an oblique basis in $\boldsymbol{R}^2$ or $\boldsymbol{R}^3$. We look at the $\boldsymbol{R}^2$ case. Let $\boldsymbol{v}$ and $\boldsymbol{w}$ be two vectors in the plane that do not have the same direction, that is, $\boldsymbol{v} \neq c\boldsymbol{w}$ for all $c \in \boldsymbol{R}^1$. Then $\boldsymbol{v}$ and $\boldsymbol{w}$ are linearly independent. Consider $\boldsymbol{S} = span\{\boldsymbol{v}, \boldsymbol{w}\} = \{\, a\boldsymbol{v} + b\boldsymbol{w} : a, b \in \boldsymbol{R}^1 \,\}$. First represent them as fixed vectors, $\boldsymbol{v} = (x_v, y_v)$ and $\boldsymbol{w} = (x_w, y_w)$. Suppose $\boldsymbol{u}$ is any vector, then it has a fixed representation $\boldsymbol{u} = (x_u, y_u)$. What vectors $\boldsymbol{u}$ in the plane are also in $\boldsymbol{S}$?

If we suppose $\{\boldsymbol{e_1}, \boldsymbol{e_2}\}$ is an orthonormal basis of $\boldsymbol{R}^2$ with $\boldsymbol{v} = x_v\boldsymbol{e_1} + y_v\boldsymbol{e_2}$ and $\boldsymbol{w} = x_w\boldsymbol{e_1} + y_w\boldsymbol{e_2}$ then $[\boldsymbol{v}\ \boldsymbol{w}] = [\boldsymbol{e_1}\ \boldsymbol{e_2}]A$ where

$$A = \begin{bmatrix} x_v & x_w \\ y_v & y_w \end{bmatrix}.$$

Suppose an arbitrary element $\boldsymbol{u} \in \boldsymbol{R}^2$ can be written $\boldsymbol{u} = x_u\boldsymbol{e_1} + y_u\boldsymbol{e_2}$. We want to know when there exists a_u, $b_u \in \boldsymbol{R}^1$ such that $\boldsymbol{u} = a_u\boldsymbol{v} + b_u\boldsymbol{w}$. That is, when there is a solution to

$$\begin{bmatrix} \boldsymbol{e_1} & \boldsymbol{e_2} \end{bmatrix} \begin{bmatrix} x_u \\ y_u \end{bmatrix} = \boldsymbol{u} = \begin{bmatrix} \boldsymbol{v} & \boldsymbol{w} \end{bmatrix} \begin{bmatrix} a_u \\ b_u \end{bmatrix}.$$

But since $[\boldsymbol{v}\ \boldsymbol{w}] = [\boldsymbol{e_1}\ \boldsymbol{e_2}]A$, this question is equivalent to asking when

$$A \begin{bmatrix} a_u \\ b_u \end{bmatrix} = \begin{bmatrix} x_u \\ y_u \end{bmatrix}$$

can be solved for unknowns a_u, b_u.

By Theorem 1.28, if $\det A \neq 0$, A^{-1} exists and the system can be solved.

Setting $[a_u\ b_u]^T = A^{-1}[x_u\ y_u]^T$ solves the system. Thus, for any $\boldsymbol{u}$, there exists $a_u, b_u \in \boldsymbol{R}^1$ such that $\boldsymbol{u} = a_u\boldsymbol{v} + b_u\boldsymbol{w}$. The pair (a_u, b_u) are the *coordinate values* in the oblique v-w coordinate system for the vector $\boldsymbol{u}$.

Consider vectors $\boldsymbol{v_1}$ and $\boldsymbol{v_2}$. Suppose they are placed so that their tails meet at the point O. Let θ denote the angle between the vectors.

We use the law of cosines to find the particular realization for the inner product in $\boldsymbol{R}^3$.

$$\|\boldsymbol{v_2} - \boldsymbol{v_1}\|^2 = \|\boldsymbol{v_2}\|^2 + \|\boldsymbol{v_1}\|^2 - 2\|\boldsymbol{v_2}\|\,\|\boldsymbol{v_1}\|\cos\theta.$$

Expanding the left side one gets:

$$\begin{aligned} \|\boldsymbol{v_2} - \boldsymbol{v_1}\|^2 &= \langle \boldsymbol{v_2} - \boldsymbol{v_1}, \boldsymbol{v_2} - \boldsymbol{v_1} \rangle \\ &= \langle \boldsymbol{v_2}, \boldsymbol{v_2} \rangle - \langle \boldsymbol{v_2}, \boldsymbol{v_1} \rangle - \langle \boldsymbol{v_1}, \boldsymbol{v_2} \rangle + \langle \boldsymbol{v_1}, \boldsymbol{v_1} \rangle \\ &= \|\boldsymbol{v_2}\|^2 - 2\langle \boldsymbol{v_1}, \boldsymbol{v_2} \rangle + \|\boldsymbol{v_1}\|^2. \end{aligned}$$

Thus one has

$$\|\boldsymbol{v_2}\|^2 - 2\langle \boldsymbol{v_1}, \boldsymbol{v_2} \rangle + \|\boldsymbol{v_1}\|^2 = \|\boldsymbol{v_2}\|^2 + \|\boldsymbol{v_1}\|^2 - 2\|\boldsymbol{v_2}\|\,\|\boldsymbol{v_1}\|\cos\theta$$

and

$$\langle \boldsymbol{v_1}, \boldsymbol{v_2} \rangle = \|\boldsymbol{v_1}\|\,\|\boldsymbol{v_2}\|\cos\theta.$$

By Corollary 1.15, $\langle \boldsymbol{v_1}, \boldsymbol{v_2} \rangle = x_1x_2 + y_1y_2 + z_1z_2$. Is there a geometric interpretation to this view of the inner product? Using the result gives the following theorem:

Theorem 1.16. *The* directed length *of the projection of* $\boldsymbol{v_1}$ *onto the direction of* $\boldsymbol{v_2}$ *is* $\langle \boldsymbol{v_1}, \boldsymbol{v_2} \rangle / \|\boldsymbol{v_2}\|$.

Proof: Place the vectors so that their tails meet at point O. Drop a perpendicular from the head of $\boldsymbol{v_1}$ onto the direction of $\boldsymbol{v_2}$. From trigonometry it is known that, if θ is the angle between the vectors, then the directed length of $\boldsymbol{v_1}$ in the direction of $\boldsymbol{v_2}$ is given by the projection of

$$\begin{aligned} \boldsymbol{v_1} &= \|\boldsymbol{v_1}\| \cos\theta \\ &= \frac{\langle \boldsymbol{v_1}, \boldsymbol{v_2} \rangle}{\|\boldsymbol{v_2}\|}. \end{aligned}$$

Note that if the angle between the vectors is greater than 90 degrees, then the directed length of the projection is considered negative. ■

We can use the simple geometric knowledge that three points determine a plane, or two vectors determine a plane to give us more complicated information.

Several operations have been defined on elements in arbitrary vector spaces. Reviewing them, we see,

- Addition: $\boldsymbol{V} \times \boldsymbol{V} \to \boldsymbol{V}$;
- Scalar multiplication: $\boldsymbol{R}^1 \times \boldsymbol{V} \to \boldsymbol{V}$;
- Inner product: $\boldsymbol{V} \times \boldsymbol{V} \to \boldsymbol{R}^1$.

Another operation, the *cross product* can be defined for the special case when the vector space is $\boldsymbol{R}^3$. The cross product of $\boldsymbol{v}$ and $\boldsymbol{w}$, $\boldsymbol{v} \times \boldsymbol{w}$, can intuitively be defined as a vector that is perpendicular to both $\boldsymbol{v}$ and $\boldsymbol{w}$, with orientation prescribed by the *right hand rule* and a magnitude prescribed by a rule which depends on the magnitude of $\boldsymbol{v}$, the magnitude of $\boldsymbol{w}$ and the angle between them. That is, point the right hand in the direction of $\boldsymbol{v}$. Then move it to the direction of $\boldsymbol{w}$ in a continuous rotational movement. The direction that the right thumb points is the direction of the cross product vector. That leaves one degree of freedom. That freedom can be restricted by requiring that $\boldsymbol{e_1} \times \boldsymbol{e_2} = \boldsymbol{e_3}$.

The formula for the coefficients can be derived from the knowledge that if $\boldsymbol{u} = \boldsymbol{v} \times \boldsymbol{w}$ then $\langle \boldsymbol{u}, \boldsymbol{v} \rangle = 0$ and $\langle \boldsymbol{u}, \boldsymbol{w} \rangle = 0$, since $\boldsymbol{u}$ is perpendicular to both $\boldsymbol{v}$ and $\boldsymbol{w}$.

If we let $\boldsymbol{e_1}$ be the unit vector in the x direction, $\boldsymbol{e_2}$ be the unit vector in the y direction, and $\boldsymbol{e_3}$ be the unit vector in the z direction, then if $\boldsymbol{u} = (x_u, y_u, z_u)$, $\boldsymbol{v} = (x_v, y_v, z_v)$, and $\boldsymbol{w} = (x_w, y_w, z_w)$, then $\boldsymbol{i} = x_i\boldsymbol{e_1} + y_i\boldsymbol{e_2} + z_i\boldsymbol{e_3}$, $i \in \{u, v, w\}$.

The inner product equations now can be written

$$\begin{aligned} \langle \boldsymbol{v}, \boldsymbol{u} \rangle &= x_v x_u + y_v y_u + z_v z_u &= 0, \\ \langle \boldsymbol{w}, \boldsymbol{u} \rangle &= x_w x_u + y_w y_u + z_w z_u &= 0. \end{aligned}$$

Since the system consists of two linear equations in three unknowns, it still has an undefined degree of freedom. Rewrite the equations in terms of x_u as known and both y_u and z_u as unknown:

$$\begin{aligned} -x_v x_u &= y_v y_u + z_v z_u, \\ -x_w x_u &= y_w y_u + z_w z_u. \end{aligned}$$

Then solve for y_u and z_u to get

$$\begin{aligned} y_u &= x_u \frac{z_v x_w - x_v z_w}{y_v z_w - z_v y_w}, \\ z_u &= x_u \frac{y_w x_v - y_v x_w}{y_v z_w - z_v y_w}. \end{aligned}$$

so

$$\begin{aligned} \boldsymbol{u} &= \frac{x_u}{y_v z_w - z_v y_w} \left(y_v z_w - z_v y_w, z_v x_w - x_v z_w, y_w x_v - y_v x_w \right) \\ &= c \left(y_v z_w - z_v y_w, z_v x_w - x_v z_w, y_w x_v - y_v x_w \right). \end{aligned}$$

Now, the last degree of freedom is set by the magnitude and "right hand rule" orientation requirement. Since c must be the same constant for all $\boldsymbol{v}, \boldsymbol{w} \in \boldsymbol{R}^3$, one can choose simple cases to determine c. Let $\boldsymbol{v} = \boldsymbol{e_1}$ and $\boldsymbol{w} = \boldsymbol{e_2}$. Since $\boldsymbol{u} = \boldsymbol{e_3} = c(0, 0, 1)$, c must equal 1.

Definition 1.17. *For $\boldsymbol{v}, \boldsymbol{w} \in \boldsymbol{R}^3$, the* cross product *operator defines a new vector, $\boldsymbol{v} \times \boldsymbol{w}$, as $\boldsymbol{v} \times \boldsymbol{w} = (y_v z_w - z_v y_w, z_v x_w - x_v z_w, y_w x_v - y_v x_w)$. The cross product is not a commutative operation.*

Lemma 1.18. $\|\boldsymbol{v} \times \boldsymbol{w}\|^2 = \|\boldsymbol{v}\|^2 \|\boldsymbol{w}\|^2 - \|(\boldsymbol{v}, \boldsymbol{w})\|^2$.

Proof: It is left as an exercise. ∎

Theorem 1.19. $\|\boldsymbol{v} \times \boldsymbol{w}\| = \|\boldsymbol{v}\| \, \|\boldsymbol{w}\| \, |\sin\theta|$.

Proof: Since $\|(\boldsymbol{v}, \boldsymbol{w})\|^2 = \|\boldsymbol{v}\|^2 \, \|\boldsymbol{w}\|^2 \cos^2\theta$, Lemma 1.18 gives,

$$\begin{aligned}\|\boldsymbol{v}\times\boldsymbol{w}\|^2 &= \|\boldsymbol{v}\|^2\|\boldsymbol{w}\|^2 - \|\boldsymbol{v}\|^2\|\boldsymbol{w}\|^2\cos^2\theta \\ &= \|\boldsymbol{v}\|^2\|\boldsymbol{w}\|^2(1-\cos^2\theta) \\ &= \|\boldsymbol{v}\|^2\|\boldsymbol{w}\|^2(\sin^2\theta).\end{aligned}$$

Since $0 \le \sin\theta$ for $0 \le \theta \le \pi$, $\|\boldsymbol{v}\times\boldsymbol{w}\| = \|\boldsymbol{v}\|\,\|\boldsymbol{w}\|\sin\theta$. ■

Thus, the cross product of two vectors $\boldsymbol{v_1}$ and $\boldsymbol{v_2}$ is the vector with magnitude equal to the area of the parallelogram described when the tails of the vectors meet at a point O, with direction perpendicular to the plane defined by $\boldsymbol{v_1}$ and $\boldsymbol{v_2}$. The orientation of this perpendicular is determined by the right hand rule. The following properties are a direct consequence of the definition:

$$\begin{aligned}\boldsymbol{u}\times\boldsymbol{v} &= -\boldsymbol{v}\times\boldsymbol{u} \\ \boldsymbol{u}\times\boldsymbol{u} &= 0 \\ \boldsymbol{u}\times(\boldsymbol{v}\times\boldsymbol{w}) &\ne (\boldsymbol{u}\times\boldsymbol{v})\times\boldsymbol{w}.\end{aligned}$$

The example $(\boldsymbol{e_1}\times\boldsymbol{e_1})\times\boldsymbol{e_2} =$ undefined, but $\boldsymbol{e_1}\times(\boldsymbol{e_1}\times\boldsymbol{e_2}) = \boldsymbol{e_1}\times\boldsymbol{e_3} = -\boldsymbol{e_2}$.

To compute the cross product for $\boldsymbol{v_i} = (x_i, y_i, z_i)$, $i = 1, 2$, determinants are commonly used:

$$\boldsymbol{v} = \boldsymbol{v_1}\times\boldsymbol{v_2} = \begin{vmatrix} \boldsymbol{e_1} & \boldsymbol{e_2} & \boldsymbol{e_3} \\ x_1 & y_1 & z_1 \\ x_2 & y_2 & z_2 \end{vmatrix}.$$

We define the *triple scalar product* of three vectors $\boldsymbol{u}$, $\boldsymbol{v}$, $\boldsymbol{w}$ as

$$\langle\boldsymbol{u},(\boldsymbol{v}\times\boldsymbol{w})\rangle = \boldsymbol{u}\cdot(\boldsymbol{v}\times\boldsymbol{w}) = \begin{vmatrix} \boldsymbol{u_x} & \boldsymbol{u_y} & \boldsymbol{u_z} \\ \boldsymbol{v_x} & \boldsymbol{v_y} & \boldsymbol{v_z} \\ \boldsymbol{w_x} & \boldsymbol{w_y} & \boldsymbol{w_z} \end{vmatrix}.$$

Note that $\langle\boldsymbol{u},\boldsymbol{v}\times\boldsymbol{w}\rangle = \langle\boldsymbol{v}\times\boldsymbol{w},\boldsymbol{u}\rangle = -\langle\boldsymbol{v},\boldsymbol{u}\times\boldsymbol{w}\rangle$.

Finally, the *triple vector product* can be decomposed in terms of inner products as

$$\boldsymbol{u}\times(\boldsymbol{v}\times\boldsymbol{w}) = (\boldsymbol{u},\boldsymbol{w})\boldsymbol{v} - (\boldsymbol{u},\boldsymbol{v})\boldsymbol{w}.$$

1.2 Linear Transformations

Definition 1.20. *Suppose $\boldsymbol{X}$ and $\boldsymbol{Y}$ are vector spaces and T is a function, $T : \boldsymbol{X} \to \boldsymbol{Y}$ such that*

$$T(r_1\boldsymbol{x_1} + r_2\boldsymbol{x_2}) = r_1T(\boldsymbol{x_1}) + r_2T(\boldsymbol{x_2}),$$

then T is called a linear transformation *from $\boldsymbol{X}$ to $\boldsymbol{Y}$, or a* linear operator.

Example 1.21. The following are examples of linear transformations:

1. $P : \mathbb{R}^3 \to \mathbb{R}^3$ by $P\big((x, y, z)\big) = (x, y, 0)$. Then P is called the *orthogonal projection* from $\mathbb{R}^3$ to the x-y plane. Orthogonal projections to the x-z plane and y-z plane are defined analogously.

   This is a linear transformation since

$$\begin{aligned} P\Big(r_1(x_1, y_1, z_1) + r_2(x_2, y_2, z_2)\Big) &= (r_1x_1 + r_2x_2, r_1y_1 + r_2y_2, 0) \\ &= (r_1x_1, r_1y_1, 0) + (r_2x_2, r_2y_2, 0) \\ &= r_1(x_1, y_1, 0) + r_2(x_2, y_2, 0) \\ &= r_1P\Big((x_1, y_1, z_1)\Big) + r_2P\Big((x_2, y_2, z_2)\Big). \end{aligned}$$

   The proof of the properties about P relies on the vector space properties of $\mathbb{R}^2$ and $\mathbb{R}^3$. Such proofs are typical of showing an operator is a linear transformation.

2. $D[f] = f'$, the derivative operator.

3. $I[f] = \int_a^b f(t)dt$.

4. $T_{j,x^*} : C^{(n)}[a, b] \to \mathbb{R}^1$ by $T_{j,x^*}[f] = f^{(j)}(x^*)$, for $j = 0, \ldots, n$: point evaluation of the jth derivative at a particular point x^*.

5. $T : C^{(n+1)}[a, b] \to P_n$, the polynomials of degree less than or equal to n, by $T[f] = f(x_0) + f'(x_0)(x - x_0) + \frac{f^{(2)}(x_0)}{2!}(x - x_0)^2 + \cdots + \frac{f^{(n)}(x_0)}{n!}(x - x_0)^n$.

   Note that T takes a function to a function and we can write $\big(T[f]\big)(x)$ or $T[f](x)$ to evaluate that function at a point x. □

Definition 1.22. *A linear transformation whose range is* $\mathbb{R}^1$ *is called a* linear functional.

Definition 1.23. *Suppose S and T are two linear transformations from $\mathbf{V}$ to $\mathbf{W}$ and $r \in \mathbb{R}^1$. Define addition by $(S + T)(v) = S(v) + T(v)$ and define scalar multiplication as $(rS)(v) = r\big(S(v)\big)$. Problem 11 shows that $(S + T)$ and rS are both linear transformations, and that the set of linear transformations from $\mathbf{V}$ to $\mathbf{W}$, $L(\mathbf{V}, \mathbf{W})$, is a vector space with these operations defined.*

1.3 Review of Matrix Properties

Definition 1.24. *Consider a rectangular array of numbers arranged as follows:*

$$A = (a_{ij}) = \begin{bmatrix} a_{1,1} & a_{1,2} & \dots & a_{1,n} \\ a_{2,1} & a_{2,2} & \dots & a_{2,n} \\ & & \vdots & \\ a_{m,1} & a_{m,2} & \dots & a_{m,n} \end{bmatrix}$$

where $\boldsymbol{a_j} = [\ a_{1,j} \quad a_{2,j} \quad \dots \quad a_{m,j}\ ]^T$, *a vector in* $\boldsymbol{R}^m$. *The array is called an* $m \times n$ matrix, *and* $\boldsymbol{M_{m,n}}$ *denotes the set of all matrices with n vectors from* $\boldsymbol{R}^m$. *If* $m = n$ *we say that the matrices in* $\boldsymbol{M_{n,n}} = \boldsymbol{M_n}$ *are square.*

Definition 1.25. *The* transpose *of* A *is* $(a_{j,i})$ *and is denoted by* A^T.

Denote by $\boldsymbol{e_j}$ the column vector $(0,\dots,0,\delta_{i,j},0,\dots,0)^T$ consisting of all zeros except a 1 in the jth position.

If A and B are two matrices in $\boldsymbol{M_{m,n}}$ we can define the following operations:

1. Addition: $+ : \boldsymbol{M_{m,n}} \times \boldsymbol{M_{m,n}} \rightarrow \boldsymbol{M_{m,n}}$. Denote by $A + B$ the matrix equal to $(a_{i,j} + b_{i,j})$, the matrix whose elements are the sum elementwise of elements of A and B.

2. Scalar Multiplication: $\boldsymbol{R}^1 \times \boldsymbol{M_{m,n}} \rightarrow \boldsymbol{M_{m,n}}$. If $r \in \boldsymbol{R}^1$ then $(rA) = (ra_{i,j})$, the matrix whose elements are multiplied elementwise by the scalar r.

3. Zero element: Let $Z = (0)$ be the matrix such that $z_{i,j} = 0$, for all i, j. Then $A + Z = A$ for all matrices A, and Z is the identity under addition. It can be shown easily that the rest of the properties hold for $\boldsymbol{M_{m,n}}$ to be called a vector space.

4. Define another operation: $\cdot : \boldsymbol{M_{m,n}} \times \boldsymbol{M_{n,k}} \rightarrow \boldsymbol{M_{m,k}}$. For $A \in \boldsymbol{M_{m,n}}$ and $B \in \boldsymbol{M_{n,k}}$ $C \in \boldsymbol{M_{m,k}}$ is defined by $C = AB = (c_{i,j})$ where $c_{i,j} = \sum_{p=1}^{n} a_{i,p} b_{p,j}$. The matrix C is called the product of A and B. Note that the dimensions must match to be able to multiply A and B.

Suppose $m = n = p$ and let $I = (\boldsymbol{e}_1, \dots, \boldsymbol{e}_n)$, then $AI = A = IA$, and I is called a multiplicative identity. Suppose I_2 is another multiplicative identity. $I = II_2 = I_2$; the left side occurring since I_2 is an identity, and the right side occurring since I is an identity. This proves that the multiplicative identity is unique.

Definition 1.26. *If $A \in \boldsymbol{M_n}$, and there exists a matrix B such that $AB = BA = I$, then B is called the* inverse *to A (conversely, A is the inverse to B).*

If A and B are in $\boldsymbol{M_n}$, does $AB \stackrel{?}{=} BA$? Namely is matrix multiplication a commutative operation? The answer is negative, in general. Let

$$A = \begin{bmatrix} 1 & 3 \\ 1 & 1 \end{bmatrix} \quad \text{and} \quad B = \begin{bmatrix} 1 & 3 \\ 1 & 1 \end{bmatrix}$$

then

$$AB = \begin{bmatrix} 4 & 3 \\ 2 & 1 \end{bmatrix} \quad \text{and} \quad BA = \begin{bmatrix} 1 & 3 \\ 2 & 4 \end{bmatrix}.$$

Definition 1.27. *Let $A = [\boldsymbol{a}_1, \ldots, \boldsymbol{a}_n] \in M_n$. The* determinant *is a functional $|\ | : \boldsymbol{M_n} \to \mathbb{R}^1$ written $|A| = D(\boldsymbol{a}_1, \ldots, \boldsymbol{a}_n)$ which is completely defined by the following conditions:*

1. $D(\boldsymbol{a}_1, \ldots, \boldsymbol{a}_i, \ldots, \boldsymbol{a}_n) = D(\boldsymbol{a}_1, \ldots, \boldsymbol{a}_i + \boldsymbol{a}_j, \ldots, \boldsymbol{a}_n)$, *for $i \neq j$;*
2. $D(\boldsymbol{a}_1, \ldots, r\boldsymbol{a}_i, \ldots, \boldsymbol{a}_n) = rD(\boldsymbol{a}_1, \ldots, \boldsymbol{a}_i, \ldots, \boldsymbol{a}_n)$;
3. $D(\boldsymbol{e}_1, \ldots, \boldsymbol{e}_i, \ldots, \boldsymbol{e}_n) = 1$.

Define the i-j^{th} *cofactor* as $A^*_{i,j} = (-1)^{i+j}|A_{i,j}|$ where $A_{i,j}$ is the $(n-1) \times (n-1)$ matrix that omits the ith row and the jth column of A.

Properties of the determinant:

1. $|A| = \left|A^T\right|$;
2. Define B as A with any two rows interchanged. Then, $|A| = -|B|$;
3. If $\boldsymbol{a}_i = \boldsymbol{a}_j$ for $i \neq j$, then $|A| = 0$;
4. $D(\boldsymbol{a}_1, \ldots, \boldsymbol{a}_i, \ldots, \boldsymbol{a}_n) = D(\boldsymbol{a}_1, \ldots, \boldsymbol{a}_i + r\boldsymbol{a}_j, \ldots, \boldsymbol{a}_n)$, for $i \neq j$;
5. $D(\boldsymbol{a}_1, \ldots, \boldsymbol{a}_i, \ldots, \boldsymbol{a}_n) + D(\boldsymbol{a}_1, \ldots, \boldsymbol{a}'_i, \ldots, \boldsymbol{a}_n) =$
   $D(\boldsymbol{a}_1, \ldots, \boldsymbol{a}_i + \boldsymbol{a}'_i, \ldots, \boldsymbol{a}_n)$;
6. Fix j, then $|A| = \sum_{i=1}^n a_{i,j}A^*_{i,j}$ or fix i, then $|A| = \sum_{j=1}^n a_{i,j}A^*_{i,j}$.

Since property (1) holds, the properties (2) through (5) hold also when the term *row* is substituted for *column* and *column* for *row*.

There is a geometric interpretation of the determinant as well. Consider either the rows or the columns of a matrix as vectors in $\mathbb{R}^n$. Then the determinant is the area of the n-dimensional hyper-parallelopiped generated by those vector *edges.*

For linear equations $\sum_{j=1}^{n} a_{i,j}x_j = b_i$ for $i = 1, \ldots, n$, where the $a_{i,j}$, and b_i are known, and the x_j, $j = 1, \ldots, n$ are unknown, it is necessary, sometimes, to determine if the system has no solutions, a unique solution, or many solutions. And, if there is a unique solution, how it can be found. The above system of equations can readily be posed as a matrix problem:

$$\begin{bmatrix} a_{1,1} & a_{1,2} & \cdots & a_{1,n} \\ a_{2,1} & a_{2,2} & \cdots & a_{2,n} \\ & & \vdots & \\ a_{n,1} & a_{n,2} & \cdots & a_{n,n} \end{bmatrix} \begin{bmatrix} x_1 \\ x_2 \\ \vdots \\ x_n \end{bmatrix} = \begin{bmatrix} b_1 \\ b_2 \\ \vdots \\ b_n \end{bmatrix}$$

$$A \qquad\qquad X \quad = \quad B$$

where $A = (a_{i,j})$, is the $n \times n$ matrix of coefficients, X is the $n \times 1$ matrix of unknowns, and B is the $n \times 1$ matrix of equation values.

The solution to this problem is given by $A^{-1}AX = IX = X = A^{-1}B$, when such an inverse exists. Does the inverse exist? These questions are answered by

Theorem 1.28. Cramer's Rule. *If $|A| \neq 0$ then the solution to the above system has a unique solution given by*

$$x_r = \frac{\sum_{i=1}^{n} A^*{}_{i,r} b_i}{|A|}, \qquad r = 1, \ldots, n$$

or, the homogeneous system with $b_j = 0$, $j = 1, \ldots, n$ possesses a nontrivial solution if and only if $|A| = 0$.

Proof: We shall treat existence only, here. If the $\boldsymbol{a_j}$ are considered as n column vectors over $\mathbb{R}^n$, then the system of equations can be written

$$x_1\boldsymbol{a_1} + x_2\boldsymbol{a_2} + \cdots + x_n\boldsymbol{a_n} = \boldsymbol{b}.$$

The homogeneous case is defined as the case when $\boldsymbol{b} = 0$. A solution to the nonhomogeneous system is equivalent to $\boldsymbol{b}$ being in the span of $\boldsymbol{a_j}$, $j = 1, \ldots, n$. A nontrivial solution to the homogeneous case is equivalent to the vectors $\boldsymbol{a_j}$, $j = 1, \ldots, n$, being dependent.

If $\boldsymbol{a_j} = \mathbf{0}$ for any j then clearly every vector in the span can be written in an infinite number of ways, so the vectors are dependent. Hence, $\boldsymbol{a_j} \neq \mathbf{0}$, $j = 1, \ldots, n$. Now, consider the homogeneous problem and suppose it has a nontrivial solution $\boldsymbol{x} = (x_1, \ldots, x_n)$. Suppose, without loss of generality, that $x_1 \neq 0$. Then, it is true that

$$\boldsymbol{a_1} + c_2'\boldsymbol{a_2} + \cdots + c_n'\boldsymbol{a_n} = 0$$

for new scalars c_i', $i = 2, \ldots, n$, not all zero. But

$$\begin{aligned} D(\boldsymbol{a}_1, \ldots, \boldsymbol{a}_n) &= D(\boldsymbol{a_1} + c_2'\boldsymbol{a_2} + \cdots + c_n'\boldsymbol{a_n}, \boldsymbol{a}_2, \ldots, \boldsymbol{a}_n) \\ &= D(\mathbf{0}, \boldsymbol{a}_2, \ldots, \boldsymbol{a}_n) \\ &= 0D(\mathbf{0}, \boldsymbol{a}_2, \ldots, \boldsymbol{a}_n) \\ &= 0. \end{aligned}$$

Thus, if the columns are dependent vectors, then the determinant is zero. So if the homogeneous system has a nontrivial solution, the determinant is zero.

Now, what vectors $\boldsymbol{b}$ can be written as unique linear combinations of the vectors $\boldsymbol{a}_1, \ldots, \boldsymbol{a}_n$? We shall show the result using a proof by induction on the size of the system of equations. First, if $n = 1$, namely, $a_{1,1}x_1 = b_1$, for $b_1 \neq 0$, has a solution if and only if $|A| = a_{1,1} \neq 0$. If $b_1 = 0$ then a nontrivial solution results if and only if $|A| = 0$. Now suppose it is true that if there are k equations in k unknowns, for all $k < n$, the theorem is true, and consider the $n \times n$ case.

If the set of n column vectors is independent over an n dimensional space and hence forms a basis,

$$\boldsymbol{e_1} = r_1\boldsymbol{a_1} + r_2\boldsymbol{a_2} + \cdots + r_n\boldsymbol{a_n},$$

for some new collection of coefficients r_i. Then, without loss of generality, suppose $r_1 \neq 0$:

$$\begin{aligned} r_1 D(\boldsymbol{a}_1, \ldots, \boldsymbol{a}_n) &= D(r_1\boldsymbol{a}_1, \ldots, \boldsymbol{a}_n) \\ &= D(r_1\boldsymbol{a_1} + r_2\boldsymbol{a_2} + \cdots + r_n\boldsymbol{a_n}, \boldsymbol{a}_2, \ldots, \boldsymbol{a}_n) \\ &= D(\boldsymbol{e}_1, \boldsymbol{a}_2, \ldots, \boldsymbol{a}_n) \\ &= \sum_{j=1}^{n} \boldsymbol{e}_{1,j} {A^*}_{1,j} \\ &= {A^*}_{1,1}. \end{aligned}$$

$A_{1,1}^*$ is a $n-1 \times n-1$ determinant. $|A| = 0$ only if $A_{1,1}^* = 0$. Let $\boldsymbol{a}_j^*$ denote the jth column vector of A with the first element omitted. Then

$A^*_{1,1} = D(\boldsymbol{a}^*_2, \ldots, \boldsymbol{a}^*_n)$. If $A^*_{1,1} = 0$ then by the induction hypothesis, there exist scalars $x_2, \ldots, x_n$ not all zero, making an $n-1$ vector $\boldsymbol{x}^*$, such that

$$x_2\boldsymbol{a}^*_2 + \cdots + x_n\boldsymbol{a}^*_n = 0.$$

Let $x_1 = -(x_2a_{1,2} + \cdots + x_na_{1,n})$. Then

$$x_1\boldsymbol{e}_1 + x_2\boldsymbol{a}_2 + \cdots + x_n\boldsymbol{a}_n = 0,$$

which means that $\boldsymbol{e}_1, \boldsymbol{a}_2, \ldots, \boldsymbol{a}_n$ is a dependent set. But since $-r_1\boldsymbol{a}_1 = -\boldsymbol{e}_1 + r_2\boldsymbol{a}_2 + \cdots + r_n\boldsymbol{a}_n$, that implies that $\boldsymbol{a}_1, \ldots, \boldsymbol{a}_n$ is a dependent set. Thus, if $\boldsymbol{a}_1, \ldots, \boldsymbol{a}_n$ are independent, then $|A| \neq 0$. We already showed that if they are dependent then $|A| = 0$.

That contradicts the hypothesis. Hence $A^*_{1,1} \neq 0$ and $|A| \neq 0$. This shows that linear independence of the columns is equivalent to a nonzero determinant and implies a basis of $\boldsymbol{R}^n$ which means that the column vector on the right can be written as a linear combination of the columns on the left.

The constructive part of the proof appears in many advanced calculus books. ■

1.4 Barycentric Coordinates

One mechanism for escaping coordinate system dependence is to develop a method for specifying arbitrary points in the plane, or in $\boldsymbol{R}^3$ as combinations of points (vectors) which have some meaning to the problem at hand. Section 3.7 develops an application of this for representing conic sections. We here develop the simplest forms and properties for barycentric coordinates.

We know that $n+1$ points P_i, $i = 0, \ldots, n$, in $\boldsymbol{R}^n$ can be used to form the vectors $\boldsymbol{r_i} = P_i - P_0$, $i = 1, \ldots, n$. If the vectors $\{\boldsymbol{r_i}\}$ form a basis of $\boldsymbol{R}^n$, then the points $\{P_i\}$ are said to be in *general position.*

Suppose there are two points P_0 and P_1 and a point T on the line through P_0 and P_1. If $P_0 \neq P_1$, then these points determine a line which can be thought of as a transformation of $\boldsymbol{R}^1$. Clearly $\boldsymbol{r_1} = P_1 - P_0$ forms a basis for this one-dimensional subspace, and hence all points on the line can be written as a combination of P_0 and P_1.

Considering $\boldsymbol{R}^1$, suppose P_0, P_1, and T are just real numbers, with $P_0 < T$ and $P_0 < P_1$. Then considered as vectors,

$$\frac{T - P_0}{\|T - P_0\|} = \frac{P_1 - P_0}{\|P_1 - P_0\|},$$

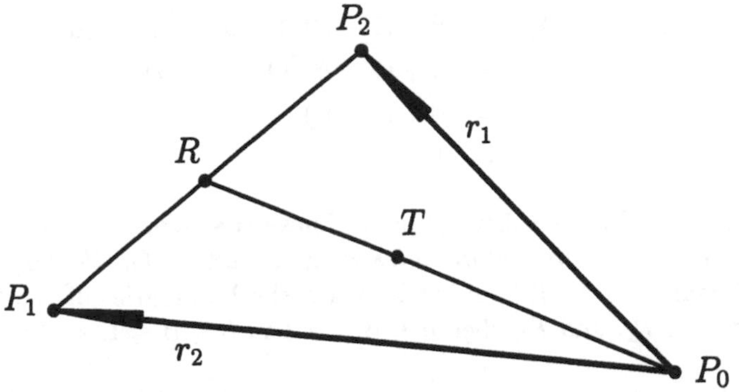

Figure 1.2. Finding barycentric coordinates.

which, setting $\lambda = \|T - P_0\| \;/\; \|P_1 - P_0\|$ simplifies to

$$T = (1 - \lambda)P_0 + \lambda P_1. \tag{1.1}$$

When $T < P_0$, Equation 1.1 holds true if $\lambda = -\|T - P_0\| \;/\; \|P_1 - P_0\|$. Here, the magnitudes of the vectors are just their absolute values. However, equation 1.1 holds true for finding the barycentric coordinates of a point in a one-dimensional subspace of $\mathbb{R}^n$with respect to two other points. Hence T is a convex combination of P_0 and P_1. Then $(1 - \lambda)$ and λ are called the *barycentric coordinates* of T with respect to P_0 and P_1. If $P_0 < P_1 < T$, then the ratio λ above is greater than 1. It is still true that $T = (1-\lambda)P_0 + \lambda P_1$, however, the coefficient of P_0 is now a negative number. Analogously, if $T < P_0$, λ is negative. If P_0 and P_1 are two points in $\mathbb{R}^n$, the exact same results hold since there exists a translation followed by a rotation which will take the line through P_0 and P_1 into the x-axis.

Now, in $\mathbb{R}^2$, suppose the three points P_0, P_1, and P_2 are in general position. Further, suppose that the point T is in the interior of the triangle formed by the three points. Draw a line from one of the points, say P_0, through the point T until it intersects the edge P_1P_2. Call that point R. (See Figure 1.2.)

We suppose that $P_i = (x_i, y_i)$, $T = (x_t, y_t)$, and $R = (x_r, y_r)$. Since R is on the line connecting P_1 and P_2, there exists a real number α such that $R = (1 - \alpha)P_1 + \alpha P_2$. Now, since T is on the line segment connecting R and P_0, there exists a real number β, such that $T = (1 - \beta)R + \beta P_0$. Putting the two equations together,

$$\begin{aligned} T &= (1 - \beta)\big[(1 - \alpha)P_1 + \alpha P_2\big] + \beta P_0 \\ &= \beta P_0 + (1 - \beta)(1 - \alpha)P_1 + (1 - \beta)\alpha P_2 \\ &= \lambda_0 P_0 + \lambda_1 P_1 + \lambda_2 P_2; \end{aligned}$$

$$\begin{aligned}\lambda_0+\lambda_1+\lambda_2 &= \beta+(1-\beta)(1-\alpha)+(1-\beta)\alpha\\ &= \beta+(1-\beta)\big[(1-\alpha+\alpha)\big]\\ &= \beta+(1-\beta)\\ &= 1.\end{aligned}$$

Thus, the sum of the coefficients is 1. These coefficients, λ_i, $i = 0, 1, 2$, are called the *barycentric coordinates* with respect to P_0, P_1, and P_2 and depend linearly on T. If T is inside or on the boundaries of the triangle formed by P_0, P_1, and P_2, then $0 \le \alpha \le 1$ and $0 \le \beta \le 1$, so $0 \le \lambda_i \le 1$, for $i = 0, \ldots, 2$.

We now want to give a geometric interpretation of barycentric coordinates as ratios of areas.

Since R is also on the line through P_0 and T, $R = \delta T + (1-\delta)P_0$ as well as $R = (1-\alpha)P_1 + \alpha P_2$. We first set the two equations equal,

$$P_0 - P_1 = \alpha(P_2 - P_1) + \delta(P_0 - T),$$

and then break the x and y components of this new vector equation apart.

$$\begin{aligned}x_0 - x_1 &= \alpha(x_2 - x_1) + \delta(x_0 - x_t),\\ y_0 - y_1 &= \alpha(y_2 - y_1) + \delta(y_0 - y_t).\end{aligned}$$

Using Cramer's Rule to solve this system yields

$$\begin{aligned}\alpha &= \frac{\begin{vmatrix} x_0 - x_1 & x_0 - x_t \\ y_0 - y_1 & y_0 - y_t \end{vmatrix}}{\begin{vmatrix} x_2 - x_1 & x_0 - x_t \\ y_2 - y_1 & y_0 - y_t \end{vmatrix}}\\ &= \frac{\|(P_0 - P_1) \times (P_0 - T)\|}{\|(P_2 - P_1) \times (P_0 - T)\|}\end{aligned}$$

with the natural extended notion of the cross products over vectors in $\boldsymbol{R}^3$, considering vector $V = (v_x, v_y)$ as $V = (v_x, v_y, 0)$, and

$$\begin{aligned}\delta &= \frac{\begin{vmatrix} x_2 - x_1 & x_0 - x_1 \\ y_2 - y_1 & y_0 - y_1 \end{vmatrix}}{\begin{vmatrix} x_2 - x_1 & x_0 - x_t \\ y_2 - y_1 & y_0 - y_t \end{vmatrix}}\\ &= \frac{\|(P_2 - P_1) \times (P_0 - P_1)\|}{\|(P_2 - P_1) \times (P_0 - T)\|}.\end{aligned}$$

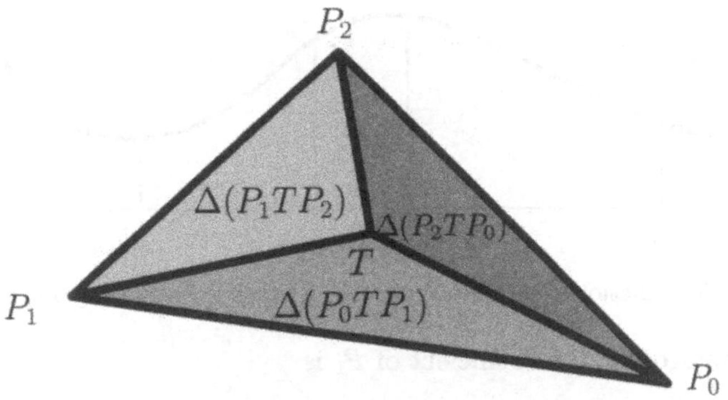

Figure 1.3. Barycentric coordinates as ratios of areas.

Since $R = P_0 + \delta(T - P_0)$, $\delta > 0$, and

$$T = \frac{1}{\delta}(R - P_0) + P_0,$$

then

$$T = \left(1 - \frac{1}{\delta}\right) P_0 + \frac{1-\alpha}{\delta} P_1 + \frac{\alpha}{\delta} P_2.$$

We finish by solving for these three coefficients in terms of the coordinates of the points. We will use the property, shown in the exercises relating cross products to areas of related parallelograms and triangles.

The coefficient of P_2 is

$$\begin{aligned}\frac{\alpha}{\delta} &= \frac{\|(P_0 - P_1) \times (P_0 - T)\|}{\|(P_2 - P_1) \times (P_0 - P_1)\|} \\ &= \frac{\text{area } \Delta(P_0TP_1)}{\text{area } \Delta(P_0P_1P_2)}.\end{aligned}$$

Applying the same equations, but reverting to the determinant values gives the coefficient of P_0 as

$$\begin{aligned}1 - \frac{1}{\delta} &= 1 - \frac{\|(P_2 - P_1) \times (P_0 - T)\|}{\|(P_2 - P_1) \times (P_0 - P_1)\|} \\ &= \frac{\|(P_2 - P_1) \times (T - P_1)\|}{\|(P_2 - P_1) \times (P_0 - P_1)\|} \\ &= \frac{\text{area } \Delta(P_1TP_2)}{\text{area } \Delta(P_0P_1P_2)}.\end{aligned}$$

We have used the identity $P_2 - P_1 = (P_2 - T) + (T - P_1)$ to get to the final

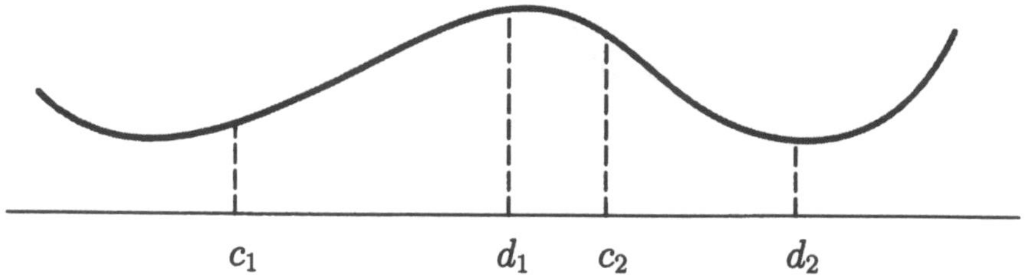

Figure 1.4. The function is increasing on (c_1, d_1) and is decreasing on (c_2, d_2).

identity. Analogously, the coefficient of P_1 is

$$\frac{1-\alpha}{\delta} = \frac{\text{area } \Delta(P_0 T P_2)}{\text{area } \Delta(P_0 P_1 P_2)}.$$

Thus, it is shown that the barycentric coordinates for a point within a triangle are the ratios of the area of the subtriangle opposite the vertex to the area of the whole triangle.

The analogous result is true for a point within a tetrahedron (in $\mathbb{R}^3$), that is, the four barycentric coordinates are the ratios of the volume of the *opposite* subtetrahedron to the volume of the whole tetrahedron.

1.5 Functions

Definition 1.29. *A function $f(x)$ is called* increasing (non-decreasing) *on an interval (c, d) if for all $u, v \in (c, d)$, $u < v$ implies $f(u) < f(v)$ $(f(u) \leq f(v))$.*

Definition 1.30. *A function $f(x)$ is called* decreasing (non-increasing) *on an interval (c, d) if for all $u, v \in (c, d)$, $u < v$ implies $f(u) > f(v)$ $(f(u) \geq f(v))$.*

Theorem 1.31. *Suppose $f(x) \in C^{(1)}(c, d)$. If $f'(x) > 0$ for $x \in (c, d)$, then $f(x)$ is increasing on (c, d). If $f'(x) < 0$ for $x \in (c, d)$, then $f(x)$ is decreasing on (c, d).*

Definition 1.32. *A* (local) maximum *for a function $f \in C^{(0)}$ occurs at a point x_0 if there exists $\epsilon > 0$ so that $f(x_0) \geq f(x)$ for all $x \neq x_0$ such that $|x - x_0| < \epsilon$. A* (local) minimum *to f is defined analogously.*

Definition 1.33. *The* extremal points *of a function are the ordered abscissa-ordinate pairs at which maxima or minima occur.*

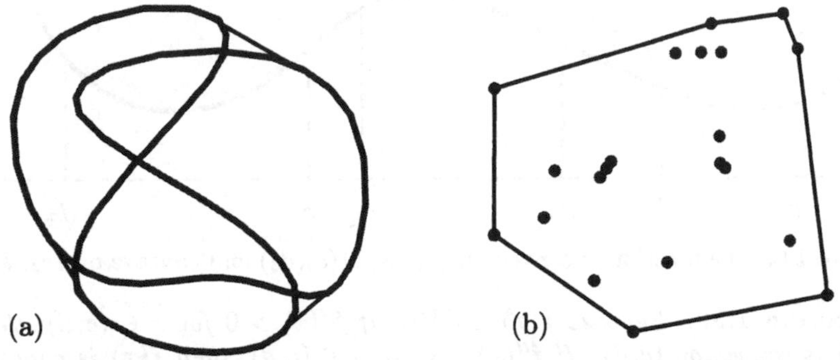

Figure 1.5. The convex hulls of (a) a continuous curve; (b) a discrete set of points.

Lemma 1.34. *Suppose a function f is piecewise $C^{(1)}$. The extremal points of a function f might occur for only the following values of x:*

- *$x = a$ and $x = b$, that is the interval endpoints,*
- *values of x for which $f'(x) = 0$,*
- *values of x for which $f'(x)$ does not exist.*

Definition 1.35. *A subset of $\mathbb{R}^3$, X is called* convex *if for all x_1, $x_2 \in X$, $(1-t)x_1 + tx_2 \in X$, for $t \in [0,1]$. That is the line segment connecting x_1 and x_2 lies entirely within the set X.*

Definition 1.36. *The* convex hull *of a set X is the smallest convex set containing X.*

If the set X is a finite set of points, the convex hull can be found by finding the line segment connecting each pair of points in the set (an n^2 operation count for the naive algorithm), and then finding out which ones form the boundary. There are more efficient algorithms, $O(n \log n)$, for finding convex hulls[68, 27].

Definition 1.37. *A function $f(x)$ is called* convex *on $[c,d]$ if for all $u, v \in [c,d]$, $f(\frac{u+v}{2}) \leq \frac{f(u)+f(v)}{2}$.*

Definition 1.38. *A function $f(x)$ is called* concave *on (c,d) if for all $u, v \in (c,d)$, $f(\frac{u+v}{2}) \geq \frac{f(u)+f(v)}{2}$.*

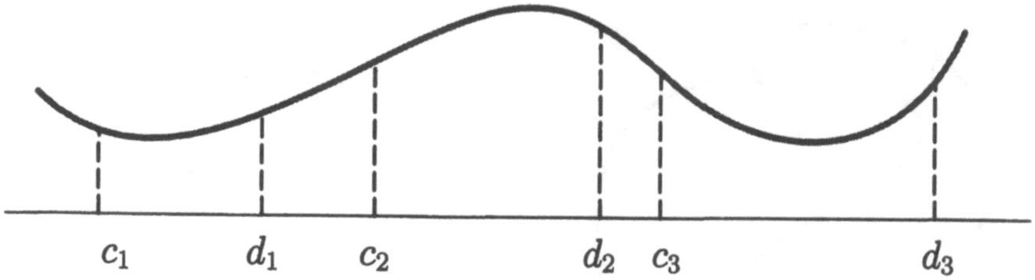

Figure 1.6. The function is convex on $(c_1, d_1) \cup (c_3, d_3)$ and concave on (c_2, d_2).

Theorem 1.39. *Suppose $f(x) \in C^{(2)}$. If $f''(x) > 0$ for $x \in (c, d)$, then $f(x)$ is convex on (c, d). If $f''(x) < 0$ for $x \in (c, d)$, then $f(x)$ is concave on (c, d).*

The implications of these results are that the signs and the values of the first and second derivatives yield important information about the shape of the curve.

1.5.1 Equations of Lines

In the plane, the implicit equation for a line is $ax + by + c = 0$. Using inner product notation yields $\langle(a, b, c), (x, y, 1)\rangle = 0$. Given a slope m which is not infinite, and a point (x_1, y_1) on the line, one has $(y - y_1) = m(x - x_1)$. It can be written more generally as, $a(x - x_1) + b(y - y_1) = \langle(a, b), (x - x_1, y - y_1)\rangle = 0$. To use parametric equations, a line can be represented using a direction vector $\boldsymbol{m}$ and one point (vector from origin) $\boldsymbol{p}$ that the line passes through. Thus $L(t) = t\boldsymbol{m} + \boldsymbol{p}$. Or one may use two points, $\boldsymbol{p_1}$ and $\boldsymbol{p_2}$, to yield $L(t) = t(\boldsymbol{p_2} - \boldsymbol{p_1}) + \boldsymbol{p_1} = (1 - t)\boldsymbol{p_1} + t\boldsymbol{p_2}$. This last form is called the *blending function* formulation. The derivations for these parametric lines occur in the introductory section on parametric functions.

What is the equation of a vector perpendicular to a line in $\mathbb{R}^2$? A unit vector that is perpendicular to a line, curve, or surface is called a *normal*. Whenever $b \neq 0$, the slope of the line $ax + by + c = 0$ is $-a/b$. It is clear that a perpendicular will have slope b/a (whenever $a \neq 0$). The vector (a, b) has that direction and, hence a vector with direction (a, b), having any length and any position, is perpendicular to the original line. We can look at this problem in a slightly different way.

Any point on the line $ax+by+c = 0$ must satisfy $\langle(a, b, c), (x, y, 1)\rangle = 0$. Now consider the situation where the line goes through the origin, that is, when $c = 0$. Then we can write $\langle(a, b), (x, y)\rangle = 0$, so the vector (a, b), is perpendicular to all points on the line. The case for $c \neq 0$ results simply in translates of the line but effects nether the slope nor the normal.

Example 1.40. Find the equation of a line through two points, $P_1 = (x_1, y_1)$ and $P_2 = (x_2, y_2)$.

The equation of a line is $ax + by + c = 0$ where we must determine a, b, and c.

$$P_1 \text{ on the line means } \langle (a,b,c),(x_1,y_1,1)\rangle = 0, \text{ and}$$
$$P_2 \text{ on the line means } \langle (a,b,c),(x_2,y_2,1)\rangle = 0.$$

This means that considered as "3-space vectors" the vectors $(x_1, y_1, 1)$ and $(x_2, y_2, 1)$ must both be perpendicular to the vector (a, b, c). Thus, we can set $(a, b, c) = (x_1, y_1, 1) \times (x_2, y_2, 1)$. Furthermore, any multiple of (a, b, c) also works! □

Example 1.41. Find the intersection point between the two lines $a_1x + b_1y + c_1 = 0$ and $a_2x + b_2y + c_2 = 0$.
Denote the intersection point by $I = (x_I, y_I)$.

$a_1x_I + b_1y_I + c_1 = 0 = \langle (a_1,b_1,c_1),(x_I,y_I,1)\rangle$ since I is on the first line, and
$a_2x_I + b_2y_I + c_2 = 0 = \langle (a_2,b_2,c_2),(x_I,y_I,1)\rangle$ since I is on the second line.

Using this "modified" three space notation, the "point" $(x_I, y_I, 1)$ must be orthogonal to both (a_1, b_1, c_1) and (a_2, b_2, c_2). Hence, it lies along the "three space vector" $Q = (a_1, b_1, c_1) \times (a_2, b_2, c_2)$ which is perpendicular to both. The third coordinate of Q is $a_1b_2 - a_2b_1$, however, not 1.

To solve this let $Q' = Q/(a_1b_2 - a_2b_1)$. Q' has its third coordinate equal to 1, $(a_1, b_1, c_1) \cdot Q' = 0$, and $(a_2, b_2, c_2) \cdot Q' = 0$. Thus the x-coordinate of Q' is x_I, and the y-coordinate of Q' is y_I. □

Example 1.42. Find the distance from a point Q to a parametric line $L(t) = \boldsymbol{p} + t\boldsymbol{d}$.

Finding the distance is equivalent to finding the magnitude of the vector perpendicular to the line through the point.

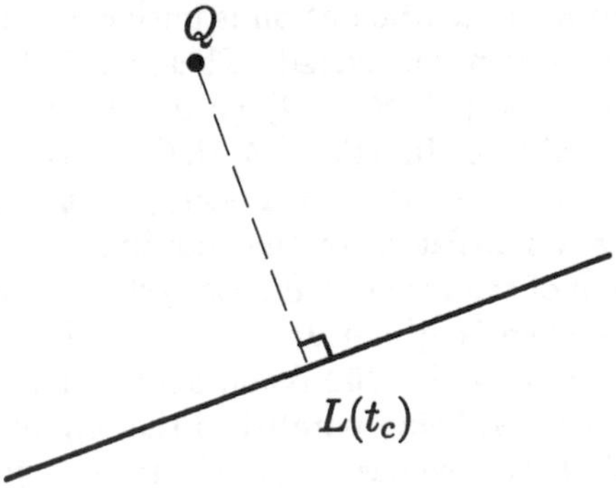

Let $L(t_c)$ denote the point on the line closest to Q. We wish to discover the value of t_c. Consider $L(t_c) - \boldsymbol{Q}$. This is a vector perpendicular to the line L through the point Q. Since it is perpendicular to L,

$$0 = \big(L(t_c) - \boldsymbol{Q}\big) \cdot \boldsymbol{d} = (\boldsymbol{p}, \boldsymbol{d}) + t_c(\boldsymbol{d}, \boldsymbol{d}) - (\boldsymbol{Q}, \boldsymbol{d}).$$

So,

$$(\boldsymbol{Q}, \boldsymbol{d}) = (\boldsymbol{p}, \boldsymbol{d}) + t_c(\boldsymbol{d}, \boldsymbol{d}),$$

and

$$t_c = \frac{(\boldsymbol{Q}, \boldsymbol{d}) - (\boldsymbol{p}, \boldsymbol{d})}{(\boldsymbol{d}, \boldsymbol{d})}.$$

The point $L(t_c)$ is now known, Q is known, so the distance from the point to the line is just the distance between these two points. □

1.5.2 Equations of Planes

The explicit equation for a plane not perpendicular to the x-y plane is $ax + by + d = z$. The implicit equation, which can be used for any plane in $\boldsymbol{R}^3$ is $Ax+By+Cz+D = 0$. While it seems as if there are four degrees of freedom, that is not true. The same plane is specified by $(rA)x + (rB)y + (rC)z + (rD) = 0$ as is specified by $Ax + By + Cz + D = 0$, for any $r \neq 0$. Since a plane has three degrees of freedom, any three independent constraints specify a unique plane. Some of the more commonly used specifications are: three points, one point and one "normal" vector, and two direction vectors and one point.

The inner product formulation developed for specifying a line can be generalized to specify planar characteristics. If (x, y, z) is on the plane, then $\langle (A, B, C, D), (x, y, z, 1) \rangle = 0$. If the plane goes through the origin, $D = 0$, and the vector (A, B, C) is perpendicular to all points in the plane. If the plane is simply translated, its orientation is unchanged, so the vectors perpendicular to it will remain unchanged. Thus, (A, B, C) is perpendicular to the plane with equation $\langle (A, B, C, D), (x, y, z, 1) \rangle = 0$.

Suppose a perpendicular direction (A, B, C) to the plane is specified. Given a point $P_1 = (x_1, y_1, z_1)$, let $x' = x - x_1$, $y' = y - y_1$, and $z' = z - z_1$. This can be seen as a translation of the coordinate axes which puts the point P_1 at the origin of the new coordinate system. The equation for the plane in that new system is $A'x' + B'y' + C'z' + D' = 0$. If that point is on the plane, $D' = 0$ by the discussion above. Further, a translation does not change directions, being parallel to the original axes, so $A' = A$, $B' = B$ and $C' = C$. Thus, one has $\langle (A, B, C), (x - x_1, y - y_1, z - z_1) \rangle = 0$

for the equation of the plane. This is modified to $\langle (A,B,C),(x,y,z)\rangle - \langle (A,B,C),(x_1,y_1,z_1)\rangle = 0$, and D is known.

To find the plane equation when two direction vectors $\boldsymbol{v}$, $\boldsymbol{w}$ for the plane, and a point in the plane are specified, one can find the normal (perpendicular) direction $(A,B,C) = \boldsymbol{v}\times\boldsymbol{w}$. This problem then reduces to the previous case.

If one wants to specify a plane in a parametric formulation, one needs two direction vectors and a point in the plane. Consider $P(s,t) = P_0 + s\boldsymbol{u} + t\boldsymbol{v}$. The points specified as the "head" of these vectors are on the surface: P_0, $P_0+\boldsymbol{u}$, and $P_0+\boldsymbol{v}$. If $P_0 = 0$, it is clear that this is simply a plane spanned by $\boldsymbol{u}$ and $\boldsymbol{v}$. Hence, $P(s,t)$ is a plane translated away from the origin.

If three points P_0, P_1, and P_2 are specified, one can set $\boldsymbol{u} = P_1 - P_0$, and $\boldsymbol{v} = P_2 - P_0$.

Example 1.43. Find the angle between two planes.

It will be shown in Exercise 5 that finding the angle, θ, between two planes is equivalent to finding the angle between the two plane normals, $\boldsymbol{n_1}$ and $\boldsymbol{n_2}$. Hence, the solution is

$$\theta = \cos^{-1}\left(\frac{\boldsymbol{n_1}\cdot\boldsymbol{n_2}}{\|\boldsymbol{n_1}\|\,\|\boldsymbol{n_2}\|}\right).$$

□

Example 1.44. What is the distance from a point R to a plane?

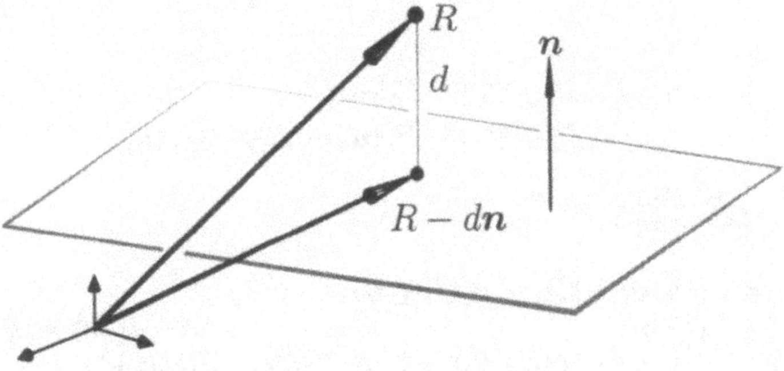

Let $\boldsymbol{n}$ be a unit normal to the plane, and $P_0 = (x_0, y_0, z_0)$ be any point in the plane. The plane equation, then is,

$$\begin{aligned} 0 &= \boldsymbol{n}\cdot(x-x_0, y-y_0, z-z_0) \\ &= \boldsymbol{n}\cdot\big[(x,y,z)-(x_0,y_0,z_0)\big]. \end{aligned}$$

Suppose d is the unknown distance from R to the plane. Then, $\boldsymbol{R} - d\boldsymbol{n}$ is a vector from the origin whose head is a point in the plane and

$$\begin{aligned} 0 &= \boldsymbol{n} \cdot [\boldsymbol{R} - d\boldsymbol{n} - \boldsymbol{P}_0] \\ &= \boldsymbol{n} \cdot \boldsymbol{R} - \boldsymbol{n} \cdot \boldsymbol{P}_0 - d\boldsymbol{n} \cdot \boldsymbol{n} \\ &= \boldsymbol{n} \cdot \boldsymbol{R} - \boldsymbol{n} \cdot \boldsymbol{P}_0 - d, \end{aligned}$$

so $d = \boldsymbol{n} \cdot \boldsymbol{R} - \boldsymbol{n} \cdot \boldsymbol{P}_0$. □

Example 1.45. Find the common perpendicular to two 3-space skew lines L_1 and L_2.

Suppose L_1 has direction vector $\boldsymbol{u_1}$ through point P_1 and L_2 has direction vector $\boldsymbol{u_2}$ through point P_2.

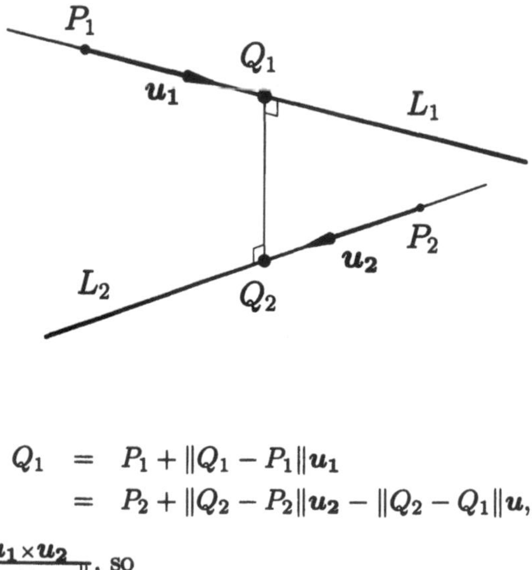

Then

$$\begin{aligned} Q_1 &= P_1 + \|Q_1 - P_1\|\boldsymbol{u_1} \\ &= P_2 + \|Q_2 - P_2\|\boldsymbol{u_2} - \|Q_2 - Q_1\|\boldsymbol{u}, \end{aligned}$$

where $\boldsymbol{u} = \frac{\boldsymbol{u_1} \times \boldsymbol{u_2}}{\|\boldsymbol{u_1} \times \boldsymbol{u_2}\|}$, so

$$\begin{aligned} \langle Q_1, \boldsymbol{u} \rangle &= \langle P_1, \boldsymbol{u} \rangle + \|Q_1 - P_1\| \langle \boldsymbol{u_1}, \boldsymbol{u} \rangle \\ &= \langle P_1, \boldsymbol{u} \rangle && \text{since } \langle \boldsymbol{u_1}, \boldsymbol{u} \rangle = 0, \text{ and} \\ &= \langle P_2, \boldsymbol{u} \rangle + \|Q_2 - P_2\| \langle \boldsymbol{u_2}, \boldsymbol{u} \rangle - \|Q_2 - Q_1\| \langle \boldsymbol{u}, \boldsymbol{u} \rangle \\ &= \langle P_2, \boldsymbol{u} \rangle - \|Q_2 - Q_1\| && \text{since } \langle \boldsymbol{u_2}, \boldsymbol{u} \rangle = 0, \\ & && \text{and } \langle \boldsymbol{u}, \boldsymbol{u} \rangle = 1. \end{aligned}$$

Thus $\langle P_2 - P_1, \boldsymbol{u} \rangle = \|Q_2 - Q_1\|$, or $\|Q_2 - Q_1\| = \left\langle P_2 - P_1, \frac{\boldsymbol{u_1} \times \boldsymbol{u_2}}{\|\boldsymbol{u_1} \times \boldsymbol{u_2}\|} \right\rangle$.

□

Example 1.46. Find the common intersection $\boldsymbol{r}$ of the three planes with normals $\boldsymbol{u}$, $\boldsymbol{v}$, $\boldsymbol{w}$.

By the equations of the three planes

$$\begin{aligned}\langle \boldsymbol{r}, \boldsymbol{u} \rangle &= p_u, \\ \langle \boldsymbol{r}, \boldsymbol{v} \rangle &= p_v, \\ \langle \boldsymbol{r}, \boldsymbol{w} \rangle &= p_w.\end{aligned}$$

This yields just three linear equations in three unknowns. □

This section concludes with some food for thought. Consider $\boldsymbol{E} = \{\boldsymbol{e}_1, \boldsymbol{e}_2, \boldsymbol{e}_3\}$ such that $\langle \boldsymbol{e}_i, \boldsymbol{e}_j \rangle = \delta_{i,j}$ for $i, j = 1, 2, 3$, and ordered so that $\boldsymbol{e}_1 \times \boldsymbol{e}_2 = \boldsymbol{e}_3$, $\boldsymbol{e}_2 \times \boldsymbol{e}_3 = \boldsymbol{e}_1$, and $\boldsymbol{e}_3 \times \boldsymbol{e}_1 = \boldsymbol{e}_2$. This does not require that $\boldsymbol{e}_1$ be a unit vector in the direction of x, nor is an analogous constraint placed upon $\boldsymbol{e}_2$ or $\boldsymbol{e}_3$.

1. Is $\boldsymbol{E}$ a basis for $\boldsymbol{R}^3$?
2. For an arbitrary $\boldsymbol{v} \in \boldsymbol{R}^3$, consider $\boldsymbol{w} = \langle \boldsymbol{v}, \boldsymbol{e}_1 \rangle \boldsymbol{e}_1 + \langle \boldsymbol{v}, \boldsymbol{e}_2 \rangle \boldsymbol{e}_2 + \langle \boldsymbol{v}, \boldsymbol{e}_3 \rangle \boldsymbol{e}_3$. Is $\boldsymbol{w} = \boldsymbol{v}$?
3. If, instead, one has three unit vectors $\{\boldsymbol{v}_1, \boldsymbol{v}_2, \boldsymbol{v}_3\}$, which form a basis for $\boldsymbol{R}^3$, but are not orthonormal, and $\boldsymbol{v} \in \boldsymbol{R}^3$, is $\boldsymbol{v}$ equal to $\langle \boldsymbol{v}, \boldsymbol{v}_1 \rangle \boldsymbol{v}_1 + \langle \boldsymbol{v}, \boldsymbol{v}_2 \rangle \boldsymbol{v}_2 + \langle \boldsymbol{v}, \boldsymbol{v}_3 \rangle \boldsymbol{v}_3$?
4. Under what conditions on the basis is such a decomposition true for all elements of $\boldsymbol{R}^3$?

These problems are left as reinforcement exercises for the reader.

1.5.3 Polynomials

Definition 1.47. *For arbitrary complex numbers $a_0, \ldots, a_n$, with $a_n \neq 0$, a* polynomial, *p_n of degree n over the complex numbers is defined as a function of the form:*

$$p_n(z) = a_n z^n + a_{n-1} z^{n-1} + \cdots + a_1 z + a_0.$$

For the general complex polynomial, one has

Theorem 1.48. The Fundamental Theorem of Algebra. *If $n > 0$ and p_n is defined over the complex domain with complex coefficients and*

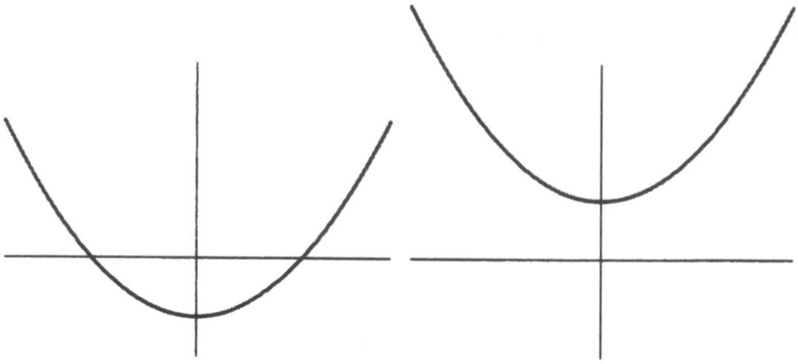

Figure 1.7. $p(x) = x^2 - 1$ $q(x) = x^2 + 1$

$a_n \neq 0$, then p_n has exactly n complex, not necessarily distinct, roots. Thus, one can write

$$p_n(z) = a_n(z - z_1)(z - z_2)\cdots(z - z_n),$$

where the values $z_1, \ldots, z_n$ are the zeros of the polynomial p_n.

If all the coefficients are real and the domain is restricted to the reals, then by Theorem 1.48 there are n roots possible, although the values may not be real.

Example 1.49. Consider the function $p(x) = x^2 - 1$. It has roots $\{1, -1\}$. The function $q(x) = x^2 + 1$, however has no real roots. Its complex roots are $\{i, -i\}$, where i is the square root of -1. □

The roots of a polynomial do not uniquely define that polynomial. For example, both $f(x) = (x - 1)$ and $g(x) = 3(x - 1)$ are straight lines that have roots at $x = 1$, but they have different slopes. In fact the coefficient of x is the slope of the line. Therefore, knowing the root and the slope uniquely defines the line. Interpolation with polynomials is based on the fundamental theorem of algebra.

An immediate corollary is

Corollary 1.50. *If a polynomial p_n of degree n vanishes (has roots) at more than n distinct points, then $p_n \equiv 0$, that is, p_n is identically zero.*

An immediate result is that the behavior of an nth degree polynomial is determined by its function values at $n+1$ points, since if two polynomials $p(x)$ and $q(x)$ of degree n agree at $n+1$ points, then their difference is a polynomial of degree n with $n+1$ zeros. The difference must, by the corollary, be identically zero, and $p(x) = q(x)$.

Within the scope of this book, the coefficients $a_0, \ldots, a_n$ and the domain (and hence the range) will be real numbers.

Definition 1.51. *The* space of polynomials *of degree n, $\mathcal{P}_n$, is defined as*

$$\mathcal{P}_n = \left\{ p(x) = a_0 + a_1 x^1 + \cdots + a_n x^n : a_i \in \boldsymbol{R},\ i = 0, \ldots,\ n \right\}.$$

Definition 1.52. *A* bivariate polynomial of degree n *is a bivariate function $f(u, v)$ such that*

$$f(u, v) = \sum_{k=0}^{n} \sum_{i=0}^{k} p_{i,k} u^i v^{k-i}.$$

A bivariate polynomial is called bilinear, biquadratic, *or* bicubic *if the highest power in each of the variables is* 1, 2, *or* 3, *respectively.*

It is clear from the definition that $\mathcal{P}_n$ is a vector space in which each polynomial is a vector. This type of space is called a function space. Further it is clear from the definition of a polynomial that the set $\{1, x, x^2, \ldots, x^n\}$ spans the space. We ask whether this set forms a basis. That is, are the functions 1, x, x^2, ..., x^n linearly independent? We shall use straightforward reasoning. Suppose they are not. Then there are coefficients c_i, $i = 0$, ..., n not all zero so that

$$0 \equiv c_0 + c_1 x + \cdots + c_n x^n.$$

That is, the polynomial on the right evaluates to zero at all values of x. Now suppose c_k is the coefficient with the lowest order subscript which is nonzero. Then differentiating both sides k times gives

$$0 \equiv k! c_k + (k+1)k \ldots 2 c_{k+1} x + \cdots + n(n-1) \cdots (n-k+1) c_n x^{n-k}.$$

Evaluating the polynomial on the right at $x = 0$ yields that $k! c_k = 0$, or that $c_k = 0$. This contradicts the hypothesis, so 0 cannot be represented as a nontrivial polynomial and the powers of x are independent functions. Thus, $\{1, x, \ldots, x^n\}$ forms a basis for the space of polynomials, and $\mathcal{P}_n$ has dimension $n + 1$, and we have shown,

Lemma 1.53. *The dimension of $\mathcal{P}_n$, the space of polynomials of degree n, is $n + 1$.*

A more concise form equation for the fundamental theorem can be derived as follows. Suppose both $p(a) = 0$ and $p'(a) = 0$. Can we tell anything more about the form of the polynomial $p(x)$? Since $p(a) = 0$, $p(x) = (x-a)p_1(x)$. But since $p'(a) = 0$, $p'(x) = (x-a)p_2(x)$, by the fundamental theorem of algebra. But from the first form on $p(x)$,

$$\begin{aligned} p'(x) &= p_1(x) + (x-a)p_1'(x) \\ &= (x-a)p_2(x). \end{aligned}$$

We see, then that $p_1(x) = (x-a)q(x)$, and $p(x) = (x-a)^2 q(x)$. Following this line of reasoning gives:

Theorem 1.54. *Suppose $p(x)$ is a polynomial of degree n with distinct real roots $x_1, \ldots, x_k$, and suppose $p^{(j)}(x_i) = 0$, $j = 0, \ldots, s_i$, $i = 1, \ldots, k$. Then*

$$p(x) = (x-x_1)^{s_1+1}(x-x_2)^{s_2+1}\cdots(x-x_k)^{s_k+1}q(x)$$

where $q(x)$ is of degree $n - k - \sum_{i=1}^{k} s_i$.

Proof: The proof is by induction. We saw above that if a is a root of both $p(x)$ and $p'(x)$, then $p(x) = (x-a)^2 q_1(x)$. Suppose we have shown that if a is a root of $p^{(j)}(x)$, $j = 0, \ldots, m$, then $p(x) = (x-a)^{m+1}q(x)$. Suppose now that in addition, a is a root of $p^{(m+1)}(x)$. We know that $p'(x) = (x-a)^{m+1}r(x)$, since a is a root of $[p']^{(j)}(x)$, $j = 0, \ldots, m$, but we also know

$$\begin{aligned} p'(x) &= (m+1)(x-a)^m q(x) + (x-a)^{m+1}q'(x) \\ &= (x-a)^m\left[(m+1)q(x) + (x-a)q'(x)\right] \\ &= (x-a)^m(x-a)r(x). \end{aligned}$$

Since the last two lines are equal, for $x \neq a$, we see that $(x-a)r(x) = (m+1)q(x) + (x-a)q'(x)$, and so $q(x) = (x-a)z(x)$, and the result is proved for a single root.

Now, if there is more than one root, we then apply the theorem to the polynomial $z(x)$ that is the remainder at a different value and get a corresponding decomposition. After applying it to all k distinct roots, the result is proved. ∎

1.5.4 Rational Functions

Definition 1.55. *A function $f(x)$ is called a* rational function *if $f(x) = p(x)/q(x)$, where both p and q are polynomials.*

Example 1.56. A simple rational function can be constructed as the quotient of two linear polynomials, that is, $f(x) = (ax+b)/(cx+d)$. □

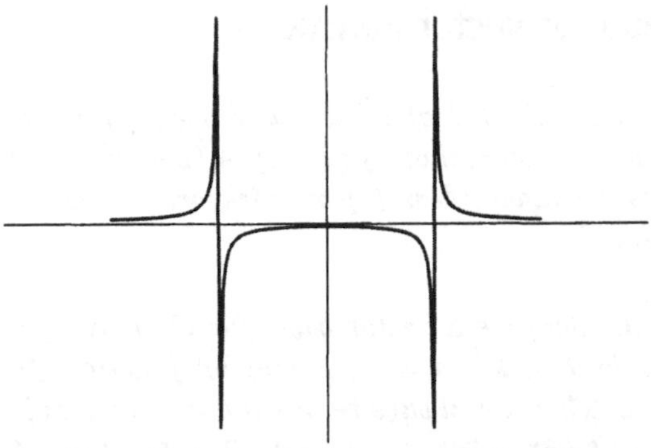

Figure 1.8. A simple rational function $(x^2+1)/(x^2-1)$.

The properties of the rational are not completely defined by the separate properties of the numerator and denominator taken separately. While the roots of the numerator may be zeros of the rational function and the roots of the denominator may be poles (infinite asymptotes) of the rational function, this may also not occur. That is, these are necessary but not sufficient conditions. If a root is common to both numerator and denominator then the number of repetitions in each may decide the final root/pole configuration.

Example 1.57. The equation

$$\frac{(x-1)(x+1)}{(x-1)(x-2)} = 0$$

has a root at $x = -1$ and a pole at $x = 2$, and is not defined at $x = 1$. However it is possible to define a function

$$f(x) = \begin{cases} \frac{(x-1)(x+1)}{(x-1)(x-2)} & x \neq 1, \\ -2 & x = 1, \end{cases}$$

which is continuous at $x = 1$.

The function $g(x) = \frac{(x-1)^2}{(x-1)(x-2)}$ has a root at $x = 1$ and a pole at $x = 2$. □

The rationals also have the feature that if $f(x) = p(x)/q(x)$, then $f(x) = ap(x)/aq(x)$ as well for all real numbers $a \neq 0$. This can lead to confusion over the number of points needed to uniquely specify a rational function.

1.6 Parametric or Vector Functions

Definition 1.58. *A subset U of $\mathbf{R}^2$ is called open if for every point $(a,b) \in U$ there exists an $\epsilon > 0$ such that, if $(x-a)^2+(y-b)^2 < \epsilon$, then $(x,y) \in U$. That is, there is a boundaryless (open) disk around each point contained entirely in the set U.*

Definition 1.59. *Suppose a vector basis for $\mathbf{R}^i$ is $\mathbf{e}^j$, $j = 1, \ldots, i$. Let U be an open subset of $\mathbf{R}^i$, $i = 1, 2$ and let function $\boldsymbol{f}(\boldsymbol{x}) : U \to \mathbf{R}^j$, where $j = 1, 2, 3$. This can always be written in vector notation as $\boldsymbol{f}(\boldsymbol{x}) = \sum_{k=1}^{j} f_k(\boldsymbol{x})\mathbf{e}^k = (f_1(\boldsymbol{x}), f_2(\boldsymbol{x}), \ldots, f_j(\boldsymbol{x}))$. The functions f_k are called the* coordinate functions *of the vector function (f).*

Definition 1.60. *A vector function $\boldsymbol{f}$ is called* continuous *at $\boldsymbol{x}^0 = (x_1^0, x_2^0)$ if for every $\epsilon > 0$ there exists $\delta > 0$ such that if $\boldsymbol{x} = (x_1, x_2)$ then $\left\|\boldsymbol{f}(\boldsymbol{x}) - \boldsymbol{f}(\boldsymbol{x}^0)\right\| < \epsilon$ whenever $\left\|\boldsymbol{x} - \boldsymbol{x}^0\right\| < \delta$, where $\left\|\boldsymbol{x} - \boldsymbol{x}^0\right\| = \sqrt{(x_1 - x_1^0)^2 + (x_2 - x_2^0)^2}$.*

It is clear that a vector function is continuous if and only if its coordinate functions are continuous.

Suppose that $\boldsymbol{f} : U \to \mathbf{R}^3$, and further, suppose that we can define functions $x_1 = \theta_1(t)$ and $x_2 = \theta_2(t)$ where $\theta_1, \theta_2 : \mathtt{I} \to U$, where $\mathtt{I}$ is an interval in $\mathbf{R}^1$. Then the function $\gamma(t) = \boldsymbol{f}(x_1, x_2) = \boldsymbol{f}\big(x_1(t), x_2(t)\big) = \boldsymbol{f}\big(\theta_1(t), \theta_2(t)\big)$ is a space curve whose image lies in the image of the function $\boldsymbol{f}$.

Definition 1.61. *Consider the set $U_{2,c} = \{(x_1, c) : (x_1, c) \in U\}$, and $U_{1,k} = \{(k, x_2) : (k, x_2) \in U\}$. Consider the space curve, $\gamma_c(x_1)$ defined by $\gamma_c(x_1) = \boldsymbol{f}(x_1, c)$, a curve in the image $\boldsymbol{f}$. Each constant c defines a distinct curve. Analogously, for a constant k, $\phi_k(x_2) = \boldsymbol{f}(k, x_2)$ defines a curve in the image $\boldsymbol{f}$, where each constant k defines a distinct curve. Each γ_c curve is has domain parallel to the x_1 axis in the $x_1 - x_2$ plane traces out a curve on the surface given by the image of $\boldsymbol{f}$. Analogously, each ϕ_k has a domain parallel to the x_2 axis and traces out a curve on the surface. The set of curves formed by the γ_c and ϕ_k on the surface is called a* curvilinear coordinate system.

Unlike curves for which derivatives can be defined only in one direction, a surface has an infinite number of curves through a point, and thus the meaning of derivative must be adapted.

Definition 1.62. *Suppose that $\boldsymbol{f}$ is defined on an open set U in $\mathbb{R}^2$, and let $\boldsymbol{x}^0$ be some point in U, with $\boldsymbol{u}^0$ a nonzero vector in U. The* directional derivative *of $\boldsymbol{f}$ at $\boldsymbol{x}^0$ in the direction of $\boldsymbol{u}^0$ is the vector*

$$D_{\boldsymbol{u}^0}\boldsymbol{f}(\boldsymbol{x}^0) = \lim_{h\to 0}\frac{\boldsymbol{f}(\boldsymbol{x}^0 + h\boldsymbol{u}^0) - \boldsymbol{f}(\boldsymbol{x}^0)}{h}$$

whenever that limit exists.

This definition has the same effect as defining a univariate derivative of a space curve of a variable h where $\gamma(h) = \boldsymbol{f}(\boldsymbol{x}^0 + h\boldsymbol{u}^0)$. This curve lies in the surface $\boldsymbol{f}$ and has as domain an open interval around zero. In general, if for each point $\boldsymbol{x} \in U$, $D_{\boldsymbol{u}^0}\boldsymbol{f}(\boldsymbol{x})$ exists, then $\boldsymbol{f}$ is said to have a directional derivative in the direction $\boldsymbol{u}^0$ in U.

In particular when $\boldsymbol{u}^0 = (1,0)$, the result is a derivative with respect to the first direction in the curvilinear coordinate system, and when $\boldsymbol{u}^0 = (0,1)$, the result is a derivative with respect to the second.

Written $\frac{\partial \boldsymbol{f}}{\partial x_1}$ and $\frac{\partial \boldsymbol{f}}{\partial x_2}$, respectively, these derivatives are called the *partial derivatives* of the function $\boldsymbol{f}$ with respect to the first and second coordinates, respectively.

The world of multivariate functions is much more complicated than that of univariate functions. Since the directional derivative depends only on the values of the function along an open line interval near a point, a function $\boldsymbol{f}$ can have derivatives in every direction at a point $\boldsymbol{x}^0$, but not be continuous at $\boldsymbol{x}^0$. Remember, the continuity depends on the actions of the function in a two-dimensional neighborhood of the point.

Theorem 1.63. *A vector function $\boldsymbol{f} = (f_1, f_2, f_3)$ is said to be of class $C^{(k)}$ if each coordinate function is of class $C^{(k)}$. A bivariate scalar function f is of class $C^{(k)}$ if all partial derivatives of order less than or equal to k exist and are continuous independent of the order of differentiation .*

Suppose that it is desired to perform a change of variables, a *reparametrization,* as it is called.

Theorem 1.64. Chain Rule. *For a parametric function $\boldsymbol{f}$, defined as $\boldsymbol{f}(x_1, x_2) = (f_1(x_1,x_2), f_2(x_1,x_2), f_3(x_1,x_2)) \in C^{(1)}$. If $x_1 = x_1(t)$ and $x_2 = x_2(t)$ and $x_1(t)$, $x_2(t) \in C^{(1)}$ then*

$$\begin{aligned}\frac{d\boldsymbol{f}}{dt} &= \frac{\partial \boldsymbol{f}}{\partial x_1}\frac{dx_1}{dt} + \frac{\partial \boldsymbol{f}}{\partial x_2}\frac{dx_2}{dt}\\ &= \left(\frac{\partial f_1}{\partial x_1}, \frac{\partial f_2}{\partial x_1}, \frac{\partial f_3}{\partial x_1}\right)\frac{dx_1}{dt} + \left(\frac{\partial f_1}{\partial x_2}, \frac{\partial f_2}{\partial x_2}, \frac{\partial f_3}{\partial x_2}\right)\frac{dx_2}{dt}\end{aligned}$$

$$= \begin{bmatrix} \frac{dx_1}{dt} & \frac{dx_2}{dt} \end{bmatrix} \begin{bmatrix} \frac{\partial f_1}{\partial x_1} & \frac{\partial f_2}{\partial x_1} & \frac{\partial f_3}{\partial x_1} \\ \frac{\partial f_1}{\partial x_2} & \frac{\partial f_2}{\partial x_2} & \frac{\partial f_3}{\partial x_2} \end{bmatrix}.$$

If $x_1 = x_1(u_1, u_2)$ *and* $x_2 = x_2(u_1, u_2)$ *and* $x_1(u_1, u_2)$, $x_2(u_1, u_2) \in C^{(1)}$ *then*

$$\begin{aligned} \frac{\partial \boldsymbol{f}}{\partial u_1} &= \frac{\partial \boldsymbol{f}}{\partial x_1}\frac{\partial x_1}{\partial u_1} + \frac{\partial \boldsymbol{f}}{\partial x_2}\frac{\partial x_2}{\partial u_1} \\ &= \left(\frac{\partial f_1}{\partial x_1}, \frac{\partial f_2}{\partial x_1}, \frac{\partial f_3}{\partial x_1}\right)\frac{\partial x_1}{\partial u_1} + \left(\frac{\partial f_1}{\partial x_2}, \frac{\partial f_2}{\partial x_2}, \frac{\partial f_3}{\partial x_2}\right)\frac{\partial x_2}{\partial u_1} \\ &= \begin{bmatrix} \frac{\partial x_1}{\partial u_1} & \frac{\partial x_2}{\partial u_1} \end{bmatrix} \begin{bmatrix} \frac{\partial f_1}{\partial x_1} & \frac{\partial f_2}{\partial x_1} & \frac{\partial f_3}{\partial x_1} \\ \frac{\partial f_1}{\partial x_2} & \frac{\partial f_2}{\partial x_2} & \frac{\partial f_3}{\partial x_2} \end{bmatrix}. \end{aligned}$$

Analogously,

$$\frac{\partial \boldsymbol{f}}{\partial u_2} = \begin{bmatrix} \frac{\partial x_1}{\partial u_2} & \frac{\partial x_2}{\partial u_2} \end{bmatrix} \begin{bmatrix} \frac{\partial f_1}{\partial x_1} & \frac{\partial f_2}{\partial x_1} & \frac{\partial f_3}{\partial x_1} \\ \frac{\partial f_1}{\partial x_2} & \frac{\partial f_2}{\partial x_2} & \frac{\partial f_3}{\partial x_2} \end{bmatrix}.$$

The fundamentally important matrix

$$\begin{bmatrix} \frac{\partial f_1}{\partial x_1} & \frac{\partial f_2}{\partial x_1} & \frac{\partial f_3}{\partial x_1} \\ \frac{\partial f_1}{\partial x_2} & \frac{\partial f_2}{\partial x_2} & \frac{\partial f_3}{\partial x_2} \end{bmatrix}$$

is called the *Jacobian matrix.* Further,

$$\begin{bmatrix} \frac{\partial \boldsymbol{f}}{\partial u_1} \\ \frac{\partial \boldsymbol{f}}{\partial u_2} \end{bmatrix} = \begin{bmatrix} \frac{\partial x_1}{\partial u_1} & \frac{\partial x_2}{\partial u_1} \\ \frac{\partial x_1}{\partial u_2} & \frac{\partial x_2}{\partial u_2} \end{bmatrix} \begin{bmatrix} \frac{\partial f_1}{\partial x_1} & \frac{\partial f_2}{\partial x_1} & \frac{\partial f_3}{\partial x_1} \\ \frac{\partial f_1}{\partial x_2} & \frac{\partial f_2}{\partial x_2} & \frac{\partial f_3}{\partial x_2} \end{bmatrix}$$

and the matrix

$$\begin{bmatrix} \frac{\partial x_1}{\partial u_1} & \frac{\partial x_2}{\partial u_1} \\ \frac{\partial x_1}{\partial u_2} & \frac{\partial x_2}{\partial u_2} \end{bmatrix}$$

is called the *Jacobian matrix of the reparametrization.* Its determinant is called the *Jacobian,* $\frac{\partial(x_1,x_2)}{\partial(u_1,u_2)}$, *of the transformation.*

1.6.1 Function Characteristics

In order to facilitate later proofs, we shall state without complete proofs some of the following theorems from calculus:

Theorem 1.65. Rolle's Theorem. *Suppose $f \in C[a,b]$ and that $f'(x)$ exists at each point of (a,b) and $f(a) = f(b)$. Then there exists a point ζ, $a < \zeta < b$ such that $f'(\zeta) = 0$.*

Proof: If $f(x) \equiv f(a)$, then f is a constant function and $f' \equiv 0$. Now, suppose there exists x such that $f(x) > f(a)$. The fact that f is continuous on $[a,b]$ implies that f achieves a maximum value, say at ζ, in (a,b). Since f' exists for all points in (a,b) that means that $f'(\zeta) = 0$. ■

Theorem 1.66. Mean Value Theorem. *Let $f \in C[a,b]$ such that $f'(x)$ exists at each point of (a,b). Then there exists a point ζ, $a < \zeta < b$ such that $(b-a)f'(\zeta) = f(b) - f(a)$.*

Proof: Consider the function $g(x) = f(x) + (b-x)\frac{f(b)-f(a)}{b-a}$. Then $g(a) = f(a) + \big(f(b) - f(a)\big) = f(b)$ and $g(b) = f(b)$. By Rolle's theorem, there exists ζ such that $g'(\zeta) = 0$. But, $g'(x) = f'(x) - \frac{f(b)-f(a)}{b-a}$ so at ζ, $f'(\zeta) = \frac{f(b)-f(a)}{b-a}$. ■

Theorem 1.67. Generalized Rolle's Theorem. *For $2 \le n$, let $f \in C[a,b]$ such that $f^{(n-1)}$ exists for each point of (a,b). If there exists $a \le x_1 < x_2 < \cdots < x_n \le b$ such that $f(x_1) = f(x_2) = \cdots = f(x_n)$, then there exists ζ, $x_1 < \zeta < x_n$ such that $f^{(n-1)}(\zeta) = 0$.*

Proof: Apply Rolle's theorem $n-1$ times. ■

Theorem 1.68. Taylor's Theorem. *For $f \in C^{(n+1)}[a,b]$ then for all x_0, $x \in [a,b]$,*

$$\begin{aligned} f(x) &= f(x_0) + f'(x_0)(x-x_0) + \frac{f''(x_0)}{2}(x-x_0)^2 + \cdots \\ &+ \frac{f^{(n)}(x_0)}{n!}(x-x_0)^n + \frac{1}{n!}\int_{x_0}^{x} f^{(n+1)}(t)(x-t)^n\,dt. \end{aligned}$$

The use of Taylor's theorem requires explicit evaluation at only one point, and other knowledge of the $(n+1)^{st}$ derivative to bound the integral. The nth degree polynomial part of the expansion is called the *Taylor* *polynomial approximation about x_0 to f of degree n* and the integral part is called the *remainder.*

Proof: From the fundamental theorem of calculus, $f(x) - f(x_0) = \int_{x_0}^{x} f'(t)\,dt$. Rewriting yields

$$f(x) = f(x_0) + \int_{x_0}^{x} f'(t)dt.$$

We integrate by parts with $u = f'(t)$, $du = f''(t)dt$ and $dv = dt$, $v = (t-x)$ to get

$$\begin{aligned} f(x) &= f(x_0) + \left[f'(t)(t-x)\right]\big|_{x_0}^{x} - \int_{x_0}^{x} (t-x)f''(t)dt \\ &= f(x_0) + f'(x_0)(x-x_0) + \int_{x_0}^{x} (x-t)f''(t)dt. \end{aligned}$$

Integrating by parts again we see

$$\int_{x_0}^{x} (x-t)f''(t)dt = f''(x_0)\frac{(x-x_0)^2}{2!} + \int_{x_0}^{x} \frac{(x-t)^2}{2} f^{(3)}(t)dt$$

so

$$f(x) = f(x_0) + f'(x_0)(x-x_0) + f''(x_0)\frac{(x-x_0)^2}{2!} + \int_{x_0}^{x} \frac{(x-t)^2}{2} f^{(3)}(t)dt.$$

Using the induction hypothesis we have

$$\begin{aligned} f(x) &= \sum_{j=0}^{n-1} \frac{f^{(j)}(x_0)}{j!}(x-x_0)^j + \int_{x_0}^{x} \frac{f^{(n)}(t)}{(n-1)!}(x-t)^{n-1}dt \\ &= \sum_{j=0}^{n-1} \frac{f^{(j)}(x_0)}{j!}(x-x_0)^j \\ &\qquad + \frac{1}{n!}\left(-(x-t)^n f^{(n)}(t)\,\big|_{t=x_0}^{x} + \int_{x_0}^{x} f^{(n+1)}(t)(x-t)^n dt\right) \\ &= \sum_{j=0}^{n} \frac{f^{(j)}(x_0)}{j!}(x-x_0)^j + \frac{1}{n!}\int_{x_0}^{x} f^{(n+1)}(t)(x-t)^n dt \\ &= \sum_{j=0}^{n} \frac{f^{(j)}(x_0)}{j!}(x-x_0)^j + \frac{1}{n!}\int_{x_0}^{b} f^{(n+1)}(t)(x-t)_+^n dt. \end{aligned}$$

Notice that the integrals in the last two lines have different beginning and ending points, and so the plus function notation is used. ■

Definition 1.69. *The function* $(x-t)_+^n$ *defined by*

$$(x-t)_+^n = \begin{cases} (x-t)^n & \text{if } x > t \\ 0 & \text{otherwise} \end{cases}$$

is called the nth *degree* plus function.

Note that by this definition, $(0)_+^0 = 0$. For $n = 0$, it is clear that $(x-t)_+^0 = 1$,if $t < x$. That is, the function is continuous and is a single polynomial, namely the constant 1 for t values less than x, and it is the constant 0 for values of t greater than or equal to x. Thus, for $n = 0$, the function is discontinuous at $x = t$, i.e., in $C^{(-1)}$. Considered as a function of t for fixed x, it is *right continuous.* That is, $\lim_{t\to x^+} (x-t)_+^0 = 0 = (x-x)_+^0$. Considered as a function of x, the function is *left continuous* since $\lim_{x\to t^-} (x-t)_+^0 = 0 = (t-t)_+^0$ but $\lim_{x\to t^+} (x-t)_+^0 = 1$.

Next consider $f(t) = (x-t)_+$, always as a function of t. For $t < x$, $f(t) = (x-t)$, a polynomial, and is continuous and differentiable. For $t > x$, $f(t) = 0$, and this is also continuous and differentiable. Now at $t = x$, the function is continuous since $0 = f(x) = \lim_{t\to x^+}(x-t)$. However, the function does not have a derivative at $t = x$, since the derivative on the left is the zero function, and the derivative on the right is 1. Thus $(x-t)_+$ is continous and continuously differentiable everywhere but at $t = x$, where the derivative is right continuous.

Now, using induction, suppose that $(x-t)_+^{n-1}$ is in $C^{(n-2)}$ at $t = x$. Then, for $t \neq x$,

$$\frac{d}{dx}(x-t)_+^n = n(x-t)_+^{n-1}.$$

Since the right hand side is $C^{(n-2)}$, then the derivative clearly exists at $t = x$, and we find:

Lemma 1.70. *For positive integers* n, $(x-t)_+^n$ *is contained in* $C^{(n-1)}$, *and as a function of* t, *its* n^{th} *derivative is continuous everywhere except* $t = x$ *where it is right continuous. As a function of* x, *its* n^{th} *derivative is continuous everywhere except at* $x = t$, *where it is left continuous.*

Exercises

1. Suppose $\boldsymbol{V}$ is a finite-dimensional vector space and $\boldsymbol{B_1}$ and $\boldsymbol{B_2}$ are two bases for $\boldsymbol{V}$. Show that $\boldsymbol{B_1}$ and $\boldsymbol{B_2}$ *must* have the same number of elements.

2. Is $C^{(0)}[a,b]$ a finite-dimensional vector space? If so prove it, if not explain why not.

3. Show that the definition of scalar multiplication of an $\boldsymbol{R}^3$ vector $\boldsymbol{v}$ by a scalar $c > 0$ makes a new vector $\boldsymbol{w}$ having the same direction as $\boldsymbol{v}$ with a magnitude c times as long. What happens when c is negative?

4. For vectors $\boldsymbol{v}, \boldsymbol{w} \in \boldsymbol{R}^3$ show that $\|\boldsymbol{v} \times \boldsymbol{w}\|^2 = \|\boldsymbol{v}\|^2 \|\boldsymbol{w}\|^2 - \|(\boldsymbol{v}, \boldsymbol{w})\|^2$.

5. If $\boldsymbol{P_1}$ and $\boldsymbol{P_2}$ are two planes with normals $\boldsymbol{n_1}$ and $\boldsymbol{n_2}$, respectively, show that finding the angle between the two planes is equivalent to finding the angle between the two plane normals.

6. Consider $\boldsymbol{E} = \{\boldsymbol{e_1}, \boldsymbol{e_2}, \boldsymbol{e_3}\}$ such that $\langle \boldsymbol{e_i}, \boldsymbol{e_j} \rangle = \delta_{i,j}$ for $i,\ j = 1,\ 2,\ 3$, and ordered so that $\boldsymbol{e_1} \times \boldsymbol{e_2} = \boldsymbol{ecd_3}$, $\boldsymbol{e_2} \times \boldsymbol{e_3} = \boldsymbol{e_1}$, and $\boldsymbol{e_3} \times \boldsymbol{e_1} = \boldsymbol{e_2}$. This does not require that $\boldsymbol{e_1}$ be a unit vector in the direction of x, nor is an analogous constraint placed on $\boldsymbol{e_2}$ or $\boldsymbol{e_3}$.

   (a) Is $\boldsymbol{E}$ a basis for $\boldsymbol{R}^3$?

   (b) For an arbitrary $\boldsymbol{v} \in \boldsymbol{R}^3$, consider $\boldsymbol{w} = \langle \boldsymbol{v}, \boldsymbol{e_1} \rangle \boldsymbol{e_1} + \langle \boldsymbol{v}, \boldsymbol{e_2} \rangle \boldsymbol{e_2} + \langle \boldsymbol{v}, \boldsymbol{e_3} \rangle \boldsymbol{e_3}$. Is $\boldsymbol{w} = \boldsymbol{v}$?

   (c) If instead one has three unit vectors $\{\boldsymbol{v_1}, \boldsymbol{v_2}, \boldsymbol{v_3}\}$, which form a basis for $\boldsymbol{R}^3$, but which is not orthonormal, and $\boldsymbol{v} \in \boldsymbol{R}^3$, if $\boldsymbol{w} = \langle \boldsymbol{v}, \boldsymbol{v_1} \rangle \boldsymbol{v_1} + \langle \boldsymbol{v}, \boldsymbol{v_2} \rangle \boldsymbol{v_2} + \langle \boldsymbol{v}, \boldsymbol{v_3} \rangle \boldsymbol{v_3}$, is $\boldsymbol{v} = \boldsymbol{w}$?

   (d) Under what conditions on the basis is such a decomposition true for all elements of $\boldsymbol{R}^3$?

   Give a proof for your response when it indicates something is true or not true all the time, and counterexamples when your response indicates that something contrary can occur.

7. Prove Lemma 1.13. That is, show that $\langle \boldsymbol{v}, \boldsymbol{w} \rangle = \sum_{i=1}^{n} r_{v,i} r_{w,i}$, for $r_{v,i}$, $r_{w,i}$, $\boldsymbol{v}$, and $\boldsymbol{w}$ defined as in the lemma.

8. Show that if $\boldsymbol{v}$ and $\boldsymbol{w}$ are vectors then $\|\boldsymbol{v} \times \boldsymbol{w}\|$ is equal to the area of the parallelogram formed with sides $\boldsymbol{v}$ and $\boldsymbol{w}$:

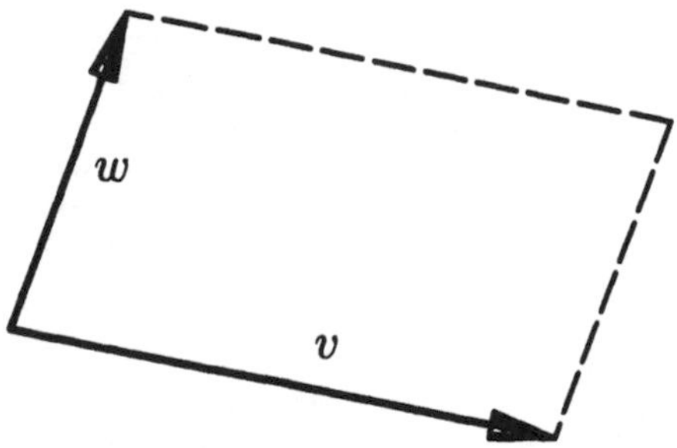

9. Show that if $\boldsymbol{u}$, $\boldsymbol{v}$, and $\boldsymbol{w}$ are vectors, then $|\langle \boldsymbol{u}, \boldsymbol{v} \times \boldsymbol{w} \rangle|$ is the volume of the parallelopiped formed with edges $\boldsymbol{u}$, $\boldsymbol{v}$, and $\boldsymbol{w}$.

10. (a) Is $T\big((x, y)\big) = (x + a, y + b)$ a linear transformation? Prove your answer.

    (b) Is $T\big((x, y, w)\big) = (x/w, y/w)$ a linear transformation? Prove your answer.

11. Suppose S and T are two linear transformations from $\boldsymbol{V}$ to $\boldsymbol{W}$ and $r \in \boldsymbol{R}$. Define $Q(v) = S(v) + T(v)$ and $R(v) = r\big(S(v)\big)$. Show that Q and R are both linear transformations. They are also written $Q = S + T$ and $R = r \cdot S$.

    Further prove that $L(\boldsymbol{V}, \boldsymbol{W})$, the set of linear transformations from $\boldsymbol{V}$ to $\boldsymbol{W}$, is a vector space with $+$ and $\cdot$ defined as above.

12. Prove that composition of two linear transformations, $S \cdot T(\boldsymbol{u}) = S(T(\boldsymbol{u}))$, is a linear transformation. Also, show by counterexample that composition of linear transformations is not a commutative operation.

13. Let $\{\boldsymbol{e}_1, \boldsymbol{e}_2, \boldsymbol{e}_3\}$ be a basis for $\boldsymbol{R}^3$. Given two vectors $\boldsymbol{v}, \boldsymbol{w} \in \boldsymbol{R}^3$, show that $\boldsymbol{v} = \boldsymbol{w}$ if and only if

$$\langle \boldsymbol{v}, \boldsymbol{e}_i \rangle = \langle \boldsymbol{w}, \boldsymbol{e}_i \rangle, \qquad \text{for i = 1,2,3.}$$

14. Let $\{\boldsymbol{u}_1, \boldsymbol{u}_2, \boldsymbol{u}_3\}$ be a basis for $\boldsymbol{R}^3$, and let $\boldsymbol{w} \in \boldsymbol{R}^3$ be arbitrary. Find and justify necessary and sufficient conditions on the basis so that

$$\boldsymbol{w} = \langle \boldsymbol{w}, \boldsymbol{u}_1 \rangle \boldsymbol{u}_1 + \langle \boldsymbol{w}, \boldsymbol{u}_2 \rangle \boldsymbol{u}_2 + \langle \boldsymbol{w}, \boldsymbol{u}_3 \rangle \boldsymbol{u}_3.$$

II

Curves

2

Representation Issues

In the various branches of mathematics, science, and engineering which use curves and surfaces, there are many formulations which occur in every day use. Each has attributes which might be desirable for different parts of the modelling or analysis process; each has drawbacks. In this chapter we investigate some of the more well-known choices. Later chapters might involve blends in which some work is initially defined in one representation and then converted to another. The interrelationship possible among them affects the conversions possible.

When one represents a curve or surface as part of the modelling process, this is a different problem than representing a curve derived from many physical experiments, or even from the clear cut mandate present in numerical analysis to fit a curve with certain abscissas and ordinates. The problems inherent in modelling are many. One problem is that there is frequently no natural coordinate system for the situation. Certain symmetries of one part of the object to be modelled might dictate one particular choice of direction to represent the axes, while symmetries of another part of the same object might dictate another totally different choice of direction. For one part of the problem a certain choice of units might be natural, but for another part a totally different scale might be appropriate.

Another part of the design might involve fitting a curve or surface to certain predetermined points or to a certain predetermined shape in space. Since a shape is not determined by the particular orientation of the coordinate system, it is desirable that the choice of coordinate system not affect the shape of the result.

The choice of a representation scheme must allow for all the considerations which will enter into the designer's specific problem. The designer usually is not well versed in the mathematical and computational issues involved in the various representations. Even after the choice of approximating function space is made, there is always the choice of which particular basis to use. This choice is extremely important since issues of geometric intuition, design ease, computation accuracy, speed of computation, and computer memory storage needs can be affected by the particular choice. Throughout the text there will be discussions about alternative basis schemes for the same function space which will reflect some of these issues. In this chapter, however, we discuss the issue on a much broader scale.

There are initially two choices of representation to make — between a *closed* form and *other*. There are basically three classes of closed form representation for curves and surfaces. They follow the cases for the line and plane discussed earlier and are called the explicit form, the implicit form and the parametric form. All of the closed forms have formulations whereby one can analyze the continuity, derivatives, and geometrical aspects of the curve.

2.1 The *Other* Representation — Data Set Representation

Another form for representing a curve or surface is a data set of ordered points. This form is extremely widespread since it can arise from direct measurements from a physical model, measurements from physical experiments, and mathematical analyses of desired properties.

Unfortunately, since a full curve specification would require an uncountable number of points, what is available is a discretized set of points from the curve. Usually there is some rule by which the values to be discretized are chosen, and almost always there is noise.

This form of representation is basically at the bottom of the complexity hierarchy. It is the most general form, but unfortunately the least useful. Since all that is present is a (perhaps) vast collection of points, their spacing and density affects the quality of the representation. But, the more points that are present, the more space that is needed to store the representation. This trait, while a definite drawback, is not enough to give the method an unfavorable status for design. However, usually if a fine enough data set is derived for an assumed computational or analytical purpose, then frequently another analysis is desired shortly thereafter which requires either finer data, or data specified at different locations, or even information which is impossible to calculate such as derivatives of various orders. This representational form, then, does not lend itself to mathematical analysis. Further, it is not amenable to uses in the area of Computer Aided Geometric Design. If the designer must position each point to be on the surface, the method becomes extremely tedious. Further, how is the designer to know if he has a sufficient number of points?

If analyses must be done on the model, the same difficulties as performing mathematical processes arise. Most computer graphics visualization techniques for objects use polyhedral, planar-sided, faceted objects, as their database. While the data representation can initially seem ideal for this purpose, it is not. If the viewpoint is taken to be close up, small facets will appear on what perhaps should be a smooth surface. If the viewpoint is taken from far away, too much work must be done computing shading for each of many little facets. Lighting and reflectivity models require computation of surface normals, a generalization of the normal for a plane, which can only be approximated by faceted models. Further problems arise. However, this method has been widely used historically whenever there is a lack of a more appropriate higher-level representation.

2.2 Explicit Formulations

Definition 2.1. *A curve formulation is called an* explicit function *if for a given formula of one variable $y = f(x)$, the set of ordered pairs $\{(x, f(x))\}$, for all x in the domain, is a complete description of the curve, and for each first element of the ordered pair, x, there is a unique second element, $f(x)$.*

Familiar examples are the slope-intercept form for a straight line, $y = ax + b$ where a and b are constants, the parabola of form $y = ax^2 + bx + c$, with a, b, and c constants, as well as $y = sinx$, $y = cosx$, $y = e^x$, etc. A complete circle or a lemniscate, however, cannot be represented as an explicit function since for a single x value, there are 2 values of y.

The sketching of this type of curve is easily accomplished. If orthogonal xy-axes are chosen, then for each x the appropriate y value can be calculated and a point sketched. Further, all derivatives that might be necessary for object analysis are immediately available. For example, for a curve $y = f(x)$, the equation of the tangent line to the curve at a point (a, b), where $b = f(a)$ is given by $(y - b)/(x - a) = f'(a)$. Thus a major advantage that this form has is its ease of use and ease in computation. Further, most theory and methods in numerical analysis have been derived for use with this formulation.

When curve or surface fitting is needed and the data fits this formulation type in a well-conditioned way, it is certainly the easiest and best

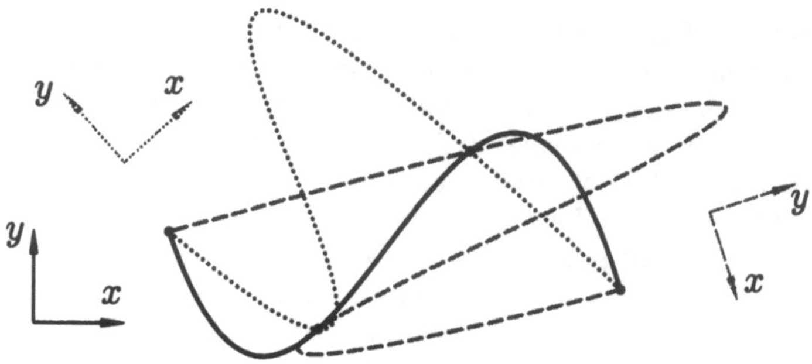

Figure 2.1. Interpolant in three different coordinate systems.

representation to use. Unfortunately, for general use in computer aided geometric design, the above is not generally the case. In particular if it is desired to fit a curve (or surface) to points in space, there is no intrinsic coordinate axis. Figure 2.1 illustrates the problem for a simple four point polynomial interpolation in the plane. Interpolation results in a different curve in each of the three coordinate systems.

Another example shows the difficulties of using an explicit form to do curve approximation.

Example 2.2. Suppose one has n data points, $\{(x_i, y_i)\}_{i=1}^{n}$, but the curve that must represent the data, $f(x)$, has only m (linear) degrees of freedom, where $m < n$. The least squares form of approximation chooses for the answer the function of the form $f(x)$ which has coefficients which minimize the error E, defined as

$$E = \sum_{i=1}^{n} \left| f(x_i) - y_i \right|^2.$$

For example, let $f(x) = \sum_{j=0}^{k} a_j x^j$. Then, for all l between 1 and k,

$$\begin{aligned} 0 &= \frac{\partial E}{\partial a_l} \\ &= \frac{\partial}{\partial a_l} \sum_{i=1}^{n} \left(\sum_{j=1}^{k} a_j x_i^j - y_i \right)^2 \\ &= 2 \sum_{i=1}^{n} \left(\sum_{j=1}^{k} a_j x_i^j - y_i \right) x_i^l, \end{aligned}$$

and we end up with k linear equations in a_j to solve.

If the data is degenerate and can be represented exactly by a function like $f(x)$, then this method will find that function and the error will be zero. Otherwise, note that what is minimized is not the absolute distance between the data points and the curve (a very hard quantity to evaluate), but only y-value differences are minimized.

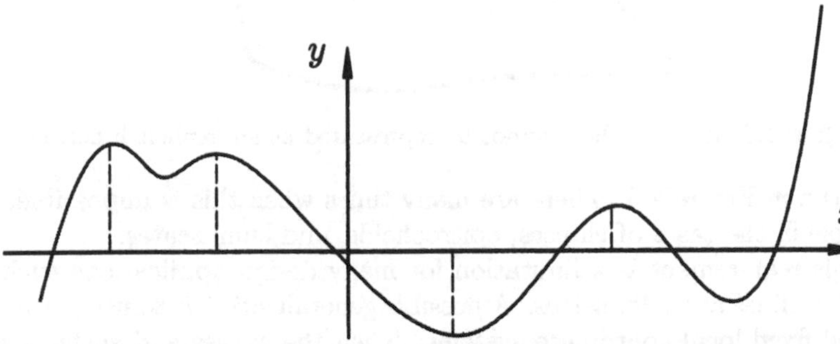

Since only the vertical distance is minimized, if a different coordinate system is used, then the meaning of vertical is changed, and the minimization process gives a different answer. Figure 2.2 shows three different least squares fit curves fit to the same data, but in different coordinate systems. □

The *shape* of the resulting curve is totally dependent on the particular choice of coordinate system. Explicit function fitting fails to satisfy shape invariance under simple rigid movements of the coordinate system. The geometry of the data is not intrinsic to the representation.

There are many curves which cannot be represented as a single explicit function, even if it is allowed to transform coordinate systems to find a *best*,

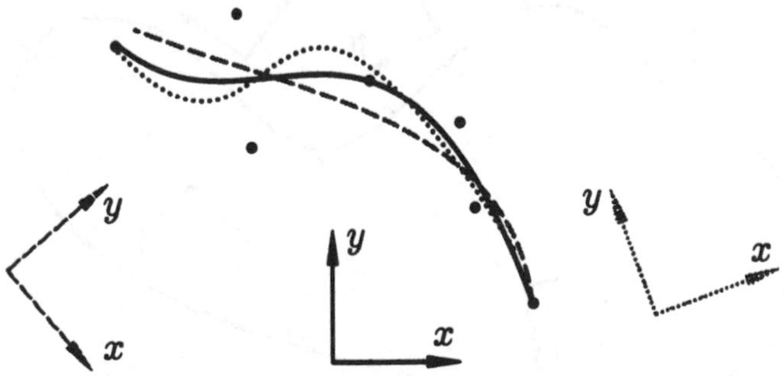

Figure 2.2. Cubic least squares approximations to the same data in three rotated coordinate systems.

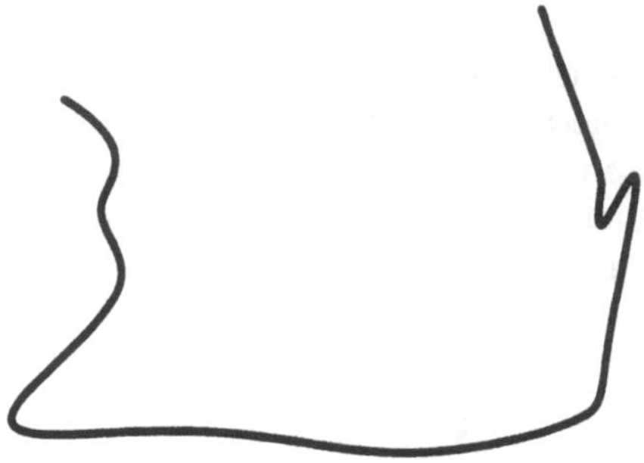

Figure 2.3. A curve that cannot be represented as an explicit function.

as shown in Figure 2.3. There are many times when this is impossible, for example in the cases of ellipses, epitrochoids, and lemniscates.

This requirement is a limitation for many design applications such as closed outlines and silhouettes. A possible generalization is to use piecewise defined fixed local coordinate systems. Then the curves and surfaces can be partitioned into piecewise explicit formulations, as in Figure 2.4

Still, there is not shape invariance in the representation. The particular curves are totally dependent here on the specific breakdown into the local coordinate systems. There is not a single unique pattern which works, nor a single strategy for deciding where to change the coordinate system. There is considerable judgment in the choice of position and orientation of each local coordinate system as well. Further, someone must keep track of the different coordinate systems, the domain ranges within each system, and the function formulation within each coordinate system. This method is used by some computer aided design modelers. It is a first step towards a more general capability than that present in the simple explicit formulation.

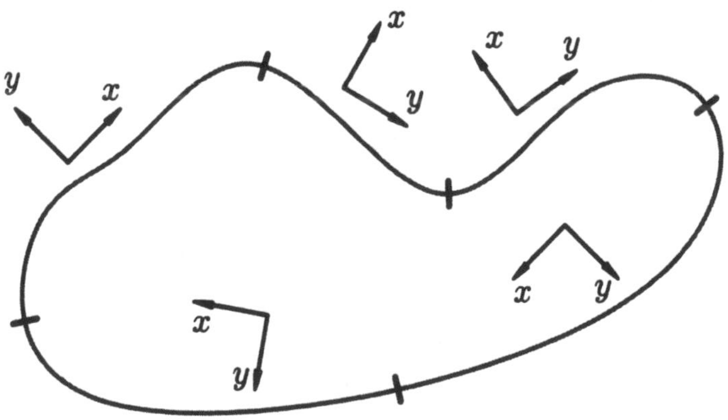

Figure 2.4. Example of piecewise explicit function in local coordinate systems.

Unfortunately, the explicit representation for a curve can be used only if the curve is planar. Since many of the defining curves in design are space curves which do not lie in planes, the explicit formulation is not useful in those cases.

2.3 Implicit Representation

Definition 2.3. *The* implicit curve *(surface) is the solution set to an equation of the form $f(x, y) = 0$ (or $f(x, y, z) = 0$ for surfaces).*

The equations for a line in the plane $ax + by + c = 0$ and a plane in three space $ax + by + cz + d = 0$ are the most common examples of implicit equations, and certainly the easiest to use. This is because the equations are linear and they can be solved for a unique dependent variable quite easily. The circle with center at (a, b) and radius r has formula $(x-a)^2+(y-b)^2 = r^2$ and is also well known. The circle cannot be represented as a single explicit function $y = f(x)$ in any rectangular coordinate system, although it can be done in a piecewise manner.

The form is called *implicit* because for a fixed value of x, only certain values of y (possibly none) will satisfy the equation and hence the graph of the curve is implied by the form $f(x, y) = 0$. In fact if one considers the explicit surface equation $f(x, y) = z$, one sees that the graph of the implicit equation is just the roots of the equation $z = f(x, y)$.

Definition 2.4. *An* implicit bivariate polynomial *of degree n has the form*

$$0 = \sum_{i=0}^{n} \sum_{j=0}^{j=i} a_{i-j,j} x^{i-j} y^j.$$

The number of terms in such an equation is the sum of the series $1+2+\cdots+(n+1) = (n+1)(n+2)/2$. But, any non-zero scalar multiple of all the coefficients yields the same curve so the number of independent conditions is $(n+1)(n+2)/2 - 1 = n(n+3)/2$. That is, there are $n(n+3)/2$ degrees of freedom. In the special case where $n = 2$, there are $2*5/2 = 5$ degrees of freedom. Therefore 5 x-y pairs will uniquely specify any implicit equation of degree 2. Of course there are the special degenerate cases when three or more points are collinear.

Let us develop this equation for the simplest case of all, the case for $n = 1$. The equation in this case is $a_{0,0} + a_{1,0}x + a_{0,1}y = 0$, or as is more commonly written, $ax + by + c = 0$. If $b \neq 0$, this reduces to the common slope-intercept form of $y = -(a/b)x - (c/b)$. The implicit form has the advantage that it treats all orientations the same. There is no favored (or unfavored) slope direction. Further, any scalar multiple of that equation will represent the same line, since the right side equals zero. Hence, the point (x_p, y_p) is on the straight line if and only if $ax_p + by_p + c = 0$. This can be likened to the inner product of generalized vector $(x_p, y_p, 1)$ with the vector (a, b, c). The point represented by the generalized vector is on the line if and only if the inner product of the generalized vector with (a, b, c) is 0, i.e., the generalized vector is perpendicular to the vector representing the line.

This same approach can be tried for quadratic implicit equations. Unfortunately the form is no longer so simple, although it is separable into either of the two following forms:

$$\begin{bmatrix} x^2 & x & 1 \end{bmatrix} \begin{bmatrix} A & 0 & 0 \\ D & B & 0 \\ F & E & C \end{bmatrix} \begin{bmatrix} 1 \\ y \\ y^2 \end{bmatrix} = 0;$$

$$\begin{bmatrix} x & y & 1 \end{bmatrix} \begin{bmatrix} A & B & D \\ 0 & C & E \\ 0 & 0 & F \end{bmatrix} \begin{bmatrix} x \\ y \\ 1 \end{bmatrix} = 0.$$

The second form has $[x\ y\ 1]$ occurring which is reminiscent of the linear case. Unfortunately, it requires two matrix multiplications to test if the point is on the curve. For higher degree implicit equations, no such decomposition, using $[x\ y\ 1]$ is feasible. Thus, even for the simple case of implicit functions which are implicit polynomials, it is very difficult to determine when a point lies on the curve.

The relations between these curves can be very important. One of the most commonly occurring requirements is to know at what points two curves intersect. In conjunction with that, the question must be answered—how many intersection points can there be between two implicit planar polynomial curves?

Theorem 2.5. Bezout's Theorem. *A curve of nth degree and a curve of mth degree intersect in at most $n * m$ points.*

Thus, two quadratics can intersect in 4 points, but a quadratic and a linear equation can intersect in at most 2 points.

It is more difficult to graph a curve specified in implicit form. If one has a point (x, y) on the curve, then one can determine all derivative values at that point. This can lead to numerical estimator techniques for finding approximate points on the curve. Unfortunately, the further one gets from

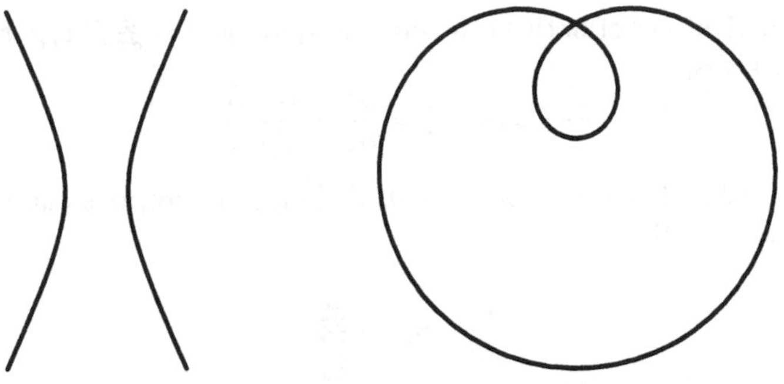

Figure 2.5. The hyperbola in (a) with implicit equation $x^2/2 - y^2/9 - 1 = 0$. Another implicit curve is shown in (b). How to specify part of the inner loop?

the original point on the curve or if the curve is rapidly changing, this method can quickly result in much error, and trouble occurs at singular points.

The immediate question arises: Why bother with implicit functions if they are so hard to evaluate? It turns out that they also have many positive attributes. Many shapes can be defined as zero sets (roots) of surfaces in this way. That is if $f(x, y)$ and $g(x, y)$ are two explicit surfaces, then the implicit equation $f(x, y) - g(x, y) = 0$ defines the domain points of their intersection. They can arise as the result of previous analysis of physical constraints. In addition, they can easily be used to define curves with disconnected parts, branched parts, or loops.

Unfortunately, this form is not intuitive. All conics have the same form, but of course do not look the same. Also, if only a portion of the curve is wanted, it can be difficult to ascertain and specify which portion is needed.

Finally, implicit curves, like explicit curves, are intrinsically planar curves. One can use piecewise implicit equations and also have local coordinate systems to allow generalizations. However, one must once again carry all the information. More often the existence of the implicit equation is a given. Perhaps the design results as the solution to another problem. Thus far, it is the most general form discussed and is used in the TIPS [63, 65] geometric modelling system.

2.3.1 Differentiating Implicit Functions

Finding the derivative to an explicit function is quite straightforward; for an implicit function, the chain rule must be invoked. Suppose for a small piece of the curve that y can be solved as a function of x, that is, $0 \equiv f\big(x, y(x)\big)$.

First the derivatives of both sides must be equal, so $0 = \frac{d}{dx}f\left(x, y(x)\right)$. The chain rule gives,

$$\frac{d}{dx}f\left(x, y(x)\right) = \frac{\partial f}{\partial x} + \frac{\partial f}{\partial y}\frac{dy}{dx}.$$

Now, the lefthand side is identically 0. If $\frac{\partial f}{\partial y}$ is not zero, one can solve for the tangent direction.

$$\frac{dy}{dx} = -\frac{\frac{\partial f}{\partial x}}{\frac{\partial f}{\partial y}}. \tag{2.1}$$

If we want to find the equation for the tangent line at a particular point (a, b) on the curve, we first denote the partial derivative of f with respect to x at (a, b) by $\frac{\partial f}{\partial x}(a, b)$, and the partial derivative of f with respect to y at (a, b) by $\frac{\partial f}{\partial y}(a, b)$. We see that

$$\frac{y-b}{x-a} = -\frac{\frac{\partial f}{\partial x}(a, b)}{\frac{\partial f}{\partial y}(a, b)}$$

or the equation for the tangent line is

$$\frac{\partial f}{\partial y}(a, b)(y-b) + \frac{\partial f}{\partial x}(a, b)(x-a) = 0. \tag{2.2}$$

This last form works even when $\frac{\partial f}{\partial y} = 0$. However if both $\frac{\partial f}{\partial x} = 0$ and $\frac{\partial f}{\partial y} = 0$, then the point is a singular point and a more detailed analysis is required to find one or more tangent directions.

2.3.2 Graphing Implicit Functions

Graphing implicit functions is a problem of root finding; that is, given an implicit function $f(x, y) = 0$ and a value x, one wants to solve for a root of f *as a function of* y. Clearly, this cannot always be done. Further, which root one converges to will be dependent on the initial guess. Further, for each point to be graphed, a rootfinding iterative technique is required. The ideas presented here are derived from a paper by Jordan *et al.* [43]. The goal of the work was to find an algorithm analogous to simple line drawing algorithms for drawing an implicit function on a raster device.

It is assumed that the function $f(x, y)$ is known, and that the point (a, b) satisfies the equation, that is $f(a, b) = 0$. Since the roots of the surface $z = f(x, y)$ form the implicit curve, the idea is to perform a crude Taylor approximation to the surface and move in the direction which gives the

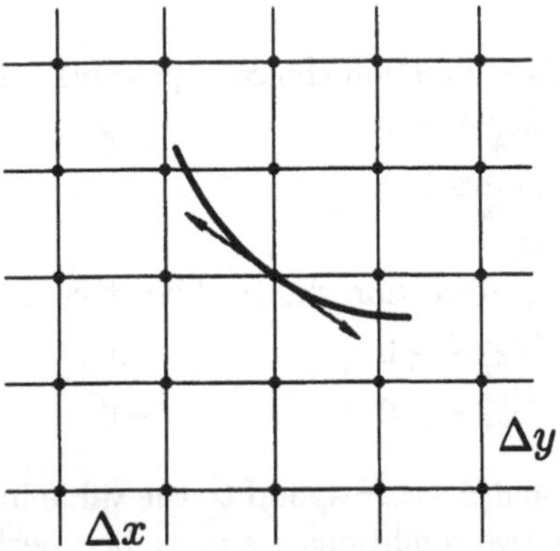

Figure 2.6. The grid and next point decision.

best approximation to 0. It is assumed that $\frac{\partial f}{\partial x}$ and $\frac{\partial f}{\partial y}$ can be determined in closed form, even if they are not fast to evaluate. Now if ξ, $\psi > 0$ are fixed, they can be used to make $\mathbb{R}^2$ into a grid with mesh points at $(a, b) + (i\xi, j\psi)$, where i and j are arbitrary integers.

When the curve is being graphed, one of the grid points $(a, b) + (i\xi, j\psi)$, $i, j \in \{-1, 0, 1\}$, i and j not both zero, must be selected as the next point to be drawn. The decision of which point to draw, and why is of interest. Note that it is highly unlikely that any of those points will lie exactly on the curve, so this is an approximation. However, since the granularity of the grid is arbitrary, it can be a fine approximation. Using Equation 2.1, the tangent vector is

$$\left(-\frac{\partial f}{\partial y}(a, b), \frac{\partial f}{\partial x}(a, b)\right) \quad \text{or} \quad \left(\frac{\partial f}{\partial y}(a, b), -\frac{\partial f}{\partial x}(a, b)\right).$$

These mark the direction of the tangent vector and differ by one vector being the negative of the other. One must select one of these directions, by multiplying either the first coordinate by -1 or the second coordinate by -1. Suppose one calls the direction obtained by multiplying the first coordinate the positive direction. Note, however, that, if the implicit function $g(x, y) = -f(x, y) = 0$ is considered it has the exact same graph, but its positive direction is the negative direction of f. The choice of the next point is then lessened from the eight possible to three, the three around the quadrant in which the tangent vector (with chosen direction) lies. They are the three points given by $(a + \Delta x, b)$, $(a + \Delta x, b + \Delta y)$, and $(a, b + \Delta y)$, where Δx and Δy are determined by the tables below, which corresponds to the slope of the tangent, the choice of f or g above, and the direction chosen to be positive:

For Δx:

partial sign/direction choice	positive	negative
$\frac{\partial f(a,b)}{\partial y} \geq 0$	$-\xi$	ξ
$\frac{\partial f(a,b)}{\partial y} < 0$	ξ	$-\xi$

and for Δy:

partial sign/direction choice	positive	negative
$\frac{\partial f(a,b)}{\partial x} \geq 0$	ψ	$-\psi$
$\frac{\partial f(a,b)}{\partial x} < 0$	$-\psi$	ψ

The values of Δx and Δy correspond to the value in the respective table when the appropriate two conditions are met. It is perhaps easiest to think of this selection process of which point to draw in the context of the surface $z = f(x, y)$. Evaluating the surface one gets values $z_x = f(a + \Delta x, b)$, $z_y = f(a, b + \Delta y)$, and $z_{xy} = f(a + \Delta x, b + \Delta y)$. Clearly, the best choice of the new coordinates will be the pair with $|z|$ value closest to zero. A consistent choice must be made for the three possible tie cases.

Clearly, after the initial point, no point will really lie on the curve. However, at every point a test is made for the closest possibility to the true curve. The tangent line is also only an approximation, after the initial point. The partial derivatives used, however, come from points close in some sense to the real curve. If the surface is slowly changing then this approximation will be close and the method gives reasonable values. See Figure 2.7. A method for terminating the algorithm is needed. If a stop point is given as well as a start point, the program can test for proximity as a stopping criterion. Thus, a closed curve can be drawn completely by requiring it to be drawn twice, giving the same start and stop points each time but the opposite tangent direction.

A method for faster evaluation uses a Taylor expansion for the bivariate function f to evaluate not only the three new function values, but also after the choice of the new point is made, approximations to the partial derivatives of the new point. If ξ and ψ are 1, these evaluations are relatively quick. However, unless a high order Taylor series is used or a low order implicit polynomial equation, such as a conic section, is used, all of the new values are approximations. If the surfaces from which the implicit curves are taken are slowly changing, the error does not compound too quickly and the method gives good results.

As a general algorithm, this method is sensitive to changes in the sizes of ξ and ψ. If the implicit function crosses itself, the method can get confused (Figure 2.7). And if both $\frac{\partial f}{\partial x}$ and $\frac{\partial f}{\partial y}$ vanish at a point, the algorithm falls apart! However, since this does not arise for second degree implicit polynomials, the graphs look quite good.

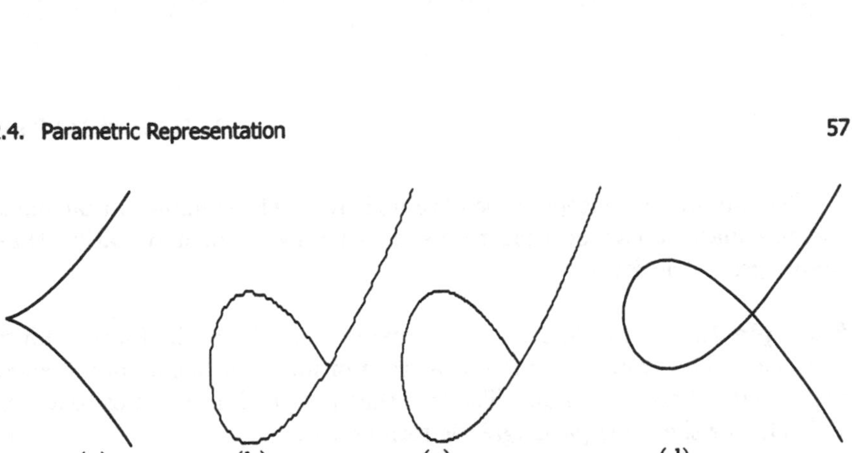

Figure 2.7. A piece of an implicit function. Self intersections might result in missing segments as is demonstrated in (a), (b) and (c); (b) and (c) also show different sampling resolutions of the algorithm; (d) shows the complete curve.

2.4 Parametric Representation

Definition 2.6. *In the* parametric representation *of a curve or surface, each coordinate is given as a function of one or two parameters, i.e.,* $\big(x(t), y(t), z(t)\big)$ *represents a space curve and* $\big(x(u,v), y(u,v), z(u,v)\big)$ *represents a surface.*

If the curve is in the x-y plane one can write $\big(x(t), y(t)\big)$. The common domain of all the *coordinate functions* for a single curve depends on the particular curve and *parameterization*. Such a representation is also called a *vector representation* of a curve or surface.

Example 2.7. An easy method of creating a parametric function is to let $x = t$ and $y = f(t)$, where $y = f(x)$ is some explicit function. An exactly analogous method can be used to create parametric surfaces from the explicit formulation. □

Example 2.8. An example of a parametric function most students of calculus see is the ideal vector description of the position of a projectile as a function of time. It is presented frequently as a planar example with the particle starting out a time $t = 0$ with some initial vector velocity, $\boldsymbol{v}_0$, and initial position at $\boldsymbol{p}_0$. The student is told that only gravity acts on the particle with vector acceleration $\boldsymbol{a}$, and then told to find the equation marking position. This requires integrating the function so that $\boldsymbol{v} = \boldsymbol{v}_0 + \boldsymbol{a}t$ and the position is the second integral $\boldsymbol{p}(t) = \frac{1}{2}\boldsymbol{a}t^2 + \boldsymbol{v}_0 t + \boldsymbol{p}_0$. This is a parametric equation since at any time the coordinate positions are $(\frac{1}{2}\boldsymbol{a}_x t^2 + \boldsymbol{v}_{0,x} t + \boldsymbol{p}_{0,x}, \frac{1}{2}\boldsymbol{a}_y t^2 + \boldsymbol{v}_{0,y} t + \boldsymbol{p}_{0,y})$, if it is presumed to be in the x-y plane. Clearly, the position is a function of time. This particular equation can be solved to find the explicit relationship for y as a function of x. That cannot always be done. □

The important concept to be obtained from this example is that parametric equations can be used to glean other pieces of information than simply geometric form.

Example 2.9. In particular note the example of the path of a pendulum of a clock. The graph, or curve, is a trace of its path, but it cannot show the velocity of the pendulum. The fact that the pendulum comes to a stop and retraces the same path again is totally lost. □

Example 2.10. Suppose a milling machine is to cut a particular path through some stock. In this case the path of the tip marks the curve in space. However, the parameterization of that path can decide much more. If the tip is moved too fast, it can lead to bad cuts or a broken tip. If it is moved too slow, it is a waste of time. If it is moved at a nonuniform rate then the cut path may be smoother in some parts than others. Paths that have high curvature in some parts while slow curvature in others are particularly prone to problems. The density of the material being cut also affects the speed at which the path can properly be cut. □

Example 2.11. Another common example is the speed of traversal of a car over a fixed road or a train over train tracks. The position of the vehicle as it moves along the track as a function of time is a parameterization of the curve which is the track. The speed is a one-dimensional quantity, but the direction can be backward as well as forward, as in the pendulum case. □

Definition 2.12. *The* arc length parameterization *of a curve is a parameterization that assigns the position of the curve as a function of the length of the curve measured from a specific point.*

Example 2.13. The arc length parameterization of a circle. The position is measured from the point $(0, R)$ on the curve. If a is the angle, measured clockwise, of a radius from the origin to the desired position,

$$\frac{a}{2\pi} = \frac{s}{2\pi R}$$

so

$$x = R\sin a = R\sin(s/R) \qquad \text{and} \qquad y = R\cos a = R\cos(s/R)$$

where s measures the arc length. For a circle, the arc length is the shorter length of its circumference between two specific points. □

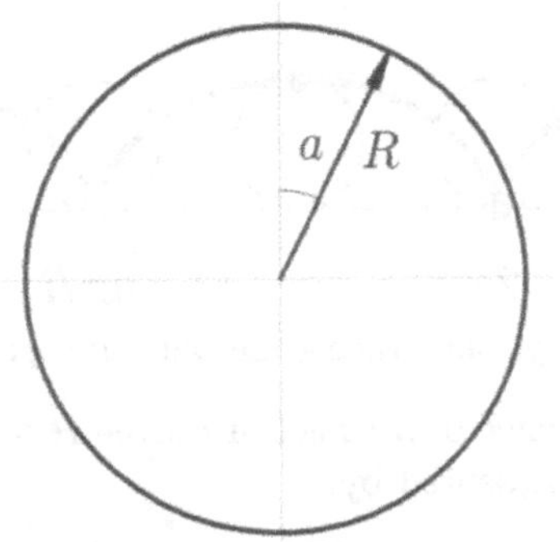

Figure 2.8. A circle in arc length parameterization.

Example 2.14. The curves in Figure 2.9 do not resemble each other, however their parametric forms look quite close.

The curve $(x, y) = (\sin\phi, \sin 2\phi)$ using trigonometric relations can be modified to: $\sin 2\phi = 2\sin\phi\cos\phi$ and $\cos\phi = \sqrt{1-\sin^2\phi}$ so $y = 2x\sqrt{1-x^2}$.

Now, the curve $(x, y) = (\sin\phi, \cos 2\phi)$ which looks almost the same can again be shown to be $y = (1 - x^2)/2$. □

The family of curves of the form

$$(x(t), y(t)) = (cos(\omega_x t + \psi_x), cos(\omega_y t + \psi_y)),$$

that includes the curves of this example, is known as the family of *Lissajous* curves.

Example 2.15. The cycloid is a curve which marks the path of a fixed

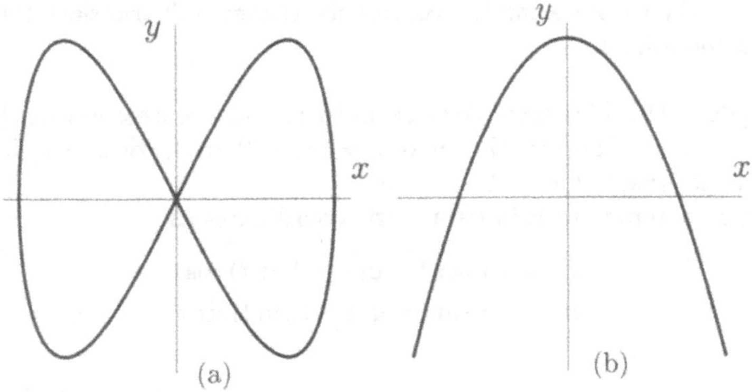

Figure 2.9. Graphs of (a) $(x, y) = (\sin\phi, \sin 2\phi)$, $\phi \in [0, 2\pi]$ and (b) $(x, y) = (\sin\phi, \cos 2\phi)$, $\phi \in [-\frac{\pi}{2}, \frac{\pi}{2}]$.

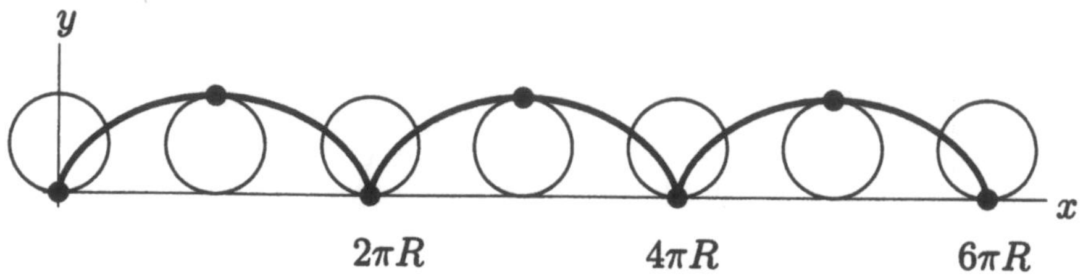

Figure 2.10. A cycloid: tracing the path of a point on a wheel.

point on the circumference of a wheel of radius R during a rotation of the wheel. This can be represented by:

$$\begin{aligned} x &= R(\phi - \sin\phi) \\ y &= R(1 - \cos\phi). \end{aligned}$$

This form can be turned into the explicit formulation

$$x = R\cos^{-1}\left(\frac{R-y}{R}\right) - \sqrt{2Ry - y^2}.$$

While y cannot easily be represented as a function of x, the other way is possible. □

A parametric representation for a curve or surface is not unique. There are many different parameterizations which lead to the same curve (or surface). In Chapter 4 different parameterizations and their properties are explored in detail.

The parametric form is axis independent. If a curve can be represented as $\sum \boldsymbol{p}_i c_i(t)$ where the $\boldsymbol{p}_i$ are vector coefficients to scalar parametric functions c_i, then a rotation or scale of the coordinate system will rotate, or scale the $\boldsymbol{p}_i$, and the shape of the curve will not be affected. If in addition, $\sum c_i = 1$, then translating the coordinate system will translate the curve without modification.

Example 2.16. This example is useful to compare representations for just a planar curve. Consider the curve $r = a(1 - 2\sin t)$, for $a > 0$, in polar coordinates. (See Figure 2.11.)

One may represent it in parametric coordinates as

$$\begin{aligned} x &= r\cos t = a(1 - 2\sin t)\cos t \\ y &= r\sin t = a(1 - 2\sin t)\sin t \end{aligned}$$

or

$$\begin{aligned} x &= a(\cos t - \sin 2t) \\ y &= a\big(\sin t + (\cos 2t - 1)\big). \end{aligned}$$

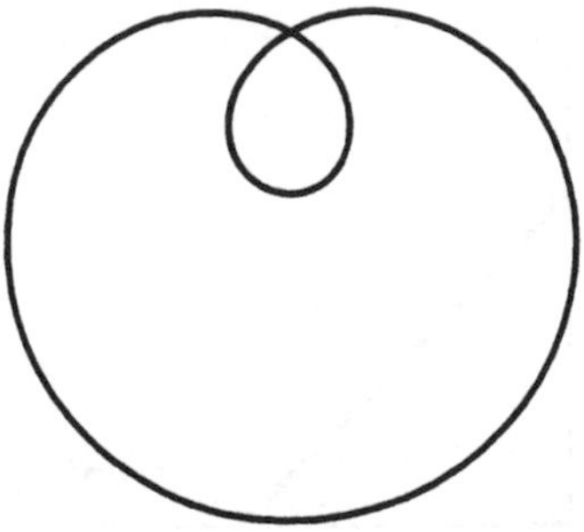

Figure 2.11. $r = a(1 - 2\sin t)$.

This representation easily can be graphed as shown in Figure 2.11. To find an implicit equation requires using trigonometric relations and much inventiveness, but results in

$$\sqrt{x^2 + y^2} = a\left(1 - \frac{2y}{\sqrt{x^2 + y^2}}\right),$$

so

$$\left(x^2 + y^2 + 2ay\right)^2 = a^2\left(x^2 + y^2\right).$$

This implicit equation is not separable, and so cannot be turned into explicit formulation in a straightforward manner. From the graph, one would not expect it to be so. A visual estimate requires at least four separate equations.

If one wanted to use a portion of this curve for design, how could that be accomplished? In the implicit form, it is quite hard, since it is difficult to determine any point in any branch. Even if a section could be identified, how could one keep track of it? For a specific x there might be 0, 1, 2, 3, or 4 points on the curve that correspond. Starting and ending values of points are again problematic. In the parametric form, this curve is well defined and a section is easily determined. One could easily use a parameter range, say of $a \le t \le b$ to give the appropriate segment of the curve. □

Unfortunately, the parametric representation has its own difficulties. The parametric form is less intuitive to use than the explicit form. Remember, it can be used to represent the position as a function of time, or the position as a function of distance along the curve. This raises another facet. A parametric parameterization is not unique. While the design or analysis to be performed might use this to advantage, it becomes a much more difficult problem to tell if the graphs of two curves are the same even if they might be traced out in a different order.

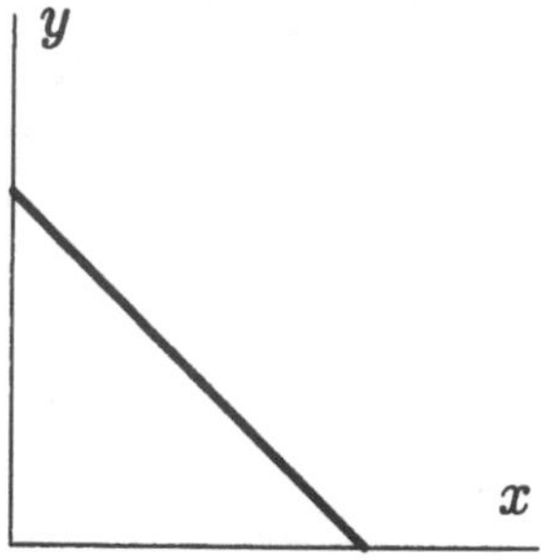

Figure 2.12. The curve $(\cos^2 \pi t, \sin^2 \pi t)$.

Example 2.17. Consider the curve $(\cos^2 \pi t, \sin^2 \pi t)$ in Figure 2.12 which has the form of a trigonometric curve. Using the trigonometric identity, $\cos^2 a + \sin^2 a = 1$ for all real values a, one finds the curve is identical to the *line* $x + y = 1$, (in coordinates $(x, 1-x)$) where $0 \leq x$ and $0 \leq y$. □

Thus, a particular curve can have infinitely many different parameterizations. The identification of equivalence classes of parameterizations is discussed later. Another difficulty with the use of parametric formulations occurs in parametric data fitting. If data is to be used for parametric interpolation or least squares, some parameter values must be assigned to the data. If the data can be assigned an x parameterization which corresponds to the explicit formulation, then that is sometimes done. This cannot be done if the data does not correspond to an explicitly ordered interpretation. Sometimes data values have been taken with uniform intervals of time between them. This time is sometimes used as a parameterization, more generally called a uniform parameterization. Finally, a different approach is used that creates a parameterization, called the chord length parameterization, that corresponds to an approximation of the arc length parameterization. The reasoning in this last method is that the chord length approximates a length of a smooth curve between the points. The following example illustrates the difference in the parameterizations.

Example 2.18. Interpolate three points by a quadratic polynomial. Data: $(0,0)$, $(1,1)$, $(3,0)$, in that order.

Parameter Values			
Parameterization:	1	2	3
Point			
$(0,0)$	0	0	0
$(1,1)$	1	1	$\sqrt{2}$
$(3,0)$	3	2	$\sqrt{2}+\sqrt{5}$

Thus Parameterization 1 imitates the behavior of an explicit function while Parameterization 2 is called the uniform spacing parameterization. Parameterization 3 is called the chord length parameterization and approximates an arc length parameterization.

In general, the strategy in parametric interpolation is to find the unique polynomials $p_x(t)$ and $p_y(t)$ so that $p_x(t_i) = x_i$ and $p_y(t_i) = y_i$, for all i. In this particular case it corresponds to having

$$\begin{aligned} p_x(t) &= a_x t^2 + b_x t + c_x, \\ p_y(t) &= a_y t^2 + b_y t + c_y. \end{aligned}$$

We study each of the above three parameterizations and note the effects on the final curve.

Since Parameterization 1 is the same as the explicit case, and because interpolation has a unique result, $p_x(t) = t$. To find $p_y(t)$, we see that

$$\begin{aligned} p_y(0) &= 0 = a_y 0^2 + b_y 0 + c_y, \\ p_y(1) &= 1 = a_y 1^2 + b_y 1 + c_y, \\ p_y(3) &= 0 = a_y 3^2 + b_y 3 + c_y. \end{aligned}$$

This can be rewritten as

$$\begin{bmatrix} 0 & 0 & 1 \\ 1 & 1 & 1 \\ 9 & 3 & 1 \end{bmatrix} \begin{bmatrix} a_y \\ b_y \\ c_y \end{bmatrix} = \begin{bmatrix} 0 \\ 1 \\ 0 \end{bmatrix}.$$

The solution to that system of equations then yields the values for the coefficients. It is clear that $c_y = 0$ from the first row. We are left with

$$\begin{aligned} a_y + b_y &= 1, \\ 3a_y + b_y &= 0. \end{aligned}$$

Since $a_y = 1 - b_y$, $0 = 3(1 - b_y) + b_y = 3 - 2b_y$ and $b_y = 3/2$, so $a_y = -1/2$, and $p_y(t) = (-1/2)t^2 + (3/2)t$.

Parameterization 2 must solve two systems of linear equations. Each system has different right sides, but the same matrix. Since $p_x(0) = 0 = p_y(0)$, $c_x = c_y = 0$. Now,

$$\begin{bmatrix} 1 & 1 \\ 4 & 2 \end{bmatrix} \begin{bmatrix} a_x & a_y \\ b_x & b_y \end{bmatrix} = \begin{bmatrix} 1 & 1 \\ 3 & 0 \end{bmatrix}.$$

Solving yields $\left((1/2)(t^2 + t), -t^2 + 2t\right)$.

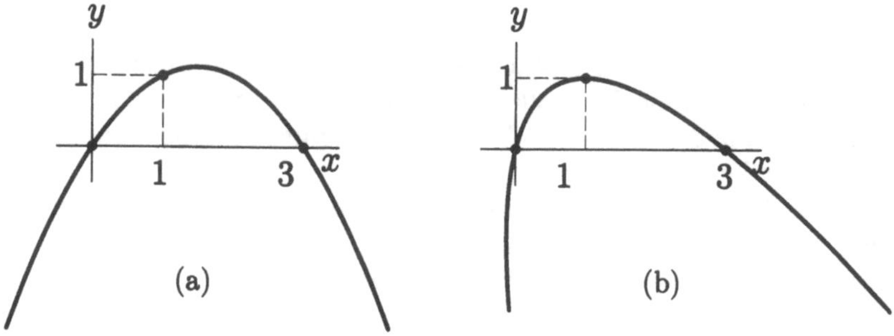

Figure 2.13. (a) Explicit parameterization; (b) uniform parameterization.

This particular form can be solved to find an implicit form for the curve using the following techniques.

$$\begin{aligned} 2x &= t^2 + t \\ y &= -t^2 + 2t \end{aligned}$$

so

$$t = \frac{2x + y}{3} \quad \text{and} \quad \left(\frac{2x + y}{3}\right)^2 + \frac{2x + y}{3} = 2x$$

and finally,

$$4x^2 + 4xy + y^2 - 12x + 3y = 0.$$

We shall later show that this is a skewed parabola. Unfortunately, solving for an implicit form cannot be done for general parametric equations, and is not easily done even for slightly more complicated cases.

Parameterization 3 is the chord length parameterization to the data and once again has $p_x(0) = 0 = p_y(0)$ so $c_x = 0 = c_y$. The other equations can be represented as a system in the same style as above:

$$\begin{bmatrix} 2 & \sqrt{2} \\ 7 + 2 * \sqrt{10} & \sqrt{2} + \sqrt{5} \end{bmatrix} \begin{bmatrix} a_x & a_y \\ b_x & b_y \end{bmatrix} = \begin{bmatrix} 1 & 1 \\ 3 & 0 \end{bmatrix}.$$

Solving this gives

$$\begin{aligned} (p_x(t), p_y(t)) &= \frac{1}{2 * \sqrt{5} - 5 * \sqrt{2}} \Big((2 * \sqrt{2} - \sqrt{5})t^2 + (1 + 2 * \sqrt{10})t, \\ &\qquad -(9 * \sqrt{2} + 6 * \sqrt{5})t^2 + 2(7 + 2 * \sqrt{10})t \Big). \end{aligned}$$

Figure 2.14(b) shows all three interpolants superimposed on the same figure to clarify the relationships of the interpolants.

From just this simple case, it is clear that the particular choice of parameterization is an important ingredient in the output of the final parametric interpolant. □

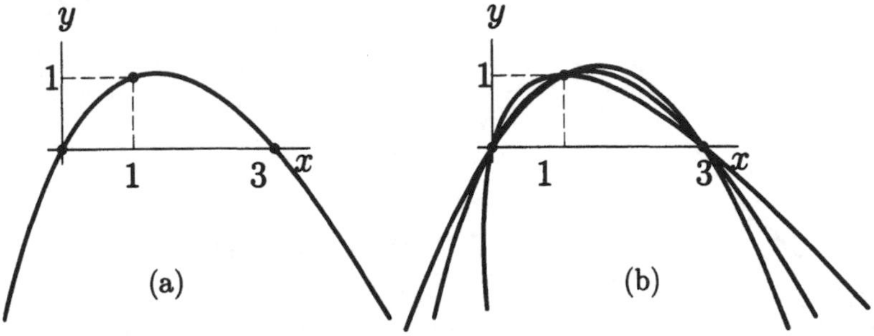

Figure 2.14. (a) Chord length parameterization; (b) all three parameterizations superimposed.

The derivative of a parametric curve is a another parametric curve, a vector quantity. An explicit or implicit curve or surface derivative gives direction information. For a parametric formulation, the derivative is also a vector and gives a *magnitude* as well as a direction. The magnitude corresponds to the speed of traversal along the curve. The larger the tangent, the faster the speed. For a plane curve the ratio of the y-component to the x-component gives the slope of the tangent vector.

Many researchers, developers, and systems in computer aided geometric design use parametric curves and surfaces. When used in *ab initio* design, the flexibility inherent in the parametric form is relied upon. When curves and surfaces must be fit, however, one must determine a parameterization to the pre-existing data. This process is difficult, but when no systems constraints dictate otherwise, chord length approximations to the data produce curves that seem reasonable. A final reason to use parametric curve formulation is that when one must have space curves, only parametric curves can be used, or one must represent the curve as an intersection of two explicit surfaces. This last form is equivalent to finding the zeros of a surface, i.e., an implicit curve formulation.

As a last example of the generalization and differences between parametric and explicit, we look at some properties of the curve as a function of parameterization.

It is known that if $f(x)$ is an explicit curve, and if $f'(x)$ exists and is continuous everywhere in the domain, then the curve $f(x)$ has no cusps. Is the same true for a parametric curve? That is, does a continuous γ' (tangent vector) assure that the curve γ has no cusps? We look at some examples. (See Figure 2.15.) The curve

$$\gamma(t) = (x(t), y(t)) = (t^2, t^3)$$

has as implicit form

$$x^3 - y^2 = 0.$$

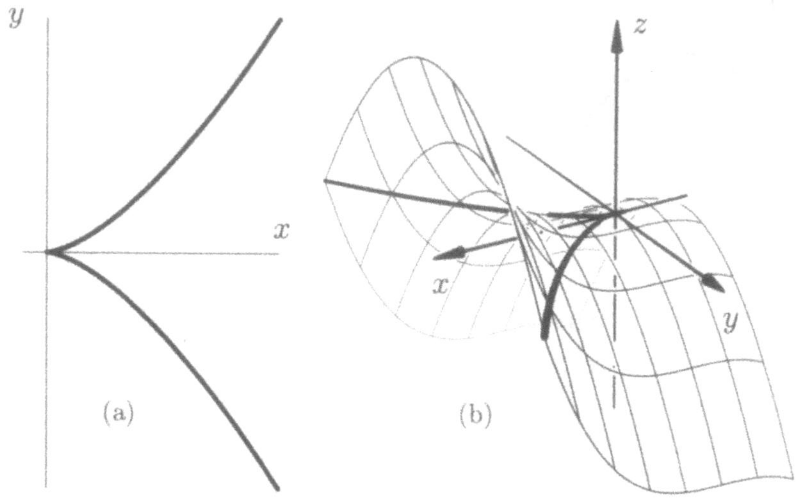

Figure 2.15. (a) Graph of (t^2, t^3); (b) explicit surface $z = x^3 - y^2$ and its zero set.

It has derivative

$$\gamma'(t) = (x'(t), y'(t)) = (2t, 3t^2)$$

which is continuous and well defined, but γ has a cusp at $t = 0$.

Note that $\gamma' = (0,0)$ at $t = 0$. Remember that for a planar curve the ratio of the y-derivative to the x-derivative gives the direction of the explicit tangent. If the tangent vanishes, at just a point, that ratio cannot be determined. A cusp can occur only where the tangent vanishes. However, if the tangent vanishes, there need not be a cusp, as for the case $(x(t), y(t)) = (t^2, t^4)$, which has explicit form $y = x^2$ for positive x.

Next we ask if a parametrically defined curve that has no cusps can have a discontinuous tangent function. Consider the curve

$$(x(t), y(t)) = \begin{cases} (t, t), & \text{for } t < 1, \\ (t^2, t^2), & \text{for } t \geq 1. \end{cases}$$

The graph of this curve is the straight line $y = x$. The derivative of the curve is $(1, 1)$ for $t < 1$ and $(2t, 2t)$ for $t \geq 1$. The directions of the two pieces are the same at $t = 1$, but the *magnitudes* are different!

It is true that a curve parametrized by arc length has no cusps if and only if the tangent vector is continuous. This will be treated in greater detail in the section on differential geometry. However, another result less restrictive on the parameterization is known. A parametric curve is free of any cusps if the tangent function is continuous and never vanishes. Conversely, a curve has a cusp if the tangent function is discontinuous for every parameterization for which the tangent vector is nonvanishing over the curve.

2.5 Shape Approximation

The term *shape approximation* means different things to mathematicians, designers, computer scientists, the layman, and others. By shape approximation we mean fidelity to the shape of the original, or *primitive* curve that exists either in closed form (by equations), in data form, or in ideal form (in the mind of the designer). Further, the approximation should be *close* to the original, if possible. Some factors which are used to characterize the idea of shape approximation include:

- Point precision: Is it desirable to require that the approximation pass through certain points on the curve?
- Derivative continuity: Will requiring first, second, or arbitrary degree continuity, frequently called smoothness, of the approximation provide shape fidelity?
- Closeness measure: Will requiring only that the approximation be close to the original function positions in some prespecified ways insure shape fidelity?
- Generalized closeness: Will requiring the approximation to be close in position and also in certain derivatives insure shape fidelity?
- Is it always possible to obtain a good fit? How many degrees of freedom will be necessary in order to get a good fit?

As we shall show later, polynomial interpolation can produce undulations. One may counter that by trying to, as is done commonly, *use more points.* Unfortunately, this strategy frequently fails. It is true that for each function $f \in C^0$, and each n, there exist sequences of points $\left\{x_j^n\right\}_{j=0}^n$ which are the defining points for a sequence of interpolants p_n to f and for which

$$\lim_{n\to\infty} \|p_n - f\|_{L^\infty} = 0.$$

Thus, for any degree, a good set of interpolating abscissa can be found. However, the shape of f is not preserved at each step. Also, the interpolation points vary completely for different n, so straightforwardly adding

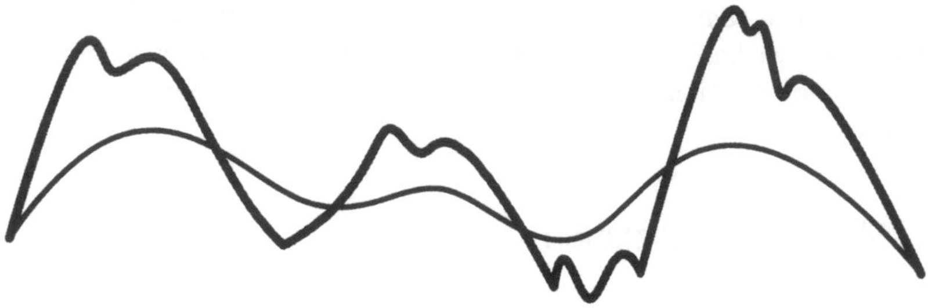

Figure 2.16. A variation diminishing approximation

more points is not guaranteed to give both interpolation and shape. In fact, it might be desirable to require, somehow, that the approximant undulate *no more* than the original. While it may be desirable to require position exactness sometimes, it is known, however, that position interpolation with polynomials *cannot* guarantee shape fidelity although eventual convergence for high enough n can occur.

A major problem of polynomial interpolation is that it is not possible to guarantee any fidelity to the shape of the original curve (or to the piecewise linear approximation to the original data). However, it is possible to characterize approximation schemes which do preserve shape in the following definition.

Definition 2.19. *We introduce the notion of* variation diminishing. *If $a(x)$ is an approximation to a function $f(x)$, a is called variation diminishing if a straight line can intersect $a(x)$ no more often than it intersects $f(x)$.*

Derivative continuity is a form of mathematical smoothness. Consider, however, that polynomials are in $C^{(\infty)}$, and they certainly can undulate. Requiring the approximant to be of a very high continuity class by itself will not guarantee fidelity to a curve of high continuity. We know that piecewise linear interpolants can converge to a curve given enough points to interpolate at some reasonable spacing, no matter how high or low the degree of continuity of the original. It is true, however, that it is possible to see second derivative discontinuities in surfaces when they are shaded realistically. The old saying that the Detroit automobile designer can look down the hood of the model and see second derivative discontinuities is true! People with less finely developed senses can also see these problems on suitably shaded images. Derivative continuity then certainly may be of interest in any shape with aesthetic requirements, and perhaps in shapes on which one might want to perform mathematical analyses.

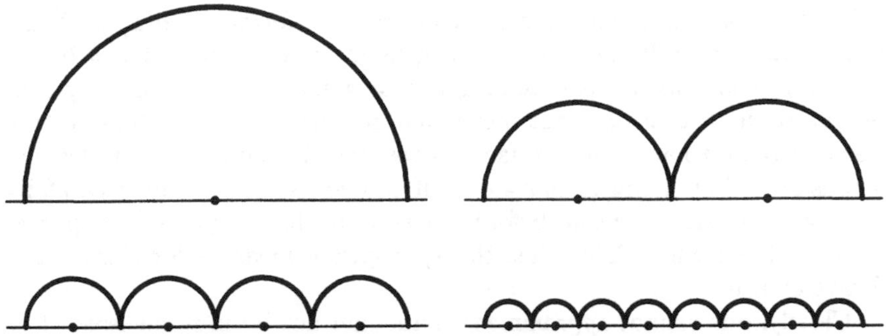

Figure 2.17. The function f_i, semicircles of radius $1/2^i$, $i = 0, \ldots, 3$.

Results relating intervals on which a curve is monotonic and for which curvature can be determined to derivative properties show that characteristics of the first and second derivatives yield important information about the shape of the curve. But it is not clear how they can be used to attain *shape approximation*.

One cannot approximate a feature of a function without being in some sense close to that function. The appropriate meaning for a particular problem is sometimes difficult to ascertain. It is important to be sure that the right closeness measure is used as this proof that $\pi = 2$ illustrates. As we know, π really equals $3.14159\ldots$, so this result is rather extraordinary. The logic of that argument follows. Consider the semicircle with the representation $f(x) = \sqrt{1-(x-1)^2}$. Its center is at the point $(1,0)$ and it has radius 1.

Its arc length can be determined to be π since the circumference of a circle is πd where d is the diameter. Thus, the circumference of a semicircle is $\pi r = \pi$ for this example. We also let $d(x) = 0$, for $x \in [0,2]$ represent the diameter function. We shall construct a sequence of piecewise circular functions which approximate $d(x)$. Let $f_0(x) = f(x)$, and let

$$f_1(x) = \begin{cases} \sqrt{1/2-(x-1/2)^2}, & x \in [0,1), \\ \sqrt{1/2-(x-3/2)^2}, & x \in [1,2]. \end{cases}$$

In general, let $f_n(x) = \sqrt{\frac{1}{2^n} - \left(x - \frac{2k+1}{2^n}\right)^2}$ when $x \in \left[\frac{2k}{2^n}, \frac{2(k+1)}{2^n}\right)$, for $k = 0, \ldots, 2^n - 1$.

Then each f_n has arc length the sum of all arc lengths of its piecewise semicircles which is π. Further,

$$\sup\left|f_n(x) - d(x)\right| \le \frac{1}{2^n}.$$

It is clear that the sequence $f_n(x)$ converges uniformly to $d(x) \equiv 0$, i.e. $\lim_{n\to\infty} \equiv 0$ on $[0,2]$. Thus, the arclengths must converge and $\pi = 2$!

The problem with this reasoning is that while $f_n(x)$ is uniformly converging to $d(x)$, that does not mean that the arc lengths of the sequence are converging to anything. In fact, as we saw, the sum of the arc lengths is constant. The moral of this story, of course, is that we must carefully investigate the desired result before choosing the desired process of approximation. Even when doing this, the appropriate measure for close is not always known!

Clearly, one cannot approximate shape and not be in some sense close to the original. One must ask if that sense is at all appropriate, and if so, is it a sufficient measure for the problem at hand? Least squares is a method of approximation which minimizes the sum of the squares of the distances between the approximant and the original, at certain points, so the approximant is close in that sense. Here one is assessing *closeness* only at certain discrete points, so the behavior of the original at other locations is unknown. Again, positional closeness is not enough.

Exercises

1. Given two points $P_1 = (1,0)$ and $P_2 = (0,1)$, each of the following four equations is a different parameterization for the line passing through those points:

   (a) $\gamma(t) = (x(t), y(t)) = (cos^2\pi t, sin^2\pi t)$,

   (b) $\gamma(t) = tP_1 + (1-t)P_2$,

   (c) $\gamma(t) = t^2P_1 + (1-t^2)P_2$,

   (d) $\gamma(t) = \frac{1}{1+t^2}P_1 + \frac{t^2}{1+t^2}P_2$

   For each parameterization:

   (a) What values of t are valid?

   (b) What range of t keeps the curves between P_1 and P_2?

   (c) Graph $\gamma(t)$ by plotting some points with various values of t. Label each point with its values of x, y, and t.

   (d) Suppose we were to plot the curve on a graphics screen by taking a nondecreasing series of t values t_i, and drawing straight line segments between $\gamma(t_i)$ and $\gamma(t_{i+1})$. What advantages or disadvantages are there for the given parameterization?

2. The curves below are given in parametric form. Find an implicit form for each curve. Sketch each curve.
   A. $x = t^2 + 1$ $y = t(t+1)$
   B. $x = \sec t$ $y = \tan t$
   C. $x = 3\cos t$ $y = 2\sin t$
   D. $x = \frac{t^2-1}{t^2+1}$ $y = \frac{2t}{t^2+1}$

3. The curves below are in implicit form. Find a parameterization for each curve. Sketch each curve.
   A. $x^2 - y^3 = 0$
   B. $x^2 + 4y^2 - 6x + 16y + 24 = 0$
   C. $3x^2 - 24x - y + 50 = 0$

4. Programming Exercise: Develop a menu-driven, interactive graphics program to draw planar curves that are specified implicitly. Use the method developed in Section 2.3.2. Provide several examples of implicit functions to be drawn.

3

Conic Sections

3.1 The Use of Conic Sections in Design

Conic sections have long been used in design. It is sometimes necessary to have an exact circle or part of a circle, while at other times only a smooth curve is necessary. The circle, ellipse, parabola, hyperbola, and the degenerate forms of lines and points give a variety of forms. Further, since, as we shall show, the general conic has five degrees of freedom, a quartic explicit polynomial would be necessary to fit a portion of the curve. A conic section has many easily found representations and it is easy to compute points along a curve with relatively fast, accurate methods. Perhaps more important than the capability to compute the particular curve is the fact that conic sections have many geometric properties which are desirable in design. One such property is that a conic can have no inflection points. Thus, one can be certain that no extraneous inflection points will be introduced when using conics, so designed curves will not have unintentional ripples. Unfortunately they also have limitations. If higher numbers of degrees of freedom are desired, piecewise methods must be used. One cannot have higher order continuity than, usually $C^{(1)}$, or rarely $C^{(2)}$. The lack of inflection points can be just as much a hinderance as an aid in certain applications. Also, the conic section is intrinsically a planar curve which means many pieces must be used to approximate a space curve. Despite all of these deficiencies, they have been used widely and continue to be used. For these reasons one should have an understanding of them.

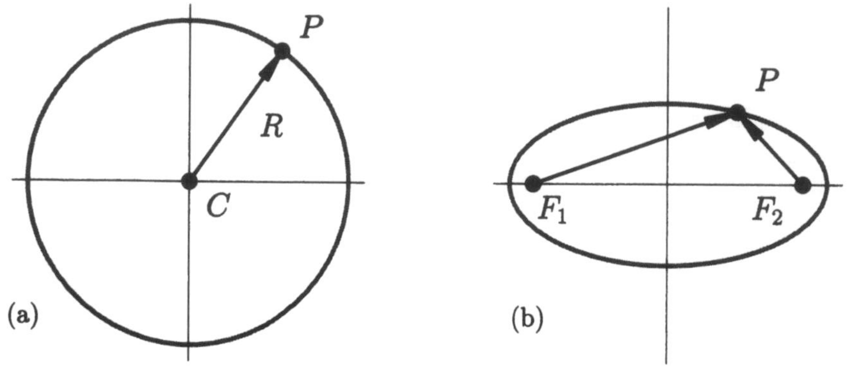

Figure 3.1. (a) Circle centered at C; (b) the ellipse.

3.2 Locus of Points Definitions

Several planar curve shapes which we will later show to be conic sections are defined as "The set of all points which..." In the material below we give those definitions and find mathematical equations which represent the same curves. Since the definitions do not imply a coordinate system, we shall derive the forms using the simplest coordinate system that satisfies the definition requirements.

3.2.1 The Circle

Definition 3.1. *Given a point C in a plane and a number $R \geq 0$, the* circle *with center C and radius R is defined as the set of all points in the plane at distance R from the point C. In set notation we write,*

$$\{P = (x, y) : \|P - C\| = R\}.$$

If $C = (x_c, y_c)$, we use the definition of Euclidean distance to see that (x, y) is a point on the circle if and only if $(x - x_c)^2 + (y - y_c)^2 = R^2$.

If $R = 0$, the circle degenerates to the point (x_c, y_c).

3.2.2 The Ellipse

Definition 3.2. *Given two points, F_1 and F_2 called the* foci *and a number $K \geq \|F_2 - F_1\|$, an* ellipse *is defined as the set of all points whose distances from the foci sum to K. That is,*

$$\{P = (x, y) : \|P - F_1\| + \|P - F_2\| = K\}.$$

We see that a circle is a kind of ellipse where $F_1 = F_2 = C$, and $K = 2R$. Also the straight line segment connecting the two points F_1 and F_2 is an ellipse since if $P = (1-t)F_1 + tF_2$ and $0 \le t \le 1$, then

$$\|P - F_1\| = t\|F_2 - F_1\|$$

and

$$\|P - F_2\| = (1-t)\|F_2 - F_1\|,$$

so

$$\begin{aligned} \|P - F_1\| + \|P - F_2\| &= t\|F_2 - F_1\| + (1-t)\|F_2 - F_1\| \\ &= \|F_2 - F_1\| \\ &= K. \end{aligned}$$

For ease in computation we shall use $F_1 = (-c, 0)$ and $F_2 = (c, 0)$, where $c > 0$. See Figure 3.1 (b).

By the definition,

$$K = \sqrt{(x+c)^2 + y^2} + \sqrt{(x-c)^2 + y^2},$$

so we will remove the radicals in two steps:

$$x^2 + 2xc + c^2 + y^2 = K^2 - 2K\sqrt{(x-c)^2 + y^2} + x^2 - 2xc + c^2 + y^2$$

so

$$4xc - K^2 = -2K\sqrt{(x-c)^2 + y^2}.$$

Squaring both sides again,

$$\begin{aligned} 16x^2c^2 - 8K^2xc + K^4 &= 4K^2x^2 - 8K^2xc + 4K^2c^2 + 4K^2y^2; \\ K^4 - 4K^2c^2 &= 4(K^2 - 4c^2)x^2 + 4K^2y^2; \\ K^2\left(K^2 - 4c^2\right) &= 4(K^2 - 4c^2)x^2 + 4K^2y^2. \end{aligned} \tag{3.1}$$

There are three cases to deal with here. Case 1 occurs when $K^2 - 4c^2 > 0$; Case 2 is when $K^2 - 4c^2 < 0$; and Case 3 is when $K^2 - 4c^2 = 0$.

Case 3 results when $2c = K$. Since $2c$ is the distance separating the foci, $K = 2c$ results in the line segment connecting the two foci, which has already been discussed above. When $2c > K$, then case 2 results. The set of points which can satisfy this condition is empty, since for any point P, $2c = \|F_1 - F_2\| = \|F_1 - P + P - F_2\| \le \|F_1 - P\| + \|F_2 - P\| = K$. Hence, K cannot be less than $2c$.

Thus, only Case 1 is left, that is $K^2 - 4c^2 > 0$. Continuing from Equation 3.1,

$$1 = \frac{4x^2}{K^2} + \frac{4y^2}{K^2 - 4c^2}.$$

Letting $a = K/2$ and $b^2 = a^2 - c^2$, the following equation results:

$$1 = \frac{x^2}{a^2} + \frac{y^2}{b^2}.$$

We see that the points $(-a, 0)$ and $(a, 0)$ are on the ellipse; there are no points on the ellipse with $|x| > a$; and the ellipse is symmetric about the x-axis. Analogously, the points $(0, -b)$ and $(0, b)$ are on the ellipse and there are no points on the ellipse with $|y| > b$, and the ellipse is symmetric about the y-axis. The axis containing the foci is called the *major axis* and the axis through the center orthogonal to the major axis is called the *minor axis.* If the points F_1 and F_2 had been chosen on the y-axis, then the same equation would result, except $b > a$. If F_1 and F_2 are chosen to be translations of either of the above cases, then the origin would translate to the average of F_1 and F_2, so, if $C = (F_1 + F_2)/2 = (c_x, c_y)$, then the translated ellipse can be written

$$1 = \frac{(x - c_x)^2}{a^2} + \frac{(y - c_y)^2}{b^2}.$$

If the two foci lie along a line which is not parallel to either axis, of the form $y = mx + b$, more complex equations result which we shall treat later.

3.2.3 The Hyperbola

Definition 3.3. *Given two points, F_1 and F_2 called the* foci, *and a number $K \neq 0$ a* hyperbola *is defined as the set of all points the difference of whose distances from the foci is K. That is,*

$$\{ P = (x, y) : \|P - F_1\| - \|P - F_2\| = \pm K \}.$$

If $K = 0$, then this hyperbola is the set of points equidistant from both F_1 and F_2, the *bisector line.*

For ease in computation, just as for ellipses, we shall use $F_1 = (-c, 0)$ and $F_2 = (c, 0)$, where $c > 0$. See Figure 3.2.

The definition of hyperbola is

$$\pm K = \sqrt{(x + c)^2 + y^2} - \sqrt{(x - c)^2 + y^2}.$$

We remove the radical in two steps in an analogous method to the ellipse.

$$\begin{aligned} x^2 + 2xc + c^2 + y^2 &= K^2 \pm 2K\sqrt{(x - c)^2 + y^2} + x^2 - 2xc + c^2 + y^2, \\ 4xc - K^2 &= \pm 2K\sqrt{(x - c)^2 + y^2}, \end{aligned}$$

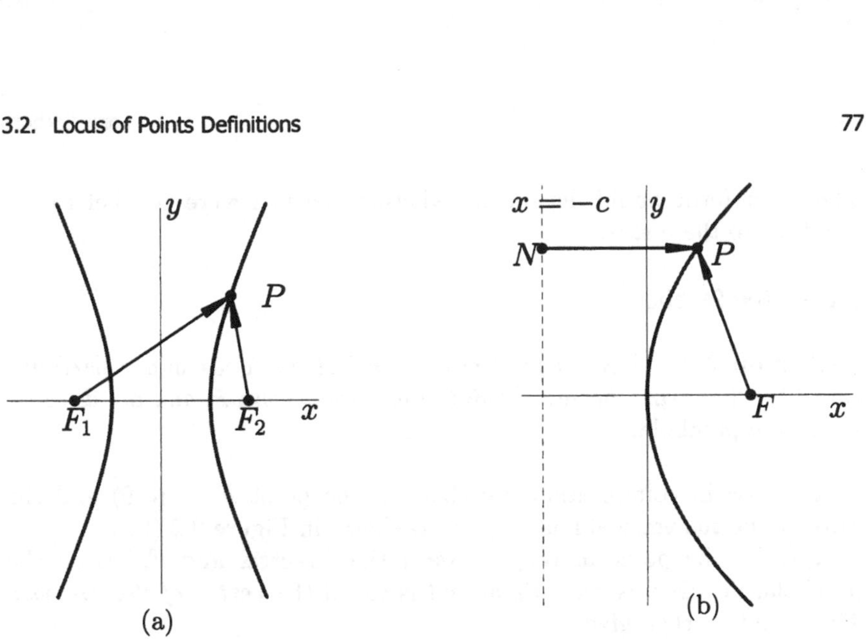

Figure 3.2. (a) The hyperbola; (b) the parabola.

$$
\begin{aligned}
16x^2c^2 - 8K^2xc + K^4 &= 4K^2x^2 - 8K^2xc + 4K^2c^2 + 4K^2y^2,\\
K^4 - 4K^2c^2 &= 4(K^2 - 4c^2)x^2 + 4K^2y^2,\\
K^2\left(K^2 - 4c^2\right) &= 4(K^2 - 4c^2)x^2 + 4K^2y^2.
\end{aligned}
$$

Suppose $K = \pm 2c$. Then the answer reduces to $y^2 = 0$, or the line $y = 0$. However, the points $(x, 0)$, $-c < x < c$ cannot satisfy this equation since the difference of the distances is always less than $2c$. Thus the points $(x, 0)$ for $x \leq -c$ and $x \geq c$ satisfy the requirement for the locus of points.

Now either $K^2 > 4c^2$ or $K^2 < 4c^2$. For any point P, $2c = \|F_1 - F_2\| = \|F_1 - P + P - F_2\| \geq \max\{\|F_1 - P\| - \|F_2 - P\|, \|F_2 - P\| - \|F_1 - P\|\} = K$, so K^2 cannot be larger than $4c^2$. Thus, $K^2 < 4c^2$. Let $a = K/2$ and $b^2 = c^2 - a^2$, then

$$1 = \frac{x^2}{a^2} - \frac{y^2}{b^2}, \tag{3.2}$$

and the denominator of y^2 in Equation 3.2 must be positive.

The points $(-a, 0)$ and $(a, 0)$ are on the hyperbola, and $a^2 + b^2 = c^2$, where c is the distance of the foci from the center axis. We see that the hyperbola is symmetric about the x- and y- axes, as was the ellipse.

Again we see that if the points F_1 and F_2 had been chosen on the y-axis, then an analogous equation would result with the roles of x and y interchanged.

If they are chosen so that $C = (F_1 + F_2)/2 = (c_x, c_y)$, then the equation for the hyperbola would be either

$$1 = \frac{(x - c_x)^2}{a^2} - \frac{(y - c_y)^2}{b^2} \quad \text{or} \quad 1 = \frac{(y - c_y)^2}{a^2} - \frac{(x - c_x)^2}{b^2}.$$

The exact form would depend on whether the foci were parallel to the x-axis or to the y-axis.

3.2.4 The Parabola

Definition 3.4. *Given a fixed point F called the* focus *and a fixed line called the* directrix, *the set of points equidistant from F and the directrix is called a* parabola.

For ease in computation, we shall let the point $F = (c, 0)$ and the directrix be the vertical line $x = -c$, as shown in Figure 3.2 (b) .

Clearly, the point midway between the directrix and F lies on the parabola, in this case the origin, and is called the *vertex of the parabola.* The definition then gives,

$$\begin{aligned} \|P - N\| &= \|P - F\|, \\ x + c &= \sqrt{(x-c)^2 + y^2}. \end{aligned}$$

Simplify by squaring both sides.

$$\begin{aligned} x^2 + 2xc + c^2 &= x^2 - 2xc + c^2 + y^2, \\ 4xc &= y^2. \end{aligned}$$

Again, we see that if the focus is placed on the y-axis and the directrix is a horizontal line, the roles of x and y are interchanged.

3.3 Conic Sections

We have defined several types of curves and derived their mathematical equations, but we have yet to present why they have the name *conic sections.* We shall rectify that now.

Consider a cone with apex at the origin and axis at the z-axis. It is defined by the equation $z^2 = m^2(x^2 + y^2)$, where m is any positive real number. We shall consider the intersection curves of that cone with various planes. While the cone is defined to be double-sided, for the sake of clarity, the illustrations show only one side. Since the cone is radially symmetric, for our construction we can choose planes which make calculation easy without affecting the shape of the results.

First consider intersecting the cone with the plane $x = k$, a plane parallel to the y-z plane. The resulting curve has the form $z^2 = m^2(k^2 + y^2)$, or $z^2/m^2 - y^2 = k^2$. If $k = 0$, then the curve degenerates to the form

$z = \pm my$, or the crossed straight lines which form the silhouette of the cone, through the origin. (See Figure 3.6.) When $k \neq 0$, then the curve takes the form $z^2/(m^2k^2) - y^2/k^2 = 1$, which we have shown is a hyperbola with foci at $\left(k, 0, -k\sqrt{1+m^2}\right)$ and $\left(k, 0, k\sqrt{1+m^2}\right)$. Notice that the x-value is fixed, so the curve is in a plane parallel to the y-z plane. As $k \to 0$, we see that the hyperbola have sharper and sharper noses until they become the pair of crossed lines at $k = 0$.

If we intersect the cone with planes of the form $z = b$, parallel to the x-y plane, then the result is $x^2 + y^2 = (b/m)^2$, the equation of a circle with center at the origin and radius b/m. If $b = 0$, the intersection is just the apex point of the cone.

Now consider planes of the form $z = ax + b$, where $a \neq 0$. The intersection of these planes with the cone form other conic sections:

$$\begin{aligned} m^2(x^2+y^2) &= (ax+b)^2 \\ &= a^2x^2 + 2abx + b^2, \\ m^2y^2 &= (a^2 - m^2)x^2 + 2abx + b^2. \end{aligned} \tag{3.3}$$

There are three cases to consider here.

If $a = \pm m$, then the intersection of the cone and the plane is a curve $y^2 = \frac{2abx}{m^2} + \frac{b^2}{m^2}$, a parabola. If, in addition, $b = 0$, that is the origin is in the plane, then $y^2 = 0$. Then the curve is the set of points $(x, 0, ax)$. It is here that the plane is tangent to the cone. If $|a| > m$ and $b = 0$, then the curve reduces to $y = \pm x\sqrt{(a^2 - m^2)/m^2}$, a pair of lines crossing at the origin.

If $a^2 < m^2$, then we set $r^2 = m^2 - a^2$ and rewrite Equation 3.3 to get

$$\begin{aligned} m^2y^2 &= -((m^2 - a^2)x^2 - 2abx) + b^2 \\ &= -(r^2x^2 - 2abx) + b^2. \end{aligned}$$

Completing the squares gives

$$m^2y^2 = -\left(rx - \frac{ab}{r}\right)^2 + \left(\frac{ab}{r}\right)^2 + b^2$$

and

$$r^2m^2y^2 + \left(r^2x - ab\right)^2 = b^2m^2.$$

Or,

$$\frac{\left(x - \frac{ab}{m^2-a^2}\right)^2}{\frac{b^2m^2}{(m^2-a^2)^2}} + \frac{y^2}{\frac{b^2}{m^2-a^2}} = 1,$$

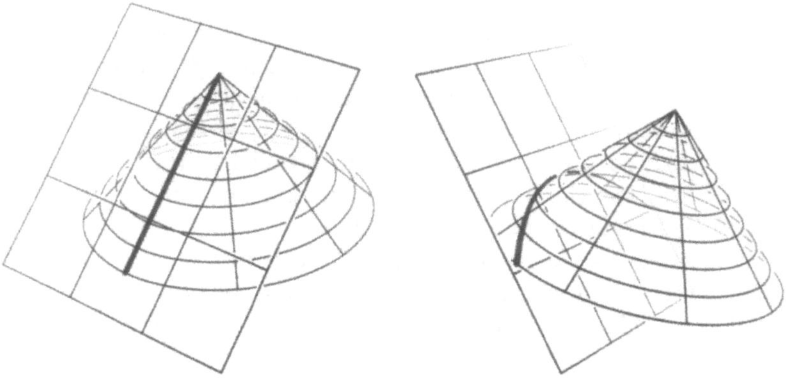

(a) (b)

Figure 3.3. (a) The straight line conic; (b) the parabolic conic.

which we have already shown is an ellipse with center at $\left(ab/(m^2-a^2), 0\right)$. The major axis is parallel to the x-axis, since the denominator of x^2 is larger than the denominator of y^2.

Now, if $a^2 > m^2$, set $r^2 = a^2 - m^2$. Rewriting Equation 3.3 and completing the squares,

$$\begin{aligned} m^2y^2 &= r^2x^2 + 2abx + b^2 \\ &= \left(rx + \frac{ab}{r}\right)^2 + b^2 - \left(\frac{ab}{r}\right)^2 \end{aligned}$$

and

$$(r^2x + ab)^2 - r^2m^2y^2 = b^2m^2.$$

If $b = 0$ then the equation immediately reduces to $x = \pm my/r$, which are crossed lines.

Otherwise, for $b \neq 0$, this can be rewritten to the standard form,

$$\frac{\left(x + \frac{ab}{a^2-m^2}\right)^2}{\frac{b^2m^2}{(a^2-m^2)^2}} - \frac{y^2}{\frac{b^2}{a^2-m^2}} = 1,$$

and this curve is a hyperbola with center at $\left(ab/(a^2-m^2), 0\right)$, and major axis along the x-axis.

Now we have seen two ways to get hyperbolas as conic sections.

We see that circles, points, lines, crossed pairs of lines, ellipses, and hyperbolas are all conic sections.

One can derive any one of the nondegenerate forms from any other through a perspective transformation, which is commonly used in computer

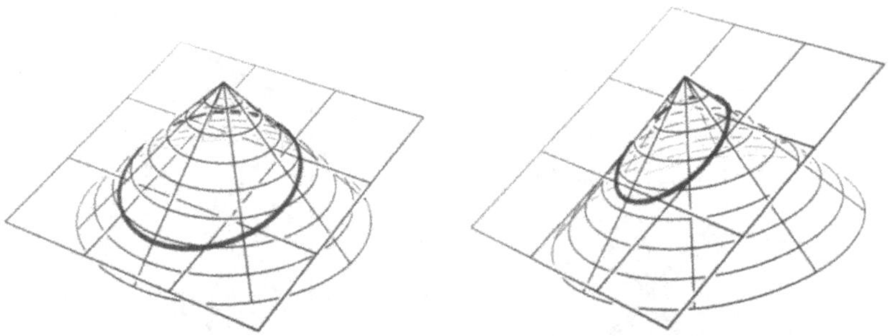

Figure 3.4. Two different ellipses resulting from intersections with different planes.

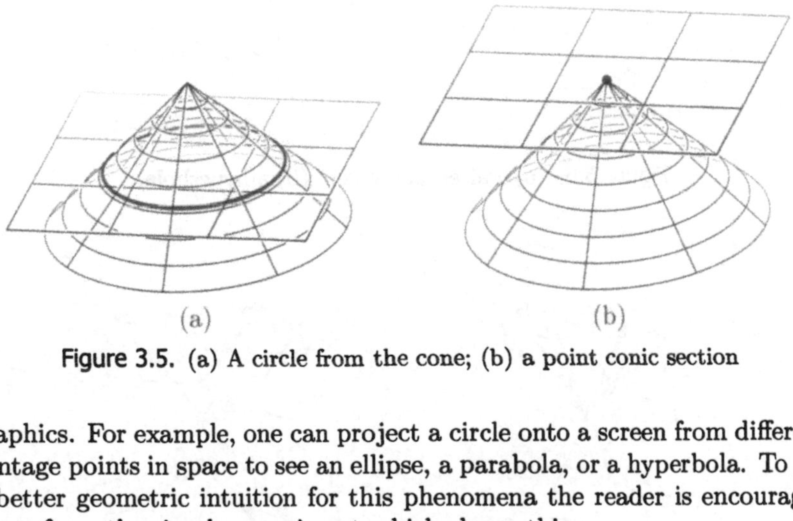

(a) (b)

Figure 3.5. (a) A circle from the cone; (b) a point conic section

graphics. For example, one can project a circle onto a screen from different vantage points in space to see an ellipse, a parabola, or a hyperbola. To get a better geometric intuition for this phenomena the reader is encouraged to perform the simple experiment which shows this.

Example 3.5. The purpose of this example is to examine shadows of circles. The equipment needed is a large fixed screen, a rigid movable circular hoop, and a point light source, which is adequately modelled by an incandescent light bulb. The experimenter holds the hoop between the light and the screen. He then can move the hoop and observe the different curve shapes projected. The factors that will affect the shape are the distance of the hoop from the light, the angle between the hoop center and the horizontal, the angle the hoop makes with the vertical screen, and the distance of the hoop center from the shortest line connecting the light and the screen.

A more analytical discussion of this phenomenon and derivations of mathematical equivalences will follow later. □

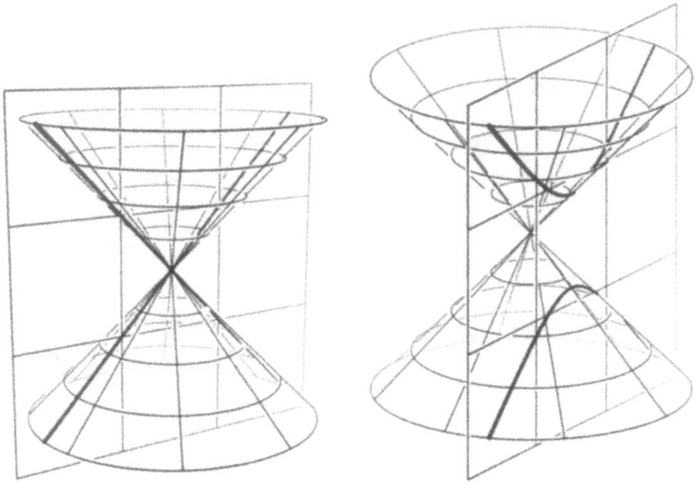

(a) (b)

Figure 3.6. (a) Paired line conic; (b) a hyperbola.

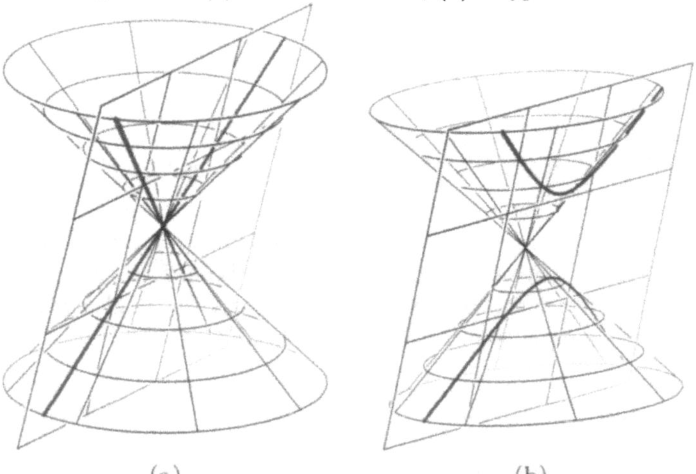

(a) (b)

Figure 3.7. (a) Slanted crossed lines; (b) slanted hyperbola.

3.4 Implicit Quadratic Functions as Conics

We saw in previous sections that all conic sections which had axes parallel to the x- or y-axes can be represented as second degree implicit functions. Now the converse is shown, namely, that all implicit quadratic functions

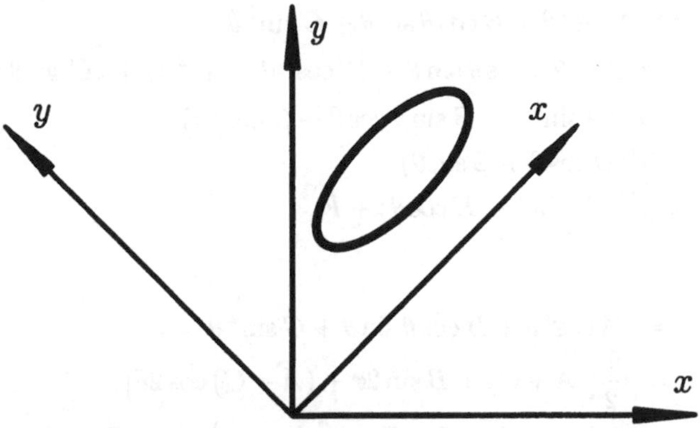

Figure 3.8. Original and rotated (in gray) coordinate systems.

are just rotated conic sections. Thus, all the specification and construction techniques which apply to implicit quadratic polynomials are just methods to construct ellipses, hyperbola, circles, and parabolas.

We start with the arbitrary implicit quadratic,

$$Ax^2 + Bxy + Cy^2 + Dx + Ey + F = 0,$$

and determine if there is a change of basis from (x, y) to (x', y') so that the same graph is drawn by the curve

$$A'x'^2 + C'y'^2 + D'x' + E'y' + F' = 0$$

in the new coordinate system. First, suppose that a simple rotation will do, where we can write (x, y) in terms of the new coordinates (x', y') and the rotation angle θ.

That can be done by writing

$$\begin{aligned} x &= x' \cos\theta - y' \sin\theta, \\ y &= x' \sin\theta + y' \cos\theta, \end{aligned}$$

and substituting into the original form. As yet it is unknown what value θ should be. The result of the substitution is

$$\begin{aligned} 0 = \; & A(x' \cos\theta - y' \sin\theta)^2 + B(x' \cos\theta - y' \sin\theta)(x' \sin\theta + y' \cos\theta) \\ & + C(x' \sin\theta + y' \cos\theta)^2 + D(x' \cos\theta - y' \sin\theta) \\ & + E(x' \sin\theta + y' \cos\theta) + F. \end{aligned}$$

If that expansion is multiplied out and regrouped according to coefficients of x'^2, $x'y'$, y'^2, x', y', and 1, the equation takes the form

$$\begin{aligned}0 &= x'^2\left(A\cos^2\theta + B\cos\theta\sin\theta + C\sin^2\theta\right)\\ &\quad + x'y'\left[-2A\cos\theta\sin\theta + B\left(\cos^2\theta - \sin^2\theta\right) + 2C\sin\theta\cos\theta\right]\\ &\quad + y'^2\left(A\sin^2\theta - B\sin\theta\cos\theta + C\cos^2\theta\right)\\ &\quad + x'(D\cos\theta + E\sin\theta)\\ &\quad + y'(-D\sin\theta + E\cos\theta) + F.\end{aligned}$$

Set

$$\begin{aligned}A' &= A\cos^2\theta + B\cos\theta\sin\theta + C\sin^2\theta\\ &= \frac{1}{2}\left[(A+C) + B\sin 2\theta + (A-C)\cos 2\theta\right];\\ B' &= -2A\cos\theta\sin\theta + B\left(\cos^2\theta - \sin^2\theta\right) + 2C\sin\theta\cos\theta\\ &= B\cos 2\theta - (A-C)\sin 2\theta;\\ C' &= A\sin^2\theta - B\sin\theta\cos\theta + C\cos^2\theta\\ &= \frac{1}{2}\left[(A+C) - B\sin 2\theta - (A-C)\cos 2\theta\right];\\ D' &= D\cos\theta + E\sin\theta;\\ E' &= -D\sin\theta + E\cos\theta;\\ F' &= F.\end{aligned}$$

If the coefficient of $x'y'$ is zero, then the equation can be written as a sum of squares (each a polynomial) plus a constant term.

Thus, the primed coefficients are the new coefficients of the conic in the new coordinate system. But, this whole effort is invoked to see if there is any rotated coordinate system in which we can write the implicit quadratic for which $B' = 0$.

If we set $B' = 0$, we can try to solve to see if there is an angle θ through which the coordinate system can be rotated:

$$B' = 0 = B\cos 2\theta - (A-C)\sin 2\theta.$$

If $B = 0$ then set $\theta = 0$ since no rotation is necessary. Now assume that $B \neq 0$. If $A = C$, then $\cos 2\theta = 0$, and $\theta = \pi/4$ satisfies the equation. Finally if $A \neq C$, setting $\theta = \frac{1}{2}\tan^{-1}(B/(A-C))$ will satisfy the equation.

Example 3.6. Find the primed coefficients in the rotated coordinate system for

$$\begin{aligned}f_1(x,y) &= \frac{1}{2}\left[x^2 + xy + y^2 - x - y\right] = 0,\\ g_1(x,y) &= \frac{1}{2}\left[x^2 - xy + y^2 + x + y - 2\right] = 0.\end{aligned}$$

Since $A = C$ for both f_1 and g_1, the cosine of 2θ must be zero; hence, if we set $\theta = \pi/4$ the cross term will disappear from both:

$$\sin(\pi/4) = \cos(\pi/4) = \sqrt{2}/2, \qquad \sin 2(\pi/4) = 1, \qquad \cos 2(\pi/4) = 0.$$

Thus,

$$\begin{array}{lll} A_1' & = & 1/2(2 + 1 * 1) = 3/2 \\ B_1' & = & 0 \\ C_1' & = & 1/2(2 - 1 * 1) = 1/2 \\ D_1' & = & -2(\sqrt{2}/2) \\ E_1' & = & 0 \\ F_1' & = & 0 \end{array} \qquad \begin{array}{lll} A_2' & = & 1/2(2 + (-1) * 1) = 1/2 \\ B_2' & = & 0 \\ C_2' & = & 1/2(2 - (-1) * 1) = 3/2 \\ D_2' & = & 2(\sqrt{2}/2) \\ E_2' & = & 0 \\ F_2' & = & -2 \end{array}$$

The new equations are

$$\begin{aligned} f_1(x', y') &= \frac{\left(x' - \frac{\sqrt{2}}{3}\right)^2}{1/3} + y'^2 - 2/3 = 0, \\ g_1(x', y') &= \left(x' + \sqrt{2}\right)^2 + \frac{(y')^2}{1/3} - 6 = 0. \end{aligned}$$

□

Since the shape of a curve is not affected by rotations of the coordinate system, and we can find θ for which $B' = 0$, let us assume that $B = 0$ in the original curve, that is,

$$Ax^2 + Cy^2 + Dx + Ey + F = 0.$$

If $A = C = 0$, then the curve is the implicit line equation. If only one of them is zero, that is, if $A = 0$ or $C = 0$, it is evident that the curve is a parabola from the simplification that occurs in the general equation. So now let us assume, without loss of generality that $AC \neq 0$. Use the method of completing the squares to rewrite the equation and suppose $Ax^2 + Cy^2 + Dx + Ey + F \equiv A(x + a)^2 + C(y + b)^2 + G$, where a, b, and G are not yet known. Then,

$$(D - 2aA)x + (E - 2bC)y + \left(F - Aa^2 - Cb^2 - G\right) \equiv 0.$$

This requires that

$$\begin{aligned} 2aA = D \quad &\text{yielding} \quad a = \frac{D}{2A}, \\ 2bC = E \quad &\text{yielding} \quad b = \frac{E}{2C}, \end{aligned}$$

and

$$G = F - Aa^2 - Cb^2.$$

Substituting for a and b,

$$\begin{aligned} Aa^2 &= A\left(\frac{D}{2A}\right)^2 = \frac{D^2}{4A}, \\ Cb^2 &= C\left(\frac{E}{2C}\right)^2 = \frac{E^2}{4C}. \end{aligned}$$

This results in

$$\begin{aligned} 0 &= A(x+a)^2 + C(y+b)^2 + G \\ &= A\left(x+\frac{D}{2A}\right)^2 + C\left(y+\frac{E}{2C}\right)^2 + \left(F - \frac{D^2}{4A} - \frac{E^2}{4C}\right) \\ &= A\left(x+\frac{D}{2A}\right)^2 + C\left(y+\frac{E}{2C}\right)^2 + \frac{4ACF - CD^2 - AE^2}{4AC}. \end{aligned}$$

Since $AC \neq 0$, one can consider the following two cases:

Case I: $AC > 0$, that is $A > 0$, $C > 0$ or $A < 0$, $C < 0$,

Case II: $AC < 0$, that is $A > 0$, $C < 0$ or $A < 0$, $C > 0$.

For Case I,

$$A\left(x+\frac{D}{2A}\right)^2 + C\left(y+\frac{E}{2C}\right)^2 = -\frac{4ACF - CD^2 - AE^2}{4AC}.$$

If $A > 0$ and $C > 0$, then this is the equation of an ellipse if the right hand side is positive. If the right hand side is zero, then only a single point can satisfy the equation. Otherwise the graph of the curve is empty. That is, it can be satisfied for no points in the plane. If $A < 0$ and $C < 0$, then the right hand side must be negative in order to solve for the ellipse.

For Case II, if $A > 0$ but $C < 0$,

$$|A|\left(x+\frac{D}{2A}\right)^2 - |C|\left(y+\frac{E}{2C}\right)^2 = \frac{4ACF - CD^2 - AE^2}{4|AC|}.$$

This is clearly a hyperbola regardless of the sign of the right hand side. An analogous result holds for $A < 0$ and $C > 0$. If the right hand side equals zero, then the hyperbola degenerates to two crossed lines.

Thus, we see that in the special case that $B = 0$, the sign of AC determines whether the curve is a parabola, a hyperbola, or an ellipse. We ask if that can be done in the more general case in which $B \neq 0$.

Suppose that we have the original equation with no information as to whether or not B is zero, and a rotated equation through an arbitrary angle θ, with primed coefficients. Let us consider the quantities $B'^2 - 4A'C'$ in terms of the unprimed coefficients.

$$\begin{aligned} B'^2 - 4A'C' = & \left(B^2\cos^2 2\theta - 2B(A-C)\cos 2\theta \sin 2\theta + (A-C)^2 \sin^2 2\theta\right) \\ & - 4M_1M_2 \end{aligned}$$

where

$$\begin{aligned} M_1 &= \frac{(A+C) + B\sin 2\theta + (A-C)\cos 2\theta}{2}, \\ M_2 &= \frac{(A+C) - B\sin 2\theta - (A-C)\cos 2\theta}{2}. \end{aligned}$$

Now,

$$\begin{aligned} 4M_1M_2 &= (A+C)^2 - (A+C)B\sin 2\theta - (A+C)(A-C)\cos 2\theta \\ &\quad + (A+C)B\sin 2\theta - B^2\sin^2 2\theta - B(A-C)\sin 2\theta\cos 2\theta \\ &\quad + (A-C)(A+C)\cos 2\theta - B(A-C)\cos 2\theta \sin 2\theta \\ &\quad - (A-C)^2\cos^2 2\theta \\ &= (A+C)^2 - B^2\sin^2 2\theta - 2B(A-C)\cos 2\theta\sin 2\theta \\ &\quad - (A-C)^2\cos^2 2\theta. \end{aligned}$$

Substituting in the original equation for $4M_1M_2$ gives

$$\begin{aligned} B'^2 - 4A'C' &= B^2 + (A-C)^2 - (A+C)^2 \\ &= B^2 - 4AC. \end{aligned}$$

Thus, the quantity $B^2 - 4AC$ is invariant under rotations.

Definition 3.7. *For a quadratic equation:* $Ax^2+Bxy+Cy^2+Dx+Ey+F$, *the quantity* $B^2 - 4AC$ *is called the* discriminant.

We have just shown that

Theorem 3.8. *The discriminant is invariant under rotations.*

Thus, we can study an arbitrary implicit quadratic equation in any rotated coordinate system and apply that knowledge to the original system.

Theorem 3.9. *Every implicit quadratic is a conic section and*

$$\textit{if } B^2 - 4AC \begin{cases} < 0, & \textit{the curve is an } \text{ellipse}, \\ = 0, & \textit{the curve is a } \text{parabola}, \\ > 0, & \textit{the curve is a } \text{hyperbola}. \end{cases}$$

Proof: Since the discriminant is invariant under rotations, rotate through the special angle θ so that $B' = 0$ in the new rotated coordinate system. Then, $B^2 - 4AC = B'^2 - 4A'C' = -4A'C'$. ∎

We have already shown that curves of this type are either a parabola, an ellipse, or a hyperbola depending on the sign of that product. But since the shape is not changed by rotation, that means that *every* implicit quadratic is a conic section.

This results in

Theorem 3.10. *An implicit function $f(x, y) = 0$ is a conic section if and only if f is a second degree polynomial in x and y.*

Thus, all the construction techniques developed for implicit quadratics really uniquely specify conic sections. In Example 3.12 we use blending function formulation in combination with discriminants.

Section 2.3 discussed implicit functions in general, and quadratic implicit functions in particular. It was shown that quadratic implicit functions have five degrees of freedom; i.e., five independent parameters. By the equivalence of implicit quadratics and conics, conics have five independent parameters. We shall explore methods for specifying the five degrees of freedom of an arbitrary implicit second degree polynomial.

3.5 5 Point Construction

Given the equation $Ax^2 + Bxy + Cy^2 + Dx + Ey + F = 0$, we wish to solve for the parameters A, B, C, D, E, and F which makes the quadratic pass through the points $P_i = (x_i, y_i)$, $i = 1, \ldots, 5$. Using straightforward substution gives the following five homogeneous equations in six unknowns:

$$\begin{aligned}
A{x_1}^2 + Bx_1y_1 + C{y_1}^2 + Dx_1 + Ey_1 + F &= 0 \\
A{x_2}^2 + Bx_2y_2 + C{y_2}^2 + Dx_2 + Ey_2 + F &= 0 \\
A{x_3}^2 + Bx_3y_3 + C{y_3}^2 + Dx_3 + Ey_3 + F &= 0 \\
A{x_4}^2 + Bx_4y_4 + C{y_4}^2 + Dx_4 + Ey_4 + F &= 0 \\
A{x_5}^2 + Bx_5y_5 + C{y_5}^2 + Dx_5 + Ey_5 + F &= 0
\end{aligned}$$

Or, written in matrix format, as

$$\begin{bmatrix} {x_1}^2 & x_1y_1 & {y_1}^2 & x_1 & y_1 & 1 \\ {x_2}^2 & x_2y_2 & {y_2}^2 & x_2 & y_2 & 1 \\ {x_3}^2 & x_3y_3 & {y_3}^2 & x_3 & y_3 & 1 \\ {x_4}^2 & x_4y_4 & {y_4}^2 & x_4 & y_4 & 1 \\ {x_5}^2 & x_5y_5 & {y_5}^2 & x_5 & y_5 & 1 \end{bmatrix} \begin{bmatrix} A \\ B \\ C \\ D \\ E \\ F \end{bmatrix} = \begin{bmatrix} 0 \\ 0 \\ 0 \\ 0 \\ 0 \end{bmatrix}.$$

It can be shown that this system yields a second degree conic, up to a nonzero scalar multiple, as long as no more than three points are colinear, but there is no geometric intuition to be gained. A constraint must be placed to obtain a solution. Examples of frequently placed constraints are setting the coefficient of the highest power of x to 1, or normalizing the coefficients so the sum of the squares of the coefficients is 1. We now proceed to derive a solution based on constructive geometric techniques.

Consider the four lines illustrated in Figure 3.9

L_1 passing through P_1 and P_2,
L_2 passing through P_3 and P_4,
L_3 passing through P_2 and P_3,
L_4 passing through P_4 and P_1.

Each point lies on two of the lines. For example, P_2 lies at the unique intersection of the two lines, L_1 and L_3.

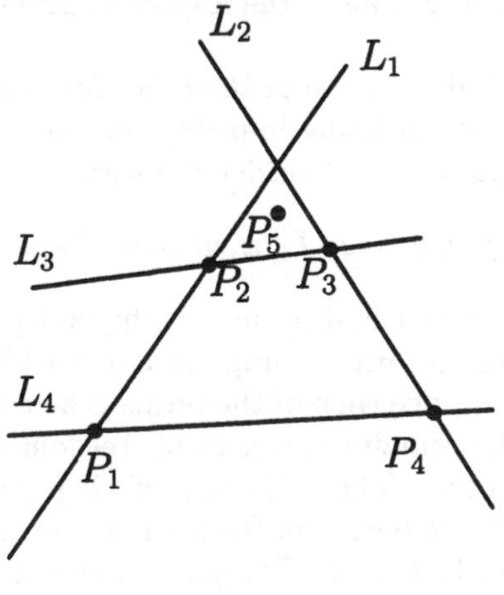

Figure 3.9. Constructing an implicit quadratic from five points.

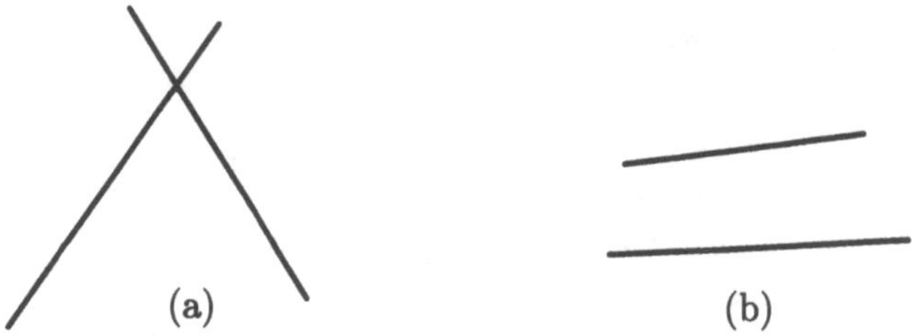

Figure 3.10. (a) $L_1L_2 = 0$; (b) $L_3L_4 = 0$.

We shall use the notation $L_i(x, y) = a_ix + b_iy + c_i$. While that is the equation of a plane in three space, a point (x, y) is on the line if and only if $L_i(x, y) = 0$. If $P_j = (x_j, y_j)$, we shall write $L_i(P_j)$ to mean $L_i(x_j, y_j)$.

Then we have,

$$\begin{aligned} L_1(P_j) &= 0 \qquad \text{for } j = 1, 2, \\ L_2(P_j) &= 0 \qquad \text{for } j = 3, 4, \\ L_3(P_j) &= 0 \qquad \text{for } j = 2, 3, \\ L_4(P_j) &= 0 \qquad \text{for } j = 4, 1. \end{aligned}$$

We also use the notation $(L_iL_j)(x, y) = L_i(x, y) * L_j(x, y) = (a_ix + b_iy + c_i)(a_jx + b_jy + c_j)$. Thus, it is a quadratic equation in x and y.

The products L_1L_2 and L_3L_4 both have the property that $L_1L_2(P_j) = 0$ and $L_3L_4(P_j) = 0$ for $j = 1, 2, 3, 4$. Considered as implicit quadratics either the function $[L_1L_2](x, y) = 0$ or the function $[L_3L_4](x, y) = 0$ could be an interpolating implicit quadratic for the first four points. Their graphs are each the unions of the graphs of the respective generating lines as shown in Figure 3.10.

If the fifth point P_5 lies on one of the other four lines, then in fact one of these curves is the interpolating implicit quadratic. Otherwise we must find a method of specification. Consider the surface

$$z = f(x, y) = L_1L_2(x, y) + cL_3L_4(x, y).$$

This is a quadric surface in x and y since the highest power of any product of variables is 2. Its roots form an implicit curve which includes the four points P_i, $i = 1, 2, 3, 4$, since each of the terms is zero at those four points. The parameter c is the remaining degree of freedom that can be used to determine a unique curve. The collection of all possible curves through those four points, with one degree of freedom is referred to as the *pencil of conics* through P_i, $i = 1, 2, 3, 4$. The particular element of the pencil is determined by the value c, which can be determined by the one remaining available point, P_5, sometimes called the *shoulder point.*

What must the value of c be to insure that $f(P_5) = 0$ for some arbitrarily positioned fifth point in the plane? Direct evaluation at P_5 gives

$$f(P_5) = L_1L_2(P_5) + cL_3L_4(P_5) = 0.$$

Solving for c yields

$$c = -\frac{L_1(P_5)L_2(P_5)}{L_3(P_5)L_4(P_5)}.$$

If P_5 lies on L_1 or L_2, then $c = 0$, and the result is L_1L_2. Unfortunately this specification was asymmetric, for if P_5 lies on L_3 or L_4, $c = \infty$. Really, the result should be L_3L_4. Thus we are given the unique quadratic through these points.

This method has used the well known and understood properties and geometries of lines in order to generate a higher order implicit equation. We continue with this geometric approach in the following modification of the derivation.

3.6 Using Tangents

We look at how the positions of the first four points affect the representation. Let P_2 move towards P_1. In the limit, that is if P_2 coalesces into P_1, the specification of the line L_1 is now in question. This can be rectified by requiring that L_1 remain constant. That is, the line specified by P_1 and P_2 is fixed during the whole limiting process. Thus, P_2 must slide down L_1 toward P_1. The line L_1 is then a chord on the curve. As the point P_2 changes, the curve changes until, when P_2 is close to P_1, L_1 is close to being tangent to the curve. (See Figure 3.11(a).) Finally, in the limit, when P_2 coalesces to P_1, L_1 is a tangent to the curve. Thus, a new pencil of quadratic rationals is defined as being tangent to L_1 and passing through P_1, P_3, and P_4. Also, the interpolation of 5 points has now turned into the interpolation of 4 points and a tangent.

In an analogous way, now move P_3 towards P_4 along L_2. This is shown in Figure 3.11(b). L_2 now defines a pencil of quadratics through the points P_1 and P_4 and also tangent to L_1 and L_2.

Call T the point of intersection of L_1 and L_2. Thus in the limit L_3 moves to L_4 and the equation for $f(x, y)$ becomes

$$f(x, y) = L_1L_2 + c{L_3}^2.$$

This is a one parameter family all of whose zero sets go through the points P_1 and P_4 and all of which are tangent to L_1 at P_1 and tangent to L_2 at P_4.

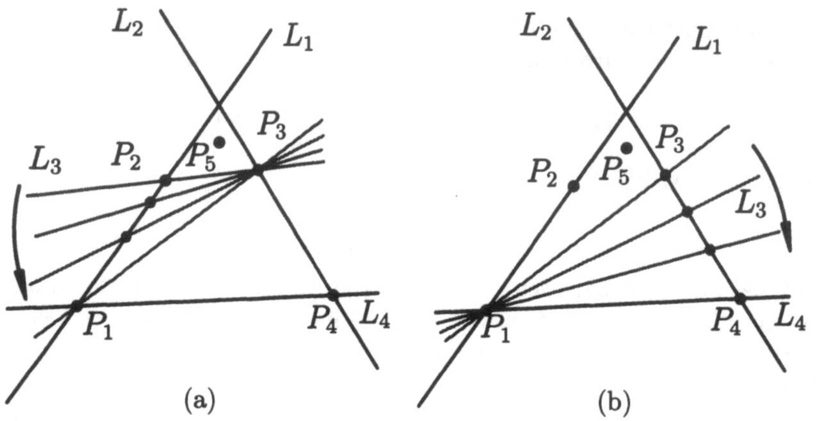

Figure 3.11. Rotating L_3: (a) P_2 converging to P_1; (b) P_3 converging to P_4.

If P_5 is inside the triangle formed by the lines L_1, L_2, and L_3 the section obtained is always a continuous curve through the points P_1 and P_4 which is tangent to L_1 and L_2. If P_5 is outside the crossed lines, the resulting curve is a hyperbola.

Many different specifications for the shoulder point (last interpolation point) lead to the same curve. A method for specifying a unique conic curve in this formulation is called a ρ *conic* (a *rho-conic*). We consider the line segment between T and M where $M = (P_1 + P_4)/2$, the midpoint of the line segment between P_1 and P_4. (See Figure 3.13.) Any point along that line may be used as the unique specification of the shoulder point P_5, and given P_5, a ratio called $\rho = \|P_5 - M\|/\|T - M\|$ is a unique scalar which

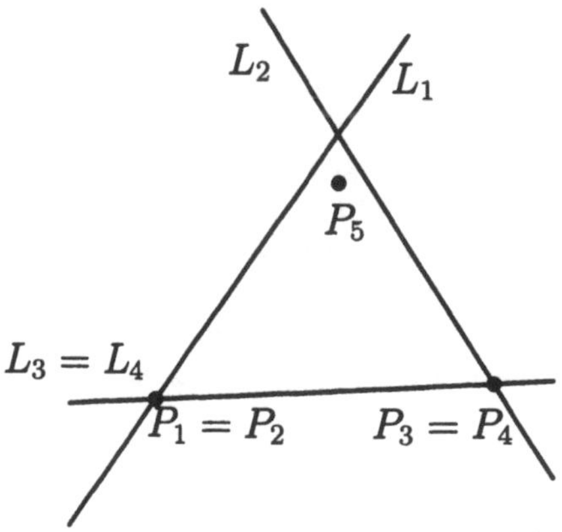

Figure 3.12. Final configuration.

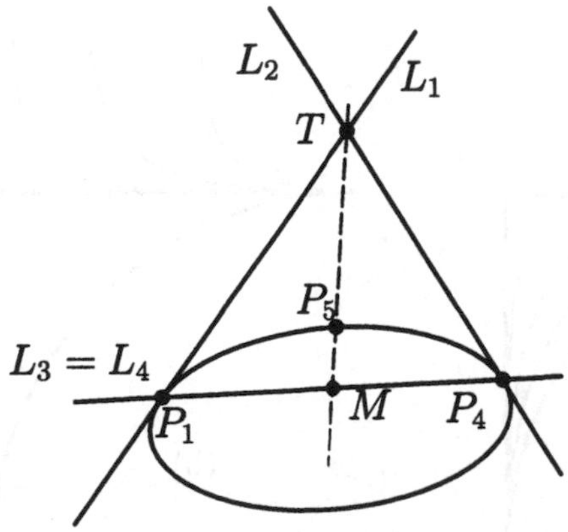

Figure 3.13. P_5 such that $\rho = \|P_5 - M\| / \|T - M\|$ is on an ellipse for $\rho < 1/2$.

specifies the shoulder point and hence the conic. Conversely, given a value of ρ, $0 \leq \rho \leq 1$, define $P_5 = (1 - \rho)M + \rho T$ to be the shoulder point. This is a unique specification for all continuous conics in the triangle, including the limiting degenerate cases. When $\rho = 1/2$, the so called ***proportional curve*** results, which will be shown to be a parabola. If $\rho < 1/2$, then the resulting curve is an ellipse, and if $\rho > 1/2$, the curve is a hyperbola.

3.6.1 Blending Formulation Revisited

To write conics as a real linear blend with respect to the blending parameter, and to only allow parameter values between zero and one, we shall slightly revise our formulation.

Suppose that C_1 and C_2 are two elements of the pencil of conics through P_1 and P_2, L_1, and L_2. Form $f(x, y) = (1 - \lambda)C_1(x, y) + \lambda C_2(x, y) = 0$. When $\lambda = 0$, $f(x, y) = C_1(x, y) = 0$, and when $\lambda = 1$, $f(x, y) = C_2(x, y) = 0$. If we restrict λ to lie between 0 and 1, $f(x, y)$ is a blend of the two curves. As Figure 3.14 shows, for any two conics, there are two orientations which can be blended. One seems more intuitive, but the other is equally valid mathematically.

Example 3.11. Consider the blending functions:

$$\begin{array}{rclll} C_1 & = & x^2 + y^2 - 1 = 0, & & \text{a unit circle at the origin, and} \\ C_2 & = & (1 - x)(1 - y) = 0, & & \text{the pair of straight lines } x = 1,\ y = 1, \end{array}$$

and set

$$f_1(x, y) = (1 - \lambda)\left(x^2 + y^2 - 1\right) + \lambda(1 - x)(1 - y) = 0 \quad \text{and}$$

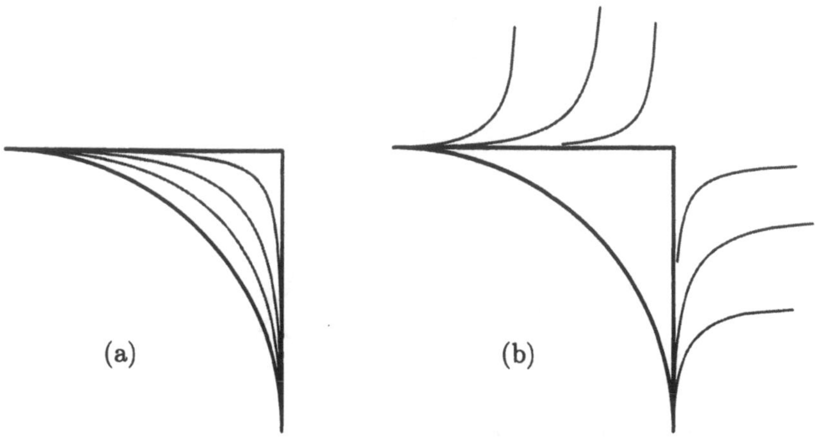

Figure 3.14. Blends (in gray) of the circle and square: (a) correct orientation; (b) wrong orientation.

$$g_1(x,y) \quad = \quad (1-\lambda)\big(x^2+y^2-1\big)-\lambda(1-x)(1-y)=0.$$

For $\lambda = 0$, $f_1(x,y)$ is the circle C_1 and for $\lambda = 1$, $f_1(x,y)$ is the pair of straight lines represented by C_2. Figure 3.14 shows several different values for λ. Figure 3.14(a) shows the blending over $g_1(x,y)$ for several λ over the same range. Let us restrict λ between 0 and 1. For example, let us set $\lambda = 1/2$ and determine the respective curves.

$$\begin{aligned} f_1(x,y) &= \frac{1}{2}\big[x^2+y^2-1+(1-x)(1-y)\big]=0 \\ &= \frac{1}{2}\big[x^2+xy+y^2-x-y\big]=0; \\ g_1(x,y) &= \frac{1}{2}\big[x^2+y^2-1-(1-x)(1-y)\big]=0 \\ &= \frac{1}{2}\big[x^2-xy+y^2+x+y-2\big]=0. \end{aligned}$$

Clearly these functions cannot be the same functions. Now, the point $\left(1/2, \frac{1+\sqrt{5}}{4}\right)$ is on f_1, while the point $\left(1/2, \frac{-1+\sqrt{21}}{4}\right)$ is on g_1. The function which is the correct blend of the circle and the pair of lines will be between them there. Since the point $(1/2, \frac{\sqrt{3}}{2})$ is on the circle, and the point $(1/2, 1)$ is on the line conic, f_1 is below the circle at that x−value, but g_1 is between, as we want. The correct blend is then g_1.

We see then that although $(1-x)(1-y)=0$ and $(x-1)(1-y)=0$ describe the same curve, when we use the curve in the blending formulation, the sign change is significant. □

Example 3.12. Consider the blending functions of Example 3.11. $C_1 = x^2+y^2-1=0$, which is a unit circle at the origin, and $C_2 = (x-1)(y-1) = 0$, which is the pair of straight lines $x = 1$, $y = 1$, can be blended as $f(x,y) = (1-\lambda)(x^2+y^2-1) + \lambda(x-1)(y-1)$.

Further, for values of $\lambda \in (0,1)$ the arc of the curve must lie between the circle C_1 and the lines C_2. But, what conic section is the curve for intermediate values of λ?

Multiplying the functions out yields

$$f(x,y) = (1-\lambda)x^2 + \lambda xy + (1-\lambda)y^2 + \cdots$$

which has a discriminant

$$\lambda^2 - 4(1-\lambda)(1-\lambda) = -(3\lambda-6)(\lambda-2/3)$$

which is zero at $\lambda = 2$ or $\lambda = 2/3$.

Thus, $0 < \lambda < 2/3$ yields an ellipse, $\lambda = 2/3$ yields a parabola, and $2/3 < \lambda < 1$ yields a hyperbola. □

3.7 Conic Arcs as Rational Functions

We will now take another look at deriving properties and formulas for conic sections, or arcs of conic sections, and derive a rational function formulation for an arc of a conic section from strictly geometrical considerations.

In this formulation we specify the five conditions as: two positions to interpolate, P_1 and P_2, two tangents, (one at each of the interpolation positions), and one last interpolation point, the *shoulder point*. This is the same as the specification in a previous blending function derivation. We do not allow the degenerate case; that is, L_1 and L_2 cannot be the same line.

Suppose the two tangent lines, L_1 and L_2 intersect at a point T. The two vectors $(P_1 - T)$ and $(P_2 - T)$ span the two-dimensional vector space in $\mathbb{R}^2$, since P_1, T, and P_2 cannot all lie on the same line. Further L_1 is the line through T and P_1, and L_2 is the line through T and P_2. We adopt shorthand and represent $(u,v) = T + u(P_1 - T) + v(P_2 - T)$. This coordinate system is a parametric coordinate system relying only on the three points P_1, T, and P_2. We need not ever know the x-y coordinates for P_1, T and P_2 in the rest of the derivation. We are concerned only with the representation of points in this intrinsic coordinate representation.

Conversion can be made to the x-y system by writing

$$(x,y) = \big(T_x + u(P_{1,x} - T_x) + v(P_{2,x} - T_x), T_y + u(P_{1,y} - T_y) + v(P_{2,y} - T_y)\big).$$

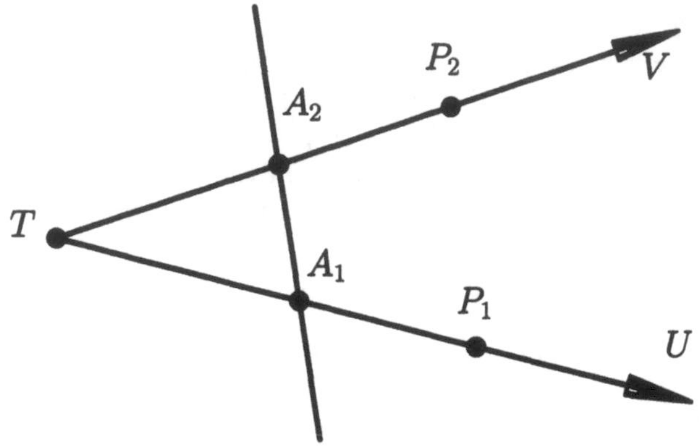

Figure 3.15. The intrinsic coordinate system to the data.

So, any point in the plane can be represented as a sum of scalar multiples of the two direction vectors, translated from the origin point T. The intersections of bivariate functions $L_1(u,v) = v$ and $L_2(u,v) = u$ with the u-v plane result in straight lines: $L_1(u,v) = v = 0$ and $L_2(u,v) = u = 0$. The graphs are $\{\,(u,v) : v = 0\,\}$ and $\{\,(u,v) : u = 0\,\}$, respectively, which are represented parametrically as $T + u(P_1 - T) = (1-u)T + uP_1$ and $T + v(P_2 - T) = (1-v)T + vP_2$, respectively. Further, the point P_1 has coordinates $(1,0)$ in the u-v coordinate system, and P_2 has coordinates $(0,1)$.

Now, for any arbitrary point on the curve, P_c, we can consider where its tangent line intersects L_1 and L_2. Call those points A_1 and A_2, respectively. The line L_3 through $P_1 = (1,0)$ and $P_2 = (0,1)$ has equation $v = 1 - u$ or $u+v-1 = 0$. It has parametric equation $L_3^t(u) = T + u(P_1 - T) + (1-u)(P_2 - T)$ in x-y coordinates. A point lies on it if it has coordinates $(u, 1-u)$.

Consider the implicit equations $L_1(u,v)L_2(u,v) + c{L_3}^2(u,v) = 0 = uv + c(u+v-1)^2$, the one parameter family of conics with interpolation points and tangents meeting the above stated specifications. We would like to determine the rational parametric equation for that portion of the curve which lies in the triangle enclosed by L_1, L_2 and L_3. Since that must lie in the first quadrant of the u-v coordinate system, the following conditions must apply to the choice of c. Since $u \geq 0$ and $v \geq 0$, the first term must be positive. Since $(u+v-1)^2$ must also be positive, the only way that this equation can sum to zero is if c is negative. Assume that $0 \leq \lambda \leq 1$, and set $c = -\lambda/(1-\lambda)$; c now can take on all negative values, so

$$C(u,v) = (1-\lambda)uv - \lambda(u+v-1)^2 = 0.$$

If $0 \leq \lambda \leq 1$, and $0 \leq u$, v with $u + v \leq 1$, the points (u,v) which are on the curve C will be inside the triangle formed by P_1, T, and P_2.

We now ask some questions about the tangent line at P_c, an arbitrary point on the desired conic. By Equation 2.2 , the equation of the tangent to the curve $C(u, v)$ at P_c is

$$\tfrac{\partial C}{\partial u}(u_c, v_c)(u - u_c) + \tfrac{\partial C}{\partial v}(u_c, v_c)(v - v_c) = 0.$$

This derives from the equation for the tangent line for a general implicit equation. For this particular problem, $\frac{\partial C}{\partial u}$ and $\frac{\partial C}{\partial v}$ at P_c are

$$\begin{aligned} \tfrac{\partial C}{\partial u}(u_c, v_c) &= (1-\lambda)v_c - 2\lambda(u_c + v_c - 1), \\ \tfrac{\partial C}{\partial v}(u_c, v_c) &= (1-\lambda)u_c - 2\lambda(u_c + v_c - 1), \end{aligned}$$

giving a tangent line equation of

$$\big[(1-\lambda)v_c - 2\lambda(u_c+v_c-1)\big](u-u_c) + \big[(1-\lambda)u_c - 2\lambda(u_c+v_c-1)\big](v-v_c) = 0.$$

We may solve for the u coordinate of A_1 and the v coordinate of A_2 using the above equation for the tangent line at P_c, since A_1 and A_2 are both on that line. But A_1 is on L_1 which means it has a v coordinate of 0. Analogously, A_2 must have a u coordinate of 0.

Solving the tangent line equation for the u coefficient of A_1 by substituting in $v = 0$ gives

$$u_1 = \frac{\big[(1-\lambda)(v_c) \quad 2\lambda(u_c + v_c - 1)\big]u_c + \big[(1-\lambda)(u_c) - 2\lambda(u_c + v_c - 1)\big]v_c}{(1-\lambda)(v_c) - 2\lambda(u_c + v_c - 1)}.$$

Now, expanding the numerator of that equation yields

$$\begin{aligned} &= 2(1-\lambda)u_c v_c - 2\lambda(u_c + v_c - 1)(u_c + v_c) \\ &= 2C(u_c, v_c) - 2\lambda(u_c + v_c - 1) \\ &= -2\lambda(u_c + v_c - 1), \end{aligned}$$

since $C(u_c, v_c) = 0$ because (u_c, v_c) is a point on the curve. Thus,

$$u_1 = \frac{-2\lambda(u_c + v_c - 1)}{(1-\lambda)v_c - 2\lambda(u_c + v_c - 1)}$$

and

$$A_1 = \left(\frac{-2\lambda(u_c + v_c - 1)}{(1-\lambda)v_c - 2\lambda(u_c + v_c - 1)}, 0\right).$$

We derived above that a point (u_c, v_c) with $u_c, v_c \geq 0$ must also have $u_c + v_c \leq 1$ to satisfy the equation C and be on the curve, so the numerator and second term of the denominator in the solution for u_1 must both be positive. We have shown that $0 \leq u_1 \leq 1$, and that A_1 will be between T and P_1. Similarly A_2 occurs when $u = 0$. Solving for v in the tangent line equation yields

$$A_2 = \left(0, \frac{-2\lambda(u_c + v_c - 1)}{(1-\lambda)u_c - 2\lambda(u_c + v_c - 1)}\right).$$

Again, $0 \leq v_2 \leq 1$ so A_2 is between T and P_2.

The ratio $r_1 = \|A_1 - T\| \,/\, \|P_1 - A_1\|$ then is

$$\begin{aligned} r_1 &= \frac{u_1}{1 - u_1} \\ &= \frac{-2\lambda(u_c + v_c - 1)}{(1-\lambda)v_c}, \end{aligned}$$

and the ratio $r_2 = \|A_2 - T\| \,/\, \|P_2 - A_2\|$ then is

$$\begin{aligned} r_2 &= \frac{v_2}{1 - v_2} \\ &= \frac{-2\lambda(u_c + v_c - 1)}{(1-\lambda)u_c}. \end{aligned}$$

The product of these ratios is

$$r_1 r_2 = \frac{4\lambda^2(u_c + v_c - 1)^2}{(1-\lambda)^2 u_c v_c},$$

but since $C(u_c, v_c) = 0$ this implies $(1-\lambda)u_c v_c = \lambda(u_c + v_c - 1)^2$. Rearranging factors yields

$$\frac{1-\lambda}{\lambda} = \frac{(u_c + v_c - 1)^2}{u_c v_c},$$

and so

$$r_1 r_2 = \frac{4\lambda^2}{(1-\lambda)^2}\frac{1-\lambda}{\lambda} = \frac{4\lambda}{1-\lambda}.$$

Notice that the right hand side of the above equation is a constant with respect to u_c and v_c. That is, although the values of both r_1 and r_2 are dependent on the values of u_c and v_c, the product $r_1 r_2$ is dependent only on the value of λ. That is, the product value is constant for a particular conic. This result means that the product of those ratios is constant for all points on the curve, and hence that product uniquely determines the curve.

Theorem 3.13. *If P_1, T, P_2, A_1 and A_2 are as above, then the product of the ratios*

$$\frac{\|TA_1\|\,\|TA_2\|}{\|A_1P_1\|\,\|A_2P_2\|}$$

is a constant for the whole conic section.

Example 3.14. Suppose we are given a blending constant λ, and a value u_1. Find the corresponding tangent line.

Since $A_1 = (1-u_1)T + u_1P_1$, the intersection of the tangent line with the u-axis of a particular point on the curve, we need only find A_2 to be able to draw the tangent line. However, $A_2 = (1-v_2)T + v_2P_2$; after v_2 is known, A_2 will also be known.

We must first solve for r_2 and then can easily obtain v_2, since by definition, $r_2 = v_2/(1-v_2)$, and

$$v_2 = \frac{r_2}{1+r_2}. \tag{3.4}$$

Now, $r_1r_2 = 4\lambda/(1-\lambda)$ and $r_1 = u_1/(1-u_1)$, so

$$r_2 = \frac{4\lambda}{1-\lambda}\frac{1}{r_1} = \frac{4\lambda}{1-\lambda}\frac{1-u_1}{u_1}.$$

For example, when $\lambda = 1/3$ and $u_1 = 3/10$,

$$\begin{aligned}
A_1 &= \frac{7}{10}T + \frac{3}{10}P_1,\\
r_1 &= \frac{3/10}{1-3/10} = 3/7,\\
r_2 &= \frac{4/3}{1-1/3}\frac{7}{3} = 14/3,\\
v_2 &= \frac{14/3}{1+14/3} = 14/17, \quad \text{and}\\
A_2 &= \frac{3}{17}T + \frac{14}{17}P_2.
\end{aligned}$$

□

Example 3.15. An example of approximations to conics appears in many children's art projects, art fairs, and homes — what we have termed string art. This example and frequently occuring phenomenon is based on the property of conics just proved above.

Given two intersecting lines and a constant greater than zero, say K, we can find the envelope of the unique conic which goes through two points on those lines and is tangent to those lines with constant product of $r_1r_2 = K$,

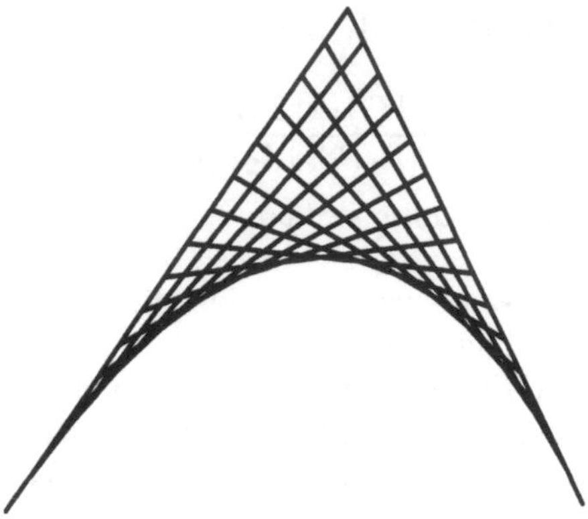

Figure 3.16. Conics drawn by envelope of tangent lines.

by drawing the tangent lines. The children's projects put yarn on nails at locations which correspond to drawing a finite number of the tangent lines at discrete and preselected locations. □

We first assumed a point $P_c = (u_c, v_c)$ on the curve and then derived values for r_1 and r_2 in terms of them, namely,

$$r_1 = \frac{-2\lambda(u_c + v_c - 1)}{(1-\lambda)v_c} \quad \text{and} \quad r_2 = \frac{-2\lambda(u_c + v_c - 1)}{(1-\lambda)u_c}.$$

We also have $r_1 = u_1/(1 - u_1)$ and $r_2 = v_2/(1 - v_2)$, where the tangent line from a point on the curve intersects L_1 at $(u_1, 0)$ and L_2 at $(0, v_2)$. For a given fixed constant, if u_1 is specified, then v_2 must be known, since $u_1 v_2/\big[(1 - u_1)(1 - v_2)\big]$ is a constant. Thus, we now try to find the points (u_c, v_c) in the curve in terms of the variables r_1 and r_2.

Solving the above equations and considering them as equations in the two unknowns u_c and v_c produces

$$\begin{aligned} r_1(1-\lambda)v_c &= -2\lambda(u_c + v_c - 1), \\ r_2(1-\lambda)u_c &= -2\lambda(u_c + v_c - 1). \end{aligned}$$

We can simplify this to

$$\begin{aligned} 2\lambda &= 2\lambda u_c + \big[r_1(1-\lambda) + 2\lambda\big]v_c, \\ 2\lambda &= \big[r_2(1-\lambda) + 2\lambda\big]u_c + 2\lambda v_c. \end{aligned}$$

The solution to the system is

$$u_c = \frac{r_1}{r_1 + r_2 + 2} \quad \text{and} \quad v_c = \frac{r_2}{r_1 + r_2 + 2}.$$

Since any point in the plane can be written as a (u, v) pair, so can any point on the curve. Let γ denote the curve. Referring to the x-y coordinate system, every point on the curve can then be written

$$\begin{aligned} P_c &= T + u_c(P_1 - T) + v_c(P_2 - T) \\ &= T + \frac{r_1}{r_1 + r_2 + 2}(P_1 - T) + \frac{r_2}{r_1 + r_2 + 2}(P_2 - T). \end{aligned}$$

We expand and combine like terms to get

$$P_c = \frac{r_1 P_1 + 2T + r_2 P_2}{r_1 + r_2 + 2}. \tag{3.5}$$

Example 3.16. We consider parameterizing the curve as a function of a variable t where $0 \le t \le 1$, by setting $u_1 = 1 - t$ and $v_2 = t$. Since, by definition,

$$\begin{aligned} r_1 &= \frac{u_1}{1 - u_1}, \\ r_2 &= \frac{v_2}{1 - v_2}, \end{aligned}$$

we can substitute in the current definitions for u_1 and v_2 to get

$$\begin{aligned} r_1 &= \frac{1 - t}{t}, \\ r_2 &= \frac{t}{1 - t}. \end{aligned}$$

Then,

$$r_1 r_2 = \frac{1 - t}{t}\frac{t}{1 - t} = 1,$$

so this parameterization is possible only when $1 = 4\lambda/(1 - \lambda)$,or, when $\lambda = 0.2$. We seek to find an equation for P_c as a function of t.

$$\begin{aligned} P_c &= \frac{r_1 P_1 + 2T + r_2 P_2}{r_1 + 2 + r_2} \\ &= \frac{\frac{1-t}{t} P_1 + 2T + \frac{t}{1-t} P_2}{\frac{1-t}{t} + 2 + \frac{t}{1-t}} \\ &= \frac{\frac{(1-t)^2 P_1 + 2t(1-t)T + t^2 P_2}{t(1-t)}}{\frac{(1-t)^2 + 2t(1-t) + t^2}{t(1-t)}} \\ &= \frac{(1 - t)^2 P_1 + 2t(1 - t)T + t^2 P_2}{(1 - t)^2 + 2t(1 - t) + t^2} \end{aligned}$$

$$\begin{aligned} &= (1-t)^2 P_1 + 2t(1-t)T + t^2 P_2 &(3.6)\\ &= (1-t)\left((1-t)P_1 + tT\right) + t\left((1-t)T + tP_2\right) &(3.7)\end{aligned}$$

Equation 3.6 follows since

$$1 = ((1-t)^2 + 2t(1-t) + t^2 = ((1-t)+t)^2 .$$

By definition, and using current definitions of u_1 and v_2,

$$\begin{aligned} A_1 &= u_1 P_1 + (1-u_1)T \\ &= (1-t)P_1 + tT; &(3.8)\\ A_2 &= (1-v_2)T + v_2 P_2 \\ &= (1-t)T + tP_2. &(3.9)\end{aligned}$$

Equation 3.7 can be rewritten in terms of A_1 and A_2. Starting with Equation 3.7,

$$\begin{aligned} P_c &= (1-t)\left((1-t)P_1 + tT\right) + t\left((1-t)T + tP_2\right) \\ &= (1-t)A_1 + tA_2 &(3.10)\end{aligned}$$

Thus, for the case when $r_1 r_2 = 1$, t can be used to create a constructive algorithm. □

We have shown,

Lemma 3.17. *For three points P_0, P_1, P_2, when $r_1 r_2 = 1$, then for each value t, $0 \le t \le 1$, set*

$$\begin{aligned} P_1^{[1]} &= (1-t)P_0 + tP_1; \\ P_2^{[2]} &= (1-t)P_1 + tP_2. \end{aligned}$$

Then

$$\begin{aligned} \gamma(u) &= (1-t)^2 P_0 + 2t(1-t)P_1 + t^2 P_2 \\ &= (1-t)P_1^{[1]} + tP_2^{[1]}. \end{aligned}$$

Now, for the general case, when $r_1 r_2 \neq 1$, it is necessary to find representations of r_1 and r_2 as different parametric functions of the same variable, say t, over the same interval, say (a, b). Hence we must find continuous $f_1(t)$ and $f_2(t)$, such that $r_1 = f_1(t)$ and $r_2 = f_2(t)$, for $t \in (a, b)$. Thus, both f_1 and f_2 must map (a, b) onto $(0, \infty)$, and $f_1(t)f_2(t) = 4\lambda/(1-\lambda)$, for every value of t. We would prefer that f_1 (and hence f_2) be a bijection onto its range, that is, each r_1 is the image of only one value of t. That means that f_1 and f_2 must be monotone functions. Further, if f_1 increases, f_2 must decrease because of the inverse relationship. One cannot choose polynomials for f_1 and f_2 since the image space is unbounded. From the original definition of r_1 and r_2, we have candidates if $(a, b) = (0, 1)$. A scaling and translation of these linear rational functions are the simplest functions with some flexibility that we can choose. So we set

$$f_1(t) = \frac{w_1(b-t)}{w(t-a)} \qquad \text{and} \qquad f_2(t) = \frac{w_2(t-a)}{w(b-t)}l.$$

Then

$$\begin{aligned} K &= \frac{4\lambda}{1-\lambda} \\ &= r_1(t)r_2(t) \\ &= f_1(t)f_2(t) \\ &= \frac{w_1(b-t)}{w(t-a)}\frac{w_2(t-a)}{w(b-t)} \\ &= \frac{w_1w_2}{w^2}. \end{aligned}$$

Any three nonzero values for w_1, w, and w_2, such that w_1w_2/w^2 is the constant above will provide f_1 and f_2 which satisfy the constraints imposed by the geometry.

Rewriting γ in this new notation yields

$$\begin{aligned} \gamma(t) &= \frac{\frac{w_1(b-t)}{w(t-a)}P_1 + 2T + \frac{w_2(t-a)}{w(b-t)}P_2}{\frac{w_1(b-t)}{w(t-a)} + 2 + \frac{w_2(t-a)}{w(b-t)}} \\ &= \frac{w_1(b-t)^2P_1 + 2w(t-a)(b-t)T + w_2(t-a)^2P_2}{w_1(b-t)^2 + 2w(t-a)(b-t) + w_2(t-a)^2}. \qquad (3.11) \end{aligned}$$

From this we see that any conic section can be written as a quadratic rational parametric equation, and that any rational quadratic parametric function with a common denominator is a conic section.

Example 3.18. Given points P_1, T, and P_2, and constant $K > 0$, find a rational parametric form, $\gamma(t)$, for the arc, with domain $[0, 1]$, where $\gamma(0) = P_1$, $\gamma(1) = P_2$, the direction of $\gamma'(0)$ is the same as the direction of $(T - P_1)$ and the direction of $\gamma'(1)$ is the same as the direction of $(T - P_2)$, and $K = \frac{4\lambda}{(1-\lambda)}$.

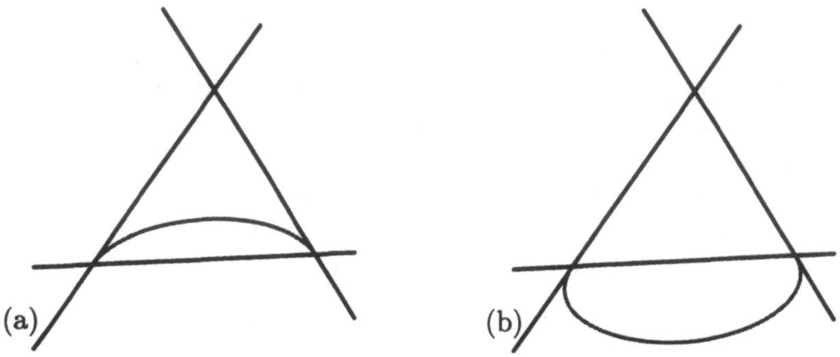

Figure 3.17. (a) Rational arc (in gray) in polygon; (b) the other part of the conic (in gray).

A common solution for this problem is to set $w_1 = w_2 = 1$, and to let $w^2 = 1/K$, that is $w = (\sqrt{K})^{-1}$. Direct substitution into the above formula gives

$$\gamma(t) = \frac{(1-t)^2 P_1 + 2(\sqrt{K})^{-1} t(1-t)T + t^2 P_2}{(1-t)^2 + 2(\sqrt{K})^{-1} t(1-t) + t^2}.$$

While this solution is simple and appealing, numerically, it may have somewhat undesirable side effects. This occurs when K is very small, or very large. □

Since $\lambda \in (0,1)$, the ratio $w_1 w_2 / w^2$ must be positive. However, individual values may be negative. If $w < 0$, the ratio is unchanged, but the part of the conic derived as t goes from 0 to 1 is the one outside the triangle. If $w_1 < 0$, then $w_2 < 0$, and this same *other* part of the conic is derived, as shown in Figure 3.17.

Example 3.19. We find the rational expansion for an arc of a circle. Suppose the circle has radius r and the arc of the circle is α. Now, at any point on the circle the radius is orthogonal to the tangent direction. Draw the line from C, the center of the circle to T, and consider the point on the circle through which that line passes, P_c. Since this is a circular arc, $\|T - P_1\| = \|T - P_2\|$, and $C - T$ bisects the angle α. In addition, since $A_1 - P_c$ is tangent to the circle at P_c it also is orthogonal to a radius, and $\|A_1 - P_1\| = \|A_1 - P_c\|$. Analogously, $\|A_2 - P_2\| = \|A_2 - P_c\|$.

Since $A_1 - A_2$ is the tangent to the circle at P_c, it is parallel to the line through P_1 and P_2, and the radius through P_c is orthogonal to the line from A_1 to A_2. Since the angle $P_iTC = \pi/2 - \alpha/2$, angle $TA_iP_c = \alpha/2$, and $\|A_i - P_c\| = \|T - A_i\| \cos(\alpha/2)$, for $i = 1, 2$. See Figure 3.18.

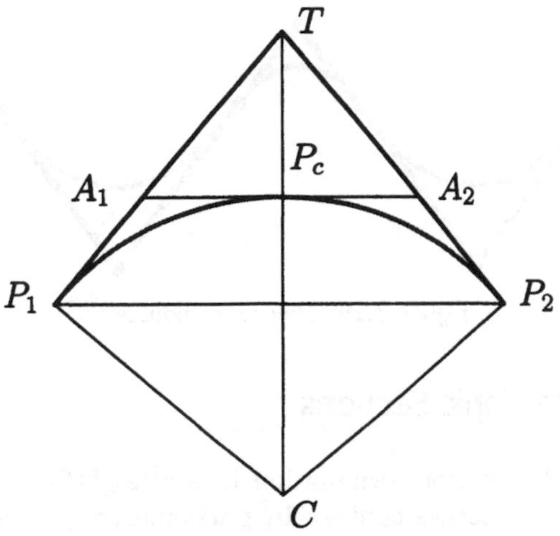

Figure 3.18. Circular arc (in gray).

$$\begin{aligned} r_i &= \frac{\|T - A_i\|}{\|A_i - P_i\|}, \qquad i = 1,2 \\ &= \frac{\|T - A_i\|}{\|A_i - P_c\|} \\ &= \frac{\|T - A_i\|}{\|T - A_i\| \cos(\alpha/2)} \\ &= \frac{1}{\cos(\alpha/2)}. \end{aligned}$$

Thus,

$$K = r_1 r_2 = \frac{1}{\cos^2(\alpha/2)}.$$

If α is close to π, then this product is very large since $\cos(\alpha/2)$ is very close to zero. Hence, if $0 < \alpha < \pi$ we can use this representation, but if α is close to π, floating point inaccuracies will affect the stability and accuracy of the numerical evaluation method. One can let $w_1 = w_2 = 1$ and let $w = \cos(\alpha/2)$. □

Methods for evaluating points on rational curves will be discussed later in a broader context.

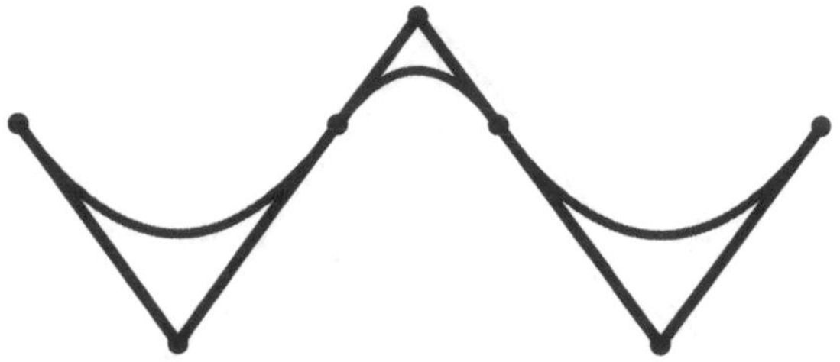

Figure 3.19. Piecewise conics.

3.8 Piecewise Conic Sections

Using the rational function formulation it is straightforward to write the equations for the piecewise conics. In particular, if one has information arranged so that the P's are interpolation points and the T's are where the tangents intersect, then the situation is pictured in Figure 3.19. With constants K_1, K_2, ..., all not zero, one can make parametric conics:

$$\begin{aligned}
\gamma_1(t) &= \frac{(1-t)^2P_1 + 2\left(\sqrt{K_1}\right)^{-1} t(1-t)T_1 + t^2P_2}{(1-t)^2 + 2\left(\sqrt{K_1}\right)^{-1} t(1-t) + t^2};\\
\gamma_2(t) &= \frac{(2-t)^2P_2 + 2\left(\sqrt{K_2}\right)^{-1} (t-1)(2-t)T_2 + (t-1)^2P_3}{(2-t)^2 + 2\left(\sqrt{K_2}\right)^{-1} (t-1)(2-t) + (t-1)^2};\\
&\vdots\\
\gamma_m(t) &= \frac{(2-t)^2P_m + 2\left(\sqrt{K_m}\right)^{-1} (t-1)(2-t)T_m + (t-1)^2P_{m+1}}{(2-t)^2 + 2\left(\sqrt{K_m}\right)^{-1} (t-1)(2-t) + (t-1)^2}.
\end{aligned}$$

If T_i, P_i, T_{i+1} are colinear then the pieces will meet with tangent direction continuity although these pieces will not in general have parametric tangent continuity. The parametric representation is used since it only requires that each trio, P_i, T_i, and P_{i+1}, be in a plane, which is automatically fulfilled. The x-y plane is not necessary, i.e., P_3 need not be in the same plane as P_1, T_1, and P_2. P_4 can be in yet another plane.

Example 3.20. The constructive technique developed in Example 3.16 and Lemma 3.17 can be extended to piecewise polygons when each piecewise curve is a parabola ($K = 1$), and all of the w values are set to 1. In that case, the algorithm is piecewise applied to the ordered triple of points $\{P_i, T_{i+1}, P_{i+1}\}$. As u varies from 0 to 1, the curve traces out from P_i to P_{i+1}. The same algorithm can be applied in parallel to all triples for drawing all the pieces simultaneously. □

3.9 Using Homogeneous Coordinates to Represent Conics

By Equation 3.11, the rational parametric form of the conic section is

$$\gamma(t) = \frac{w_1(b-t)^2 P_1 + 2w(t-a)(b-t)T + w_2(t-a)^2 P_2}{w_1(b-t)^2 + 2w(t-a)(b-t) + w_2(t-a)^2}.$$

Let $\theta_1(t) = (b-t)^2$, $\theta(t) = 2(t-a)(b-t)$, and $\theta_2(t) = (t-a)^2$.
Then, the equation is rewritten

$$\begin{aligned}\gamma(t) &= \frac{w_1 P_1 \theta_1(t) + wT\theta(t) + w_2 P_2 \theta_2(t)}{w_1\theta_1(t) + w\theta(t) + w_2\theta_2(t)} \\ &\equiv (w_1P_1, w_1)\theta_1(t) + (wT, w)\theta(t) + (w_2P_2, w_2)\theta_2(t).\end{aligned}$$

This notation looks as if the function has become a parametric polynomial in one extra dimension, but looks are deceiving. For $K > 0$, since $K = w_1w_2/w^2$, the functions

$$(P_1, 1)\theta_1(t) + \left(\left(\sqrt{K}\right)^{-1}T, \left(\sqrt{K}\right)^{-1}\right)\theta(t) + (P_2, 1)\theta_2(t),$$

$$(KP_1, K)\theta_1(t) + (T, 1)\theta(t) + (P_2, 1)\theta_2(t),$$

and

$$(2P_1, 2)\theta_1(t) + (2T, 2)\theta(t) + (2KP_2, 2K)\theta_2(t)$$

all represent the same curve! Since in regular Euclidean space, even parametric equations have to behave according to different rules, this is clearly not Euclidean Space. In face, the coordinate system has been embedded in the *homogeneous coordinate system* of the *projective plane*. Different curves in this space project onto the same rational quadratic curve in $\mathbb{R}^3$. This representation shall be worked for the user's benefit later, but caution must always be paid for the anomalies of the projective plane.

3.9.1 Extending the Constructive Algorithm to All Conics

In Example 3.16, a constructive algorithm was developed for the case in which $K = 1$, that is, the parabola. That approach does not extend to arbitrary conics using the normal Euclidean point representation. Let us consider the general case and suppose the ordered geometric points are $\{P_0, P_1, P_2\}$ and the corresponding coefficients in the denominator are $\{w_0, w_1, w_2\}$. Then, $\gamma(t) = n(t)/d(t)$ where

$$\begin{aligned}n(t) &= w_0P_0(b-t)^2 + w_1P_1 2(b-t)(t-a) + w_2P_2(t-a)^2, \\ d(t) &= w_0(b-t)^2 + w_1 2(b-t)(t-a) + w_2(t-a)^2.\end{aligned}$$

Note that Lemma 3.17 cannot be used directly. However, set $Q_i = w_i P_i$. Then,

$$\begin{aligned} n(t) &= (b-t)\left((b-t)Q_0 + (t-a)Q_1\right) + (t-a)\left((b-t)Q_1 + (t-a)Q_2\right) \\ &= (b-t)Q_1^{[1]} + (t-a)Q_2^{[1]}, \end{aligned}$$

where

$$\begin{aligned} Q_1^{[1]} &= (b-t)Q_0 + (t-a)Q_1, \\ Q_2^{[1]} &= (b-t)Q_1 + (t-a)Q_2, \end{aligned}$$

and

$$\begin{aligned} d(t) &= (b-t)\left((b-t)w_0 + (t-a)w_1\right) + (t-a)\left((b-t)w_1 + (t-a)w_2\right) \\ &= (b-t)w_1^{[1]} + (t-a)w_2^{[1]}, \end{aligned}$$

where

$$\begin{aligned} w_1^{[1]} &= (b-t)w_0 + (t-a)w_1, \\ w_2^{[1]} &= (b-t)w_1 + (t-a)w_2. \end{aligned}$$

These two constructions can be compressed into one by using the homogeneous coordinate formulation

$$H_i = (Q_i, w_i), \qquad i = 0, 1, 2.$$

Then,

$$\begin{aligned} H_i^{[1]} &= (b-t)H_{i-1} + (t-a)H_i \\ &= ((b-t)Q_{i-1}, (b-t)w_{i-1}) + ((t-a)Q_i, (t-a)w_i) \\ &= \left(Q_i^{[1]}, w_i^{[1]}\right). \end{aligned}$$

And,

$$\begin{aligned} (n(t), d(t)) &= \left((b-t)Q_1^{[1]}, (b-t)w_1^{[1]}\right) + \left((t-a)Q_2^{[1]}, (t-a)w_2^{[1]}\right) \\ &= (b-t)H_1^{[1]} + (t-a)H_2^{[1]}. \end{aligned}$$

Thus, we have developed a constructive algorithm for all conics using the homogeneous curve formulation.

Algorithm 3.21. *Given geometric points $\{P_0, P_1, P_2\}$, and corresponding weights, $\{w_0, w_1, w_2\}$, such that $w_0 w_2 / w_1^2 = K$, the conic section through point P_0 with tangent $P_1 - P_0$ and through P_2 with tangent $P_2 - P_1$ and such that $K = 4\lambda(1-\lambda)$, can be evaluated over an interval (a, b).*

Step 1: Set $H_i = (w_i P_i, w_i)$*, for* $i = 0, \ldots, 2$.

Step 2: Set $H_i^{[1]} = (b - t)H_{i-1} + (t - a)H_i$*, for* $i = 1, \ldots, 2$.

Step 3: Set $(N, D) = (b - t)H_1^{[1]} + (t - a)H_2^{[1]}$.

Step 4: Then $\gamma(t) = N/D$.

Beware! Trying to visualize the homogenous constructive algorithm in regular $\mathbb{R}^3$ space can lead to inappropriate conclusions and incorrect computations.

3.9.2 The Homogeneous Coordinate System and the Projective Plane

We define an equivalence relation on $\mathbb{R}^3 - \{0\}$, that is punctured Euclidean space. Two points (x_1, y_1, z_1) and (x_2, y_2, z_2) are related if there exists a real number $r \neq 0$ such that $(x_1, y_1, z_1) = r(x_2, y_2, z_2) = (rx_2, ry_2, rz_2)$. Thus, the unit sphere contains two points in each equivalence class, the points at opposite ends of a diameter. The identification of these points cannot be shown in 3D. The set of equivalence classes is called the *projective plane.* The ideas of distance and separation cannot have the traditional meaning, as was illustrated in the examples of Section 3.7. Usually an element of the representation is picked with $r = 1$. That is, we identify $[x, y, z, 1]$ directly with $\mathbb{R}^3$, while $[rx, ry, rz, r]$, $r \neq 0$, represents all other members of the equivalence class. The limiting case in each direction is an infinity in that direction, represented by $[x, y, z, 0]$.

To give some ideas of the identification, we can derive the equivalence relation one dimension lower, from $\mathbb{R}^2$ to the *projective line.* The punctured plane is used. Given a point, P, a line is drawn through P and the origin. As a first step, P is identified with the point on the unit circle which also lies on the line through P and the origin.

As in the case of the projective plane, after performing this identification, there are still two representatives for each equivalence class on the unit circle. They are the opposite ends of each diameter. The identification cannot be performed in $\mathbb{R}^2$. We must twist part of the circle through three space. This identification can be performed by (see Figure 3.20) squeezing the diameter of the x-axis to identify $(-1, 0)$ and $(1, 0)$ and twisting the top semicircle. Continue the twisting until the new circle, coming from the top semicircle can just be projected into the plane onto what was originally the bottom semicircle. Hence, the projective line is a circle.

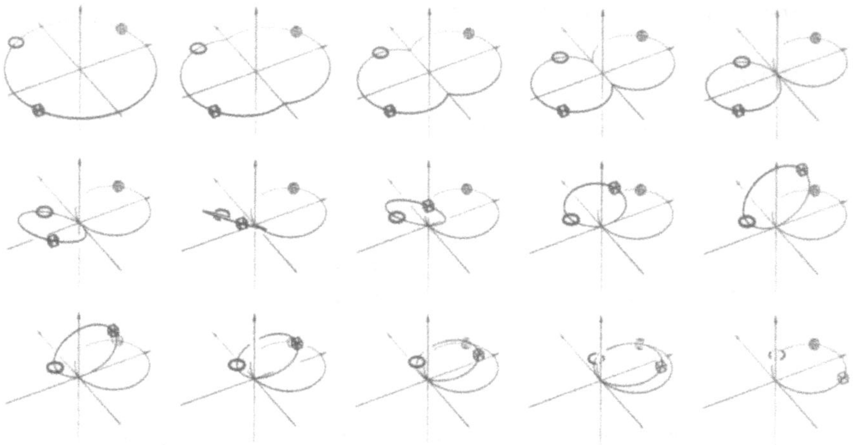

Figure 3.20. Partial Identification. Sequence goes from top left to bottom right. Note the sphere, torus, and cube markers.

Exercises

1. For a three-dimensional result, use two straight sticks of wood and glue them at some specific intersecting angle. Select a constant $k > 0$. The result of this is to find an approximate envelope of the unique conic which goes through two specified points on the sticks and is tangent to the sticks with a constant k. Use $r_1 r_2 = k$, where r_1 and r_2 are defined as in Equation 3.4.

2. Consider the four points $(0,0)$, $(0,1)$, $(2,0)$, and $(3,1)$. What is the formula for the family of implicit conics passing through these points. Calculate the value of the last unknown to force interpolation through an arbitrary fifth point, (x_5, y_5).

3. For $P_1 = (0,0)$, $P_2 = (1,2)$, and $P_3 = (3,0)$, what is the formula for the family of implicit conics passing through P_1 and P_3, with tangent directions P_1P_2 and P_2P_3? Calculate the value of the last degree of freedom which would force the conic to interpolate through an arbitrary point Q.

4. Given three points P_1, T, and P_2, and the blending constant λ, determine the discriminant of the resulting conic. For what values is the curve an ellipse? a parabola? a hyperbola? If you want to describe this curve arc in the rational parametric form, what values of w_1,

w_2, and w can be used? Select a triple and tell why you think this particular one would be useful.

5. If we remove the constraint that $u \geq 0$ and $v \geq 0$ in the derivation of the conic sections as rational functions, then the blend value c could be either positive or negative, so that allows λ to take on any real value. How does that potentially affect the values u_1 and v_2? What about $r_1 r_2$? If $\frac{4\lambda}{1-\lambda} < 0$, how does that affect the choice of w_1, w_2, and w? Which part of the curve is drawn, for $a \leq u \leq b$?

6. Suppose L_1 and L_2 intersect at a point P, and you are given discrete evenly spaced pegs along each line. If you can draw only approximations to tangent lines by connecting the dots how do you do it? How good of an approximation to a real conic curve do you get? How does the spacing and the number of points change the quality of the approximation? Are there any values of $K = \frac{\lambda}{1-\lambda}$ for which you can get a real conic using this approximation?

7. Programming Exercise: Develop an interactive program to draw conic curves specified by the tangent form of the blending formulation. One way to draw this conic is to draw its envelope by displaying various segments of the tangent lines, via string art method described. You should:

   (a) Allow the user to pick three points, P_1, P_2, and T graphically.

   (b) Display the lines L_1 and L_2.

   (c) Allow the user to specify the last degree of freedom, by one of the two common ways: 1) specifying the K parameter, or 2) interpolation through the fifth point.

4

Differential Geometry for Space Curves

4.1 More on Parameterizations

We have learned that for parametric curves, the parameterization gives information about rates of traversal which are important in many practical applications such as specifying the feed rate of numerically controlled cutting machines. The particular parameterization of a curve can make it easier or more difficult to obtain information about the geometry of the curve. The importance of the parameterization leads us to ask if there are equivalence classes of transformations. That is, is there some way of characterizing different parameterizations for a curve that will allow decomposition into equivalence classes with well-defined, well-behaved transformations between the equivalence classes?

Definition 4.1. *Let I_1 and I_2 be intervals of $\mathbb{R}^1$. If $\gamma(t) : I_1$ into $\mathbb{R}^3$ and $p : I_2$ into/onto I_1, γ and p are* composable functions *and $\gamma(p(u)) : I_2$ into $\mathbb{R}^3$ is a* reparameterization *of γ. This is also called a* change of parameter *from t to u.*

Both $\gamma(t)$ and $\gamma(p(u))$ have the same graph, and so are different representations of the same curve.

Definition 4.2. *γ is a* regular parametric representation *if for all $t \in I$,*

1. *$\gamma(t) \in C^{(1)}$, and*
2. *$\gamma'(t) \neq 0$.*

Definition 4.3. $p = p(u)$ *is called an* allowable change of parameter *if, for all* $u \in I_2$, $p(u) \in C^{(1)}$ *and* $p'(u) \neq 0$.

The two definitions can be used to show:

Lemma 4.4. *The result of reparametrizing a regular parametric representation using an allowable change of parameter is another regular parametric representation representing the same curve or, a part of the same curve.*

Proof: Suppose that γ is a regular parametric representation such that $\gamma : I_1 \to \mathbb{R}^3$, and that there exists an allowable change of parameters $p : I_2 \to I_1$ such that $\beta(u) = \gamma(p(u))$.

Since $p'(u)$ is continuous and nonzero, by Theorem 1.31, p must be an increasing or a decreasing function. That means that it is an injection (a one-to-one function). Further, β must have as its graph part of the same image as γ since it is just γ evaluated at some subset of I_1.

By thc chain rule, $\beta'(u) = \gamma'(p(u))p'(u)$. Since both γ' and p' are continuous and nonzero, β' is continuous and nonzero throughout I_2. Thus, β is a regular parametric representation. ■

Example 4.5. Consider $\gamma(t) = (t, 1-t)$, for $t \in [a, 1-a]$, for any a such that $0 < a < 1/2$. $\gamma'(t) = (1, -1)$, and γ is a regular representation.

Now, let $t = p(u) = \cos^2 u$ for $u \in [b, \pi/2 - c]$ where b and c are such that $cos^2 b = a$ and $\cos^2(\pi/2 - c) = 1 - a$. Then $p'(u) = -2\cos u \sin u$ which is greater than 0 for $u \in [b, \pi/2 - c]$. Thus $\gamma(p(u)) = \beta(u) = (\cos^2 u, \sin^2 u)$ is also a regular parametric representation of the same geometric shape. □

Definition 4.6. *A regular representation* $\gamma(t)$, $t \in I_1$, *is related to another regular representation* $\phi(u)$, $u \in I_2$, *if there exists an allowable change of parameter* $t = p(u)$ *such that* 1) $p(I_2) = I_1$, *and* 2) $\gamma(p(u)) = \phi(u)$.

Lemma 4.7. *The above relation is an equivalence relation on all regular parametric representations.*

Proof: This is left as an exercise for the reader in Exercise 2. ■

Definition 4.8. *Each equivalence class of the relation,* "allowable change of parameter", *defined in Definition 4.3 is called a* regular curve.

Note that every element in the equivalence class has the same graph and so sensibly can be grouped together as collectively representing a single curve.

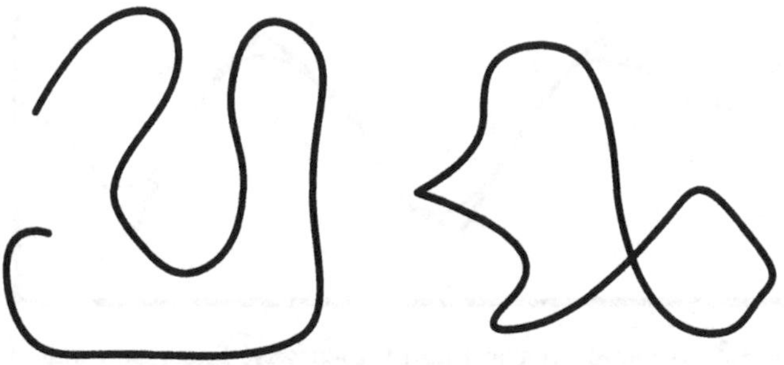

Figure 4.1. (a) Example of a simple curve; (b) a complex curve.

Definition 4.9. *A* simple curve *is a curve that does not cross itself. More analytically,* $\gamma(t)$ *is a simple curve if it is an injection, i.e., for all* t_1, $t_2 \in I$, $t_1 \neq t_2$ *implies that* $\gamma(t_1) \neq \gamma(t_2)$.

4.2 Arc Length Parameterization

Given an arbitrary regular curve, we would like to know the length of the curve between two distinct parameter values t_a and t_b. That is, how far would a bug walking along the curve walk if it started at $\gamma(t_a)$ and ended at $\gamma(t_b)$?

In the case where the parameterization is equivalent to an explicit curve, $\beta(t) = (t, f(t))$, we might consider approximating the curve as a sequence of piecewise constants and then computing the length of that approximant. Figure 4.2 demonstrates that the length of the piecewise constant approximant is then $t_b - t_a$, for every curve!

This occurs because the vertical distance is being ignored. We might take the horizontal and vertical components and add them together, but a shorter and closer approximation is obtained by taking the hypotenuse of

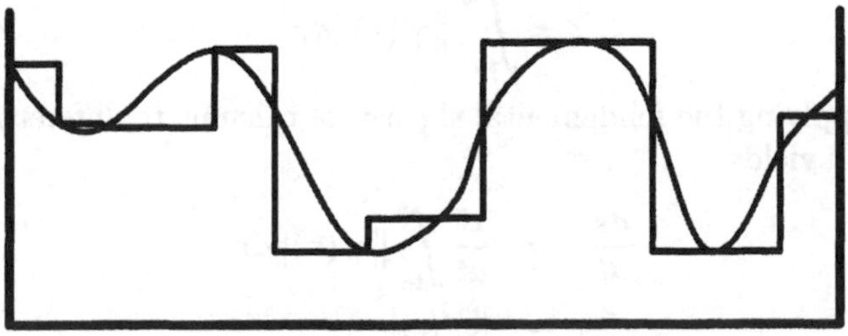

Figure 4.2. A curve (in gray) and a piecewise constant approximation.

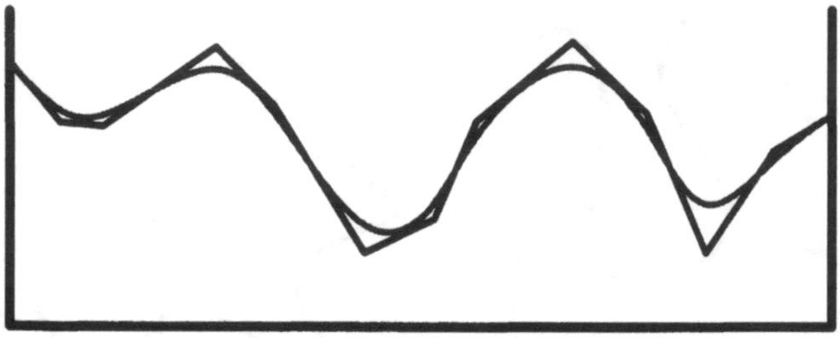

Figure 4.3. A curve (in gray) and a piecewise linear approximation.

each of the right triangles and adding them together. While the meaning of horizontal and vertical change with every rotation of the coordinate system, the hypotenuse is attached to the curve and is invariant.

For a general regular curve $\gamma(t)$ from an interval I into $\mathbb{R}^3$, we can approximate the curve by using piecewise linear segments to interpolate along the curve. Suppose $n+1$ points of interpolation are used, including the left endpoint, t_0, and up to some point $t = t_n \in I$, $t_a = t_0 < t_1 < \ldots < t_n = t_b$. Letting s denote the arc length we get

$$\begin{aligned} s &\approx \sum_{i=0}^{n-1} \left\| \gamma(t_{i+1}) - \gamma(t_i) \right\| \\ &= \sum_{i=0}^{n-1} \left\| \tfrac{\gamma(t_{i+1}) - \gamma(t_i)}{t_{i+1} - t_i} \right\| (t_{i+1} - t_i), \end{aligned}$$

where

$$\left\| \gamma(t_{i+1}) - \gamma(t_i) \right\| = \sqrt{\left(x(t_{i+1}) - x(t_i)\right)^2 + \left(y(t_{i+1}) - y(t_i)\right)^2 + \left(z(t_{i+1}) - z(t_i)\right)^2}.$$

But this is just a Riemann sum of a well-behaved function.

As $\lim_n \max_{i\,=\,0,\ldots,\,n} |t_{i+1} - t_i| \longrightarrow 0$, two things happen: first, the quantity in the quotient approaches the derivative, and second, the sum approaches the integral to yield

$$s = \int_{t_a}^{t_b} \left\| \gamma'(\tau) \right\| d\tau.$$

Directly applying the fundamental theorm of calculus to differentiate with respect to t yields

$$\begin{aligned} \frac{ds}{dt} &= \frac{d}{dt} \int_{t_0}^{t} \left\| \gamma'(\tau) \right\| d\tau \\ &= \left\| \gamma'(t) \right\| \\ &= \left\| \boldsymbol{v}_\gamma(t) \right\|. \end{aligned}$$

Since $\gamma'(t) = \boldsymbol{v}_\gamma(t)$, the velocity for the curve γ at t, then $\|\gamma'(t)\|$ is just the magnitude, or speed.

Now, $s = s(t)$ is a strictly increasing function, since it is the integral of a positive-valued function, which means it is invertible. That is, there exists some function q such that, $t = q(s)$. Further, since $\gamma(t)$ is regular, $\|\gamma'(t)\| > 0$, so $\frac{ds}{dt}$ is nonzero, and hence $\frac{dt}{ds}$ exists and is nonzero.

So for any regular representation, a reparameterization by arc length is an allowable representation, and is in the same equivalence class, yielding the same curve.

Suppose now that $\gamma(s)$ is parametrized by arc length. Then, as shown above, $\frac{ds}{ds} = 1 = \|\gamma'(s)\|$. Conversely, suppose $\gamma(t)$ is a uniform speed curve with speed 1, that is $\|\gamma'(t)\| = 1$ for all t. What are the consequences of such a parameterization?

$$s = \int_{t=0}^{t} \|\gamma'(t)\| dt = t.$$

Thus, we have shown that γ has an arc length parameterization if and only if $\|\gamma'\| = 1$.

Example 4.10. A model train moves at a fixed speed along the track. Digital plotters move at different speeds in different directions at different times so are not fixed speed devices. □

Unfortunately computing the arc length parameterization is not always simple. In fact, for many applications in CAGD it is impractical to try to use it. However, the question arises whether two curves are the same curve, just in different parameterizations. There are two scalar functions that completely characterize a parametric curve of a smooth enough class. These scalar valued functions are known as the curvature and the torsion.

Theorem 4.11. *Let $k(s)$ and $f(s)$ be continuous scalar valued functions on $a \le s \le b$, and $0 < k(s)$. There exists exactly one regular curve $\gamma(s)$ for which $k(s)$ is the curvature and $f(s)$ is the torsion. The position of the curve in space is, however, free. These functions are called the* intrinsic equations *of the curve.*

In the following sections, we derive the geometry of parametric curves, derive definitions for the above two scalar curves, and examine their local effects on the curve.

4.3 Intrinsic Vectors of a Space Curve

We first derive the formulations of these vectors in terms of an arc length parameterization. Later, we shall generalize to an arbitrary regular representation, which is more complicated but is the case that occurs in practice.

For this section we shall assume with no further statement that

- s is always the arc length parameter;
- $\gamma : I$ to $\mathbb{R}^3$;
- γ is parametrized by its arc length, that is, $\gamma(s)$, and $\|\gamma'(s)\| = 1$.

Definition 4.12. *$T = T(s) = \gamma'(s)$ is called the* unit tangent *to the curve γ since $\|T(s)\| = 1$ for all s.*

Definition 4.13. *$T' = T'(s) = \gamma''(s)$ is called the* curvature vector.

Definition 4.14. *$\kappa = \kappa(s) = \|T'(s)\|$ is called the* curvature *of γ at the point $\gamma(s)$. Note that $0 \leq \kappa$. For $\kappa > 0$, we define another unit length vector N as, $N = N(s) = T'/\kappa$. N is called the* normal vector *or the* unit normal.

Rewriting, we see $T'(s) = \kappa(s)N(s)$. If $\kappa = 0$, N is clearly not defined. Actually, it can be shown that if $\kappa(s) = 0$ over an interval, then γ is a straight line over that interval.

Definition 4.15. *The scalar value $1/\kappa$ is called the* radius of curvature.

The motivation and geometric intuition behind this term is derived in Section 4.4.

Several properties of inner products will be useful in our consideration of the curve geometry.

Lemma 4.16. *Let $\beta(t)$ and $\alpha(t)$ be regular parametric representations. Then*

$$\frac{d}{dt}\langle \alpha(t), \beta(t)\rangle = \langle \alpha'(t), \beta(t)\rangle + \langle \alpha(t), \beta'(t)\rangle.$$

Proof: If $\alpha(t) = (x_\alpha, y_\alpha, z_\alpha)$ and $\beta(t) = (x_\beta, y_\beta, z_\beta)$ then $\langle \alpha(t), \beta(t)\rangle = x_\alpha x_\beta + y_\alpha y_\beta + z_\alpha z_\beta$. Using the product rule for differentiating functions,

$$\begin{aligned}\frac{d}{dt}\langle\alpha(t),\beta(t)\rangle &= x'_\alpha x_\beta + y'_\alpha y_\beta + z'_\alpha z_\beta + x_\alpha x'_\beta + y_\alpha y'_\beta + z_\alpha z'_\beta \\ &= \langle\alpha'(t),\beta(t)\rangle + \langle\alpha(t),\beta'(t)\rangle.\end{aligned}$$ ■

Lemma 4.17. *If $\|\alpha(t)\| = 1$ for all values of t, then $\langle\alpha(t),\alpha'(t)\rangle = 0$, that is, α is perpendicular to its derivative α'.*

Proof: $1 = \|\alpha(t)\|^2 = \langle\alpha(t),\alpha(t)\rangle$, so

$$\begin{aligned}\frac{d}{dt}(1) \quad = 0 = \quad & \frac{d}{dt}\langle\alpha(t),\alpha(t)\rangle \\ = \quad & 2\langle\alpha(t),\alpha'(t)\rangle.\end{aligned}$$ ■

Using these properties we show:

Theorem 4.18. $\langle T, N\rangle = 0$. *That is, the unit tangent and the unit normal are perpendicular to each other for all s.*

Proof: Since γ is in the arc length parameterization, for all s,

$$\|T\|^2 = 1 = \langle T(s), T(s)\rangle.$$

we obtain from Lemma 4.17 and Definition 4.14,

$$0 = 2\langle T(s), T'(s)\rangle = 2\kappa(s)\langle T(s), N(s)\rangle.$$ ■

Definition 4.19. *Finally we define the* binormal, $B(s) = T(s) \times N(s)$.

Note that since T and N are orthonormal, the triple $\{T, N, B\}$ forms a *right-handed* orthonormal set.

Definition 4.20. *The triple $\{T, N, B\}$ forms an orthonormal basis for $\mathbb{R}^3$ which changes for each point on the curve. It is called the* moving trihedron.

Definition 4.21. *The plane through the curve $\gamma(s)$ orthogonal to the curve normal N is called the* rectifying plane. *It has direction vectors T and B.*

Definition 4.22. *The plane through the curve $\gamma(s)$ with direction vectors T and N (and hence normal B) is called the* osculating plane.

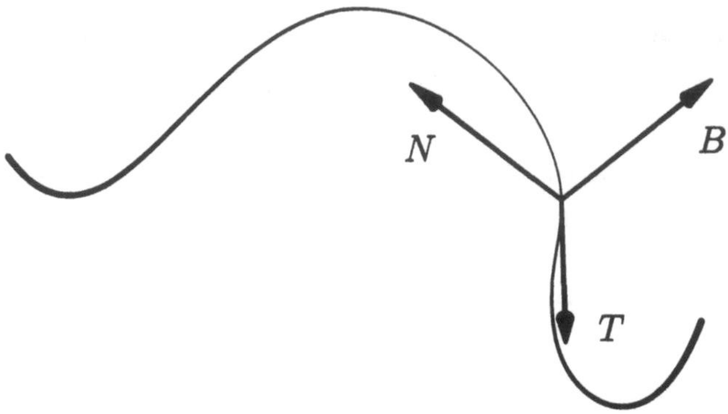

Figure 4.4. A moving trihedron on a space curve.

Example 4.23. For arbitrary a, b with $a > 0$, let $c = \sqrt{a^2 + b^2}$ and define

$$\gamma(s) = \left(a\cos(s/c), a\sin(s/c), \frac{bs}{c}\right).$$

We see that $\sqrt{\left(a\cos(s/c)\right)^2 + \left(a\sin(s/c)\right)^2} = \sqrt{a^2} = a$, so the x and y coordinate values lie on a circle of constant radius for all values of s. This implies that the curve lies on a cylinder. Further, since the z- component is linear in the parameter, this is just a helix. We investigate its properties.

$$\begin{aligned}
\gamma'(s) &= \left(-\frac{a}{c}\sin(s/c), \frac{a}{c}\cos(s/c), \frac{b}{c}\right), \\
\|\gamma'(s)\| &= \sqrt{\left(-a/c\sin(s/c)\right)^2 + \left(a/c\cos(s/c)\right)^2 + (b/c)^2} \\
&= \sqrt{(-a/c)^2 + (b/c)^2} \\
&= \frac{\sqrt{a^2 + b^2}}{c} \\
&= 1.
\end{aligned}$$

Thus, $\gamma(s)$ has an arc length parameterization and

$$\begin{aligned}
T(s) &= \left(-\frac{a}{c}\sin(s/c), \frac{a}{c}\cos(s/c), \frac{b}{c}\right); \\
\gamma''(s) = T'(s) &= \left(-\frac{a}{c^2}\cos(s/c), -\frac{a}{c^2}\sin(s/c), 0\right),
\end{aligned}$$

so

$$\begin{aligned}\kappa(s) = \|T'(s)\| &= \sqrt{\left(-\frac{a}{c^2}\cos(s/c)\right)^2 + \left(-\frac{a}{c^2}\sin(s/c)\right)^2} \\ &= \frac{\sqrt{a^2}}{c^2} \\ &= \frac{a}{c^2}.\end{aligned}$$

Thus, this curve has constant curvature. Since $T'(s) = \kappa N(s)$, $N(s) = (-\cos(s/c), -\sin(s/c), 0)$. This means that for any a and b, N always points in towards the axis of the cylinder on which the curve lies.

The binormal B is computed as follows.

$$\begin{aligned}B = T \times N &= \left(-\frac{a}{c}\sin(s/c), \frac{a}{c}\cos(s/c), \frac{b}{c}\right) \times (-\cos(s/c), -\sin(s/c), 0) \\ &= \left(\frac{b}{c}\sin(s/c), -\frac{b}{c}\cos(s/c), \frac{a}{c}\right).\end{aligned}$$

Other interesting items are: What is the pitch of the helix? What is the vertical thread spacing? □

4.4 Frenet Equations

The moving trihedron forms a basis for $\mathbb{R}^3$ at each point on the curve, which is intrinsic to the representation (not to the geometry!). The direction and motion of the curve is well defined by the trihedron. However, its rate of change of direction and twist is well defined by the rates of changes of these quantities. If we can write the derivatives in terms of simple combinations of the originals, we can use this information to study the shape of the curve.

We know that $\gamma' = T$ and $T' = \kappa N$. What is B'?

Since $\|B\| = \|T \times N\| = \|T\| \, \|N\| \sin \pi/2 = 1$ for all s, it follows from Lemma 4.17 that $(B, B') = 0$. This means that B' can be written in terms of only T and N. Since $B = T \times N$, $(B, T) = 0$ then implies

$$\begin{aligned}0 &= \frac{d}{ds}\langle B(s), T(s)\rangle \\ &= \langle B', T\rangle + \langle B, T'\rangle \\ &= \langle B', T\rangle + \kappa(s)\langle B(s), N(s)\rangle \\ &= \langle B', T\rangle.\end{aligned}$$

Hence, B' is perpendicular to T also, and hence must lie in the N direction only. Select the scalar so that $B' = -\tau(s)N(s)$. The negative sign is a convention.

Definition 4.24. *The quantity τ given by $B' = -\tau(s)N(s)$ is called the* torsion *of the curve.*

The torsion is a measure of the nonplanarity of the curve. The larger the torsion, the faster the osculating plane is changing; the smaller the torsion magnitude, the slower the osculating plane is changing. Note that for a plane curve, $B' = 0$ for all s since the plane containing N and T never changes. Thus,the direction of B does not change and hence $\tau(s) = 0$ for a plane curve.

Finally we ask what is N' in terms of $\{T, N, B\}$?

Since $N = B \times T$ we can use the properties of differentiating cross products to obtain N'.

$$\begin{aligned} N' &= B' \times T + B \times T' \\ &= (-\tau N) \times T + B \times (\kappa N) \\ &= -\tau(N \times T) + \kappa(B \times N) \\ &= \tau B - \kappa T \\ &= -\kappa T + \tau B. \end{aligned}$$

The decomposition is then $N'(s) = -\kappa(s)T(s) + \tau(s)B(s)$.

We have derived,

Theorem 4.25. Frenet Equations. *For an arc length parametrized curve $\gamma(s)$ with unit tangent $T(s)$, normal $N(s)$, binormal $B(s)$, curvature $\kappa(s)$, and torsion $\tau(s)$,*

$$\begin{aligned} T'(s) &= && \kappa(s)N(s) && \\ N'(s) &= -\kappa(s)T(s) && && +\tau(s)B(s) \\ B'(s) &= && -\tau(s)N(s). && \end{aligned}$$

This can be written in matrix form as

$$\begin{bmatrix} T'(s) \\ N'(s) \\ B'(s) \end{bmatrix} = \begin{bmatrix} 0 & \kappa(s) & 0 \\ -\kappa(s) & 0 & \tau(s) \\ 0 & -\tau(s) & 0 \end{bmatrix} \begin{bmatrix} T(s) \\ N(s) \\ B(s) \end{bmatrix}.$$

Example 4.26. For the helix above,

$$\begin{aligned} B'(s) &= \left(\frac{b}{c^2}\cos(s/c), \frac{b}{c^2}\sin(s/c), 0\right) \\ &= -\tau(s)(-\cos(s/c), -\sin(s/c), 0); \end{aligned}$$

$\tau = b/c^2$, the torsion, is a constant for this curve. □

Lemma 4.27. *If γ is an arc length parametrized curve such that τ is identically 0 and such that κ is a constant, then γ is a circle.*

Proof: Define $\alpha = \gamma + (1/\kappa)N$. Then, since $\tau = 0$,

$$\begin{aligned} \alpha' &= \gamma' + \left(\frac{1}{\kappa}\right)' N + \left(\frac{1}{\kappa}\right) N' \\ &= T + 0N + \left(\frac{1}{\kappa}\right)[-\kappa T + \tau B] \\ &= T - T \\ &= 0. \end{aligned}$$

Thus α' is identically 0 and, hence, $\alpha = \mathbf{c}$, a constant. Now, $\|\alpha - \gamma\| = \|\mathbf{c} - \gamma\| = \|(1/\kappa)N\| = |1/\kappa|$. Thus all points on γ are the same distance, $1/\kappa$, from a specified point, $\mathbf{c}$. This meets the definition for γ to be a circle. Note that a constant radius of curvature alone is insufficient since a spiral meets that condition. ■

The orthonormal triple $\{T, N, B\}$ is also known as the *Frenet frame field.* It can be shown that the osculating plane contains *more* of the curve locally than any other plane and is a kind of local *best fit,* an orthogonal projection of the curve onto that plane will give the least distortion locally.

Consider the second order Taylor expansion of γ about s_0, denoted as γ_T.

$$\begin{aligned} \gamma_T(s) &= \gamma(s_0) + \gamma'(s_0)(s - s_0) + \gamma''(s_0)(s - s_0)^2/2 \\ &= \gamma(s_0) + T(s_0)(s - s_0) + \kappa(s_0)N(s_0)(s - s_0)^2/2. \end{aligned}$$

Thus the second order Taylor series expansion for the arc length parametrized curve γ lies in the osculating plane and has the same position, first and second derivative values as γ at $s = s_0$. What happens when the third order term is added?

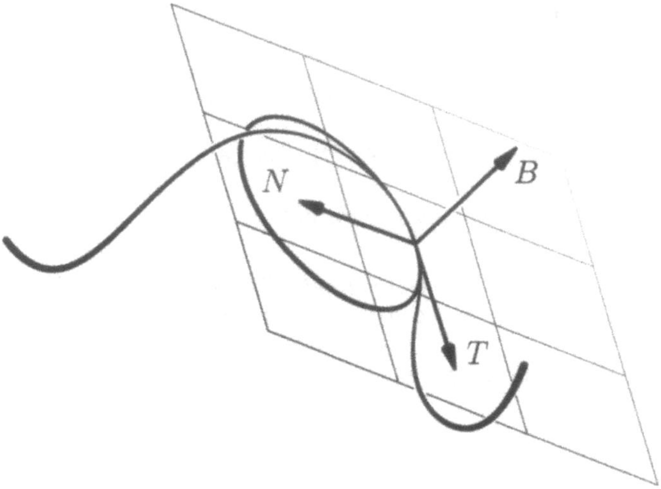

Figure 4.5. A space curve with osculating plane and osculating circle.

$$\begin{aligned}\gamma'''(s_0) &= -\kappa^2 T + \frac{d\kappa}{ds}(s_0)N + \kappa\tau B \quad \text{and} \\ \gamma_T(s) &= \gamma(s_0) + T(s_0)\left[(s-s_0) - \kappa^2(s-s_0)^3/6\right] \\ &\quad + N(s_0)\left[\kappa(s)(s-s_0)^2/2 + \frac{d\kappa}{ds}(s_0)(s-s_0)^3/6\right] \\ &\quad + B\left[\kappa\tau(s-s_0)^3/6\right].\end{aligned}$$

If the torsion is slowly changing, then one can decompose the curve to a good approximation using the Frenet frame, the torsion and the curvature. If $\|\tau\|$ is small enough, then the second degree Taylor approximation gives a very good planar fit to the curve.

We shall investigate the importance of this expansion. Consider the parametric form: $\mathbf{c}(s) = \boldsymbol{k} + \big(r\cos((s-s_0)/r)\big)\boldsymbol{k}_1 + \big(r\sin((s-s_0)/r)\big)\boldsymbol{k}_2$, where $\boldsymbol{k}_1$ and $\boldsymbol{k}_2$ are two direction vectors. If $(\boldsymbol{k}_1, \boldsymbol{k}_2) = \delta_{ij}$ then $\|\mathbf{c}(s) - \boldsymbol{k}\| = r$, and $\mathbf{c}$ is a circle. In order to represent a circle in this way, one must have $\boldsymbol{k}$, the center, r the radius, and the two orthonormal direction vectors. Set

$$\begin{aligned}\boldsymbol{k}_1 &= -N(s_0); \\ \boldsymbol{k}_2 &= T(s_0); \\ r &= 1/\kappa(s_0); \\ \boldsymbol{k} &= \gamma(s_0) + \frac{1}{\kappa}N.\end{aligned}$$

In the equations above, $T(s_0)$ is the unit tangent of γ at $s = s_0$, $N(s_0)$ is the unit normal to γ at $s = s_0$, and $\kappa(s_0)$ is the curvature of γ at $s = s_0$. Since T and N are orthonormal, the requirement for c to be a circle is met. Then,

$$\begin{aligned}
\mathbf{c}(s_0) &= \boldsymbol{k} + r\boldsymbol{k_1} \\
&= \gamma(s_0);
\end{aligned}$$

$$\begin{aligned}
\mathbf{c}'(s_0) &= \boldsymbol{k_2} \\
&= T(s_0) \\
&= \gamma'(s_0);
\end{aligned}$$

$$\begin{aligned}
\mathbf{c}''(s_0) &= -\frac{1}{r}\boldsymbol{k_1} \\
&= \kappa(s_0)N(s_0) \\
&= \gamma''(s_0).
\end{aligned}$$

Thus, the position and first two derivatives of the circle match the position and first two derivatives of the curve γ at $s = s_0$.

Definition 4.28. *Two curves γ_1 and γ_2 have contact of degree p at s_0 if $\gamma_1^{(m)}(s_0) = \gamma_2^{(m)}(s_0)$ for $m = 0, \ldots, p$.*

However,

$$\begin{aligned}
\mathbf{c}^{(3)}(s_0) &= -\frac{1}{r^2}\boldsymbol{k_2} \\
&= \kappa^2(s_0)T(s_0); \\
\gamma^{(3)}(s_0) &= \kappa^2(s_0)T(s_0) + \frac{d\kappa}{ds}(s_0)N(s_0) + \kappa(s_0)\tau(s_0)B(s_0).
\end{aligned}$$

Thus, the third derivative of the circle is already completely defined and cannot match γ unless γ is planar and has constant curvature, i.e., γ is already a circle. Thus, the highest order contact with a circle possible is second order contact. It also has the same contact with the second order Taylor series.

Definition 4.29. *The second degree Taylor expansion at $s = s_0$ for a curve γ, with arc length parameterization, matches the position, unit tangent, curvature and unit normal to γ at $s = s_0$. This parabola is called the* osculating parabola.

Definition 4.30. *A circle defined so that it passes through a point on a curve and matches the curve tangent and curve normal direction with radius defined as* $1/\kappa$ *is called the* osculating circle *to the curve at that point.*

The last question of great concern is based on the name of this section. Can two distinct curves have the same torsion and curvature? If not, is knowledge of the torsion and curvature equations as functions of arc length enough to specify a curve uniquely? What are the necessary and sufficient conditions? It turns out that:

Theorem 4.31. *Let* $k(s)$ *and* $f(s)$ *be continuous scalar valued functions on* $a \leq s \leq b$, *and* $0 < k(s)$. *There exists exactly one regular curve* $\gamma(s)$ *for which* $k(s)$ *is the curvature and* $f(s)$ *is the torsion. The position of the curve in space is, however, free. These functions are called the* intrinsic equations *of the curve.*

The proof of this theorem requires solving systems of differential equations, and is best left to the many differential geometry books that devote more space to this subject. Although theory guarantees that it can be done, in practice it is typically an intractable computational problem.

The importance of this theorem cannot be overstated. Since one can characterize, that is, uniquely define and construct the shape of a three space curve by two scalar functions, one might be tempted to use these quantities as design parameters. That is, specify the curvature and torsion and then reconstruct the curve. This is not a easy procedure, and becomes more difficult since most people do not have geometric intuition about shape directly from two scalar functions.

However, if one is given two distinct parametric representations, one can tell if they define the same regular curve by deriving the curvature and torsion equations (in arc length parameterization). If those two equations are the same, then the curves are the same. If one pair is not identical, the curves are distinct.

The difficulty in obtaining an arc length parameterization from an arbitrary regular parameterization can most easily be shown by example. In Section 3.7 it was shown that any conic arc can be represented by many different rational quadratics. In Section 2.13 the arc length parameterization for a circle was determined in terms of sines and cosines. The process of finding the arc length parameterization of a circular arc from a rational quadratic parametric form must transform from rationals to trignometric functions. The thought is altogether overwhelming!

4.5 Frenet Equations for Non-Arc Length Parameterizations

For practical purposes, most of the time the user has only a non-arc length parameterization available. As we shall see later, it is frequently, in fact, undesirable to obtain an arc length parameterization. One wants, therefore, to obtain the calculations for $\{T, N, B\}$ as functions of the given parameterization, and obtain from that the Frenet equations for arbitrary regular representations.

Given $\beta(t)$ a regular curve, we review some properties. Since $s = \int_{t_0}^{t} \|\beta'(w)\| dw$, we know several things. First, since β is regular, s is an increasing function of t. That is, there is always a bijection from some closed interval (for t) to another closed interval (for s). Hence one can write t as a function of s, $t = p(s)$, for some function p. Also $\frac{ds}{dt} = \|\beta'(t)\|$, an application of the fundamental theorem of calculus. Thus, for any regular representation, one can transform to the arc length parameterization with an allowable transformation and remain in the equivalency class. Suppose $\gamma(s)$ is the arc length parameterization for $\beta(t)$. Then for any point on the curve the moving trihedron is known, the curvature is known, and the torsion is known. Now, for a non-arc length parametrized curve $\beta(t)$, since $t = p(s)$,

$$\frac{dT_\beta}{dt} = \frac{dT_\beta}{ds}\frac{ds}{dt} = \kappa N_\beta \|\beta'(t)\|; \tag{4.1}$$

$$\frac{dN_\beta}{dt} = \frac{dN_\beta}{ds}\frac{ds}{dt} = (-\kappa_\beta T_\beta + \tau_\beta B_\beta)\|\beta'(t)\|; \tag{4.2}$$

$$\frac{dB_\beta}{dt} = \frac{dB_\beta}{ds}\frac{ds}{dt} = -\tau_\beta N_\beta \|\beta'(t)\|. \tag{4.3}$$

4.6 Intrinsic Functions on Arbitrary Parameterizations

Having found the rates of change for $\{T, N, B\}$ of a curve β with respect to arbitrary parameterization, let us now find $\{T, N, B\}$ for arbitrary parameterization. We begin by finding the most straightforward.

$$\begin{aligned} T &= \frac{d\beta(t)}{ds} = \frac{d\beta(t)}{dt}\frac{dt}{ds} \\ &= \frac{\beta'(t)}{\left(\frac{ds}{dt}\right)} = \frac{\beta'(t)}{\|\beta'(t)\|}, \end{aligned} \tag{4.4}$$

so

$$T_\beta(t) = \frac{\beta'(t)}{\|\beta'(t)\|},$$

and

$$\beta'(t) = \|\beta'(t)\| T_\beta(t).$$

We shall leave out the subscript β, since it is easily understood. Now we would like to find N and B as functions of β' and β''.

$$\begin{aligned} \beta''(t) = \frac{d\beta'(t)}{dt} &= \left(\frac{d\|\beta'(t)\|}{dt}\right) T(t) + \left(\frac{dT(t)}{dt}\right)\|\beta'(t)\| \\ &= \left(\frac{d\|\beta'(t)\|}{dt}\right) T(t) + (\kappa(t)N(t))\|\beta'(t)\|^2. \end{aligned} \quad (4.5)$$

Thus, if β is not in an arc length parameterization, its second derivative, β'', is no longer in the N direction, although it remains in the osculating plane. Thus $\beta' \times \beta''$ either points in the direction of B or the direction of $-B$. We will try to discover which.

$$\begin{aligned} \beta' \times \beta'' &= (\|\beta'\|T) \times \left[\left(\frac{d\|\beta'(t)\|}{dt}\right) T(t) + \kappa(t)N(t)\|\beta'(t)\|^2\right] \\ &= \|\beta'\| \left(\frac{d\|\beta'(t)\|}{dt}\right)(T \times T) + \|\beta'\|^3 \kappa(t)(T \times N). \end{aligned}$$

Now, $T \times N = B$, by definition of B, and $T \times T = 0$ which gives

$$\beta' \times \beta'' = \|\beta'\|^3 \kappa(t) B. \quad (4.6)$$

Since κ and $\|\beta'\|$ are nonnegative, $\beta' \times \beta''$ points in the direction of B, and

$$B = \frac{\beta' \times \beta''}{\|\beta' \times \beta''\|}, \quad (4.7)$$

the unit vector in that direction. However,

$$\|\beta' \times \beta''\| = \|\beta'\|^3 \kappa(t) \|B\| = \|\beta'\|^3 \kappa(t).$$

Regrouping yields

$$\kappa(t) = \frac{\|\beta' \times \beta''\|}{\|\beta'\|^3}. \quad (4.8)$$

Since $N = B \times T$ we can immediately find N as a function of β.

$$\begin{aligned} N &= \frac{\beta'(t) \times \beta''(t)}{\|\beta'(t) \times \beta''(t)\|} \times \frac{\beta'(t)}{\|\beta'(t)\|} \\ &= \frac{(\beta'(t) \times \beta''(t)) \times \beta'(t)}{\|\beta'(t)\| \, \|\beta'(t) \times \beta''(t)\|} \\ &= \frac{\langle\beta'(t), \beta'(t)\rangle\beta''(t)}{\|\beta'(t)\| \, \|\beta'(t) \times \beta''(t)\|} - \frac{\langle\beta'(t), \beta''(t)\rangle\beta'(t)}{\|\beta'(t)\| \, \|\beta'(t) \times \beta''(t)\|}. \end{aligned} \tag{4.9}$$

We must still calculate the torsion. Unfortunately, to calculate τ we must find β''' in the Frenet frame. We continue to use the tools developed earlier and write

$$\beta''' = \langle\beta''', T\rangle T + \langle\beta''', N\rangle N + \langle\beta''', B\rangle B.$$

We know β'' from Equation 4.5 so we can immediately differentiate both sides with respect to t.

$$\begin{aligned} \beta''' &= \left(\frac{d\|\beta'\|}{dt} T + \kappa\|\beta'\|^2 N\right)' \\ &= \left(\frac{d\|\beta'\|}{dt} T\right)' + \left(\kappa\|\beta'\|^2 N\right)' \\ &= \frac{d^2\|\beta'\|}{dt^2} T + \frac{d\|\beta'\|}{dt}\frac{dT}{dt} + \left(\kappa\|\beta'\|^2\right)' N + \left(\kappa\|\beta'\|^2\right)\frac{dN}{dt} \\ &= \frac{d^2\|\beta'\|}{dt^2} T + \frac{d\|\beta'\|}{dt}\|\beta'\|\kappa N + \left(\kappa\|\beta'\|^2\right)' N \\ &\quad + \left(\kappa\|\beta'\|^2\right)\|\beta'\|\left(-\kappa T + \tau B\right) \\ &= \left(\frac{d^2\|\beta'\|}{dt^2} - \kappa^2\|\beta'\|^3\right) T + \left(\kappa\frac{d\|\beta'\|}{dt}\|\beta'\| + \left(\kappa\|\beta'\|^2\right)'\right) N \\ &\quad + \kappa\tau\|\beta'\|^3 B. \end{aligned}$$

By uniqueness of representation,

$$\langle\beta''', B\rangle = \kappa\tau\|\beta'\|^3,$$

so

$$\begin{aligned} \tau &= \frac{\langle\beta''', B\rangle}{\kappa\|\beta'\|^3} \\ &= \frac{\left\langle\beta''', \frac{\beta' \times \beta''}{\|\beta' \times \beta''\|}\right\rangle}{\|\beta' \times \beta''\|} \\ &= \frac{\langle\beta''', \beta' \times \beta''\rangle}{\|\beta' \times \beta''\|^2}. \end{aligned} \tag{4.10}$$

Example 4.32. Calculate the moving trihedron values as functions of t. Also calculate the curvature and torsion for

$$\beta(t) = \left(3t - t^3, 3t^2, 3t + t^3\right).$$

We must first show that the curve parameterization is not an arc length parameterization.

$$\beta'(t) = \left(3 - 3t^2, 6t, 3 + 3t^2\right) \qquad \text{and} \qquad \|\beta'\| \not\equiv 1.$$

Now find the derivatives to use in subsequent calculations.

$$\begin{aligned}
\beta'(t) &= 3\left(1 - t^2, 2t, 1 + t^2\right); \\
\beta''(t) &= 6(-t, 1, t); \\
\beta'''(t) &= 6(-1, 0, 1).
\end{aligned}$$

Using the derived equations:

$$\langle \beta', \beta' \rangle = 9\left(\left(1 - t^2\right)^2 + 4t^2 + \left(1 + t^2\right)^2\right) = 18\left(1 + t^2\right)^2$$

and

$$\|\beta'\| = \left(1 + t^2\right)\sqrt{18}.$$

Therefore,

$$T = \frac{\left(1 - t^2, 2t, 1 + t^2\right)}{\sqrt{2}\,(1 + t^2)}.$$

To calculate the other quantities we need the cross product

$$\beta' \times \beta'' = 18\begin{vmatrix} \mathbf{e_1} & \mathbf{e_2} & \mathbf{e_3} \\ (1 - t^2) & 2t & 1 + t^2 \\ -t & 1 & t \end{vmatrix},$$

$$\begin{aligned}
\|\beta' \times \beta''\|^2 &= 18^2\left[\left(-1 + t^2\right)^2 + 4t^2 + \left(1 + t^2\right)^2\right] \\
&= 2 \cdot 18^2\left(1 + t^2\right)^2.
\end{aligned}$$

Thus,

$$B = \frac{\left(-1 + t^2, -2t, 1 + t^2\right)}{\sqrt{2}\,(1 + t^2)}.$$

Now,

$$\begin{aligned}
\langle \beta' \times \beta'', \beta''' \rangle &= 6 \cdot 18\left(\left(1 - t^2\right) + 0 + \left(1 + t^2\right)\right) \\
&= 6 \cdot 18 \cdot 2,
\end{aligned}$$

and

$$\begin{aligned}
\tau &= \frac{\langle \beta' \times \beta'', \beta''' \rangle}{\|\beta' \times \beta''\|^2} \\
&= \frac{12 \cdot 18}{2 \cdot 18^2 \left(1+t^2\right)^2} \\
&= \frac{1}{3\left(1+t^2\right)^2}.
\end{aligned}$$

The remaining unknowns are $N = B \times T$ which can be calculated, and

$$\begin{aligned}
\kappa &= \frac{\|\beta' \times \beta''\|}{\|\beta'\|^3} \\
&= \frac{18\sqrt{2}\left(1+t^2\right)}{\left(1+t^2\right)^3 18^{3/2}} \\
&= \frac{1}{3\left(1+t^2\right)^2}.
\end{aligned}$$

□

Example 4.33. Suppose $\gamma(t) = (t, f(t), 0)$, where $f(t)$ is a $C^{(2)}$function. We compute the curvature of γ. Since

$$\begin{aligned}
\gamma'(t) &= (1, f'(t), 0), \\
\gamma''(t) &= (0, f''(t), 0),
\end{aligned}$$

by Equation 4.8,

$$\begin{aligned}
\kappa(t) &= \frac{\|\gamma' \times \gamma''\|}{\|\gamma'\|^3} \\
&= \frac{\|f''\|}{\left(\sqrt{1+\left(f'(t)\right)^2}\right)^3}. \qquad (4.11)
\end{aligned}$$

□

4.7 Piecing together Parametric Curves

Suppose it is necessary to either move two parametric curves in space until they form a compound curve, or suppose that a compound curve already exists. For design or analysis reasons it may be necessary to know if the compound curve is tangent and curvature continuous. We derive necessary and sufficient conditions.

If γ_1 and γ_2 are two regular parametric curves defined on the interval $t \in [0,1]$, let

$$\gamma(t) = \begin{cases} \gamma_1(t), & t \in [0,1) \\ \gamma_2(t-1), & t \in [1,2]. \end{cases} \tag{4.12}$$

We investigate when γ is $C^{(0)}$, $C^{(1)}$, $C^{(2)}$, or curvature continuous at $t = 1$.

Definition 4.34. *The term* geometrically $C^{(n)}$ *will mean that a curve is* $C^{(n)}$ *when represented in the arc length parameterization.*

Geometrically $C^{(1)}$ then means that the unit tangent vector is continuous. We shall say that the curve is curvature continuous if, in addition to having a continuous tangent vector, the normal vector as well as the scalar curvature are also changing continuously. This will be equivalent to being $C^{(2)}$ in the arc length parameterization. Remember, the osculating plane is defined by the tangent and the normal vector, so this ensures a continuously changing osculating plane. The term $C^{(1)}$ and $C^{(2)}$ will be used in the usual manner.

$C^{(0)}$is easy to check, since γ can be $C^{(0)}$at $t = 1$ if and only if $\gamma_1(1) = \gamma_2(0)$, so we move to geometrically $C^{(1)}$. In order for this to occur,

$$\boldsymbol{T_1} = \frac{\gamma_1'(1)}{\|\gamma_1'(1)\|} = \frac{\gamma_2'(0)}{\|\gamma_2'(0)\|} = \boldsymbol{T_2}.$$

Let us denote that vector T_1, the tangent vector of γ at $t = 1$, and let $c_1 = \|\gamma_1'(1)\|$, $c_2 = \|\gamma_2'(0)\|$, and $k_1 = c_1/c_2$. We have

$$\gamma_1'(1) = k_1\gamma_2'(0). \tag{4.13}$$

As a consequence of Equation 4.13, the parametric derivatives can abruptly change magnitude at $t = 1$, but no change of direction is allowed. If the magnitudes are the same, that is, if $k_1 = 1$, then the compound curve is $C^{(1)}$at $t = 1$.

For the curve to be curvature continuous, the unit tangent vector and the normal vector must be continuous. But then it follows that the binormal is also continuous. In fact, continuity of any two of these implies continuity of the third. From Equation 4.6 we know that

$$\kappa(t)B(t) = \frac{\gamma'(t) \times \gamma''(t)}{\|\gamma'(t)\|^3}.$$

A necessary and sufficient condition for curvature continuity at $t = 1$ is that $\kappa_1(1)\boldsymbol{B_1}(1) = \kappa_2(0)\boldsymbol{B_2}(0)$. Now,

$$T_1 = \frac{\gamma'(1)}{\|\gamma'(1)\|} = \frac{\gamma_1'(1)}{\|\gamma_1'(1)\|} = \frac{\gamma_2'(0)}{\|\gamma_2'(0)\|};$$

$$\frac{\gamma_1'(1)\times\gamma_1''(1)}{\|\gamma_1'(1)\|^3} = \frac{\gamma_1'(1)}{\|\gamma_1'(1)\|}\times\frac{\gamma_1''(1)}{\|\gamma_1'(1)\|^2} = T_1\times\frac{\gamma_1''(1)}{\|\gamma_1'(1)\|^2};$$

$$\frac{\gamma_2'(0)\times\gamma_2''(0)}{\|\gamma_2'(0)\|^3} = \frac{\gamma_2'(0)}{\|\gamma_2'(0)\|}\times\frac{\gamma_2''(0)}{\|\gamma_2'(0)\|^2} = T_1\times\frac{\gamma_2''(0)}{\|\gamma_2'(0)\|^2}.$$

Setting them equal, then requires that

$$T_1\times\frac{\gamma_1''(1)}{\|\gamma_1'(1)\|^2} = T_1\times\frac{\gamma_2''(0)}{\|\gamma_2'(0)\|^2}$$

and

$$\begin{aligned}
0 &= T_1\times\left(\frac{\gamma_1''(1)}{\|\gamma_1'(1)\|^2}-\frac{\gamma_2''(0)}{\|\gamma_2'(0)\|^2}\right)\\
&= T_1\times\left(\gamma_1''(1)-\frac{\|\gamma_1'(1)\|^2}{\|\gamma_2'(0)\|^2}\gamma_2''(0)\right)\\
&= T_1\times\left(\gamma_1''(1)-(k_1)^2\gamma_2''(0)\right).
\end{aligned}$$

The cross product being zero means that $(\gamma_1''(1)-(k_1)^2\gamma_2''(0))$ is in the same direction as T_1, and so there exists a constant c_3 such that

$$c_3\gamma_2'(0) = \gamma_1''(1)-(k_1)^2\gamma_2''(0)$$

from which we have

$$\gamma_1''(1) = c_3\gamma_2'(0)+(k_1)^2\gamma_2''(0). \tag{4.14}$$

For every real value c_3, curvature continuity is maintained whenever this condition is upheld. Clearly, if a composite curve already exists it is then possible to test if it is curvature continuous by solving for k_1 and c_3. Conversely, it is possible to construct such a curve by prescribing the vector T_1 and the constant c_3 and k_1. Now $\gamma_2''(0)$ and $\gamma_2'(0)$ form the osculating plane, so it is clear that the osculating plane is preserved.

Example 4.35. Suppose $\gamma_1(t) = (1,-5)(1-t) + (1,0)t$ and $\gamma_2(t) = (\cos\pi t, \sin\pi t)$, and we form the compound γ as described above.

$\gamma_1'(1) = (0,5)$ and $\gamma_2'(0) = (0,\pi)$. Since the directions are the same, but the magnitudes differ, the compound curve is not $C^{(1)}$, but it is geometrically $C^{(1)}$; $k_1 = 5/\pi$.

$\gamma_1''(1) = (0,0)$ and $\gamma_2''(0) = (-\pi^2, 0)$, so the compound curve is not $C^{(2)}$at $t = 1$. Further,

$$\begin{aligned} c_3\gamma_2'(0) + (k_1)^2\gamma_2''(0) &= c_3(0,\pi) + \frac{25}{\pi^2}(-\pi^2, 0) \\ &= (-25, \pi c_3). \end{aligned}$$

Since the above quantity cannot equal (0,0) (which equals $\gamma_1''(1)$), the compound curve cannot be curvature continuous. □

Many engineering drawings and older models are done with constructions consisting of lines and circular arcs. The above example shows that it is impossible to make a curve designed that way curvature continuous.

4.8 Exercises

1. Consider the curve $r = 2\cos u - 1$, $u \in [0, 2\pi]$, represented in polar coordinates.

   (a) A corresponding parameterization is $x(u) = \cos u(2\cos u - 1)$ and $y(u) = \sin u(2\cos u - 1)$.

   (b) Modify that to get $t = u-1$. Then $I_2 = [-1, 2\pi-1]$, with $x(t) = \cos(t+1)(2\cos(t+1)-1)$ and $y(t) = \sin(t+1)(2\cos(t+1)-1))$ as the representation.

   (c) Let

   $$u = \begin{cases} t & \text{for } t \in [0, \pi/3], \\ -t + 2\pi & \text{for } t \in (\pi/3, 5\pi/3), \\ t & \text{for } t \in [5\pi/3, 2\pi]. \end{cases}$$

   In what directions is the curve traversed for a, b, and c? Are they in the same equivalence class?

2. An operation $\sim$ on a set of elements S is called an *equivalence relation* if it has the following three properties for elements x, y, $z \in S$:

   (a) $x \sim x$ (reflexivity and read as "x is related to itself").

(b) $x \sim y$ implies $y \sim x$ (commutativity).

(c) $x \sim y$ and $y \sim z$ implies $x \sim z$ (transitivity).

Real number equality is an equivalence relation, as is modulo arithmetic. Floating point equality, $=_f$, is not an equivalence relation since $x =_f y$ and $y =_f z$ does not imply that $x =_f z$. This implication can fail because of error accumulation.

Prove Lemma 4.7 by showing:

(a) The identity function $p(u) = u$ is an allowable change of parameter where $I_1 = I_2$.

(b) If p is an allowable change of parameter from I_1 to I_2, then p^{-1} exists and is an allowable change of parameter from I_2 to I_1.

(c) If p is an allowable change of parameter from I_1 to I_2 and q is an allowable change of parameter from I_2 to I_3 then $q(p)$ is an allowable change of parameter from I_1 to I_3.

Prove and use these three points to complete the proof of the lemma.

3. Suppose an arc of a circle in rational parametric form for $t \in [0,1]$ is pieced together with a straight line, parametrized for $t \in [1,2]$, and then another arc in rational parametric form, parametrized for $t \in [2,3]$ is added in a different plane from the first arc, but all composed to form a smooth looking curve. This is a very traditional form of design with arcs and lines.

   Find the form of the unit tangent vector for such a curve. Is this composite curve curvature continuous?

4. Suppose

$$\gamma(t) = \begin{cases} (1-t)^3 P_0 + 3(1-t)^2 t P_1 + 3(1-t)t^2 P_2 & \\ \quad + t^3 P_3 & \text{for } 0 \le t < 1 \\ (2-t)^3 P_3 + 3(2-t)^2 (t-1) P_4 & \\ \quad + 3(2-t)(t-1)^2 P_5 + (t-1)^3 P_6 & \text{for } 1 \le t \le 2 \end{cases}$$

   Under what conditions is the tangent vector of γ continuous for all t? Under what conditions is the curvature vector continuous for all t? When is the Frenet frame changing continuously? You may specify your conditions as requirements on the vector coefficients P_i. Are there any degrees of freedom left?

5. Given a regular curve $\alpha(t)$, the evolute of α is defined as

$$\alpha^*(t) = \alpha(t) + \frac{1}{\kappa(t)} N(t).$$

What is the geometrical meaning of α^*? For points (x, y) in the plane, define $A(x, y) = (-y, x)$. If α is a planar curve, show

$$\alpha^* = \alpha + \frac{(\alpha', \alpha')}{(\alpha'', A(\alpha'))} A(\alpha').$$

If $\alpha(t) = (a \cos t/c, a \sin t/c, bt/c)$, where $c^2 = a^2 + b^2$, what is the evolute of α? What is the evolute of its evolute?

6. Programming Exercise: Develop a self-documenting interactive graphics program that will help to visualize the moving Frenet frame, curvature, and torsion on various space curves for arbitrary parameterizations.

5

Bézier Curves and Bernstein Approximation

Chapter 3 developed a parametric rational quadratic representation for conics and in Example 3.16, a constructive algorithm was developed for evaluating a special case conic section. Later in Section 3.9.1, the algorithm is generalized to evaluate all conic sections constructively. Recall that attractive properties of the geometric approach for conics included:

1. the curve interpolated the first point and the last point;
2. the curve was tangent to the *sides* of the polygon formed by the three points that determine it;
3. the rational parametric form used the defining three points for coefficients in the rational parametric formulation derived.

Property 3 means that no computationally costly process is needed to derive the analytical coefficients from the geometric solution.

In this chapter, that constructive process is extended to create curves from ordered sets of $n+1$ points, for arbitrary n. We also investigate which of the above three characteristics are retained by curves constructed with the generalized formulation, and also if any other shape characteristics can be determined.

This leads us to the formulation of the Bézier curve, presentation of its properties, and corresponding proofs. We also show its relationship to the Bernstein curve approximation method. This relationship allows us to use approximation and convergence properties from the Bernstein method to understand the behavior of the Bézier curve.

5.1 Constructive Evaluation Curves

An arbitrary collection of $n+1$ points in $\mathbb{R}^3$, $\{P_i\}_{i=0}^n$, is used to define the constructive process and the resulting curve γ. Without loss of generality, we initially define the curve over the parametric interval $[0,1]$. The piecewise linear curve formed by the line segments through P_i and P_{i+1}, for $i = 0, \ldots, n-1$ is called the curve's *control polygon*, or sometimes simply the *polygon*. Since the subscripts start with zero, it has $n+1$ vertices.

Guided by the constructive process from Lemma 3.17, we shall extend this to a polygon of arbitrary size.

Algorithm 5.1. *Given points P_i, $i = 0, \ldots, n$, the goal is to determine a curve $\gamma(t)$, for all values $t \in [0,1]$.*

Step 1: Select a value $t \in [0,1]$. This value remains fixed for the remaining three steps.

Step 2: For $i = 0, \ldots, n$, set $P_i^{[0]}(t) = P_i$.

Step 3: For $j = 1, \ldots, n$, recursively define

$$P_i^{[j]}(t) = (1-t)P_{i-1}^{[j-1]}(t) + tP_i^{[j-1]}(t), \qquad \text{for } i = j, \ldots, n.$$

Step 4: $\gamma(t) = P_n^{[n]}(t)$.

The ordered original points form the $zero^{th}$ level. Figures 5.1–5.4 show recursive sequences of polygons for two distinct t values on the curve. Figure 5.5 shows two complete nested sequences. Note that the sequence of constructive polygons differ for the two values at all but the starting polygon.

Thus, a point on the curve is straightforward to compute as a linear combination of points on polygons with an ever decreasing number of sides.

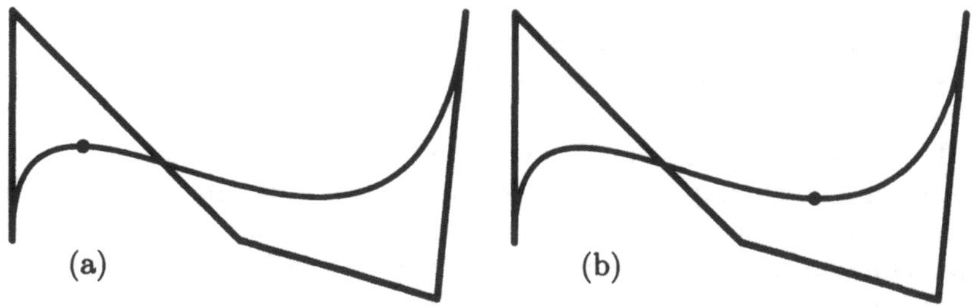

Figure 5.1. Control polygon and curve, also the $j = 0$ polygon, evaluated at (a) $t = 0.25$; (b) $t = 0.625$.

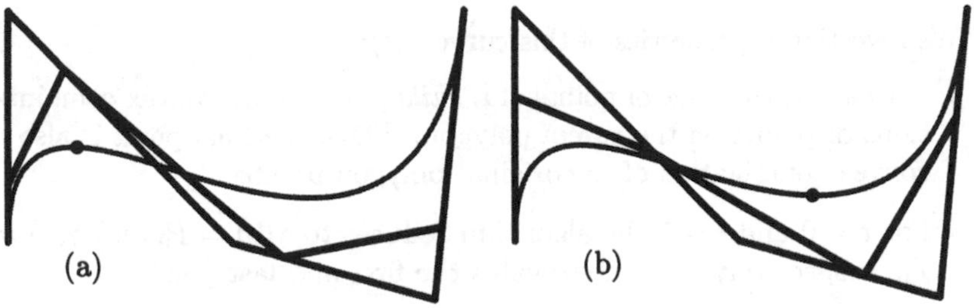

Figure 5.2. The $j = 1$ polygons for (a) $t = 0.25$; (b) $t = 0.625$.

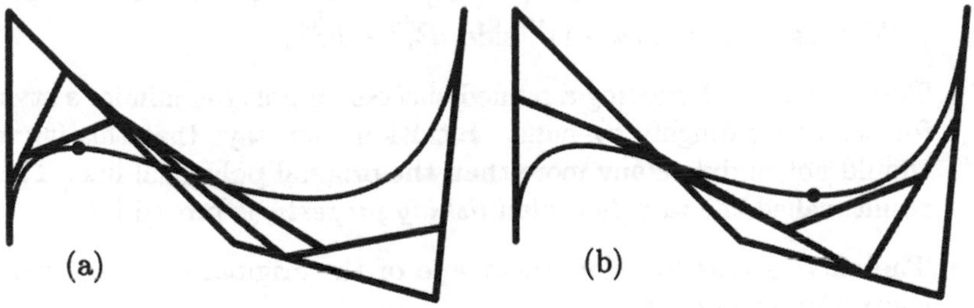

Figure 5.3. The $j = 2$ polygons for (a) $t = 0.25$; (b) $t = 0.625$.

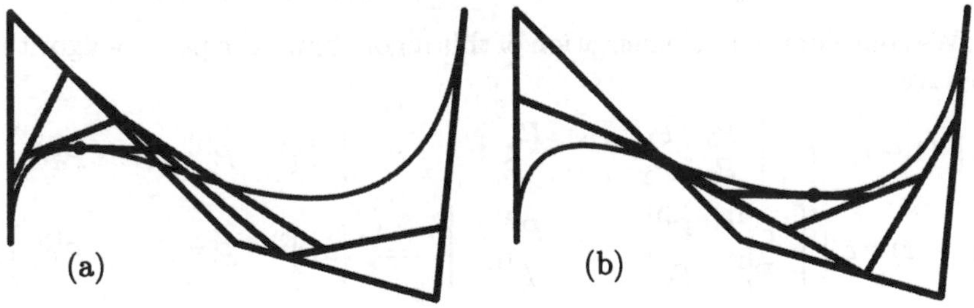

Figure 5.4. The $j = 3$ polygons for (a) $t = 0.25$; (b) $t = 0.625$.

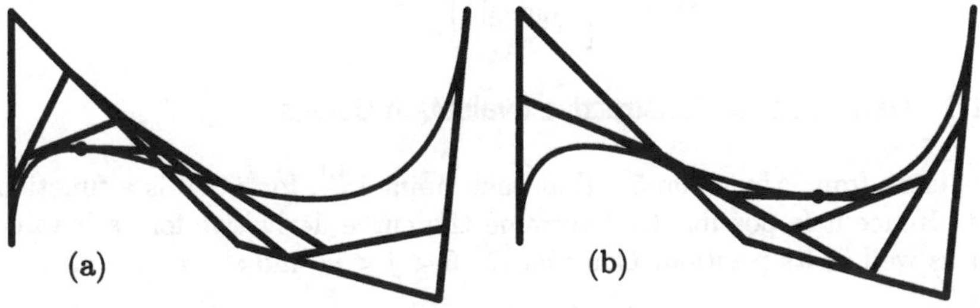

Figure 5.5. The whole sequence of polygons for (a) $t = 0.25$; (b) $t = 0.625$.

We investigate properties of this curve $\gamma(t)$:

- For each t, the control points of a *child* polygon are convex combinations of points on the *parent* polygon. Thus, the final point is also a convex combination of the original polygon points.
- For $t = 0$ and $t = 1$ the algorithm reduces to $\gamma(0) = P_0$ and $\gamma(1) = P_n$, respectively, so γ interpolates the first and last points.
- Intuitively we see that as t gets close to zero, the 0th side of the control polygon approximates its behavior, that is, the curve looks tangent to the $zero^{th}$ side, $P_1^{[0]} - P_0^{[0]}$. Similarly, at $t = 1$ the curve looks tangent to the $(n-1)^{st}$ side, $P_n^{[0]} - P_{n-1}^{[0]}$.
- The operation of creating a refined succession polygon mimics a style for sketching roughly by hand. Intuition then says that this curve should not undulate any more than the original polygonal line. This result, called the *variation diminishing property*, is proved later.
- The curve seems to follow the shape of the original piecwise linear control polygon curve.
- The behavior of the curve is easily controlled. If point P_k, is moved, it is quite easy to recalculate the resulting modified curve.

We consider the implementation of this algorithm. As a pseudo-algorithm we have

$$\begin{bmatrix} (1-t) & t \end{bmatrix} \begin{bmatrix} P_0 & P_1 & \dots & P_{n-1} \\ P_1 & P_2 & \dots & P_n \end{bmatrix} = \begin{bmatrix} P_1^{[1]} & P_2^{[1]} & \dots & P_n^{[1]} \end{bmatrix};$$

$$\begin{bmatrix} (1-t) & t \end{bmatrix} \begin{bmatrix} P_1^{[1]} & P_2^{[1]} & \dots & P_{n-1}^{[1]} \\ P_2^{[1]} & P_3^{[1]} & \dots & P_n^{[1]} \end{bmatrix} = \begin{bmatrix} P_2^{[2]} & P_3^{[2]} & \dots & P_n^{[2]} \end{bmatrix};$$

$$\vdots$$

$$\begin{bmatrix} (1-t) & t \end{bmatrix} \begin{bmatrix} P_{n-1}^{[n-1]} \\ P_n^{[n-1]} \end{bmatrix} = P_n^{[n]}.$$

5.1.1 Derivatives of Constructive Evaluation Curves

It is clear from Algorithm 5.1 that each point $P_i^{[j]}$, for $j > 0$ is a function of t. Hence it is possible to determine the curve derivative for each value of t as well as its position. Consider for $0 < j < n$, and $i = j, \dots, n$,

$$\frac{d}{dt}P_i^{[j]}(t) = P_i^{[j-1]}(t) - P_{i-1}^{[j-1]}(t) + (1-t)\frac{d}{dt}P_{i-1}^{[j-1]}(t) + t\frac{d}{dt}P_i^{[j-1]}(t), \quad (5.1)$$

which shows that the derivative can be recursively determined. Now, since $P_i^{[0]}(t) = P_i$

$$\begin{aligned}
\frac{d}{dt}P_i^{[0]}(t) &= 0 \\
\frac{d}{dt}P_i^{[1]}(t) &= P_i^{[0]}(t) - P_{i-1}^{[0]}(t) \\
&= P_i - P_{i-1}. \qquad (5.2)
\end{aligned}$$

We show that

Lemma 5.2. *The derivative of the constructive curve γ defined in Algorithm 5.1 is given by*

$$\gamma'(t) = n\left(P_n^{[n-1]}(t) - P_{n-1}^{[n-1]}(t)\right). \qquad (5.3)$$

Proof: We use an induction argument, and show that

$$\frac{d}{dt}P_i^{[j]}(t) = j\left(P_i^{[j-1]}(t) - P_{i-1}^{[j-1]}(t)\right).$$

Then, setting $j = n$ proves the lemma. By Equation 5.2, the assertion is true for $j = 1$. Now suppose it is true that

$$\frac{d}{dt}P_i^{[j-1]}(t) = (j-1)\left(P_i^{[j-2]}(t) - P_{i-1}^{[j-2]}(t)\right), \qquad i = j-1, \ldots, n.$$

Then,

$$\begin{aligned}
\frac{d}{dt}P_i^{[j]}(t) &= P_i^{[j-1]}(t) - P_{i-1}^{[j-1]}(t) \\
&+ \quad (1-t)\frac{d}{dt}P_{i-1}^{[j-1]}(t) + t\frac{d}{dt}P_i^{[j-1]}(t) \\
&= P_i^{[j-1]}(t) - P_{i-1}^{[j-1]}(t) \\
&\quad +(1-t)(j-1)\left(P_{i-1}^{[j-2]}(t) - P_{i-2}^{[j-2]}(t)\right) \\
&\quad +t(j-1)\left(P_i^{[j-2]}(t) - P_{i-1}^{[j-2]}(t)\right) \\
&= P_i^{[j-1]}(t) - P_{i-1}^{[j-1]}(t) \\
&\quad +(j-1)\Big(\left((1-t)P_{i-1}^{[j-2]}(t) + tP_i^{[j-2]}(t)\right) \\
&\quad - \left((1-t)P_{i-2}^{[j-2]}(t) + tP_{i-1}^{[j-2]}(t)\right)\Big)
\end{aligned}$$

$$= P_i^{[j-1]}(t) - P_{i-1}^{[j-1]}(t) + (j-1)\left(P_i^{[j-1]}(t) - P_{i-1}^{[j-1]}(t)\right)$$
$$= j\left(P_i^{[j-1]}(t) - P_{i-1}^{[j-1]}(t)\right),$$

and the result is proved. ■

We end this section with a pseudocode implementation of Algorithm 5.1, augmented to include computation of the derivative using Lemma 5.2.

```
//Evaluates a Bezier curve and derivative at t = t̂
procedure Bez_eval(P, n, t, γ,γp)
    //P denotes the control polygon for the curve
    //n gives the degree of the curve (n + 1 control points)
    for i = 0 to n
        Qi <- Pi
    for i = 1 to n − 1
        for j = n to i
            Qj = (1 − t)Qj−1 + tQj
    γp <- n(Qn − Qn−1)
    γ <- (1 − t)Qn−1 + tQn
```

5.2 Bézier Curves

While this formulation is intuitive and easy to compute, unlike the case for the conics, we do not yet have the ability to perform certain computations nor the capability of analyzing all the characteristics of the curves. For example, how can we be sure that we have drawn enough points to show the real shape of the curve? How can we define surfaces with similar straightforward computational algorithms and analogous properties?

We would like to understand this curve and if possible write it in closed form. It would be desirable to keep the characteristic that was shown for the conics, namely that the parametric formulation can be written as an analytical blend of the original points, i.e., we would like to find scalar functions $\theta_{i,n}$, $i = 0, \ldots, n$, which *do not depend on the points* P_i, $i = 0, \ldots, n$ *in any way*, such that $\gamma(t) = \sum_{i=0}^{n} P_i\theta_{i,n}(t)$. If we can find these functions, then questions of form can be answered.

We consider the evaluation graph for a point on the curve. All of the original polygon points are at the bottom leaves of the graph. Lines are drawn connecting them to the first order iteration polygons, and the scalars multiplying them are noted. This process is continued, with one less node

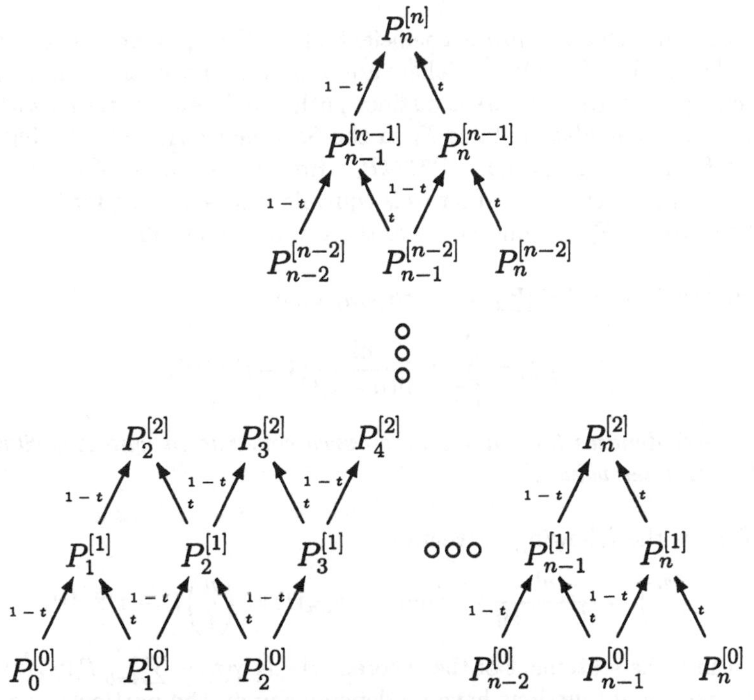

Figure 5.6. Constructive curve evaluation tree.

at each level until the root node represents the function value. Figure 5.6 shows this graph.

We use this graph to percolate the multipliers of each original polygon point to the top. Notice that P_0 appears only as a dependency for $P_1^{[1]}$, which appears only as a dependency for $P_2^{[2]}$, and so forth. And each of these points (with a subscript of i at the ith iteration level) is multiplied by $(1-t)$. Hence, P_0 percolates to the top along exactly one path, and the product of the multipliers along that path is $(1-t)^n$. Flattening out the recursive evaluation, we see that P_0 appears in exactly one term, with coefficient $(1-t)^n$.

Next, consider P_1. Both $P_1^{[1]}$ and $P_2^{[1]}$ depend on P_1. And, in general, only $P_{j+1}^{[j+1]}$ and $P_{j+2}^{[j+1]}$ depend on $P_{j+1}^{[j]}$, and in those terms $P_{j+1}^{[j]}$ appears with a multiplier of t and $(1-t)$, respectively. Hence, following a path up the tree for P_1 we see that we can move from $P_j^{[j-1]}$ to $P_{j+1}^{[j]}$, $j = 1, \ldots, k$, through k steps of the iteration, then from $P_{k+1}^{[k]}$ to $P_{k+1}^{[k+1]}$, and then from $P_{j-1}^{[j-1]}$ to $P_j^{[j]}$, $j = k+2, \ldots, n$. We have already seen that the first k and the last $(n-k-1)$ iterations have multipliers of $(1-t)$, and the step where the subscript changes has a coefficient of t. The polynomial associated with that path is $(1-t)^{n-1}t$. There are n distinct places that the subscript can change and so there are n distinct paths, and hence n terms with that polynomial that also contains P_1 in the flattened form. Combining them in the final form gives $n(1-t)^{n-1}tP_1$. Hence the search of the flattened form for P_i as coefficients of terms is equivalent to searching for the distinct paths in which P_i can bubble up to the top of the graph.

Theorem 5.3. *If $\{P_i\}_{i=0}^{n}$ is a polygon, then*

$$\gamma(t) = \sum_{i=0}^{n} P_i \frac{n!}{i!(n-i)!}(1-t)^{n-i}t^i,$$

where $\gamma(t)$ denotes the point on the curve constructed with Algorithm 5.1 with parameter value t.

We use the following conventions.

$$\binom{n}{i} = \frac{n!}{i!(n-i)!} \quad \text{and} \quad \theta_{i,n}(t) = \binom{n}{i}(1-t)^{n-i}t^i.$$

Rewriting, the statement of the theorem yields $\gamma(t) = \sum_{i=0}^{n} P_i\theta_{i,n}(t)$. It is clear that the θ-functions have no dependency on the particular polygons chosen, but depend only on the number of control points. Hence, once this result is shown, we will have a closed form representation of the constructive curve from Algorithm 5.1 in a blending function formulation, where the vertices of the polygon constitute the data.

Before we prove this result, we prove a property about the blending functions.

Theorem 5.4.

$$\theta_{k,n}(t) = \begin{cases} 1, & \text{for } n = 0, \\ t\,\theta_{k-1,n-1}(t) + (1-t)\theta_{k,n-1}(t), & \text{for } 0 < k < n, \\ (1-t)\theta_{0,n-1}(t), & \text{for } n > 0, k = 0, \\ t\,\theta_{n-1,n-1}(t), & \text{for } k = n > 0. \end{cases} \tag{5.4}$$

Proof: The cases for $n = 0$ and $n > 0$, $k = 0$ or $k = n$ are left as Exercise 5. Using the definition of

$$\binom{n}{k} = \frac{n!}{k!(n-k)!},$$

we see that

$$\begin{aligned}\binom{n}{k} &= \frac{n}{k}\frac{(n-1)!}{(k-1)!((n-1)-(k-1))!} \\ &= \frac{n}{n-k}\frac{(n-1)!}{k!((n-1)-k)!}.\end{aligned}$$

These combinatorial results yield for $0 < k < n$,

$$\begin{aligned}\theta_{k,n}(t) &= \binom{n}{k}t^k(1-t)^{n-k} \\ &= \frac{n}{k}\frac{(n-1)!}{(k-1)!((n-1)-(k-1))!}t^k(1-t)^{n-k} \\ &= \frac{n}{k}t\frac{(n-1)!}{(k-1)!((n-1)-(k-1))!}t^{k-1}(1-t)^{(n-1)-(k-1)} \\ &= \frac{n}{k}t\,\theta_{k-1,n-1}(t) \qquad (5.5)\end{aligned}$$

and

$$\frac{k}{n}\theta_{k,n}(t) = t\,\theta_{k-1,n-1}(t).$$

Similarly,

$$\begin{aligned}\theta_{k,n}(t) &= \binom{n}{k}t^k(1-t)^{n-k} \\ &= \frac{n}{n-k}\frac{(n-1)!}{k!((n-1)-k)!}t^k(1-t)^{n-k} \\ &= \frac{n}{n-k}(1-t)\frac{(n-1)!}{k!((n-1)-k)!}t^k(1-t)^{(n-1)-k} \\ &= \frac{n}{n-k}(1-t)\theta_{k,n-1}(t) \qquad (5.6)\end{aligned}$$

and

$$\frac{n-k}{n}\,\theta_{k,n}(t) = (1-t)\theta_{k,n-1}(t).$$

Putting both together yields

$$\begin{aligned}\theta_{k,n}(t) &= \frac{k}{n}\,\theta_{k,n}(t) + \frac{n-k}{n}\theta_{k,n}(t) \\ &= t\theta_{k-1,n-1}(t) + (1-t)\theta_{k,n-1}(t).\end{aligned}$$ ■

We now return to the proof of Theorem 5.3.

Proof: Denote by $\alpha(t)$ the right hand side of the result. That is,

$$\alpha(t) \stackrel{def}{=} \sum_{i=0}^{n} P_i\theta_{i,n}(t).$$

Using this definition and Theorem 5.4 we see:

$$\begin{aligned}
\alpha(t) &= \sum_{i=0}^{n} P_i\theta_{i,n}(t) \\
&= P_0\theta_{0,n}(t) + \sum_{i=1}^{n-1} P_i\theta_{i,n}(t) + P_n\theta_{n,n}(t) \\
&= P_0(1-t)\theta_{0,n-1}(t) \\
&\quad + \sum_{i=1}^{n-1} P_i\left(t\,\theta_{i-1,n-1}(t) + (1-t)\theta_{i,n-1}(t)\right) + P_n t\,\theta_{n-1,n-1}(t).
\end{aligned}$$

Regrouping by all terms that have t as a factor and those with $(1-t)$ as a factor, and using Algorithm 5.1 yields

$$\begin{aligned}
\alpha(t) &= (1-t)\sum_{i=0}^{n-1} P_i\theta_{i,n-1}(t) + t\sum_{i=1}^{n} P_i\theta_{i-1,n-1}(t) \\
&= (1-t)\sum_{i=0}^{n-1} P_i\theta_{i,n-1}(t) + t\sum_{i=0}^{n-1} P_{i+1}\theta_{i,n-1}(t) \\
&= \sum_{i=0}^{n-1}\left[(1-t)P_i + tP_{i+1}\right]\theta_{i,n-1}(t) \\
&= \sum_{i=0}^{n-1} P_{i+1}^{[1]}\theta_{i,n-1}(t).
\end{aligned}$$

Repeated applications of the same process give

$$\alpha(t) = \sum_{i=0}^{n-j} P_{j+i}^{[j]}\theta_{i,n-j}(t).$$

Hence, after n applications of the process, the result is $\alpha(t) = P_n^{[n]}\theta_{0,0}(t) = P_n^{[n]} = \gamma(t)$. We have arrived at a closed formulation with the desired characteristics. ■

Definition 5.5. *The curve created by the constructive Algorithm 5.1 whose closed form is given in Theorem 5.3 is called the* Bézier curve. *Algorithm 5.1 is referred to as either the* constructive Bézier algorithm *or the* de Casteljau algorithm.

Definition 5.6. *The functions*

$$\theta_{i,n}(t) = \binom{n}{i}(1-t)^{n-i}t^i,$$

for $i = 0, \ldots, n$, *for any nonnegative integer* n, *are called the* Bernstein basis functions *or the* Bernstein blending functions.

Definition 5.7. *For* $t \in [a, b]$, *the functions*

$$\begin{aligned} \theta_{i,n}(a, b; t) &= \theta_{i,n}\left(\frac{t-a}{b-a}\right) \\ &= \binom{n}{i}\frac{(b-t)^{n-i}(t-a)^i}{(b-a)^n}, \end{aligned}$$

for $i = 0, \ldots, n$, *where* n *is any nonnegative integer, are called the* generalized Bernstein basis functions *or the* generalized Bernstein blending functions.

We see that the generalized Bernstein blending functions are just a linear scaling of the interval $[a, b]$ into the interval $[0, 1]$, and $\theta'_{i,n}(a, b; t) = \frac{1}{b-a}\theta'_{i,n}\left(\frac{t-a}{b-a}\right)$.

A natural question to investigate is whether these blending functions do indeed form a basis for the space of polynomials of degree n. We shall now investigate their properties.

5.2.1 Properties of Bernstein Blending Functions

What are some of the properties of these basis functions which make them interesting?

Theorem 5.8. $\theta_{i,n}(t) > 0$ *for* $t \in (0, 1)$, *for all* $n \geq 0$ *and* $i = 0, \ldots, n$.

Proof: For $t \in (0, 1)$, $t > 0$ and $(1 - t) > 0$. $\theta_{i,n}(t)$ is then just the product of n positive factors, and is hence positive. ■

Theorem 5.9. $\sum_{i=0}^{n} \theta_{i,n}(t) \equiv 1$, *for* $t \in [0, 1]$.

Proof: This follows from the binomial theorem in which

$$(a+b)^n = \sum_{i=0}^{n} \binom{n}{i} a^{n-i} b^i.$$

Letting $a = (1-t)$ and $b = t$ we have

$$((1-t)+t)^n = 1^n = 1 = \sum_{i=0}^{n} \binom{n}{i} (1-t)^{n-i} t^i = \sum_{i=0}^{n} \theta_{i,n}(t).$$

■

Theorem 5.10. *The Bernstein blending functions are unimodal with maximum for the ith blending function of degree n at $t = i/n$.*

Proof: The term unimodal means that it has one extremal point over the domain, that is it has exactly one maximum or minimum. Since $\theta_{i,n}(t)$ is a polynomial for all i and n, it is continuously differentiable.

Since $\theta_{0,n}(t) = (1-t)^n$ and $\theta_{n,n}(t) = t^n$, the behavior of these functions is well known, and it is clear that they are always decreasing or increasing, respectively, on the interval $(0,1)$. Hence they have only one maximum which occurs at the appropriate endpoint.

Now let us consider $\theta_{i,n}(t)$ for $i = 1, \ldots, n-1$. Since $\theta'_{i,n}(t)$ is continuous, the extremal points of $\theta_{i,n}$ can occur only at $t = 0$ or $t = 1$ or at a zero of $\theta'_{i,n}(t)$ by Corollary 1.34. We solve for the zeros of the derivative to find the extremal points inside $(0,1)$ for $\theta_{i,n}$.

$$\begin{aligned}
\theta'_{i,n}(t) &= \binom{n}{i} \left[i t^{i-1}(1-t)^{n-i} - (n-i) t^i (1-t)^{n-1-i} \right] \\
&= \binom{n}{i} t^{i-1}(1-t)^{n-1-i} \left[i(1-t) - (n-i)t \right] \\
&= \binom{n}{i} t^{i-1}(1-t)^{n-1-i} [i - nt].
\end{aligned}$$

Setting $\theta'_{i,n}(t) = 0$ and solving for possible extremal abscissa gives

$$t_e = 0 \quad \text{or} \quad t_e = 1 \quad \text{or} \quad t_e = \frac{i}{n}.$$

Now, $\theta_{i,n}(0) = 0 = \theta_{i,n}(1)$ for $i = 1, \ldots, n-1$. Since $\theta_{i,n}$ is not identically zero, there is only one maximum and it occurs at $t_e = i/n$. ■

Example 5.11. We study the cubic Bernstein blending functions at the various maximal points.

For $t = 0$, $\theta_{0,3}(0) = 1$, and it is the only blending function that is nonzero. Similarly, for $t = 1$, $\theta_{3,3}(1) = 1$ is the only blending function that is nonzero.

For $t = 1/3$, the next maximal point,

$$\begin{aligned}\theta_{0,3}(1/3) &= 8/27,\\ \theta_{1,3}(1/3) &= 4/9,\\ \theta_{2,3}(1/3) &= 2/9,\\ \theta_{3,3}(1/3) &= 1/27.\end{aligned}$$

For $t = 2/3$, the next maximal point,

$$\begin{aligned}\theta_{0,3}(2/3) &= 1/27,\\ \theta_{1,3}(2/3) &= 2/9,\\ \theta_{2,3}(2/3) &= 4/9,\\ \theta_{3,3}(2/3) &= 8/27.\end{aligned}$$

P_1 will have the single largest influence at $t = 1/3$, and P_2 will have the single largest influence at $t = 2/3$, but the other polygon points contribute over half of the weighting. □

Theorem 5.12. *The Bernstein blending functions of degree n form a basis for the polynomials of degree n.*

Proof: Suppose $\{\theta_{i,n}(t)\}_{i=0}^{n}$ are linearly dependent. Then,

$$\begin{aligned}0 &\equiv \sum_{i=0}^{n} c_i\theta_{i,n}(t)\\ &\equiv \sum_{i=0}^{n} c_i\frac{n!}{i!(n-i)!}t^i(1-t)^{n-i} &(5.7)\\ &\equiv \sum_{i=0}^{n} c_i\frac{n!}{i!(n-i)!}t^i\sum_{j=0}^{n-i}\frac{(n-i)!}{j!(n-i-j)!}(-t)^{n-i-j} &(5.8)\\ &\equiv \sum_{i=0}^{n} c_i\frac{n!}{i!(n-i)!}\sum_{j=0}^{n-i}(-1)^{n-i-j}\frac{(n-i)!}{j!(n-i-j)!}t^{n-j}\\ &\equiv \sum_{j=0}^{n} t^{n-j}\sum_{i=0}^{n-j}(-1)^{n-i-j}\frac{n!}{i!(n-i)!}\frac{(n-i)!}{j!(n-i-j)!}c_i\\ &\equiv \sum_{j=0}^{n} t^{n-j}\sum_{i=0}^{n-j} a_{ij}c_i, &(5.9)\end{aligned}$$

where a_{ij} is some non-zero coefficient that is a function of i and j. Equation 5.8 follows from Equation 5.7 by applying the binomial theorem to the $(1-t)^{n-i}$ factor in each term.

Since this polynomial is identically equal to zero, each coefficient of each power of t must be zero. Hence,

$$0 = \sum_{i=0}^{n-j} a_{ij} c_i,$$

for $j = n,\ n-1, \ldots,\ 0$. When $j = n$ there is just one term in the sum, so $c_0 = 0$. When $j = n-1$, there are two terms: one for c_0 and one for c_1. But since $c_0 = 0$, $c_1 = 0$ also. Continuing in this manner using straightforward back substitution, we see that as j is lowered in value by one, $c_{n-j} = 0$. The conclusion results that there is exactly one way to write zero as a sum of these functions; therefore they are independent. Because there are $n+1$ independent Bernstein blending functions of degree n, they form a basis for the space of all polynomials of degree n. ■

5.2.2 Curve Properties

Corollary 5.13. *The Bézier curve lies in the convex hull of the points* $\{P_i\}_{i=0}^{n}$.

Proof: The proof follows immediately from Theorems 5.8 and 5.9. ■

Later in this chapter we shall introduce the Bernstein approximation to an arbitrary continuous function and show it has the variation diminishing property. We shall also show that we can consider the Bézier curve a Bernstein approximation to its control polygon. The following theorem is a corollary to such a result.

Theorem 5.14. *Considered as an approximation to its control, the Bézier curve polygon is variation diminishing.*

Some implications of this theorem are that the Bézier curve has no more undulations than the polygon and the undulations which exist in the curve must occur roughly near the undulations in the polygon. The convex hull property and the variation diminishing property together give the user a very strong sense of where the curve should be. This reinforces the user's intuition and helps make this scheme very popular to use.

Theorem 5.10 is indicative of the strength of the Bernstein functions as blending functions. Here, the data appear as coefficients for the blending functions, but more than one blending function is nonzero for all points in the interior of the unit interval, so the data cannot be interpolated. To pass through some point P_j, the coefficient of P_j must be 1, and all other contributions to the sum must vanish.

However, the influence of the ith basis function is strongest at $t = i/n$ and hence in evaluating the Bézier curve, the largest weight is given to the ith polygon point at $t = i/n$. However, the ith point need not contribute the majority of the influence to the curve as Example 5.11 shows.

5.2.3 Derivative Evaluation

In Lemma 5.2 we found the derivative in terms of the constructive algorithm. We now move towards finding derivatives of all orders in closed form as other Bézier curves.

Since $\theta_{i,n}(t) = \binom{n}{i} t^i (1-t)^{n-i}$,

$$\theta'_{i,n}(t) = \binom{n}{i} \left[it^{i-1}(1-t)^{n-i} - (n-i)t^i(1-t)^{n-i-1} \right].$$

Now

$$\binom{n}{i} i = \frac{n!i}{i!(n-i)!} = n\frac{(n-1)!}{(i-1)!(n-i)!} = n\binom{n-1}{i-1},$$

and

$$\binom{n}{i}(n-i) = \frac{n!(n-i)}{i!(n-i)!} = n\frac{(n-1)!}{(i)!(n-1-i)!} = n\binom{n-1}{i}.$$

Substituting yields

$$\begin{aligned} \theta'_{i,n}(t) &= n\binom{n-1}{i-1} t^{i-1}(1-t)^{n-i} - n\binom{n-1}{i} t^i(1-t)^{n-i-1} \\ &= n\left[\theta_{i-1,n-1}(t) - \theta_{i,n-1}(t)\right]. \end{aligned} \tag{5.10}$$

Clearly, $i = 0$ and $i = n$ are handled differently. For $i = 0$,

$$\begin{aligned} \theta'_{0,n}(t) &= \binom{n}{0}(-1)n(1-t)^{n-1} \\ &= -n\theta_{0,n-1}(t). \end{aligned}$$

The analogous result holds for $\theta'_{n,n}(t)$. Thus, we have developed the following theorem:

Theorem 5.15. *When $\theta_{i,n}(t)$ denotes the ith Bernstein blending function of degree n then*

$$\theta'_{i,n}(t) = n\big[\theta_{i-1,n-1}(t) - \theta_{i,n-1}(t)\big], \qquad \textit{for } i = 0, \ldots, n,$$

where $\theta_{-1,n-1}(t) \equiv 0$ and $\theta_{n,n-1}(t) \equiv 0$.

This gives a method of recursively obtaining the derivatives of the blending functions and thereby calculating the derivatives of the curves since $\gamma(t) = \sum_{i=0}^{n} P_i\theta_{i,n}(t)$ implies

$$\begin{aligned} \gamma'(t) &= \sum_{i=0}^{n} P_i\theta'_{i,n}(t) \\ &= \sum_{i=0}^{n} P_i n\left(\theta_{i-1,n-1}(t) - \theta_{i,n-1}(t)\right) \\ &= \sum_{i=0}^{n-1} n\left(P_{i+1} - P_i\right)\theta_{i,n-1}(t). \end{aligned}$$

We have shown

Corollary 5.16. *If $\gamma(t) = \sum_{i=0}^{n} P_i\theta_{i,n}(t)$, then*

$$\gamma'(t) = \sum_{i=0}^{n-1} Q_i\theta_{i,n-1}(t),$$

where $Q_i = n(P_{i+1} - P_i)$.

Corollary 5.17. *If $\gamma(t) = \sum_{i=0}^{n} P_i\theta_{i,n}(t)$, then*

$$\gamma^{(j)}(t) = \sum_{i=0}^{n-j} Q_{j,i}\theta_{i,n-j}(t),$$

where

$$Q_{j+1,i} = \begin{cases} P_i, & j = -1 \\ (n-j)\left[Q_{j,i+1} - Q_{j,i}\right], & j = 0, \ldots, n-1. \end{cases} \tag{5.11}$$

Proofs of these corollaries follow naturally from previous results and are left to the reader. See Exercises 1 and 2.

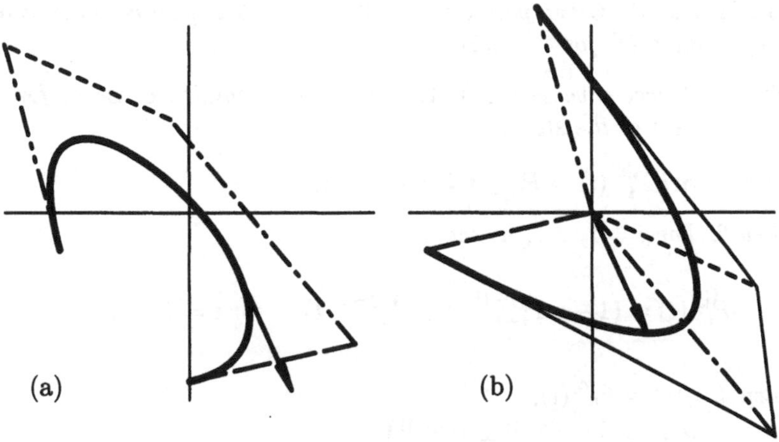

Figure 5.7. (a) A Bézier curve; (b) its hodograph.

The curve $\delta(t) = \gamma'(t)/n = \sum_{i=0}^{n-1}(P_{i+1} - P_i)\theta_{i,n-1}(t)$ is called the *hodograph* of $\gamma(t)$. (See Figure 5.7.)

We can continue this approach, and show its equivalence to Lemma 5.2 by setting $L_i = P_{i+1}$ and $R_i = P_i$ to get

$$\begin{aligned}\gamma'(t) &= \sum_{i=0}^{n-1} n\,(P_{i+1} - P_i)\,\theta_{i,n-1}(t) \\ &= n\left(\sum_{i=0}^{n-1} P_{i+1}\theta_{i,n-1}(t) - \sum_{i=0}^{n-1} P_i\theta_{i,n-1}(t)\right) \\ &= n\left(\sum_{i=0}^{n-1} L_i\theta_{i,n-1}(t) - \sum_{i=0}^{n-1} R_i\theta_{i,n-1}(t)\right).\end{aligned}$$

By Algorithm 5.1,

$$\begin{aligned}L_{n-1}^{[n-1]} &= \sum_{i=0}^{n-1} L_i\theta_{i,n-1}(t) &= P_n^{[n-1]}, \quad \text{and} \\ R_{n-1}^{[n-1]} &= \sum_{i=0}^{n-1} R_i\theta_{i,n-1}(t) &= P_{n-1}^{[n-1]}.\end{aligned}$$

This proves Lemma 5.2 in a totally different way. Further, these results can be used to create an algorithm for simultaneous curve and derivative evaluation.

Algorithm 5.18. *Given points P_i, $i = 0, \ldots, n$. The goal is to determine both $\gamma(t)$, and $\gamma'(t)$ for $t \in [0,1]$.*

Step 1: Select a value $t \in [0,1]$. This value remains constant for the rest of the steps.

Step 2: Set $P_i^{[0]}(t) = P_i$, for $i = 0, \ldots, n$.

Step 3: For $j = 1, \ldots, n$, set

$$P_i^{[j]}(t) = (1-t)P_{i-1}^{[j-1]}(t) + tP_i^{[j-1]}(t), \quad \textit{for } i = j, \ldots, n.$$

Step 4: $\gamma(t) = P_n^{[n]}(t)$;
$$\gamma'(t) = n\left(P_n^{[n-1]} - P_{n-1}^{[n-1]}\right).$$

5.2.4 Midpoint Subdivision

Sometimes it happens that one part of a curve is exactly the shape that is needed, but the other part needs modifying. Since the Bézier curve is global, it is not possible to move any control point without changing the whole curve. However, it is possible to *subdivide* the curve. That is, to develop a Bézier curve representation for just part of the curve. This is the opposite problem to that of *piecing* curves together.

In this section we develop a representation for the part of a cubic Bézier curve on the interval $[0, 1/2]$. Exercise 7 requests that the reader develop an analogous formulation for the interval $[1/2, 1]$. In Theorem 16.12 we develop formulations for subdivision at an arbitrary point and arbitrary degree.

Let $\gamma(t) = \sum_{i=0}^{3} P_i\theta_{i,3}(t)$ be a Bézier curve and $\beta(t) = \sum_{i=0}^{3} Q_i\theta_{i,3}(2t)$ be a Bézier representation for the same curve for $t \in [0, 1/2]$. We seek to determine the coefficients of β in terms of those for γ.

Since $\gamma^{(j)}(0) = \beta^{(j)}(0)$ for $j = 0, \ldots, 3$, this implies that

$$\begin{aligned} P_0 &= Q_0; && (5.12)\\ 3(P_1 - P_0) &= \frac{3}{1/2}(Q_1 - Q_0); && (5.13)\\ 3*2(P_2 - 2P_1 + P_0) &= \frac{3*2}{(1/2)^2}(Q_2 - 2Q_1 + Q_0); && (5.14)\\ 3*2(P_3 - 3P_2 + 3P_1 - P_0) &= \frac{3*2}{(1/2)^3}(Q_3 - 3Q_2 + 3Q_1 - Q_0). && (5.15) \end{aligned}$$

By Equations 5.12 and 5.13,

$$\begin{aligned} Q_1 &= Q_0 + (1/2)(P_1 - P_0) \\ &= P_0 + (1/2)(P_1 - P_0) \\ &= (1/2)(P_0 + P_1) \\ &= P_1^{[1]}. \end{aligned}$$

The last equality uses Step 3 of Algorithm 5.1 with t = 1/2. Solving Equation 5.14 and 5.15 similarly yields

$$\begin{aligned} Q_0 &= P_0; \\ Q_1 &= P_1^{[1]}; \\ Q_2 &= P_2^{[2]}; \\ Q_3 &= P_3^{[3]}. \end{aligned}$$

The points $P_j^{[i]}$ are those generated by Algorithm 5.1.

5.3 Bernstein Approximation

It is time to consider whether Bézier approximation has its roots in any classical mathematics. That is, is there some known mathematical method on which the Bézier curve can be theoretically founded?

Definition 5.19. *For a function $f \in C^{(m)}[0,1]$, the nth Bernstein polynomial approximation is*

$$B_n(f;x) = \sum_{k=0}^{n} f\left(\frac{k}{n}\right) \theta_{k,n}(x).$$

Represent the polygon line defining the shape of the Bézier curve as a piecewise linear parametric function f with parameter value i/n at P_i. The associated Bézier curve is the parametric Bernstein approximation of degree n to f. It is easy to see that $B_n(f;0) = f(0)$ and $B_n(f;1) = f(1)$, so interpolation at the end points is achieved.

Definition 5.20. *A sequence of functions $\{s_j(x)\}_j$ is said to* converge uniformly *to a function $s(x)$ on an interval I if for every $\epsilon > 0$ there exists a natural number N such that, for all $j > N$, $|s_j(x) - s(x)| < \epsilon$ for all x.*

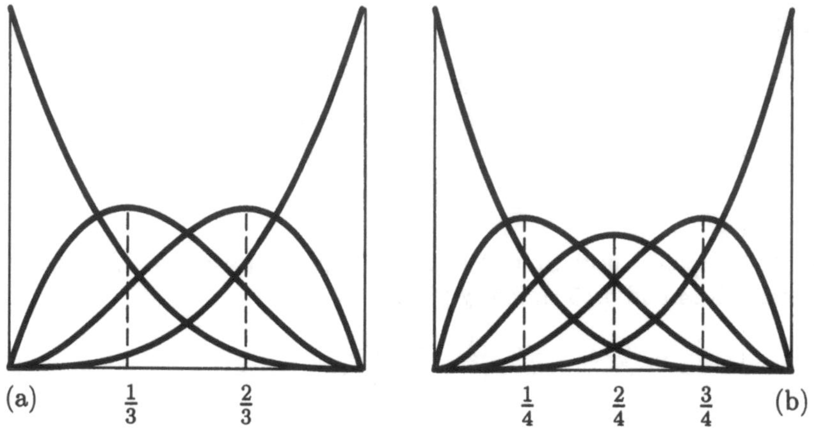

Figure 5.8. Bernstein basis functions: (a) $n = 3$; (b) $n = 4$.

Theorem 5.21. *If $f(x)$ is bounded on $[0, 1]$, then $\lim_{n\to\infty} B_n(f; x) = f(x)$ at any point x where f is continuous. If $f \in C^{(0)}[0, 1]$ then it converges uniformly. Further, if $f \in C^{(m)}[0, 1]$ then $\lim_{n\to\infty} B_n^{(m)}(f; x) = f^{(m)}(x)$ uniformly on $[0, 1]$.*

Thus the Bernstein approximation converges to the function and also to all its derivatives simultaneously, whenever the derivatives exist. This extraordinary convergence property guarantees that a Bernstein approximation of sufficiently high degree provides fidelity to position, tangency, curvature, and torsion simultaneously over the total domain.

The very property of *simultaneous convergence* which makes the Bernstein approximation so interesting to us comes at the cost of slow convergence in position. For numerical approximation problems needing position only, without regard to overall shape, this method is much too slow!

Example 5.22. Let $f(x) = 1$. Since $f(k/n) = 1$ for all n and k,

$$\begin{aligned} B_n(1; x) &= \sum_{k=0}^{n} \binom{n}{k} x^k (1-x)^{n-k} \\ &= \big(x + (1-x)\big)^n \\ &= 1. \end{aligned}$$

Hence, the Bernstein approximation reproduces constant functions exactly. □

Example 5.23. Let $f(x) = x$. $f(k/n) = k/n$ for all n and k and

$$\begin{aligned}
B_n(x;x) &= \sum_{k=0}^{n} \frac{k}{n}\binom{n}{k} x^k(1-x)^{n-k} \\
&= \sum_{k=1}^{n} \frac{k}{n}\binom{n}{k} x^k(1-x)^{n-k} \\
&= \sum_{k=1}^{n} \binom{n-1}{k-1} x^k(1-x)^{n-k} \\
&= x\sum_{k=1}^{n} \binom{n-1}{k-1} x^{k-1}(1-x)^{(n-1)-(k-1)} \\
&= x\sum_{k=0}^{n-1} \binom{n-1}{k} x^k(1-x)^{(n-1)-k} \\
&= x.
\end{aligned}$$

The last step follows from the previous example. Thus, linear functions are also reproduced exactly. □

Example 5.24. Let $f(x) = x^2$. $f(k/n) = (k/n)^2$ for all n and k and

$$B_n(x^2;x) = \sum_{k=0}^{n} \left(\frac{k}{n}\right)^2 \binom{n}{k} x^k(1-x)^{n-k}. \tag{5.16}$$

But, for $n \geq 2$,

$$\begin{aligned}
x^2 &= x^2\sum_{r=0}^{n-2} \binom{n-2}{r} x^r(1-x)^{n-2-r} \\
&= \sum_{r=0}^{n-2} \binom{n-2}{r} x^{r+2}(1-x)^{n-2-r} \\
&= \sum_{r=0}^{n-2} \frac{(r+2)(r+1)}{n(n-1)} \binom{n}{r+2} x^{r+2}(1-x)^{n-(r+2)} \\
&= \sum_{k=2}^{n} \frac{k(k-1)}{n(n-1)} \binom{n}{k} x^k(1-x)^{n-k} \\
&= \sum_{k=0}^{n} \frac{k(k-1)}{n(n-1)} \binom{n}{k} x^k(1-x)^{n-k}.
\end{aligned} \tag{5.17}$$

Since $\frac{k(k-1)}{n(n-1)} = 0$ for $k = 0$ and $k = 1$, Equation 5.17 is true. Equations 5.16 and 5.17 are not equivalent, so $B_n(x^2; x) \not\equiv x^2$. We can solve for the error in approximating x^2 with $B_n(x^2; x)$:

$$\begin{aligned}
\left|B_n(x^2; x) - x^2\right| &= \sum_{k=0}^{n} \left|\frac{k^2}{n^2} - \frac{k(k-1)}{n(n-1)}\right| \theta_{k,n}(t) \\
&= \sum_{k=0}^{n} \left|\frac{k}{n}\left(\frac{k}{n} - \frac{(k-1)}{n-1}\right)\right| \theta_{k,n}(t) \\
&= \sum_{k=0}^{n} \left|\frac{k}{n^2}\left(\frac{n-k}{n-1}\right)\right| \theta_{k,n}(t) \\
&\le \sum_{k=0}^{n} \frac{1}{n} \theta_{k,n}(t). \\
&\le 1/n.
\end{aligned}$$

Hence, we see that as $n \to \infty$, the error gets small uniformly. □

Example 5.25. Three points in space, $P_i = (x_i, y_i, z_i)$, $i = 0, 1, 2$, define a Bézier curve with the following representation:

$$\begin{aligned}
\gamma(t) &= \left(\sum_{i=0}^{2} x_i \theta_{i,2}(t), \sum_{i=0}^{2} y_i \theta_{i,2}(t), \sum_{i=0}^{2} z_i \theta_{i,2}(t)\right) \\
&= ((1-t)^2 x_0 + 2t(1-t)x_1 + t^2 x_2, \\
&\quad (1-t)^2 y_0 + 2t(1-t)y_1 + t^2 y_2, \\
&\quad (1-t)^2 z_0 + 2t(1-t)z_1 + t^2 z_2).
\end{aligned}$$

This is exactly the form derived for the rational conic (in Equation 3.11) when $w_0 = w_1 = w_2 = 1$. Thus, we see that the form used for the rational conic is compatible with the Bézier curve. In fact, when one represents a rational conic in this parametric form using "extended" points, it is called the *rational Bézier Curve.*, and is written

$$\gamma(t) = \frac{\left(\sum_{i=0}^{2} w_i x_i \theta_{i,2}(t), \sum_{i=0}^{2} w_i y_i \theta_{i,2}(t), \sum_{i=0}^{2} w_i z_i \theta_{i,2}(t)\right)}{\sum_{i=0}^{2} w_i \theta_{i,2}(t)}.$$

□

We show that Bernstein approximation is a variation diminishing approximation to its primitive, and consequently the Bézier curve has no more undulations than its control polygon.

Lemma 5.26. *Suppose $s(x) = ax + b$ is an arbitrary straight line and $B_n(f;x)$ is the nth degree Bernstein approximation to f. Then the number of times $B_n(f;x)$ crosses s is exactly the same as the number of sign changes of the Bernstein approximation to $f - s$.*

Proof: Since, by Examples 5.23 and 5.24,

$$\begin{aligned} B_n(f-s;x) &= \sum_{i=0}^{n} (f(i/n) - s(i/n))\,\theta_{i,n}(x) \\ &= \sum_{i=0}^{n} f(i/n)\theta_{i,n}(x) - \sum_{i=0}^{n} s(i/n)\theta_{i,n}(x) \\ &= \sum_{i=0}^{n} f(i/n)\theta_{i,n}(x) - s(x). \end{aligned}$$

This shows that counting zeros is equivalent to counting intersections. ■

Denote by $Z(a_0, a_1, \ldots, a_n)$ the number of sign changes of the sequence $a_0, \ldots, a_n$, and $Z[f]$ the number of sign changes of an arbitrary continuous function f.

Theorem 5.27. *For an arbitrary continuous function f, over the interval* $[0,1]$, $Z[B_n(f;x)] \le Z(f(0), f(1/n), \ldots, f(n/n))$.

Proof: For $x \in (0,1)$,

$$\begin{aligned} Z[B_n(f;x)] &= Z\left[\frac{B_n(f;x)}{(1-x)^n}\right] \\ &= Z\left[\sum_{i=0}^{n} f(i/n)\binom{n}{i}\left(\frac{x}{1-x}\right)^i\right] \\ &= Z\left[\sum_{i=0}^{n} f(i/n)\binom{n}{i} z^i\right] \\ &\le Z\left(f(0), f(1/n), f(2/n), \ldots, f(n/n)\right), \end{aligned}$$

where $z = \left(\frac{x}{1-x}\right)$ can take on all positive real values and $\binom{n}{i}$ is positive throughout. The inequality in the last line follows from Descartes rule of signs. ■

Definition 5.28. *If S is a finite set, the* cardinality of S, *$card(S)$, is the number of elements of S.*

Now, supppose that $f = (x(t), y(t)) = \sum_{j=0}^{n}(x_j, y_j)\theta_{j,n}(t)$ is a planar parametric Bézier curve, and $ax + by + c = 0$ is the implicit equation of an arbitrary line. Then the number of intersection points is $card\{t : ax(t) + by(t) + c = 0\}$. Now $ax(t) + by(t) + c = \sum_{j=0}^{n}(ax_j + by_j + c)\theta_{j,n}(t)$. Thus by Theorem 5.27,

$$Z[\sum_{j=0}^{n}(ax_j + by_j + c)\theta_{j,n}(t)] \leq Z\,(ax_0 + by_0 + c, ax_1 + by_1 + c, \ldots, ax_n + by_n + c)\,.$$

But this last expression is just the number of times that the line crosses the edges of the control polygon.

The classic definition of variation diminishing is defined for curves in the plane. The space curves prevalent in computer aided geometric design do not conform to that requirement. A generalization of the definition keeps the idea and properties when the curve is a space curve.

Definition 5.29. *If $f(t)$ is a parametric curve in $\mathbb{R}^3$ and $V[f](t)$ is a* parametric variation diminishing *approximation, then a arbitrary plane can intersect $V[f]$ no more often then it intersects $f(t)$.*

5.4 Interpolation Using the Bernstein Blending Functions

The problem here is to interpolate $n+1$ points, V_i, $i = 0, \ldots, n$, with a parametric polynomial of degree n, and to represent this unique interpolant in the Bernstein blending function basis.

Since the data points have no parameter values, we must begin by assigning a parameter value to each of them. Since we want the ith blending function to exert maximal control at the ith interpolation point, and the ith blending function has a maximum at i/n, we assign the parameter value $t_i = i/n$ to V_i and seek to find $\gamma(t)$ where $\gamma(i/n) = V_i, i = 0, \ldots, n$. That means control points P_j, $j = 0, \ldots, n$ must be found so that $\gamma(i/n) = \sum_{j=0}^{n} P_j\theta_{j,n}(i/n) = V_i, i = 0, \ldots, n$. This represents a system of $n+1$ equations in $n+1$ unknowns. Writing this in matrix form,

$$\left[\ \theta_{j,n}(i/n)\ \right]\left[\ P_0 \quad P_1 \quad \ldots \quad P_n\ \right]^T = \left[\ V_0 \quad V_1 \quad \ldots \quad V_n\ \right]^T,$$

or

$$\Theta \boldsymbol{P} = \boldsymbol{V}.$$

We show in Chapter 10 that a unique interpolating polynomial exists as long as the abscissa values are all distinct. Since the Bernstein blending functions form a basis for $\mathcal{P}_n$, the above system can be solved for a unique solution.

Example 5.30. In this example we use the Bézier formulation to interpolate the ordered points: (0,2), (3,2), (5,4), (1,0).

Let $V_0 = (0,2)$, $V_1 = (3,2)$, $V_2 = (5,4)$, $V_3 = (1,0)$. Referring to Example 5.11, we assign parameter values of the data to correspond to the maximal points of the blending functions. That is, V_i is assigned parameter value i/n, where $n = 3$ in this example. This is called *nodal interpolation.* Directly substituting into the formula:

$$\begin{bmatrix} 1 & 0 & 0 & 0 \\ 8/27 & 4/9 & 2/9 & 1/27 \\ 1/27 & 2/9 & 4/9 & 8/27 \\ 0 & 0 & 0 & 1 \end{bmatrix} \begin{bmatrix} P_0 \\ P_1 \\ P_2 \\ P_3 \end{bmatrix} = \begin{bmatrix} V_0 \\ V_1 \\ V_2 \\ V_3 \end{bmatrix}.$$

The solution requires solving the system of equations for x and y coefficients. □

5.4.1 Comparing Bézier Curves and Interpolation

Figures 5.9 and 5.10 illustrate the differences between the approximations to the same data via the methods of interpolation and Bézier approximation. To reiterate, the Bézier method is variation diminishing, which means that the Bézier curve can have no more undulations than the polygon connecting the original data. Figure 5.9 (a) shows the original points while Figure 5.9 (b) has the data sequentially connected by linear segments. This curve is frequently called the control polygon or the control net.

Figure 5.10 (a) shows a piecewise linear curve interpolating the data, and also the polynomial interpolant to the data. One can see the undulation of the curve around the data, which is an inherent problem for any interpolation scheme.

The Bézier curve in Figure 5.10 (b) has the same degree as the polynomial interpolant. In this example, the smoothing nature of this type of curve becomes evident. While the Bézier curve is much smoother than

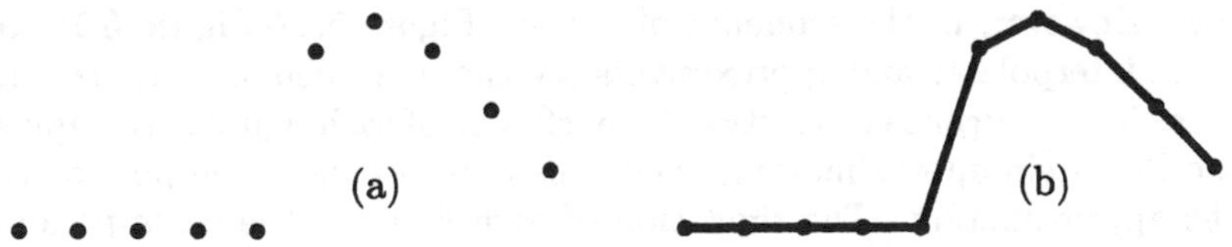

Figure 5.9. (a) Original data; (b) data connected piecewise linearly.

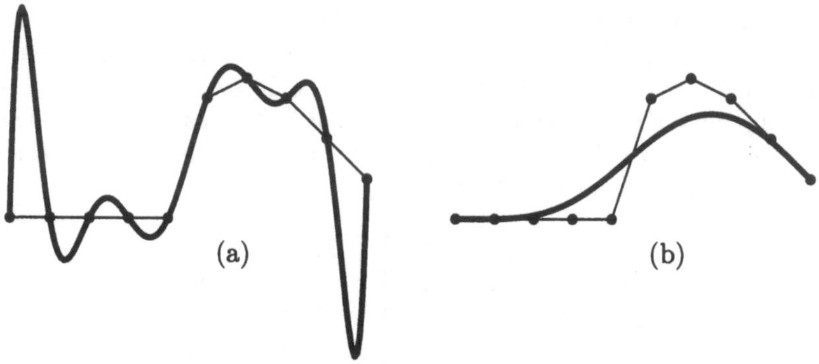

Figure 5.10. (a) Polynomial interpolant; (b) Bézier curve.

the interpolant, the approximation problem may require that some intermediate points be interpolated. Figure 5.11 uses a piecewise cubic Bézier approach to solve this problem, an approach discussed more fully later.

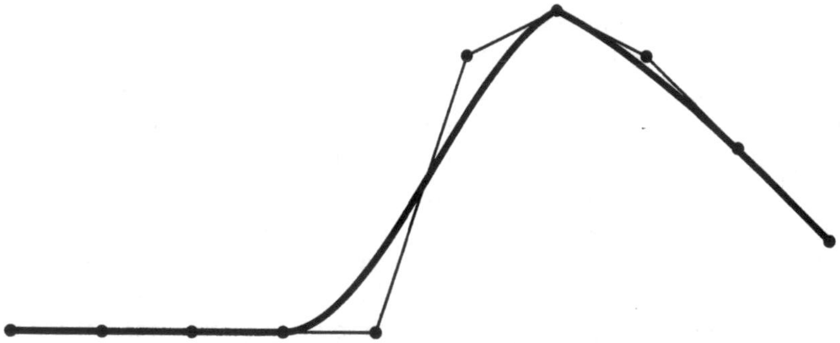

Figure 5.11. Polygon and piecewise cubic Bézier curve.

5.4.2 Comparing Bernstein Approximation to Functions and Interpolation

Both interpolation and shape preserving approximation have a role in CAGD. However, in the sequence of figures, Figure 5.12–Figure 5.15, respective interpolants and approximants are chosen to demonstrate the differences in convergence properties. The left side of each figure is the appropriate Bernstein approximation, where the crosses denote the points used in the approximation. The right side of each figure is the corresponding interpolation, where the crosses denote the points used in the interpolation. The original curve is reproduced in each figure for comparison.

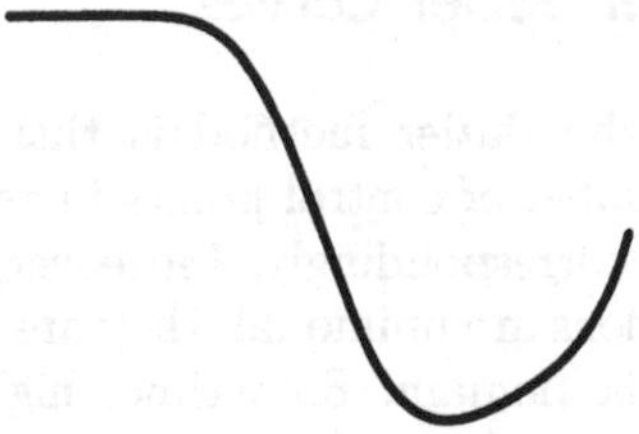

Figure 5.12. Original curve. Examine the sequence Figure 5.13–Figure 5.15.

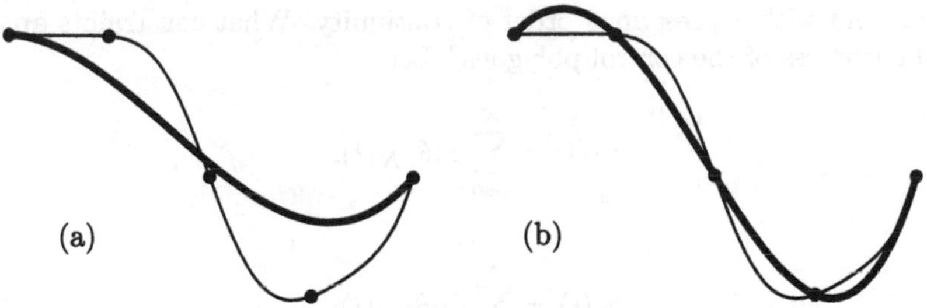

Figure 5.13. Quartic comparison: (a) Bernstein approximation; (b) interpolation.

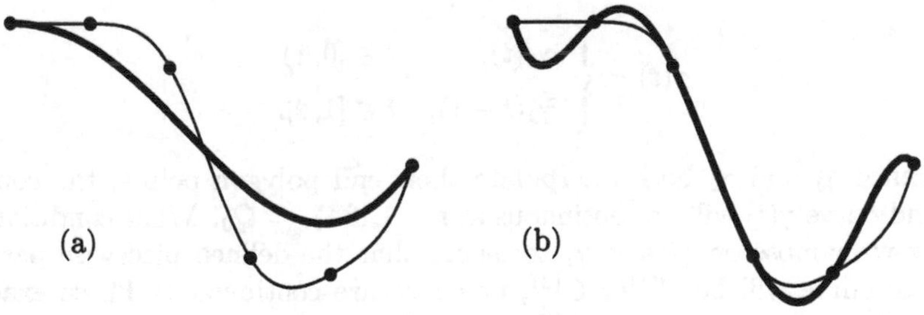

Figure 5.14. Quintic comparison: (a) Bernstein approximation; (b) interpolation.

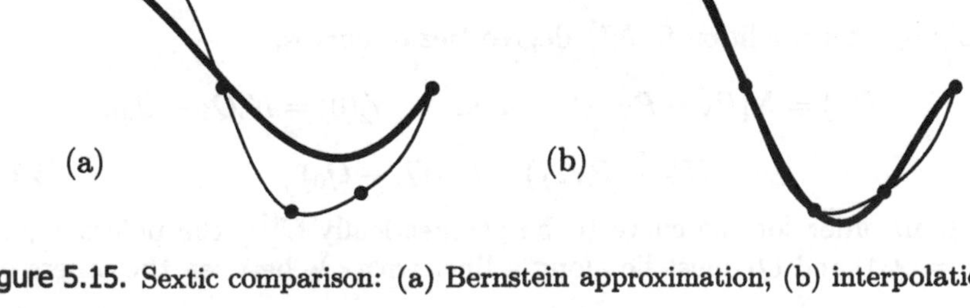

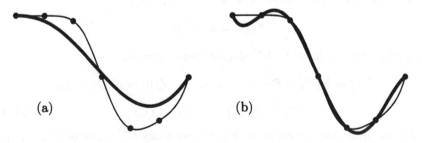

Figure 5.15. Sextic comparison: (a) Bernstein approximation; (b) interpolation.

5.5 Piecing together Bézier Curves

A serious drawback of the Bézier method is that the curve is a single polynomial, so as the number of control points increases, the degree of the approximation increases correspondingly. Moreover, the method is global. Although the basis functions are unimodal, they are polynomials and hence nonzero everywhere in the domain. Sometimes higher order continuity is *traded off* and given up to get lower degree computation. Piecing together Bézier curves while maintaining some derivative continuity allows this.

Suppose it is desired that two Bézier curves of the same degree be joined at one end with a prescribed order of continuity. What constraints apply to the vertices of the control polygons? Let

$$\gamma_1(t) = \sum_{i=0}^{N} P_i\theta_{i,N}(t),$$

and

$$\gamma_2(t) = \sum_{i=0}^{N} Q_i\theta_{i,N}(t),$$

and denote by $\gamma(t)$ the compound curve where

$$\gamma(t) = \begin{cases} \gamma_1(t), & t \in [0,1) \\ \gamma_2(t-1), & t \in [1,2]. \end{cases}$$

Since γ_1 and γ_2 both interpolate their end polygon points, the compound curve $\gamma(t)$ will be continuous at $t = 1$ if $P_N = Q_0$. What conditions must we impose on γ_1 and γ_2 to insure that the defined piecewise parametric curve will be $C^{(1)}$or $C^{(2)}$, or curvature continuous? Those exact conditions were investigated in Section 4.7, and we will use them.

Equation 4.13 displayed the condition that was developed for geometric $C^{(1)}$. Namely, that there exists $k_1 > 0$ such that

$$\gamma_1'(1) = k_1\gamma_2'(0).$$

Applying this condition to N^{th} degree Bézier curves,

$$\gamma_1'(1) = N[P_N - P_{N-1}] \qquad \text{and} \qquad \gamma_2'(0) = N[Q_1 - Q_0];$$

$$(P_N - P_{N-1}) = k_1\left(Q_1 - Q_0\right). \tag{5.18}$$

Thus, in order for the curve to be geometrically $C^{(1)}$, the points P_{N-1}, $P_N\,(= Q_0)$, and Q_1 must lie along a line, with Q_0 between the others. If the curve is to be $C^{(1)}$ then $\gamma_1'(1) = \gamma_2'(0)$ so that Q_0 must be the midpoint of the line segment connecting P_{N-1} and Q_1.

The curvature continuous condition of Equation 4.14, that there exists a constant c_3 such that

$$\gamma_1''(1) = c_3\gamma_2'(0) + (k_1)^2\gamma_2''(0),$$

has implications for the Bézier curves. Notice that if $k_1 = 1$ and $c_3 = 0$, then the curve is $C^{(2)}$.

$$\begin{aligned}\gamma_1''(1) &= N(N-1)(P_N - 2P_{N-1} + P_{N-2})\\ \gamma_2''(0) &= N(N-1)(Q_0 - 2Q_1 + Q_2).\end{aligned}$$

Applying Equation 4.14 to the compound Bézier curve gives

$$\begin{aligned}N(N-1)(P_N - 2P_{N-1} + P_{N-2}) &= c_3N(Q_1 - Q_0) + (k_1)^2N(N-1)(Q_0 - 2Q_1 + Q_2)\\ (N-1)(P_N - 2P_{N-1} + P_{N-2}) &= c_3(Q_1 - Q_0) + (k_1)^2(N-1)(Q_0 - 2Q_1 + Q_2)\\ (P_{N-2} - 2P_{N-1} + P_N) &= (k_1)^2(Q_0 - 2Q_1 + Q_2) + \frac{c_3}{N-1}(Q_1 - Q_0).\end{aligned}$$

Since

$$P_N = Q_0 \quad \text{and} \quad (P_N - P_{N-1}) = k_1(Q_1 - Q_0),$$

substituting for P_N and P_{N-1} yields

$$P_{N-2} - P_{N-1} = \left(k_1 - \frac{c_3}{N-1} + (k_1)^2\right)(Q_0 - Q_1) + (k_1)^2(Q_2 - Q_1). \quad (5.19)$$

But we can rewrite $P_{N-1} = Q_1 + (1 + k_1)(Q_0 - Q_1)$, so $P_{N-2} = Q_1 + a(Q_0 - Q_1) + b(Q_2 - Q_1)$, for certain constants a, b. Hence, P_{N-2} is in the plane spanned by Q_0, Q_1, and Q_2, which we have shown also contains P_{N-1}.

Thus, for the compound curve to be curvature continuous, it is required that P_{N-2}, P_{N-1}, $P_N = Q_0$, Q_1, and Q_2 be coplanar, and that the middle three be collinear. We see that the problem of piecing together curve segments forces the loss of several degrees of freedom. Notice that since k_1 is known, specifying $c_2 = \|\gamma_2'(1)\|$, then also $c_1 = \|\gamma_1'(1)\|$, since $c_1 = k_1c_2$. Now P_{N-1} is completely specified, since $P_N = Q_0$ and Q_1 are known. Since c_3 and Q_2 are also given, P_{N-2} is also completely specified. Thus, only P_i, $i = 0, \ldots, N-3$ have complete freedom now.

Under what conditions would one want to *piece together* curves? In many practical situations, that is just what occurs. This approach is in fact a sort of blend between Bézier and piecewise Hermite (see Section 10.3.1). It is important to remember that the Bézier method is still a single polynomial, and hence a global scheme. Sometimes it is desirable to isolate

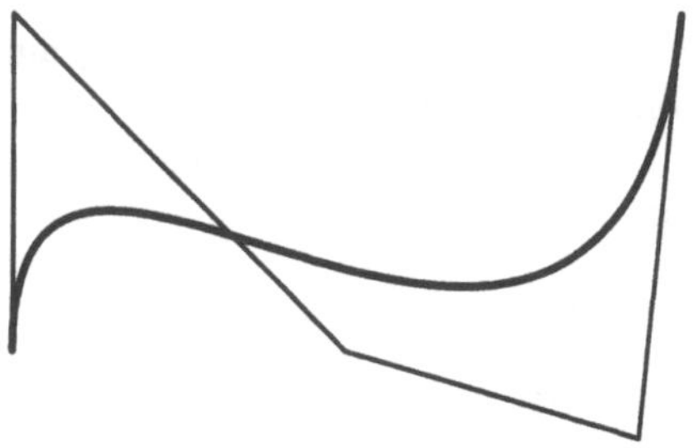

Figure 5.16. Original polygon with curve.

the curve elements to make them local. Other times, when the number of sides of the polygon gets very large, the degree of the polynomial becomes high and computational accuracy problems result. This is one way to try to avoid such difficulties.

5.6 Adding Flexibility — Degree Raising

We have seen that the Bézier method allows for manipulation of the polygon points and hence editing (modifying) the curve. Composite curves may be developed by piecing the Bézier polynomials together. A further problem is encountered, however, when one discovers that one has used the method quite successfully to gain a reasonable approximation of the desired shape, but now would like to make some finer changes — and cannot because there are not enough degrees of freedom. The solution is, of course, to add degrees of freedom. This can be accomplished in several ways. One such way is to break the whole Bézier curve and develop a formulation that allows it to be treated as two, or more, Bézier curves that have been joined together to form the composite curve. This process is called *subdivision*. Example 5.2.4 has developed an algorithm for subdividing cubic Bézier curves at their midpoint. It will be treated in full generality in a later chapter.

The other method of adding degrees of freedom to a polynomial curve is to represent the curve as a higher degree polynomial. In other words given an nth degree polynomial $p_n(x) = \sum_{i=0}^{n} a_i x^i$, represent it as an mth degree polynomial, q_m, with $m > n$ in the following way. Let $q_m(x) = \sum_{i=0}^{m} b_i x^i$ where $b_i = a_i, i = 0, \ldots, n$, and $b_i = 0, i = n+1, \ldots, m$. One can then change all the values of the coefficients to modify the polynomial. We must then ask a similar question when using the Bernstein basis with the Bézier curve. How can we write $\gamma_p(t) = \sum_{i=0}^{n} P_i \theta_{i,n}(t)$ as an $(n+1)^{st}$ degree polynomial? i.e., $\gamma_q(t) = \sum_{i=0}^{n+1} Q_i \theta_{i,n+1}(t)$. Further, what is the relationship between the original polygon $\{P_i\}$ and the new polygon $\{Q_i\}$?

From Equations 5.5 and 5.6,

$$\frac{k}{n+1}\theta_{k,n+1} = t\,\theta_{k-1,n}(t) \quad \text{and} \quad \frac{n+1-k}{n+1}\theta_{k,n+1} = (1-t)\theta_{k,n}(t).$$

The first equality can be modified by substituting $k+1$ wherever k appears to yield

$$\frac{k+1}{n+1}\theta_{k+1,n+1} = t\,\theta_{k,n}(t).$$

Adding this last equation to the second equality above yields

$$\theta_{k,n}(t) = \frac{k+1}{n+1}\theta_{k+1,n+1}(t) + \frac{n+1-k}{n+1}\theta_{k,n+1}(t).$$

Substituting this into the curve equation we have:

$$\begin{aligned}
\gamma_p(t) &= \sum_{i=0}^{n} P_i\theta_{i,n}(t) \\
&= \sum_{i=0}^{n} P_i\left[\frac{i+1}{n+1}\theta_{i+1,n+1}(t) + \frac{n+1-i}{n+1}\theta_{i,n+1}(t)\right] \\
&= \sum_{i=0}^{n+1}\left[\frac{iP_{i-1}+(n+1-i)P_i}{n+1}\right]\theta_{i,n+1}(t).
\end{aligned}$$

This yields the following theorem:

Theorem 5.31. *A Bézier curve $\gamma(t) = \sum_{i=0}^{n} P_i\theta_{i,n}(t)$ can be represented as*

$$\gamma(t) = \sum_{i=0}^{n+1} Q_i\theta_{i,n+1}(t)$$

where $Q_i = \frac{iP_{i-1}+(n+1-i)P_i}{n+1}$.

Observe that when $i = 0$, P_{-1} appears, it is multiplied by 0. Analogously, P_{n+1} is multiplied by 0. Hence their appearance is symbolic only, and so can be set to 0.

The new higher degree representation for the original curve has a control polygon with more points, and hence more design handles. The designer can now choose to modify the curve by moving the new control points. Figure 5.17 illustrates degree raising for Bézier curves. Note that the degree-raised curve is always identical, but the defining control polygon changes as the degree changes.

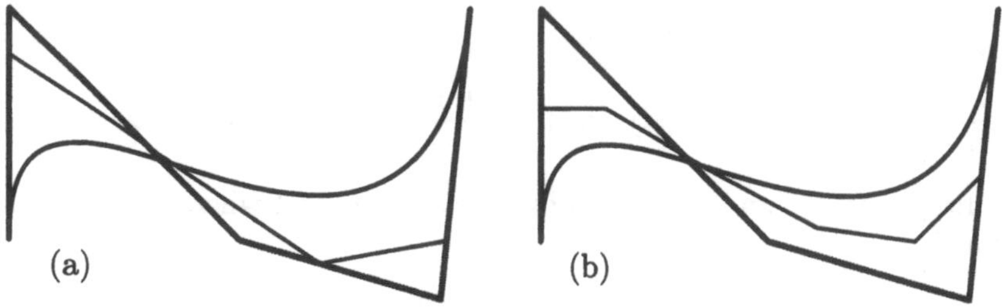

Figure 5.17. (a) Single application of degree raising; (b) multiple applications.

Exercises

1. Prove Corollary 5.16.
2. Using Exercise 1, prove Corollary 5.17.
3. When one is piecing together Bézier curves, one might want to consider torsion continuity as well as tangent direction and curvature continuity. Study this problem as it affects the degrees of freedom allowed to Q_3.
4. How would Equations 5.18 and 5.19 be changed if the two Bézier curves had different degrees? Derive the new equations.
5. Finish the proof of Theorem 5.4.
6. Prove by constructive argument that, if γ is defined as in Example 5.2.4 and $\rho(t) = \sum_{i=0}^{3} R_i\theta_{i,3}(2t-1)$ is a representation for the same curve for $t \in [1/2, 1]$, then

$$\begin{aligned} R_0 &= P_3^{[3]}; \\ R_1 &= P_3^{[2]}; \\ R_2 &= P_3^{[1]}; \\ R_3 &= P_3^{[0]}. \end{aligned}$$

7. In a manner analogous to that in Example 5.2.4, develop a Bézier curve type of representation for the part of a cubic Bézier curve on the interval $[1/2, 1]$.
8. Programming Exercise: Write a program to interactively design Bézier curves. Evaluate the points on the curves to be displayed using Algorithm 5.1 . Your program should be able to:

- Graphically enter and then display n-sided polygons by entering the control points.
- Turn off the display of the polygons and show only the curves.
- Edit a polygon defining a curve and redraw the curve. This requires the ability to pick a vertex of the polygon and move it to a new location without altering the rest of the polygon.
- Optionally, provide the facility for adding or deleting points, either interior or end points), from an existing polygon.
- Be able to display and edit multiple curves together on the screen.
- Delete a curve from the display.

9. Programming Exercise: For a given parametric function, $f(t) = (x(t), y(t))$, defined over $[0,1]$, write a program to compute and display

   - the function $f(t)$;
   - the Bernstein approximation of degree n;
   - the piecewise cubic Bernstein approximation to f, requiring values of f at $f(i/3)$, $i = 0, ..., 3n$;
   - the cubic interpolating spline to the data $f(i/n)$, $i = 0, \ldots, n$, $f'(0)$, $f'(1)$.
     The user should be able to specify n.

10. Using the variation diminishing property, the convex hull property, and the midpoint Bézier subdivision algorithm, develop an arbitrarily accurate algorithm to determine if two Bézier curves intersect (within $\epsilon > 0$).

6

B-Spline Curves

Chapter 5 explored the types of curves that are generated by a direct generalization of the constructive process first developed for conics in Example 3.16. The resulting curves are the Bézier curves. Positive features of the Bézier form are the variation diminishing property, the convex hull property, and the straightforward constructive algorithm. These properties allow us to think of the curve as a smoothed version of the control polygon. One positive feature of the Bézier form, that it results in a single polynomial curve, can also be a negative feature since

- when the ordered point set defining the curve gets large, the degree of the curve is high, and
- all polynomial bases, including the Bernstein basis, must be nonzero over the domain except at a finite number of points. Thus, changing the coefficient of any basis function modifies the whole curve. For a Bézier curve this means that moving any one control point changes the whole curve. The global effect is lessened because of the particular features of the Bernstein basis, but it is there.

In this chapter we seek to examine the constraints on overcoming the above negative features while retaining positive features of the Bézier curve design method. That is, we are seeking to retain the characteristics that allow the curve to be considered as a smoothed version of the control polygon. We investigate developing a representation that is piecewise polynomial and allows specification of degree of smoothness across the joins of the polynomial pieces. The techniques discussed in Section 5.5 are bottom

Figure 6.1. (a) Control polygon and first generalization; (b) discontinuous generalization.

up and require significant effort on the part of the user. Instead we seek methods that, based on the specification of the polynomial degree and the specification of smoothness across the joins of the pieces, will automatically create a constructive algorithm that generalizes the Bézier constructive algorithm. For each such method determined, we seek

- to find a closed form expression for the curve in terms of blending functions whose coefficients are the vertices of the control polygon, but which are not dependent on the particular control polygon (but are dependent on specified polynomial degree and join smoothness conditions). It is possible to think of this as separating the geometric properties of the curve from the parametric properties of the curve;
- to determine which characteristics and properties of the Bézier curves have been retained.

6.1 Constructive Piecewise Curves

For example, suppose we consider generalizing the linear Bézier curve, a line segment defined by a control polygon with two points. If a control polygon with three points is created, then the only possible generalization is a $C^{(0)}$ piecewise linear curve. Now if the control polygon has four points there are multiple generalizations. There could be a single $C^{(0)}$ piecewise linear curve with three pieces, graphically the same as the control polygon, as shown in Figure 6.1 (a). Or the curve could be disconnected ($C^{(-1)}$) and consist of two separate line segments, as shown in Figure 6.1 (b).

Consider the issues in generalizing the quadratic Bézier curve, a parabolic segment defined by a control polygon with three points. If a control polygon with four points is created, the extension method discussed in Section 3.8 is not applicable. Five points are necessary for that extension, and the resulting curve is guaranteed to be $C^{(0)}$, but only can be made $C^{(1)}$if the three middle vertices lie along a line. Thus, derivative smoothness across the joins cannot be specified as a separate degree of freedom. It is totally dependent on the geometry.

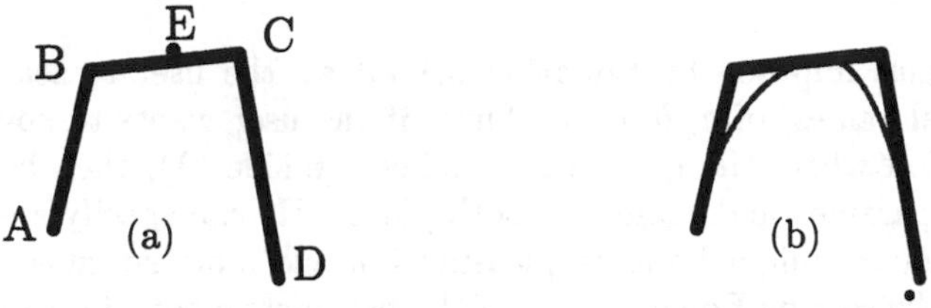

Figure 6.2. (a) Control polygon with breakpoint, E; (b) desired curve.

Now consider the specification that an arbitrary control polygon with four points, points A, B, C, and D, as shown in Figure 6.2, be used to generate a piecewise quadratic curve whose pieces meet with $C^{(1)}$ smoothness across the joins. The obvious generalization cannot be applied at all, much less guarantee the smoothness, so we explore a different strategy for generating such a curve. We specify a point $E = (1-\lambda)B + \lambda C$ on the line segment between B and C (so $0 < \lambda < 1$). This process is used to create two quadratic segments, one with control polygon A, B, E, and the other with control polygon E, C, D. Since the three points B, E, and C are on a line and properly ordered, the resulting curve will look $C^{(1)}$, as in Figure 6.2. However, it will not be $C^{(1)}$unless the magnitudes of the derivatives are adjusted to be equal. That affects the curve, and so we now develop a method for determining how to make the resulting composite curve $C^{(1)}$.

In order to develop this method, we use the generalized functions introduced in Definition 5.7 to represent the Bézier curve with polygon A, B, E on the interval $[a, b]$. It has derivative $2(E-B)/(b-a)$ at parameter value b, the upper endpoint of its interval. If the Bézier curve with polygon E, C, D is over the interval $[b, c]$, it has derivative $2(C-E)/(c-b)$ at parameter value b, the lower endpoint of its interval. If the composite curve is to be $C^{(1)}$, the two derivative values must be the same so,

$$(E-B)/(b-a) = (C-E)/(c-b).$$

Now since $E = (1-\lambda)B + \lambda C$,

$$\begin{aligned} E - B &= ((1-\lambda)B + \lambda C) - B \\ &= \lambda(C - B); \\ C - E &= C - ((1-\lambda)B + \lambda C) \\ &= (1-\lambda)(C - B). \end{aligned}$$

Thus, $\lambda(C-B)/(b-a) = (1-\lambda)(C-B)/(c-b)$, so

$$\lambda(c-b) = (1-\lambda)(b-a). \tag{6.1}$$

The constraint imposed by Equation 6.1 allows the user to set three of the four real values, of a, b, c, λ. Thus, if the user wants to specify the exact location where the quadratic touches the side (λ), then he cannot specify the parametric domains for both pieces. He can specify any pair of them. Symmetry might be more pleasing if a and c are specified and b is determined by solving Equation 6.1. If the parametric domains ($a < b < c$) are to be specified, then the place where the curve touches the line BC is determined from the equation (λ). If the user does not care to specify any of the values, the system could supply defaults. Reasonable defaults might be $a = 0$, $b = 1$, $c = 2$, with $\lambda = 1/2$, the value that satisfies Equation 6.1.

Notice that the above construction has no dependence on A or D. This enables the user to construct piecewise quadratic $C^{(1)}$curves from control polygons with $n + 1$ vertices, for any n. This can be done in a general way, for an arbitrary polygon $\{P_i\}_{i=0}^{n}$, by specifying $\boldsymbol{u} = \{u_i\}_{i=1}^{n}$, $u_i < u_{i+1}$. The composite curve domain is the interval $[u_1, u_n]$, with subinterval $[u_i, u_{i+1}]$ serving as the domain of the ith quadratic piece, resulting in $n - 1$ different quadratic segments. Then λ_i, between the ith and $(i+1)^{st}$ segment, $i = 1, \ldots, n-2$ is determined by solving Equation 6.1, setting $a = u_i$, $b = u_{i+1}$, and $c = u_{i+2}$.

While we have just evaluated the constructive curve algorithm with generalized Bernstein functions for each of the quadratic pieces, we have not yet determined a new constructive algorithm. Once again we return to the simple case with just two quadratic segments.

If a, b, and c are specified, then Equation 6.1 yields $\lambda = (b-a)/(c-a)$. To make this task easier, set $Q_0 = A$, $Q_1 = B$, $Q_2 = E$, $Q_3 = E$, $Q_4 = C$, $Q_5 = D$. The constructive algorithm for the curve with generalized Bernstein functions becomes, for $t \in [a, b]$,

$$
\begin{aligned}
P_1^{[1]} &= \frac{b-t}{b-a}Q_0 + \frac{t-a}{b-a}Q_1; \\
P_2^{[1]} &= \frac{b-t}{b-a}Q_1 + \frac{t-a}{b-a}Q_2; \\
P_2^{[2]} &= \frac{b-t}{b-a}P_1^{[1]} + \frac{t-a}{b-a}P_2^{[1]},
\end{aligned}
$$

and the curve point is $P_2^{[2]}$ for t. By definition of E, $Q_2 = ((c-b)/(c-a))Q_1 + ((b-a)/(c-a))Q_4$, so

$$
\begin{aligned}
P_2^{[1]} &= \frac{b-t}{b-a}Q_1 + \frac{t-a}{b-a}Q_2 \\
&= \frac{b-t}{b-a}Q_1 + \frac{t-a}{b-a}\left(\frac{c-b}{c-a}Q_1 + \frac{b-a}{c-a}Q_4\right)
\end{aligned}
$$

$$
\begin{aligned}
&= \left(\frac{b-t}{b-a} + \frac{t-a}{b-a}\frac{c-b}{c-a}\right) Q_1 + \frac{t-a}{c-a} Q_4 \\
&= \frac{c-t}{c-a} Q_1 + \frac{t-a}{c-a} Q_4.
\end{aligned}
$$

Analogously, for $t \in [b, c]$,

$$
\begin{aligned}
P_2^{[1]} &= \frac{c-t}{c-b} Q_3 + \frac{t-b}{c-b} Q_4; \\
P_3^{[1]} &= \frac{c-t}{c-b} Q_4 + \frac{t-b}{c-b} Q_5; \\
P_3^{[2]} &= \frac{c-t}{c-b} P_2^{[1]} + \frac{t-b}{c-b} P_3^{[1]},
\end{aligned}
$$

and $P_3^{[2]}$ is the curve point at t. Again, rewriting Q_3 in terms of original control points,

$$
\begin{aligned}
P_2^{[1]} &= \frac{c-t}{c-b} Q_3 + \frac{t-b}{c-b} Q_4 \\
&= \frac{c-t}{c-b}\left(\frac{c-b}{c-a} Q_1 + \frac{b-a}{c-a} Q_4\right) + \frac{t-b}{c-b} Q_4 \\
&= \frac{c-t}{c-a} Q_1 + \left(\frac{c-t}{c-b}\frac{b-a}{c-a} + \frac{t-b}{c-b}\right) Q_4 \\
&= \frac{c-t}{c-a} Q_1 + \frac{t-a}{c-a} Q_4.
\end{aligned}
$$

Notice that the final equation for computing $P_2^{[1]}$ is the same whichever interval t is in. Thus, the algorithm can be stated as

Algorithm 6.1. *Given a control polygon* $\{P_i\}_{i=0}^3$ *and values a,b,c,*
Step 1: Set $P_i^{[0]} = P_i, i = 0, \ldots, 3.$
Step 2: for any value $t \in (a, c)$

$$
\begin{aligned}
P_1^{[1]} &= \frac{b-t}{b-a} P_0^{[0]} + \frac{t-a}{b-a} P_1^{[0]}; \\
P_2^{[1]} &= \frac{c-t}{c-a} P_1^{[0]} + \frac{t-a}{c-a} P_2^{[0]}; \\
P_3^{[1]} &= \frac{c-t}{c-b} P_2^{[0]} + \frac{t-b}{c-b} P_3^{[0]}.
\end{aligned}
$$

Step 3: the second level then gives the appropriate curve value

$$\gamma(t) = \begin{cases} P_2^{[2]} & t \in [a,b] \\ P_3^{[2]} & t \in [b,c] \end{cases}$$

where

$$P_2^{[2]} = \frac{b-t}{b-a}P_1^{[1]} + \frac{t-a}{b-a}P_2^{[1]}; \tag{6.2}$$

$$P_3^{[2]} = \frac{c-t}{c-b}P_2^{[1]} + \frac{t-b}{c-b}P_3^{[1]}. \tag{6.3}$$

We have found a constructive algorithm which uses only the original control points. However it has been derived for piecewise quadratic curves only. Now we seek to address our second goal, to find a closed form expression for the curve in terms of blending functions whose coefficients are the original control points. Let

$$Q_i^{[0]} = \begin{cases} Q_i, & i = 0, \dots, 2, \\ Q_{i-1}, & i = 3, \dots, 5, \end{cases}$$

and define

$$\zeta_i^{[0]}(t) = \begin{cases} \theta_{i,2}(a,b;t) & a \le t \le b,\ i = 0, \dots, 2 \\ 0 & b \le t \le c, i = 0, \dots, 2 \\ 0 & a \le t \le b, i = 3, \dots, 5 \\ \theta_{i-3,2}(b,c;t) & b \le t \le c,\ i = 3, \dots, 5. \end{cases}$$

The curve can be written as

$$\gamma(t) = \sum_{i=0}^{5} Q_i^{[0]} \zeta_i^{[0]}(t).$$

Figure 6.3 shows $\zeta_i^{[0]}(t)$, $i = 0, \dots, 5$.

However, this form represents E twice. The next step is to remove that redundancy by setting

$$Q_i^{[1]} = \begin{cases} Q_i^{[0]}, & i = 0, \dots, 2 \\ Q_{i+1}^{[0]}, & i = 3, \dots, 4 \end{cases}$$

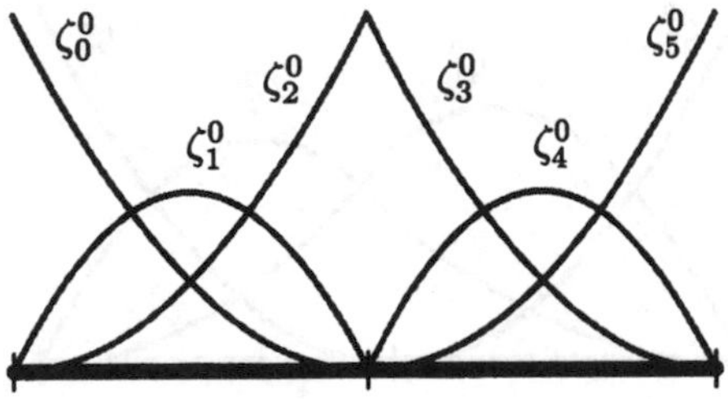

Figure 6.3. Original blending functions.

and

$$\zeta_i^{[1]}(t) = \begin{cases} \zeta_i^{[0]}(t), & i = 0, 1, \\ \zeta_2^{[0]}(t) + \zeta_3^{[0]}(t), & i = 2, \\ \zeta_{i+1}^{[0]}(t), & i = 3, 4. \end{cases}$$

Figure 6.4 shows $\zeta_i^{[1]}(t)$, $i = 0, \ldots, 4$. Now since $Q_2^{[1]} = (1-\lambda)Q_1^{[1]} + \lambda Q_3^{[1]}$, and $\lambda = (b-a)/(c-a)$,

$$\begin{aligned} \gamma(t) &= \sum_{i=0}^{4} Q_i^{[1]}\zeta_i^{[1]}(t) \\ &= Q_0^{[1]}\zeta_0^{[1]}(t) + Q_1^{[1]}\zeta_1^{[1]}(t) + ((1-\lambda)Q_1^{[1]} + \lambda Q_3^{[1]})\zeta_2^{[1]}(t) \\ &\quad + Q_3^{[1]}\zeta_3^{[1]}(t) + Q_4^{[1]}\zeta_4^{[1]}(t) \\ &= Q_0^{[1]}\zeta_0^{[1]}(t) + Q_1^{[1]}\zeta_1^{[1]}(t) + (Q_1^{[1]}(1-\lambda)\zeta_2^{[1]}(t) + Q_3^{[1]}\lambda\zeta_2^{[1]}(t)) \\ &\quad + Q_3^{[1]}\zeta_3^{[1]}(t) + Q_4^{[1]}\zeta_4^{[1]}(t) \\ &= Q_0^{[1]}\zeta_0^{[1]}(t) + Q_1^{[1]}(\zeta_1^{[1]}(t) + (1-\lambda)\zeta_2^{[1]}(t)) \\ &\quad + Q_3^{[1]}(\lambda\zeta_2^{[1]}(t) + \zeta_3^{[1]}(t)) + Q_4^{[1]}\zeta_4^{[1]}(t). \end{aligned}$$

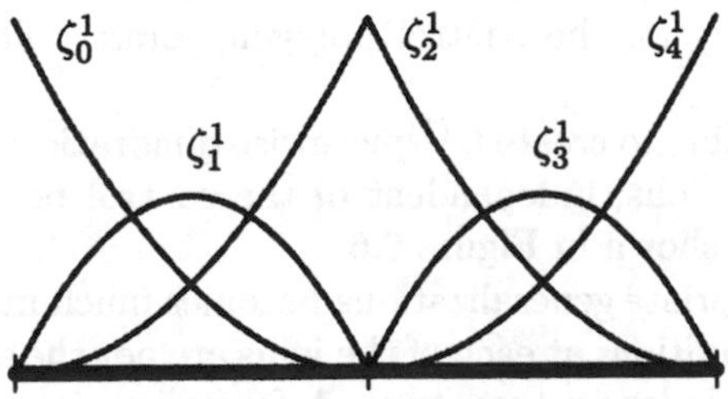

Figure 6.4. Blending functions after first reduction.

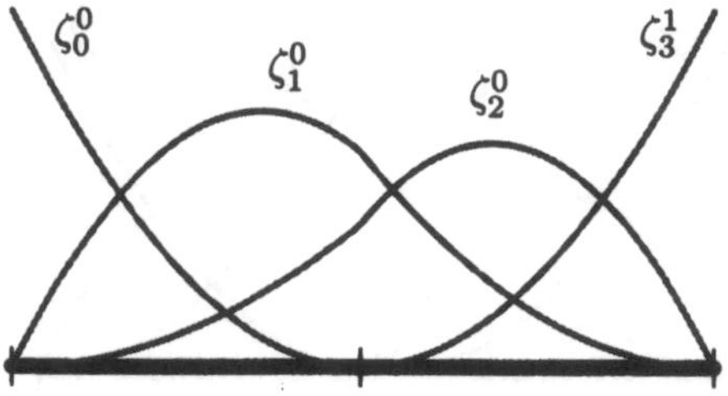

Figure 6.5. Final blending functions; $\lambda = 0.4$.

Using these results we set

$$Q_i^{[2]} = \begin{cases} Q_i^{[1]}, & i = 0, 1, \\ Q_{i+1}^{[1]}, & i = 2, 3, \end{cases}$$

and

$$\zeta_i^{[2]}(t) = \begin{cases} \zeta_i^{[1]}(t) & i = 0, \\ \zeta_1^{[1]}(t) + (1-\lambda)\zeta_2^{[1]}(t), & i = 1, \\ \lambda\zeta_2^{[1]}(t) + \zeta_3^{[1]}(t), & i = 2, \\ \zeta_4^{[1]}(t) & i = 3, \end{cases}$$

or simplifying further,

$$\zeta_i^{[2]}(t) = \begin{cases} \zeta_i^{[1]}(t) & i = 0, \\ \zeta_1^{[1]}(t) + \frac{c-b}{c-a}\zeta_2^{[1]}(t), & i = 1, \\ \frac{b-a}{c-a}\zeta_2^{[1]}(t) + \zeta_3^{[1]}(t), & i = 2, \\ \zeta_4^{[1]}(t) & i = 3. \end{cases}$$

The functions $\zeta_i^{[2]}(t), i = 0, \ldots, 3$ are $C^{(1)}$ on curve domain ($[a, c]$), and have no dependency on the control polygon points. They are shown in Figure 6.5.

Thus, it is possible to create $C^{(1)}$piecewise quadratic curves. The closed form blending functions, independent of the control polygon, for a curve with six pieces are shown in Figure 6.6.

Deriving appropriate generalizations becomes much more difficult when the smoothness conditions at each of the joins are not the same, or when the polynomial degree is larger than two. A formalism is necessary in which all the constructive processes and closed form blending functions can be

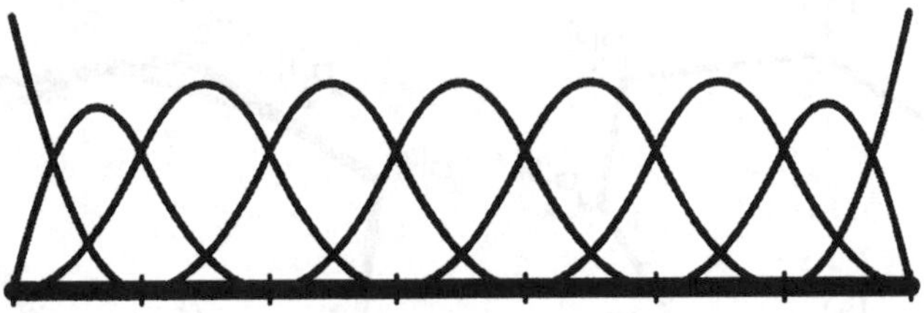

Figure 6.6. Quadratic blending functions with six subintervals.

represented, but is still compact. That formalism is called the *B-spline representation.*

Definition 6.2. *A sequence* $\boldsymbol{u} = \{u_i\}_{i=0}^{s}$ *of distinct real values is called a* breakpoint sequence, *and* $\boldsymbol{m} = \{m_i\}_{i=0}^{s}$ *an associated sequence of positive integer values, one for each element of* $\boldsymbol{u}$ *is called the* multiplicity vector. *A nondecreasing sequence of real numbers* $\mathbf{t} = \{t_j\}$ *such that* $m_i = card\{j : t_j = u_i\}$ *is called a* knot vector.

Definition 6.3. *The number of elements in* $\mathbf{t}$*, a knot vector, is called its* length *and is designated* $card(\boldsymbol{t})$.

Set $N+1 = \sum_{i=0}^{s} m_i$, that is, $\boldsymbol{t}$ has $N+1$ elements. The breakpoint sequence is so named because the interval $u_i \le t < u_{i+1}$ defines one piecewise polynomial span; $u_{i+1} \le t < u_{i+2}$ defines the next. m_{i+1} is related to continuity with which the pieces meet, as will be shown later.

We give an algorithmic definition of a curve.

Algorithm 6.4. Recursive B-Spline Algorithm. *To define a piecewise polynomial curve of degree* κ*, called* $\gamma(t)$*, we require the domain be* $[t_\kappa, t_{N-\kappa})$*, where* $\{t_i\}_{i=0}^{N}$ *is as defined in Definition 6.2.*

Step 1: Define $P_i^{[0]} = P_i$.

Step 2: For a given $t \ge t_\kappa$*, find* J *such that* $t \in [t_J, t_{J+1})$.

Step 3: For $p = 1, \ldots, \kappa$

for $i = J - \kappa + p, \ldots, J$*, let*

$$P_i^{[p]} = \frac{t - t_i}{t_{i+\kappa-(p-1)} - t_i} P_i^{[p-1]} + \frac{t_{i+\kappa-(p-1)} - t}{t_{i+\kappa-(p-1)} - t_i} P_{i-1}^{[p-1]}.$$

Step 4: Then, $\gamma(t) = P_J^{[\kappa]}$.

Once the interval is found, step two is exactly the same as the first step of the constructive Bézier algorithm. We see that the next step is

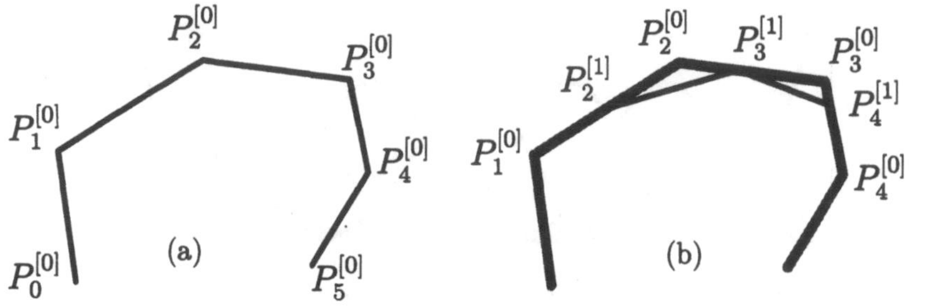

Figure 6.7. (a) Original polygon; (b) $P_i^{[0]}$ and $P_i^{[1]}$.

the modification. $P_i^{[p]}$ is always a convex combination of $P_{i-1}^{[p-1]}$ and $P_i^{[p-1]}$ which depends on the value of t, but it is a different combination for each value of i, p, and κ. We see, however, that if the degree is cubic, then there are exactly four points involved in the *zero*th iteration, three points in the first iteration, two points in the second iteration, and one point, the answer, in the third iteration, just like the Bézier case. This is true no matter how many points there are in the original polygon. However, we see that the particular values of t that are chosen affect the algorithm. For a polygon with $n+1$ points, if we want a curve of degree κ, then we must have a $\boldsymbol{t}$ vector defined which has exactly $n+2+\kappa$ elements in it. The restriction on the curve that occurs by using only $\kappa+1$ points to evaluate this curve of degree κ over a particular interval causes the defined curve to be local. Clearly, it is the same polynomial for all t in the open interval. The next interval will use different points to define the polynomial, although some may be the same. The multiplicity gives the ***skip*** information; that is, it tells how many polygon points are the same in two adjacent spans. We shall leave verification of the continuity that is embedded to Chapter 7. However, intuitively we see that if $\kappa - p + 1$ points are in common, then the polynomial pieces must meet with continuity $C^{(\kappa-p)}$.

Example 6.5. $\kappa = 3$ and $\boldsymbol{t} = \{0,0,0,0,1,4,5,5,5,5\}$. To evaluate the curve at $t = 2$ using the algorithm, $J = 4$ since $t_4 = 1 \le 2 < t_5 = 4$.

$$P_i^{[p]} = \frac{t - t_i}{t_{i+4-p} - t_i} P_i^{[p-1]} + \frac{t_{i+4-p} - t}{t_{i+4-p} - t_i} P_{i-1}^{[p-1]}, \qquad i = 4,\ 3, \ldots,\ 1+p.$$

For $p = 1$,

$$\begin{aligned} P_4^{[1]} &= \frac{t - t_4}{t_{8-1} - t_4} P_4^{[1-1]} + \frac{t_{8-1} - t}{t_{8-1} - t_4} P_3^{[1-1]} \\ &= \frac{2-1}{5-1} P_4^{[0]} + \frac{5-2}{5-1} P_3^{[0]} \\ &= \frac{1}{4} P_4^{[0]} + \frac{3}{4} P_3^{[0]}; \end{aligned}$$

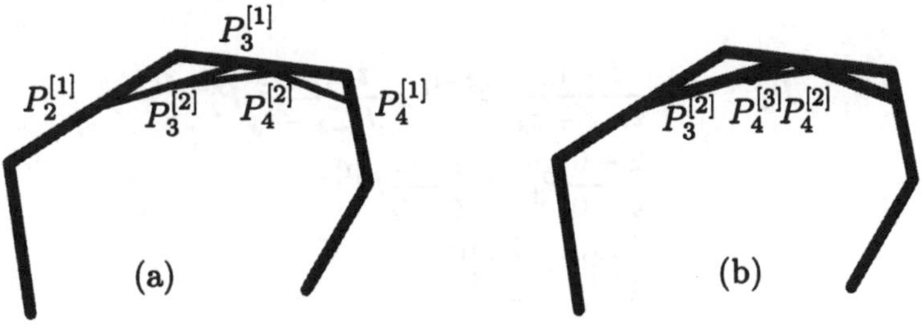

Figure 6.8. (a) $P_i^{[1]}$ and $P_i^{[2]}$; (b) $P_i^{[2]}$ and $P_i^{[3]}$.

$$\begin{aligned} P_3^{[1]} &= \frac{t-t_3}{t_{7-1}-t_3}P_3^{[1-1]} + \frac{t_{7-1}-t}{t_{7-1}-t_3}P_2^{[1-1]} \\ &= \frac{2-0}{5-0}P_3^{[0]} + \frac{5-2}{5-0}P_2^{[0]} \\ &= \frac{2}{5}P_3^{[0]} + \frac{3}{5}P_2^{[0]}; \\ P_2^{[1]} &= \frac{t-t_2}{t_{6-1}-t_2}P_2^{[1-1]} + \frac{t_{6-1}-t}{t_{6-1}-t_2}P_1^{[1-1]} \\ &= \frac{2-0}{4-0}P_2^{[0]} + \frac{4-2}{4-0}P_1^{[0]} \\ &= \frac{1}{2}P_2^{[0]} + \frac{1}{2}P_1^{[0]}. \end{aligned}$$

For $p = 2$,

$$\begin{aligned} P_4^{[2]} &= \frac{t-t_4}{t_{8-2}-t_4}P_4^{[2-1]} + \frac{t_{8-2}-t}{t_{8-2}-t_4}P_3^{[2-1]} \\ &= \frac{2-1}{5-1}P_4^{[1]} + \frac{5-2}{5-1}P_3^{[1]} \\ &= \frac{1}{4}P_4^{[1]} + \frac{3}{4}P_3^{[1]}; \end{aligned}$$

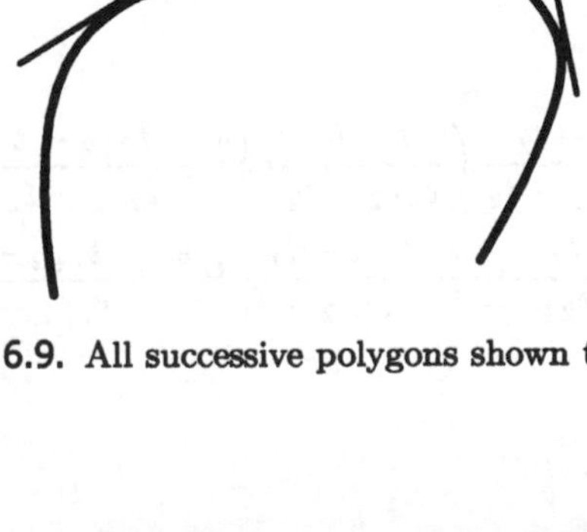

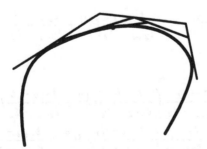

Figure 6.9. All successive polygons shown together.

$$\begin{aligned} P_3^{[2]} &= \frac{t-t_3}{t_{7-2}-t_3}P_3^{[2-1]} + \frac{t_{7-2}-t}{t_{7-2}-t_3}P_2^{[2-1]} \\ &= \frac{2-0}{4-0}P_3^{[1]} + \frac{4-2}{4-0}P_2^{[1]} \\ &= \frac{1}{2}P_3^{[1]} + \frac{1}{2}P_2^{[1]}. \end{aligned}$$

For $p = 3$,

$$\begin{aligned} P_4^{[3]} &= \frac{t-t_4}{t_{8-3}-t_4}P_4^{[3-1]} + \frac{t_{8-3}-t}{t_{8-3}-t_4}P_3^{[3-1]} \\ &= \frac{2-1}{4-1}P_4^{[2]} + \frac{4-2}{4-1}P_3^{[2]} \\ &= \frac{1}{3}P_4^{[2]} + \frac{2}{3}P_3^{[2]}. \end{aligned}$$

Finally, $\gamma(2) = P_4^{[3]}$. Note that these recursively defined polygons are totally dependent on the value of t used. □

Example 6.6. Let $\boldsymbol{t} = \{0, 0, 0, 1, 3, 4, 4, 4\}$. We consider the algorithm for $\kappa = 2$, quadratic curves, with control polygon P_0, P_1, P_2, P_3, P_4, and find the closed form solution.

The domain will be $[0, 4)$.
For $t \in [0, 1)$, J = 2.

1. $P_i^0 = P_i$, for $i = 0, 1, 2$;

2. $$\begin{aligned} P_2^{[1]} &= \frac{t-t_2}{t_{2+2}-t_2}P_2^{[0]} + \frac{t_{2+2}-t}{t_{2+2}-t_2}P_{2-1}^{[0]}; \\ P_1^{[1]} &= \frac{t-t_1}{t_{1+2}-t_1}P_1^{[0]} + \frac{t_{1+2}-t}{t_{1+2}-t_1}P_{1-1}^{[0]}. \end{aligned}$$

3. $$P_2^{[2]} = \frac{t-t_2}{t_{2+1}-t_2}P_2^{[1]} + \frac{t_{2+1}-t}{t_{2+1}-t_2}P_{2-1}^{[1]}.$$

Thus for $t \in [0, 1)$,

$$\begin{aligned} \gamma(t) &= \frac{t-t_2}{t_{2+1}-t_2}\left(\frac{t-t_2}{t_{2+2}-t_2}P_2^{[0]} + \frac{t_{2+2}-t}{t_{2+2}-t_2}P_{2-1}^{[0]}\right) \\ &\quad + \frac{t_{2+1}-t}{t_{2+1}-t_2}\left(\frac{t-t_1}{t_{1+2}-t_1}P_1^{[0]} + \frac{t_{1+2}-t}{t_{1+2}-t_1}P_{1-1}^{[0]}\right). \end{aligned}$$

For $t \in [1,3)$, J = 3.

1. $P_i^0 = P_i$, for $i = 1, 2, 3$

2.
$$\begin{aligned} P_3^{[1]} &= \frac{t - t_3}{t_{3+2} - t_3} P_3^{[0]} + \frac{t_{3+2} - t}{t_{3+2} - t_3} P_{3-1}^{[0]}; \\ P_2^{[1]} &= \frac{t - t_2}{t_{2+2} - t_2} P_2^{[0]} + \frac{t_{2+2} - t}{t_{2+2} - t_2} P_{2-1}^{[0]}. \end{aligned}$$

3.
$$P_3^{[2]} = \frac{t - t_3}{t_{3+1} - t_3} P_3^{[1]} + \frac{t_{3+1} - t}{t_{3+1} - t_3} P_{3-1}^{[1]}.$$

Thus, for $t \in [1,3)$,

$$\begin{aligned} \gamma(t) &= \frac{t - t_3}{t_{3+1} - t_3} \left(\frac{t - t_3}{t_{3+2} - t_3} P_3^{[0]} + \frac{t_{3+2} - t}{t_{3+2} - t_3} P_{3-1}^{[0]} \right) \\ &\quad + \frac{t_{3+1} - t}{t_{3+1} - t_3} \left(\frac{t - t_2}{t_{2+2} - t_2} P_2^{[0]} + \frac{t_{2+2} - t}{t_{2+2} - t_2} P_{2-1}^{[0]} \right). \end{aligned}$$

The result for $t \in [3,4)$ can be similarly obtained. □

Like Bézier curves, this new curve form will not, in general, interpolate the control polygon. Looking at generating Algorithm 6.4, we see that the curve will have the form $\gamma(t) = \sum P_i \mathcal{B}_{i,\kappa}(t)$, where each $\mathcal{B}_{i,\kappa}(t)$ is a piecewise polynomial, and exactly $(\kappa + 1)$ of them are nonzero on each open interval between knots.

6.2 B-Spline Blending Functions

We seek to find a collection of functions $\{\mathcal{B}_{i,\kappa}(t)\}_{i=0}^{n}$ with a nice recursive form, a generalization of the Bernstein/Bézier blending functions, the same way that the curve is a generalization. Since κ denotes the degree of the piecewise polynomials, we see that we would like Algorithm 6.4 to fit the paradigm

$$\begin{aligned} \gamma(t) &= \sum i P_i \mathcal{B}_{i,\kappa}(t) \\ &= \sum i P_i^{[j]} \mathcal{B}_{i,\kappa-j}(t) \qquad j = 0, \ldots, \kappa. \end{aligned}$$

For a given value of t, there are only $(\kappa+1)$ values of $P_i^{[0]}$, $i = J - \kappa, \ldots, J$ defined, so that means

$$\begin{aligned}\gamma(t) &= \sum_{i=J-\kappa+j}^{J} P_i^{[j]} \mathcal{B}_{i,\kappa-j}(t), \qquad j = 0, \ldots, \kappa \\ &= \sum \left(\frac{t-t_i}{t_{i+\kappa-j+1}-t_i} P_i^{[j-1]} + \frac{t_{i+\kappa-j+1}-t}{t_{i+\kappa-j+1}-t_i} P_{i-1}^{[j-1]} \right) \mathcal{B}_{i,\kappa-j}(t), \\ &\qquad\qquad j = 1, \ldots, \kappa \\ &= \sum P_i^{[j-1]} \left(\frac{t-t_i}{t_{i+\kappa-j+1}-t_i} \mathcal{B}_{i,\kappa-j}(t) \right. \\ &\qquad \left. + \frac{t_{i+1+\kappa-j+1}-t}{t_{i+1+\kappa-j+1}-t_{i+1}} \mathcal{B}_{i+1,\kappa-j}(t) \right).\end{aligned}$$

But since we would like $\gamma(t) = \sum P_i^{[j-1]} \mathcal{B}_{i,\kappa-(j-1)+1}(t)$, a function to satisfy our requirements would have recursive form

$$\mathcal{B}_{i,\kappa-j+1)}(t) = \left(\frac{t-t_i}{t_{i+\kappa-j+1)}-t_i} \mathcal{B}_{i,\kappa-j}(t) + \frac{t_{i+\kappa-j+2}-t}{t_{i+\kappa-j+2}-t_{i+1}} \mathcal{B}_{i+1,\kappa-j}(t) \right).$$

Letting $r = \kappa - j$ gives

$$\mathcal{B}_{i,r+1}(t) = \left(\frac{t-t_i}{t_{i+r+1}-t_i} \mathcal{B}_{i,r}(t) + \frac{t_{i+r+2}-t}{t_{i+r+2}-t_{i+1}} \mathcal{B}_{i+1,r}(t) \right).$$

Now we shall define collections of piecewise polynomials in a recursive formulation which have this exact characteristic, and show that the curve obtained is the same as that in Algorithm 6.4. Later we shall show that these functions are independent and form a basis for the linear space that they span. This will give us uniqueness of representation which is helpful.

Definition 6.7. *Let $t_0 \le t_1 \le \cdots \le t_N$ be a sequence of real numbers. For $\kappa = 0, \ldots, N-1$, and $i = 0, \ldots, N-\kappa-1$, define the ith (normalized) B-spline of degree κ and order $k = (\kappa + 1)$ as*

$$\mathcal{B}_{i,1}(t) = \begin{cases} 1 & \textit{for } t_i \le t < t_{i+1} \\ 0 & \textit{otherwise} \end{cases}$$

and for $\kappa > 0$,

$$\mathcal{B}_{i,\kappa}(t) = \begin{cases} \frac{(t-t_i)}{t_{i+\kappa}-t_i} \mathcal{B}_{i,\kappa-1}(t) + \frac{(t_{i+1+\kappa}-t)}{t_{i+1+\kappa}-t_{i+1}} \mathcal{B}_{i+1,\kappa-1}(t), & t_i < t_{i+1+\kappa}, \\ 0 & \textit{otherwise.} \end{cases}$$

Notice that the second subscript is the order which is always one more than the degree. Although the concept of polynomial degree is more familiar, B-splines were introduced using the order as the second subscript.

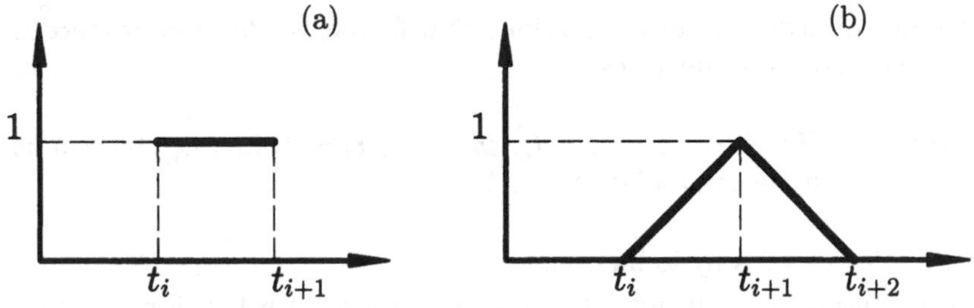

Figure 6.10. (a) B-spline of degree 0; (b) B-spline of degree 1.

Some recent authors have chosen to use the degree as the second subscript. While this is correct, it can be confusing for those new to concepts in B-splines. For this reason, we adopt the notation that B and N are used to denote B-splines using the order as the second subscript while $\mathcal{B}$ and $\mathcal{N}$ are used to designate B-splines for which the second subscript is the polynomial degree. Thus, $\mathcal{B}_{i,\kappa} = B_{i,\kappa+1}$. For the rest of this chapter, we will use the degree, $\mathcal{B}$, formulation. Thereafter, we will use both degree and order notation.

This definition needs a bit of explanation. We consider the case $\kappa = 0$. If $t_i = t_{i+1}$, then it is impossible to find any values in the half open interval. The convention that has been adopted, then, is to set it equal to zero, the limiting value from both above and below. Now, suppose one wants to calculate $\mathcal{B}_{i,\kappa}(t)$, and $t_i = \cdots = t_{i+\kappa} < t_{i+\kappa+1}$. In this instance, the denominator in the first term is zero, and it would seem that the recursive definition fails. However, in this case, $\mathcal{B}_{i,\kappa-1}(t)$ is identically zero also, by definition. The convention has been adopted that such a term in the recursive formulation will be set equal to zero. Hence, for the example above the recursive definition becomes, $\mathcal{B}_{i,\kappa}(t) = \frac{(t_{i+1+\kappa}-t)}{t_{i+1+\kappa}-t_{i+1}}\mathcal{B}_{i+1,\kappa-1}(t)$.

Example 6.8. Find the form of $\mathcal{B}_{i,1}(t)$. To solve this we first note that $\mathcal{B}_{i,1}(t)$ depends only on $\mathcal{B}_{i,0}(t)$ and $\mathcal{B}_{i+1,0}(t)$. Suppose $t_i < t_{i+1} < t_{i+2}$. Then $\mathcal{B}_{i,0}(t) \neq 0$ if $t \in [t_i, t_{i+1})$, and $\mathcal{B}_{i+1,0}(t) \neq 0$ if $t \in [t_{i+1}, t_{i+2})$. Hence, if $\mathcal{B}_{i,1}(t)$ is nonzero, then $t \in [t_i, t_{i+2})$. Now,

$$\begin{aligned}\mathcal{B}_{i,1}(t) &= \frac{(t-t_i)}{t_{i+1}-t_i}\mathcal{B}_{i,0}(t) + \frac{(t_{i+2}-t)}{t_{i+2}-t_{i+1}}\mathcal{B}_{i+1,0}(t) \\ &= \begin{cases} \frac{(t-t_i)}{t_{i+1}-t_i} & \text{for } t_i \le t < t_{i+1} \\ \frac{(t_{i+2}-t)}{t_{i+2}-t_{i+1}} & \text{for } t_{i+1} \le t < t_{i+2}. \end{cases}\end{aligned}$$

$\mathcal{B}_{i,1}(t_{i+1}) = 1$, and $\mathcal{B}_{i,1}(t)$ is a piecewise linear function which is nonzero over two knot spans. It immediately follows that $\mathcal{B}_{i,\kappa}(t)$ is a piecewise polynomial of degree κ. In the sections that follow, we develop properties of these collections of functions. □

Lemma 6.9. *If $t \geq t_{i+1+\kappa}$ or $t < t_i$, then $\mathcal{B}_{i,\kappa}(t) = 0$, i.e., $\mathcal{B}_{i,\kappa}(t)$ can be nonzero only on the interval $[t_i, t_{i+\kappa+1})$.*

Proof: The proof is by induction.

By definition, it is true for for $\kappa = 0$. Now assume it is true for B-splines of degree $\kappa - 1$. We shall show it is true for B-splines of degree κ.

By the induction hypothesis, $\mathcal{B}_{i,\kappa-1}(t)$ is zero for $t < t_i$ and $t \geq t_{i+\kappa}$, and $\mathcal{B}_{i+1,\kappa-1}(t)$ is zero for $t < t_{i+1}$ and $t \geq t_{i+\kappa+1}$, since they are both B-splines of degree $\kappa - 1$. Putting both of these together in the recursive definition, we find that the first term might be nonzero for $t \in [t_i, t_{i+\kappa})$ and the second term might be nonzero for $t \in [t_{i+1}, t_{i+1+\kappa})$. Thus, $\mathcal{B}_{i,\kappa}(t)$ can be nonzero only in the union of these two sets, that is, $\mathcal{B}_{i,\kappa}(t)$ can be nonzero only for $t \in [t_i, t_{i+1+\kappa})$. ■

Lemma 6.10. $\mathcal{B}_{i,\kappa}(t) > 0$ *for* $t \in (t_i, t_{i+\kappa+1})$.

This lemma says that $\mathcal{B}_{i,\kappa}(t)$ is positive on the open interval from t_i to $t_{i+\kappa+1}$, is zero at $t_{i+\kappa+1}$, and can be zero or nonzero at t_i, depending on the function. We shall see separate instances below.

Proof: By definition, $\mathcal{B}_{i,0}(t) > 0$ for $t \in (t_i, t_{i+1})$, and so the result is true for $\kappa = 0$.

Now assume it is true for B-splines of degree $\kappa - 1$. We shall show it is true for B-splines of degree κ.

By the induction hypothesis, $\mathcal{B}_{i,\kappa-1}(t) > 0$, for $t \in (t_i, t_{i+\kappa})$, and $\mathcal{B}_{i+1,\kappa-1}(t) > 0$ for $t \in (t_{i+1}, t_{i+1+\kappa})$, since they are both B-splines of degree $\kappa - 1$. Putting both of these together in the recursive definition, we find that the first term is positive for $t > t_i$, and the second term is positive for $t < t_{i+\kappa+1}$. Hence, $\mathcal{B}_{i,\kappa}(t)$ is positive on the intersection of these values of t, that is $\mathcal{B}_{i,\kappa}(t) > 0$ for $t \in (t_i, t_{i+\kappa+1})$.

Now, since $\mathcal{B}_{j,0}(t_{j+1}) = 0$ for all j, we have shown the second assertion for $\kappa = 0$. Assume the induction hypothesis, that $\mathcal{B}_{j,\kappa-1}(t_{j+\kappa}) = 0$ for all j. Then,

$$\begin{aligned}\mathcal{B}_{i,\kappa}(t_{i+\kappa+1}) &= \frac{(t_{i+\kappa+1} - t_i)}{t_{i+\kappa} - t_i}\mathcal{B}_{i,\kappa-1}(t_{i+\kappa+1}) \\ &\quad + \frac{(t_{i+1+\kappa} - t_{i+1+\kappa})}{t_{i+1+\kappa} - t_{i+1}}\mathcal{B}_{i+1,\kappa-1}(t_{i+1+\kappa}).\end{aligned}$$

For the last point, note that if $t_i = t_{i+1}$, $\mathcal{B}_{i,1}(t_i) = \mathcal{B}_{i,1}(t_{i+1}) = 1$, but if $t_i < t_{i+1}$, $\mathcal{B}_{i,1}(t_i) = 0$. ∎

Definition 6.11. *A function which is nonzero over only a finite interval of the real line is called a* local function.

An added result of this lemma is that B-splines are local for all i and κ. Assume that $t_j < t_{j+1}$. Which functions can be nonzero over (t_j, t_{j+1})? From the previous lemma, we know that the support of $\mathcal{B}_{i,\kappa}(t)$ must be contained in $[t_i, t_{i+\kappa+1})$, so if $t_i \leq t_j < t_{i+\kappa+1}$, that is, $t_j \in \{t_i, t_{i+1}, \ldots, t_{i+\kappa}\}$, then $\mathcal{B}_{i,\kappa}$ is nonzero in the interval (t_j, t_{j+1}). For $i = j,\ j-1, \ldots,\ j-\kappa$, $\mathcal{B}_{j,\kappa}(t)$ is nonzero in $[t_j, t_{j+1})$, and we have shown that $\kappa + 1$ functions are nonzero on any single open interval.

Corollary 6.12. *If $t \in (t_j, t_{j+1})$ and $i \in \{j-\kappa,\ j-\kappa+1, \ldots,\ j\}$, then $\mathcal{B}_{i,\kappa}(t) > 0$.*

6.2.1 Proof of Equivalence of Constructive Algorithm with Curve Form

We prove the equivalence relationship between the curve generation algorithm definition, Algorithm 6.4, and the curve formed using the polygon vertices as coefficients to the blending functions defined in Definition 6.7.

Theorem 6.13. *If $\gamma(t)$ is the curve defined by Algorithm 6.4 and $\alpha(t) = \sum P_i\mathcal{B}_{i,\kappa}(t)$, where the $\mathcal{B}_{i,\kappa}(t)$ are defined in Definition 6.7, then $\gamma(t) = \alpha(t)$.*

Proof: For $t \in [t_J, t_{J+1})$,

$$\begin{aligned} \alpha(t) &= \sum_{m=0}^{n} P_m\mathcal{B}_{m,\kappa}(t) \\ &= \sum_{m=J-\kappa}^{J} P_m\mathcal{B}_{m,\kappa}(t), \end{aligned}$$

since only nonzero B-splines contribute. Now, using the recursive definition

$$
\begin{aligned}
\alpha(t) &= \sum_{m=J-\kappa}^{J} P_m \left(\frac{(t_{m+\kappa+1} - t)}{t_{m+\kappa+1} - t_{m+1}} \mathcal{B}_{m+1,\kappa-1}(t) + \frac{(t - t_m)}{t_{m+\kappa} - t_m} \mathcal{B}_{m,\kappa-1}(t) \right) \\
&= \sum_{m=J-\kappa}^{J} P_m \frac{(t_{m+\kappa+1} - t)}{t_{m+\kappa+1} - t_{m+1}} \mathcal{B}_{m+1,\kappa-1}(t) \\
&\qquad + \sum_{m=J-\kappa}^{J} P_m \frac{(t - t_m)}{t_{m+\kappa} - t_m} \mathcal{B}_{m,\kappa-1}(t) \\
&= \sum_{m=J+1-\kappa}^{J+1} P_{m-1} \frac{(t_{m+\kappa} - t)}{t_{m+\kappa} - t_m} \mathcal{B}_{m,\kappa-1}(t) \\
&\qquad + \sum_{m=J-\kappa}^{J} P_m \frac{(t - t_m)}{t_{m+\kappa} - t_m} \mathcal{B}_{m,\kappa-1}(t).
\end{aligned}
$$

Now, $t \in [t_J, t_{J+1})$, and so $\mathcal{B}_{J+1,\kappa-1}(t) = 0$ (this appears as a term in the first summation). The support of $\mathcal{B}_{J-\kappa,\kappa-1}$ is in $[t_{J-\kappa}, t_{J-\kappa+\kappa}) = [t_{J-\kappa}, t_J)$, so the particular value of t of interest is outside the support of $\mathcal{B}_{J-\kappa,\kappa-1}$, so $\mathcal{B}_{J-\kappa,\kappa-1}(t) = 0$ (this appears in the second summation). We are left with

$$
\begin{aligned}
\alpha(t) &= \sum_{m=J+1-\kappa}^{J} P_{m-1} \frac{(t_{m+\kappa} - t)}{t_{m+\kappa} - t_m} \mathcal{B}_{m,\kappa-1}(t) \\
&\qquad + \sum_{m=J+1-\kappa}^{J} P_m \frac{(t - t_m)}{t_{m+\kappa} - t_m} \mathcal{B}_{m,\kappa-1}(t) \\
&= \sum_{m=J+1-\kappa}^{J} \left(\frac{(t_{m+\kappa} - t)}{t_{m+\kappa} - t_m} P_{m-1} + \frac{(t - t_m)}{t_{m+\kappa} - t_m} P_m \right) \mathcal{B}_{m,\kappa-1}(t) \\
\alpha(t) &= \sum_{m=J+1-\kappa}^{J} P_m^{[1]} \mathcal{B}_{m,\kappa-1}(t).
\end{aligned}
$$

The higher order iterates follow by induction, see Exercise 1, and we are left with

$$
\begin{aligned}
\alpha(t) &= \sum_{m=J}^{J} P_m^{[\kappa]} \mathcal{B}_{m,0}(t) \\
&= P_J^{[\kappa]} \\
&= \gamma(t).
\end{aligned}
$$

■

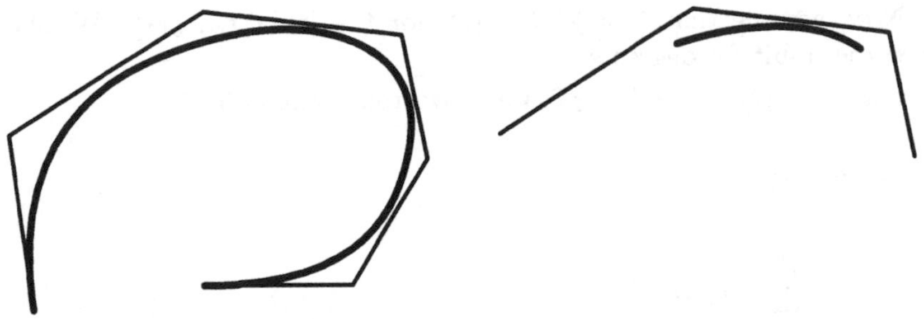

Figure 6.11. A curve and associated subcontrol polygon for an interior interval.

We see that this construction is very stable since at each step the next level polygon results from convex combinations of adjacent vertices of the present polygon.

Corollary 6.14. *For $t_i \leq t < t_{i+1}$, the B-spline curve is a convex combination of $P_{i-\kappa}, \ldots, P_i$. This is called the* convex hull property.

6.3 More Properties of B-Splines

Definition 6.15. *Suppose $\boldsymbol{t}$ is a knot vector with $N+1$ elements. Then there are exactly $N-\kappa$ B-spline functions of degree κ defined over the sequence $\boldsymbol{t}$. Let $k = \kappa + 1$ be the order. $S_{k,\boldsymbol{t}} = \{\sum P_i B_{i,k}(t) : P_i \in R\}$, the collection of all linear combinations. Analogous notation for the degree formulation gives $\mathcal{S}_{\kappa,\boldsymbol{t}} = \{\sum P_i \mathcal{B}_{i,\kappa}(t) : P_i \in R\}$. Remember $\mathcal{S}_{\kappa,\boldsymbol{t}} = S_{k,\boldsymbol{t}}$. So, $S(\mathcal{S})$ is called the* space of B-splines of order k (degree κ) over the knot vector $\boldsymbol{t}$.

Lemma 6.16. *For a knot vector $\boldsymbol{t} = \{t_i\}_{i=0}^{N}$, and for $t \in [t_\kappa, t_{N-\kappa})$,*

$$\sum_{i=0}^{N-(\kappa+1)} \mathcal{B}_{i,\kappa}(t) = 1, \qquad \textit{for all } \kappa \geq 0.$$

Further, for $t < t_\kappa$ or $t > t_{N-\kappa}$, $\sum_{i=0}^{N-(\kappa+1)} \mathcal{B}_{i,\kappa}(t) < 1$.

Proof: Again the proof is by induction on the degree κ.

In the proof below, let $j \in \{\kappa, \ldots, N-(\kappa+1)\}$. For $\kappa = 0$, and $t \in [t_j, t_{j+1})$ only one B-spline is nonzero, and by definition, it has value 1, so the sum over all of them has value 1.

Now, assume that $1 = \sum \mathcal{B}_{i,\kappa-1}(t)$ for $t \in [t_{\kappa-1}, t_{N-\kappa+1}]$. We shall show the result for degree κ.

For $t \in [t_j, t_{j+1})$, and $\kappa \geq 0$, we know from Lemma 6.12

$$\begin{aligned}
\sum_i \mathcal{B}_{i,\kappa}(t) \\
&= \sum_{i=j-\kappa}^{j} \mathcal{B}_{i,\kappa}(t) \\
&= \sum_{i=j-\kappa}^{j} \left(\frac{t - t_i}{t_{i+\kappa} - t_i} \mathcal{B}_{i,\kappa-1}(t) + \frac{t_{i+1+\kappa} - t}{t_{i+1+\kappa} - t_{i+1}} \mathcal{B}_{i+1,\kappa-1}(t) \right) \\
&= \sum_{i=j-\kappa}^{j} \frac{t - t_i}{t_{i+\kappa} - t_i} \mathcal{B}_{i,\kappa-1}(t) + \sum_{i=j-\kappa}^{j} \frac{t_{i+1+\kappa} - t}{t_{i+1+\kappa} - t_{i+1}} \mathcal{B}_{i+1,\kappa-1}(t) \\
&= \sum_{i=j-\kappa+1}^{j} \frac{t - t_i}{t_{i+\kappa} - t_i} \mathcal{B}_{i,\kappa-1}(t) + \sum_{i=j-\kappa}^{j-1} \frac{t_{i+1+\kappa} - t}{t_{i+1+\kappa} - t_{i+1}} \mathcal{B}_{i+1,\kappa-1}(t).
\end{aligned}$$

These changes in the summations follow from Lemma 6.9.

$$\begin{aligned}
&= \sum_{i=j+1-\kappa}^{j} \frac{t - t_i}{t_{i+\kappa} - t_i} \mathcal{B}_{i,\kappa-1}(t) + \sum_{i=j+1-\kappa}^{j} \frac{t_{i+\kappa} - t}{t_{i+\kappa} - t_i} \mathcal{B}_{i,\kappa-1}(t) \\
&= \sum_{i=j+1-\kappa}^{j} \left(\frac{t - t_i}{t_{i+\kappa} - t_i} + \frac{t_{i+\kappa} - t}{t_{i+\kappa} - t_i} \right) \mathcal{B}_{i,\kappa-1}(t) \\
&= \sum_{i=j+1-\kappa}^{j} \mathcal{B}_{i,\kappa-1}(t) \\
&= \sum_{i=0}^{N-\kappa} \mathcal{B}_{i,\kappa-1}(t) \\
&= 1, \qquad \text{by induction hypothesis.}
\end{aligned}$$

The result is then shown. We now show that $1 > \sum_{i=0}^{N-(\kappa+1)} \mathcal{B}_{i,\kappa}(t)$, for $t < t_\kappa$. Select values $t_{-\kappa} \leq \ldots \leq t_{-1} < t_0$ and consider the augmented knot vector $\{t_{-\kappa}, \ldots, t_{-1}, t_0, t_1, \ldots, t_N\}$. There are κ extra B-splines of degree κ over this knot collection. $\mathcal{B}_{i,\kappa}(t)$, $i = 0, \ldots, N - (\kappa + 1)$, are identical to the B-splines of degree κ defined over the original knot vector. We know that each of these B-splines is positive in its support and that

$$
\begin{aligned}
1 &= \sum_{i=-\kappa}^{N-(\kappa+1)} \mathcal{B}_{i,\kappa}(t), \qquad \text{for } t \in [t_0, t_{N-(\kappa+1)}) \\
&= \sum_{i=-\kappa}^{0} \mathcal{B}_{i,\kappa}(t) + \sum_{i=0}^{N-(\kappa+1)} \mathcal{B}_{i,\kappa}(t).
\end{aligned}
$$

For $t \in [t_0, t_\kappa)$, at least one of the B-splines in the first sum is nonzero, and so the sum of the B-splines in the second sum must be less than 1.

Analogous methods prove the result for $t \geq t_{N-\kappa}$. ∎

Theorem 6.17. *If $\mathcal{B}_{i,\kappa}(t)$ is a B-spline of degree κ defined by the knot vector* **t***, when the derivative exists (i.e., between knots and sometimes at knots) it is defined by*

$$
\mathcal{B}'_{i,\kappa}(t) = (\kappa)\left(\frac{\mathcal{B}_{i,\kappa-1}(t)}{t_{i+\kappa} - t_i} - \frac{\mathcal{B}_{i+1,\kappa-1}(t)}{t_{i+1+\kappa} - t_{i+1}}\right).
$$

Proof: Once again, the proof is by induction on the degree. For $\kappa = 0$,

$$
\mathcal{B}_{i,0}(t) = \begin{cases} 1 & \text{for } t_i \leq t < t_{i+1} \\ 0 & \text{otherwise,} \end{cases}
$$

so it is clear that if $\mathcal{B}_{i,0}(t)$ is nonzero anywhere, it is discontinuous at $t = t_i$ and $t = t_{i+1}$, but $\mathcal{B}'_{i,0}(t) = 0$ for $t \in (t_i, t_{i+1})$.

Now, consider $\kappa = 1$. If $t_i < t_{i+2}$,

$$
\mathcal{B}_{i,1}(t) = \frac{(t - t_i)}{t_{i+1} - t_i}\mathcal{B}_{i,0}(t) + \frac{(t_{i+2} - t)}{t_{i+2} - t_{i+1}}\mathcal{B}_{i+1,0}(t),
$$

so for $t \notin \{t_i, t_{i+1}, t_{i+2}\}$,

$$
\begin{aligned}
\mathcal{B}'_{i,1}(t) &= \frac{1}{t_{i+1} - t_i}\mathcal{B}_{i,0}(t) - \frac{1}{t_{i+2} - t_{i+1}}\mathcal{B}_{i+1,0}(t) \\
&\quad + \frac{(t - t_i)}{t_{i+1} - t_i}\mathcal{B}'_{i,0}(t) + \frac{(t_{i+2} - t)}{t_{i+2} - t_{i+1}}\mathcal{B}'_{i+1,0}(t) \\
&= \frac{1}{t_{i+1} - t_i}\mathcal{B}_{i,0}(t) - \frac{1}{t_{i+2} - t_{i+1}}\mathcal{B}_{i+1,0}(t),
\end{aligned}
$$

which is the conclusion of the theorem.

Now we shall assume the conclusion is true for degree $\kappa - 1$, that is, for $t \notin \{t_j, \ldots, t_{j+\kappa}\}$, $\mathcal{B}'_{j,\kappa-1}(t) = (\kappa - 1)\left(\frac{\mathcal{B}_{j,\kappa-2}(t)}{t_{j+\kappa-1}-t_j} - \frac{\mathcal{B}_{j+1,\kappa-2}(t)}{t_{j+\kappa}-t_{j+1}}\right)$, and prove it to be true for degree κ:

$$\begin{aligned}\mathcal{B}'_{i,\kappa}(t) &= \frac{\mathcal{B}_{i,\kappa-1}(t)}{t_{i+\kappa} - t_i} - \frac{\mathcal{B}_{i+1,\kappa-1}(t)}{t_{i+1+\kappa} - t_{i+1}} \\ &\quad + \frac{t - t_i}{t_{i+\kappa} - t_i}\mathcal{B}'_{i,\kappa-1}(t) + \frac{t_{i+1+\kappa} - t}{t_{i+1+\kappa} - t_{i+1}}\mathcal{B}'_{i+1,\kappa-1}(t).\end{aligned}$$

Now,

$$\begin{aligned}&\frac{t - t_i}{t_{i+\kappa} - t_i}\mathcal{B}'_{i,\kappa-1}(t) + \frac{t_{i+1+\kappa} - t}{t_{i+1+\kappa} - t_{i+1}}\mathcal{B}'_{i+1,\kappa-1}(t) \\ &\quad = \frac{t - t_i}{t_{i+\kappa} - t_i}(\kappa - 1)\left(\frac{1}{t_{i+\kappa-1} - t_i}\mathcal{B}_{i,\kappa-2}(t) - \frac{1}{t_{i+\kappa} - t_{i+1}}\mathcal{B}_{i+1,\kappa-2}(t)\right) \\ &\quad\quad + \frac{t_{i+1+\kappa} - t}{t_{i+1+\kappa} - t_{i+1}}(\kappa - 1)\left(\frac{1}{t_{i+\kappa} - t_{i+1}}\mathcal{B}_{i+1,\kappa-2}(t)\right. \\ &\quad\quad\quad \left. - \frac{1}{t_{i+1+\kappa} - t_{i+2}}\mathcal{B}_{i+2,\kappa-2}(t)\right).\end{aligned}$$

This substitution is true since we are assuming the induction hypothesis for the theorem is true. Continuing,

$$\begin{aligned}&= (\kappa - 1)\left(\frac{1}{t_{i+\kappa} - t_i}\left(\frac{t - t_i}{t_{i+\kappa-1} - t_i}\mathcal{B}_{i,\kappa-2}(t) + \frac{t_{i+\kappa} - t}{t_{i+\kappa} - t_{i+1}}\mathcal{B}_{i+1,\kappa-2}(t)\right)\right. \\ &\quad \left. - \frac{1}{t_{i+\kappa} - t_{i+1}}\left(\frac{t_{i+\kappa} - t}{t_{i+\kappa} - t_i} + \frac{t - t_i}{t_{i+\kappa} - t_i} - \frac{t_{i+1+\kappa} - t}{t_{i+1+\kappa} - t_{i+1}}\right)\mathcal{B}_{i+1,\kappa-2}(t)\right) \\ &\quad -(\kappa - 1)\frac{1}{t_{i+1+\kappa} - t_{i+1}}\frac{t_{i+1+\kappa} - t}{t_{i+1+\kappa} - t_{i+2}}\mathcal{B}_{i+2,\kappa-2} \\ &= (\kappa - 1)\left(\frac{1}{t_{i+\kappa} - t_i}\mathcal{B}_{i,\kappa-1}(t)\right. \\ &\quad \left. - \frac{1}{t_{i+1+\kappa} - t_{i+1}}\left(\frac{t - t_{i+1}}{t_{i+\kappa} - t_{i+1}}\mathcal{B}_{i+1,\kappa-2}(t) + \frac{t_{i+1+\kappa} - t}{t_{i+1+\kappa} - t_{i+2}}\mathcal{B}_{i+2,\kappa-2}(t)\right)\right) \\ &= (\kappa - 1)\left(\frac{1}{t_{i+\kappa} - t_i}\mathcal{B}_{i,\kappa-1}(t) - \frac{1}{t_{i+1+\kappa} - t_{i+1}}\mathcal{B}_{i+1,\kappa-1}(t)\right).\end{aligned}$$

Substituting this into the derivative equation,

$$
\begin{aligned}
&\mathcal{B}'_{i,\kappa}(t) \\
&= \frac{\mathcal{B}_{i,\kappa-1}(t)}{t_{i+\kappa}-t_i} - \frac{\mathcal{B}_{i+1,\kappa-1}(t)}{t_{i+1+\kappa}-t_{i+1}} + \frac{t-t_i}{t_{i+\kappa}-t_i}\mathcal{B}'_{i,\kappa-1}(t) \\
&\quad + \frac{t_{i+1+\kappa}-t}{t_{i+1+\kappa}-t_{i+1}}\mathcal{B}'_{i+1,\kappa-1}(t) \\
&= \frac{\mathcal{B}_{i,\kappa-1}(t)}{t_{i+\kappa}-t_i} - \frac{\mathcal{B}_{i+1,\kappa-1}(t)}{t_{i+1+\kappa}-t_{i+1}} + (\kappa-1)\left(\frac{\mathcal{B}_{i,\kappa-1}(t)}{t_{i+\kappa}-t_i} - \frac{\mathcal{B}_{i+1,\kappa-1}(t)}{t_{i+1+\kappa}-t_{i+1}}\right) \\
&= (\kappa)\left(\frac{1}{t_{i+\kappa}-t_i}\mathcal{B}_{i,\kappa-1}(t) - \frac{1}{t_{i+1+\kappa}-t_{i+1}}\mathcal{B}_{i+1,\kappa-1}(t)\right),
\end{aligned}
$$

and the proof is complete. ■

Theorem 6.18. *For a B-spline curve $\gamma(t) = \sum_{i=0}^{n} P_i\mathcal{B}_{i,\kappa}(t)$, the jth derivative is given by*

$$\frac{d^j}{dt^j}(\gamma(t)) = \sum_i Q_{j,i}\mathcal{B}_{i,\kappa-j}(t)$$

where

$$Q_{j,i} = \begin{cases} P_i & \text{for } j=0 \\ \dfrac{Q_{j-1,i}-Q_{j-1,i-1}}{\frac{t_{i+\kappa-j+1}-t_i}{\kappa-j+1}}, & \text{for } j>0. \end{cases}$$

Proof: The theorem is clearly true for $j=0$. Assume true for all lower order derivatives and prove the theorem is true for the jth derivative. That is

$$
\begin{aligned}
\frac{d^j}{dt^j}\gamma(t) &= \frac{d}{dt}\left(\frac{d^{j-1}}{dt^{j-1}}\gamma(t)\right) \\
&= \frac{d}{dt}\left(\sum Q_{j-1,i}\mathcal{B}_{i,\kappa-j+2-1}(t)\right) \\
&= \sum Q_{j-1,i}\frac{d}{dt}\mathcal{B}_{i,\kappa-j+2-1}(t) \\
&= \sum Q_{j-1,i}(\kappa-j+1)H_{i,\kappa,j}(t) \\
H_{i,\kappa-j}(t) &= \frac{\mathcal{B}_{i,\kappa-j}}{t_{i+\kappa-j+1}-t_i} - \frac{\mathcal{B}_{i+1,\kappa-j}}{t_{i+\kappa-j+2}-t_{i+1}}.
\end{aligned}
$$

The rest of this proof follows from expanding the sumation and reordering, leaving out the zero valued functions (if over a finite knot sequence) and appears as Exercise 5. ■

Corollary 6.19. *Suppose the conditions of Theorem 6.18 apply. Then for* $t_J \le t < t_{J+1}, i = J - (\kappa - 1) + p, \ldots, J,$ *and* $p = 1, \ldots, \kappa - 1,$

$$Q_{1,i}^{[p]} = \frac{1}{\kappa}\frac{P_i^{[p]} - P_{i-1}^{[p]}}{t_{i+\kappa-p} - t_i}.$$

This corollary follows from straightforwardly applying Algorithm 6.4 to the derivative function and then substituting for the definition of $Q_{1,i}$. Finally,

Corollary 6.20. *It is possible to evaluate both position and derivative with a single application of Algorithm 6.4, since*

$$Q_{1,J}^{[\kappa-1]} = \frac{1}{\kappa}\frac{P_J^{[\kappa-1]} - P_{J-1}^{[\kappa-1]}}{t_{i+1} - t_i}.$$

Corollary 6.21. *If* $\boldsymbol{\tau}$ *is a knot vector of length* $n+\kappa+2$, $n+1 \ge \kappa \ge 0$ *and* t *is defined with length* $n + \kappa + 4$, $t_{-1} < \tau_0$, $\tau_n < t_{n+1}$, $\tau_{n+\kappa+1} < t_{n+\kappa+2}$, *and* $t_i = \tau_i, i = 0, \ldots, n + \kappa + 1$. *Then,* $\mathcal{B}_{i,\kappa} \in \mathcal{S}_{\kappa,\boldsymbol{\tau}}$ *has an indefinite integral given by*

$$\int_{-\infty}^{t} \mathcal{B}_{i,\kappa}(u)du = \begin{cases} 0, & t < t_i \\ \sum_{j=i}^{n} \frac{t_{i+1+\kappa} - t_i}{\kappa+1}\mathcal{B}_{j,\kappa+1,\boldsymbol{t}}(t) & t_\kappa \le t < t_{n+1} \\ \frac{t_{i+1+\kappa} - t_i}{\kappa+1} & t \ge t_{n+1}. \end{cases}$$

Proof: By Theorem 6.18, for $\gamma \in \mathcal{S}_{\kappa+1,\boldsymbol{t}}$,

$$\begin{aligned} \gamma'(t) &= \left(\sum_{j=-1}^{n} c_j \mathcal{B}_{j,\kappa+1,\boldsymbol{t}}(t)\right) \\ &= \sum_{j=0}^{n} (\kappa+1)\frac{c_j - c_{j-1}}{t_{j+\kappa+1} - t_j}\mathcal{B}_{j,\kappa,\boldsymbol{\tau}}(t). \end{aligned}$$

If $\mathcal{B}_{i,\kappa}(t) = \gamma'(t)$, then

$$\frac{(\kappa+1)(c_j - c_{j-1})}{t_{j+\kappa+1} - t_j} = \begin{cases} 0 & \text{for } j < i, \\ 1 & \text{for } j = i, \\ 0 & \text{for } j > i. \end{cases}$$

Thus, $c_j = c_{j-1}, j < i$. But $\gamma(t) = 0$, $t < t_i$, implies that $c_j = 0, j < i$. Since $c_i - c_{i-1} = (t_{i+\kappa+1} - t_i)/(\kappa+1)$ and $c_{i-1} = 0$, $c_i = (t_{i+1+\kappa} - t_i)/(\kappa+1)$. Finally, since $c_j = c_{j-1}, j > i$, then $c_j = c_i$, and the theorem is proved. ■

Corollary 6.22. *If $\boldsymbol{\tau}$ is a knot vector of length $n + \kappa + 2$, $n+1 \geq \kappa > 0$ and $\boldsymbol{t}$ is defined with length $n + \kappa + 4$, $t_{-1} < \tau_0$ and $\tau_n < t_{n+1}$, and $t_i = \tau_i, i = 0, \ldots, n$. Then, $\sum_{i=0}^{n} P_i \mathcal{B}_{i,\kappa}(t) \in \mathcal{S}_{\kappa,\boldsymbol{\tau}}$ has an indefinite integral given by*

$$\int_{-\infty}^{t} \sum_{i=0}^{n} P_i \mathcal{B}_{i,\kappa}(u)du = \sum_{i=0}^{n} \left(\sum_{j=0}^{i} \frac{\tau_{j+1+\kappa} - \tau_j}{(\kappa+1)} P_j \right) \mathcal{B}_{i,\kappa+1}(t).$$

Proof: By Theorem 6.18, for $\gamma \in \mathcal{S}_{\kappa+1,\boldsymbol{t}}$,

$$\begin{aligned} \gamma'(t) &= \left(\sum_{i=-1}^{n} c_i \mathcal{B}_{i,\kappa+1,\boldsymbol{t}}(t) \right) \\ &= \sum_{i=0}^{n} (\kappa+1) \frac{c_i - c_{i-1}}{t_{i+\kappa+1} - t_i} \mathcal{B}_{i,\kappa,\boldsymbol{\tau}}(t) \\ &= \sum_{i=0}^{n} P_i \mathcal{B}_{i,\kappa}(t). \end{aligned}$$

So

$$P_i = \frac{(\kappa+1)(c_i - c_{i-1})}{t_{i+\kappa+1} - t_i}$$

and

$$c_i = c_{i-1} + \frac{t_{i+\kappa+1} - t_i}{(\kappa+1)} P_i.$$

Letting $c_{-1} = 0$, then $c_0 = \frac{\tau_{\kappa+1} - \tau_0}{(\kappa+1)} P_0$. More generally,

$$c_i = \sum_{j=0}^{i} \frac{\tau_{j+\kappa+1} - \tau_j}{(\kappa+1)} P_j.$$

■

While the behavior of the derivatives of the B-spline has been determined except at the knots, behavior at those points is very important. We now seek to determine the continuity classes of the B-splines.

Definition 6.23. *If $\boldsymbol{u}$ and $\boldsymbol{m}$ are defined as in Definition 6.2, then define $\mathcal{PP}_{\kappa,\boldsymbol{u},\boldsymbol{m}} = \{f(t) : f(t)|_{(u_i,u_{i+1})}$ a polynomial of degree less than or equal* *to κ and $f \in C^{(\kappa-m_i)}$ at $u_i\}$. When the space of piecewise polynomials over a restricted interval, say $[a,b]$, is needed we write $\mathcal{PP}_{\kappa,\mathbf{u},\mathbf{m}}[a,b]$.*

$\mathcal{PP}_{\kappa,\mathbf{u},\mathbf{m}}$ is the collection of all piecewise polynomials of degree κ such that the pieces meet at the values u_i with continuity $\kappa - m_i$. Then it is easy to show that $\mathcal{PP}_{\kappa,\mathbf{u},\mathbf{m}}$ is a vector space.

Theorem 6.24. *If $\{t_j\}$ is a knot sequence developed from $\{u_i\}$ and $\{m_i\}$, as in Definition 6.2, then $\mathcal{B}_{i,\kappa} \in \mathcal{PP}_{\kappa,\mathbf{u},\mathbf{m}}[t_\kappa, t_{N-\kappa}]$.*

Proof: Since $\mathcal{B}_{i,0}$ is not continuous at either t_i or t_{i+1} the result is immediately true for degree 0, since we let $C^{(-1)}$ designate discontinuous functions. If m is any negative integer, it is understood that $C^{(m)}$ stands for $C^{(-1)}$, since we do not consider degrees of discontinuity.

If $t_i = t_{i+\kappa+1}$, then $\mathcal{B}_{i,\kappa}(t) = 0$ for all t, and the theorem is trivially true.

The rest of the proof is done by induction on the degree. By Theorem 6.17, except at the knot points,

$$\mathcal{B}'_{i,\kappa}(t) = (\kappa)\left[\frac{\mathcal{B}_{i,\kappa-1}(t)}{t_{i+\kappa} - t_i} - \frac{\mathcal{B}_{i+1,\kappa-1}(t)}{t_{i+1+\kappa} - t_{i+1}}\right].$$

Now, suppose $u_{i_p} = t_j \in \{t_i, \ldots, t_{i+\kappa+1}\}$. If $t_i < t_{i+\kappa}$ and $t_{i+1} < t_{i+\kappa+1}$, then $\mathcal{B}'_{i,\kappa}$ is a linear combination of functions of class $C^{((\kappa)-1-m_{i_p})}$, so $\mathcal{B}_{i,\kappa}$, its integral, must be of class $C^{(\kappa-m_{i_p})}$.

If $t_i = t_{i+\kappa} < t_{i+\kappa+1}$, then $\mathcal{B}_{i,\kappa-1}(t) \equiv 0$, and $m_{i_p} \geq \kappa+1$, so $\mathcal{B}_{i+1,\kappa-1} \in C^{(-1)}$ at t_κ. Further

$$\mathcal{B}_{i,\kappa}(t) = \frac{t_{i+\kappa+1} - t}{t_{i+\kappa+1} - t_{i+1}}\mathcal{B}_{i+1,\kappa-1}(t).$$

Since $lim_{t<t_{i+\kappa}, t\to t_{i+\kappa}}\mathcal{B}_{i,\kappa}(t) = 0 = lim_{t<t_{i+\kappa}, t\to t_{i+\kappa}}\mathcal{B}_{i+1,\kappa-1}(t)$, however $\mathcal{B}_{i,\kappa}(t_{i+\kappa}) = \mathcal{B}_{i+1,\kappa-1}(t_{i+\kappa})$, $\mathcal{B}_{i,\kappa} \in C^{(-1)}$ at $t_j = u_{i_p}$. Since $\kappa - m_{i_p} \leq -1$, the conclusion is shown.

When $t_i < u_{i_p} = t_{i+1} = t_{i+\kappa+1}$, an analogous proof shows the conclusion. ■

Corollary 6.25. *Given knot vector $\{t_i, t_{i+1}, \ldots, t_{i+\kappa+1}\}$ which has distinct increasing values $\{u_{i_1}, \ldots, u_{i_s}\}$, each repeated with multiplicity $\{\mu_{i_1}, \ldots, \mu_{i_s}\}$, then $\mathcal{B}_{i,\kappa}(t)$ is contained in $C^{(\kappa-\mu_{i_j})}$ at u_{i_j}.*

The proof is left as an exercise for the reader.

Theorem 6.26. *$\mathcal{B}_{i,\kappa}$ is unimodal. That is, for $t \in (t_i, t_{i+\kappa+1})$, there exists exactly one extremal point.*

Proof: If $t_{i+1} = t_{i+\kappa}$,

$$\mathcal{B}'_{i,\kappa}(t) = \begin{cases} (\kappa)\frac{\mathcal{B}_{i,\kappa-1}}{t_{i+\kappa}-t_i}, & t \in (t_i, t_{i+1}) \\ -(\kappa)\frac{\mathcal{B}_{i+1,\kappa-1}}{t_{i+\kappa+1}-t_{i+1}}, & t \in (t_{i+1}, t_{i+\kappa+1}). \end{cases}$$

Thus, $\mathcal{B}'_{i,\kappa}$ cannot have any zeros since neither $\mathcal{B}_{i+1,\kappa-1}$ nor $\mathcal{B}_{i,\kappa-1}$ can have zeros on the specified disjoint domains. Hence, possible extremal points are $\{t_i, t_{i+1}, t_{i+\kappa+1}\}$. If $t_i = t_{i+1}$, then $\mathcal{B}_{i,\kappa-1}$ is identically zero, and $\mathcal{B}_{i,\kappa}(t)$ is decreasing. Analogously if $t_{i+1} = t_{i+\kappa+1}$, we can show that $\mathcal{B}_{i,\kappa}$ is always increasing. Finally, if $t_i < t_{i+1} = t_{i+\kappa} < t_{i+\kappa+1}$, then $\mathcal{B}_{i,\kappa} \in C^{(\kappa-1)}$ at t_i and $t_{i+\kappa+1}$ and hence equals zero at those points. Thus, there is only one maximum and it occurs at t_{i+1}.

Now, if $t_{i+1} < t_{i+\kappa}$, $\mathcal{B}'_{i,\kappa}$ exists and is continuous everywhere in the open interval $(t_i, t_{i+\kappa+1})$. Using simple results from calculus, the extremal points of $\mathcal{B}_{i,\kappa}$ occur at t_i, $t_{i+\kappa+1}$, at the zeros of $\mathcal{B}'_{i,\kappa}$ and nowhere else. Since $\mathcal{B}_{i,\kappa}(t_i) = \mathcal{B}_{i,\kappa}(t_{i+\kappa+1}) = 0$, Rolle's theorem tells us that there is at least one value of $t \in (t_i, t_{i+\kappa+1})$ for which $\mathcal{B}'_{i,\kappa} = 0$, and so there is at least one local maximum. The converse direction which proves at most one zero for the derivative, is based on theorems which count roots of spline curves and appear in [76]. The two results give exactly one root for the derivative, and so exactly one maximum for $\mathcal{B}_{i,\kappa}(t)$.

It is shown that for all possible cases, the conclusion of the Theorem holds true. ∎

Example 6.27. Let $\kappa = 2$ and find the value of τ_i, the parameter value where the maximum of $\mathcal{B}_{i,3-1}(t)$ occurs. First, suppose $t_i < t_{i+1} < t_{i+2} < t_{i+3}$. Then,

$$\mathcal{B}'_{i,2}(t) = 2\left(\frac{\mathcal{B}_{i,2-1}}{t_{i+2}-t_i} - \frac{\mathcal{B}_{i+1,2-1}}{t_{i+3}-t_{i+1}}\right)$$

and,

$$\mathcal{B}'_{i,2}(t) = 2\begin{cases} \frac{(t-t_i)}{(t_{i+1}-t_i)(t_{i+2}-t_i)}, & \text{for } t \in [t_i, t_{i+1}] \\ \frac{(t_{i+2}-t)}{(t_{i+2}-t_{i+1})(t_{i+2}-t_i)} - \frac{(t-t_{i+1})}{(t_{i+2}-t_{i+1})(t_{i+3}-t_{i+1})}, & \text{for } t \in [t_{i+1}, t_{i+2}] \\ -\frac{(t_{i+3}-t)}{(t_{i+3}-t_{i+2})(t_{i+3}-t_{i+1})}, & \text{for } t \in [t_{i+2}, t_{i+3}]. \end{cases}$$

Since $\mathcal{B}'_{i,2}$ has zeros neither on (t_i, t_{i+1}) nor on (t_{i+2}, t_{i+3}), it must have a zero on (t_{i+1}, t_{i+2}).

$$\frac{(t_{i+2} - \tau)}{(t_{i+2} - t_{i+1})(t_{i+2} - t_i)} - \frac{(\tau - t_{i+1})}{(t_{i+2} - t_{i+1})(t_{i+3} - t_{i+1})} = 0$$

so

$$\begin{aligned} \frac{(t_{i+2} - \tau)}{(t_{i+2} - t_i)} &= \frac{(\tau - t_{i+1})}{(t_{i+3} - t_{i+1})}, \\ \frac{\tau}{(t_{i+3} - t_{i+1})} + \frac{\tau}{(t_{i+2} - t_i)} &= \frac{t_{i+2}}{(t_{i+2} - t_i)} + \frac{t_{i+1}}{(t_{i+3} - t_{i+1})}, \\ \frac{\tau(t_{i+3} - t_{i+1} + t_{i+2} - t_i)}{(t_{i+3} - t_{i+1})(t_{i+2} - t_i)} &= \frac{t_{i+2}(t_{i+3} - t_{i+1}) + t_{i+1}(t_{i+2} - t_i)}{(t_{i+2} - t_i)(t_{i+3} - t_{i+1})}, \\ \tau &= \frac{t_{i+2}(t_{i+3} - t_{i+1}) + t_{i+1}(t_{i+2} - t_i)}{(t_{i+3} - t_{i+1} + t_{i+2} - t_i)}. \end{aligned}$$

If we define $\beta_i = t_{i+2} - t_i$ then this yields

$$\tau = \frac{\beta_{i+1} t_{i+2} + \beta_i t_{i+1}}{\beta_{i+1} + \beta_i},$$

and τ is a convex combination of t_{i+1} and t_{i+2}. Note that in the case that $\beta_i = \beta_{i+1}$, τ is half way between t_{i+1} and t_{i+2}. A further special case that falls within this category is when the knots are uniformly spaced.

Now, if $t_i < t_{i+1} = t_{i+2} < t_{i+3}$, we know from Theorem 6.26 that the maximum occurs at t_{i+1}, or

$$\tau = \frac{\beta_{i+1} t_{i+2} + \beta_i t_{i+1}}{\beta_{i+1} + \beta_i} = t_{i+1} = t_{i+2}.$$

Further, if $t_i = t_{i+1} = t_{i+2}$, then from Theorem 6.26, the maximum occurs at t_i, and fits the convex combination result as well. The analogous result holds for $t_{i+1} = t_{i+2} = t_{i+3}$. Now, if $t_i = t_{i+1} < t_{i+2}$,

$$\mathcal{B}'_{i,2}(t) = 2 \begin{cases} \left(\frac{(t_{i+2} - t)}{(t_{i+2} - t_{i+1})(t_{i+2} - t_i)} - \frac{(t - t_{i+1})}{(t_{i+2} - t_{i+1})(t_{i+3} - t_{i+1})} \right) & t \in (t_{i+1}, t_{i+2}] \\ \frac{(t_{i+3} - t)}{(t_{i+3} - t_{i+2})(t_{i+3} - t_{i+1})}, & t \in [t_{i+2}, t_{i+3}). \end{cases}$$

This result yields that the maximum must still occur on (t_{i+1}, t_{i+2}), as will the condition that $t_{i+1} < t_{i+2} = t_{i+3}$. Similar results hold for degree 3 basis functions since the solution requires solving quadratic equations for their zeros and determining if the zeros lie in the correct interval. Since the manipulations are very complicated, they are omitted. □

6.3.1 Continuity at the Curve Ends

Now, $\gamma(t) = \sum_{i=0}^{N-(\kappa+1)} P_i \mathcal{B}_{i,\kappa}(t)$, where $\{t_0, t_1, \ldots, t_N\}$ is the defining knot set. Since the B-splines of degree κ sum to 1 for $t \in [t_\kappa, t_{N-\kappa})$, the half open interval would seem to be the domain of the space of this collection of functions. We consider continuity properties at the ends.

Suppose $t_\kappa = u_{j_\kappa}$. If $t_\kappa = t_{\kappa+1}$, $\mathcal{B}_{0,\kappa}$ will not contribute to the function γ. Hence, we suppose $t_\kappa < t_{\kappa+1}$ so the first span of the function $\gamma(t)$ will not be degenerate. Now, $t_{\kappa+1-m_{j_\kappa}} = \cdots = t_\kappa$, so assume $m_{j_\kappa} < \kappa + 1$. Then at $t = t_\kappa$ the function is of continuity class $C^{(\kappa - m_{j_\kappa})}$, which is at least continuous. However, if $m_{j_\kappa} = \kappa + 1$, that is if $t_0 = t_1 = \cdots = t_\kappa < t_{\kappa+1}$, then the function is discontinuous. Because of the definition of the B-splines as being right continuous,

$$
\begin{aligned}
\mathcal{B}_{0,\kappa}(t) &= \lim_{t \to t_\kappa^+} \mathcal{B}_{0,\kappa}(t) \\
&= \lim_{t \to t_\kappa^+} \left(\frac{t_{\kappa+1} - t}{t_{\kappa+1} - t_0} \right)^\kappa \mathcal{B}_{\kappa,0}(t) \\
&= 1.
\end{aligned}
$$

This means that the behavior of $\mathcal{B}_{i,\kappa}(t)$, $i = 0, \ldots,$ at least looks continuous from the inside of the interval.

Now, let us consider the upper endpoint. We assume that $t_{N-(\kappa+1)} < t_{N-\kappa}$ for the same reasons as at the lower endpoint. Now, we consider the case where $\mathcal{B}_{N-(\kappa+1),\kappa}$ is discontinuous at $t = t_{N-\kappa}$. Here, $t_{N-\kappa} = \cdots = t_N$, and the knot has multiplicity $\kappa+1$. Here, as we have seen, on the interior of the interval $\mathcal{B}_{N-(\kappa+1),\kappa}(t) = \left(\frac{t - t_{N-(\kappa+1)}}{t_{N-\kappa} - t_{N-(\kappa+1)}} \right)^\kappa \mathcal{B}_{N-(\kappa+1),0}(t)$, so

$$
\begin{aligned}
\lim_{t \to t_{N-\kappa}^-} \mathcal{B}_{N-(\kappa+1),\kappa}(t) &= \lim_{t \to t_{N-\kappa}^-} \left(\frac{t - t_{N-(\kappa+1)}}{t_{N-\kappa} - t_{N-(\kappa+1)}} \right)^\kappa \mathcal{B}_{N-(\kappa+1),0}(t) \\
&= 1,
\end{aligned}
$$

but $\mathcal{B}_{N-(\kappa+1),\kappa}(t_{N-\kappa}) = 0$, from the recursive definition. Thus, at the upper endpoint, the last function does not look continuous from the inside of the domain. That is unfortunate, since we would like it to look continuous. There is a solution, however, since we can modify the definition of the last B-spline at the upper endpoint of the domain to make it conform to desired practice in curve approximation and design.

Definition 6.28. *Given a finite knot sequence* $\{t_j\}_{j=0}^{N}$*, suppose that* $J = \max\{j : x_j < x_{j+1}, j = 0, \ldots, N-1\}$*. Then we define*

$$\mathcal{B}_{i,0}(t) = \begin{cases} 1 & \text{for } t_i \le t < t_{i+1}, i < J, \\ 1 & \text{for } t_J \le t \le t_{J+1}, i = J \\ 0 & \text{otherwise.} \end{cases}$$

This then affects only the last nontrivial blending function at any level, making it look continuous from the inside.

The implications are then that $\mathcal{B}_{i,\kappa}(t_{N-\kappa}) = 1$ using this new modified definition. This definition in no way changes the definition of any of the B-splines which are continuous at $t_{N-\kappa}$. By using the value of the limit to define the value of the function at t_{n+1}, only that last discontinuous value is changed.

Theorem 6.29. *If* $\mathbf{u} = \{a, b\}$ *and* $\mathbf{m} = \{\kappa+1, \kappa+1\}$ *let the knot vector be* $\mathbf{t} = \{t_0 = t_1 = \cdots = t_\kappa < t_{\kappa+1} = \cdots = t_{2\kappa+1}\}$*, where* $t_0 = a$ *and* $t_{\kappa+1} = b$*. Then for* $t \in [a, b]$*,* $p = 1, \ldots, \kappa$*, and* $i = 0, \ldots, p$*,*

$$\mathcal{B}_{i+\kappa-p,p}(t) = \theta_{i,p}\left(\frac{t-a}{b-a}\right).$$

This further shows that for this special choice of knot vectors, the B-splines reduce to the Bernstein/Bézier blending functions.

Proof: For $p = 0$, $\mathcal{B}_{i+\kappa,0}$ has support in $[t_{i+\kappa}, t_{i+\kappa+1}]$. But $t_{i+\kappa} < t_{i+\kappa+1}$ if and only if $i = 0$. Thus

$$\mathcal{B}_{\kappa,0}(t) \equiv 1 \equiv \theta_{0,0}\left(\frac{t-a}{b-a}\right),$$

and the case for $p = 0$ is proved. Now, assume the conclusion is true for $0 < p-1 < \kappa$, $i = 0, \ldots, p-1$, and show it is true for p^{th} degree splines.

$$\begin{aligned} \mathcal{B}_{i+\kappa-p,p}(t) &= \frac{t - t_{i+\kappa-p}}{t_{i+\kappa-p+p} - t_{i+k-p}} \mathcal{B}_{i+\kappa-p,p-1}(t) \\ &\quad + \frac{t_{i+\kappa-p+1+p} - t}{t_{i+\kappa-p+1+p} - t_{i+\kappa-p+1}} \mathcal{B}_{i+\kappa-p+1,p-1}(t) \\ &= \frac{t - t_{i+\kappa-p}}{t_{i+\kappa} - t_{i+\kappa-p}} \mathcal{B}_{i+\kappa-p,p-1}(t) \\ &\quad + \frac{t_{i+\kappa+1} - t}{t_{i+\kappa+1} - t_{i+\kappa+1-p}} \mathcal{B}_{i+\kappa+1-p,p-1}(t). \end{aligned}$$

If $i = 0$, then $\mathcal{B}_{\kappa-p,p-1}$ has support in $[t_{\kappa-p}, t_\kappa] = [a, a]$, and hence is identically zero, thus,

$$\mathcal{B}_{\kappa-p,p}(t) = \frac{t_{\kappa+1} - t}{t_{\kappa+1} - t_{\kappa+1-p}} \mathcal{B}_{\kappa+1-p,p-1}(t).$$

Since $p - 1 < \kappa$,

$$\begin{aligned} &= \frac{b-t}{b-a}\theta_{0,p-1}\left(\frac{t-a}{b-a}\right) \\ &= \theta_{0,p}\left(\frac{t-a}{b-a}\right) \end{aligned}$$

as desired.

Since $\mathcal{B}_{i+\kappa+1-p,p-1}$ has support in $[t_{i+\kappa-p+1}, t_{i+\kappa+1}]$, if $i = p$, $supp(\mathcal{B}_{\kappa+1,p-1}) = [t_{\kappa+1}, t_{\kappa+1+p-1}] = [b, b]$. Hence $\mathcal{B}_{\kappa+1,p-1} \equiv 0$, so

$$\begin{aligned} \mathcal{B}_{\kappa,p}(t) &= \frac{t - t_\kappa}{t_{\kappa+p} - t_\kappa}\mathcal{B}_{\kappa,p-1}(t) \\ &= \frac{t-a}{b-a}\mathcal{B}_{(p-1)+\kappa-p}(t) \\ &= \frac{t-a}{b-a}\theta_{p-1,p-1}(t) \\ &= \theta_{p,p}(t). \end{aligned}$$

This shows the result for $i = 0$ and $i = p$. Now, for $0 < i < p$,

$$\begin{aligned} &\mathcal{B}_{i+\kappa-p,p}(t) \\ &= \frac{t - t_{i+\kappa-p}}{t_{i+\kappa} - t_{i+\kappa-p}}\mathcal{B}_{i+\kappa-p,p-1}(t) \\ &\quad + \frac{t_{i+\kappa+1} - t}{t_{i+\kappa+1} - t_{i+\kappa+1-p}}\mathcal{B}_{i+\kappa+1-p,p-1}(t) \\ &= \frac{t-a}{b-a}\theta_{i-1,p-1}\left(\frac{t-a}{b-a}\right) + \frac{b-t}{b-a}\theta_{i,p-1}\left(\frac{t-a}{b-a}\right) \\ &= \theta_{i,p}\left(\frac{t-a}{b-a}\right). \end{aligned}$$

■

Thus, the Bernstein/Bézier blending functions are a special case of B-spline blending functions. The Bézier curve is then also a special case of a curve determined by using B-spline blending functions.

Corollary 6.30. *For $\boldsymbol{\tau} = \{\tau_0 = \tau_1 = \cdots = \tau_\kappa < \tau_{\kappa+1} = \cdots = \tau_{2\kappa+1}\}$, $S_{\kappa,\boldsymbol{\tau}} = P_\kappa[\tau_\kappa, \tau_{\kappa+1}]$, and, for this special choice of knot vectors, the B-spline curve reduces to the Bézier curve.*

Proof: Theorem 6.29 showed that under certain conditions, the B-spline functions defining $\mathcal{S}_{\kappa,\boldsymbol{\mathcal{T}}}$ are the Bernstein/Bézier blending functions of degree κ. Since the Bernstein/Bézier blending functions form a basis for $P_\kappa[\tau_\kappa, \tau_{\kappa+1}]$ clearly P_κ is contained in $\mathcal{S}_{\kappa,\boldsymbol{\mathcal{T}}}$. However, since $\mathcal{S}_{\kappa,\boldsymbol{\mathcal{T}}}$ contains only polynomials of degree κ over $[\tau_\kappa, \tau_{\kappa+1}]$, it is contained in P_κ. Hence the two spaces are equal.

Thus, we see that the Bézier curve is a special case of the B-spline curve.

■

Theorem 6.31. *If* $\boldsymbol{\mathcal{T}} = \{\tau_0 = \tau_1 = \cdots = \tau_\kappa < \tau_{\kappa+1} = \cdots = \tau_{2\kappa+1}\}$*, then the B-spline blending functions defining* $\mathcal{S}_{\kappa,\boldsymbol{\mathcal{T}}}$ *are linearly independent and form a basis for* $\mathcal{S}_{\kappa,\boldsymbol{\mathcal{T}}}$*.*

Proof: By Corollary 6.30, $\mathcal{S}_{\kappa,\boldsymbol{\mathcal{T}}} = P_\kappa[\tau_\kappa, \tau_{\kappa+1}]$, so the set of $\kappa+1$ functions $\{\mathcal{B}_{i,\kappa}\}_{i=0}^{\kappa}$ span the vector space $P_\kappa[\tau_\kappa, \tau_{\kappa+1}]$, of dimension $\kappa+1$. Hence they are linearly independent. ■

The unimodality of the Bernstein blending functions also follows from the unimodality of all B-splines, regardless of the specific configuration of the knot vector. However, we have additionally already determined in Theorem 5.10, $\theta_{i,\kappa}(t)$ has its maximum at $(b-a)\,i/\kappa$ for arbitrary degree κ.

Exercises

1. Complete the proof of Theorem 6.13.

2. Give a proof for Corollary 6.25.

3. Given the knot vector $\boldsymbol{t} = \{0, 0, 1, 2\}$, express the quadratic B-spline defined on those knots as a piecewise polynomial. Evaluate the left and right hand limits of the function, and its derivatives at the knots to find what continuity constraints the B-spline satisfies. How does your answer on the continuity constraints compare with what you would expect from the theory?

4. Consider the knot vector $\boldsymbol{t} = \{0, 0, 0, 0, 1, 2, 2.5, 3, 3, 4, 5, 5, 5, 6, 6, 6, 6\}$, and the cubic B-spline basis functions it defines. How many cubic B-splines are defined? What knots define each basis function? In general, given a specified degree and knot vector, how many basis functions are defined?

5. Complete the proof for Theorem 6.18.

6. Programming Exercise: Develop a program for interactive curve design for B-spline curves with arbitrary control polygons, orders, and knot vectors. You should allow the user to specify polygons graphically or to read them in from a file. Allow the user to insert new control points into the spline. Think carefully about the insertion points to be chosen by the user, and how that effects the knot vector. Allow the user to specify (and change) order, and knot vector for each control polygon. Consider how to make the specification and changes easy for the user. Uniform knot vectors (with open end conditions) should be generated automatically when requested.

7

Linear Spaces of B-Splines

In this chapter we study curves which are linear combinations of B-spline basis functions. We shall derive the properties of these curves. In addition, we develop the property which underlies most subdivision curve and surface algorithms as a property of uniform B-spline curves. We shall use the notation of the previous chapter and briefly review the most important definitions. Let $\boldsymbol{u} = \{u_i\}$ denote a collection of distinct values, and $\boldsymbol{m} = \{m_i\}$ denote a set of positive integer values, one for each element of $\boldsymbol{u}$. Define a nondecreasing sequence of real numbers $\boldsymbol{t} = \{t_j\}$ such that $m_i = \text{card}\{t_j : t_j = u_i\}$, and $t_j \leq t_{j+1}$. $\boldsymbol{u}$ and $\boldsymbol{t}$ can be infinite or finite.

Suppose that $N + 1 = \sum_{i=0}^{s} m_i$, that is, $\boldsymbol{t}$ has $N + 1$ elements. Then there are exactly $N - \kappa$ basis functions of degree κ. The linear combinations of these basis functions can be written

$$\gamma(t) = \sum_{i=0}^{n} P_i \mathcal{B}_{i,\kappa}(t),$$

where $n = N - \kappa - 1$. While it is somewhat confusing, it is a current convention to call the functions $\mathcal{B}_{i,\kappa}$ B-splines while calling linear combinations of them B-spline curves or B-spline expansions.

Lemma 7.1. *Consider $\mathcal{S}_{\kappa,\boldsymbol{t}} = \{\sum P_i \mathcal{B}_{i,\kappa}(t) : P_i \in \mathbb{R}^q\}$, the collection of all linear combinations of the modified normalized B-splines of degree κ over the knot vector $\boldsymbol{t}$ with coefficients in $\mathbb{R}^1$ (for $q = 1$), $\mathbb{R}^2$ (for $q = 2$), etc.*

Then, $\mathcal{S}_{\kappa,\boldsymbol{t}}$ is a vector space.

The proof is left to the reader.

By Corollary 6.12 if $t_j < t < t_{j+1}$, then $\mathcal{B}_{i,\kappa}(t) > 0$ if and only if $i = j - \kappa, \ldots, j$. Now if $j < \kappa$, and t_0 is the first element of the knot vector, fewer than $\kappa + 1$ blending functions can be nonzero. Since every polynomial of degree κ can be represented with $\kappa + 1$ basis functions, for $j < \kappa$, it is impossible to represent all polynomials of degree κ over (t_j, t_{j+1}) using these B-splines. In order to have any hope of such a representation, there must be at least $\kappa+1$ nonzero functions available. Also, on $[t_0, t_\kappa)$ the B-splines no longer sum to 1. On the interval (t_j, t_{j+1}), for $\kappa \leq j < N - \kappa$, there are $\kappa + 1$ nonzero B-spline functions available. The domain is usually considered to start at t_κ, and for analogous reasons, it terminates at t_{n+1}. We shall use the modified B-splines to allow the interval domain to be closed on both sides. For $\gamma(t) = \sum P_i \mathcal{B}_{i,\kappa}(t)$, the modified value at t_{n+1} is $\lim_{t \to t_{n+1}^-} \gamma(t)$. This is equal to the unmodified $\gamma(t_{n+1})$ whenever the multiplicity of t_{n+1} is less than $\kappa + 1$ since then all $\mathcal{B}_{i,\kappa} \in C^{(0)}$ at t_{n+1}. If the multiplicity at t_{n+1} is $\kappa+1$, then on the open interval ending at that value, all basis functions except one have a limit of zero, and the curve value interpolates the polygon point corresponding to that blending function.

Further, we want $t_\kappa < t_{\kappa+1}$, $t_n < t_{n+1}$, and $t_i < t_{i+\kappa+1}$ for all i. The reasons for this are simple. If $t_\kappa = t_{\kappa+1}$, then the support of $\mathcal{B}_{0,\kappa}$ is not contained in the domain for the functions of $\mathcal{S}_\kappa$, and any coefficient of it does not contribute to the curve. The other end has the analogous situation. Also, if $t_i = t_{i+\kappa+1}$ for a knot sequence interior to the domain, $\mathcal{B}_{i,\kappa} \equiv 0$, and contributes nothing to the curve. We require that all B-spline blending functions of degreee κ defined over $\boldsymbol{t}$ are nonzero over some part of the domain.

Example 7.2. Let us compare the knot vector $\{0, 0, 0, 0, 1, 2, 2, 2, 2\}$ with the knot vector $\{0, 1, 2, 3, 4, 5, 6, 7, 8\}$. If $\kappa = 3$ and $n = 4$, there are five B-spline blending functions which can be used to make curves in both cases.

For the first knot vector, the domain of the curve is considered to extend from $t_3 = 0$ to $t_5 = 2$. In this case, $\mathcal{B}_{4,3}$ is $C^{(-1)}$ at $t = 2$, and the modified functional value is 1. $\mathcal{B}_{i,3}(t)$, for $i = 0, 1, 2, 3$, all have both a function value and a limit equal to zero at t_5. Hence $\gamma(t_5) = \sum_{i=0}^4 P_i \mathcal{B}_{i,3}(t_5) = P_4$.

For the second knot vector, the domain of the curve is considered to extend from $t_3 = 3$ to $t_5 = 5$. In this case all the basis functions are contained in $C^{(2)}$ and hence have both function value and limit equal to the same value at t_5. Note that $\mathcal{B}_{4,3}(t_5) > 0$, $\mathcal{B}_{3,3}(t_5) > 0$, $\mathcal{B}_{2,3}(t_5) > 0$, and $\mathcal{B}_{1,3}(t_5) = 0$, as is expected. Hence, $\gamma(t_5) = \sum_{i=0}^4 P_i \mathcal{B}_{i,3}(t_5) = \sum_{i=2}^4 P_i \mathcal{B}_{i,3}(t_5)$. □

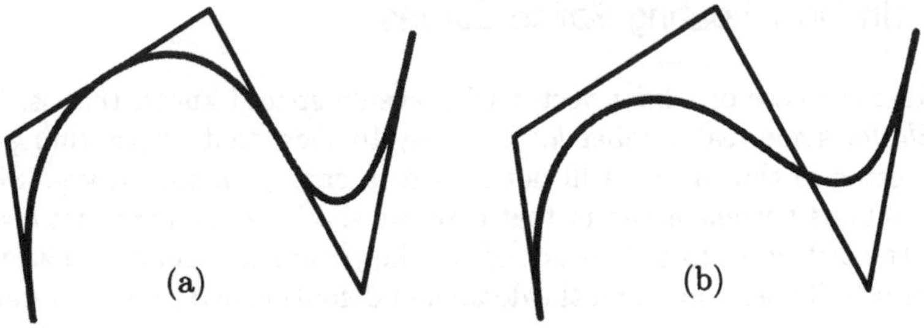

Figure 7.1. (a) Degree 2, $\boldsymbol{t} = \{0,0,0,1,2,3,3,3\}$;
(b) degree 3, $\boldsymbol{t} = \{0,0,0,0,1.5,3,3,3,3\}$.

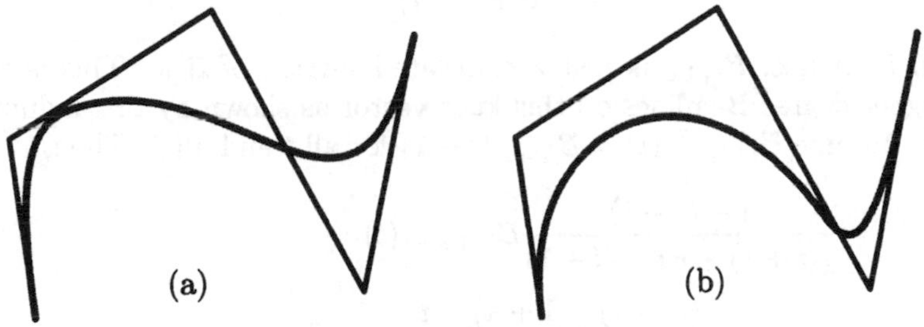

Figure 7.2. Degree 3: (a) $\boldsymbol{t} = \{0,0,0,0,.5,3,3,3,3\}$;
(b) $\boldsymbol{t} - \{0,0,0,0,2.5,3,3,3,3\}$.

For a given knot vector $\boldsymbol{t}$ and degree κ it is important to know which of the B-splines are nonzero over some portion of the domain of $\mathcal{S}_{\kappa,\boldsymbol{t}}$. We consider all the cases. $\mathcal{B}_{j,\kappa}$, $j = 0, \ldots, \kappa$, are all nonzero on $(t_\kappa, t_{\kappa+1})$, and so they are counted. When $t_i < t_{i+1}$, $\mathcal{B}_{j,\kappa}$ is nonzero for $j = i, i-1, \ldots, i-\kappa$. Finally, $\mathcal{B}_{n,\kappa}$ is the last completely specified κ^{th} degree B-spline. So there are $n+1$ distinct B-splines of degree κ which can be nonzero over the interval $[t_\kappa, t_{n+1})$.

When the curve is considered as a parametric function, and the coefficients $\{P_i\}$ are points in $\boldsymbol{R}^3$, these coefficients give geometric insight into the position and shape of the curve. We shall investigate computational, geometric, and theoretical properties of these curves. Additionally we shall determine exactly what types of curves are contained in the space $\mathcal{S}_{\kappa,\boldsymbol{t}}$.

Example 7.3. Figures 7.1 and 7.2 show examples of B-spline curves with different knot vectors and different degrees, but created using the same polygon. This shows the increased flexibility over the Bézier curve. □

7.1 Uniform Floating Spline Curves

Consider the case of a knot vector with evenly spaced knots, that is, $t_i = t_0 + ih$, for some real number h. It is easy to show that transforming the knot vector so that $h = 1$ will not lead to a change in the curves. Since computations become easier in that case, we shall assume here that $h = 1$. Since translating t_0 to 0 does not affect the parametric curve, we assume that $t_0 = 0$. Thus, $t_i = i$, and the domain of definition of $\mathcal{S}_{\kappa,\mathbf{t}}$ is the interval $[\kappa, n+1]$. Now,

$$\begin{aligned}\mathcal{B}_{i+1,0}(t) &= \begin{cases} 1 & t \in [i+1, i+2) \\ 0 & t \notin [i+1, i+2) \end{cases} \\ &= \mathcal{B}_{i,0}(t-1),\end{aligned}$$

for all i. Hence, $\mathcal{B}_{i+1,0}$ is just a translated version of $\mathcal{B}_{i,0}$. This is true for higher degree B-splines on this knot vector as shown by this inductive proof. Assume $\mathcal{B}_{j+1,\kappa-1}(t) = \mathcal{B}_{j,\kappa-1}(t-1)$ for all t and all j. Then,

$$\begin{aligned}\mathcal{B}_{i+1,\kappa}(t) = &\frac{t-(i+1)}{((i+1)+\kappa)-(i+1)}\mathcal{B}_{i+1,\kappa-1}(t) \\ &+ \frac{((i+1)+1+\kappa)-t}{((i+1)+1+\kappa)-((i+1)+1)}\mathcal{B}_{(i+1)+1,\kappa-1}(t) \\ = &\frac{(t-1)-i}{(i+\kappa)-i}\mathcal{B}_{i+1,\kappa-1}(t) + \frac{(i+\kappa+1)-(t-1)}{(i+\kappa+1)-(i+1)}\mathcal{B}_{(i+1)+1,\kappa-1}(t) \\ = &\frac{(t-1)-i}{(i+\kappa)-i}\mathcal{B}_{i,\kappa-1}(t-1) + \frac{i+\kappa+1-(t-1)}{(i+\kappa+1)-(i+1)}\mathcal{B}_{i+1,\kappa-1}(t-1),\end{aligned}$$

by the induction hypothesis, so

$$\mathcal{B}_{i+1,\kappa}(t) = \mathcal{B}_{i,\kappa}(t-1).$$

We have shown,

Theorem 7.4. *For* $\mathbf{t}$ *defined so that* $t_i = i$,

$$\mathcal{B}_{i+1,\kappa}(t) = \mathcal{B}_{i,\kappa}(t-1), \qquad \textit{for all } t.$$

Corollary 7.5. *If* $t_i = i$ *for all elements of knot vector* $\mathbf{t}$, *the B-splines of degree* κ *are translates of each other, i.e.,*

$$\mathcal{B}_{i+j,\kappa}(t) = \mathcal{B}_{i,\kappa}(t-j),$$

for all t, *and for all* i, j, *and* κ *for which the B-splines are defined.*

Corollary 7.6. *Let t be defined so that $t_i = i$. Then for all j and m,*

$$\mathcal{B}_{j,\kappa}(t) = \mathcal{B}_{m,\kappa}(t - (j - m)).$$

Immediate questions about spaces of B-splines formed with arbitrary knot vectors are

- If $\boldsymbol{t}$ and $\boldsymbol{\tau}$ are different knot vectors, are there any relationships between $\mathcal{S}_{\kappa,\boldsymbol{t}}$ and $\mathcal{S}_{\kappa,\boldsymbol{\tau}}$?
- How do B-spline curves compare with the Bézier curves and other representations for polynomials?
- Are B-splines linearly independent in $\mathcal{S}_{\kappa,\boldsymbol{t}}$?
- What curves can be represented by linear combinations of these B-splines?
- What knot collections are suitable for use?

In the rest of this chapter we investigate these topics more fully.

7.2 Spline Space Hierarchy

Suppose $\boldsymbol{t}$ and $\boldsymbol{\tau}$ are two knot vectors defining spline spaces $\mathcal{S}_{\kappa,\boldsymbol{t}}$ and $\mathcal{S}_{\kappa,\boldsymbol{\tau}}$, with card$(\boldsymbol{t}) = n + \kappa + 2$ and card$(\boldsymbol{\tau}) = m + \kappa + 2$, $m, n > \kappa$. We consider function spaces with the same domain, for which $\mathcal{S}_{\kappa,\boldsymbol{\tau}}$ is a subspace. First, $t_\kappa = \tau_\kappa$ and $t_{n+1} = \tau_{m+1}$, so the domains are the same. Let $\boldsymbol{u}$ and $\boldsymbol{\mu}$ be the breakpoint vectors for $\boldsymbol{t}$ and $\boldsymbol{\tau}$, respectively, and $\boldsymbol{m}$ and $\boldsymbol{\nu}$ be the multiplicity vectors, for $\boldsymbol{t}$ and $\boldsymbol{\tau}$, respectively. Suppose $\hat{t}$ is in the domain of arbitrary functions f and g. If $f(t) \in C^{(n)}$ at $\hat{t}$, and $g(t)$ is smoother at $\hat{t}$ than f, then $g \in C^{(m)}$ at $\hat{t}$, where $m > n$. If $\mathcal{S}_{\kappa,\boldsymbol{\tau}}$ is smoother at u_i than $\mathcal{S}_{\kappa,\boldsymbol{t}}$, then the multiplicity of u_i as an element of $\boldsymbol{\mu}$ is lower, and perhaps, could even be zero (i.e., u_i is not an element in $\boldsymbol{\mu}$). Using Theorem 6.24, we could hypothesize that $\mathcal{S}_{\kappa,\boldsymbol{\tau}} \subset \mathcal{S}_{\kappa,\boldsymbol{t}}$. However, if there is a μ_j which is not present in $\boldsymbol{u}$ or has lower multiplicity in $\boldsymbol{u}$ than in $\boldsymbol{\mu}$, then we might hypothesize the opposite. Both cannot be true unless the two vectors are identical, although they can share subspaces. However, to obtain containment, we show

Theorem 7.7. *Suppose $\boldsymbol{\tau}$ and $\boldsymbol{t}$ are knot vectors, with lengths $n + \kappa + 2$ and $m + \kappa + 2$, respectively, $m > n \geq \kappa$. If $\boldsymbol{t}$ contains the elements of $\boldsymbol{\tau}$ with the same or greater multiplicity, then $\mathcal{S}_{\kappa,\boldsymbol{\tau}} \subset \mathcal{S}_{\kappa,\boldsymbol{t}}$.*

In Chapter 16 we give a constructive proof of this result by actually deriving the coefficients of the B-spline basis for $\mathcal{S}_{\kappa,\boldsymbol{\tau}}$ as linear combinations of the B-spline basis for $\mathcal{S}_{\kappa,\boldsymbol{t}}$. By using the constructive result, originally shown in [13], for the restricted case of $\boldsymbol{t}$ having just one more element than $\boldsymbol{\tau}$, we can show existence of the general case. However, the general algorithms presented in Chapter 16 allow for $\boldsymbol{t}$ to be arbitrary, as long as it has $\boldsymbol{\tau}$ as a knot vector subset and do not require the iterative approach necessary with the restricted case. This particular proof follows the style of that in [5].

Lemma 7.8. *Let $\boldsymbol{\tau}$ be an arbitrary knot vector with corresponding B-spline functions $\mathcal{B}_{i,\kappa}(t)$. Let $\hat{t}$ be a real value such that $\tau_J \leq \hat{t} < \tau_{J+1}$, and form a new knot vector $\mathbf{t} = \boldsymbol{\tau} \cup \{\hat{t}\}$, with corresponding B-spline functions $\mathcal{N}_{j,\kappa}(t)$. Then*

$$\mathcal{B}_{i,\kappa}(t) = \begin{cases} \mathcal{N}_{i,\kappa}(t) & i \leq J-\kappa-1 \\ \frac{\hat{t}-t_i}{t_{i+\kappa+1}-t_i}\mathcal{N}_{i,\kappa}(t) + \frac{t_{i+\kappa+2}-\hat{t}}{t_{i+\kappa+2}-t_{i+1}}\mathcal{N}_{i+1,\kappa}(t) & J-\kappa \leq i \leq J \\ \mathcal{N}_{i+1,\kappa}(t) & J+1 \leq i. \end{cases}$$

Proof: The recursive relation of the B-spline functions is defined as,

$$\mathcal{B}_{i,\kappa}(t) = \begin{cases} \begin{cases} 1 & t \in [t_i, t_{i+1}) \\ 0 & t \notin [t_i, t_{i+1}) \end{cases} & \kappa = 0 \\ \frac{t-\tau_i}{\tau_{i+\kappa}-\tau_i}\mathcal{B}_{i,\kappa-1}(t) + \frac{\tau_{i+\kappa+1}-t}{\tau_{i+\kappa+1}-\tau_{i+1}}\mathcal{B}_{i+1,\kappa-1}(t) & \kappa > 0. \end{cases}$$

This proof is inductive, so first let $\kappa = 0$. Then

$$\begin{aligned} \mathcal{B}_{J,0}(t) &= \mathcal{N}_{J,0}(t) + \mathcal{N}_{J+1,0}(t) \\ &= \frac{t_{J+1}-t_J}{t_{J+1}-t_J}\mathcal{N}_{J,0}(t) + \frac{t_{J+2}-t_{J+1}}{t_{J+2}-t_{J+1}}\mathcal{N}_{J+1,0}(t) \\ &= \frac{\hat{t}-t_J}{t_{J+1}-t_J}\mathcal{N}_{J,0}(t) + \frac{t_{J+2}-\hat{t}}{t_{J+2}-t_{J+1}}\mathcal{N}_{J+1,0}(t). \end{aligned}$$

Thus,

$$\mathcal{B}_{i,0}(t) = \begin{cases} \mathcal{N}_{i,0}(t) & i \leq J-1 \\ \frac{\hat{t}-t_i}{t_{i+1}-t_i}\mathcal{N}_{i,0}(t) + \frac{t_{i+2}-\hat{t}}{t_{i+2}-t_{i+1}}\mathcal{N}_{i+1,0}(t) & J \leq i \leq J \\ \mathcal{N}_{i+1,0}(t) & J+1 \leq i. \end{cases}$$

Now assume the conclusion holds for $\kappa - 1$, that is

$$\mathcal{B}_{i,\kappa-1}(t) = \begin{cases} \mathcal{N}_{i,\kappa-1}(t) & i \le J - \kappa \\ \frac{\hat{t}-t_i}{t_{i+\kappa}-t_i}\mathcal{N}_{i,\kappa-1}(t) & \\ \quad +\frac{t_{i+\kappa+1}-\hat{t}}{t_{i+\kappa+1}-t_{i+1}}\mathcal{N}_{i+1,\kappa-1}(t) & J-\kappa+1 \le i \le J \\ \mathcal{N}_{i+1,\kappa-1}(t) & J+1 \le i. \end{cases}$$

Now, for $J - \kappa < i < J$, using the recursive definition, the above substitutions, and renaming the elements of $\boldsymbol{\tau}$ as elements of $\boldsymbol{t}$,

$$\begin{aligned} \mathcal{B}_{i,\kappa} &= \frac{t-\tau_i}{\tau_{i+\kappa}-\tau_i}\mathcal{B}_{i,\kappa-1} + \frac{\tau_{i+\kappa+1}-t}{\tau_{i+\kappa+1}-\tau_{i+1}}\mathcal{B}_{i+1,\kappa-1} \\ &= \frac{t-t_i}{t_{i+\kappa+1}-t_i}\left(\frac{\hat{t}-t_i}{t_{i+\kappa}-t_i}\mathcal{N}_{i,\kappa-1} + \frac{t_{i+\kappa+1}-\hat{t}}{t_{i+\kappa+1}-t_{i+1}}\mathcal{N}_{i+1,\kappa-1}\right) \\ &\quad + \frac{t_{i+\kappa+2}-t}{t_{i+\kappa+2}-t_{i+1}}\left(\frac{\hat{t}-t_{i+1}}{t_{i+\kappa+1}-t_{i+1}}\mathcal{N}_{i+1,\kappa-1}\right. \\ &\qquad\qquad \left. + \frac{t_{i+\kappa+2}-\hat{t}}{t_{i+\kappa+2}-t_{i+2}}\mathcal{N}_{i+2,\kappa-1}\right). \end{aligned} \tag{7.1}$$

Adding and subtracting

$$\left(\frac{\hat{t}-t_i}{t_{i+\kappa+1}-t_i}\,\frac{t_{i+\kappa+1}-t}{t_{i+\kappa+1}-t_{i+1}} + \frac{t_{i+\kappa+2}-\hat{t}}{t_{i+\kappa+2}-t_{i+1}}\,\frac{t-t_{i+1}}{t_{i+\kappa+1}-t_{i+1}}\right)\mathcal{N}_{i+1,\kappa-1},$$

from Equation 7.1 yields,

$$\begin{aligned} &\mathcal{B}_{i,\kappa} \\ &= \frac{\hat{t}-t_i}{t_{i+\kappa+1}-t_i}\left(\frac{t-t_i}{t_{i+\kappa}-t_i}\mathcal{N}_{i,\kappa-1} + \frac{t_{i+\kappa+1}-t}{t_{i+\kappa+1}-t_{i+1}}\mathcal{N}_{i+1,\kappa-1}\right) \\ &\quad + \frac{t_{i+\kappa+2}-\hat{t}}{t_{i+\kappa+2}-t_{i+1}}\left(\frac{t-t_{i+1}}{t_{i+\kappa+1}-t_{i+1}}\mathcal{N}_{i+1,\kappa-1} + \frac{t_{i+\kappa+2}-t}{t_{i+\kappa+2}-t_{i+2}}\mathcal{N}_{i+2,\kappa-1}\right) \\ &\quad + \left(-\frac{\hat{t}-t_i}{t_{i+\kappa+1}-t_i}\,\frac{t_{i+\kappa+1}-t}{t_{i+\kappa+1}-t_{i+1}} - \frac{t_{i+\kappa+2}-\hat{t}}{t_{i+\kappa+2}-t_{i+1}}\,\frac{t-t_{i+1}}{t_{i+\kappa+1}-t_{i+1}}\right. \\ &\quad \left. + \frac{t-t_i}{t_{i+\kappa+1}-t_i}\,\frac{t_{i+\kappa+1}-\hat{t}}{t_{i+\kappa+1}-t_{i+1}} + \frac{t_{i+\kappa+2}-t}{t_{i+\kappa+2}-t_{i+1}}\,\frac{\hat{t}-t_{i+1}}{t_{i+\kappa+1}-t_{i+1}}\right)\mathcal{N}_{i+1,\kappa-1}. \end{aligned} \tag{7.2}$$

Figure 7.3. A sequence of refinements of a cubic curve (left). The curve is shown in gray and the previous control polygon in the sequence is shown in light gray.

The coefficient of $\mathcal{N}_{i+1,\kappa-1}$ in parenthesis in Equation 7.2 is zero so

$$\begin{aligned}\mathcal{B}_{i,\kappa} &= \frac{\hat{t}-t_i}{t_{i+\kappa+1}-t_i}\left(\frac{t-t_i}{t_{i+\kappa}-t_i}\mathcal{N}_{i,\kappa-1}+\frac{t_{i+\kappa+1}-t}{t_{i+\kappa+1}-t_{i+1}}\mathcal{N}_{i+1,\kappa-1}\right)\\ &\quad+\frac{t_{i+\kappa+2}-\hat{t}}{t_{i+\kappa+2}-t_{i+1}}\left(\frac{t-t_{i+1}}{t_{i+\kappa+1}-t_{i+1}}\mathcal{N}_{i+1,\kappa-1}+\frac{t_{i+\kappa+2}-t}{t_{i+\kappa+2}-t_{i+2}}\mathcal{N}_{i+2,\kappa-1}\right)\\ &= \frac{\hat{t}-t_i}{t_{i+\kappa+1}-t_i}\mathcal{N}_{i,\kappa}+\frac{t_{i+\kappa+2}-\hat{t}}{t_{i+\kappa+2}-t_{i+1}}\mathcal{N}_{i+1,\kappa}. \qquad (7.3)\end{aligned}$$

The cases $i = J$ and $i = J-\kappa$, and the limiting cases, $i < J-\kappa$ and $J < i$, are left as an exercise. ■

Corollary 7.9. *For $\boldsymbol{\tau}$ and $\mathbf{t}$ as in Lemma 7.8, then*

$$\sum_i P_i\mathcal{B}_{i,\kappa}(t) = \sum_j D_j\mathcal{N}_{j,\kappa}(t),$$

where

$$D_j = \begin{cases} P_j & \text{for } j \le J-\kappa\\ \frac{\hat{t}-t_j}{t_{j+\kappa+1}-t_j}P_j + \frac{t_{j+\kappa+1}-\hat{t}}{t_{j+\kappa+1}-t_j}P_{j-1}, & \text{for } J-\kappa+1 \le j \le J\\ P_{j-1}, & \text{for } J+1 \le j.\end{cases}$$

Corollary 7.10. *Let $\boldsymbol{\tau}$ be an arbitrary knot vector with corresponding spline space $\mathcal{S}_{\kappa,\boldsymbol{\tau}}$. Let $\hat{t}$ be a real value such that $\tau_0 \le \hat{t} < \tau_{n+\kappa+1}$, and form a new knot vector $\mathbf{t} = \boldsymbol{\tau} \cup \{\hat{t}\}$, with corresponding spline space $\mathcal{S}_{\kappa,\mathbf{t}}$. If $\hat{t} \notin [t_\kappa, t_{n+1})$, then consider $\mathcal{S}_{\kappa,\mathbf{t}}$ only over the same domain as $\mathcal{S}_{\kappa,\boldsymbol{\tau}}$. Then $\mathcal{S}_{\kappa,\boldsymbol{\tau}} \subset \mathcal{S}_{\kappa,\mathbf{t}}$.*

Definition 7.11. *Consider two knot vectors $\boldsymbol{\tau}$ and $\mathbf{t}$ where $\boldsymbol{\tau}$ is a subset of $\mathbf{t}$, that is $\mathbf{t} = \{t_i\} = \boldsymbol{\tau} \cup \boldsymbol{\eta}$ where $\boldsymbol{\eta} = \{\eta_i\}_{i=0}^q$ for some nondecreasing sequence $\boldsymbol{\eta}$. $\boldsymbol{\eta}$ may contain the same value multiple times. Then $\mathbf{t}$ is called* a refinement *of $\boldsymbol{\tau}$.*

Theorem 7.12. *Suppose $\boldsymbol{\tau}$ is an arbitrary knot vector, $\tau_i < \tau_{i+\kappa+1}$ for all i, with corresponding spline space $\mathcal{S}_{\kappa,\boldsymbol{\tau}}$, and $\mathbf{t}$ is any refinement of $\boldsymbol{\tau}$ such that $t_j < t_{j+\kappa+1}$, for all j, with corresponding spline space $\mathcal{S}_{\kappa,\boldsymbol{t}}$. Then $\mathcal{S}_{\kappa,\boldsymbol{\tau}} \subset \mathcal{S}_{\kappa,\boldsymbol{t}}$.*

Proof: Let $\hat{t}_1$ be any element in $\boldsymbol{t}$ that is not in $\boldsymbol{\tau}$, and let $\boldsymbol{t}_1$ be a knot vector formed as $\boldsymbol{t}_1 = \boldsymbol{\tau} \cup \{\hat{t}_1\}$, with associated spline space $\mathcal{S}_{\kappa,\boldsymbol{t}_1}$. By Corollary 7.10, $\mathcal{S}_{\kappa,\boldsymbol{\tau}}$ is a subspace of $\mathcal{S}_{\kappa,\boldsymbol{t}_1}$. Continuing in this way, we let $\hat{t}_i$ be any element in $\boldsymbol{t}$ that is not in $\boldsymbol{t}_{i-1}$, and form the knot vector $\boldsymbol{t}_i = \boldsymbol{t}_{i-1} \cup \{\hat{t}_i\}$, having associated B-spline space $\mathcal{S}_{\kappa,\boldsymbol{t}_i}$. Each time Corollary 7.10 shows that $\mathcal{S}_{\kappa,\boldsymbol{t}_{i-1}}$ is a subspace of $\mathcal{S}_{\kappa,\boldsymbol{t}_i}$. This process must terminate after card($\boldsymbol{t}$) - card($\boldsymbol{\tau}$) steps. Containment forms a total ordering on the set of spaces, with $\mathcal{S}_{\kappa,\boldsymbol{\tau}}$ the smallest element in the ordering, and $\mathcal{S}_{\kappa,\boldsymbol{t}_{\mathrm{card}(\boldsymbol{t})-\mathrm{card}(\boldsymbol{\tau})}} = \mathcal{S}_{\kappa,\boldsymbol{t}}$ is the largest element. ■

7.2.1 Uniform Subdivison Curves

Let Δ be a constant, and consider $\tau_i = 2i\Delta$ and $\boldsymbol{t}$ defined so that

$$t_j = \begin{cases} \tau_i = 2i\Delta & \text{if } j = 2i \\ (2i+1)\Delta & \text{if } j = 2i+1. \end{cases}$$

Let $\mathcal{B}_{i,\kappa}(t) = \mathcal{B}_{i,\kappa,\boldsymbol{\tau}}(t)$ and $\mathcal{N}_{j,\kappa}(t) = \mathcal{B}_{j,\kappa,\boldsymbol{t}}(t)$.

Then since

$$\mathcal{B}_{i,0}(t) = \begin{cases} 1 & \text{if } t \in [\tau_i, \tau_{i+1}) \\ 0 & \text{otherwise,} \end{cases}$$

$\mathcal{B}_{i,0}(t) = \mathcal{N}_{2i,0}(t) + \mathcal{N}_{2i+1,0}(t)$.

If $\gamma_0(t) = \sum_i P_i \mathcal{B}_{i,0}(t)$,

$$\begin{aligned} \gamma_0(t) &= \sum_i P_i(\mathcal{N}_{2i,0}(t) + \mathcal{N}_{2i+1,0}(t)) \\ &= \sum_j Q_j \mathcal{N}_{j,0}(t); \\ Q_j &= P_i \qquad \text{for } j = 2i \text{ and } j = 2i+1. \end{aligned}$$

Next, consider $\kappa = 1$.

$$\begin{aligned}
\mathcal{B}_{i,1}(t) &= \frac{t-\tau_i}{2\Delta}\mathcal{B}_{i,0}(t) + \frac{\tau_{i+2}-t}{2\Delta}\mathcal{B}_{i+1,0}(t) \\
&= \frac{t-t_{2i}}{2\Delta}\left(\mathcal{N}_{2i,0}+\mathcal{N}_{2i+1,0}\right) + \frac{t_{2(i+2)}-t}{2\Delta}\left(\mathcal{N}_{2(i+1),0}+\mathcal{N}_{2(i+1)+1,0}\right) \\
&= \frac{1}{2}\left(\frac{(t-t_{2i})}{\Delta}\mathcal{N}_{2i,0} + \frac{t_{2i+2}-t}{\Delta}\mathcal{N}_{2i+1,0}\right) \\
&\quad + \left(\frac{t-t_{2i+1}}{\Delta}\mathcal{N}_{2i+1,0} + \frac{t_{2i+3}-t}{\Delta}\mathcal{N}_{2i+2,0}\right) \\
&\quad + \frac{1}{2}\left(\frac{t-t_{2i+2}}{\Delta}\mathcal{N}_{2i+2,0} + \frac{t_{2i+4}-t}{\Delta}\mathcal{N}_{2i+3,0}\right) \\
&= \frac{1}{2}\mathcal{N}_{2i,1}(t) + \mathcal{N}_{2i+1,1}(t) + \frac{1}{2}\mathcal{N}_{2(i+1),1}(t).
\end{aligned}$$

Thus

$$\begin{aligned}
\gamma_1(t) &= \sum P_i\mathcal{B}_{i,1}(t) \\
&= \sum P_i\left(\frac{1}{2}\mathcal{N}_{2i,1}(t) + \mathcal{N}_{2i+1,1}(t) + \frac{1}{2}\mathcal{N}_{2(i+1),1}(t)\right) \\
&= \sum\left(\frac{1}{2}P_i + \frac{1}{2}P_{i-1}\right)\mathcal{N}_{2i,1}(t) + P_i\mathcal{N}_{2i+1,1}(t) \\
&= \sum_j Q_j\mathcal{N}_{j,1}(t)
\end{aligned}$$

$$Q_j = \begin{cases} \frac{1}{2}(P_{i-1}+P_i), & j = 2i \\ P_i, & j = 2i+1. \end{cases}$$

Let $\boldsymbol{\tau}$ and $\boldsymbol{t}$ be defined as above and consider $\kappa = 2$.

$$\begin{aligned}
&\mathcal{B}_{i,2}(t) \\
&= \frac{t-\tau_i}{2(2\Delta)}\mathcal{B}_{i,1}(t) + \frac{\tau_{i+3}-t}{2(2\Delta)}\mathcal{B}_{i+1,1}(t) \\
&= \frac{t-\tau_i}{2(2\Delta)}\left(\frac{1}{2}\mathcal{N}_{2i,1} + \mathcal{N}_{2i+1,1} + \frac{1}{2}\mathcal{N}_{2i+2,1}\right) \\
&\quad + \frac{\tau_{i+3}-t}{2(2\Delta)}\left(\frac{1}{2}\mathcal{N}_{2i+2,1} + \mathcal{N}_{2i+3,1} + \frac{1}{2}\mathcal{N}_{2i+4,1}\right) && (7.4) \\
&= \frac{1}{4}\frac{t-t_{2i}}{2\Delta}\mathcal{N}_{2i,1} + \frac{1}{2}\frac{t-t_{2i}}{2\Delta}\mathcal{N}_{2i+1,1} + \frac{1}{4}\frac{t-t_{2i}}{2\Delta}\mathcal{N}_{2i+2,1} + \frac{1}{4}\frac{t_{2i+6}-t}{2\Delta}\mathcal{N}_{2(i+1),1} \\
&\quad + \frac{1}{2}\frac{t_{2i+6}-t}{2\Delta}\mathcal{N}_{2(i+1)+1,1} + \frac{1}{4}\frac{t_{2i+6}-t}{2\Delta}\mathcal{N}_{2(i+2),1} && (7.5)
\end{aligned}$$

Since

$$\begin{aligned}\mathcal{N}_{2i,2}(t) &= \frac{t-t_{2i}}{2\Delta}\mathcal{N}_{2i,1}(t)+\frac{t_{2i+3}-t}{2\Delta}\mathcal{N}_{2i+1,1}(t)\\ \mathcal{N}_{2i+3,2}(t) &= \frac{t-t_{2i+3}}{2\Delta}\mathcal{N}_{2i+3,1}(t)+\frac{t_{2i+6}-t}{2\Delta}\mathcal{N}_{2i+4,1}(t)\end{aligned}$$

we first solve

$$\begin{aligned}&\mathcal{B}_{i,2}-\frac{1}{4}\mathcal{N}_{2i,2}-\frac{1}{4}\mathcal{N}_{2i+3,2}\\ &= \frac{1}{4}\frac{3t-2t_{2i}-t_{2i+3}}{2\Delta}\mathcal{N}_{2i+1,1}+\frac{1}{4}\frac{t_{2i+6}-t}{2\Delta}\mathcal{N}_{2i+2,1}\\ &\quad+\frac{1}{4}\frac{t-t_{2i}}{2\Delta}\mathcal{N}_{2i+2,1}+\frac{1}{4}\frac{2t_{2i+6}+t_{2i+3}-3t}{2\Delta}\mathcal{N}_{2i+3,1}\\ &= \frac{1}{4}\frac{3t-2*2i\Delta-(2i+3)\Delta}{2\Delta}\mathcal{N}_{2i+1,1}+\frac{1}{4}\frac{(2i+6)\Delta-2i\Delta}{2\Delta}\mathcal{N}_{2i+2,1}\\ &\quad+\frac{1}{4}\frac{2*(2i+6)\Delta+(2i+3)\Delta-3t}{2\Delta}\mathcal{N}_{2i+3,1}\\ &= \frac{1}{4}\frac{3(t-(2i+1)\Delta)}{2\Delta}\mathcal{N}_{2i+1,1}+\frac{1}{4}\frac{6\Delta}{2\Delta}\mathcal{N}_{2i+2,1}+\frac{1}{4}\frac{3((2i+5)\Delta-t)}{2\Delta}\mathcal{N}_{2i+3,1}\\ &= \frac{3}{4}\frac{(t-t_{2i+1})}{2\Delta}\mathcal{N}_{2i+1,1}+\frac{3}{4}\frac{2\Delta}{2\Delta}\mathcal{N}_{2i+2,1}+\frac{3}{4}\frac{t_{2i+5}-t}{2\Delta}\mathcal{N}_{2i+3,1}\\ &= \frac{3}{4}\left(\mathcal{N}_{2i+1,2}+\mathcal{N}_{2i+2,2}\right).\end{aligned}$$

So,

$$\mathcal{B}_{i,2}(t)=\frac{1}{4}\mathcal{N}_{2i,2}(t)+\frac{3}{4}\mathcal{N}_{2i+1,2}(t)+\frac{3}{4}\mathcal{N}_{2i+2,2}(t)+\frac{1}{4}\mathcal{N}_{2i+3,2}(t).$$

Rewriting γ_2 in terms of the refined knot vector,

$$\begin{aligned}\gamma_2(t) &= \sum_{i=0}^{n}P_i\mathcal{B}_{i,2}(t)\\ &= \sum_{i=0}^{n}P_i\left(\frac{1}{4}\mathcal{N}_{2i,2}(t)+\frac{3}{4}\mathcal{N}_{2i+1,2}(t)+\frac{3}{4}\mathcal{N}_{2i+2,2}(t)+\frac{1}{4}\mathcal{N}_{2i+3,2}(t)\right)\\ &= \sum_{i=1}^{n}\left(\frac{1}{4}P_i+\frac{3}{4}P_{i-1}\right)\mathcal{N}_{2i,2}(t)+\left(\frac{3}{4}P_i+\frac{1}{4}P_{i-1}\right)\mathcal{N}_{2i+1,2}(t)\\ &= \sum_{j=2}^{2n+1}Q_j\mathcal{N}_{j,2}(t);\end{aligned}$$

$$Q_j = \begin{cases} \frac{1}{4}P_{i-2} + \frac{3}{4}P_{i-1} & j = 2i-1 \\ \frac{3}{4}P_{i-1} + \frac{1}{4}P_i & j = 2i. \end{cases}$$

Since the curve over the refined knot vector uses only a subset of the possible curve domain and only a subset of the possible B-splines, it is possible to write it as a B-spline curve over a smaller floating knot vector. That knot vector must include in its domain the domain of $\mathcal{S}_{2,\mathcal{T}}$. In this case, call the knot vector $\hat{\mathbf{t}}$. Then, since the curve uses $\mathcal{N}_{2,2}$ through $\mathcal{N}_{2n+1,2}$, this modified refined knot vector has $(2n+3)$ elements and is defined as

$$\hat{t}_j = t_{j+2} \qquad j = 0, \ldots, 2n+2,$$

with control points defined as

$$\hat{Q}_j = Q_{j+2} \qquad j = 0, \ldots, 2n-1.$$

So the refined curve can be written as $\gamma(t) = \sum_{j=0}^{2n-1} \hat{Q}_j \mathcal{B}_{j,2,\hat{\mathbf{t}}}(t)$.

Notice that this is an edge directed algorithm. That is, the child of an edge is an edge, and the child of a vertex is an edge.

$$\begin{aligned} [P_{i-1}, P_i] &\rightarrow [Q_{2i}, Q_{2i+1}] \\ P_i &\rightarrow [Q_{2i+1}, Q_{2(i+1)}]. \end{aligned}$$

(See Figure 7.4.)

We leave it as an exercise (see Exercise 3) to show that

$$\mathcal{B}_{i,3}(t) = \frac{1}{8}\mathcal{N}_{2i,3} + \frac{1}{2}\mathcal{N}_{2i+1,3} + \frac{3}{4}\mathcal{N}_{2i+2,3} + \frac{1}{2}\mathcal{N}_{2i+3,3} + \frac{1}{8}\mathcal{N}_{2i+4,3}$$

Figure 7.4. Quadratic uniform refinement. Original is on the left and two stages of uniform refinements are shown in the middle and right. Curve is shown in gray and previous control polygon in light gray.

and

$$\begin{aligned}\gamma_3(t) &= \sum_{i=0}^{n} P_i \mathcal{B}_{i,3}(t)\\ &= \sum_{j=3}^{2n+1} Q_j \mathcal{N}_{j,3}(t);\\ Q_j &= \begin{cases} \frac{P_{i-2}+P_{i-1}}{2} & j = 2i-1\\ \frac{P_{i-2}}{8} + \frac{3}{4}P_{i-1} + \frac{P_i}{8} & j = 2i. \end{cases}\end{aligned}$$

Once again, because the curve over the refined knot vector uses only a subset of the possible curve domain and only a subset of the possible B-splines, it is possible to write it as a B-spline curve over a smaller floating knot vector. That knot vector must include in its domain the domain of $\mathcal{S}_{3,\mathcal{T}}$. In this case, call the knot vector $\hat{\mathbf{t}}$. Then, since the curve uses $\mathcal{N}_{3,3}$ through $\mathcal{N}_{2n+1,3}$, this modified refined knot vector has $(2n+3)$ elements and is defined as

$$\hat{t}_j = t_{j+3} \qquad j = 0, \ldots, 2n+2,$$

with control points defined as

$$\hat{Q}_j = Q_{j+3} \qquad j = 0, \ldots, 2n-2.$$

So the refined curve can be written as $\gamma(t) = \sum_{j=0}^{2n-2} \hat{Q}_j \mathcal{B}_{j,3,\hat{\mathbf{t}}}(t)$.

Notice that this algorithm is vertex oriented. That is, a child of an edge is a vertex, and a child of a vertex is a vertex.

$$\begin{aligned}[P_{i-2}, P_{i-1}] &\rightarrow Q_{2i-1},\\ P_{i-1} &\rightarrow Q_{2i}.\end{aligned}$$

(See Figure 7.5.)

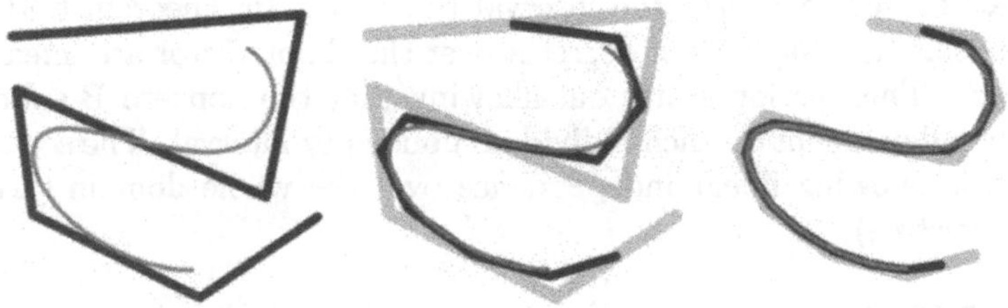

Figure 7.5. Cubic uniform refinement. Original is on the left and two stages of uniform refinements are shown in the middle and right. Curve is shown in gray and previous control polygon in light gray.

Since the derivations in this section require only that the knots be evenly spaced, a sequence of control polygons can be derived, for example, for the cubic case, as follows. For $\gamma(t) = \sum_i P_i \mathcal{B}_{i,3}(t)$, then for $\boldsymbol{t_0} = \boldsymbol{\tau}$, let $\boldsymbol{t_p}$ designate the uniform refinement of $\hat{\boldsymbol{t}}_{\boldsymbol{p-1}}$, and $\hat{\boldsymbol{t}}_{\boldsymbol{p}}$ the modified refined knot vector:

$$\gamma(t) = \sum_j \hat{Q}_j^{[p]} \mathcal{B}_{j,3,\hat{\boldsymbol{t}}_{\boldsymbol{p}}}(t),$$

where

$$\hat{Q}_{j-3}^{[p]} = Q_j^{[p]} = \begin{cases} P_j & \text{for } p = 0 \text{ and all } j \\ \frac{\hat{Q}_{i-2}^{[p-1]} + \hat{Q}_{i-1}^{[p-1]}}{2} & \text{for } p > 0 \text{ and } j = 2i-1 \\ \frac{\hat{Q}_{i-2}^{[p-1]} + 6\hat{Q}_{i-1}^{[p-1]} + \hat{Q}_{i}^{[p-1]}}{8} & \text{for } p > 0 \text{ and } j = 2i. \end{cases}$$

Call $\alpha_{[p]}$ the function whose image is the polyline defined by $\{\hat{Q}_j^{[p]}\}$, and consider the sequence of functions $\{\alpha_{[p]}\}$, as p varies. As p increases, line segments defining $\alpha_{[p]}$ get shorter, and the exterior angles between the sides get smaller, so the resulting curve looks smoother. Further, visually it appears that these polygons approach the curve. In Chapter 16 we show that this sequence of piecewise linear curves converges uniformly to γ. Each polyline in the sequence is called a *subdivision curve* of its predecessor.

7.3 Linear Independence of B-Splines

Given

$$\boldsymbol{\tau} = \{\tau_0 = \tau_1 = \cdots = \tau_\kappa < \tau_{\kappa+1} = \cdots = \tau_{2\kappa+1}\}$$

Corollary 6.30 showed that

$$\mathcal{S}_{\kappa,\boldsymbol{\tau}} = P_\kappa[\tau_\kappa, \tau_{\kappa+1}],$$

polynomials of degree κ over the interval $[\tau_\kappa, \tau_{\kappa+1}]$. In Theorem 6.31 it was shown that the B-splines of degree κ over that knot vector are linearly independent. This section starts out showing that the nonzero B-splines are linearly independent on each individual nonempty interval. The section culminates in showing linear independence over the whole domain (arbitrary knot vectors).

Theorem 7.13. *Suppose $\boldsymbol{\tau}$ is a knot vector with $2*\kappa+2$ elements such that $\tau_\kappa < \tau_{\kappa+1}$. Then the set of functions $\{\mathcal{B}_{i,r}(t)\}_{i=\kappa-r}^{\kappa}$ is linearly independent for $t \in [\tau_\kappa, \tau_{\kappa+1})$, for $r = 0, \ldots, \kappa$.*

Proof: Assume $t \in [\tau_\kappa, \tau_{\kappa+1})$. We prove the result by induction on r. First consider $r = 0$. Since $\mathcal{B}_{\kappa,0}(t) \equiv 1$, $c\,\mathcal{B}_{\kappa,0}(t) \equiv 0$ if and only if $c = 0$.

Now for $0 < r \leq \kappa$, assume $\{\mathcal{B}_{i,r-1}(t)\}_{i=\kappa-r+1}^{\kappa}$ is linearly independent. Now, suppose

$$0 \equiv \sum_{j=\kappa-r}^{\kappa} c_j\,\mathcal{B}_{j,r}(t). \tag{7.6}$$

Then the derivative of the zero function is also zero, so

$$\begin{aligned} 0 &\equiv \sum_{j=\kappa-r}^{\kappa} c_j\,\mathcal{B}'_{j,r}(t) \\ &\equiv \sum_{j=\kappa-r+1}^{\kappa} \frac{c_j - c_{j-1}}{\tau_{j+r} - \tau_j}\,\mathcal{B}_{j,r-1}(t). \end{aligned}$$

The linear independence of the B-splines of degree r then gives that

$$\begin{aligned} 0 &\equiv \frac{c_j - c_{j-1}}{\tau_{j+r} - \tau_j}, \\ &\equiv c_j - c_{j-1}, \qquad \text{for } j = \kappa - r + 1, \ldots, \kappa, \end{aligned}$$

so

$$c_\kappa = c_{\kappa-1} = \ldots = c_{\kappa+1-r} = c_{\kappa-r}.$$

Equation 7.6 then becomes

$$\begin{aligned} 0 &\equiv c_\kappa \sum_{j=\kappa-r}^{\kappa} \mathcal{B}_{j,r}(t). \\ &= c_\kappa. \end{aligned}$$

Thus,

$$0 = c_\kappa = c_{\kappa-1} = \ldots = c_{\kappa-r+1} = c_{\kappa-r},$$

so we have shown the desired result. ■

Remember that $PP_{\kappa,\boldsymbol{u},\boldsymbol{m}}$ is the space of piecewise polynomials of degree κ and continuity $C^{\kappa-m_j}$ at breakpoint u_j. Theorem 6.24 showed that $\mathcal{B}_{i,\kappa} \in PP_{\kappa,\boldsymbol{u},\boldsymbol{m}}$, for all i and κ, so $\mathcal{S}_{\kappa,\boldsymbol{t}}$ is contained in $PP_{\kappa,\boldsymbol{u},\boldsymbol{m}}$. We now show that all polynomials of degree κ on $[t_\kappa, t_{n+1})$ are in $\mathcal{S}_{\kappa,\boldsymbol{t}}$.

Let $\boldsymbol{t}$ be an arbitrary knot vector. Define $\boldsymbol{\tau}$ to have the same domain as $\boldsymbol{t}$, but without any internal knots. Finally, let $\mathcal{S}_{\kappa,\boldsymbol{t}}$ and $\mathcal{S}_{\kappa,\boldsymbol{\tau}}$ be their respective spline spaces of degree κ. If $\boldsymbol{t}$ has $n + \kappa + 2$ elements, $n > \kappa$, then $\boldsymbol{\tau}$ has $2\,\kappa + 2$ elements. By the theorems above, $\mathcal{S}_{\kappa,\boldsymbol{\tau}} = P_\kappa[\tau_\kappa, \tau_{\kappa+1})$, the space of polynomials of degree κ. By Theorem 7.12,

$$P_\kappa[\tau_\kappa, \tau_{\kappa+1}) = \mathcal{S}_{\kappa,\boldsymbol{\tau}} \subset \mathcal{S}_{\kappa,\boldsymbol{t}}.$$

Theorem 7.14. *If $\boldsymbol{t}$ is a knot vector with $n+\kappa+2$ elements, $n \geq \kappa$, such that $t_\kappa < t_{n+1}$, then P_κ is contained in $\mathcal{S}_{\kappa,\boldsymbol{t}}$.*

Example 7.15. Given a knot vector $\boldsymbol{t}$, an integer $\kappa > 0$, such that $t_i < t_{i+\kappa+1}$, where $\boldsymbol{t}$ has $n+\kappa+2$ elements, $n > \kappa$, then Theorem 7.14 proves that there exist real coefficients, $c_{\kappa,i}, i = 0, \ldots, n$, such that

$$t \equiv \sum_{i=0}^{n+\kappa-r} c_{r,i}\, \mathcal{B}_{i,r}(t) \qquad \text{for } t \in [t_r, t_{n+\kappa-r+1}),\ r = 1, \ldots, \kappa.$$

Unfortunately, the theorem does not show how to find such coefficients. We do so here using induction.

Let $r = 1$. Since $\mathcal{B}_{i,1}(t_{j+1}) = \delta_{i,j}$,

$$\begin{aligned} t_{j+1} &= \sum_{i=0}^{n+\kappa-1} c_{2,i}\, \mathcal{B}_{i,1}(t_{j+1}) \\ &= \sum_{i=0}^{n+\kappa-1} c_{2,i}\, \delta_{i,j} \\ &= c_{2,j}. \end{aligned}$$

Thus, $t \equiv \sum_{i=0}^{n+\kappa-1} t_{i+1}\, \mathcal{B}_{i,1}(t)$. Now, define $t^*_{i,r} = (\sum_{q=1}^{r} t_{i+q})/r$, and suppose it is true that

$$t \equiv \sum_{i=0}^{n+\kappa+1-r} t^*_{i,r-1}\, \mathcal{B}_{i,r-1}(t) \qquad t \in [t_{r-1}, t_{n+\kappa-r+2}).$$

Let $t \in [t_r, t_{n+\kappa-r+1})$. Since the support of $\mathcal{B}_{0,r-1}$ is $[t_0, t_r)$, $\mathcal{B}_{0,r-1}$ cannot contribute over the new shortened domain. Similarly, the support of $\mathcal{B}_{n+\kappa+1-r,r-1}$ is $[t_{n+\kappa+1-r}, [t_{n+\kappa+1})$, so $\mathcal{B}_{n+\kappa+1-r,r-1}$ does not contribute. Thus, for $t \in [t_r, t_{n+\kappa-r+1})$,

$$t = \sum_{i=1}^{n+\kappa-r} t^*_{i,r-1} \mathcal{B}_{i,r-1}.$$

Now,

$$\sum_{i=0}^{n+\kappa-r} t^*_{i,r}\, \mathcal{B}_{i,r}(t) = \frac{1}{r} \sum_{i=0}^{n+\kappa-r} \left(\sum_{q=1}^{r} t_{i+q} \right) \mathcal{B}_{i,r}(t)$$

$$
\begin{aligned}
&= \frac{1}{r}\sum_{i=0}^{n+\kappa-r}\left(\sum_{q=1}^{r} t_{i+q}\right)\left(\frac{t-t_i}{t_{i+r}-t_i}\mathcal{B}_{i,r-1}(t)+\frac{t_{i+r+1}-t}{t_{i+r+1}-t_{i+1}}\mathcal{B}_{i+1,r-1}(t)\right)\\
&= \frac{1}{r}\sum_{i=0}^{n+\kappa-r}\left(\sum_{q=1}^{r} t_{i+q}\right)\frac{t-t_i}{t_{i+r}-t_i}\mathcal{B}_{i,r-1}(t)\\
&\quad +\frac{1}{r}\sum_{i=0}^{n+\kappa-r}\left(\sum_{q=1}^{r} t_{i+q}\right)\frac{t_{i+r+1}-t}{t_{i+r+1}-t_{i+1}}\mathcal{B}_{i+1,r-1}(t)\\
&= \frac{1}{r}\sum_{i=1}^{n+\kappa-r}\left(\sum_{q=1}^{r} t_{i+q}\right)\frac{t-t_i}{t_{i+r}-t_i}\mathcal{B}_{i,r-1}(t)\\
&\quad +\frac{1}{r}\sum_{i=1}^{n+\kappa-r}\left(\sum_{q=1}^{r} t_{i-1+q}\right)\frac{t_{i+r}-t}{t_{i+r}-t_i}\mathcal{B}_{i,r-1}(t)\\
&= \frac{1}{r}\sum_{i=1}^{n+\kappa-r} t\left(\frac{\sum_{q=1}^{r} t_{i+q}-\sum_{q=0}^{r-1} t_{i+q}}{t_{i+r}-t_i}\right)\mathcal{B}_{i,r-1}(t)\\
&\quad +\frac{1}{r}\sum_{i=1}^{n+\kappa-r}\left(\sum_{q=1}^{r-1} t_{i+q}\right)\mathcal{B}_{i,r-1}(t)\\
&= \sum_{i=1}^{n+\kappa-r} t\left(\frac{t_{i+r}-t_i}{t_{i+r}-t_i}\right)\mathcal{B}_{i,r-1}(t)+\frac{r-1}{r}\sum_{i=1}^{n+\kappa-r}\left(\frac{1}{r-1}\sum_{q=1}^{r-1} t_{i+q}\right)\mathcal{B}_{i,r-1}(t)\\
&= \frac{t}{r}+\frac{r-1}{r}t\\
&= t.
\end{aligned}
$$

Hence we have shown that

$$
t \equiv \sum_{i=0}^{n+\kappa-r} t_{i,r}^{*}\,\mathcal{B}_{i,r}(t) \qquad t \in [t_r, t_{n+k-r}),\ r = 1,\ldots,\ \kappa
$$

where

$$
t_{i,r}^{*} = \frac{1}{r}\sum_{q=1}^{r} t_{i+q}. \qquad \square
$$

Example 7.16. Now suppose it is necessary to draw the curve using a plotting package that works for parametric curves only. How could one plot the explicit B-spline, $f(x)$, of degree κ with knots $\{x_0, \ldots, x_{n+\kappa+1}\}$, written, $f(x) = \sum_{i=0}^{n} f_i \mathcal{B}_{i,\kappa}(x)$?

We let $x_i^* = \frac{1}{\kappa}\sum_{r=1}^{\kappa-1} x_{i+r}$, and let $P_i = (x_i^*, f_i)$. Then

$$\begin{aligned}
\gamma(x) &= \sum_{i=0}^{n} P_i\mathcal{B}_{i,\kappa}(x) \\
&= \sum_{i=0}^{n} (x_i^*, f_i)\mathcal{B}_{i,\kappa}(x) \\
&= \left(\sum_{i=0}^{n} x_i^*\mathcal{B}_{i,\kappa}(x), \sum_{i=0}^{n} f_i\mathcal{B}_{i,\kappa}(x)\right) \\
&= \left(x, \sum_{i=0}^{n} f_i\mathcal{B}_{i,\kappa}(x)\right) \\
&= (x, f(x)).
\end{aligned}$$

□

By definition $\mathcal{S}_{\kappa,\boldsymbol{t}}$ is spanned by the B-splines. However, it is reasonable to ask if there is a subset of these functions which, in the general case, forms a basis. Suppose $\boldsymbol{t}$ has $N+1$ elements, $n = N - \kappa + 1$, and each distinct value has multiplicity $\kappa + 1$ or fewer. If $\mathcal{B}_{i,\kappa}(t), i = 0, \ldots, n$, are linearly independent, then they form a basis for $\mathcal{S}_{\kappa,\boldsymbol{t}}$, and each function in that space is represented uniquely.

Theorem 7.17. *If $\kappa \geq 0$ is an integer and $\boldsymbol{t}$ is a knot vector with $n+\kappa+2$ elements, $\kappa \leq n$, then $\{\mathcal{B}_{i,\kappa,\boldsymbol{t}}, i = 0, \ldots, n\}$ is linearly independent in $\mathcal{S}_{\kappa,\boldsymbol{t}}$ and hence forms a basis.*

Proof: Suppose $0 \equiv \sum_{i=0}^{n} c_i\mathcal{B}_{i,\kappa}(t)$. For $t \in [t_\kappa, t_{\kappa+1})$, the B-splines are independent, and hence $0 \equiv \sum_{i=0}^{\kappa} c_i\mathcal{B}_{i,\kappa}(t)$ if and only if $c_i = 0$, $i = 0, \ldots, \kappa$. Thus $0 \equiv \sum_{i=\kappa+1}^{n} c_i\mathcal{B}_{i,\kappa}(t)$. But for $t \in [t_{\kappa+1}, t_{\kappa+2})$, the non-zero B-splines are also independent, and $0 \equiv \sum_{i=1}^{\kappa+1} c_i\mathcal{B}_{i,\kappa}(t)$ if and only if $c_i = 0$, $i = 1, \ldots, \kappa+1$. Hence, in general, for $t \in [t_j, t_{j+1})$ the non-zero B-splines are independent, and $0 \equiv \sum_{i=j-\kappa+2}^{j} c_i\mathcal{B}_{i,\kappa}(t)$ if and only if $c_i = 0$, $i = j - \kappa + 2, \ldots, j$. But each B-spline appears in a sum over at least one nontrivial interval since each multiplicity is $\kappa + 1$ or less, and so $c_i = 0$, $i = 0, \ldots, n$, and the B-splines are linearly independent.

Since the B-splines also span $\mathcal{S}_{\kappa,\boldsymbol{t}}$, we have shown that they form a basis. ■

We end this section by showing the relationship of $PP_{\kappa,\boldsymbol{u},\boldsymbol{m}}$ to $\mathcal{S}_{\kappa,\boldsymbol{t}}$.

Theorem 7.18. $\mathcal{S}_{\kappa,\boldsymbol{t}} = PP_{\kappa,\boldsymbol{u},\boldsymbol{m}}$.

Proof: It has already been shown in Theorem 6.24 that $\mathcal{S}_{\kappa,\boldsymbol{t}} \subset PP_{\kappa,\boldsymbol{u},\boldsymbol{m}}$, so it is necessary only to show the converse.

Suppose $\boldsymbol{u}$ and $\boldsymbol{m}$ are the breakpoint and multiplicity vectors, respectively, each of length $q+1$, for a knot vector $\boldsymbol{t}$, of length $n+\kappa+2$, as presented in Definition 6.2. Suppose a and b are integers such that $u_a = t_\kappa$ and $u_b = t_{n+1}$. Then $\kappa+1 = \sum_{i=0}^{a} m_i = \sum_{i=b}^{q} m_i$.

Let $\gamma(t) \in PP\kappa, \boldsymbol{u}, \boldsymbol{m}$. For $t \in [u_a, u_{a+1}) = [t_\kappa, t_{\kappa+1})$, $\gamma(t)$ is a polynomial of degree κ and so can be represented exactly by a linear combination of $\{\mathcal{B}_{i,\kappa}(t)\}_{i=0}^{\kappa}$, those B-splines in $\mathcal{S}_{\kappa,\boldsymbol{t}}$ whose support contains $[t_\kappa, t_{\kappa+1})$. Suppose $p_a(t)$ is that function. Then $r_a(t) = \gamma(t) - p_a(t) \in PP_{\kappa,\boldsymbol{u},\boldsymbol{m}}$, and $r_a(t) \equiv 0$, for $t \in [u_a, u_{a+1}) = [t_\kappa, t_{\kappa+1})$. Analogously, $p_{a+i}(t) \in \mathcal{S}_{\kappa,\boldsymbol{t}}$ represents $r_{a+i-1}(t)$ for $t \in [u_{a+i}, u_{a+i+1}) = [t_{\kappa+\sum_{j=a+1}^{a+i} m_j}, t_{\kappa+1+\sum_{j=a+1}^{a+i} m_j})$, and $r_{a+i}(t) = r_{a+i-1}(t) - p_{a+i}(t)$.

It is important to characterize the behavior of $p_{a+1}(t)$ on $[u_a, u_{a+1})$, so we must determine, in particular, if the support of $p_{a+1}(t)$ includes $[u_a, u_{a+1})$. Since $\gamma(t), p_a(t) \in C^{(\kappa-m_{a+1})}$ at u_{a+1}, so is r_a. Since $r_a(t) \equiv 0$ on $[u_a, u_{a+1})$, and is in $C^{(\kappa-m_{a+1})}$ at u_{a+1}, and p_{a+1} represents r_a exactly on $[u_{a+1}, u_{a+2})$, $p_{a+1}^{(j)}(u_{a+1}) = 0, j = 0, \ldots, \kappa - m_{a+1}$. Now, $p_{a+1}(t) = \sum_{i=m_{a+1}}^{n} c_{1,i}\mathcal{B}_{i,\kappa}(t)$, $\mathcal{B}_{i,\kappa}, i = 0, \ldots, m_{a+1}-1$ have support in $[u_0, u_{a+1})$, and cannot contribute to reconstructing r_a on $[u_{a+1}, u_{a+2})$. Thus, since ${\mathcal{B}_{i,\kappa}}^{(j)}(u_{a+1}) = 0$, for $j = 0, \ldots, \kappa - m_{a+1}$, and $i = \kappa+1, \ldots, n$,

$$
\begin{aligned}
p_{a+1}^{(j)}(u_{a+1}) &= 0 \\
&= \sum_{i=m_{a+1}}^{n} c_{1,i}{\mathcal{B}_{i,\kappa}}^{(j)}(u_{a+1}) \\
&= \sum_{i=m_{a+1}}^{\kappa} c_{1,i}{\mathcal{B}_{i,\kappa}}^{(j)}(u_{a+1}).
\end{aligned}
$$

The system of linear equations above has $\kappa+1-m_{a+1}$ independent, homogeneous equations in $\kappa+1-m_{a+1}$ unknowns. Then the only solution is

$$c_{1,i} = 0, \qquad i = m_{a+1}, \ldots, \kappa.$$

Thus $p_{a+1} \equiv 0$ on $[u_a, u_{a+1})$, so $r_{a+1}(t) \equiv 0$ on $[u_a, u_{a+2})$. An induction argument is now applicable (see Exercise 2) to show that $p_i \equiv 0$ on $[u_a, u_i)$, and so $r_i(t) \equiv 0$ on $[u_a, u_{i+1})$, $i = a+1, \ldots, b-1$. We have shown that for $q \in PP_{\kappa,\boldsymbol{u},\boldsymbol{m}}$,

$$q(t) = \sum_{i=a}^{b-1} p_i(t)$$

where $p_i \in \mathcal{S}_{\kappa,\boldsymbol{t}}$. ■

7.4 A Second Look

A different approach was used originally to arrive at Theorem 7.14. We present the approach, using proofs inductive on the degree. This first theorem is due to Marsden [59].

Theorem 7.19. Marsden's Identity. *For any $y \in \boldsymbol{R}$ and $t \in [t_\kappa, t_{n+1}]$,*

$$(y-t)^\kappa = \sum_{j=0}^{n} \Phi_{j,\kappa}(y)\mathcal{B}_{j,\kappa}(t),$$

where

$$\Phi_{j,\kappa}(y) = \begin{cases} 1, & \text{for } \kappa = 0, \\ \prod_{r=1}^{\kappa}(y - t_{j+r}) = (y - t_{j+\kappa})\Phi_{j,\kappa-1}, & \text{for } \kappa > 0. \end{cases} \tag{7.7}$$

Proof: For $\kappa = 0$, the equation takes the form $1 = \sum_{j=0}^{n} \mathcal{B}_{j,0}(t)$, which we have already shown to be true. Induction is used for the rest of the theorem. Assume the hypothesis is true for $\kappa - 1$, that is, $(y-t)^{\kappa-1} = \sum_{j=1}^{n} \Phi_{j,\kappa-1}(y)\mathcal{B}_{j,\kappa-1}(t)$ for $t \in [t_\kappa, t_{n+1}]$. We shall show it true for κ.

If we define $Q_r(y,t) = \sum_{j=0}^{n} \Phi_{j,r}(y)\mathcal{B}_{j,r}(t)$, then we know $Q_{\kappa-1}(y,t) = (y-t)^{\kappa-1}$ and we wish to show $Q_\kappa(y,t) = (y-t)^\kappa$. Using the recursive definition of the B-spline functions,

$$Q_\kappa(y,t) = \sum_{j=0}^{n} \Phi_{j,\kappa}(y)\left[(t-t_j)\frac{\mathcal{B}_{j,\kappa-1}(t)}{t_{j+\kappa}-t_j} + (t_{j+\kappa+1}-t)\frac{\mathcal{B}_{j+1,\kappa-1}}{t_{j+\kappa+1}-t_{j+1}}\right].$$

Now denote the support by *supp*.

Then $supp(\mathcal{B}_{0,\kappa-1}) = [t_0, t_\kappa]$ and $supp(\mathcal{B}_{n+1,\kappa-1}) = [t_{n+1}, t_{n+\kappa+1}]$. Since $t \in [t_\kappa, t_{n+1})$, $\mathcal{B}_{0,\kappa-1}(t) = \mathcal{B}_{n+1,\kappa-1}(t) = 0$. Hence rearranging the terms in the sum we have

$$Q_\kappa(y,t) = \sum_{j=1}^{n} \zeta_{j,\kappa}(y,t)\frac{\mathcal{B}_{j,\kappa-1}(t)}{t_{j+\kappa}-t_j},$$

where

$$\zeta_{j,\kappa}(y,t) = \Phi_{j,\kappa}(y)(t-t_j) + \Phi_{j-1,\kappa}(y)(t_{j+\kappa}-t).$$

Now, since $\Phi_{j,\kappa}(y) = (y-t_{j+\kappa})\Phi_{j,\kappa-1}(y)$ and

$$\Phi_{j-1,\kappa}(y) = \prod_{r=1}^{\kappa}(y-t_{j-1+r}) = \prod_{r=0}^{\kappa-1}(y-t_{j+r}) = (y-t_j)\Phi_{j,\kappa-1}(y),$$

then

$$\begin{aligned}\zeta_{j,\kappa}(y,t) &= \Phi_{j,\kappa-1}(y)\left[(y-t_{j+\kappa})(t-t_j)+(y-t_j)(t_{j+\kappa}-t)\right] \\ &= (y-t)(t_{j+\kappa}-t_j)\Phi_{j,\kappa-1}(y).\end{aligned}$$

Thus

$$\begin{aligned}Q_\kappa(y,t) &= (y-t)\sum_{j=1}^{n}\Phi_{j,\kappa-1}(y)(t_{j+\kappa}-t_j)\frac{\mathcal{B}_{j,\kappa-1}(t)}{t_{j+\kappa}-t_j} \\ &= (y-t)\sum_{j=1}^{n}\Phi_{j,\kappa-1}(y)\mathcal{B}_{j,\kappa-1}(t) \\ &= (y-t)Q_{\kappa-1}(y,t).\end{aligned}$$

By the induction hypothesis on $Q_{\kappa-1}$,

$$\begin{aligned}Q_\kappa(y,t) &= (y-t)(y-t)^{\kappa-1} \\ &= (y-t)^\kappa,\end{aligned}$$

which proves the hypothesis. ■

Corollary 7.20. *t^i, $i=0,\ldots,\kappa$ are contained in $\mathcal{S}_{\kappa,\mathbf{t}}$, so P_κ is contained in $\mathcal{S}_{\kappa,\mathbf{t}}$.*

Proof: Let D_y^i denote the $i-^{th}$ derivative with respect to y. Using the result from Theorem 7.19,

$$\begin{aligned}D_y^i\left[(y-t)^\kappa\right] &= \kappa\cdots(\kappa-i+1)(y-t)^{\kappa-i}, \qquad i=1,\ldots,\kappa, \\ &= \sum_{j=0}^{n} D_y^i\left[\Phi_{j,\kappa}(y)\right]\mathcal{B}_{j,\kappa}(t).\end{aligned}$$

Evaluate that function at $y=0$ to get

$$\kappa\cdots(\kappa-i+1)(-t)^{\kappa-i}=\sum_{j=0}^{n} D_y^i\left[\Phi_{j,\kappa}(y)\right]\Big|_{y=0}\mathcal{B}_{j,\kappa}(t). \tag{7.8}$$

Since $\Phi_{j,\kappa}$ is a polynomial of degree κ, this makes perfect sense, and we have all polynomials of degree less than or equal to κ are contained in $\mathcal{S}_{\kappa,\mathbf{t}}$. ■

Example 7.21. We find the coefficients, $c_{i,\kappa}$, so that $t=\sum c_{i,\kappa}\mathcal{B}_{i,\kappa}(t)$. To apply Corollary 7.20 using Theorem 7.19, it is necessary to find the

$(\kappa - 1)^{st}$ derivative. Consider the function Φ expanded as a power basis in y.

$$\begin{aligned}\Phi_{j,\kappa}(y) &= \prod_{r=1}^{\kappa}(y - t_{j+r}) \\ &= y^{\kappa} - \left(\sum_{r=1}^{\kappa} t_{j+r}\right) y^{\kappa-1} + \text{lower degree powers of } y.\end{aligned}$$

Taking the $(\kappa - 1)$st derivative gives

$$D_y^{\kappa-1}\Phi_{j,\kappa}(y) = (\kappa)\cdots(\kappa - (\kappa - 2))y - (\kappa - 1)!\left(\sum_{r=1}^{\kappa} t_{j+r}\right). \tag{7.9}$$

By Equation 7.8 and Equation 7.9,

$$\begin{aligned}(\kappa)\cdots(2)(-t)^{\kappa-(\kappa-1)} &= \sum_{j=0}^{n} D_y^{\kappa-1}\left[\Phi_{j,\kappa}(y)\right]\big|_{y=0}\mathcal{B}_{j,\kappa}(t) \\ &= \sum_{j=0}^{n}(\kappa - 1)!\left(-\sum_{r=1}^{\kappa} t_{j+r}\right)\mathcal{B}_{j,\kappa}(t).\end{aligned}$$

Simplifying,

$$t = \sum_{j=0}^{n}\frac{\sum_{r=1}^{\kappa} t_{j+r}}{\kappa}\mathcal{B}_{j,\kappa}(t)$$

and the problem is solved by setting $c_{j,\kappa} = \frac{1}{\kappa}\sum_{r=1}^{\kappa} t_{j+r}$. For any degree, κ, the straight line $at + b$ can be written

$$at + b = \sum_{j=0}^{n}(ac_{j,\kappa} + b)\mathcal{B}_{j,\kappa}(t).$$

□

Example 7.22. In this example the coefficients, $\zeta_{i,\kappa}$, so that $t^2 = \sum \zeta_{i,\kappa}\mathcal{B}_{i,\kappa}(t)$ are determined. To apply Corollary 7.20 using Theorem 7.19, it is necessary to find the $(\kappa - 2)$ derivative. Again, consider the function Φ expanded as a power basis in y.

$$\begin{aligned}\Phi_{j,\kappa}(y) &= \prod_{r=1}^{\kappa}(y - t_{j+r}) \\ &= y^{\kappa} - \left(\sum_{r=1}^{\kappa} t_{j+r}\right) y^{\kappa-1} + \left(\sum_{r=1}^{\kappa-1}\sum_{q=r+1}^{\kappa} t_{j+r}t_{j+q}\right) y^{\kappa-2} \\ &\quad + \text{lower degree powers of } y.\end{aligned}$$

Taking the $(\kappa-2)^{nd}$ derivative gives

$$\begin{aligned} D_y^{\kappa-2}\Phi_{j,\kappa}(y) &= (\kappa)\cdots(\kappa-(\kappa-3))\,y^2-(\kappa-1)!\left(\sum_{r=1}^{\kappa}t_{j+r}\right)y \\ &\quad +(\kappa-2)!\left(\sum_{r=1}^{\kappa-1}\sum_{q=r+1}^{\kappa}t_{j+r}t_{j+q}\right). \qquad (7.10) \end{aligned}$$

By Equation 7.8 and Equation 7.10,

$$\begin{aligned} (\kappa)\cdots(3)(-t)^{\kappa-(\kappa-2)} &= \sum_{j=0}^{n} D_y^{\kappa-2}\left[\Phi_{j,\kappa}(y)\right]\Big|_{y=0}\mathcal{B}_{j,\kappa}(t) \\ &= \sum_{j=0}^{n}(\kappa-2)!\left(\sum_{r=1}^{\kappa-1}\sum_{q=r+1}^{\kappa}t_{j+r}t_{j+q}\right)\mathcal{B}_{j,\kappa}(t). \end{aligned}$$

Simplifying,

$$t^2=\sum_{j=0}^{n}\frac{\sum_{r=1}^{\kappa-1}\sum_{q=r+1}^{\kappa}t_{j+r}t_{j+q}}{\binom{\kappa}{2}}\mathcal{B}_{j,\kappa}(t)$$

and the problem is solved by setting $\zeta_{j,\kappa}=\frac{1}{\binom{\kappa}{2}}\sum_{r=1}^{\kappa-1}\sum_{q=r+1}^{\kappa}t_{j+r}t_{j+q}$. □

Example 7.23. The cubic B-spline representation for the functions $1, t, t^2$, and t^3 using the knot vector $\{-1,-1,-1,-1,0,1,4,4,4,4\}$ is found using Marsden's identity. Then these results as well as those in Examples 7.21 and 7.22 are used to find the B-spline representation for at^3+bt^2+ct+d over the same knot vector.

The domain is $[-1,4]$, $\kappa=3$, $n=5$, and

$$(0-t)^3=\sum_{j=0}^{5}\Phi_{j,3}(0)\mathcal{B}_{j,3}(t), \qquad \text{where } \Phi_{j,3}(0)=\prod_{r=1}^{3}(0-t_{j+r}),$$

so

$$\begin{aligned} t^3 &= \sum_{j=0}^{5}(-1)^3\Phi_{j,3}(0)\mathcal{B}_{j,3}(t), \\ &= \sum_{j=0}^{5}\left(\prod_{r=1}^{3}t_{j+r}\right)\mathcal{B}_{j,3}(t), \\ D_y\Phi_{j,3}(y) &= \sum_{p=1}^{3}\prod_{r=1,r\neq p}^{3}(y-t_{j+r}). \end{aligned}$$

By Example 7.22,

$$t^2 = \sum_{j=0}^{5} \frac{1}{3} \left[\sum_{p=1}^{3} \prod_{r=1, r \neq p}^{3} (0 - t_{j+r}) \right] \mathcal{B}_{j,3}(t).$$

From Example 7.21,

$$t = \sum_{j=0}^{5} \frac{\sum_{r=1}^{3} t_{j+r}}{3} \mathcal{B}_{j,\kappa}(t),$$

and from the fundamental properties on normalized B-splines

$$1 = \sum_{j=0}^{5} \mathcal{B}_{j,3}(t).$$

Finally, we can write any cubic polynomial over this domain by this appropriate linear combination:

$$\begin{aligned} & at^3 + bt^2 + ct + d \\ = \; & a \sum_{j=0}^{5} \left(\prod_{r=1}^{3} t_{j+r} \right) \mathcal{B}_{j,3}(t) + b \sum_{j=0}^{5} \frac{1}{3} \left(\sum_{p=1}^{3} \prod_{r=1, r\neq p}^{3} t_{j+r} \right) \mathcal{B}_{j,3}(t) \\ & + c \sum_{j=0}^{5} \frac{\sum_{r=1}^{3} t_{j+r}}{3} \mathcal{B}_{j,3}(t) + d \sum_{j=0}^{5} \mathcal{B}_{j,3}(t) \\ = \; & \sum_{j=0}^{5} \left(a \prod_{r=1}^{3} t_{j+r} + b \frac{\sum_{p=1}^{3} \prod_{r=1, r\neq p}^{3} t_{j+r}}{3} + c \frac{\sum_{r=1}^{3} t_{j+r}}{3} + d \right) \mathcal{B}_{j,3}(t). \end{aligned}$$

□

Exercises

1. Prove Lemma 7.1.
2. Use induction to finish the proof of Theorem 7.18.
3. Let $\boldsymbol{\tau}$ and $\boldsymbol{t}$ be defined as in Section 7.2.1, with $\{\mathcal{B}_{i,\kappa}\}$ and $\{\mathcal{N}_{j,\kappa}\}$ the basis functions of degree κ over $\boldsymbol{\tau}$ and $\boldsymbol{t}$, respectively. Show that

$$\mathcal{B}_{i,3}(t) = \frac{1}{8}\mathcal{N}_{2i,3} + \frac{1}{2}\mathcal{N}_{2i+1,3} + \frac{3}{4}\mathcal{N}_{2i+2,3} + \frac{1}{2}\mathcal{N}_{2i+3,3} + \frac{1}{8}\mathcal{N}_{2i+4,3}$$

and

$$\begin{aligned}\gamma_3(t) &= \sum P_i \mathcal{B}_{i,3}(t) \\ &= \sum Q_j \mathcal{N}_{j,3}(t), \\ Q_j &= \begin{cases} \frac{P_{i-2}}{8} + \frac{3}{4}P_{i-1} + \frac{P_i}{8} & j = 2i \\ \frac{P_{i-1}+P_i}{2} & j = 2i+1. \end{cases}\end{aligned}$$

4. Complete the proof of Lemma 7.8 by proving the conclusion for $i = J - \kappa, J$ and limiting cases $i < J - \kappa$ and $J < i$. (Hint: It may be helpful to use the fact that $t_{J+1} = \hat{t}$.)

8

Choosing a B-Spline Space

Once the general theory and the general algorithms are known, they must be interpreted and made practical for use in design. While the possible spaces of curves are very general, there is no theoretical method for determining which ones are feasible in particular design situations. For this reason we discuss each of the criteria for specifying spline curves and consider their effects on the resulting curves. These issues include the effect of knot vector selection, including end conditions and multiple knots, control polygon specification, and degree specification. We also discuss the interrelationship of these criteria. Multiple vertices are sometimes used to cause interpolation at the end of the curve, but the curve obtained is not the same curve as the curve obtained with an open knot vector. We discuss and compare the resulting curves.

8.1 Knot Patterns

As we have already shown, the Bézier curve of degree κ is a special case of the B-spline curve of degree κ when the knot vector $\{\tau_i\}$ configuration is $\tau_i = 0$, $i = 0, \ldots, \kappa$ and $\tau_i = 1$, $i = \kappa + 1, \ldots, 2\kappa + 1$. It has been shown that the Bézier curve interpolates position and has the tangents $\gamma'(0) = (\kappa)(P_1 - P_0)$ and $\gamma'(1) = (\kappa)(P_\kappa - P_{\kappa-1})$ at the ends. We now show that

Theorem 8.1. *Let* $\{t_i\}_{i=0}^{n+\kappa+1}$ *be a knot sequence with corresponding B-splines* $\{\mathcal{B}_{i,\kappa}(t)\}_{i=0}^{n}$, *and* $\gamma(t) = \sum_{i=0}^{n} P_i \mathcal{B}_{i,\kappa}(t)$, $t \in [t_\kappa, t_{n+1}]$. *Suppose*

$t_i < t_{i+\kappa+1}$ for all i so the basis functions of degree κ are not degenerate. Suppose there exists a J such that $t_J = \cdots = t_{J+\kappa-1}$. Then

1. *If $t_{J-1} < t_J < t_{J+\kappa}$, the curve is $C^{(0)}$ and interpolates P_J.*
2. *If $t_{J-1} = t_J$ the curve is $C^{(-1)}$. It has $\lim_{t \to t_J^-} \gamma(t) = P_{J-2}$, and $\lim_{t \to t_J^+} \gamma(t) = P_{J-1}$.*

   If instead, $t_{J+\kappa} = t_J$, then the curve is also $C^{(-1)}$, but, $\lim_{t \to t_J^-} \gamma(t) = P_{J-1}$, and $\lim_{t \to t_J^+} \gamma(t) = P_J$.
3. *If $t_\kappa \le t_{J-1} < t_J$, then $\lim_{t \to t_J^-} \gamma'(t) = \left(\frac{\kappa}{t_J - t_{J-1}}\right)(P_J - P_{J-1})$.*
4. *If $t_{J+\kappa-1} < t_{J+\kappa} \le t_{n+1}$, then $\lim_{t \to t_J^+} \gamma'(t) = \left(\frac{\kappa}{t_{J+\kappa} - t_J}\right)(P_{J+1} - P_J)$.*

Proof: It is known that there are at most $\kappa + 1$ basis functions nonzero at any parameter value, and the domain of $\mathcal{B}_{i,\kappa}(t)$ is $[t_i, t_{i+\kappa+1})$.

For item 1: Since $t_{J-1} < t_J < t_{J+\kappa}$, the highest multiplicity possible for t_J is κ and so all the basis functions are at least $C^{(0)}$ at t_J. Since $t_{J+\kappa-1} \le t_J < t_{J+\kappa}$,

$$\begin{aligned}\gamma(t_J) &= \sum_{i=J-1}^{J+\kappa-1} P_i \mathcal{B}_{i,\kappa}(t_J) \\ &= \sum_{i=0}^{\kappa} P_{J-1+i} \mathcal{B}_{J-1+i,\kappa}(t_J).\end{aligned}$$

Now, $\mathcal{B}_{J-1+i,\kappa}$ has $(\kappa - i)$ copies of t_J in its determining knot set and is in $C^{(i)}$ at t_J by Corollary 6.25. Thus, $\mathcal{B}_{J-1+i,\kappa}(t_J) = 0$ for $i = 1, \ldots, \kappa$, since they each meet the zero function continuously. Now again,

$$1 = \sum_{i=0}^{\kappa} \mathcal{B}_{J-1+i,\kappa}(t_J) = \mathcal{B}_{J-1,\kappa}(t_J),$$

so $\gamma(t_J) = P_{J-1}$.

For item 2: If $t_{J-1} = t_J = \cdots = t_{J+\kappa-1}$, then $\mathcal{B}_{J-1,\kappa}$ has domain $[t_{J-1+\kappa}, t_{J+\kappa})$ and is right continuous. Further it has a $(\kappa+1)$-tuple knot so $\mathcal{B}_{J-1,\kappa} \in C^{(\kappa-(\kappa+1))} = C^{(-1)}$ at $t_{J+\kappa-1}$. Each $\mathcal{B}_{J-1+j,\kappa}(t)$, $j = 1, \ldots, \kappa$, has a $(\kappa - j + 1)$-tuple knot at $t_{J+\kappa-1}$ and hence is in $C^{(j-1)}$ at $t_{J+\kappa-1}$. Hence, they are continuously joined with the zero function on the left of t_{J-1}, and so must have function value 0. Since they are the only possible functions which are nonzero at $t_{J+\kappa-1}$, $\mathcal{B}_{J-1,\kappa}(t_J)=1$ and $\gamma(t_J)=P_{J-1}=\lim_{t\to t_J^+}\gamma(t)$. Now, for $t_{J-2}<t<t_{J-1}$, $\gamma(t)=\sum_{i=J-2-\kappa}^{J-2}P_i\mathcal{B}_{i,\kappa}(t)$. Since $\mathcal{B}_{J-2-j,\kappa}$ is defined by the knots $\{t_{J-2-j},\ldots,t_{J-2-j+\kappa+1}\}$, for $j=1$, $\ldots$, κ, $\mathcal{B}_{J-2-j,\kappa}$ is $C^{(j-1)}$ at t_{J-1}. Further, since $j>0$, the function continuously joins with the zero function at t_{J-1}. Thus, $\lim_{t\to t_J^-}\mathcal{B}_{J-2-j,\kappa}(t)=0$. But $1=\sum_{j=0}^{\kappa}\mathcal{B}_{J-2+j,\kappa}(t)$, so

$$\begin{aligned} 1 &= \lim_{t\to t_J^-} 1 \\ &= \sum_{j=0}^{\kappa}\lim_{t\to t_J^-}\mathcal{B}_{J-2-j,\kappa}(t) \\ &= \lim_{t\to t_J^-}\mathcal{B}_{J-2,\kappa}(t). \end{aligned}$$

And hence, $\lim_{t\to t_J^-}\gamma(t)=P_{J-2}$. The second part of item 1 is the same proof shifted over one subscript.

The results from item 1 and item 2 can be used to prove items 3 and 4 by simply considering the function γ' wherever γ is used in items 1 and 2, and noting that one must also replace κ by $(\kappa-1)$. ∎

Corollary 8.2. *For* $\mathbf{t}=\{t_i\}_{i=0}^{n+\kappa+1}$ *such that* $t_0=\cdots=t_\kappa$, *and* $t_{n+1}=\cdots=t_{n+\kappa+1}$ *the curve*

$$\gamma(t)=\sum_{i=0}^{i=n}P_i\mathcal{B}_{i,\kappa}(t)$$

interpolates P_0 *and* P_n *and is tangent to* (P_1-P_0) *and* (P_n-P_{n-1}) *at* $t=t_\kappa$ *and* $t=t_{n+1}$, *respectively.*

Definition 8.3. *The B-spline curve defined in Corollary 8.2 is called the* open B-spline curve.

Definition 8.4. *A B-spline curve of degree* κ *defined over* $[t_\kappa,t_{n+1})$ *for which the multiplicities necessary for interpolation do not apply at either end is called a* floating B-spline curve.

Definition 8.5. *A* periodic B-spline curve *is a curve defined with parametric domain of a unit circle. The mapping from the real line is made by the function* circle : $\boldsymbol{R}\to\{\text{unit circle}\}$ *such that*

$$\text{circle}(t)=\exp\left(2\pi i\frac{t-t_0}{t_{n+1}-t_0}\right).$$

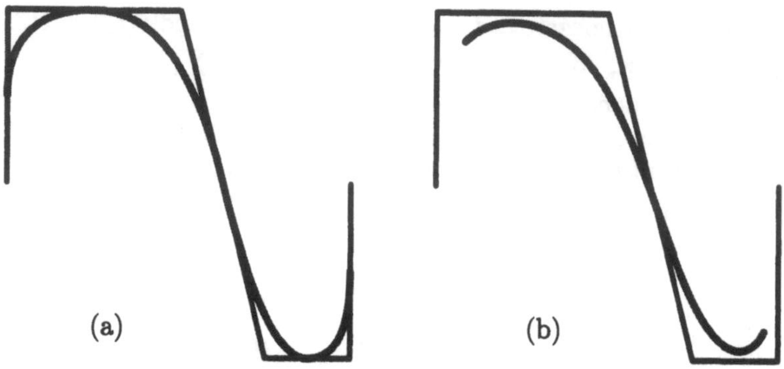

Figure 8.1. Floating B-spline curves: (a) degree 2, t ={0,1,2,3,4,5,6,7,8}; (b) degree 3, t={0,1,2,3,4,5,6,7,8,9}

There are $n+1$ *distinct basis functions over this domain. We say that* $u = v \bmod a$ *if* $u = v + ma$ *where* m *is integer valued. This mapping to the circle is equivalent to setting* $a = t_{n+1} - t_0$ *and applying the modular function.*

There are certain conflicts between the definition of periodic spline and the aperiodic (open, floating) splines.

Definition 8.6. *For* $\Delta_i = t_{i+1} - t_i$,

If $\Delta_i = \Delta_{i+1}$, *for all* i, *then all B-spline curves using* $\mathbf{t}$ *as a knot vector are called* uniform floating *or* uniform periodic B-spline curve, *depending on whether the modulus identification is made.*

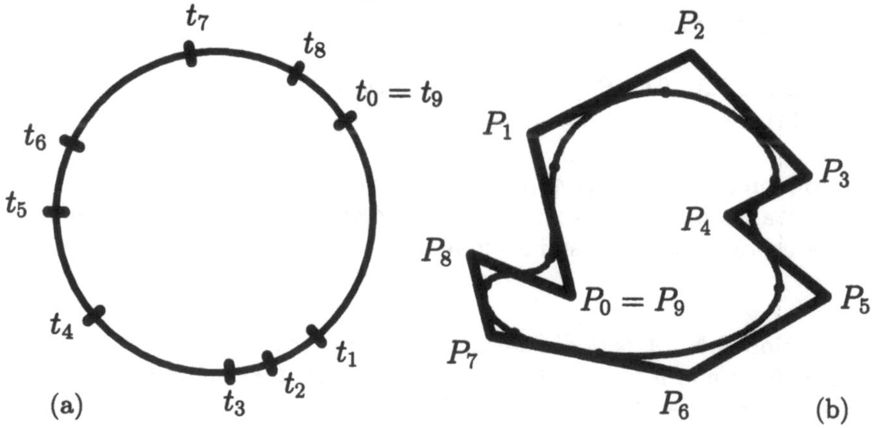

Figure 8.2. (a) Mapping to unit circle, t ={}; (b) A cubic curve using the basis functions in (a).

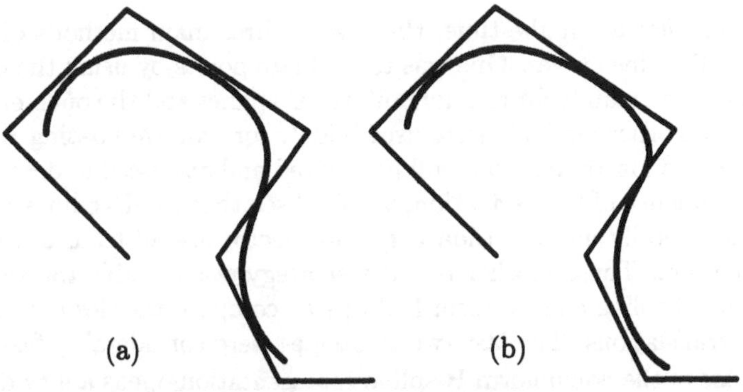

Figure 8.3. Floating cubic B-splines: (a) uniform knot vector; (b) nonuniform, $t = \{0,2,4,5,6,7,8,12,14,17,19\}$.

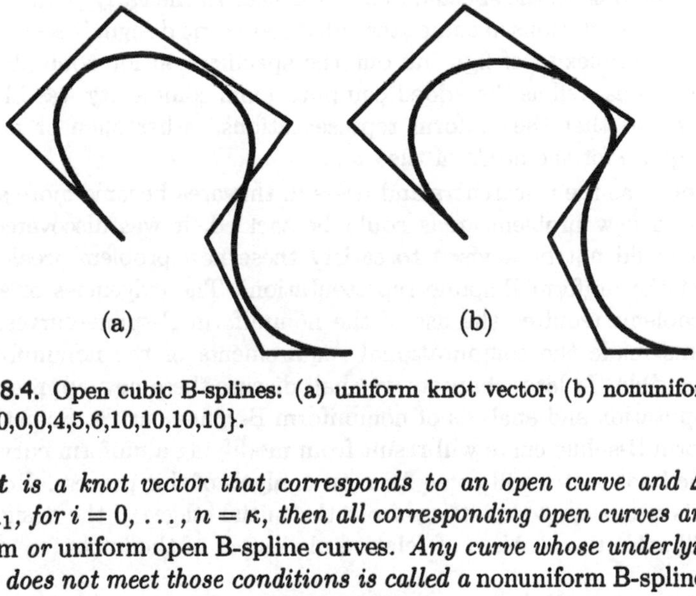

Figure 8.4. Open cubic B-splines: (a) uniform knot vector; (b) nonuniform, $t = \{0,0,0,0,4,5,6,10,10,10,10\}$.

If **t** *is a knot vector that corresponds to an open curve and* $\Delta_{\kappa+i} = \Delta_{\kappa+i+1}$, *for* $i = 0, \ldots, n = \kappa$, *then all corresponding open curves are called* uniform *or* uniform open B-spline curves. *Any curve whose underlying knot vector does not meet those conditions is called a* nonuniform B-spline curve.

Thus we see that for unique specification of the B-spline curve, two quantities are needed: a control polygon and a knot vector. Modifying either one can change the resulting curve.

8.1.1 Uniform vs. Nonuniform Knot Vectors

When the *B-spline method* of curve design was first introduced by Riesenfeld [71] there was not much distinction made over the utility of uniform versus nonuniform. At the time, there were three main methods of calculating the B-spline curves. One was to evaluate points by using the deBoor algorithm; used mainly for the nonuniform B-splines and the open end conditions. Two other methods were used widely for uniform floating B-spline curves. One, a matrix method, will be derived and discussed in detail later. As a consequence of the derivation, we shall see that the B-splines that result from a totally uniform (floating) knot vector are all translations of a single function. Thus, another popular strategy was to table the values of that single B-spline and perform lookups to compute the desired nonzero B-spline translations. The last two strategies were considerably faster.

The use of the nonuniform B-spline representation opens a new door on complexity in the use of B-splines. Suddenly the choice is no longer between open and periodic, but between relative knot spacing, multiplicities, etc.; the same control polygon can yield an infinite number of different B-spline curves depending on the specific knot sequence. In the early years of using B-spline representations in computer aided geometric design, it seemed that the added complexity of figuring out the specification for nonuniform B-spline curves as well as the added computational complexity was ill worth the effort and that the uniform representations, either open or floating, were adequate for the needs of the area.

However, as the researchers and users in this area became more sophisticated and new problem areas could be tackled, it was discovered that methods could not be devised to satisfy these new problem areas which just used the uniform B-spline representation. The exigencies of solving these problems required the use of the nonuniform B-spline curves. New algorithms made the computational requirements of the nonuniform B-splines feasible. In later chapters we shall discuss these new approaches to the computation and analysis of nonuniform B-spline curves. Sometimes a nonuniform B-spline curve will result from modifying a uniform curve, and hence the knot vector will come from an analysis of this process. However, when it is necessary to use them from the initial stages of the design, one must still solve the problem of relative placement of the knots.

8.1.2 End Conditions for B-Spline Curves

It has been shown that the multiplicity of a knot affects the curve very strongly. Two types of curves have been named that in reality have only one distinctive difference: the way the ends of the domains are defined and the behavior of the curves near those end conditions. A third type, the periodic B-spline is fundamentally different. When Riesenfeld first introduced the use of B-splines in design, the desire for faster computation

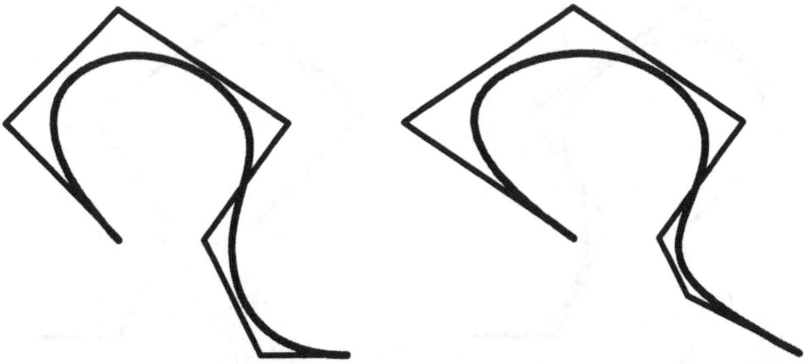

Figure 8.5. Open B-spline curves from identical polygons, except at P_1 and P_{n-1}.

led to the use of uniform, floating cubic B-splines. In general, that reason is no longer valid, but each of these conditions can be useful and will be discussed.

Note that if $\boldsymbol{t}$ and $\boldsymbol{\tau}$ are two knot vectors, each with $n+\kappa+2$ elements, for which $t_i = \tau_i$, $i = \kappa, \ldots, n+1$, then the spaces $\mathcal{S}_{\kappa,\boldsymbol{t}}$ and $\mathcal{S}_{\kappa,\boldsymbol{\tau}}$ prescribe piecewise polynomials having exactly the same domain, exactly the same interior knot values and multiplicities, and hence, and exactly the same piecewise polynomials, except for the end conditions. However, the B-splines which define the space are different.

8.1.2.1 Using open B-spline curves. The determining characteristics of the open B-spline curve are that the endpoints of the control polygon are interpolated and the curve is tangent to the first and last side of the control polygon, for all degree open B-spline curves. It is a direct generalization of the Bézier curve near the ends of the control polygon. In fact, if $\{P_0, \ldots, P_n\}$ are the vertices of the control polygon, P_1 and P_{n-1} are used most strongly to control the direction and magnitude of the tangent vectors at the ends. Since $t_0 = \cdots = t_\kappa < t_{\kappa+1}$, and $t_n < t_{n+1} = \cdots = t_{n+\kappa+1}$, the domain of $\mathcal{S}_{\kappa,\boldsymbol{t}}$ is $[t_\kappa, t_{n+1}] = [t_0, t_{n+\kappa+1}]$. Namely, the whole support of each B-spline function is included in the domain of the space of *open* B-splines.

There are many times when it is necessary to control the exact points at which the curve begins and ends. To do this, the open curve is used. One such application is in interpolation. Another is when it is necessary to combine two B-spline curves into one sequential representation with one knot vector. It is always possible to represent a B-spline curve defined over the domain $[t_\kappa, t_{n+1}]$ as an open B-spline curve defined over the same domain. The knot vectors of the original and the open form would be exactly the same for the κth through the $(n+1)^{st}$ element of the knot sequences,

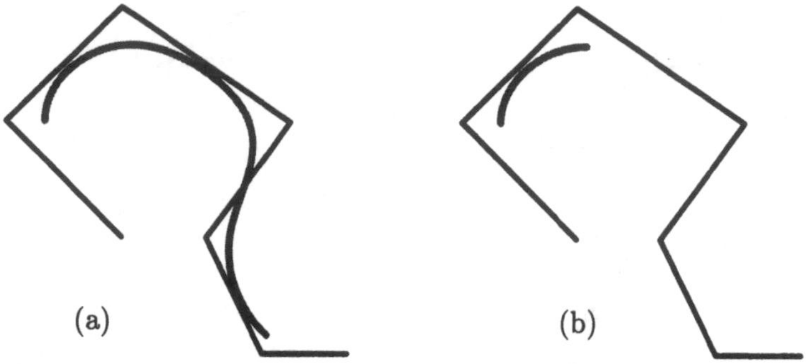

Figure 8.6. (a) A floating cubic B-spline; (b) a curve segment.

but would differ at the ends. Such a case can arise when one originally has floating or mixed end conditions, and one would like to have more information on the specific behavior of the curve. Graphing, or rendering the curve can provide such a motivation, as can the necessity for modification of the curve, or performance of analysis.

Since this is just a basis transformation, one could accomplish this using matrix methods to solve for the open end polygon points that result from the old ones and the old and new knot vectors. This method involved solving systems of equations and was quite tedious. Another technique which accomplishes this in a constructive geometric way is called *subdivision* or *refinement*, and will be discussed later.

8.1.2.2 Using floating B-spline curves. The term *floating* is used to intuitively give information about the end conditions of this type of B-spline curve. Historically this curve has had only simple knots ($m_i = 1$, for all i). In addition, all the knots have been uniformly spaced. The wedding of these two constraints arose from the availability of easily understood, fast computational methods, which we shall also discuss. In theory, however, these constraints are unnecessary. All that is necessary is that the open end conditions not hold; i.e., the knot vector specification does not cause the curve to interpolate the first and last vertex of the control polygon.

In this case, the exact position of the end of the curve is not a point on the control polygon. In fact, the curve floats near the polygon and responds when a vertex is moved, but no intuitive grasp of exactly where the curve begins and where it ends is implicit from viewing the mathematical representation. Tangency conditions at the ends are also not known.

Since $t_0 < t_1 < \cdots < t_\kappa < t_{\kappa+1} < \cdots < t_n < \cdots < t_{n+\kappa+1}$, the domain of $S_{\kappa,\mathbf{t}}$, $[t_\kappa, t_{n+1}]$ overlaps only part of the domains of 2κ of the B-splines. Thus, care must be taken to choose only the appropriate values of t. They do not automatically follow, as in the case of open end conditions.

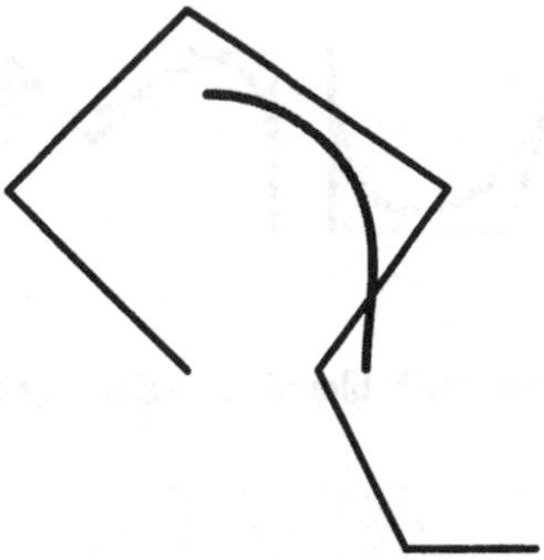

Figure 8.7. A quintic floating B-spline on same control polygon.

8.1.2.3 **Using mixed end conditions.** One can recommend the use of mixed end conditions for the following case. A specific curve that can be represented as a B-spline curve is shown below on the left in Figure 8.8. On the right in Figure 8.8 is another such curve. The instructions to the designer are "we need a single curve which traces out a path which includes the curve on the left and the curve on the right." This new single curve should blend smoothly between the two original curves. However, we don't exactly know how to specify this blend region, but do something nice.

There is not a well-known successful strategy for doing this. We propose one here using the mixed end conditions as an intermediate step.

Suppose knot vector $\boldsymbol{\tau_1} = \{ \tau_{1,i} : i = 0, \ldots, n_1 + \kappa + 1 \}$ and control polygon $\{ Q_{1,i} : i = 0, \ldots, n_1 \}$ correspond to the curve on the left, $\gamma_1(t)$, where $\tau_{1,0} = \tau_{1,1} = \cdots = \tau_{1,\kappa-1}$ and $\tau_{1,n_1+1} < \tau_{1,n_1+2} < \cdots < \tau_{1,n_1+\kappa+1}$. Further suppose $\gamma_2(t)$ has control polygon $\{ Q_{2,j} : j = 0, \ldots, n_2 \}$ and knot vector $\boldsymbol{\tau_2} = \{ \tau_{2,j} : j = 0, \ldots, n_2 + \kappa + 1 \}$, such that $\tau_{2,0} < \tau_{2,1} < \cdots < \tau_{2,\kappa-1}$ and $\tau_{2,n_2+1} = \tau_{2,n_2+2} = \cdots = \tau_{2,n_2+\kappa+1}$. We add the additional constraint that $\tau_{1,n_1+1+j} = \tau_{2,j}$, $j = 0, \ldots, \kappa$.

One approach to solving this problem forms a new curve $\gamma(t)$ defined with control polygon

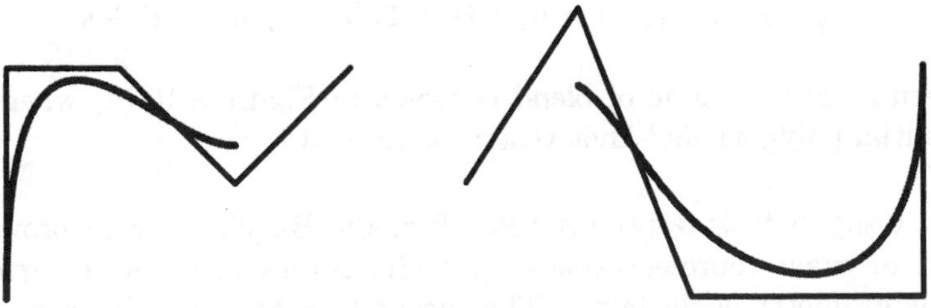

Figure 8.8. Curves to be blended.

Figure 8.9. Compound curve with blending region: (a) nice fit; (b) undulating blend.

$$P_j = \begin{cases} Q_{1,j}, & j = 0, \ldots, n_1, \\ Q_{2,j-n_1-1}, & j = n_1+1, \ldots, n_1+n_2+1, \end{cases}$$

and knot vector

$$t_j = \begin{cases} \tau_{1,j}, & j = 0, \ldots, n_1+\kappa+1, \\ \tau_{2,j-n_1-1}, & j = n_1+\kappa+2, \ldots, n_1+n_2+\kappa+2. \end{cases}$$

This new curve will then have reproduced γ_1 and γ_2 exactly and will have a blending region which has κ new polynomial segments, joined to the original curves with $C^{(\kappa-1)}$ continuity.

Sometimes, as in Figure 8.9(b), the blend region undulates. A method which can be used to effect a smoother blend is to let

$$P_j = \begin{cases} Q_{1,j}, & j = 0, \ldots, n_1-1, \\ Q_{2,j-n_1+1}, & j = n_1, \ldots, n_1+n_2-1, \end{cases}$$

used with knot vector

$$t_j = \begin{cases} \tau_{1,j}, & j = 0, \ldots, n_1+(\kappa+1)/2-1, \\ \tau_{2,j-n_1+1}, & j = n_1+(\kappa+1)/2, \ldots, n_1+n_2+\kappa. \end{cases}$$

An example of that style of blend is shown in Figure 8.10 (a) where the same initial polygons and knot vectors were used.

8.1.2.4 Using periodic B-spline curves. Periodic B-spline curves provide a method of exact representation of periodic curves in a way compatible with the geometry of the curve. This means that the periodic knot vector and periodic control polygon will not have distinguished points. When the

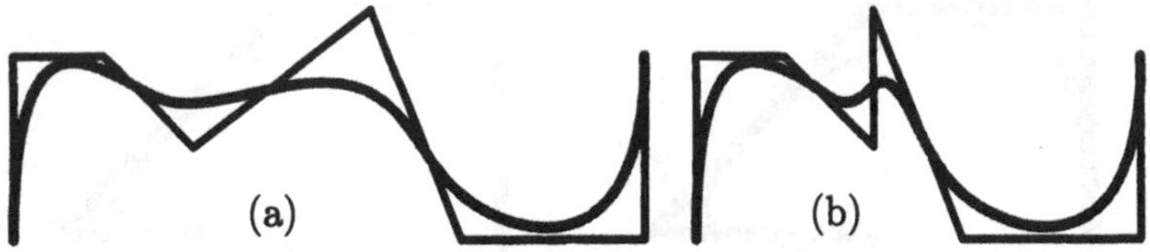

Figure 8.10. Blended region curve with second method: (a) nice fit; (b) undulating blend.

knot vector is uniform, any control point can be considered the first control point, so the control polygon can be counted from that control point. If the knot vector is nonuniform, the knot vector must be reordered if there is any change made in which control point is considered the first.

In order to actually compute periodic splines one usually develops a corresponding floating knot vector and control polygon. However, since one can start the periodic curves at any place, the periodic and floating curves are intrinsically different. Below we shall give a concrete example with a floating curve at two such different starting places.

Suppose the periodic knot vector is $\boldsymbol{\tau} = \{\tau_0, \ldots, \tau_{n+1}\}$. To implement periodic B-spline curves one can make a floating knot vector, $\boldsymbol{t}$, on domain $[\tau_0, \tau_{n+1})$ subject to the condition that $t_{\kappa+i} = \tau_i$, $i = 0, \ldots, n+1$. The values of $t_0, \ldots, t_{\kappa-1}$, and $t_{n+2}, \ldots, t_{n+\kappa+1}$ must be determined to fulfill the *modulo* condition. That is, they cannot add knots when they are wound onto the unit circle using the modulo operator.

Thus, $t_{n+1+i} = \tau_i + (\tau_{n+1} - \tau_0)$, $i = 0, \ldots, \kappa$, and $t_i = \tau_{n-\kappa+1+i} - (\tau_{n+1} - \tau_0)$, $i = 0, \ldots, \kappa$. If $\kappa \geq n$, then the knot vectors will just wrap around the circle one or more times. The control polygon must also be unwrapped. There, choice of the implementer becomes a large factor.

Consider periodic knot vector $\{0, 1, 3, 4, 5, 7, 9, 10\}$ for a cubic curve. Since there are eight values in the knot vector, $n = 6$, and $10 \equiv 0 \bmod (10-0)$. There are seven periodic B-splines and seven coefficients in the coefficient control polygon, $P_0, \ldots, P_6$. The floating knot vector which accompanies this knot vector is $\boldsymbol{t} = \{-5, -3, -1, 0, 1, 3, 4, 5, 7, 9, 10, 11, 13, 14\}$ with floating polygon $\boldsymbol{V}$ where

$$
\begin{aligned}
V_0 &= P_6 \\
V_{i+1} &= P_i, \qquad i = 0, \ldots, 6 \\
V_8 &= P_0 \\
V_9 &= P_1.
\end{aligned}
$$

So, $\boldsymbol{V} = \{P_6, P_0, P_1, \ldots, P_6, P_0, P_1\}$. There are other appropriate choices for the accompanying floating polygon since there is no theoretical reason

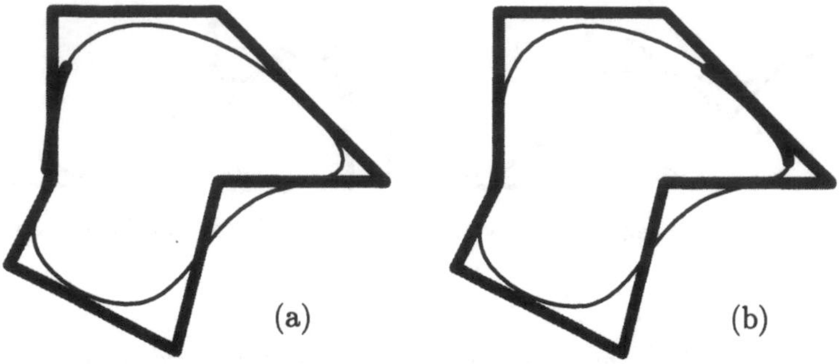

Figure 8.11. First segment for unrolled periodic polygon: (a) I; (b) II.

for choosing one over the other nor well-established convention. The particular one selected draws the first of the floating segments based on the vertices P_6, P_0, P_1, and P_2 and is shown in Figure 8.11 (a).

If instead $t_3 = 3$, then $\boldsymbol{t} = \{-1, 0, 1, 3, 4, 5, 7, 9, 10, 11, 13, 14, 15, 17\}$, and the control polygon becomes $\{P_1, P_2, P_3, P_4, P_5, P_6, P_0, P_1, P_2, P_3\}$. Now, the domain of this floating curve is $[3, 13)$, and the first section of the curve is shown in Figure 8.11 (b).

While other choices of correspondences between unrolling the knot vector and unrolling a control polygon are possible, unless the periodic vector is uniform, each will result in a different final curve. Caution and consistency must be the watchwords if periodic curves are implemented in this manner.

8.1.3 Computing Uniform Floating B-Splines

We shall discuss this for uniform floating cubics which have had the most widespread use. The ideas have been generalized to higher degree uniform floating splines [72]. Other methods for drawing curves with this knot vector are discussed in Section 7.2.

By Corollary 7.5 there is only one distinct B-spline. Thus, we need determine the values of just one function everywhere in its support, *or we can determine the values of all functions nonzero over one interval,* $t_i \le t < t_{i+1}$.

Using matrix notation, and letting the knot vector and the control polygon indices both start at -3, for $t_j \le t < t_{j+1}$, $0 \le j < n$, gives

$$\gamma(t) = \begin{bmatrix} \mathcal{B}_{j-3,3}(t) & \mathcal{B}_{j-2,3}(t) & \mathcal{B}_{j-1,3}(t) & \mathcal{B}_{j,3}(t) \end{bmatrix} \begin{bmatrix} P_{j-3} \\ P_{j-2} \\ P_{j-1} \\ P_j \end{bmatrix}.$$

Now,

$$\begin{aligned}\mathcal{B}_{j-3,3}(t) &= \mathcal{B}_{-3,3}(t-j)\\ \mathcal{B}_{j-2,3}(t) &= \mathcal{B}_{-2,3}(t-j)\\ \mathcal{B}_{j-1,3}(t) &= \mathcal{B}_{-1,3}(t-j)\\ \mathcal{B}_{j,3}(t) &= \mathcal{B}_{0,3}(t-j).\end{aligned}$$

But since $t_j = j \le t < t_{j+1} = j+1$, $0 \le t-j < 1$. Rewriting in this new format, for $0 \le t < 1$,

$$\gamma(t+j) = \begin{bmatrix} \mathcal{B}_{-3,3}(t) & \mathcal{B}_{-2,3}(t) & \mathcal{B}_{-1,3}(t) & \mathcal{B}_{0,3}(t) \end{bmatrix} \begin{bmatrix} P_{j-3} \\ P_{j-2} \\ P_{j-1} \\ P_j \end{bmatrix}.$$

From these equations, it is clear that knowing the equations of $\mathcal{B}_{-3,3}(t)$, $\mathcal{B}_{-2,3}(t)$, $\mathcal{B}_{-1,3}(t)$, and $\mathcal{B}_{0,3}(t)$ for $t \in [0,1)$ gives the value of the curve at any location. The matrix multiplication process is really just a moving window over the polygon vertices.

For example, suppose it is desired to graph the curve with knot vector $\{-3,-2,-1,0,1,2,3,4,5,6,7\}$ and control polygon having the seven vertices, $\{P_{-3}, P_{-2}, P_{-1}, \ldots, P_3\}$. The domain of the curve is $[0,4]$.

1. Select values x_m, $m = 0, \ldots, r$, $0 = x_0 < \ldots < x_r = 1$.
2. Evaluate $\mathcal{B}_{i,3}(x_m)$, $i = -3$, -2, -1, 0, and $m = 0, \ldots, r$.
3. For $j = 0$, 1, 2, and 3 compute $\gamma(x_m + j)$ using

$$\begin{bmatrix} \mathcal{B}_{-3,3}(x_m) & \mathcal{B}_{-2,3}(x_m) & \mathcal{B}_{-1,3}(x_m) & \mathcal{B}_{0,3}(x_m) \end{bmatrix} \begin{bmatrix} P_{j-3} \\ P_{j-2} \\ P_{j-1} \\ P_j \end{bmatrix}.$$

$m = 0, \ldots, r-1$, (for $j = 0$, 1, 2,) and $m = 0, \ldots, r$ (for $j = 3$).

While this method seems very straightforward, it is necessary to compute and store $(r+1)$ values of four distinct B-splines. One could proceed with this matrix analysis one more step and write

$$\begin{aligned} B &= \begin{bmatrix} \mathcal{B}_{-3,3}(t) & \mathcal{B}_{-2,3}(t) & \mathcal{B}_{-1,3}(t) & \mathcal{B}_{0,3}(t) \end{bmatrix}\\ &= \begin{bmatrix} t^3 & t^2 & t & 1 \end{bmatrix} C \end{aligned}$$

since each of those functions is one polynomial segment over the interval $[0,1]$. We shall derive the forms of these frequently used functions. All calculations are based upon $0 \le t < 1$.

For $t \in [0,1)$, $t_0 \le t < t_1$, so

$$\begin{aligned}
\mathcal{B}_{0,3}(t) &= \frac{t-t_0}{t_3-t_0}\mathcal{B}_{0,2}(t) + \frac{t_4-t}{t_4-t_1}\mathcal{B}_{1,2}(t) \\
&= \frac{t}{3}\mathcal{B}_{0,2}(t), \qquad \text{since } t < t_1. \\
&= \frac{t}{3}\left(\frac{t-t_0}{t_2-t_0}\mathcal{B}_{0,1}(t) + \frac{t_3-t}{t_3-t_1}\mathcal{B}_{1,1}(t)\right) \\
&= \frac{t}{3}\frac{t}{2}\mathcal{B}_{0,1}(t), \qquad \text{since } t < t_1. \\
&= \frac{t}{3}\frac{t}{2}\left(\frac{t-t_0}{t_1-t_0}\mathcal{B}_{0,0}(t) + \frac{t_2-t}{t_2-t_1}\mathcal{B}_{1,0}(t)\right) \\
&= \frac{t}{3}\frac{t}{2}\frac{t}{1}\mathcal{B}_{0,0}(t) \\
&= \frac{t^3}{3!}.
\end{aligned}$$

Continuing,

$$\begin{aligned}
\mathcal{B}_{-1,3}(t) &= \frac{t-t_{-1}}{t_2-t_{-1}}\mathcal{B}_{-1,2}(t) + \frac{t_3-t}{t_3-t_0}\mathcal{B}_{0,2}(t) \\
&= \frac{t+1}{3}\mathcal{B}_{-1,2}(t) + \frac{3-t}{3}\mathcal{B}_{0,2}(t) \\
&= \frac{t+1}{3}\left(\frac{t-t_{-1}}{t_1-t_{-1}}\mathcal{B}_{-1,1}(t) + \frac{t_2-t}{t_2-t_0}\mathcal{B}_{0,1}(t)\right) \\
&\quad + \frac{3-t}{3}\left(\frac{t-t_0}{t_2-t_0}\mathcal{B}_{0,1}(t) + \frac{t_3-t}{t_3-t_1}\mathcal{B}_{1,1}(t)\right) \\
&= \frac{t+1}{3}\left(\frac{t-t_{-1}}{t_1-t_{-1}}\mathcal{B}_{-1,1}(t) + \frac{t_2-t}{t_2-t_0}\mathcal{B}_{0,1}(t)\right) \\
&\quad + \frac{3-t}{3}\left(\frac{t-t_0}{t_2-t_0}\mathcal{B}_{0,1}(t)\right) \\
&= \frac{t+1}{3}\frac{t+1}{2}\mathcal{B}_{-1,1}(t) + \left(\frac{t+1}{3}\frac{2-t}{2} + \frac{3-t}{3}\frac{t}{2}\right)\mathcal{B}_{0,1}(t) \\
&= \frac{t+1}{3}\frac{t+1}{2}\left(\frac{t-t_{-1}}{t_0-t_{-1}}\mathcal{B}_{-1,0}(t) + \frac{t_1-t}{t_1-t_0}\mathcal{B}_{0,0}(t)\right) \\
&\quad + \left(\frac{t+1}{3}\frac{2-t}{2} + \frac{3-t}{3}\frac{t}{2}\right)\left(\frac{t-t_0}{t_1-t_0}\mathcal{B}_{0,0}(t) + \frac{t_2-t}{t_2-t_1}\mathcal{B}_{1,0}(t)\right) \\
&= \frac{t+1}{3}\frac{t+1}{2}\frac{1-t}{1} + \left(\frac{t+1}{3}\frac{2-t}{2} + \frac{3-t}{3}\frac{t}{2}\right)\frac{t}{1}
\end{aligned}$$

$$
\begin{aligned}
&= \frac{(t+1)^2(1-t)+t(t+1)(2-t)+(3-t)t^2}{3!} \\
&= \frac{-3t^3+3t^2+3t+1}{3!}.
\end{aligned}
$$

Now, for $t \in [0,1)$, $0 \equiv \mathcal{B}_{-3,2}(t) \equiv \mathcal{B}_{-2,1}(t) \equiv \mathcal{B}_{-1,0}(t)$, so

$$
\begin{aligned}
\mathcal{B}_{-3,3}(t) &= \frac{t-t_{-3}}{t_0-t_{-3}}\mathcal{B}_{-3,2}(t) + \frac{t_1-t}{t_1-t_{-2}}\mathcal{B}_{-2,2}(t) \\
&= \frac{1-t}{3}\mathcal{B}_{-2,2}(t) \\
&= \frac{1-t}{3}\left(\frac{t-t_{-2}}{t_0-t_{-2}}\mathcal{B}_{-2,1}(t) + \frac{t_1-t}{t_1-t_{-1}}\mathcal{B}_{-1,1}(t)\right) \\
&= \frac{1-t}{3}\frac{1-t}{2}\mathcal{B}_{-1,1}(t) \\
&= \frac{(1-t)^2}{3!}\mathcal{B}_{-1,1}(t) \\
&= \frac{(1-t)^2}{3!}\left(\frac{t-t_{-1}}{t_0-t_{-1}}\mathcal{B}_{-1,0}(t) + \frac{t_1-t}{t_1-t_0}\mathcal{B}_{0,0}(t)\right) \\
&= \frac{(1-t)^2}{3!}\frac{1-t}{1}\mathcal{B}_{0,0}(t) \\
&= \frac{(1-t)^3}{3!}.
\end{aligned}
$$

Since $1 \equiv \sum_{i=-3}^{0} \mathcal{B}_{i,3}(t)$, finding the form of the last unknown function is straightforward.

$$
\begin{aligned}
\mathcal{B}_{-2,3}(t) &= 1 - \mathcal{B}_{-3,3}(t) - \mathcal{B}_{-1,3}(t) - \mathcal{B}_{0,3}(t) \\
&= \frac{4-6t^2+3t^3}{3!}.
\end{aligned}
$$

Filling in the C matrix,

$$
C = \begin{bmatrix} -1/6 & 1/2 & -1/2 & 1/6 \\ 1/2 & -1 & 1/2 & 0 \\ -1/2 & 0 & 1/2 & 0 \\ 1/6 & 2/3 & 1/6 & 0 \end{bmatrix}.
$$

Putting this together, for $0 \le t < 1$,

$$
\gamma(t+j) = \begin{bmatrix} t^3 & t^2 & t & 1 \end{bmatrix} \begin{bmatrix} -1/6 & 1/2 & -1/2 & 1/6 \\ 1/2 & -1 & 1/2 & 0 \\ -1/2 & 0 & 1/2 & 0 \\ 1/6 & 2/3 & 1/6 & 0 \end{bmatrix} \begin{bmatrix} P_{j-3} \\ P_{j-2} \\ P_{j-1} \\ P_j \end{bmatrix}.
$$

8.2 Analysis of Lower Degree Curves

When splines are used in design or numerical methods, cubics and quadratics are used most often. Originally the cubic spline found use in the context of complete spline interpolation. The $C^{(2)}$ continuity at all interior knots was sufficient for most mathematical analysis purposes. Further since other methods such as piecewise cubic Hermite interpolation could be cast in the framework of cubic spline theory, it seems quite general. While quadratic B-spline curves have at most $C^{(1)}$ continuity at the knots, they are particularly simple in their formulation and easy to compute. In their rational form, which will be discussed later, they allow the *rational quadratic B-spline* to represent exactly all the piecewise conic sections as parametric space curves. At the same time, new uses are being discovered for higher degree B-splines with their higher order continuity. The feasibility of computing curves with these elements has been made possible by the existence of the deBoor-Cox algorithm, the deBoor algorithm, and a different method of computing information about the curve called *refinement*, which will be discussed later in Chapter 16.

We shall start the discussion with the easiest to compute of the B-spline curves, degree $\kappa = 0$. For a knot vector $\boldsymbol{t} = \{t_0, \ldots, t_{n+1}\}$, $t_i < t_{i+1}$, we consider the curve

$$\gamma(t) = \sum_{i=0}^{n} P_i \mathcal{B}_{i,0}(t).$$

For $t_i \le t < t_{i+1}$, $i < n$, only one of the B-splines is nonzero, namely $\mathcal{B}_{i,0}$ and hence the curve is $\gamma(t) = P_i$ for those values of t. For $t_n \le t \le t_{n+1}$, only $\mathcal{B}_{n,0}$ is nonzero and $\gamma(t) = P_n$. Thus, the 0 degree B-spline curve is just a sequence of ordered points in the plane. The curve sits at the ith point for time $(t_{i+1} - t_i)$.

Next consider the degree $\kappa = 1$ curves. The non-degenerate knot vector will be of the form $\boldsymbol{t} = \{t_0, \ldots, t_{n+2}\}$. If $t_0 = t_1 < t_{n+2}$ and $t_n < t_{n+1} = t_{n+2}$, while all interior knots have multiplicity one, the curve is the open linear B-spline curve,

$$\gamma(t) = \sum_{i=0}^{n} P_i \mathcal{B}_{i,1}(t).$$

Again, for $t_j \le t < t_{j+1}$,

$$\gamma(t) = P_{j-1}\mathcal{B}_{j-1,1}(t) + P_j \mathcal{B}_{j,1}(t)$$

and

$$\gamma'(t) = P_j - P_{j-1}.$$

Thus, $\mathcal{B}_{i,1}(t_j) = \delta_{i,j-1}$, $\gamma(t_j) = P_{j-1}$, $j = 0, \ldots, n+1$, since we are using the modified B-splines, and the curve is the piecewise linear curve connecting the points P_j, $j = 0, \ldots, n$. The tangent vector of the curve is a constant on each segment, which means that not only is the graph a piecewise linear function, but it is constant speed parametrization on each segment with abrupt discontinuous changes between.

8.2.1 Quadratic B-Spline Curves

The quadratic curve, $\kappa = 2$, is a more interesting example. For this example, consider the knot vector $\boldsymbol{t} = \{t_0, \ldots, t_{n+3}\}$, such that each interior knot has multiplicity smaller than three, but allows multiplicity three at the ends. The domain of the curve is $[t_2, t_{n+1}]$. In general,

$$\gamma(t) = \sum_{i=p-2}^{p} P_i \mathcal{B}_{i,2}(t), \qquad \text{for } t_p \le t < t_{p+1}.$$

Consider the curve over the interval $[t_j, t_{j+1}]$, where $2 < j < n$, and $t_j < t_{j+1}$. Since the multiplicity at t_j and t_{j+1} is less than or equal to two, the curve is continuous at the endpoints of the interval. The curve is also $C^{(1)}$ on the whole interval if both t_j and t_{j+1} are *simple knots*, that is, each has multiplicity one in $\boldsymbol{t}$. Let us treat that case first. Since $\mathcal{B}_{j,2}(t_j) = 0$,

$$\begin{aligned} \gamma(t_j) &= \sum_{i=j-2}^{j} P_i \mathcal{B}_{i,2}(t_j) \\ &= \sum_{i=j-2}^{j-1} P_i \mathcal{B}_{i,2}(t_j), \end{aligned}$$

and, analogously, at t_{j+1}

$$\begin{aligned} \gamma(t_{j+1}) &= \sum_{i=j-1}^{j+1} P_i \mathcal{B}_{i,2}(t_{j+1}) \\ &= \sum_{i=j-1}^{j} P_i \mathcal{B}_{i,2}(t_{j+1}). \end{aligned}$$

Since $\mathcal{B}_{j-2,2}(t_j) + \mathcal{B}_{j-1,2}(t_j) = 1$ and $\mathcal{B}_{j-1,2}(t_{j+1}) + \mathcal{B}_{j,2}(t_{j+1}) = 1$, the values of γ at the respective values of t are just convex combinations of P_{j-2} and P_{j-1} and P_{j-1} and P_j, respectively.

The curve then interpolates somewhere along both sides of the polygon around the vertex P_{j-1}. Since $\mathcal{B}_{j-2,2}(t_j) = 1 - \mathcal{B}_{j-1,2}(t_j)$ and $\mathcal{B}_{j,2}(t_{j+1}) = 1 - \mathcal{B}_{j-1,2}(t_{j+1})$, if we find the values of $\mathcal{B}_{j-1,2}(t_j)$ and $\mathcal{B}_{j-1,2}(t_{j+1})$, we shall understand exactly where this interpolation takes place. Using Definition 6.7,

$$\mathcal{B}_{j-1,2}(t_j) = \frac{t_j - t_{j-1}}{t_{j-1+2} - t_{j-1}}\mathcal{B}_{j-1,1}(t_j) + \frac{t_{j-1+3} - t_j}{t_{j-1+3} - t_{j-1+1}}\mathcal{B}_{j,1}(t_j).$$

Since t_j is a simple knot, $\mathcal{B}_{j,1}$ is continuous at t_j, and $\mathcal{B}_{j,1}(t_j) = 0$. Further, $\mathcal{B}_{j-1,1}(t_j) = 1$ since it is the only nonzero B-spline at t_j, so

$$\mathcal{B}_{j-1,2}(t_j) = \frac{t_j - t_{j-1}}{t_{j+1} - t_{j-1}}.$$

Analogously,

$$\mathcal{B}_{j-1,2}(t_{j+1}) = \frac{t_{j+2} - t_{j+1}}{t_{j+2} - t_j}.$$

Thus,

$$\begin{aligned}\gamma(t_j) &= \left(1 - \frac{t_j - t_{j-1}}{t_{j+1} - t_{j-1}}\right)P_{j-2} + \frac{t_j - t_{j-1}}{t_{j+1} - t_{j-1}}P_{j-1} \\ &= \frac{t_{j+1} - t_j}{t_{j+1} - t_{j-1}}P_{j-2} + \frac{t_j - t_{j-1}}{t_{j+1} - t_{j-1}}P_{j-1}.\end{aligned}$$

Analogously,

$$\gamma(t_{j+1}) = \frac{t_{j+2} - t_{j+1}}{t_{j+2} - t_j}P_{j-1} + \frac{t_{j+1} - t_j}{t_{j+2} - t_j}P_j.$$

In the particularly simple case where $t_{j+1} - t_j = t_j - t_{j-1}$, $\gamma(t_j)$ interpolates the *midpoint* of the line between P_{j-2} and P_j.

The curve is tangent continuous at both t_j and t_{j+1}, and we would like to know the relationship of this tangent to the polygon. For $t_j \le t < t_{j+1}$,

$$\gamma'(t) = \sum_{i=j-1}^{j} \frac{2}{t_{i+2} - t_i}[P_i - P_{i-1}]\mathcal{B}_{i,1}(t)$$

so

$$\begin{aligned}\gamma'(t_j) &= \sum_{i=j-1}^{j} \frac{2}{t_{i+2} - t_i}[P_i - P_{i-1}]\mathcal{B}_{i,1}(t_j) \\ &= \frac{2}{t_{j+1} - t_{j-1}}[P_{j-1} - P_{j-2}].\end{aligned}$$

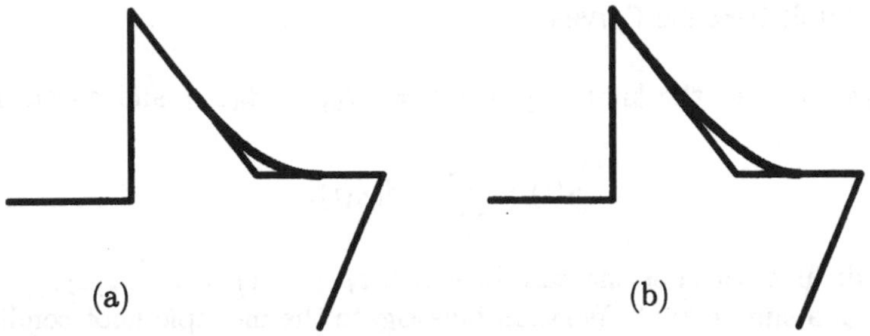

Figure 8.12. A typical quadratic segment: (a) simple knots at t_j and t_{j+1}; (b) $t_{j-1} = t_j$.

This shows that the tangent to the curve at t_j is in the direction of the side of the polygon between the points P_{j-2} and P_{j-1}. It can be similarly shown that the tangent to the curve at $t = t_{j+1}$ is in the direction of the side of the polygon between the points P_{j-1} and P_j.

Hence we have shown that if t_j and t_{j+1} are simple knots, the quadratic curve interpolates the polygon between P_{j-2} and P_{j-1}, and P_{j-1} and P_j respectively. Further, the curve is tangent to the polygon at those points with magnitude dependent on the relative spacing and scale of the knot vector. Figure 8.12 (a) shows this type of curve.

Now we consider the case in which t_j has multiplicity 2. If $t_{j-2} < t_{j-1} = t_j < t_{j+1}$, one has that $\mathcal{B}_{j-2,2}(t)$ and $\mathcal{B}_{j-1,2}(t) \in C^{(0)}$ at t_j, and $\mathcal{B}_{j,2}(t) \in C^{(1)}$ at t_j. Thus, $\mathcal{B}_{j-1,2}(t_j) = \mathcal{B}_{j,2}(t_j) = 0$, and $1 = \sum_{i=j-2}^{j} \mathcal{B}_{i,2}(t_j)$ gives that $\mathcal{B}_{j-2,2}(t_j) = 1$. Hence, $\gamma(t_j) = \gamma(t_{j-1}) = P_{j-2}$. Further, it can be shown that $\lim_{t \to t_j^+} \gamma(t)$ is tangent to the polygon side through P_{j-2} and P_{j-1}, while $\lim_{t \to t_j^+} \gamma(t)$ is tangent to the polygon side through P_{j-3} and P_{j-2}. Thus, the case of a double knot acts exactly as a limit of the single knot case. See Figure 8.12 (b).

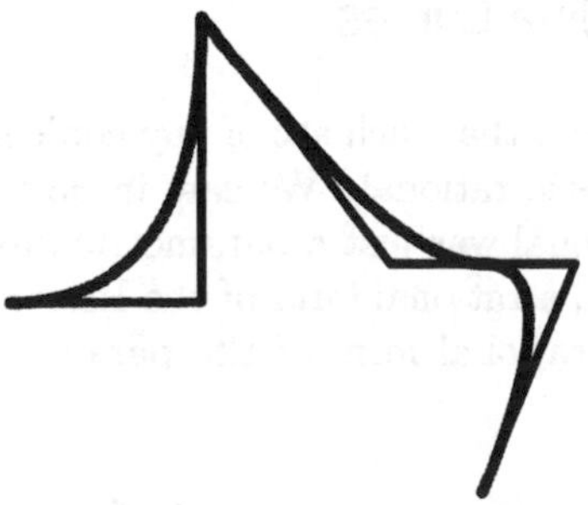

Figure 8.13. The whole quadratic curve for t ={0,0,0,2,2,3,4,4,4}.

8.2.2 Cubic B-Spline Curves

We now consider the knot sequence $t = \{t_0, \ldots, t_{n+4}\}$ and consider the curves

$$\gamma(t) = \sum_{i=0}^{n} P_i \mathcal{B}_{i,3}(t).$$

We shall first consider the case in which $t_{j-1} < t_j < t < t_{j+1}$, namely, that t_j is a simple knot. We shall investigate the multiple knot conditions afterwards. As in the case of the quadratic B-spline curve, our primary tools will be the knowledge of which B-splines will be nonzero.

Since t_j is a simple knot, all B-splines including it in their domain are $C^{(2)}$ at that point and $\mathcal{B}_{j,3}(t_j)$ must be equal to zero. Analogously, $\mathcal{B}_{j-4,3}(t_j) = 0$, so

$$\gamma(t_j) = \sum_{i=j-3}^{j-1} P_i \mathcal{B}_{i,3}(t_j),$$

and we know only that the curve is somewhere within the triangle specified by the points P_{j-3}, P_{j-2}, P_{j-1}. Clearly, the smaller that triangle becomes, the more we know about the curve at that point. Note that the curve does not interpolate the controlling polygon. In fact, the exact location depends on the relative positioning of the knot values entering into the evaluation of $\mathcal{B}_{j-3,3}(t_j)$, $\mathcal{B}_{j-2,3}(t_j)$, and $\mathcal{B}_{j-1,3}(t_j)$. Its tangent vector will be

$$\begin{aligned}\gamma'(t_j) &= c_1[P_{j-2} - P_{j-3}]\mathcal{B}_{j-2,2}(t_j) + c_2[P_{j-1} - P_{j-2}]\mathcal{B}_{j-1,2}(t_j)\\ &= -k_1[P_{j-3} - P_{j-2}] + k_2[P_{j-1} - P_{j-2}].\end{aligned}$$

Analogous results hold when t_{j+1} is a simple knot. The multiple knot conditions for the cubics will be discussed in Section 8.4.

8.3 Rational B-Spline Curves

It was shown in Chapter 3 that each arc of any conic section can be written as a parametric quadratic rational. We saw in Section 2.5 that the form of that parametric rational was just a parametric quadratic rational using the Bernstein basis, i.e., a rational form of the Bézier curve. Versprille [83] first proposed using a rational form of the parametric B-spline curve for design.

Definition 8.7. *Let* $\mathbf{t} = \{t_0, \ldots, t_{n+\kappa+1}\}$, $t_i \le t_{i+1}$, *and* $t_i < t_{i+\kappa+1}$, *for all appropriate* i, *and let* $\mathcal{B}_{0,\kappa-1}(t)$, $\ldots$, $\mathcal{B}_{n,\kappa-1}(t)$ *be the* $n + 1$ *B-splines* *defined over that knot vector. For a sequence of coefficients $\{P_0, \ldots, P_n\}$, which can be scalar or vector and a sequence of scalars, $\{w_0, \ldots, w_n\}$, the curve*

$$\gamma(t) = \frac{\sum_{i=0}^{n} w_i P_i \mathcal{B}_{i,\kappa}(t)}{\sum_{i=0}^{n} w_i \mathcal{B}_{i,\kappa}(t)}$$

is called a rational B-spline curve *of degree κ. The w_i's are sometimes called the* homogeneous coordinates *for the P_i's since there is one associated with each P_i. The determining parameters of the rational B-spline curve are the knot vector and the* homogeneous point *$H_i = (h_{x,i}, h_{y,i}, h_{z,i}, h_{w,i})$ where if $P_i = (x_i, y_i, z_i)$, $h_{x,i} = w_i x_i$, $h_{y,i} = w_i y_i$, $h_{z,i} = w_i z_i$, and $h_{w,i} = w_i$.*

When Versprille first advocated their use, he also argued that the uniform floating form was sufficient for design purposes. Since then, for the same reasons as for regular B-spline curves, it is clear that this is not true. At the same time, the same methods now useful for nonuniform B-spline curves are also useful for nonuniform rational B-spline curves.

We have already seen that for the quadratic with $n = 2$ and no internal knots, as long as the ratio $w_0 w_2 / w_1^2$ is constant, the conic section defined by it is also constant. The particular values of w_0, w_1 and w_2 are not important theoretically, although they affect the computational stability. An analogous result holds for the general quadratic spline, but here the relative spacing of the knots also enters into the ratio. Also, in this case, ratios of sequential triples of w_i's will determine the resulting conic. Hence if all the w_i's are constant, then the curve is the quadratic B-spline, with denominator a constant. The following figures show rational B-splines of different degree and different w_i's.

The preservation of the shape of the conic section as long as the ratio $w_0 w_2 / w_1^2$ is constant has an extension to arbitrary degree rational functions. Consider the rational Bézier curve,

$$\gamma(t) = \frac{\sum_{i=0}^{n} w_i P_i \mathcal{B}_i(t)}{\sum_{i=0}^{n} w_i \mathcal{B}_i(t)},$$

and let $t(r) = \frac{ra}{1+(a-1)r}$ be a linear rational change of parameter. Then,

$$\begin{aligned} \gamma(t(r)) &= \frac{\sum_{i=0}^{n} w_i P_i \mathcal{B}_i(t(r))}{\sum_{i=0}^{n} w_i \mathcal{B}_i(t(r))} \\ &= \frac{\sum_{i=0}^{n} w_i P_i \mathcal{B}_i\left(\frac{ra}{1+(a-1)r}\right)}{\sum_{i=0}^{n} w_i \mathcal{B}_i\left(\frac{ra}{1+(a-1)r}\right)} \end{aligned}$$

$$
\begin{aligned}
&= \frac{\sum_{i=0}^{n} w_i P_i \binom{n}{i} \left(\frac{ra}{1+(a-1)r}\right)^i \left(1 - \frac{ra}{1+(a-1)r}\right)^{n-i}}{\sum_{i=0}^{n} w_i \binom{n}{i} \left(\frac{ra}{1+(a-1)r}\right)^i \left(1 - \frac{ra}{1+(a-1)r}\right)^{n-i}} \\
&= \frac{\sum_{i=0}^{n} w_i P_i \binom{n}{i} \left(\frac{ra}{1+(a-1)r}\right)^i \left(\frac{1-r}{1+(a-1)r}\right)^{n-i}}{\sum_{i=0}^{n} w_i \binom{n}{i} \left(\frac{ra}{1+(a-1)r}\right)^i \left(\frac{1-r}{1+(a-1)r}\right)^{n-i}} \\
&= \frac{\frac{\sum_{i=0}^{n} w_i P_i \binom{n}{i} (ra)^i (1-r)^{n-i}}{(1+(a-1)r)^n}}{\frac{\sum_{i=0}^{n} w_i \binom{n}{i} (ra)^i (1-r)^{n-i}}{(1+(a-1)r)^n}} \\
&= \frac{\sum_{i=0}^{n} w_i a^i P_i \mathcal{B}_i(r)}{\sum_{i=0}^{n} w_i a^i \mathcal{B}_i(r)}.
\end{aligned}
$$

The change of parameter of $t(r) = \frac{ra}{1+(a-1)r}$ is also known as *Moebius transformation*. Further, a piecewise knot vector dependent on this transformation to B-spline functions is known. See [51].

As can be seen by Figure 8.14–Figure 8.16, and as was shown by Versprille, the homogenous coordinates can act in some sense as *tension* parameters, pulling the curve towards the vertices of the control polygon or repelling them. When some of the w_i's are positive and some are negative, the denominator might take a zero value. It corresponds to the point at infinity, but is not practical to use since it leads to a loss of geometric intuition and special case computations. While the behavior of the rational quadratic spline is well understood locally, and less well understood as a whole, the behavior of rational B-spline curves of higher degree is not well analyzed. Whether there is a corresponding relationship like the constant ratios in the case of the quadratic is not known. However, since the ratio-

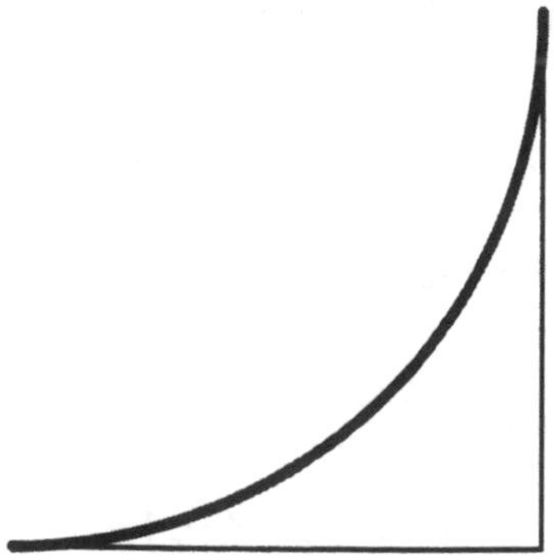

Figure 8.14. A rational quadratic spline, no internal knots.

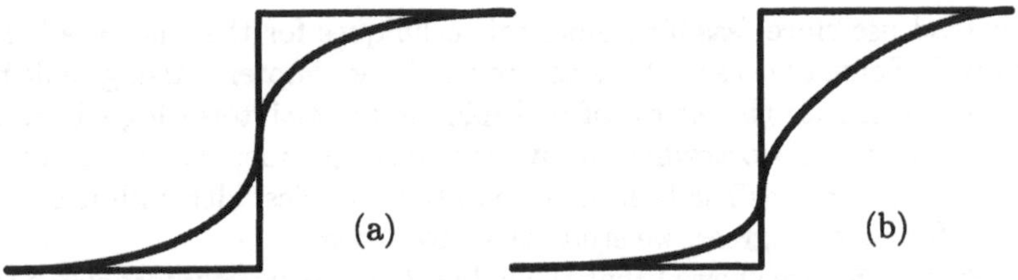

Figure 8.15. (a) A rational quadratic spline; (b) different homogeneous coordinates.

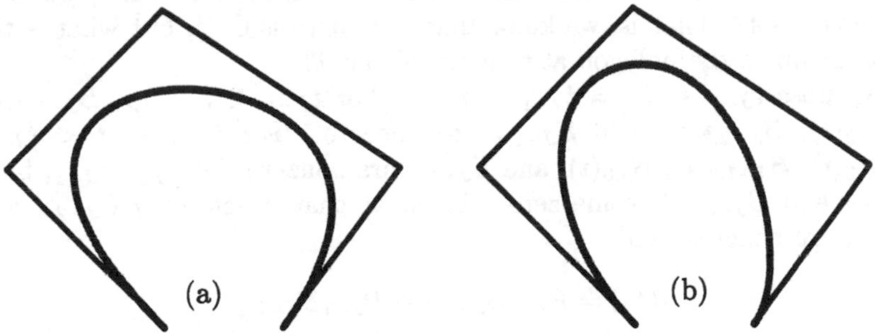

Figure 8.16. (a) A rational cubic spline; (b) different homogeneous coordinates.

nal B-spline curves form a generalization of B-spline curves, and include all conic sections (for degree greater than one), they are becoming very widely used.

8.4 Effects of Multiple Knots vs. Multiple Vertices

As we noted, one can change the curve by modifying the polygon or modifying the knot vector. The nonuniform B-splines are not all translates of each other which means that each must be calculated. While the algorithms for doing so are stable they are not necessarily fast. When B-splines first started to be used in computer aided geometric design [41, 71], many decided that the benefits of using the nonuniform open B-spline were not worth the computational costs. Further, if the uniform floating B-splines were used, all B-splines were just translations of one generic function, and could be computed and tabled once. Also, simple matrix methods could be used to evaluate points along the whole curve [72]. Hence, attempts were made to simulate the effects of multiple knot conditions by using multiple vertices [83, 84]. More recently the need for real nonuniform splines allowing multiple interior knots has been recognized. They are starting to have widespread use since feasible numerical techniques for their use are being developed. Some of these will be covered in later chapters. Using multiple vertices to simulate the effects of multiple knots has interesting effects on the curve, and it is worthwhile to investigate the geometry of their behavior. Since this process has been used mainly for cubics, although it can be adapted for other degrees, we study the cubic case.

In Section 8.1 we showed that the cubic B-spline curve interpolates the appropriate control polygon vertex whenever there is a knot with multiplicity three, and it is tangent to the sides of the polygon leaving that vertex. A knot with multiplicity four allows discontinuities. What happens with a double knot? That is, we know that the curve is $C^{(1)}$, but what is the relationship to the polygon at that double knot?

Suppose $t_{J-1} < t_J = t_{J+1} < t_{J+2}$. For $t_{J-1} < t < t_J$, $\mathcal{B}_{J-4,3}(t)$, $\mathcal{B}_{J-3,3}(t)$, $\mathcal{B}_{J-2,3}(t)$, and $\mathcal{B}_{J-1,3}$ are nonzero. For $t_{J+1} < t < t_{J+2}$, $\mathcal{B}_{J-2,3}(t)$, $\mathcal{B}_{J-1,3}(t)$, $\mathcal{B}_{J,3}(t)$, and $\mathcal{B}_{J+1,3}$ are nonzero. At $t_J = t_{J+1}$, both $\mathcal{B}_{J-4,3}$ and $\mathcal{B}_{J-3,3}$ become zero. Thus, at that point, only $\mathcal{B}_{J-2,3}$ and $\mathcal{B}_{J-1,3}$ are nonzero, and

$$\gamma(t_J) = P_{J-2}\mathcal{B}_{J-2,3} + P_{J-1}\mathcal{B}_{J-1,3}.$$

So $\gamma(t_J)$ lies on the polygon between P_{J-2} and P_{J-1}. Hence, a double knot yields the B-spline curve interpolates to a point on the side of the polygon. The exact point along that side is determined by the spacing of the knots. We know that the sum is a convex combination since each function is nonnegative and

$$\mathcal{B}_{J-2,3}(t_J) + \mathcal{B}_{J-1,3}(t_J) = 1. \tag{8.1}$$

Since the B-splines are all $C^{(1)}$ or $C^{(2)}$ at $t_J = t_{J+1}$,

$$\mathcal{B}'_{J,3}(t_J) = \mathcal{B}'_{J+1,3}(t_J) = 0,$$

and differentiating Equation 8.1, we get

$$\mathcal{B}'_{J-2,3}(t_J) + \mathcal{B}'_{J-1,3}(t_J) = 0.$$

Thus

$$\begin{aligned}
\gamma'(t) &= P_{J-2}\mathcal{B}'_{J-2,3}(t_J) + P_{J-1}\mathcal{B}'_{J-1,3}(t_J) \\
&= P_{J-2}\mathcal{B}'_{J-2,3}(t_J) + P_{J-1}\left(1 - \mathcal{B}'_{J-2,3}(t_J)\right) \\
&= (P_{J-2} - P_{J-1})\mathcal{B}'_{J-2,3}(t_J) + P_{J-1}.
\end{aligned}$$

We see that the tangent vector is a scalar in the direction from $P_{J-2} - P_{J-1}$. Thus, in cubics, a double knot means tangent continuity but second

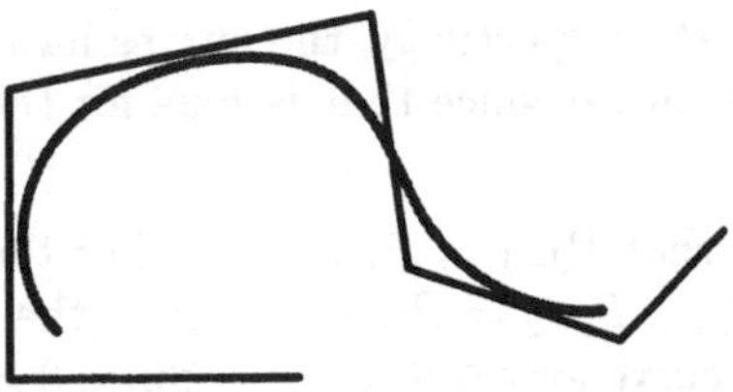

Figure 8.17. A typical floating uniform B-spline curve.

derivative discontinuity. We have already seen that it can still be curvature continuous.

Having reviewed multiple knot possibilities, we should now approach multiple vertices and uniform knots. The key to this whole area is the property that if $t_J \leq t \leq t_{J+1}$, then $\gamma(t) = \sum_{i=J-3}^{J} P_i \mathcal{B}_{i,3}(t)$. Four points define the curve, but, in this floating case, the curve does not interpolate any of the four, except possibly by the degeneracy conditions we describe. Finally, the curve lies in the convex hull of the four points. Call this curve segment $\gamma_J(t)$. The curve at $t = t_J$ is parametrically $C^{(2)}$, which means that $\gamma_{J-1}^{(j)}(t_J) = \gamma_J^{(j)}(t_J)$, $j = 0,\ 1,\ 2$. The third derivative is then not well defined at the knots. Suppose now that $P_{J-1} = P_{J-2} = P_{J-3}$. For $t \in [t_J, t_{J+1}]$, $\gamma(t)$ lies in the convex hull of P_{J-3}, P_{J-2}, P_{J-1}, and P_J, namely along the straight line between P_{J-1} and P_J. Analogously, for $t \in [t_{J-1}, t_J]$ the curve lies in the convex hull of P_{J-4}, $P_{J-3} = P_{J-2} = P_{J-1}$, which is the straight line connecting P_{J-4} and P_{J-1}.

The curve value at t_J lies in the intersection of those two convex hulls, namely the point P_{J-1} (really, the multiple vertex).

There are several distinctive properties caused by inserting a triple vertex in the cubic uniform floating curve:

1. The curve interpolates the triple polygon vertex.

2. The curve has a linear segment extending part way along the line from P_{J-1} to P_J.

3. The curve has a linear segment starting part way in the line from P_{J-4} to P_{J-1}.

4. There is a corner in the curve at t_J corresponding to the point P_{J-1}. Clearly, this representation is *not* a regular representation (Definition 4.2) for the B-spline curve.

5. The linear segments are joined to the nonlinear portions with parametric $C^{(2)}$ continuity. At t_{J-1} and t_{J+1} along the segments P_{J-4},

P_{J-1} and P_{J-1}, P_J, respectively, the curves have zero second derivative and zero curvature, since that is true for the respective straight lines.

Now suppose only that $P_{J-1} = P_{J-2}$, i.e., that the multiplicity of the vertex is two. Then P_{J-3}, $P_{J-2} = P_{J-1}$, P_J forms the control polygon and the convex hull of the curve for $t \in [t_J, t_{J+1}]$. Since $0 = \mathcal{B}_{J,3}(t_J)$, $\gamma(t_J)$ lies in the convex hull of $\{P_{J-3}, P_{J-2}, P_{J-1}\}$, that is, along the straight line between P_{J-3} and P_{J-2}. Consider $\gamma'(t)$ near t_J. It exists geometrically if and only if $\gamma'_{J-1}(t_J) = \gamma'_J(t_J)$. But,

$$\begin{aligned}
\gamma'_J(t_J) &= \sum_{i=0}^{2} Q_{J-i}\mathcal{B}_{J-i,2}(t_J) \\
&= Q_{J-1}\mathcal{B}_{J-1,2}(t_J) + Q_{J-2}\mathcal{B}_{J-2,2}(t_J) \\
&= Q_{J-2}\mathcal{B}_{J-2,2}(t_J), \qquad \text{since } Q_{J-1} = 0,
\end{aligned}$$

and

$$\begin{aligned}
\gamma'_{J-1}(t_J) &= \sum_{i=0}^{2} Q_{J-1-i}\mathcal{B}_{J-1-i,2}(t_J) \\
&= Q_{J-1}\mathcal{B}_{J-1,2}(t_J) + Q_{J-2}\mathcal{B}_{J-2,2}(t_J) \\
&= Q_{J-2}\mathcal{B}_{J-2,2}.
\end{aligned}$$

Finally,

$$Q_{J-2} = \frac{3(P_{J-2} - P_{J-3})}{t_{J-2+3} - t_{J-2}},$$

so the curve $\gamma(t)$ interpolates the side of the polygon between P_{J-2} and P_{J-3}, and is also tangent to the polygon there.

Now, the points $P_{J-2} = P_{J-1}$, P_J, P_{J+1} form the control polygon for the curve over $t \in [t_{J+1}, t_{J+2}]$. Again we know that the value for the curve at t_{J+1} occurs on the line segment from P_{J-1} to P_J. We may show in an analogous fashion that the curve is also tangent to the polygon at that point, having a tangent in the direction of $P_J - P_{J-1}$.

Hence we see the effect of a double vertex $P_{J-2} = P_{J-1}$:

1. The curve interpolates the two sides of the polygon around the double vertex, the precise location depending on knot spacing.

2. The curve is tangent to the sides of the polygon at those two points of interpolation.

Several sequences below show first the effect of coalescing the knots and then coalescing the vertices at one place from the floating original curve above.

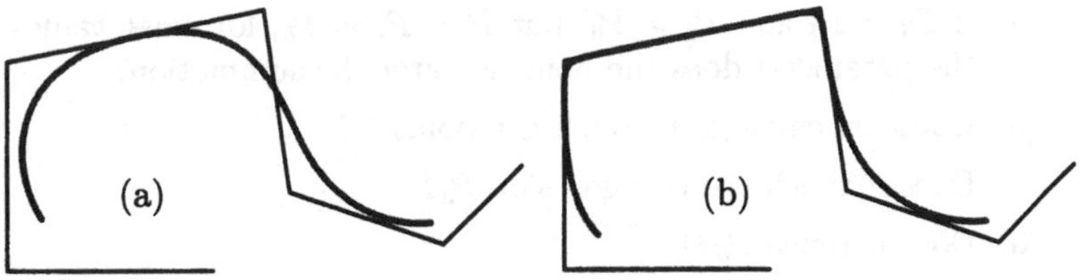

Figure 8.18. (a) A double value in the knot vector; (b) two double vertices.

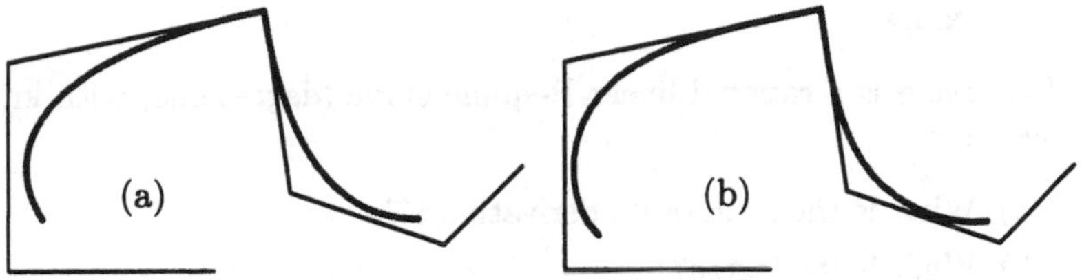

Figure 8.19. (a) A triple value in the knot vector; (b) a triple vertex.

Exercises

1. If a curve γ is to have a Frenet frame, it must have a continuous second derivative, and a regular representation (that is, $\gamma'(t) \neq 0$ for all t). This prerequisite is frequently not met for spline curves at the knots. In addition, computation of torsion assumes a continuous *third* derivative, which a cubic spline curve hardly ever has at the knots. Discuss for a general spline curve what meaning can be assigned to the Frenet frame geometry at and around the knot. What approach might you consider taking at these discontinuities?

2. Find the matrix representation for the cubic Bézier curve, i.e., find C when
$$\begin{aligned}\gamma(t) &= \left[\ \theta_{0,3}(t)\ \ \theta_{1,3}(t)\ \ \theta_{2,3}(t)\ \ \theta_{3,3}(t)\ \right]\left[\ P_0\ \ P_1\ \ P_2\ \ P_3\ \right]^T \\ &= \left[\ t^3\ \ t^2\ \ t\ \ 1\ \right] C \left[\ P_0\ \ P_1\ \ P_2\ \ P_3\ \right]^T.\end{aligned}$$

3. Find the analogous representation for the degree n Bézier curve.

4. Suppose that $\gamma(t) = \sum_{i=0}^{n} P_i \mathcal{B}_{i,3}(t)$ where the B-splines are defined over the knot vector $\boldsymbol{t} = \{-3, -2, -1, 0, 1, 2, 3, \ldots, n-3, n-2, n-1, n, n+1\}$. (Remember $t_i = i - 3$.) So γ is a uniform, floating, cubic B-spline curve.

(a) If $P_3 \neq P_4$ and $P_5 \neq P_6$, but $P = P_4 = P_5$, for what values of the parameter does the point P enter the summation?

(b) Does the curve go through the point P?

(c) Does it touch the polygon side P_3P_4?

(d) Does it touch P_5P_6?

(e) If it touches anywhere, where is it tangent there?

(f) If it touches anywhere, find the *exact* locations and parameter values.

5. Suppose γ is a rational linear B-spline curve (degree one) with knot vector τ.

   (a) What is the form of its derivative γ'?

   (b) What is its degree?

   (c) If γ is an *open* B-spline with simple internal knots, what is its continuity class?

   (d) If γ is as in 5(c),what is the knot vector for γ'? Can you find a formula for γ'?

   (e) Suppose

$$\gamma(t) = \frac{\sum_{i=0}^{6} w_i P_i \mathcal{B}_{i,1}(t)}{\sum_{i=0}^{6} w_i \mathcal{B}_{i,1}(t)},$$

   where $P_i = (x_i, y_i, z_i)$ and $\tau = \{0, 0, 1, 2, 3, 4, 5, 6, 6\}$. Find the knot vector for γ', and then find γ'.

6. Suppose γ is a rational B-spline curve of degree κ with simple internal knots . What is the form of its derivative γ'? In general, is it a rational B-spline? If so, what is its knot vector?

7. Investigate the different effects to the moving trihedron, the curvature, and the torsion when multiple vertices or multiple knots are used.

9

Data Fitting with B-Splines

Once the general theory and the general algorithms are known, they must be interpreted and made practical for use in design. While the possible spaces of curves are very general, there is no theoretical method for determining which ones are feasible in particular design situations. In this chapter approaches are presented for fitting to a pre-existing primitive — either data or function. We present interpolation techniques, including complete spline interpolation, nodal interpolation, and piecewise Hermite interpolation, all with the B-spline basis. We also include approximation techniques, such as quasi-interpolation, least squares, and the Schoenberg variation diminishing approximation. We touch on B-spline wavelets in the section on multiresolution curve decomposition and editing.

9.1 Interpolation with B-Splines

One widely used form of data fitting is interpolation. Sometimes the problem is to interpolate to positional data alone, and sometimes it is necessary to interpolate to derivatives as well as position, at least at some of the points. We can state the general problem over a spline space $\mathcal{S}_{\kappa,\mathcal{T}}$. First, however, consider a collection of $s+1$ data pairs, $\{(u_i, p_i)\}_{i=0}^{s}$. The position interpolation problem in the space $\mathcal{S}_{\kappa,\mathcal{T}}$ is to find values for coefficients c_j, such that

$$p_i = \sum_{j=0}^{n} c_j \mathcal{B}_{j,\kappa,\mathcal{T}}(u_i), \qquad i = 0, \ldots, s.$$

In general, suppose that at u_i, it is necessary to interpolate the value p_i and its first r_i derivatives. That is, the interpolation problem is to find coefficients c_j such that

$$p_i^{(q)} = \sum_{j=0}^{n} c_j \mathcal{B}_{j,\kappa,\mathcal{T}}^{(q)}(u_i), \qquad q = 0, \ldots, r_i, i = 0, \ldots, s;$$

p_i^q means this data value represents the $q-th$ derivative evaluated at u_i, and the 0th derivative is just the function value. In this general case there are $s + 1 + \sum_{i=0}^{s} r_i$ interpolation conditions.

Below we develop the system of equations for solving the cubic interpolation problems using the B-spline basis. Afterwards we shall define a kth order interpolation problem and develop the system of equations for solving it using the B-spline curve representation.

9.1.1 C^2 Cubic Interpolation at Knots

In this section we develop formulas for using nonuniform open cubic B-splines to solve the problem of interpolating a collection of points $\{(u_i, p_i)\}_{i=0}^{s}$ with a $C^{(2)}$piecewise cubic polynomial curve, with internal knot values defined at the abscissas of the data values.

To satisfy these conditions, the domain of the interpolant curve must be the interval $[u_0, u_s]$. Since cubic splines are being used, the specification that the curve be $C^{(2)}$ at the joins requires that the interior knots have multiplicity one, or be *simple knots*, occurring at u_i, $i = 1, \ldots, s-1$. These requirements lead to a unique knot vector $\boldsymbol{t} = \{t_j\}$ where $t_j = u_0$, $j = 0$, 1, 2, 3; $t_j = u_{j-3}$, $j = 4, \ldots, s+2$; and $t_j = u_s$, $j = s+3, \ldots, s+6$. Label the resulting cubic B-splines $\{\mathcal{B}_{j,3}\}$. Then, there are $s+3$ distinct B-splines defined with this knot set. Since there are $s+1$ points to be interpolated, this leaves two additional degrees of freedom.

While there are many ways to constrain the two additional degrees of freedom, several are used most frequently.

1. Sometimes, the two extra degrees are constrained by requiring that the second derivative of the interpolant evaluate to zero at the endpoints, that is, requiring that at u_0 and u_n, $p_0^{(2)} = 0$ and $p_s^{(2)} = 0$, respectively, be interpolated. This type of interpolant is called a *natural spline interpolant.*

2. Other times, no new data is interpolated, but the constraints are specified by requiring $C^{(3)}$ continuity of the interpolant at u_1 and u_{s-1}.

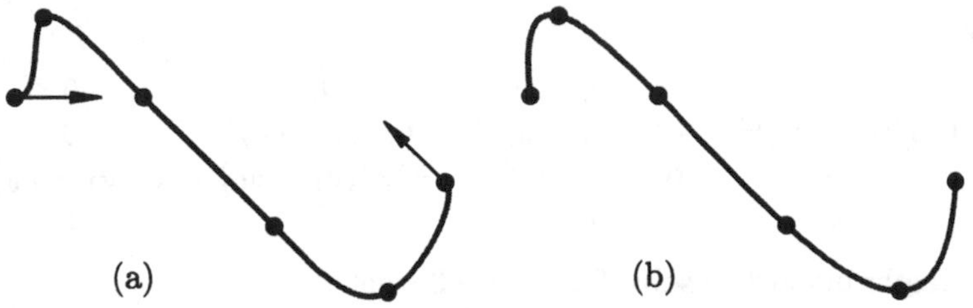

Figure 9.1. (a) Complete spline interpolation; (b) *not a knot* spline interpolation.

These conditions are equivalent to specifying a single cubic polynomial for the first three datum and the last three datum, and then requiring the $C^{(2)}$ continuity at all other internal knots. Sometimes this is referred to as the *not a knot* condition.

3. Finally, and most frequently, the two extra degrees are constrained by requiring that the interpolant satisfy interpolating conditions at the two interval endpoints, that is, requiring that the tangents at u_0 and u_n, p_0' and p_s', respectively, be interpolated. That interpolant is called the *complete cubic spline interpolant.*

Figure 9.1 shows a data set and shows the last two types of spline interpolants.

Example 9.1. We consider each of the types of interpolation for the two point problem set, $\{(u_0, p_0), (u_1, p_1)\}$. In all cases the curve domain is the interval $[u_0, u_1]$, so the knot vector is, for any $b \geq u_1$, $\boldsymbol{t} = \{u_0, u_0, u_0, u_0, u_1, u_1, u_1, b\}$. The interpolant has the form $\gamma(t) = \sum_{i=0}^{3} C_i \mathcal{B}_{i,3,\boldsymbol{t}}(t)$, with $C_i, i = 0, \dots, 3$ unknown.

Solving for the natural spline interpolant adds constraints: $\gamma''(u_0) = 0$ and $\gamma''(u_1) = 0$. All constraints can be written,

$$\begin{array}{rcccc} p_0 & = & \gamma(u_0) & = & C_0 \\ 0 & = & \gamma''(u_0) & = & 6\,(C_2 - 2C_1 + C_0)\,/(u_1 - u_0)^2 \\ 0 & = & \gamma''(u_0) & = & 6\,(C_3 - 2C_2 + C_1)\,/(u_1 - u_0)^2 \\ p_1 & = & \gamma(u_1) & = & C_3. \end{array}$$

Written in matrix notation,

$$\boldsymbol{N}\begin{bmatrix} C_0 \\ C_1 \\ C_2 \\ C_3 \end{bmatrix} = \begin{bmatrix} p_0 \\ 0 \\ 0 \\ p_1 \end{bmatrix},$$

where

$$N = \begin{bmatrix} 1 & 0 & 0 & 0 \\ 6/(u_1 - u_0)^2 & -12/(u_1 - u_0)^2 & 6/(u_1 - u_0)^2 & 0 \\ 0 & 6/(u_1 - u_0)^2 & -12/(u_1 - u_0)^2 & 6/(u_1 - u_0)^2 \\ 0 & 0 & 0 & 1 \end{bmatrix}.$$

It is straightforward to solve for the coefficients.

The "not a knot" case cannot be solved uniquely for the two point data set. Since there are no internal knots, there are no degrees of freedom to remove.

We proceed to consider complete cubic spline interpolation for the two point data set. Designate by p_0' and p_1' the tangents at u_0 and u_1, respectively. Complete cubic spline interpolating constraints yield

$$\begin{aligned} p_0 &= \gamma(u_0) &=& \quad C_0 \\ p_0' &= \gamma'(u_0) &=& \quad 3\,(C_1 - C_0)\,/(u_1 - u_0) \\ p_1' &= \gamma'(u_1) &=& \quad 3\,(C_3 - C_2)\,/(u_1 - u_0) \\ p_1 &= \gamma(u_1) &=& \quad C_3. \end{aligned}$$

Writing in matrix notation,

$$B \begin{bmatrix} C_0 \\ C_1 \\ C_2 \\ C_3 \end{bmatrix} = \begin{bmatrix} p_0 \\ p_0' \\ p_1' \\ p_1 \end{bmatrix},$$

where

$$B = \begin{bmatrix} 1 & 0 & 0 & 0 \\ -3/(u_1 - u_0) & 3/(u_1 - u_0) & 0 & 0 \\ 0 & 0 & -3/(u_1 - u_0) & 3/(u_1 - u_0) \\ 0 & 0 & 0 & 1 \end{bmatrix}. \tag{9.1}$$

Since

$$B^{-1} = \begin{bmatrix} 1 & 0 & 0 & 0 \\ 1 & (u_1 - u_0)/3 & 0 & 0 \\ 0 & 0 & -(u_1 - u_0)/3 & 1 \\ 0 & 0 & 0 & 1 \end{bmatrix}, \tag{9.2}$$

it is easily seen that

$$\begin{aligned} C_0 &= p_0, \\ C_1 &= p_0 + (u_1 - u_0)p_0'/3, \\ C_2 &= p_1 - (u_1 - u_0)p_1'/3, \qquad \text{and} \\ C_3 &= p_1, \end{aligned}$$

which solves the complete cubic spline interpolation problem for two points. This two point problem is also called *Hermite interpolation.* □

We proceed by finding the general solution to the problem of complete cubic spline interpolation. Letting $n = s + 2$, we seek the interpolant, $A(u) = \sum_{i=0}^{n} P_i\mathcal{B}_{i,3}(u)$ such that $p_j = A(u_j) = \sum_{i=0}^{n} P_i\mathcal{B}_{i,3}(u_j)$, $j = 0, \ldots, s$ and $p'_j = A'(u_j) = \sum_{i=0}^{n} P_i\mathcal{B}'_{i,3}(u_j)$, $j = 0$, $j = s$. Because open end conditions are specified, that means that $P_0 = p_0$ and $P_n = p_s$. Further, at $u_0 = t_0 = \cdots = t_3$ only $\mathcal{B}_{0,3}$ and $\mathcal{B}_{1,3}$ have nonzero derivatives. Since the sum of the derivatives of all B-splines at any point is zero, $\mathcal{B}'_{0,3}(u_0) = -\mathcal{B}'_{1,3}(u_0)$. Similarly, $\mathcal{B}'_{n-1,3}(u_s) = -\mathcal{B}'_{n,3}(u_s)$.

Finally, we analyze the functions to determine which might be nonzero at u_i, $i = 1, \ldots, s-1$. Since $t_{i+3} = u_i \le u_i < t_{i+4}$, only $\mathcal{B}_{i,3}$, $\mathcal{B}_{i+1,3}$, $\mathcal{B}_{i+2,3}$, and $\mathcal{B}_{i+3,3}$ might be nonzero. But on the interval $[t_{i+2}, t_{i+3})$, only $\mathcal{B}_{i-1,3}$, $\mathcal{B}_{i,3}$, $\mathcal{B}_{i+1,3}$, and $\mathcal{B}_{i+2,3}$ can be nonzero. Now, $\mathcal{B}_{i-1,3}$ and $\mathcal{B}_{i+3,3}$ are both $C^{(2)}$ at t_{i+3} and since each is the zero function on one side of it and nonzero on the other, $\mathcal{B}_{i-1,3}(u_i) = 0$, and $\mathcal{B}_{i+3,3}(u_i) = 0$, so $p_i = \sum_{j=i}^{i+2} P_j\mathcal{B}_{i,3}(u_i)$.

Applying all of these results to the interpolation problem yields the following system.

$$\boldsymbol{B}\begin{bmatrix} P_0 \\ P_1 \\ P_2 \\ \vdots \\ P_{N-2} \\ P_{n-1} \\ P_n \end{bmatrix} = \begin{bmatrix} p_0 \\ p'_0 \\ p_1 \\ \vdots \\ p_{s-1} \\ p'_s \\ p_s \end{bmatrix}. \tag{9.3}$$

Without any ambiguity we drop the degree subscript and expand the $\boldsymbol{B}$ matrix. Call $\beta_{i,j} = \mathcal{B}_{i,3}(u_j)$. Then,

$$\begin{bmatrix} 1 & 0 & 0 & 0 & \ldots & 0 & 0 & 0 & 0 \\ \mathcal{B}'_0(u_0) & \mathcal{B}'_1(u_0) & 0 & 0 & \ldots & 0 & 0 & 0 & 0 \\ 0 & \beta_{1,1} & \beta_{2,1} & \beta_{3,1} & \ldots & 0 & 0 & 0 & 0 \\ & & & \ddots & \ddots & \ddots & & & \\ 0 & 0 & 0 & 0 & \ldots & \beta_{n-3,s-1} & \beta_{n-2,s-1} & \beta_{n-1,s-1} & 0 \\ 0 & 0 & 0 & 0 & \ldots & 0 & 0 & \mathcal{B}'_{n-1}(u_s) & \mathcal{B}'_n(u_s) \\ 0 & 0 & 0 & 0 & \ldots & 0 & 0 & 0 & 1 \end{bmatrix}. \tag{9.4}$$

The solution of this $(n+1)\times(n+1)$ system of equations gives the coefficients of the B-splines. While this matrix can get large, it is a tridiagonal matrix for which there are many special efficient, fast, numerical techniques. In the special case where the knots are uniformly spaced, the matrix has the form

$$\begin{bmatrix} 1 & 0 & 0 & 0 & 0 & \dots & 0 & 0 & 0 & 0 & 0 \\ -a & a & 0 & 0 & 0 & \dots & 0 & 0 & 0 & 0p & 0 \\ 0 & 1/4 & 7/12 & c & 0 & \dots & 0 & 0 & 0 & 0 & 0 \\ 0 & 0 & c & d & c & \dots & 0 & 0 & 0 & 0 & 0 \\ & & & & \ddots & \ddots & \ddots & & & & \\ 0 & 0 & 0 & 0 & 0 & \dots & c & d & c & 0 & 0 \\ 0 & 0 & 0 & 0 & 0 & \dots & 0 & c & 7/12 & 1/4 & 0 \\ 0 & 0 & 0 & 0 & 0 & \dots & 0 & 0 & 0 & -a & a \\ 0 & 0 & 0 & 0 & 0 & \dots & 0 & 0 & 0 & 0 & 1 \end{bmatrix},$$

where $a = .3$, $c = 1/6$ and $d = 2/3$.

The continuity conditions are carried in the basis functions, so the user need not solve for them directly, and in fact never knows the values of the tangents at the knots unless he wishes to evaluate them.

To use this method for parametric interpolation requires that the system be solved for the coefficients in each of the three variable directions independently.

9.1.2 Higher-Order Complete Interpolation with B-Splines

There is a higher order analogy to complete cubic interpolation. We shall determine the necessary order for the type of constraints required. The data to be interpolated is:

$$\begin{aligned} &(u_i, p_i), & i &= 0, \dots, s, & &\text{where } u_j < u_{j+1}; \\ &\left(u_0, p_0^{(j)}\right), & j &= 1, \dots, m-1 & &\text{and} \\ &\left(u_s, p_s^{(j)}\right), & j &= 1, \dots, m-1. & & \end{aligned}$$

It is desired to first find κ, the lowest degree with which it might be possible to find a unique piecewise polynomial in $C^{(\kappa-1)}$ which interpolates this generalized data.

Since it is necessary to have the curve defined over $[u_0, u_s]$, $t_\kappa = u_0$. Having interior continuity in $C^{(\kappa-1)}$ means that u_i, $i = 1, \dots, s-1$, must all be simple knots, so $t_{\kappa+i} = u_i$, $i = 1, \dots, s-1$, and $t_{\kappa+s} = u_s$. Thus the spline space must have $n = \kappa + s$ degrees of freedom (dimension). Now the data specifies $(s+1) + 2(m-1)$ constraints, and hence one must have a space with dimension exactly $(s+1)+2(m-1) = s-1+2m$ in order to try to solve this problem uniquely. Thus, if κ is set equal to $2m-1$, then the dimensions of the problem are equal. Whether in fact the interpolation conditions are all independent over this space must be shown.

We now are trying to find an element in $\mathcal{S}_{\kappa,\boldsymbol{t}}$ which satisfies the complete interpolation conditions, where $\kappa = 2m-1$, $n = 2m+s-2$, and $\boldsymbol{t}$ has $t_i = u_0$, $i = 0, \ldots, \kappa$, $t_{\kappa+i} = u_i$, $i = 1, \ldots, s-1$, and $t_{\kappa+s+i} = u_s$, $i = 0, \ldots, \kappa$. These conditions make the spline have open end conditions. To that end, define

$$L_{i,j} = \begin{cases} \mathcal{B}^{(i)}_{j,\kappa}(u_0), & i = 0, \ldots, m-1, \;\; j = 0, \ldots, n \\ \mathcal{B}_{j,\kappa}(u_{i-(m-1)}), & i = m, \ldots, m+s-2, \;\; j = 0, \ldots, n \\ \mathcal{B}^{((2m+s-2-i))}_{j,\kappa}(u_s), & i = m+s-1, \ldots, 2m+s-2, \;\; j = 0, \ldots, n. \end{cases}$$

Set $\boldsymbol{L} = \left[\; L_{i,j} \;\right]$ and

$$\boldsymbol{D} = \begin{bmatrix} p_0 \\ p_0^{(1)} \\ \vdots \\ p_0^{(m-1)} \\ p_1 \\ \vdots \\ p_{i-m+1} \\ \vdots \\ p_{s-1} \\ p_s^{(m-1)} \\ \vdots \\ p_s^{(1)} \\ p_s \end{bmatrix}; \qquad \boldsymbol{P} = \begin{bmatrix} P_0 \\ P_1 \\ \vdots \\ P_{n-1} \\ P_n \end{bmatrix},$$

where $\boldsymbol{L}$ is an $(n+1)\times(n+1)$ matrix, $\boldsymbol{D}$ is an $(n+1)\times 1$ matrix, and $\boldsymbol{P}$ is an $(n+1)\times 1$ matrix. Solving the interpolation problem then amounts to solving the linear system

$$\boldsymbol{LP} = \boldsymbol{D},$$

for the vector $\boldsymbol{P}$.

$L_{0,j} = \delta_{0,j}$, since there is only one nonzero B-spline at u_0. For each i between 0 and $m-1$, there are only $(i+1)$ different B-splines with nonzero i^{th} derivatives at u_0, $\mathcal{B}^{(i)}_{j,\kappa}(u_0)$, $j = 0, \ldots, i$. Finally, the elements of each of the ith rows of the matrix, $i = 1, \ldots, m-1$, and $i = n-m+1, \ldots, n-1$, representing the B-spline values of the ith derivative at u_0 and the $(m-i)^{th}$ derivative at u_s, respectively, sum to zero since $\mathcal{B}_{j,\kappa}(u) \equiv 1$. The first m rows of the matrix have nonzero elements only on and below the main diagonal, with the following form:

$$\begin{array}{cccccc}
1 & 0 & 0 & \dots & 0 & \underbrace{0 \ \dots \ 0}_{n-m \text{ times}} \\
\mathcal{B}^{(1)}_{0,\kappa}(u_0) & \mathcal{B}^{(1)}_{1,\kappa}(u_0) & 0 & \dots & 0 & \underbrace{0 \ \dots \ 0}_{n-m \text{ times}} \\
\vdots & \vdots & \vdots & \dots & \vdots & \vdots \\
\mathcal{B}^{(m-1)}_{0,\kappa}(u_0) & \mathcal{B}^{(m-1)}_{1,\kappa}(u_0) & & \dots & \mathcal{B}^{(m-1)}_{m-1,\kappa}(u_0) & \underbrace{0 \ \dots \ 0}_{n-m \text{ times}}
\end{array}$$

Now for the $(m-1)^{st}$ to $(n-m+1)^{st}$ rows, $L_{i,j} = \mathcal{B}_{j,\kappa}(u_{i-m+1}) = \mathcal{B}_{j,\kappa}(t_{\kappa+(i-m+1)}) = \mathcal{B}_{j,\kappa}(t_{m+i})$, which is nonzero only for $j = (m+i) - \kappa$, $\dots$, $m+i-1$. Rewriting shows that $j = i-(m-1)$, $\dots$, $i+m-1$. The ith row of the matrix has only those elements nonzero in columns q for which $\mathcal{B}_{q,\kappa}(t_{i+m})$ is nonzero, that is $q = i+m-1$, $\dots$, $i-m+1$. Further, the sum of each of those rows is 1. In particular, the (i,i) element of the matrix is nonzero. Thus the ith row of the matrix looks like

$$\underbrace{0 \dots 0}_{i-m+1 \text{ times}} \quad \mathcal{B}_{i-m+1,\kappa}(t_{m+i}) \quad \mathcal{B}_{i-m+2,\kappa}(t_{m+i})$$
$$\dots \ \mathcal{B}_{i+m-1,\kappa}(t_{m+i}) \qquad \underbrace{0 \dots 0}_{n+1-m-i \text{ times}}$$

This system has a unique solution exactly when $|\boldsymbol{L}| \neq 0$, which follows from results in [76]. For a more complete treatment, see [29].

9.2 Other Positional Interpolation with B-Splines

Suppose it is necessary to fit the data set $\{(v_j, f_j)\}_{j=0}^{n}$ to an element of $\mathcal{S}_{\kappa,\boldsymbol{t}}$, where $2 \leq \kappa \leq n$, and $\boldsymbol{t}$ has length $n+\kappa+2$. In this section we develop the matrix system of equations and determine the conditions under which it is uniquely solvable.

Once again, suppose we are trying to solve for $\{P_i\}_{i=0}^{n}$ such that

$$\begin{aligned} f_j &= \gamma(v_j) \\ &= \sum_{i=0}^{n} P_i \mathcal{B}_{i,\kappa}(v_j). \end{aligned}$$

The $\boldsymbol{L}$ matrix then has elements $L_{j,i} = \mathcal{B}_{i,\kappa}(v_j)$. In the previous section, we investigated a slightly different problem. In that case we were interpolating a data set with $(n-1)$ elements of position at each of the $(n-1)$ distinct elements of associated knot vector, and both derivative values at the domain end points. The problem here is to solve for strictly positional data interpolation in a spline space whose knots *may not be* at the data parameter values.

Suppose there is some value $j = J$ such that $\mathcal{B}_{J,\kappa}(v_J) = 0$. Since $\mathcal{B}_{J,\kappa}(t)$ is nonzero on the interval $[t_J, t_{J+\kappa+1})$, then $v_J < t_J$ or $t_{J+\kappa+1} \leq v_J$. Without loss of generality, assume that $v_J < t_J$, so $\mathcal{B}_{q,\kappa}(v_j) = 0, j = 0, \ldots, J, q = J, \ldots, n$, since $v_j < v_{j+1}$, for all j. Rephrasing that, $P_j, j = J, \ldots, n$ depend on the parameter values $\{v_j\}_{j=J+1}^{n}$. In matrix analysis terms, that means that the Jth through n^{th} columns of L are linearly dependent since there are $n - J + 1$ column vectors each having the same $n - J$ elements nonzero. So the system cannot be solved uniquely. Hence, the system *may* have a solution if $\mathcal{B}_{i,\kappa}(v_i) \neq 0, i = 0, \ldots, n$, whereas if that constraint is not met, the system *cannot* have a solution. In fact Schoenberg and Whitney [75] showed

Theorem 9.2. Schoenberg-Whitney. *Let* $\mathbf{t}$ *be a knot vector,* κ *and* n *integers such that* $n > \kappa > 0$*, and suppose* $\mathbf{v}$ *is strictly increasing with* $n+1$ *elements. Then the matrix* $L = (\mathcal{B}_{i,\kappa}(v_j))$ *is invertible if and only if* $\mathcal{B}_{i,\kappa}(v_i) \neq 0, i = 0, \ldots, n$ *,i.e., if and only if* $t_i < v_i < t_{i+\kappa+1}$*, for all* i*.*

Notice that since each row of L can have at most $(\kappa + 1)$ columns nonzero, this requirement for invertibility of L, and hence solvability of the system, means that L is a banded matrix of width $(2\kappa + 1)$. As long as the conditions of Theorem 9.2 are satisfied, the system has a theoretical solution. The actual stability of the algorithms to determine the solution is affected by how close the determinant of the matrix $\boldsymbol{L}$ is to zero. Hence if the value of v_i is close to the center of the support of $\mathcal{B}_{i,\kappa}$, then the determinant is well behaved and the process of solving the system is more robust.

9.2.1 Nodal Interpolation

Suppose it is necessary to find an interpolant in $\mathcal{S}_{\kappa,\boldsymbol{t}}$, where the knot vector $\boldsymbol{t}$ is specified. In this case it is assumed that the interpolation data set consists of $\boldsymbol{f}_i \in \mathbb{R}^k$, where $k > 1$, and that no parameter values are given. The problem solution requires assigning parameter values to the data points so the problem can be solved, and then solving it.

Definition 9.3. *Given a knot vector* $\boldsymbol{t} = \{t_i\}_{i=0}^{n+\kappa+1}$*, define another vector* $\boldsymbol{t}^*$ *whose elements are*

$$t^*_{i,\kappa} = \frac{t_{i+1} + t_{i+2} + \cdots + t_{i+\kappa}}{\kappa} \qquad \textit{for } i = 0, \ldots, n.$$

The values $t^*_{i,\kappa}$ *are called the* nodes.

Now, define the parameter values for the ith data value f_i to be $v_i = t^*_{i,\kappa}$. Theorem 9.2 immediately shows that the resulting system is solvable. This is called *nodal* interpolation. The matrix equations to be solved can be written

$$\begin{bmatrix} \mathcal{B}_{0,\kappa}(t^*_0) & \mathcal{B}_{1,\kappa}(t^*_0) & \ldots \mathcal{B}_{n-1,\kappa}(t^*_0) & \mathcal{B}_{n,\kappa}(t^*_0) \\ \mathcal{B}_{0,\kappa}(t^*_1) & \mathcal{B}_{1,\kappa}(t^*_1) & \ldots \mathcal{B}_{n-1,\kappa}(t^*_1) & \mathcal{B}_{n,\kappa}(t^*_1) \\ & & \vdots & \\ \mathcal{B}_{0,\kappa}(t^*_{n-1}) & \mathcal{B}_{1,\kappa}(t^*_{n-1}) & \ldots \mathcal{B}_{n-1,\kappa}(t^*_{n-1}) & \mathcal{B}_{n,\kappa}(t^*_{n-1}) \\ \mathcal{B}_{0,\kappa}(t^*_n) & \mathcal{B}_{1,\kappa}(t^*_n) & \ldots \mathcal{B}_{n-1,\kappa}(t^*_n) & \mathcal{B}_{n,\kappa}(t^*_n) \end{bmatrix} \begin{bmatrix} P_0 \\ P_1 \\ \vdots \\ P_{n-1} \\ P_n \end{bmatrix} = \begin{bmatrix} f_0 \\ f_1 \\ \vdots \\ f_{n-1} \\ f_n \end{bmatrix}. \tag{9.5}$$

Example 9.4. Bezier Nodal Interpolation. Suppose there are $n + 1$ data points. Consider B-splines of degree $\kappa = n$ over knot vector $\boldsymbol{t}$ having value 0 with multiplicity $n + 1$ and value 1 with multiplicity $n + 1$. $\mathcal{S}_{n,\boldsymbol{t}}$ is the interpolating space.

By Theorem 6.29, the B-splines over this knot vector reduce to being the cubic Bernstein/Bézier blending functions. So the B-spline nodal interpolation reduces to the Bezier nodal interpolation. In Example 5.30 it was degree 3. In this case, the nodes are $\boldsymbol{t}^* = \{0, 1/n, 2/n, \ldots, (n-2)/n, (n-1)/n, 1\}$.

$$\begin{bmatrix} \theta_{0,n}(0) & \ldots & \theta_{n,n}(0) \\ \theta_{0,n}(1/n) & \ldots & \theta_{n,n}(1/n) \\ & \vdots & \\ \theta_{0,n}((n-1)/n) & \ldots & \theta_{n,n}((n-1)/n) \\ \theta_{0,n}(1) & \ldots & \theta_{n,n}(1) \end{bmatrix} \begin{bmatrix} P_0 \\ P_1 \\ \vdots \\ P_{n-1} \\ P_n \end{bmatrix} = \begin{bmatrix} f_0 \\ f_1 \\ \vdots \\ f_{n-1} \\ f_n \end{bmatrix}$$

□

9.2.2 Piecewise Cubic Hermite Interpolation

Suppose $\kappa = 3$, and the data has (m+1) triples composed of parametric value, position, and derivative, u_i, f_i, s_i. We construct an open knot vector $\boldsymbol{t}$ by setting

$$t_i = \begin{cases} u_0 & \text{for } i = 0, \dots, 3 \\ u_j & \text{for } i = 2j+2, 2j+3, \ j = 1, \dots, m-1 \\ u_m & \text{for } i = 2m+2, \dots, 2m+5. \end{cases}$$

The space $\mathcal{S}_{3,\boldsymbol{t}}$ is $C^{(1)}$ at each parametric data value since $\kappa = 3$ and each distinct internal knot value has multiplicity two. Finally, $\boldsymbol{t}$ has $2m+6$ elements. It is necessary to solve the system of $2m+2$ equations for the unknown coefficients.

$$\begin{aligned} \sum_{i=0}^{2m+1} P_i \mathcal{B}_i(u_j) &= f_j \qquad \text{for } j = 0, \dots, m, \\ \sum_{i=0}^{2m+1} P_i \mathcal{B}_i'(u_j) &= s_j \qquad \text{for } j = 0, \dots, m. \end{aligned}$$

In this generalized case, the L matrix members become:

$$\begin{aligned} L_{q,i} &= \mathcal{B}_i(u_j) \ \text{ for } q = 2j \\ &= \mathcal{B}_i'(u_j) \ \text{ for } q = 2j+1 \\ & \qquad\qquad\quad i = 0, \dots, 2m+1. \end{aligned}$$

Note that $\mathcal{B}_i(u_j) \neq 0$ if and only if $i = 2j, 2j+1$, and $\mathcal{B}_i'(u_j) \neq 0$ if and only if $i = 2j, 2j+1$. Thus L is a matrix which is zero except for 2×2 block matrices along the diagonal. Let

$$\mathcal{D}_j = \begin{bmatrix} \mathcal{B}_{2j}(u_j) & \mathcal{B}_{2j+1}(u_j) \\ \mathcal{B}_{2j}'(u_j) & \mathcal{B}_{2j+1}'(u_j) \end{bmatrix}$$

diagonal elements are nonzero and every other row has one element to the right of the diagonal nonzero. Hence the matrix is invertible, and the system is solvable. Figure 9.2 show data and tangents and the piecewise Hermite interpolant. The complete spline interpolant is in light gray for comparison. Remember that the complete spline interpolant does not interpolate the interior prescribed tangent vectors.

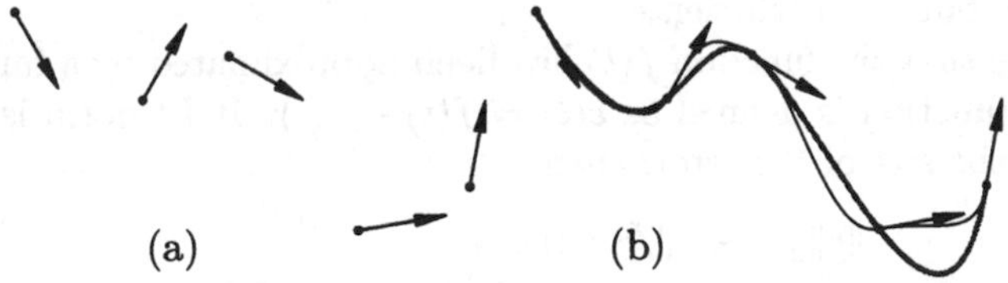

Figure 9.2. (a) Data; (b) data, piecewise Hermite interpolant (in black), and complete spline interpolant (in light gray).

9.2.3 Generalized Interpolation

Karlin and Ziegler [45] extended the Schoenberg-Whitney theorem to include general types of spline interpolation.

Theorem 9.5. *Suppose* $\boldsymbol{t}$ *is a knot vector with* $n+\kappa+2$ *elements,* $n > \kappa > 0$*, such that* $t_i < t_{i+\kappa+1}$ *for all* i*. Suppose* $\boldsymbol{s}$ *is a nondecreasing sequence of values in the domain of* $\mathcal{S}_{\kappa,\boldsymbol{t}}$ *such that* $s_j < s_{j+\kappa+1}$*,* $\boldsymbol{s}$ *has* $n+1$ *elements, and whenever*

$$max(t_i, s_j) < s_{j+1} = s_{j+2} = \ldots = s_{j+r} = t_{i+1} = \ldots = t_{i+q} < min(t_{i+q+1}, s_{j+r+1})$$

then $r + q \leq \kappa + 1$*. Suppose the real value* s_j *occurs in* $\boldsymbol{s}$ *with multiplicity* ι*, and such multiplicity is taken to mean that it is necessary to interpolate values* $f^{(i)}, i = 0, \ldots, \iota - 1$*, that is, position and the first* $\iota - 1$ *derivatives. Then there exists exactly one element of* $\mathcal{S}_{\kappa,\boldsymbol{t}}$ *(that is, the system can be solved uniquely) if and only if* $\mathcal{B}_{j,\kappa}(s_j) \neq 0$ *for all* j*.*

We refer interested parties to [28] for a proof.

9.3 B-Spline Least Squares

Suppose, as usual, that $\boldsymbol{t}$ is a knot vector with $n+\kappa+2$ elements, $n \geq \kappa$, $\mathcal{S}_{\kappa,\boldsymbol{t}}$ the related function space and $\{\mathcal{B}_{i,\kappa}(t)\}_{i=0}^{n}$ the B-spline basis over domain $[t_\kappa, t_{n+1})$. The least squares problem has two frequently used formulations: one in which it is necessary to approximate an already known function, and the other in which it is necessary to approximate (and interpolate if possible) known discrete data. In both cases, the information to be approximated is called the *primitive.*

In both formulations, it is necessary to find the coefficients of the *best* approximation in the space, where that means we need to find $\{P_i\}_{i=0}^{n}$ so that the curve $\gamma(t) = \sum_{i=0}^{n} P_i \mathcal{B}_{i,\kappa}(t)$ is the *best* approximation in $\mathcal{S}_{\kappa,\boldsymbol{t}}$ to the primitive. In what follows, we describe several meanings of *best* and also how to find the coefficients.

Suppose that the function $f(t)$ has been approximated by a function γ. The error function is defined as $e(t) = f(t) - \gamma(t)$. If L^2 norm is used to determine the size of the error, then

$$\begin{aligned} \|e\|_2 &= \|f - \gamma\|_2 \\ &= \left(\int (f(t) - \gamma(t))^2 dt \right)^{(1/2)}. \end{aligned}$$

In the case that the approximation $\gamma(t)$ is unknown, finding a *best* L^2 fit then means determining the unknown $(n+1)$ P_i's to minimize that error. Using standard techniques from calculus, extremal points (maxima and minima) occur at those values of the unknowns for which the partial derivatives (with respect to the unknowns) are simultaneously zero. However, the square root makes the problem unnecessarily complex, since the extreme points for the square are the same. Hence, it is usual to solve for the minimum of

$$\begin{aligned} E &= (\|e\|_2)^2 \\ &= (\|f-\gamma\|_2)^2 \\ &= \int (f(t)-\gamma(t))^2 dt. \end{aligned}$$

Differentiating with respect to P_j yields

$$\begin{aligned} \frac{\partial E}{\partial P_j} &= 2\int (f(t)-\gamma(t))\left(-\frac{\partial \gamma(t)}{\partial P_j}\right) dt \\ &= -2\int (f(t)-\gamma(t))\mathcal{B}_{j,\kappa}(t)dt \\ &= -2\int f(t)\mathcal{B}_{j,\kappa}(t)dt + 2\sum_{i=0}^{n} P_i \int \mathcal{B}_{i,\kappa}(t)\mathcal{B}_{j,\kappa}(t)dt. \end{aligned}$$

Setting the partials to zero gives

$$\sum_{i=0}^{n} P_i \int \mathcal{B}_{i,\kappa}(t)\mathcal{B}_{j,\kappa}(t)dt = \int f(t)\mathcal{B}_{j,\kappa}(t)dt \quad \text{for } j=0,\ldots,n. \tag{9.6}$$

Define $\mathcal{L} = (\mathcal{L}_{i,j})$, as $\mathcal{L}_{i,j} = \int \mathcal{B}_{i,\kappa}(t)\mathcal{B}_{j,\kappa}(t)dt$. Let $\mathcal{F} = [\mathcal{F}_0\ \mathcal{F}_1\ \ldots \mathcal{F}_n]^T$ be defined by $\mathcal{F}_j = \int f(t)\mathcal{B}_{j,\kappa}(t)dt$. Set $\boldsymbol{P} = [P_0\, P_1 \ldots P_n]^T$ and rewrite Equation 9.6

$$\mathcal{L}\boldsymbol{P} = \mathcal{F}. \tag{9.7}$$

In the discrete case, the function f is replaced by a data set, $\{(s_q, f_q)\}_{q=0}^{m}$, which must be approximated. The definitions for $\mathcal{L}$ and $\mathcal{F}$ become

$$\mathcal{L}_{j,i,} = \sum_{q=0}^{m} \mathcal{B}_{j,\kappa}(s_q)\mathcal{B}_{i,\kappa}(s_q), \tag{9.8}$$

$$\mathcal{F}_j = \sum_{q=0}^{m} f_q \mathcal{B}_{j,\kappa}(s_q). \tag{9.9}$$

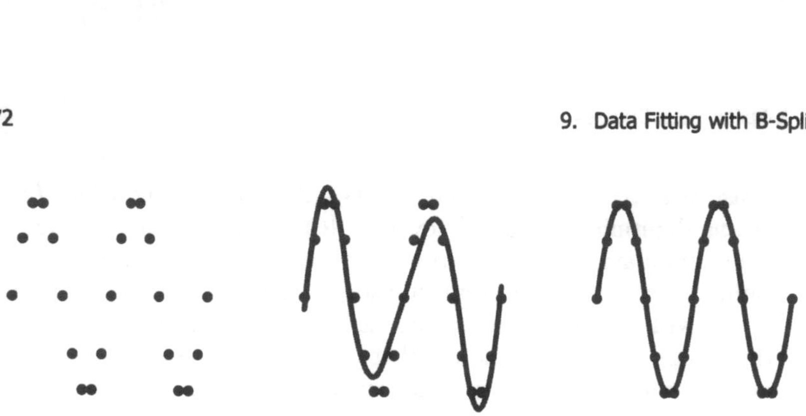

(a) (b) (c)

Figure 9.3. (a) Sampled data of a sine function; (b) least squares fitting with a quadratic B-spline function and seven control points; (c) least squares fitting with a quadratic B-spline function and 10 control points.

The matrix $\mathcal{L}$ resulting from the discrete case is invertible (and hence the system has a unique solution) only when restrictions are placed on the number, $m+1$, of data points and their parametric locations. Figure 9.3 (a) shows the data , while in (b), both data and approximation are shown. Remember, for this and other parametric data fitting schemes, the data must be assigned parametric values, which can change the fit. Figure 9.3 (c) shows the fit with a different set of parametric values assigned to the same data.

For there to be any possibility of satisfying these conditions, the data must be over the domain $[t_\kappa, t_{n+1}]$. If $m < n$, then the number of data points is less than the number of variables. In this case the problem is underspecified, so it is necessary to add enough independent linear constraint conditions, which may not be positional data fitting, so that the resulting system has rank n. In Section 9.3.1 such a scheme, called *direct manipulation*, is presented.

If $m = n$, then if the data meets the requirements of Theorem 9.2, the system is solvable and interpolates the data. If it does not, then some intervals will be overspecified, but the system will be underspecified and cannot be solved uniquely.

If $m > n$, let j be fixed but arbitrary. If $s_q < t_j$ or $s_q > t_{j+\kappa+1}$ for all q, then $\mathcal{L}_{j,i} = \sum_{q=0}^{m} \mathcal{B}_{i,\kappa}(s_q)\mathcal{B}_{j,\kappa}(s_q)$, and the row can have only zeros. Thus, the system of equations is dependent and cannot be solved uniquely. Thus, one requirement in order for a solution to exist is that there must be a subsequence of $\boldsymbol{s}, \{s_{q_j}\}_{j=0}^{n}$ such that $\mathcal{B}_{j,\kappa}(s_{q_j}) \neq 0$. Further since $\mathcal{B}_{i,\kappa}(t)\mathcal{B}_{j,\kappa}(t) \neq 0$ only for those t in the intersection of the supports of the B-splines, that is, $t \in [t_i, t_{i+\kappa+1}) \cap [t_j, t_{j+\kappa+1})$, then the only elements in the j-th row which might be nonzero are the $2\kappa + 1$ elements, $\mathcal{L}_{j,i}, i = j - \kappa, \ldots, j + \kappa$. Thus $\mathcal{L}$ is a banded matrix.

9.3.1 Direct Manipulation

Now, suppose a curve has already be specified, say by creating a control polygon or interpolating some conditions or by positional least squares. However, a frequently occuring problem is that the creator wants to adjust the position or tangent of a specific point on a curve, but also wants to keep the rest of the curve as much the same as possible. Let $\gamma(t) = \sum_{i=0}^{n} P_i \mathcal{B}_{i,\kappa,\boldsymbol{\tau}}(t)$ be the existing curve. Suppose $(\hat{s}, \gamma(\hat{s}))$ is the selected point on the curve whose position should become Q. That is, we need to find $\beta(t) = \sum_{i=0}^{n} Q_i \mathcal{B}_{i,\kappa,\boldsymbol{\tau}}(t)$, such that $\beta(\hat{s}) = Q$ and β is as *close* to γ as possible. This problem statement was first formulated for interactive design in [6]. Let $\Delta_i = Q_i - P_i$. We can restate the problem to be that of finding a curve $\Delta(t) = \sum_{i=0}^{n} \Delta_i \mathcal{B}_{i,\kappa,\boldsymbol{\tau}}(t)$, such that $\Delta(\hat{s}) = Q - \gamma(\hat{s})$ and either $E\{\Delta_i\} = \int (\Delta(t))^2 dt$ or $E\{\Delta_i\} = \sum_j (\Delta(s_j))^2$ are minimized. Δ, by this definition, is the function that is the minimal deviation from the zero function and still satisfies the constraint. Then the problem is formulated as the construction of $\Delta(t)$ as an underconstrained linear system, with the single interpolation condition, $\Delta(\hat{s}) - Q + \gamma(\hat{s}) = 0$.

To solve this problem, first find J such that $\tau_J \leq \hat{s} < \tau_{J+1}$. Then $\Delta_i = 0$, for $i < J - \kappa$ and $i > J$, so $\Delta(t) = \sum_{i=J-\kappa}^{\kappa} \Delta_i \mathcal{B}_{i,\kappa,\boldsymbol{\tau}}(t)$. Set

$$a = \sum_{i=J-\kappa}^{\kappa} \left(\mathcal{B}_{i,\kappa,\boldsymbol{\tau}}(\hat{s})\right)^2 .$$

It is easy to show that

$$\Delta_i = \frac{(Q - \gamma(\hat{s}))\mathcal{B}_{i,\kappa,\boldsymbol{\tau}}(\hat{s})}{a}, \qquad \text{for } J - \kappa \leq i \leq J \tag{9.10}$$

form a solution (see [32]). The final solution then is $\beta(t) = \gamma(t) + \Delta(t)$. The above solution is correct only when there is exactly one interpolation condition. More general direct manipulation constraints can be applied by simultaneously constraining multiple positions and tangent vectors. The multiple conditions can lead to a system that is overconstrained in some intervals and underconstrained in others. Finding solutions to the general problem can be done by an L^2 minimizing solution via, for example, singular value decomposition (SVD) or QR factorization [40]. See Figure 9.4.

Such solutions are approached by considering the interpolation conditions as a linear system, written $B\Delta = Q$, $\Delta \in \boldsymbol{R}^n$ is the vector of the control points for the difference curve $\Delta(t)$, $B \in \boldsymbol{R}^{m\times n}$ is the matrix consisting of interpolation constraints, one row for each constraint, and Q is the vector of values that the difference curve must interpolate, one for each

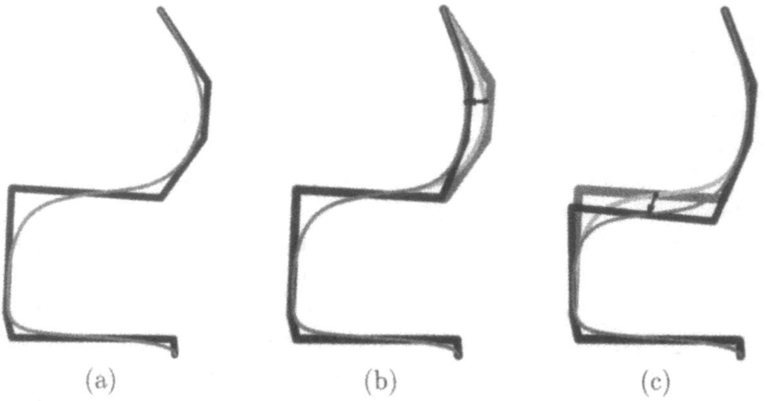

Figure 9.4. Example of direct manipulation (a) before and (b) and (c) after the changes. Originals are shown in light gray.

row of B. If each row is independent, the linear system has infinitely many solutions and is *underdetermined.* QR factorization and SVD are two methods that can compute the solution that is L^2 minimizing in this infinite set of solutions, the solution that is required in this application in order to minimize the global change of the curve. If the system is *overdetermined,* i.e., all the constraints cannot be met, the same methods still can be used to find L^2 minimizing solutions. See [40] for details of proofs and algorithms for the different cases.

9.4 Schoenberg Variation Diminishing Splines

Given a knot vector $\boldsymbol{t}$ with $n + \kappa + 2$ elements, the problem is to approximate a function f, called the *primitive*, which is continuous on the interval $[t_\kappa, t_n]$, by an element of $\mathcal{S}_{\kappa,\boldsymbol{t}}$, with a generalized Bernstein approximation. We would like to keep the shape characteristics of the Bernstein approximation.

By Example 5.22, the Bernstein approximation reproduces constants, and by Example 5.23, it reproduces linear functions. A desirable property of the generalization is to keep these characteristics. However, Example 5.24 shows that Bernstein approximation does not reproduce quadratic functions, although the sequence of approximations does converge.

Examine approximating $f(x) = x$ using Bernstein approximation when $n = \kappa$, and knot vector $\boldsymbol{t}$, has $t_i = 0$, for $i = 0, \ldots, \kappa$, and $t_i = 1$, for $i = \kappa + 1, \ldots, 2\kappa + 1$. The ith coefficient is

$$\begin{aligned} P_i &= \frac{t_{i+1}+\ldots+t_\kappa+t_{\kappa+1}+\ldots+t_{i+\kappa}}{\kappa} \\ &= \frac{t_{\kappa+1}+\ldots+t_{i+\kappa}}{\kappa} \\ &= \frac{i}{n} \\ &= f(i/n). \end{aligned}$$

We showed in Example 7.15 that an analogous assertion holds. That is, $x = \sum_{i=0}^{n} f\left(\frac{t_{i+1}+\ldots+t_{i+\kappa}}{\kappa}\right) \mathcal{B}_{i,\kappa,\mathbf{t}}(x)$. While this is rather simple for linear functions, it suggests a method for obtaining coefficients for approximations to more complex functions by simply evaluating the primitive at $n+1$ points. We shall formalize the definition of this approximation, and then proceed to prove when it converges to the primitive function.

Definition 9.6. *For f any continuous function, define*

$$V(f;t) = \sum_{i=0}^{n} f(t^*_{i,\kappa})\mathcal{B}_{i,\kappa}(t).$$

$V(f)$ is called the Schoenberg variation diminishing spline approximation to f.

The proof that the approximant defined in Definition 9.6 is variation diminishing uses material outside the scope of this book. The interested reader is referred to [44, 74, 25, 76], which use the theory of positive definite operators, or more recently [50], which uses properties of B-splines and discrete splines, for complete proofs.

To find V(f) it is necessary to evaluate only the function f. While V(f) does not interpolate those values, it does approximate them and roughly approximates the shape of the whole curve. Originally, Schoenberg [74] introduced this approximation as a data fitting scheme for discrete data points and called it a *smoothing interpolant.*

Definition 9.7. *For a knot vector $\mathbf{t}$ we call $\|\mathbf{t}\| = \max_i\{t_{i+1} - t_i\}$ the* norm *of the knot vector.*

Lemma 9.8. *$V(x^2;x) \neq x^2$, but $\left|V_{\mathbf{t}}(x^2;x) - x^2\right| \to 0$ as $\|\mathbf{t}\| \to 0$.*

Proof: By the results of Example 7.22 and the above definition,

$$V(x^2;x) - x^2 = \sum_{j=0}^{n}\left(\left(\frac{\sum_{s=1}^{\kappa} t_{j+s}}{\kappa}\right)^2 - \frac{\sum_{1\le r<q\le\kappa} t_{j+r}t_{j+q}}{\binom{\kappa}{2}}\right)\mathcal{B}_{j,\kappa}(t).$$

Let $e(x) = V(x^2; x) - x^2$, and let e_j designate the B-spline coefficients of e. Then,

$$\begin{aligned}
e_j &= \left(\frac{\sum_{s=1}^{\kappa} t_{j+s}}{\kappa}\right)^2 - \frac{\sum_{1\le r<q\le\kappa} t_{j+r}t_{j+q}}{\binom{\kappa}{2}} \\
&= \frac{1}{\kappa^2}\left(\sum_{1\le r\le\kappa} t_{j+r}^2 + 2\sum_{1\le r<s\le\kappa} t_{j+r}t_{j+s}\right) \\
&\qquad - \frac{2}{\kappa(\kappa-1)}\sum_{1\le r<s\le\kappa} t_{j+r}t_{j+s} \\
&= \frac{(\kappa-1)}{\kappa^2(\kappa-1)}\sum_{1\le r\le\kappa} t_{j+r}^2 - \frac{2}{\kappa^2(\kappa-1)}\sum_{1\le r<s\le\kappa} t_{j+r}t_{j+s} \\
&= \frac{\kappa+1}{\kappa^2(\kappa-1)}\sum_{1\le r\le\kappa} t_{j+r}^2 \\
&\qquad - \frac{2}{\kappa^2(\kappa-1)}\left(\sum_{1\le r\le\kappa} t_{j+r}^2 + \sum_{1\le r<s\le\kappa} t_{j+r}t_{j+s}\right) \\
&= \frac{\kappa+1}{\kappa^2(\kappa-1)}\sum_{1\le r\le\kappa} t_{j+r}^2 - \frac{2}{\kappa^2(\kappa-1)}\sum_{1\le r\le s\le\kappa} t_{j+r}t_{j+s}.
\end{aligned}$$

To finish and bound the coefficient, we expand the first summation further:

$$\begin{aligned}
(\kappa+1)\sum_{1\le r\le\kappa} t_{j+r}^2 &= \sum_{1\le r\le\kappa}\left(\sum_{1\le s\le r} t_{j+r}^2 + \sum_{r<s\le\kappa} t_{j+r}^2\right) + \sum_{1\le r\le\kappa} t_{j+r}^2 \\
&= \sum_{1\le s\le r\le\kappa} t_{j+r}^2 + \sum_{1\le r\le s\le\kappa} t_{j+r}^2 \\
&= \sum_{1\le r\le s\le\kappa} t_{j+s}^2 + \sum_{1\le r\le s\le\kappa} t_{j+r}^2.
\end{aligned}$$

Combining the results we get

$$\begin{aligned}
e_j &= \frac{1}{\kappa^2(\kappa-1)}\sum_{1\le r\le s\le\kappa}\left(t_{j+s}^2 + t_{j+r}^2 - 2t_{j+r}t_{j+s}\right) \\
&= \frac{1}{\kappa^2(\kappa-1)}\sum_{1\le r\le s\le\kappa}\left(t_{j+s} - t_{j+r}\right)^2. \qquad (9.11)
\end{aligned}$$

From the above calculations we see that

$$0 \le e_j \le \frac{1}{2\kappa}(t_{j+\kappa} - t_{j+1})^2 \le \frac{1}{2}\|\mathbf{t}\|^2. \qquad \blacksquare$$

In order to bound the error for an arbitrary continuous function, one more definition and property is useful.

Definition 9.9. *For a function f defined over an interval D,*

$$\omega(f;h) = \sup_{|x-y|\le h} \{|f(x)-f(y)| : x,y \in D\}$$

is called the modulus of continuity *of the function f.*

Lemma 9.10. *For any $\lambda > 0$, $\omega(f;\lambda h) \le (\lambda+1)\omega(f;h)$.*

Proof: Recall that for all integers $N > 0$,

$$\begin{aligned} |f(y)-f(x)| &= \left|\textstyle\sum_{i=1}^{N} \left(f\left(x+i\frac{y-x}{N}\right) - f\left(x+(i-1)\frac{y-x}{N}\right)\right)\right| \\ &\le \sum_{i=1}^{N} \left|\left(f\left(x+i\frac{y-x}{N}\right) - f\left(x+(i-1)\frac{y-x}{N}\right)\right)\right|. \end{aligned}$$

Let $N = \lceil \lambda \rceil$, the smallest integer greater than λ. Then

$$\begin{aligned} \omega(f;\lambda h) &\le \omega(f;Nh) \\ &\le \sup_{|y-x|\le Nh} \{|f(y)-f(x)| : x,y\in D\} \\ &\le \sup_{|y-x|\le Nh} \{\sum_{i=1}^{N} |f(x+i\tfrac{y-x}{N}) - f(x+(i-1)\tfrac{y-x}{N})| : x,y \in D\} \\ &\le \sum_{i=1}^{N} \sup_{|t|\le h} \{|f(x+t)-f(x)| : x, x+t \in D\} \\ &\le N\omega(f;h) \\ &\le (\lambda+1)\omega(f;h). \end{aligned}$$

■

We finally make our assertion about convergence of $V(f)$ to f, and find some bounds on the rate of convergence. Note that there are two parts to the theorem, and convergence of the Bernstein approximation is a consequence of one of them.

Theorem 9.11. *Suppose that f is a continuous function and $\boldsymbol{x}$ is a knot vector with $n+\kappa+2$ elements and norm $\|\boldsymbol{x}\|$. Suppose that no real value occurs in $\boldsymbol{x}$ with frequency more than κ internally and frequency $\kappa+1$*

on the ends. If $e_{\kappa,\boldsymbol{x}}(f;x) = V_{\kappa,\boldsymbol{x}}(f;x) - f(x)$ is the error between the approximation and the function f, then

$$e_{\kappa,\boldsymbol{x}}(f;x) \leq 2\omega(f;\|\boldsymbol{x}\|).$$

Proof: If $e_{\kappa,\boldsymbol{x}}(f;x) = V_{\kappa,\boldsymbol{x}}(f;x) - f(x)$ is the error between the approximation and the function f, then

$$\begin{aligned} \|e_{\kappa,\boldsymbol{x}}(f;x)\| &= \|V_{\kappa,\boldsymbol{x}}(f;x) - f(x)\| \\ &= \left\|\textstyle\sum_{i=0}^{n} \left(f(x_{i,\kappa}^*) - f(x)\right) \mathcal{B}_{i,\kappa}(x)\right\| \\ &\leq \sum_{i=0}^{n} \omega(f;\left|x_{i,\kappa}^* - x\right|)\mathcal{B}_{i,\kappa}(x). \end{aligned} \tag{9.12}$$

For any real $h > 0$ we use Lemma 9.10 to get

$$\begin{aligned} \omega(f;\left|x_{i,\kappa}^* - x\right|) &= \omega(f; \frac{\left|x_{i,\kappa}^* - x\right|}{h} h) \\ &\leq \left(\frac{\left|x_{i,\kappa}^* - x\right|}{h} + 1\right) \omega(f;h). \end{aligned} \tag{9.13}$$

Putting Equations 9.12 and 9.13 together yields

$$\begin{aligned} \|e_{\kappa,\boldsymbol{x}}(f;x)\| &\leq \omega(f;h) \sum_{i=0}^{n} \left(\frac{\left|x_{i,\kappa}^* - x\right|}{h} + 1\right) \mathcal{B}_{i,\kappa}(x) \\ &\leq \omega(f;h) + \frac{\omega(f;h)}{h} \sum_{i=0}^{n} \left|x_{i,\kappa}^* - x\right| \mathcal{B}_{i,\kappa}(x). \end{aligned}$$

Now, using Hölder's inequality,

$$\begin{aligned} &\left(\sum_{i=0}^{n} \left|x_{i,\kappa}^* - x\right| \mathcal{B}_{i,\kappa}(x)\right)^2 \\ &\quad = \left(\sum_{i=0}^{n} \left(\left|x_{i,\kappa}^* - x\right| \sqrt{\mathcal{B}_{i,\kappa}(x)}\right) \sqrt{\mathcal{B}_{i,\kappa}(x)}\right)^2 \\ &\quad \leq \sum_{i=0}^{n} \left|x_{i,\kappa}^* - x\right|^2 \mathcal{B}_{i,\kappa}(x) \sum_{i=0}^{n} \mathcal{B}_{i,\kappa}(x) \\ &\quad \leq \sum_{i=0}^{n} \left((x_{i,\kappa}^*)^2 - 2x_{i,\kappa}^* x + x^2\right) \mathcal{B}_{i,\kappa}(x) \end{aligned}$$

$$
\begin{aligned}
&\leq \sum_{i=0}^{n}(x_{i,\kappa}^*)^2 \mathcal{B}_{i,\kappa}(x) - 2x\sum_{i=0}^{n} x_{i,\kappa}^* \mathcal{B}_{i,\kappa}(x) + x^2 \sum_{i=0}^{n} \mathcal{B}_{i,\kappa}(x) \\
&\leq V_{\kappa,\boldsymbol{x}}(x^2;x) - 2x^2 + x^2 \\
&\leq V_{\kappa,\boldsymbol{x}}(x^2;x) - x^2 \\
&\leq \|\boldsymbol{x}\|^2/2.
\end{aligned}
$$

This last inequality is true by the bounds on the coefficients of the function $e(x)$, the error function of approximating x^2 by the variation diminishing approximation. Substituting this inequality into the error for f:

$$\|e_{\kappa,\boldsymbol{x}}(f;x)\| \leq \omega(f;h)\left(\frac{\|\boldsymbol{x}\|}{\sqrt{2}h} + 1\right).$$

Thus if $\|\boldsymbol{x}\|/\sqrt{2} \leq h$,

$$
\begin{aligned}
\|e_{\kappa,\boldsymbol{x}}(f;x)\| &\leq 2\omega(f;\|\boldsymbol{x}\|/\sqrt{2}) \\
&\leq 2\omega(f;\|\boldsymbol{x}\|).
\end{aligned}
$$

■

Since a continuous function over a closed interval is uniformly continuous, given an $\epsilon > 0$ there exists $h > 0$ such that $\omega(f;h) < \epsilon$. Hence, if a knot vector is selected so that $\|\boldsymbol{x}\| < h$, where h is related to ϵ as above, then the κth degree variation diminishing B-spline approximation to f is uniformly close to f, and is variation diminishing, so it maintains roughly the same shape as f.

9.5 Quasi-Interpolation

The variation diminishing spline approximation is an example of a broader class of spline approximants called *quasi-interpolants.* They have the characteristic that the coefficient of $\mathcal{B}_{i,\kappa}$ is derived as a blend of values of the primitive and various derivatives of the primitive in $[t_i, t_{i+\kappa+1})$. In this section we introduce a quasiinterpolant due to deBoor [31] that is the identity on $\mathcal{S}_{\kappa,\boldsymbol{t}}$.

Definition 9.12. *Fix $\kappa \geq 1$. Suppose $\mathbf{t} = \{t_j\}_{j=0}^{n+\kappa+1}$ is a knot vector with corresponding space $\mathcal{S}_{\kappa,\boldsymbol{t}}$, and B-splines, $\{\mathcal{B}_{i,\kappa} = \mathcal{B}_{i,\kappa,\boldsymbol{t}}, i = 0, \ldots, n\}$. Let $\boldsymbol{a} = \{a_i\}_{i=0}^{n}$ be such that $a_i \in (t_i, t_{i+\kappa+1})$. Define the linear functionals $\lambda_{i,\kappa,\boldsymbol{a}}[f] = \sum_{r=0}^{\kappa}(-1)^r \psi_{i,\kappa}^{(\kappa-r)}(a_i) f^{(r)}(a_i)$, where*

$$\psi_{j,\kappa}(y) = \begin{cases} 1, & \text{for } \kappa = 0, \\ \frac{\prod_{r=1}^{\kappa}(y - t_{j+r})}{\kappa!}, & \text{for } \kappa > 0. \end{cases} \tag{9.14}$$

The λ's are clearly dependent on the knot vector and the degree of the B-splines. They are also dependent on the vector $\boldsymbol{a}$. However, de-Boor [28, 29] has shown that the values of the functionals on the B-splines are independent of the choice of $\boldsymbol{a}$, and have an orthonormal relationship.

Theorem 9.13. *If $\{\lambda_{i,\kappa,\boldsymbol{a}}\}$ is as in Definition 9.12 then*

$$\lambda_{i,\kappa,\boldsymbol{a}}[\mathcal{B}_{j,\kappa}] = \delta_{i,j},$$

for every choice of $\boldsymbol{a}$.

Definition 9.14. *The* quasi-interpolant *of f with respect to $\boldsymbol{a}$, and $\mathcal{S}_{\kappa,\boldsymbol{t}}$ is defined as*

$$Q[f](t) = \sum_{i=0}^{n} \lambda_{i,\kappa,\boldsymbol{a}}[f]\mathcal{B}_{i,\kappa}(t),$$

where $\lambda_{i,\kappa,\boldsymbol{a}}$ is defined in Definition 9.12.

Theorem 9.15. *For $\gamma \in \mathcal{S}_{\kappa,\boldsymbol{t}}$, $Q[\gamma] = \gamma$. That is, Q acts as the identity transformation on $\mathcal{S}_{\kappa,\boldsymbol{t}}$ for every choice of $\boldsymbol{a}$.*

Proof: For $\gamma(t) = \sum_{i=0}^{n} P_i\, \mathcal{B}_{i,\kappa}$, it follows directly from Theorem 9.13 that

$$\begin{aligned}
\lambda_{j,\kappa,\boldsymbol{a}}[\gamma] &= \lambda_{j,\kappa,\boldsymbol{a}}(\sum_{i=0}^{n} P_i\, \mathcal{B}_{i,\kappa}) \\
&= \sum_{i=0}^{n} \lambda_{j,\kappa,\boldsymbol{a}}(P_i\, \mathcal{B}_{i,\kappa}) \\
&= \sum_{i=0}^{n} P_i\, \lambda_{j,\kappa,\boldsymbol{a}}(\mathcal{B}_{i,\kappa}) \\
&= \sum_{i=0}^{n} P_i\, \delta_{j,i} \\
&= P_j.
\end{aligned}$$

■

Example 9.16. For a given κ, let $\boldsymbol{\tau}$ be an arbitrary knot vector, $\tau_i < \tau_{i+\kappa+1}$ for all i, with corresponding spline basis functions $\{\mathcal{B}_{\kappa,\boldsymbol{\tau}}\}$, and let $\boldsymbol{t}$ be any refinement of $\boldsymbol{\tau}$ such that $t_j < t_{j+\kappa+1}$, for all j, with corresponding spline basis functions $\{\mathcal{N}_{\kappa,\boldsymbol{t}}\}$. By Theorem 7.12, $\mathcal{S}_{\kappa,\boldsymbol{\tau}} \subset \mathcal{S}_{\kappa,\boldsymbol{t}}$. Thus, for each i,

$$\mathcal{B}_{i,\kappa}(t) = \sum_{j} \lambda_{j,\kappa,\boldsymbol{t}_{\kappa^*}}(\mathcal{B}_{i,\kappa})\mathcal{N}_{j,\kappa}(t).$$

□

Example 9.17. Let $\kappa = 1$, and let $\boldsymbol{\tau}$, $\{\mathcal{B}_{i,1}(t)\}$, $\boldsymbol{t}$, and $\{\mathcal{N}_{j,1}(t)\}$ be defined as in Example 9.16. Then $t^*_{1,j} = t_{j+1}$, and for each i,

$$\begin{aligned}
\lambda_{j,1,\boldsymbol{t_1}^*}(\mathcal{B}_{i,\kappa}) &= \sum_{r=0}^{1}(-1)^r \psi_{i,1}^{(1-r)}(t^*_{1,j})\mathcal{B}_{i,1}^{(r)}(t^*_{1,j}) \\
&= \mathcal{B}_{i,1}(t^*_{1,j}) + \left(t^*_{1,j} - t_{j+1}\right)\mathcal{B}_{i,1}^{(1)}(t^*_{1,j}) \\
&= \mathcal{B}_{i,1}(t_{j+1}) + \left(t_{j+1} - t_{j+1}\right)\mathcal{B}_{i,1}^{(1)}(t_{j+1}) \\
&= \mathcal{B}_{i,1}(t_{j+1}).
\end{aligned}$$

Thus,

$$\mathcal{B}_{i,1}(t) = \sum_j \mathcal{B}_{i,1}(t_{j+1})\mathcal{N}_{j,1}(t). \tag{9.15}$$

□

In general, if $f(t)$ is an arbitrary function with appropriate continuity, and $Q[f](t) = \sum Q_i \mathcal{B}_{i,\kappa}(t)$, then

$$\begin{aligned}
Q_i &= \lambda_{i,\kappa,\boldsymbol{a}}[f] \\
&= \sum_{r=0}^{\kappa}(-1)^r \psi_{i,\kappa}^{(\kappa-r)}(a_i) f^{(r)}(a_i) \\
&= f(a_i)\psi_{i,\kappa}^{\kappa}(a_i) - f'(a_i)\psi_{i,\kappa}^{(\kappa-1)}(a_i) + \sum_{r=2}^{\kappa}(-1)^r \psi_{i,\kappa}^{(\kappa-r)}(a_i) f^{(r)}(a_i).
\end{aligned}$$

If ψ is expanded into powers of t,

$$\psi_{i,\kappa}(t) = \frac{1}{\kappa!}t^{\kappa} - \frac{\sum_{j=1}^{\kappa} t_{i+j}}{\kappa!}t^{\kappa-1} + \text{all lower powers of } t.$$

Then,

$$\psi_{i,\kappa}^{(\kappa)}(a_i) = 1 \quad \text{and} \quad \psi_{i,\kappa}^{(\kappa-1)}(a_i) = \left(a_i - \frac{\sum_{j=1}^{\kappa} t_{i+j}}{\kappa}\right).$$

If $a_{i,\kappa} = t^*_i = \frac{\sum_{j=1}^{\kappa} t_{i+j}}{\kappa}$, then $\psi_{i,\kappa}^{(\kappa-1)}(t^*_i) = 0$ and

$$Q_i = f(t^*_i) + \sum_{r=2}^{\kappa} \psi_{i,\kappa}^{(\kappa-r)}(t^*_i) f^{(r)}(t^*_i).$$

In general,

$$\psi_{i,\kappa}^{(r)}(t) = C_r \sum \cdots \sum_{0 \le i_1 < i_2 < \ldots < i_{\kappa-r} \le \kappa} \cdots \sum (t - t_{i+i_1}) \cdots (t - t_{i+i_{\kappa-r}}),$$

where $C_r = 2^r$ is a constant depending on r, but not the particular knot values in $\boldsymbol{t}$. If $\psi_{i,\kappa}^{(r)}(t)$ is evaluated at t_i^*, each factor in each of the terms looks like, for each fixed $j = 1, \ldots, \kappa$,

$$\begin{aligned}
|t_i^* - t_{i+j}| &= \left|\frac{\sum_{p=1}^{\kappa} t_{i+p}}{\kappa} - t_{i+j}\right| \\
&= \left|\frac{\sum_{p=1}^{\kappa}(t_{i+p} - t_{i+j})}{\kappa}\right| \\
&= \left|\frac{\sum_{p=1}^{j-1}(t_{i+p} - t_{i+j})}{\kappa} + \frac{\sum_{p=j+1}^{\kappa}(t_{i+p} - t_{i+j})}{\kappa}\right| \\
&= \left|\frac{\sum_{p=1}^{j-1}\sum_{s=p}^{j-1}(t_{i+s} - t_{i+s+1})}{\kappa} + \frac{\sum_{p=j+1}^{\kappa}\sum_{s=j}^{p-1}(t_{i+s+1} - t_{i+s})}{\kappa}\right| \\
&\leq \|\boldsymbol{t}\|\frac{\sum_{p=1}^{j-1}\sum_{s=p}^{j-1}1}{\kappa} + \|\boldsymbol{t}\|\frac{\sum_{p=j+1}^{\kappa}\sum_{s=j}^{p-1}1}{\kappa} \\
&\leq \|\boldsymbol{t}\|\frac{(j-1)j + (\kappa-j)(\kappa-j-1)}{2\kappa}.
\end{aligned}$$

Thus,

$$\left\|\psi_{i,\kappa}^{(r)}(t_i^*)\right\| \leq \|\boldsymbol{t}\|^{\kappa-r} C_r \sum \cdots \sum_{0 \leq i_1 < i_2 < \cdots < i_{\kappa-r} \leq \kappa} \cdots \sum \prod_{j=1}^{\kappa-r} \frac{(i_j - 1)i_j + (\kappa - i_j)(\kappa + 1 - i_j)}{2\kappa}$$

and

$$\left\|\psi_{i,\kappa}^{(r)}(t_i^*)\right\| \leq \|\boldsymbol{t}\|^{\kappa-r} A_{\kappa,r}, \quad \text{so} \quad \left\|\psi_{i,\kappa}^{(\kappa-r)}(t_i^*)\right\| \leq \|\boldsymbol{t}\|^{r} A_{\kappa,\kappa-r},$$

where the constant $A_{\kappa,r}$ depends on the order of the derivative and κ, but not on the knot vector $\boldsymbol{t}$ or the value of i, and is the appropriate maximum value. Set $A_\kappa = max\{A_{\kappa,r} : r = 2, \ldots, \kappa\}$, and $C \geq \left\|f^{(r)}(t)\right\|$.

$$\begin{aligned}
\|Q_i - f(t_i^*)\| &\leq A_\kappa \sum_{r=2}^{\kappa} \|\boldsymbol{t}\|^r \left\|f^{(r)}(t_i^*)\right\| \\
&\leq A_\kappa C \sum_{r=2}^{\kappa} \|\boldsymbol{t}\|^r \\
&\leq A_\kappa C \|\boldsymbol{t}\|^2 \sum_{r=0}^{\kappa-2} \|\boldsymbol{t}\|^r \\
&\leq A_\kappa C \|\boldsymbol{t}\|^2 \frac{1 - \|\boldsymbol{t}\|^{r-1}}{1 - \|\boldsymbol{t}\|}.
\end{aligned}$$

Now, when $\|\boldsymbol{t}\|$ gets small, the squared factor dominates, so

$$Q_i = f(t_i^*) + O(\|\boldsymbol{t}\|^2). \tag{9.16}$$

Theorem 9.18. *Given a knot vector* $\boldsymbol{t}$ *and an integer* $\kappa > 1$. *Let* $f \in \mathcal{S}_{\kappa,\boldsymbol{t}}$. *Suppose* $\{\boldsymbol{t}_j\}$ *is a sequence of knot vectors such that* $\boldsymbol{t}_0 = \boldsymbol{t}$, *and* $\boldsymbol{t}_j \subset \boldsymbol{t}_{j+1}$, *as knot vectors, and* $f \in \mathcal{S}_{\kappa,\boldsymbol{t}_j}$. *Let* $\boldsymbol{V}_j[f](t) = \sum V_{j,i}\mathcal{B}_{i,\kappa,j}(t)$, *be the variation diminishing spline approximation to* f *in* $\mathcal{S}_{\kappa,\boldsymbol{t}_j}$, *and let* $\boldsymbol{Q}_j[f]$ *designate the quasi-interpolant to* f *in* $\mathcal{S}_{\kappa,\boldsymbol{t}_j}$. *Then,*

$$\|f - \boldsymbol{V}_j[f]\| \le \|\boldsymbol{t}_j\|^2 A_\kappa C.$$

Proof: Since $f \in \mathcal{S}_{\kappa,\boldsymbol{t}_j}$, then $Q_j[f] = f$ for all j.

$$\begin{aligned}
\|f - \boldsymbol{V}_j[f]\| &= \|Q_j[f] - V_j[f]\| \\
&= \left\|\textstyle\sum_i Q_{j,i}\mathcal{B}_{i,\kappa,j}(t) - \sum_i V_{j,i}\mathcal{B}_{i,\kappa,j}(t)\right\| \\
&= \left\|\textstyle\sum_i (Q_{j,i} - V_{j,i})\mathcal{B}_{i,\kappa,j}(t)\right\| \\
&\le \sum_i \|Q_{j,i} - V_{j,i}\|\mathcal{B}_{i,\kappa,j}(t) \\
&\le \max_i\{\|Q_{j,i} - V_{j,i}\|\}\sum_i \mathcal{B}_{i,\kappa,j}(t) \\
&\le \max_i\{\|Q_{j,i} - V_{j,i}\|\}.
\end{aligned}$$

By the derivation resulting in Equation 9.16, then

$$\max_i\{\|Q_{j,i} - V_{j,i}\|\} \le \|\boldsymbol{t}_j\|^2 A_\kappa C$$

where A_κ and C do not depend on $\boldsymbol{t_j}$. ■

Corollary 9.19. *If the sequences of knot vectors* $\boldsymbol{t}_j$ *are chosen so that* $\|\boldsymbol{t}_j\| \to 0$ *as* $j \to \infty$, *then* $\boldsymbol{V}_j[f]$ *converges to* f *quadratically in the knot sequence norms.*

9.5.1 Variation Diminishing Splines and B-Spline Curves

In the design context, one may have a B-spline curve of the form $\gamma(t) = \sum P_i\mathcal{B}_{i,\kappa}(t)$ from $\mathcal{S}_{\kappa,\boldsymbol{t}}$, where the space is now taken with the $P_i \in \mathbb{R}^3$. If one wants to place this in the framework of approximation theory, one may view the polygon $P(t)$ as the primitive curve, with $P(t_i^*) = P_i$, and the B-spline curve $\gamma(t)$ as the approximation. Now suppose it is necessary to find the point on the curve γ that is closest to P_i. Finding the exact solution requires solving a nonlinear equation for a root. However, we observe that, from Theorem 9.18, $\|P_i - \gamma(t_i^*)\| \leq A_\kappa C$. Even though $\gamma(t_i^*)$ is not closest to P_i, it is a close first order approximation (and easily obtained!).

9.6 Multiresolution Decompostion

In this Section generalizations to the least squares and constrained spline approximation approaches from section 9.3 are presented.

Definition 9.20. *Subspaces V_1 and V_2 of an inner product vector space V are called* orthogonal *if for all $v_1 \in V_1$ and $v_2 \in V_2$ $< v_1, v_2 >= 0$.*

Definition 9.21. *If S is an inner product space with subspaces V and W such that $S = V + W = \{v + w : v \in V,\ \text{and}\ w \in W\}$, then V and W are called an* orthogonal decomposition *of S and written $S = V \oplus W$ if V and W are orthogonal.*

Lemma 9.22. *$\mathcal{S}_{\kappa,\boldsymbol{t}}$ is an inner product space, with*

$$< f, g >= \int_{t_\kappa}^{t_{n+1}} f(t)g(t)dt$$

defining the inner product, $f, g \in \mathcal{S}_{\kappa,\boldsymbol{t}}$.

The proof is left as an exercise. (See Exercise 3.)

Now we consider orthogonal decompositions of the space $\mathcal{S}_{\kappa,\boldsymbol{t}}$ which can be used for multiresolution approximation and shape editing. We approach this problem by considering a knot vector $\boldsymbol{t}_1$ which is identical to $\boldsymbol{t}$ except that it has a single knot removed from the interior. Then $\mathcal{S}_{\kappa,\boldsymbol{t}_1}$ is a subspace of $\mathcal{S}_{\kappa,\boldsymbol{t}}$ of one dimension less. Is there a subspace V_0 of $\mathcal{S}_{\kappa,\boldsymbol{t}}$ that is orthogonal to $\mathcal{S}_{\kappa,\boldsymbol{t}_1}$? If so it can have only a single dimension. Suppose it exists with element $w_0(t)$. Since $w_0 \in \mathcal{S}_{\kappa,\boldsymbol{t}}$, it can be expressed as

$$w_0(t) = \sum_{i=0}^{n} a_{0,i} \mathcal{B}_{i,\kappa,\boldsymbol{t}}(t).$$

The orthogonality constraint between $\mathcal{S}_{\kappa,\boldsymbol{t}_1}$ and $V_0 = span\{w_0\}$ imposes that, for $j = 0, \ldots, n-1$,

$$\begin{aligned} 0 &= \left\langle \mathcal{B}_{j,\kappa,\boldsymbol{t}_1}(t), w_0(t) \right\rangle \\ &= \sum_{i=0}^{n} a_{0,i} \left\langle \mathcal{B}_{j,\kappa,\boldsymbol{t}_1}(t), \mathcal{B}_{i,\kappa,\boldsymbol{t}}(t) \right\rangle. \end{aligned} \tag{9.17}$$

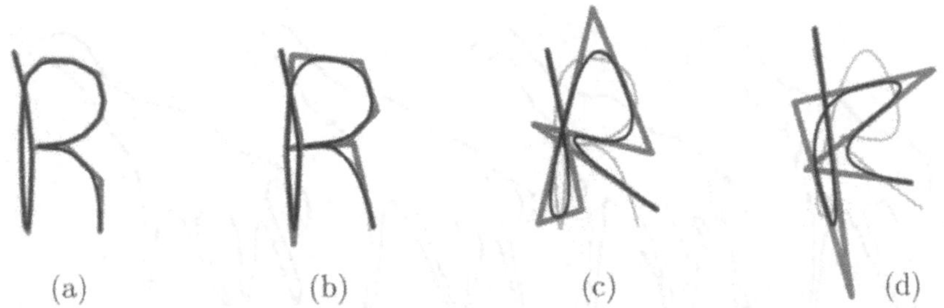

Figure 9.5. (a) Simple example of multiresolution decomposition of a quadratic B-sline curve with 22 control points; in (b) to (d), the projection of the curve into a new B-spline curve, removing 10, 15, and 16 knots, respectively, are presented. The previous curve in each stage is shown in light gray.

This creates n linear constraints in $n+1$ unknowns. The last degree of freedom on $w_0(t)$ is a scaling factor that is typically selected to normalize w_0 as $< w_0, w_0 >= 1$. Then, $\mathcal{S}_{\kappa,\boldsymbol{t}} = \mathcal{S}_{\kappa,\boldsymbol{t}_1} \oplus V_0$. This process can be applied again, this time using $\boldsymbol{t}_1$ as the full knot vector and calling the reduced knot vector $\boldsymbol{t}_2$. In this situation a function w_1 can be found that is in $\mathcal{S}_{\kappa,\boldsymbol{t}_1}$ but orthogonal to $\mathcal{S}_{\kappa,\boldsymbol{t}_2}$. Since $w_1 \in \mathcal{S}_{\kappa,\boldsymbol{t}_1}$ it is orthogonal to V_0. Form $W_1 = V_1 \oplus V_0$, then

$$\begin{aligned} \mathcal{S}_{\kappa,\boldsymbol{t}} &= \mathcal{S}_{\kappa,\boldsymbol{t}_1} \oplus V_0 \\ &= \mathcal{S}_{\kappa,\boldsymbol{t}_2} \oplus W_1. \end{aligned}$$

Continue, each time setting $W_j = V_j \oplus W_{j-1}$, so

$$\mathcal{S}_{\kappa,\boldsymbol{t}} = \mathcal{S}_{\kappa,\boldsymbol{t}_{n-\kappa}} \oplus W_{n-\kappa-1}.$$

These w functions are sometimes called *B-spline wavelets* or *B-Wavelets.*

Now, consider the curve $\gamma(t) = \sum_{i=0}^{n} P_i B_{i,\kappa,\boldsymbol{t}}(t) \in \mathcal{S}_{\kappa,\boldsymbol{t}}$. We look for the best fit to γ in $\mathcal{S}_{\kappa,\boldsymbol{t}_1}$, where *best fit* is taken in the L^2 norm sense. Such an approximation is called an *orthogonal projection.* Since the spaces $\mathcal{S}_{\kappa,\boldsymbol{t}_1}$ and W_0 are orthogonal, by finding $\gamma_1 \in \mathcal{S}_{\kappa,\boldsymbol{t}_1}$ and $\psi_0 \in W_0$ so that $\gamma(t) = \gamma_1(t) + \psi_0(t)$, $\gamma_1(\mathrm{t})$ is the best fit in $\mathcal{S}_{\kappa,\boldsymbol{t}_1}$ to $\gamma(t)$. This can be continued according to the hierarchy we have developed, each time projecting γ_j onto its best fit in $\mathcal{S}_{\kappa,\boldsymbol{t}_{j+1}}$, a space with fewer degrees of freedom. However, each time the function ψ_j, a multiple of w_j, must be found. Figure 9.5 shows a simple example of a projection into a lower dimensional space. Figure 9.5 (a) shows the original quadratic curve with 22 control points while Figure 9.5 (b) to (d) are the projections into subspace after 10, 15, and 16 interior knots were removed.

This presentation has shown the development of the multiresolution approach at its broadest; that is, projecting onto a lower dimensional subspace

Figure 9.6. (a) Original quadratic curve with 200 control points; (b)–(d) show the projection of (a) to lower resolution spaces, having 100, 50, and 25 control points, respectively.

where the dimension difference is 1. It is more often applied projecting onto lower dimensional spaces in which groups of interior knots have been removed. Frequently as many as half of the current knots are removed in a single step. Figure 9.6 is such an example. It shows a projection of a complex curve onto a subspace with just a few internal knots, a low resolution subspace. The exact support of ψ depends on which knots have been removed, the order of removal, and κ. Clearly, that approach requires algorithms and approaches that can find the functions in the B-wavelet space as well as the projected spline space rapidly and efficiently. Greater detail and discussion can be found in [26, 46].

The hierarchical decomposition of a B-spline curve is expensive to compute. Each inner product of $< f(t), g(t) >= \int f(t)g(t)dt$ requires the computation of products as well as integration of splines. Such operations are computationally intensive [60, 35]. Fortunately, one can refrain from explicitly computing the decomposition when using multiresolution in many applications, such as editing which now follows.

9.6.1 Multiresolution Curve Editing

While the local support of the B-spline basis is an important attribute for some design purposes, it also causes difficulty when it is desirable to make more global design changes. Then it may be necessary to move multiple coefficients in a coordinated way. Also, the basic representation does not allow intermixing editing changes at multiple resolutions. Consider a curve defined with a large number of control points, perhaps one with a great number of small local perturbations in shape, although having an overall smooth shape. Now, suppose the design dictates that the curve retain its

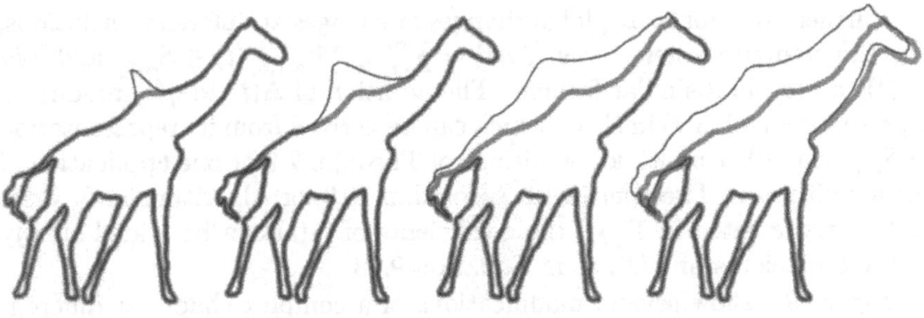

(a) (b) (c) (d)

Figure 9.7. The back of the giraffe is edited at various resolutions, resulting in an intuitive global modification of the shape. Original curve is shown in wide gray color

local attributes, but the left half should be bent, the way one bends a pipe. Performing this operation by interactively adjusting each individual control point would be prohibitive in time and lead to large inaccuracies. One approach to dealing with this is to use *shape operators* that act on regions of the curve, and one subclass of these operators relies on the multiresolution approach we have presented.

For an arbitrary knot vector, $\boldsymbol{t_0}$, where $n = card(\boldsymbol{t_0}) - (2\kappa+2)$, let $\boldsymbol{r_0}$ be the vector consisting of $\{t_{\kappa+1+2i}\}_{i=0}^{\lfloor n/2 \rfloor}$ of $\boldsymbol{t_0}$, and define $\boldsymbol{t_1} = \{\boldsymbol{t_0} \setminus \boldsymbol{r_0}\}$, that is, the knot vector $\boldsymbol{t}$ minus the elements in $\boldsymbol{r_0}$ where multiplicity counts as distinct values. Repeat this process, letting $n_j = card(\boldsymbol{t_j}) - (2\kappa+2)$ and let $\boldsymbol{r_j}$ be the vector consisting of $\{t_{j,\kappa+1+2i}\}_{i=0}^{\lfloor n_j/2 \rfloor}$ of $\boldsymbol{t_j}$, define $\boldsymbol{t_{j+1}} = \{\boldsymbol{t_j} \setminus \boldsymbol{r_j}\}$.

Then, by Theorem 7.7, $\{\mathcal{S}_{\kappa,\boldsymbol{t_j}}\}_j$ forms a hierarchy of subspaces,

$$\mathcal{S}_{\kappa,\boldsymbol{t_0}} \supset \mathcal{S}_{\kappa,\boldsymbol{t_1}} \supset \ldots \supset \mathcal{S}_{\kappa,\boldsymbol{t_i}} \supset \mathcal{S}_{\kappa,\boldsymbol{t_{i+1}}} \supset \ldots \supset \mathcal{S}_{\kappa,\boldsymbol{t_m}},$$

so that $\boldsymbol{t_m}$ has $2\kappa + 2$ elements (i.e., no interior elements). Consider the effects of performing a curve editing change on the original curve by direct manipulation (see Section 9.3.1), $\gamma(t) = \sum_{i=0}^{n} P_i \mathcal{B}_{i,\kappa,\boldsymbol{t_0}}(t)$. Suppose the user selects $\gamma(\hat{t})$ and drags it. The direct manipulation method finds $\Delta(t) \in \mathcal{S}_{\kappa,\boldsymbol{t_0}}$ as the L^2 minimizing solution to the under-constrained problem of $\Delta(\hat{t}) = 1$. The L_2 minimizing solution for this under-constrained problem with a single interpolation constraint can be achieved using SVD decomposition or via QR factorization [40]. With $\Delta(t)$, the new modified curve can be reconstructed as the user drags point $\gamma(\hat{t})$:

$$\gamma_{new}(t) = \gamma_{orig}(t) + \Delta(t) dM, \tag{9.18}$$

where dM is the motion vector of the input device.

Furthermore, for more global changes or changes at different resolutions, one can compute a new curve, $\Delta_i(t) = \sum_{j=0}^{n} P_j \mathcal{B}_{j,\kappa,\boldsymbol{t}_i}(t) \in \mathcal{S}_{\kappa,\boldsymbol{t}_i}$ and add $\Delta_i(t)$ to $\gamma(t)$, in a similar fashion. The addition of $\Delta(t)$ to $\gamma(t)$ requires a representation of $\Delta(t)$ in $\mathcal{S}_{\kappa,\boldsymbol{t}_0}$, which can be derived from its representation in $\mathcal{S}_{\kappa,\boldsymbol{t}_i}$ via either multiple repetitions of Theorem 7.9 or one application of the algorithms in Theorem 16.13, Algorithm 16.8, or Algorithm 16.9. Once $\Delta(t)$ is represented in $\mathcal{S}_{\kappa,\boldsymbol{t}_0}$, the coefficients of $\Delta(t)$ can be added simply to the coefficients of $C(t)$, as in Equation 9.18.

Figure 9.7 show several modifications of a complex shape, at different resolutions.

9.6.2 Constraints and Multiresolution Curve Editing

In geometric design, it is frequently the case that the shape should fit a previously defined point or shape in a certain area or be tangent to another shape at another location. Here, local constraints are imposed on the shape and should be maintained throughout the editing process of the freeform curve.

Restating the interpolation constraint for a curve $\gamma(t)$ to interpolate position Q_0 at parameter value s_0 equals,

$$Q_0 = \gamma(s_0) = \sum_{i=0}^{n} P_i \mathcal{B}_{i,\kappa,\boldsymbol{t}_0}(s_0),$$

which is a linear equation in the unknown coefficients of the curve. Similarly, setting interpolation of a tangent vector constraint that $\gamma'(s_1) = T_1$ reduces to

$$T_1 = \gamma'(t_1) = \sum_{i=0}^{n} P_i \mathcal{B}'_{i,\kappa,\boldsymbol{t}_0}(s_1),$$

which, again, is linear in the curve's coefficients or control points, P_i.

The ability to represent the interpolation constraints as linear constraints greatly simplifies the solution process. In order to support multiresolution curves while imposing constraints on the curve, it is necessary to guarantee that the technqiues can preserve the constraints. Thus, supposing that γ_{curr} satisfies the constraints, the minimal solution to the multiresolution change, $\Delta(t)$, must satisfy $\Delta(s_0) = 0$ for $\gamma_{new}(s_0)$ to remain at Q_0. Similarly $\Delta(t)$ must satisfy $\Delta'(s_1) = 0$ for $\gamma'_{new}(s_1)$ to remain correct.

Other linear constraints may be supported with the same ease and so, other constraints are sought that are linear or can be linearized. Here, we present two more examples, a symmetry constraint and an area constraint.

Example 9.23. A planar curve $C(t) = (x(t), y(t))$, $t \in [0,1]$ is called *symmetric with respect to the y-axis* if,

$$x(t) = -x(1-t), \qquad y(t) = y(1-t).$$

Now, suppose $C(t) = \sum_{i=0}^{n} P_i B_i(t)$, $P_i = (x_i, y_i)$, $t \in [0, 1]$ is symmetric with respect to the y-axis. The two equality constraints can be reduced to a set of equality constraints on the coefficients of $C(t)$, because the B-spline functions form a basis for the space:

$$x_i = -x_{n-i}, \qquad y_i = y_{n-i}, \qquad i = 0, \left\lceil \frac{n}{2} \right\rceil.$$

Hence, symmetry in a planar curve of $n+1$ control points can be expressed as $\left\lceil \frac{n}{2} \right\rceil$ linear constraints. □

Example 9.24. Employing the Stokes theorem from vector analysis [2], if $C(t) = (x(t), y(t))$, is a closed planar curve, the area enclosed is given by

$$\mathcal{A} = \frac{1}{2} \int (-x'(t)y(t) + x(t)y'(t))\, dt = \frac{1}{2} \int C(t) \times C'(t) dt. \qquad (9.19)$$

Now suppose $C(t) = \sum_{i=0}^{n} P_i \mathcal{B}_{i,\kappa}(t) = \sum_{i=0}^{n} (x_i, y_i) \mathcal{B}_{i,\kappa}(t)$. Let the vectors of the coefficients of the curves be $\boldsymbol{X} = (x_0, x_1, \cdots, x_n)$ and $\boldsymbol{Y} = (y_0, y_1, \cdots, y_n)$. Then,

$$\begin{aligned} 2\mathcal{A} &= \int \left(-\sum_i x_i \mathcal{B}'_{i,\kappa}(t) \sum_j y_j \mathcal{B}_{j,\kappa}(t) + \sum_i x_i \mathcal{B}_{i,\kappa}(t) \sum_j y_j \mathcal{B}'_{j,\kappa}(t) \right) dt \\ &= \sum_i x_i \sum_j y_j \int (-\mathcal{B}'_{i,\kappa}(t) \mathcal{B}_{j,\kappa}(t) + \mathcal{B}_{i,\kappa}(t) \mathcal{B}'_{j,\kappa}(t))\, dt. \end{aligned}$$

The area of a closed parametric curve could then be written in matrix form as,

$$2\mathcal{A} = \begin{bmatrix} x_0, x_1, \cdots, x_n \end{bmatrix} \begin{bmatrix} \pi_{00}, & \pi_{01}, & \cdots, & \pi_{0n} \\ \pi_{10}, & \pi_{11}, & \cdots, & \pi_{1n} \\ \vdots & \vdots & \ddots & \vdots \\ \pi_{n0}, & \pi_{n1}, & \cdots, & \pi_{nn} \end{bmatrix} \begin{bmatrix} y_0 \\ y_1 \\ \vdots \\ y_n \end{bmatrix} = \boldsymbol{X}\, \boldsymbol{\Pi}\, \boldsymbol{Y}. \qquad (9.20)$$

where, recalling the derivative Equation 6.17 of B-spline basis functions,

$$\begin{aligned} \pi_{ij} &= \kappa \int \Bigg(-\left(\frac{\mathcal{B}_{i,\kappa-1}(t)}{t_{i+\kappa} - t_i} - \frac{\mathcal{B}_{i+1,\kappa-1}(t)}{t_{i+\kappa+1} - t_{i+1}} \right) \mathcal{B}_{j,\kappa}(t) \\ &\quad + \left(\frac{\mathcal{B}_{j,\kappa-1}(t)}{t_{j+\kappa} - t_j} - \frac{\mathcal{B}_{j+1,\kappa-1}(t)}{t_{j+\kappa+1} - t_{j+1}} \right) \mathcal{B}_{i,\kappa}(t) \Bigg) dt. \end{aligned}$$

Unfortunately, it is necessary to compute the integrals of products of B-spline curves, a complex process for which there are algorithms [60, 35]. Fortunately, once the space of $C(t)$ is fixed, the coefficients of $\mathbf{\Pi} = (\pi_{i,j})$ can be evaluated and so this expensive product computation is required neither frequently nor during the process of selection and dragging a curve point. Yet, Equation 9.20 is quadratic in $\boldsymbol{X}$ and $\boldsymbol{Y}$. In order to linearize the problem, it is possible to solve for the X axis while ignoring the area constraint, and then solve for the Y axis with the area constraint and $\boldsymbol{X}$ that is known. Then, in the next iteration, or mouse event during the drag operation, the roles of X and Y are exchanged. A solution for the Y axis is now derived first while ignoring the area constraint, only to be satisfied with the solution of the X axis while $\boldsymbol{Y}$ is known [36]. □

Exercises

1. Show that Equation 9.10 finds coefficients for Δ.

2. Given the knot set $\boldsymbol{t} = \{0,0,0,1,2,3,3\}$, use quasi-interpolants to represent the function $f(t) = t^2$ as a quadratic B-spline.

3. Prove Lemma 9.22.

10

Other Polynomial Bases for Interpolation

We introduce the classical ideas of interpolation and approximation of preexisting data. While some of these ideas have been used extensively in many areas of applied mathematics, others of them were useful mainly to prove other theorems. However, the emergence of computer aided geometric design has modified accepted ideas of the usefulness of classical methods. Some of the already popular ones were used in fresh new ways to provide the underpinnings for some curve and surface modeling efforts (Coons' surfaces), while others which had been deemed not practical have become workhorses in CAGD through the reassessment of the methods (Bernstein/Bézier curves and surfaces).

This chapter presents the most traditional form of curve and surface modeling, that of polynomial interpolation. In its most widespread form, the interpolation problem addresses the passing of a curve from a preselected class of functions through a set of ordered points. These points may have been the result of experiments and exist only at discrete positions, or the closed form of the original curve, the *primitive function*, may be known but not easily computed. In either case, interpolation theory assumes the existence of such a primitive function, and then analysis can be done to test closeness of the interpolant to the primitive with a variety of analytical measures.

Early geometric modeling used this method to fit curves, and it continues to be widely used today, although it is not completely satisfactory. The theoretically *bad* cases arise frequently in practice. Nonetheless, there is a strong intuitive appeal to the idea of modeling by selecting points and then passing a curve through those points which mimic their position and the general shape. Unfortunately, the classes of functions from which one usually chooses the interpolant, mainly polynomials, do not allow for this capability. Modifications of positional interpolation, for example adding more information or using piecewise methods, have been developed to overcome the limitations on polynomial interpolation. They have had mixed success. Other classes of functions from which to choose the interpolant have been tried. In this chapter interpolation using polynomials and several piecewise generalizations are introduced. Generalizations of the interpolating conditions to be more inclusive are introduced.

In this chapter we use a *blending function* approach to solving the interpolation problem. That is, if the original data is $\{(x_i, y_i)\}_{i=0}^{n}$, we would like to find functions $\{f_i(t)\}_{i=0}^{n}$ such that the function $f(t) = \sum_{i=0}^{n} y_i f_i(t)$ is the solution to the proposed approximation problem. This is called a *blending function* formulation since the answer is a *blend* of the data. The matrix formulation would be

$$f(t) = \begin{bmatrix} f_1(t) & f_2(t) & \dots & f_n(t) \end{bmatrix} \begin{bmatrix} y_1 & y_2 & \dots & y_n \end{bmatrix}^T.$$

We can see here that if y_2 is changed, re-evaluating the curve is simply a matrix multiplication; the f-matrix does not require any re-evaluation.

10.1 Position Interpolation

We start our discussion of interpolation with the most frequently used form. Given data of the form $\{(x_i, y_i)\}_{i=0}^{n}$ it is necessary to find a unique polynomial, p, which passes through that data. That is, $p(x_i) = y_i$, $i = 0, \dots, n$.

The blending formulation approach requires polynomial blending functions $L_i(x)$, $i = 0, \dots, n$, with interpolant $p(x) = \sum_{j=0}^{n} y_j L_j(x)$ such that $p(x_i) = \sum_{j=0}^{n} y_j L_j(x_i) = y_i$. Determining a collection of $n+1$ polynomials $\{L_j(x)\}_{j=0}^{n}$ such that $L_j(x_i) = 0$ when $i \neq j$, and $L_i(x_i) = 1$, would solve the problem. Hence, we try to solve the problem of finding $n+1$ polynomials that incorporate those properties; i.e., $L_j(x_i) = \delta_{i,j}$, for $i = 0, \dots, n, j = 0, \dots, n$.

The n roots of L_i occur at all the abscissa values of the data, except x_i. The fundamental theorem of algebra applied to the root requirements gives

$$\begin{aligned} L_i(x) &= A_i(x - x_0)(x - x_1) \cdots (x - x_{i-1})(x - x_{i+1}) \cdots (x - x_n) \\ &= A_i \prod_{j=0, j \neq i}^{n} (x - x_j). \end{aligned}$$

L_i is a polynomial of degree n. Finally, the condition that $L_i(x_i) = 1$ is the normalization condition. We solve for A_i by evaluating the function.

$$\begin{aligned} L_i(x_i) &= 1 \\ &= A_i(x_i - x_0)(x_i - x_1)\cdots(x_i - x_{i-1})(x_i - x_{i+1})\cdots(x_i - x_n). \end{aligned}$$

As long as $x_0 < x_1 < \cdots < x_i < \cdots < x_n$ the value on the right side of the equation is unequal to zero and it is possible to solve for A_i and find that, for $i = 0, \ldots, n$,

$$L_i(x) = \frac{(x - x_0)(x - x_1)\cdots(x - x_{i-1})(x - x_{i+1})\cdots(x - x_n)}{(x_i - x_0)(x_i - x_1)\cdots(x_i - x_{i-1})(x_i - x_{i+1})\cdots(x_i - x_n)}.$$

This straightforward approach shows that using the blending formulation, we can solve the interpolation problem *as long as the values of the abscissas are all distinct.*

Example 10.1. Find the interpolant to the data (0, 1), (2, 4), (3, 4) (see Figure 10.1). By the fundamental theorem of algebra,

$$\begin{aligned} L_1(x) &= A_1(x - 2)(x - 3); \\ L_2(x) &= A_2(x - 0)(x - 3); \\ L_3(x) &= A_3(x - 0)(x - 2). \end{aligned}$$

Using normalization properties,

$$\begin{aligned} L_1(0) &= 1 \\ &= A_1(0 - 2)(0 - 3) \\ &= 6A_1, \end{aligned}$$

so $A_1 = 1/6$. Analogously,

$$\begin{array}{llll} L_2(2) = & 1 & = A_2(2 - 0)(2 - 3) = -2A_2; & \text{so } A_2 = -1/2; \\ L_3(3) = & 1 & = A_3(3 - 0)(3 - 2) = 3A_3; & \text{so } A_3 = 1/3. \end{array}$$

Then the final blending functions are:

$$\begin{aligned} L_1(x) &= \frac{(x - 2)(x - 3)}{6}; \\ L_2(x) &= -\frac{(x - 0)(x - 3)}{2}; \\ L_3(x) &= \frac{(x - 0)(x - 2)}{3}; \end{aligned}$$

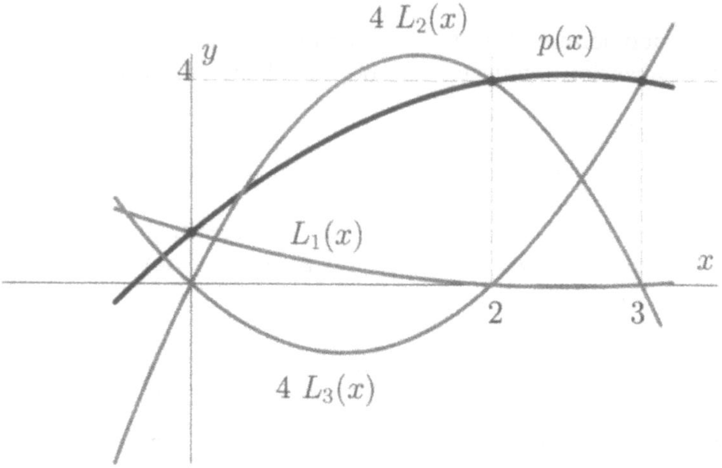

Figure 10.1. $L_1(x)$, $L_2(x)$, and $L_3(x)$ in gray.

and the interpolating polynomial is

$$\begin{aligned} p(x) &= 1L_1(x) + 4L_2(x) + 4L_3(x) \\ &= \frac{(x-2)(x-3)}{6} - \frac{4(x-0)(x-3)}{2} + \frac{4(x-0)(x-2)}{3}. \end{aligned} \quad \square$$

This is called the *Lagrange Formula.*

For each set of $n+1$ abscissa values, there are $n+1$ different polynomials of degree n which are blended together to form the interpolant. Further, any set of $n+1$ ordinate values can be used with these functions. Hence, we ask if these functions form a basis for the space of polynomials of degree n. Suppose $L_i(x)$, $i = 0, \ldots, n$ are the *Lagrange blending functions* defined by abscissa values x_j, $j = 0, \ldots, n$. To show linear independence of the set, we assume there is a set of coefficients such that

$$0 \equiv c_0 L_0(x) + c_1 L_1(x) + \cdots + c_n L_n(x).$$

Now,

$$\begin{aligned} 0 &= c_0 L_0(x_i) + c_1 L_1(x_i) + \cdots + c_n L_n(x_i) \\ &= c_i \end{aligned}$$

and each c_i must be zero. Hence this set of $n+1$ functions forms a linearly independent set and spans a subspace of $\mathcal{P}_n$, so they form a basis for $\mathcal{P}_n$. They are called the *Lagrange basis functions.* While this formulation appears straightforward and polynomials have many nice analytic properties, there are nevertheless distinct problems with polynomial interpolation.

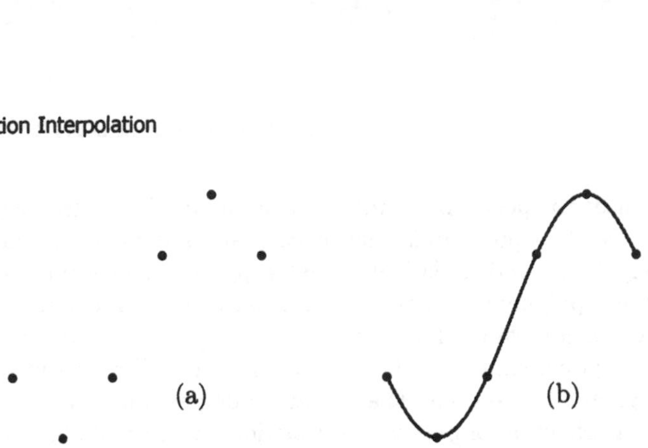

Figure 10.2. (a) Data points to be interpolated; (b) polynomial interpolant.

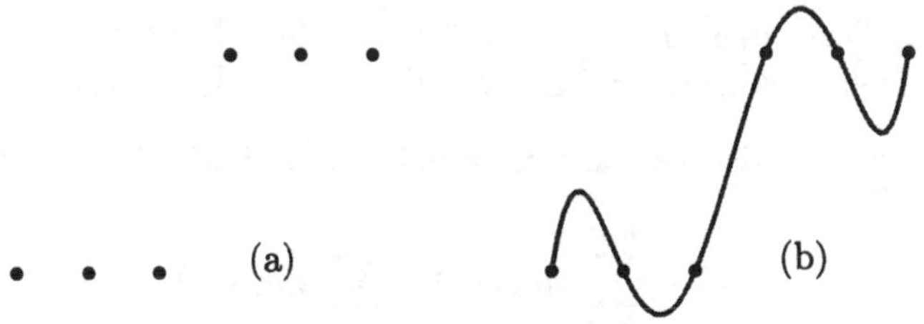

Figure 10.3. (a) Data points to be interpolated; (b) polynomial interpolant.

When many points must be interpolated, the polynomial has high degree, but high degree interpolation can lead to undesirable undulations. Large numbers of points in the data set create a greater potential for error in determining the coefficients for the power basis to form the interpolating polynomial, or for difficulty in computing the Lagrange basis functions. Further, once these functions are obtained, high degree and floating point errors compound when arriving at a specific function value at an abscissa value.

Since using only position values to constrain the interpolant can lead to undulating functions, there have been attempts to lessen these effects by incorporating shape information into the interpolating data. The shape information usually incorporated are derivative values at the position data points. Many of the methods developed are variants of the method discussed below.

10.1.1 Hermite Interpolation

This approach to incorporating shape into the interpolant uses two pieces of information at each point of interpolation. The first is the position, and the second is the required derivative value. Since the first derivative gives information on the increasing/decreasing nature of a function, it then allows the user to specify more information about the desired curve. The formulation for this problem is that at each abscissa value x_i, one is given both a height, y_i, and a derivative value y'_i, and one wants the lowest degree unique polynomial $p(x)$ such that $p(x_i) = y_i$ and $p'(x_i) = y'_i$. If one is given m abscissa values, and hence $2m$ pieces of information, the interpolating polynomial is of degree $2m-1$ or less. First we investigate if such a polynomial exists, and then what its degree must be. We continue to look for a collection of polynomial functions which will allow a *blending function* representation for the solution. That is, we seek polynomials, $h_{i,j}$, $j = 0, 1, i = 0, \ldots, m-1$, each of degree $2m-1$ such that

$$\begin{array}{lll} h_{i,0}(x_j) = \delta_{i,j}, & h'_{i,0}(x_j) = 0, & j = 0, \ldots, m-1, \; i = 0, \ldots, m-1 \\ h_{i,1}(x_j) = 0, & h'_{i,1}(x_j) = \delta_{i,j}, & j = 0, \ldots, m-1, \; i = 0, \ldots, m-1. \end{array}$$

If these functions exist we can form a polynomial $h(x)$ of degree $2m-1$,

$$h(x) = \sum_{i=0}^{m-1} y_i h_{i,0}(x) + \sum_{i=0}^{m-1} y'_i h_{i,1}(x)$$

which has the following properties:

- $h(x_j) = \sum_{i=0}^{m-1} y_i h_{i,0}(x_j) + \sum_{i=0}^{m-1} y'_i h_{i,1}(x_j) = y_j$;
- $h'(x_j) = \sum_{i=0}^{m-1} y_i h'_{i,0}(x_j) + \sum_{i=0}^{m-1} y'_i h'_{i,1}(x_j) = y'_j$.

That is, the function $h(x)$ will be the polynomial that we seek which interpolates to both the positions and the derivatives at the prescribed locations, and has the data as coefficients.

In order to find the form of these functions we use the general form of the fundamental theorem of algebra to get

$$h_{i,0}(x) = q_i(x) \prod_{j=0}^{i-1} (x - x_j)^2 \prod_{j=i+1}^{m-1} (x - x_j)^2; \tag{10.1}$$

$$h_{i,1}(x) = a_i(x - x_i) \prod_{j=0}^{i-1} (x - x_j)^2 \prod_{j=i+1}^{m-1} (x - x_j)^2. \tag{10.2}$$

Here $h_{i,0}$ has $(m-1)$ quadratic factors and is a polynomial of degree $2m-1$ so $q_i(x) = c_i x + d_i$, for some values c_i and d_i.

Example 10.2. We complete the solution for $m = 2$. For this case there are two points, x_0 and x_1, and the solutions will be cubics. For simplification we shall let $x_0 = 0$ and $x_1 = 1$.

From Equations 10.1 and 10.2, we start with

$$h_{0,0}(x) = q_0(x)(x-1)^2, \qquad h_{1,0}(x) = q_1(x)(x-0)^2,$$

and

$$h_{0,1}(x) = a_0(x-0)(x-1)^2, \qquad h_{1,1}(x) = a_1(x-0)^2(x-1).$$

Now,

$$\begin{aligned} h'_{0,1}(0) &= 1 \\ &= a_0\left[(0-1)^2 + (0-0)2(0-1)\right] \\ &= a_0. \end{aligned}$$

Similarly,

$$\begin{aligned} h'_{1,1}(1) &= 1 \\ &= a_1\left[2(1-0)(1-1) + (1-0)^2\right] \\ &= a_1. \end{aligned}$$

We use both constraints at zero to solve for $h_{0,0}(x)$.

$$\begin{aligned} h'_{0,0}(0) &= 0 \\ &= q'_0(0)(0-1)^2 + q_0(0)2(0-1) \\ &= c_0 - 2d_0; \\ h_{0,0}(0) &= 1 \\ &= (c_0 0 + d_0)(0-1)^2 \\ &= d_0. \end{aligned}$$

Solving this we have $d_0 = 1$ and $c_0 = 2$.

Finally,

$$\begin{aligned} h'_{1,0}(1) &= 0 \\ &= q'_1(1)(1-0)^2 + q_1(1)2(1-0) \\ &= 3c_1 + 2d_1; \\ h_{1,0}(1) &= 1 \\ &= (c_1 1 + d_1)(1-0)^2 \\ &= c_1 + d_1. \end{aligned}$$

Solving that system gives $c_1 = -2$ and $d_1 = 3$. The final cubic blending functions (Figure 10.4) are

$$\begin{array}{ll} h_{0,0}(x) = (2x+1)(x-1)^2, & h_{1,0}(x) = (-2x+3)x^2, \\ h_{0,1}(x) = x(x-1)^2, & h_{1,1}(x) = x^2(x-1). \end{array} \tag{10.3}$$

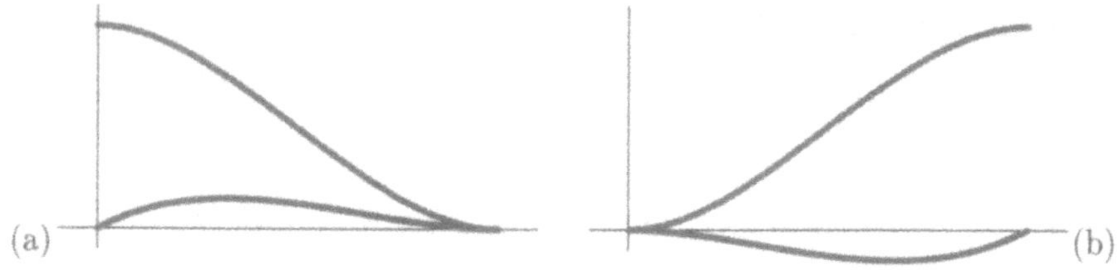

Figure 10.4. (a) Functions $h_{0,0}$ and $h_{0,1}$; (b) functions $h_{1,0}$ and $h_{1,1}$.

Suppose one wants to look at the curve, $c(x)$, which has tangent value 2 at position $(0, 4)$ and tangent value 3 at position $(1, 1)$. The equation would be

$$c(x) = 4h_{0,0}(x) + 1h_{1,0}(x) + 2h_{0,1}(x) + 3h_{1,1}(x).$$ □

Example 10.3. Suppose we want to interpolate data $(x, y, y') : (0, 1, 3)$, $(1, 0, -1)$, then (see Figure 10.5)

$$I(x) = 1h_{0,0}(x) + 0h_{1,0}(x) + 3h_{0,1}(x) + (-1)h_{1,1}(x).$$

Now, if it is decided that the tangent at the point $(0, 1)$ should be 0 instead of 3, one simply changes the third coefficient to get

$$I(x) = (1)h_{0,0}(x) + (0)h_{1,0}(x) + (0)h_{0,1}(x) + (-1)h_{1,1}(x).$$

The connection of the coefficients to the geometry is straightforward and easy to understand as well as manipulate. □

We return to solving the general case for the remaining unknown constants. Now, to find $h_{i,1}(x)$ we need only find the normalizing constant a_i:

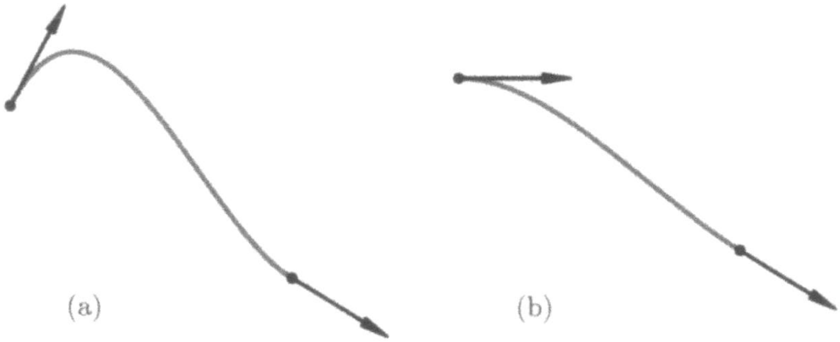

Figure 10.5. Hermite interpolants to data of Example 10.3: (a) initial data; (b) modified data.

$$
\begin{aligned}
h'_{i,1}(x) &= 2a_i \sum_{j=0, j\neq i}^{m-1} (x - x_i)(x - x_j) \prod_{r=0, r\notin\{i,j\}}^{m-1} (x - x_r)^2 \\
&\quad + a_i \prod_{r=0, r\neq i}^{m-1} (x - x_r)^2,
\end{aligned}
$$

so we evaluate at x_i to get

$$
\begin{aligned}
1 &= h'_{i,1}(x_i) \\
&= a_i \left(2 \sum_{j=0, j\neq i}^{m-1} (x_i - x_i)(x_i - x_j) \prod_{r=0, r\notin\{i,j\}}^{m-1} (x_i - x_r)^2 \right. \\
&\quad \left. + \prod_{r=0, r\neq i}^{m-1} (x_i - x_r)^2 \right) \\
&= a_i \left(\prod_{r=0, r\neq i}^{m-1} (x_i - x_r)^2 \right).
\end{aligned}
$$

This gives

$$
a_i = \frac{1}{\prod_{r=0, r\neq i}^{m-1} (x_i - x_r)^2}
$$

and

$$
h_{i,1}(x) = \frac{(x - x_i) \prod_{j=0}^{i-1} (x - x_j)^2 \prod_{j=i+1}^{m-1} (x - x_j)^2}{\prod_{r=0, r\neq i}^{m-1} (x_i - x_r)^2}.
$$

This solution is well defined as long as the abscissa values are all distinct. This is not imposing a constraint since by the definition of function, a single abscissa can have exactly one ordinate value. We are left with the problem of solving for the remaining unknowns in $h_{i,0}$ which are specified by the information at x_i. Note that $q_i(x)$ is just a linear function, $q_i(x) = c_i x + d_i$. We shall define and use some shorthand below. Let $\alpha_i = \prod_{j=0, j\neq i}^{m-1} (x_i - x_j)$ and define $\beta_i = \sum_{j=0, j\neq i}^{m-1} 1/(x_i - x_j)$:

$$
\begin{aligned}
1 &= h_{i,0}(x_i) \\
&= q_i(x_i)\alpha_i^2 \\
&= (c_i x_i + d_i)\alpha_i^2, \qquad (10.4)
\end{aligned}
$$

and

$$\begin{aligned}
0 &= h'_{i,0}(x_i) \\
&= q'_i(x_i)\alpha_i^2 \\
&\quad + 2q(x_i) \sum_{j=0,j\neq i}^{m-1} (x_i - x_j) \prod_{r=0,r\notin\{i,j\}}^{m-1} (x_i - x_r)^2 \\
&= q'_i(x_i)\alpha_i^2 + 2q(x_i)\alpha_i^2\beta_i \\
&= c_i + 2c_i x_i \beta_i + 2d_i\beta_i \\
&= (1 + 2x_i\beta_i)c_i + 2\beta_i d_i. \qquad (10.5)
\end{aligned}$$

Equations 10.4 and 10.5 together give a linear system of two equations in two unknowns:

$$\begin{aligned}
0 &= (1 + 2x_i\beta_i)c_i + 2\beta_i d_i, \\
1 &= x_i\alpha_i^2 c_i + \alpha_i^2 d_i.
\end{aligned}$$

Solving the system gives

$$c_i = -\frac{2\beta_i}{\alpha_i^2}, \qquad d_i = \frac{1 + 2x_i\beta_i}{\alpha_i^2}.$$

We see that this interpolation problem can be solved then for any ordinate values and also any abscissa values, as long as the abscissa values are distinct. We now see that we have $2m$ polynomials, the blending functions, each of degree $2m - 1$. If they are independent, then they are also a basis for the space of polynomials of degree $2m - 1$, which would mean that every polynomial of degree $2m - 1$ could be represented in this form. Then one could experiment with modifying the shape by tweaking some of the values slightly. Or, one could compare the behavior of two different polynomials of degree $2m - 1$ at these points by representing both of them in the Hermite formulation and compare the coefficients of the blending functions. One could then discover if they had the same position values at the desired abscissa and also compare the slopes at those points.

Suppose

$$0 = \sum_{i=0}^{m-1} \left[a_i h_{i,0}(x) + b_i h_{i,1}(x)\right].$$

Evaluating the equation at x_j gives

$$0 = \sum_{i=0}^{m-1} \left[a_i h_{i,0}(x_j) + b_i h_{i,1}(x_j)\right] = a_j, \qquad j = 0, \ldots, m-1.$$

We see then, that the coefficients of the $h_{j,0}$-functions must be 0. Now, evaluating the derivative at each abscissa point gives

$$0 = \sum_{i=0}^{m-1} \left[a_i h'_{i,0}(x_j) + b_i h'_{i,1}(x_j)\right] = b_j, \qquad j = 0, \ldots, m-1.$$

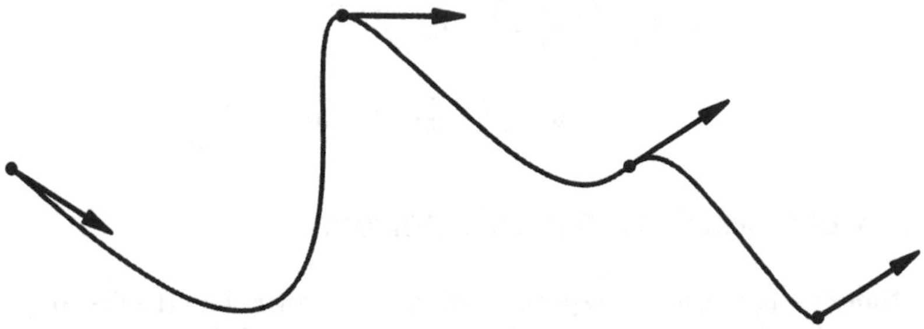

Figure 10.6. Example of Hermite interpolation with four points.

So the coefficients of the $h_{j,1}$ must be 0. Hence, the Hermite blending functions are independent and form a basis for all polynomials of degree $2m-1$.

Example 10.4. Translation of the Cubic Hermite Basis. Suppose we want to interpolate data $(x, y, y') : (1, 1, 3), (2, 0, -1)$ using a shifted version of Example 10.3. It is possible to do this by simply performing a translation to make this new interval look right, instead of working out the Hermite basis functions. Using $\theta(x) = x - 1$, $\theta(1) = 0$ and $\theta(2) = 1$; and $\theta(x) \in (0, 1)$ and is onto since θ is an increasing linear function. If $I_t(x)$ is the interpolant for this example, then

$$\begin{aligned} I_t(x) &= (1)h_{0,0}(\theta(x)) + (0)h_{1,0}(\theta(x)) + (3)h_{0,1}(\theta(x)) + (-1)h_{1,1}(\theta(x)) \\ &= h_{0,0}(x-1) + 3h_{0,1}(x-1) - h_{1,1}(x-1) \\ &= (2(x-1)+1)((x-1)-1)^2 + 3(x-1)((x-1)-1)^2 \\ &\quad - (x-1)^2((x-1)-1) \\ &= (2x-1)(x-2)^2 + 3(x-1)(x-2)^2 - (x-1)^2(x-2). \qquad \square \end{aligned}$$

Example 10.5. Translation and Scaling of Cubic Hermite Bases. In this case it is necessary to interpolate $(x, y, y') : (1, 1, 3), (4, 0, -1)$. Instead of working out the Hermite basis functions, it is possible to perform the translation and scale by $\theta(x) = (x-1)/3$. Then $\theta(1) = 0$ and $\theta(4) = 1$. $\theta(x) \in (0, 1)$ and is onto since it is an increasing linear function. Further, $dh_{i,j}(\theta(x))/dx = h'_{i,j}(\theta(x))\theta'(x)$. $\theta'(x) = 1/3$, so in order to get the derivatives to have a unit value at the endpoints, the new blending functions must be $\frac{h_{i,1}(\theta(x))}{1/3} = 3h_{i,1}(\theta(x))$. If $I_{t,s}(x)$ is the interpolant for this example, then

$$\begin{aligned} I_{t,s}(x) &= (1)h_{0,0}(\theta(x)) + (0)h_{1,0}(\theta(x)) \\ &\quad + (3)3h_{0,1}(\theta(x)) + (-1)3h_{1,1}(\theta(x)) \end{aligned}$$

$$= h_{0,0}\left(\frac{x-1}{3}\right) + 0h_{1,0}\left(\frac{x-1}{3}\right)$$
$$+ 9h_{0,1}\left(\frac{x-1}{3}\right) - 3h_{1,1}\left(\frac{x-1}{3}\right). \qquad \square$$

10.2 Generalized Hermite Interpolation

Sometimes designers need to specify derivatives higher than the first degree, or the results of experiments might contain data with higher derivatives. The generalized Hermite interpolation method is designed to solve these cases.

The problem here is to consider data of the form $\left(x_i, y_i, y_i^{(1)}, \ldots, y_i^{(m_i)}\right)$, for $i = 0, \ldots, n$, and $x_i < x_{i+1}$. There are $N = (1+m_0)+(1+m_1)+\cdots+(1+m_n)$ data values which must be interpolated, so a polynomial with degree as high as $N-1$ may be necessary. While the first derivatives prescribed the tangent direction of the curve, for example, the second derivative would prescribe the curvature of the curve, κ. Given $y = y(x)$,

$$\kappa(x) = \frac{\|y''(x)\|}{\|1+y'(x)\|^{3/2}},$$

which follows from Equation 4.11 in Example 4.33.

Proceeding along the same lines as with the Hermite polynomials and using the blending formulation, we want to find polynomials, $h_{i,j}$, $j = 0, \ldots, m_i$, $i = 0, \ldots, n$, each of degree $N-1$ so that

$$h_{i,j} = q_{i,j}(x)(x-x_i)^j \prod_{k=0,k\neq i}^{n} (x-x_k)^{1+m_k},$$

where $q_{i,j}(x)$ is a polynomial of degree $m_i - j$. This formulation uses the generalization of the fundamental theorem of algebra, but leaves us with the problem of showing the exact formulation for $q_{i,j}(x)$. The existence of the $h_{i,j}$ functions is shown by construction. That is, if they can be constructed they exist, and the problem can be solved. It is straightforward to show that the functions $h_{i,j}$, $j = 0, \ldots, m_i$, $i = 0, \ldots, n$, form a basis for the space of polynomials of degree N using a generalization of the arguments used in the Hermite case. We shall show the more general technique of finding these basis polynomials in the case where we have two points and position, first, and second derivative information specified at each. That is six constraints, and a fifth degree polynomial results. This case is interesting since if one wants to piece together curves with $C^{(2)}$ continuity, one must use quintics of this form.

Example 10.6. The point of this problem is to solve for the generalized Hermite blending functions for the two point problem when the data have the form $\left(x_i, y_i, y_i^{(1)}, y_i^{(2)}\right)$, for $i = 0, 1$. In other words, each data point prescribes the position, the tangent and the curvature of curve at the data point location. Once again we use the simplifying assumption that $x_0 = 0$ and $x_1 = 1$. The generic form yields

$$\begin{aligned} h_{0,j} &= q_{0,j}(x)x^j(x-1)^3, \\ h_{1,j} &= q_{1,j}(x)(x-1)^j x^3. \end{aligned}$$

We shall solve for the functions oriented around $x_0 = 0$, and leave the ones oriented around $x_1 = 1$ as problems for the reader. Now, from the general case $h_{0,2}(x) = q_{0,2}x^2(x-1)^3$. Applying the normalization condition,

$$\begin{aligned} 1 &= h''_{0,2}(0) \\ &= q_{0,2}\left[2(0-1)^3 + 12(0)(0-1)^2 + 6(0)^2(0-1)\right] \\ &= -2q_{0,2}. \end{aligned}$$

Since $q_{0,2} = -1/2$, $h_{0,2}(x) = -\frac{1}{2}x^2(x-1)^3$. The general case gives, $h_{0,1}(x) = q_{0,1}(x)(x-0)(x-1)^3$ so we need only solve for $q_{0,1}(x) = c_0 x + d_0$. Using the root and normalizing conditions,

$$\begin{aligned} 1 &= h'_{0,1}(0) \\ &= c_0(0)(0-1)^3 + (c_0 0 + d_0)(0-1)^3 + 3(c_0 0 + d_0)(0)(0-1)^2 \\ &= -d_0; \\ 0 &= h''_{0,1}(0) \\ &= 2c_0(0-1)^3 + (12c_0 0 + 6d_0)(0-1)^2 + 6(c_0 0 + d_0)(0)(0-1) \\ &= -2c_0 + 6d_0 \\ &= -c_0 + 3d_0. \end{aligned}$$

We have determined that $d_0 = -1$ and $c_0 = -3$. Thus, $h_{0,1}(x) = -(3x+1)x(x-1)^3$. The last one has a bit more to solve, for here, $h_{0,0}(x) = q_{0,0}(x)(x-1)^3$, where $q_{0,0}(x) = u_i x^2 + v_i x + w_i$.

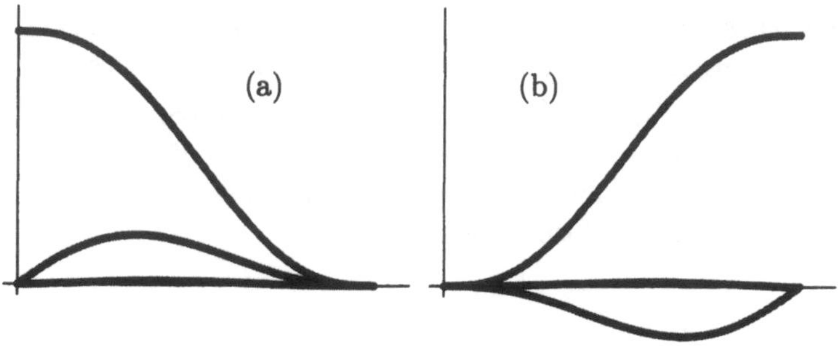

Figure 10.7. (a) Functions $h_{0,0}$, $h_{0,1}$ and $h_{0,2}$; (b) functions $h_{1,0}$, $h_{1,1}$ and $h_{1,2}$.

$$\begin{aligned}
h_{0,0}(0) &= 1 \\
&= w_i(-1); \qquad \text{so } w_i = -1; \\
h'_{0,0}(0) &= 0 \\
&= (2u_i(0) + v_i)\,(0-1)^3 + (u_i 0^2 + v_i 0 - 1)3(0-1)^2 \\
&= -v_i - 3; \qquad \text{so } v_i = -3; \\
h''_{0,0}(0) &= 0 \\
&= 2u_i(0-1)^3 + 6(0-1)^2(2u_i 0 + v_i) + 6(0-1)(u_i 0^2 + v_i 0 + w_i) \\
&= -2u_i - 12 \qquad \text{and } u_i = -6.
\end{aligned}$$

Hence $h_{0,0}(x) = (-6x^2 - 3x - 1)(x-1)^3$.

Similar derivations for the other three Hermite basis functions yield

$$\begin{aligned}
h_{1,0}(x) &= (6x^2 - 15x + 10)x^3; \\
h_{1,1}(x) &= (-3x + 4)(x-1)x^3; \\
h_{1,2}(x) &= \frac{1}{2}(x-1)^2 x^3.
\end{aligned}$$

Figure 10.7 shows these six basis functions. □

Example 10.7. Consider a generalized Hermite interpolation for the following data:

- Position: (0, 0.3), (1, 0.7),
- Tangents at each point respectively (slopes): 0, -1,
- Second derivative values: 5, 5.

The result is shown in Figure 10.8 (a). If the position and tangent constraints are kept the same, but the second derivative is constrained to have

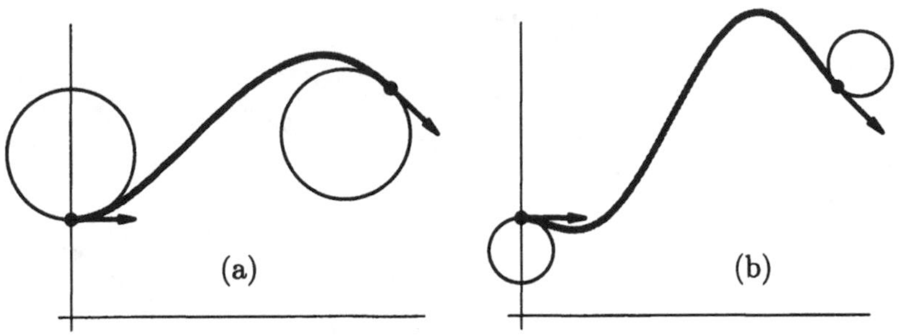

Figure 10.8. Generalized quintic Hermite interpolant to the data of Example 10.7.

- Second derivative values: -10, -10,

the resulting curve is shown in Figure 10.8 (b). This new curve interpolates the same end locations and presents similar tangents while the curvature at the end points is the opposite and with twice the magnitude. □

Example 10.8. Choosing mixed conditions gives rise to a slightly different type of example. The problem is to interpolate the following conditions:

1. Position: $(-.8, 0)$, $(-.1, .5)$, $(.5, 0)$, $(.9, -.5)$;
2. Tangents at each point respectively (slopes): -1, 0, 1, 1;
3. Second derivative values at the first and third point: 0 (i.e., flat curvature).

The position and first derivative values are the same as in Figure 10.6 with several second derivatives added. The resulting polynomial has degree six and is shown in Figure 10.9.

For comparison, the original four point Hermite and this generalized Hermite polynomial are shown in Figure 10.10. □

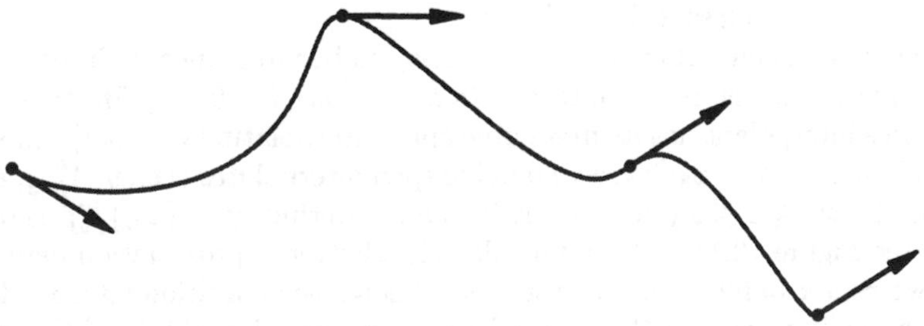

Figure 10.9. Generalized Hermite interpolant to the data of Example 10.8.

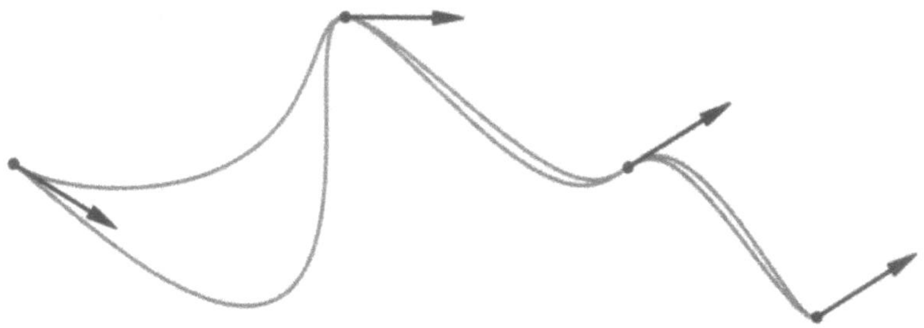

Figure 10.10. Comparison of Hermite, and generalized Hermite interpolants.

Thus, one can specify second derivatives and have the ability to completely manipulate curvature of the curve. Then for two points, instead of having a cubic polynomial, a quintic polynomial would result. For three points it would be eighth degree, etc. This can quickly get out of hand, leading to higher and higher degree polynomials with the attendent computational problems.

However, most designers cannot see third derivative discontinuities, since curvature is a continuous function if the interpolant has second derivative continuity. Also, many analysis methods do not assume more than second order continuity. In fact, frequently, tangent continuity is sufficient for many purposes. Thus, one could trade the single polynomial representation for a piecewise polynomial representation. One would give up derivative continuity above that presecribed by the method, but one could gain a fixed low order method.

10.3 Piecewise Methods

There are several commonly used methods which seek to alleviate the problem of high degree polynomials, whether obtained from many position data, or from a combination of positional and derivative information. Cubic methods may be classified as follows:

Calculate a new cubic every four spans, similar to numerical integration techniques. That is, if the data points are $(x_i, y_i), i = 0, \ldots, 3n$, then the piecewise interpolant has its first cubic span interpolating $\{(x_i, y_i)\}_{i=0}^{3}$, and is valid for $x_0 \le x \le x_3$; its second cubic span interpolates $\{(x_i, y_i)\}_{i=3}^{6}$, and is valid for $x_3 \le x \le x_6$, and so forth. Thus, continuity at $\{x_{3i}\}_{i=1}^{n-1}$ is only $C^{(0)}$ (see Figure 10.11). The interpolant is a better approximation between the two center points of each cubic. It is known that position interpolation gives a smaller error to the original function near the middle of the data to be interpolated. This method leads to an uneven distribution of error.

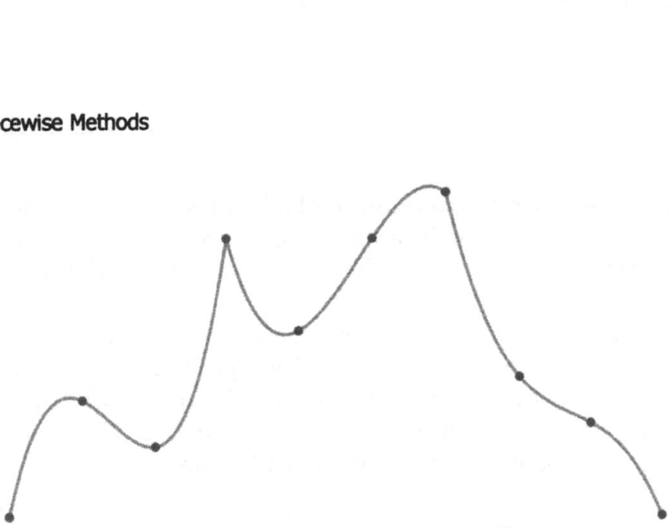

Figure 10.11. One span every four points.

Out of each cubic span of four points and three intervals between them, the center interval has the smallest error bound while the two exterior intervals have greater error. Thus, using this method, two thirds of the intervals between data points have larger error than need be.

The next method attempts to overcome this difficulty. This method uses overlapping cubics. For each interval between abscissa values, evaluate a new cubic polynomial based on the four positions around that interval, that is the cubic to be used for $x_i \leq x \leq x_{i+1}$ is the interpolant to the data $(x_j, y_j), j = i-1, \ldots, i+2$. So there is a new cubic on every interval, lots of calculation and matrix solving; and only $C^{(0)}$ continuity. However, the error function is much more tractable since only the center interval of each cubic calculated is used. This method is fairly common.

A third method calculates a new cubic for evaluation between each successive pair of data values but requires specification of tangent values

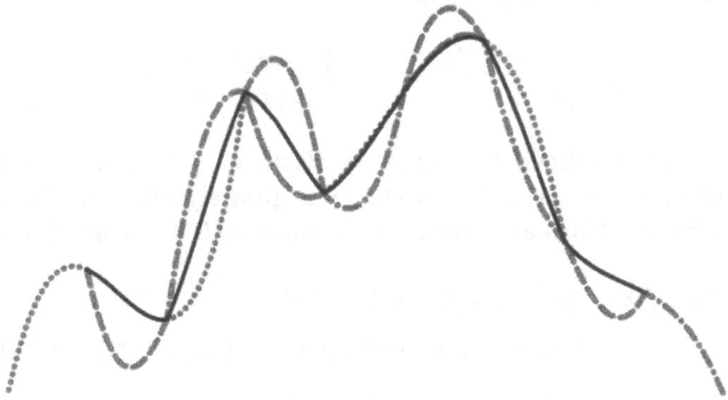

Figure 10.12. Overlapping cubics.

at each point, a piecewise Hermite method. This method is discussed in detail below. It requires specification of tangent values at every data point just as the regular hermite case. The price for the lower cubic order is that only $C^{(1)}$ continuity results.

The last method calculates a new cubic for evaluation between each successive pair of data values. It requires specification of position and requires continuity of the tangents and second derivative at the data points, but the exact derivative values must not be specified. The results of this method are called splines and are discussed in detail below.

10.3.1 Piecewise Hermite Interpolation

Even with Hermite interpolation, when the degree of the polynomial becomes large, the interpolant has wiggles which do not reflect the data. Using lower degree piecewise Hermite methods is an attempt to keep the shape preserving features of the Hermite method, and avoid the undesirable problems of high degree. Frequently, cubics are used, so we discuss that problem.

Between each pair of abscissa values a cubic Hermite polynomial is interpolated. Since the data requires it, the pieces meet with $C^{(1)}$ continuity, that is they match in position and tangent values for each data value of x. However, usually there is no higher order continuity. The specific statement is to interpolate $\{(x_i, f_i, f_i')\}_{i=0}^{N}$ with cubic polynomial pieces $P_i(x)$, $i = 0, \ldots, N-1$ such that

$$
\begin{aligned}
P_{i-1}(x_i) &= P_i(x_i) = f_i, \\
P'_{i-1}(x_i) &= P'_i(x_i) = f'_i = s_i,
\end{aligned}
$$

for values of i, $i = 1, \ldots, n-1$ and

$$
\begin{aligned}
P_0(x_0) &= f(x_0) &= f_0, \qquad & P_{n-1}(x_n) &= f(x_n) &= f_n, \\
P'_0(x_0) &= f'(x_0) &= s_0, \qquad & P'_{n-1}(x_n) &= f'(x_n) &= s_n.
\end{aligned}
$$

The interpolant is defined to be $p(x) = P_i(x)$, $x_i \le x < x_{i+1}$, where each $P_i(x)$ is the linear combination of the appropriately translated and scaled cubic Hermite functions as derived in Example 10.5, to be $g_i(x) = \frac{x - x_i}{x_{i+1} - x_i}$:

$$
\begin{aligned}
P_i(x) = {} & f_i h_{0,0}(g_i(x)) + f_{i+1} h_{1,0}(g_i(x)) \\
& + s_i(x_{i+1} - x_i) h_{0,1}(g_i(x)) + s_{i+1}(x_{i+1} - x_i) h_{1,1}(g_i(x)).
\end{aligned}
$$

Why should anyone choose (generalized) Hermite or piecewise (generalized) Hermite interpolation over polynomial interpolation? The tangent

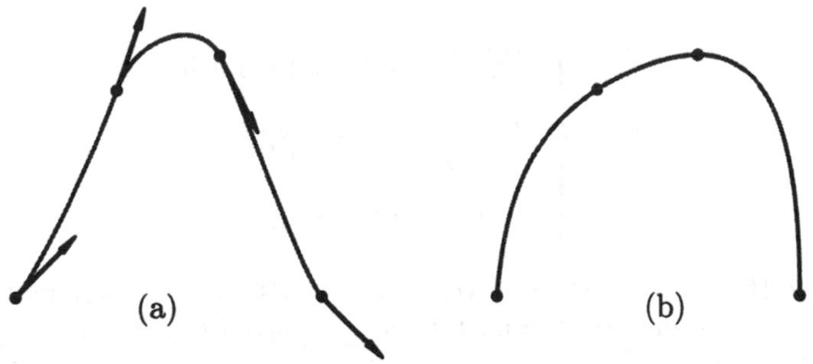

Figure 10.13. (a) Piecewise cubic Hermite interpolant; (b) single Hermite interpolant to the data.

specification gives additional degrees of freedom. One can use this to iron out the wiggles in the interpolant. The piecewise capability lets one iron out the wiggles and other problems caused by potentially high degree interpolants. There is a price, however, for this additional flexibility. The user must specify values for the additional needed information, the derivative, and higher derivative values. Usually this information is unavailable, and some heuristic or ad hoc procedure must be used to guess at a value of the tangent. Even when the user has some idea for the tangent value, he usually has no idea what the second or higher derivative values should be.

10.3.1.1 A constructive approach to smooth piecewise interpolation. This method uses only positional data, low order polynomial pieces, and a local blending to try to smooth out the undulations caused by interpolation. Suppose it is desired to interpolate parametrically (or explicitly) data of the form (t_i, p_i), $i = 1, \ldots, n$ and there are available two extra points, one at the beginning, (t_0, p_0), and one at the end, (t_{n+1}, p_{n+1}). The additional points will help with local smoothing. Let $q_i(t)$ be the quadratic function that interpolates (t_{i-1}, p_{i-1}), (t_i, p_i), and (t_{i+1}, p_{i+1}). Then,

$$q_i(t) = p_{i-1} + (t - t_{i-1})\frac{p_i - p_{i-1}}{t_i - t_{i-1}} + (t - t_{i-1})(t - t_i)\frac{\frac{p_{i+1}-p_i}{t_{i+1}-t_i} - \frac{p_i - p_{i-1}}{t_i - t_{i-1}}}{t_{i+1} - t_{i-1}}$$

and

$$q_i'(t_i) = \frac{t_i - t_{i-1}}{t_{i+1} - t_{i-1}}\frac{p_{i+1} - p_i}{t_{i+1} - t_i} + \frac{t_{i+1} - t_i}{t_{i+1} - t_{i-1}}\frac{p_i - p_{i-1}}{t_i - t_{i-1}}.$$

Interpolation error is, in general, smallest in the center of the interval of interpolation, that is, for $q_i(t)$ it would be near t_i. Hence, the new interpolant should use the information from q_i around t_i. The difficulty is figuring a way to accomplish this. The most straightforward way is to define

$$l_i(t) = \begin{cases} \frac{t-t_{i-1}}{t_i-t_{i-1}}, & \text{for } t \in [t_{i-1}, t_i], \\ \frac{t_{i+1}-t}{t_{i+1}-t_i}, & \text{for } t \in [t_i, t_{i+1}], \\ 0, & \text{otherwise.} \end{cases}$$

Then $q_i(t)l_i(t) = 0$ for $t \notin (t_{i-1}, t_{i+1})$, $q_i(t_i)l_i(t_i) = p_i$, and it is a piecewise cubic function. Define $I(t) = \sum_{i=1}^{n} q_i(t)l_i(t)$. For $t \in (t_j, t_{j+1})$, $I(t) = \sum_j^{j+1} q_i(t)l_i(t)$, since any interval occurs in the support of exactly two of the piecewise linear functions. Also $I(t_j) = q_j(t_j)l_j(t_j) = q_j(t_j)$, for all j, so $I(t)$ is a piecewise continuous cubic interpolant to the data. Further, this formulation blends the $q_i(t)$'s in such a way that the weight of q_i, is strongest near t_i.

Does $I(t)$ have a derivative at each t_j, and if so, is that derivative continuous? For $t \in (t_{j-1}, t_j)$,

$$I(t) = \sum_{j-1}^{j} q_i(t)l_i(t),$$

and

$$I'(t) = \sum_{j-1}^{j} \left[q_i'(t)l_i(t) + q_i(t)l_i'(t)\right].$$

But,

$$l_j'(t) = \frac{1}{(t_j - t_{j-1})} \quad \text{and} \quad l_{j-1}'(t) = \frac{-1}{(t_j - t_{j-1})}.$$

So

$$I'(t) = \sum_{j-1}^{j} \left[q_i'(t)l_i(t)\right] + \frac{q_j(t) - q_{j-1}(t)}{t_j - t_{j-1}}$$

and

$$\begin{aligned} \lim_{t \to t_j^-} I'(t) &= \sum_{j-1}^{j} \left[q_i'(t_j)l_i(t_j)\right] + \frac{q_j(t_j) - q_{j-1}(t_j)}{t_j - t_{j-1}} \\ &= q_j'(t_j) + \frac{p_j - p_j}{t_j - t_{j-1}} \\ &= q_j'(t_j). \end{aligned}$$

Analogously, $\lim_{t \to t_j^+} I'(t) = q_j'(t_j)$. Thus, since the right and left derivatives exist and are equal, the piecewise cubic interpolant $I(t)$ is $C^{(1)}$ with derivative equal to $q_j'(t_j)$ at t_j.

Does this curve have second derivatives? For $t \in (t_{j-1}, t_j)$,

$$\begin{aligned} I''(t) &= \sum_{j-1}^{j} [q_i''(t)l_i(t) + 2q_i'(t)l_i'(t) + q_i(t)l_i''(t)] \\ &= \sum_{j-1}^{j} [q_i''(t)l_i(t) + 2q_i'(t)l_i'(t)], \end{aligned}$$

since the l's are linear in the open interval and have second derivative zero. Now, going through the limiting process for the second derivative gives

$$\lim_{t \to t_j^-} I''(t) = q_j''(t_j) + 2\frac{q_j'(t_j) - q_{j-1}'(t_j)}{t_j - t_{j-1}}$$

and analogously,

$$\lim_{t \to t_j^+} I''(t) = q_j''(t_j) + 2\frac{q_{j+1}'(t_j) - q_j'(t_j)}{t_{j+1} - t_j}.$$

The two limits are equal, and the second derivative exists and is continuous, if and only if

$$\frac{q_j'(t_j) - q_{j-1}'(t_j)}{t_j - t_{j-1}} = \frac{q_{j+1}'(t_j) - q_j'(t_j)}{t_{j+1} - t_j}.$$

This requirement is, in general, not met.

Hence this curve form is $C^{(1)}$. It is in fact identical to the piecewise cubic Hermite interpolant using the tangent value of the local quadratic interpolant, $q_j'(t_j)$ at t_j. That is, piecewise cubic Hermite interpolation is applied to the data $(t_j, p_j, q_j'(t_j)), j = 1, \ldots, n$. This method is useful when it is necessary that the interpolant be a piecewise polynomial of low degree, such as quadratic or cubic and no derivative information is known. It supplies tangent information from the local quadratic interpolant to be used in piecewise Hermite interpolation of the data. The method of estimating tangent values at data abscissa t_i by evaluating the tangents of the quadratic interpolating polynomial to the position data at t_{i-1}, t_i, t_{i+1}, and then interpolating using piecewise cubic Hermite functions using those tangent values with the position information is called *cubic Bessel interpolation.*

Once the form of the derivative for the quadratic functions q_i at t_i is derived, it is necessary to evaluate these n derivatives. If the t_i's are evenly spaced, the local cubic Hermite basis functions can be computed once and tabled for graphic and other calculations. If the tangents are not known for piecewise cubic Hermite interpolation, one might want to use these tangent values as an initial guess in an iterative fitting scheme.

10.4 Parametric Extensions

All the explicit forms developed in the preceding sections can be modified to be parametric methods by simply inserting a vector coefficient value instead of a scalar coefficient value and simply solving the system three times for the coefficients, one for each dimension. But identical methods using identical data with different parameterizations lead to different curves being defined, all of which will still interpolate the data. Further, solving for the coefficients can require much work. The Hermite methods pose special difficulties since, now the derivative that must be specified is a vector and has magnitude as well as direction. If two sets of data have the same position points and the same derivative directions, but different magnitudes at the respective points, the resulting interpolants can be startlingly different. We have seen in the section on geometry that a curve can look smooth and be smooth in the arc length parameterization without being smooth in the given parameterization. Therefore, in parametric applications of the piecewise Hermite form, one frequently matches the tangent direction at the joins and uses the extra scalar degree of freedom to set the relative magnitudes of these tangent vectors.

The blending formulation shows up quite strongly here, for one can modify the actual physical data locations without modifying their associated parameter values. For example, the Lagrange form for a space curve would be (t_i, x_i, y_i, z_i), where one would solve the Lagrange problem for the data sets (t_i, x_i), (t_i, y_i), and (t_i, z_i) independently, obtaining x as a polynomial function of t, y as a polynomial function of t, and z as a polynomial function of t. Each of the three polynomial functions use the same blending functions since their parameter space abscissa values are all the same. Thus, the parametric interpolant would have the form

$$\begin{aligned} p(t) &= \left(\sum_{i=1}^{n} x_i L_i(t), \sum_{i=1}^{n} y_i L_i(t), \sum_{i=1}^{n} z_i L_i(t)\right), \\ &= \sum_{i=1}^{n} (x_i, y_i, z_i) L_i(t), \\ &= \sum_{i=1}^{n} D_i L_i(t), \end{aligned}$$

using vector space properties of linearity and scalar multiplication, and setting $D_i = (x_i, y_i, z_i), i = 1, \ldots, n$. One can evaluate the curve at many points by determining parameter values t_j, $j = 0, \ldots, k$, at which one wants the interpolant evaluated, and then evaluating and tabling these values. Then,

$$\begin{bmatrix} p(t_0) \\ p(t_1) \\ \vdots \\ p(t_{n-1}) \\ \vdots \\ p(t_{k-1}) \\ p(t_k) \end{bmatrix} = \begin{bmatrix} L_1(t_0) & L_2(t_0) & \dots & L_n(t_0) \\ L_1(t_1) & L_2(t_1) & \dots & L_n(t_1) \\ & & \vdots & \\ L_1(t_{n-1}) & L_2(t_{n-1}) & \dots & L_n(t_{n-1}) \\ & & \vdots & \\ L_1(t_{k-1}) & L_2(t_{k-1}) & \dots & L_n(t_{k-1}) \\ L_1(t_k) & L_2(t_k) & \dots & L_n(t_k) \end{bmatrix} \begin{bmatrix} D_1 \\ D_2 \\ \vdots \\ D_n \end{bmatrix}.$$

If D_j is moved, but the parameter value associated with it is unchanged, re-evaluating the curve is a simple matrix multiplication, since it is unnecessary either to redetermine or to re-evaluate the blending functions.

A more frequently used form of this is the parametric cubic Hermite form. In this case exactly the same reasoning is applied. If P_0 and S_0 are the position and tangent vectors, respectively, at $t = 0$, and P_1 and S_1 are the position and tangent vectors, respectively at $t = 1$, the cubic Hermite interpolant is defined as

$$h(t) = \begin{bmatrix} h_{0,0}(t) & h_{0,1}(t) & h_{1,0}(t) & h_{1,1}(t) \end{bmatrix} \begin{bmatrix} P_0 & S_0 & P_1 & S_1 \end{bmatrix}^T.$$

Once again the values of the blending functions may be tabled to obtain

$$\begin{bmatrix} h(u_0) \\ h(u_1) \\ h(u_2) \\ h(u_3) \\ h(u_4) \\ \vdots \\ h(u_k) \end{bmatrix} = \begin{bmatrix} h_{0,0}(u_0) & h_{0,1}(u_0) & h_{1,0}(u_0) & h_{1,1}(u_0) \\ h_{0,0}(u_1) & h_{0,1}(u_1) & h_{1,0}(u_1) & h_{1,1}(u_1) \\ h_{0,0}(u_2) & h_{0,1}(u_2) & h_{1,0}(u_2) & h_{1,1}(u_2) \\ h_{0,0}(u_3) & h_{0,1}(u_3) & h_{1,0}(u_3) & h_{1,1}(u_3) \\ h_{0,0}(u_4) & h_{0,1}(u_4) & h_{1,0}(u_4) & h_{1,1}(u_4) \\ & & \vdots & \\ h_{0,0}(u_k) & h_{0,1}(u_k) & h_{1,0}(u_k) & h_{1,1}(u_k) \end{bmatrix} \begin{bmatrix} P_0 \\ S_0 \\ P_1 \\ S_1 \end{bmatrix}.$$

Call the $(k+1) \times 4$ matrix of blending function values $\boldsymbol{H}$. Now suppose one has position and tangent data (P_i, S_i), $i = 0, \ldots, n$, and one wants

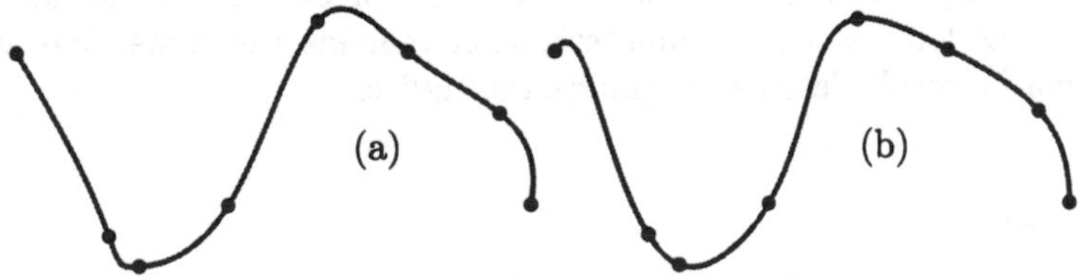

Figure 10.14. Polynomial interpolants: (a) uniform parameterization; (b) nonuniform parameterization.

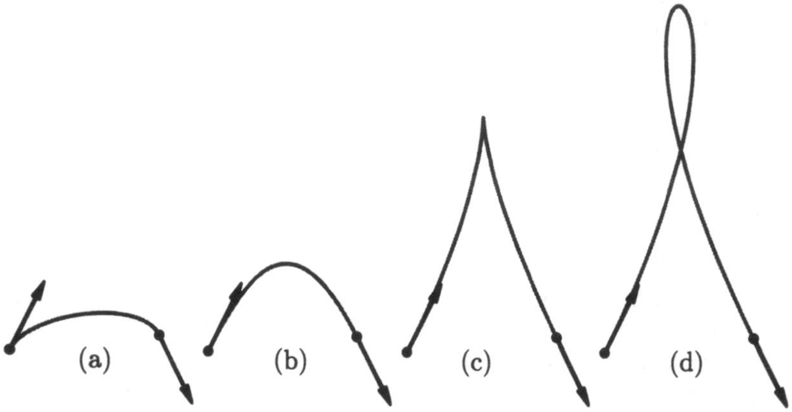

Figure 10.15. Examples of parametric cubic Hermite interpolation with different tangent magnitudes.

to evaluate the parametric piecewise cubic Hermite function for this data using a uniform parametric spacing of the data. Using the above matrix formulation, one can draw the 0th cubic piece immediately.

Now, let

$$\boldsymbol{V}_i = \begin{bmatrix} P_i & S_i & P_{i+1} & S_{i+1} \end{bmatrix}^T.$$

Using translation properties developed in Example 10.5, we see that the curve on the ith segment is

$$h(t) = \begin{bmatrix} h_{0,0}(t-i) & h_{0,1}(t-i) & h_{1,0}(t-i) & h_{1,1}(t-i) \end{bmatrix} \boldsymbol{V}_i.$$

It immediately follows that

$$\begin{bmatrix} h(i+u_0) & h(i+u_1) & \ldots & h(i+u_k) \end{bmatrix}^T = \boldsymbol{H}\boldsymbol{V}_i.$$

Using this method, the whole curve can be drawn from the table of values that was developed for the single interval.

Now, if a position or tangent is modified, or even if several data are modified, as long as the parameterization remains the same, the whole curve can be recalculated as n matrix multiplies.

Exercises

1. Finish Example 10.6 by finding the quintic generalized Hermite blending functions at the upper endpoint, $t = 1$.

2. Suppose $P_0, S_0, P_1, S_1, P_2, S_2$ are three points and their corresponding tangent vectors. Use the parametric extension of the piecewise uniform cubic Hermite form and determine the tangent vector, the curvature, and the torsion at each parametric value along the curve. Does the Frenet frame vary continuously for each value of t?.

3. Find the cubic Hermite blending functions for two arbitrary abscissa by generalizing the method of Example 10.5.

4. Programming Exercise: Implement a program to design parametric piecewise cubic curves using the cubic Hermite interpolation method. Your program should be capable of the following:

   (a) The user should be able to graphically select n points that the curve will pass through.

   (b) The user should be able to *move*, *delete*, or *add* a single point after the initial entry of the n points.

   (c) The user should be able to graphically specify the (vector) derivatives at each data value, including magnitude and direction, and to select and change those data values randomly.

   (d) Display the piecewise cubic Hermite curve.

11

Other Derivations of B-Splines

We have discussed a constructive approach to the introduction of B-splines using a recursive definition. We shall introduce several approaches below that stem from very different views.

11.1 Higher-Dimensional Volumetric Projections

This view of splines was first noted by Curry and Schoenberg [25] and later expanded by deBoor [28].

Select $n > 0$. Define $\boldsymbol{P} : \mathbb{R}^n \to \boldsymbol{R}$ by $\boldsymbol{P}(t_1, \ldots, t_n) = t_1$, the projection operator on the first coordinate. Given $\{x_0, \ldots, x_n\}$, in $\boldsymbol{R}$, not all the same value, choose points P_i in $\mathbb{R}^n$ such that $\boldsymbol{P}[P_i] = x_i$ and $\{P_0, \ldots, P_n\}$ form the vertices of an arbitrary n-simplex, σ. It is easily shown that this can be done by a constructive example. Note, however that the simplex obtained is only one of the infinite number possible. For the purposes of example, suppose $x_0 \neq x_1$, and choose $P_0 = (x_0, 0, \ldots, 0)$, $P_1 = (x_1, 0, \ldots, 0)$. Let $P_i = (x_i, 0, \ldots, \delta_{i,j}, 0, \ldots, 0)$, $i = 2, \ldots, n$, where $\delta_{i,j} = 1$ if $i = j$ and $\delta_{i,j} = 0$ otherwise; i.e., if $n = 3$, $P_0 = (x_0, 0, 0)$, $P_1 = (x_1, 0, 0)$, $P_2 = (x_2, 1, 0)$, $P_3 = (x_3, 0, 1)$. It can be shown that $P_i - P_0$, $i = 1, \ldots, n$, considered as vectors, form a basis for $\mathbb{R}^n$, and thus $\{P_i\}$, $i = 0, \ldots, n$ form the vertices of an n-simplex.

Define $\sigma(x) = \{P \in \sigma; \boldsymbol{P}(P) = x\}$. Then $\sigma(x)$ is an $n - 1$ dimensional subset of σ. Set

$$M_n(x) = \frac{vol_{n-1}(\sigma(x))}{vol_n(\sigma)}.$$

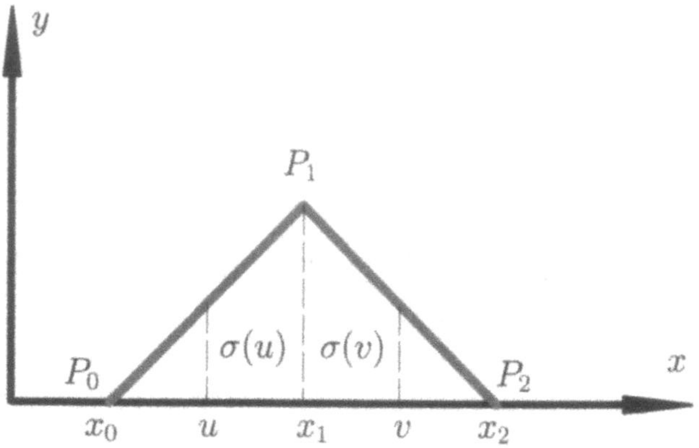

Figure 11.1. The generic two simplex to find the linear B-spline of Example 11.1.

Clearly M_n depends on n and $x_0, \ldots, x_n$. But at first glance it appears that it also depends on the particular simplex σ used. It can be shown that this is not the case. That is, M_n is independent of the choice of σ as long as $\boldsymbol{P}(\{P_0, \ldots, P_n\}) = \{x_0, \ldots, x_n\}$.

Example 11.1. Let $n = 2$ and $\{x_0, x_1, x_2\}$ be all distinct. Choose $P_0 = (x_0, 0)$, $P_1 = (x_1, 1)$, $P_2 = (x_2, 0)$. Then σ is the triangle enclosed by those three points. The area enclosed by σ is

$$vol_2(\sigma) = \frac{1}{2}(base * height) = \frac{1}{2}(x_2 - x_0) * 1.$$

Consider $x \in \mathbb{R}^1$; $vol_1(\sigma(x))$ is simply the length of the vertical line through x which is contained in σ. If $x < x_0$, or $x > x_2$, this line segment contains no points and thus has zero length. Now suppose $x_0 < x < x_1$. The length of the line segment is the height from the x-axis to the line through P_0 and P_1. This line has equation $height = \frac{1}{x_1 - x_0}(x - x_0)$. Thus

$$\begin{aligned} M_2(x) &= \frac{\frac{1}{x_1 - x_0}(x - x_0)}{\frac{1}{2}(x_2 - x_0)} \\ &= \frac{2}{(x_1 - x_0)(x_2 - x_0)}(x - x_0). \end{aligned}$$

Now suppose $x_1 < x < x_2$. The length of the line segment is the height from the x-axis to the line through P_1 and P_2. This line has equation $height = \frac{1}{x_1 - x_2}(x - x_2) = \frac{1}{x_2 - x_1}(x_2 - x)$, and again

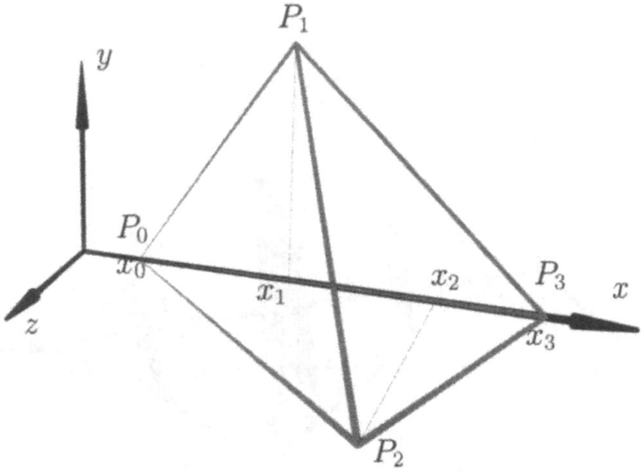

Figure 11.2. Tetrahedron to form quadratic B-spline of Example 11.2.

$$M_2(x) = \frac{\frac{1}{x_2-x_1}(x_2 - x)}{\frac{1}{2}(x_2 - x_0)} = \frac{2}{(x_2 - x_1)(x_2 - x_0)}(x_2 - x).$$

But this shows that $M_2(x) = M_2(x; x_0, x_1, x_2)$. □

Example 11.2. Suppose $n = 3$, and $\{x_0, x_1, x_2, x_3\}$ are all distinct. Choose $P_0 = (x_0, 0, 0)$, $P_1 = (x_1, 1, 0)$, $P_2 = (x_2, 0, 1)$, and $P_3 = (x_3, 0, 0)$. Then σ is the tetrahedron enclosed by those four points. Further $vol_3(\sigma) = 1/6(x_3 - x_0)\,[1/6(base * height * depth)]$. Consider $x \in \mathbf{R}^1$; $vol_{n-1}(\sigma(x))$ is the area of the vertical plane through x which is contained in σ. We must investigate the shape of this polygon. If $x < x_0$ or $x > x_3$, this polygon contains no points and thus has zero area. Now suppose $x_0 < x < x_1$. The shape of this polygon is a triangle with

$$height(y) = \frac{x - x_0}{x_1 - x_0} \quad \text{and} \quad base(z) = \frac{x - x_0}{x_2 - x_0}.$$

The area is then

$$\frac{1}{2}yz = \frac{(x - x_0)^2}{(x_1 - x_0)(x_2 - x_0)}.$$

This yields

$$M_3(x) = \frac{3(x - x_0)^2}{(x_1 - x_0)(x_2 - x_0)(x_3 - x_0)}, \qquad \text{for } x_0 < x < x_1.$$

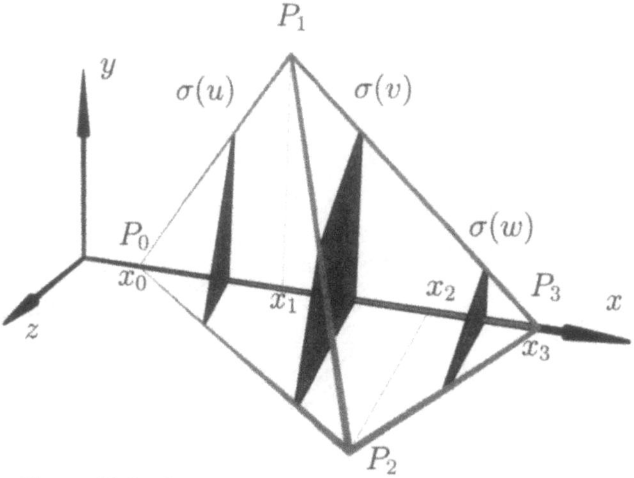

Figure 11.3. Quadrilateral cross section of tetrahedron.

Analogously, for $x_2 < x < x_3$, the polygon is a triangle with

$$\begin{aligned} height(y) &= \frac{x_3 - x}{x_3 - x_1}, \\ base(z) &= \frac{x_3 - x}{x_3 - x_2}, \\ M_3(x) &= \frac{3(x_3 - x)^2}{(x_3 - x_0)(x_3 - x_2)(x_3 - x_1)}. \end{aligned}$$

The situation is not so simple when $x_1 < x < x_2$. In that region, the polygon is a quadrilateral as shown in Figure 11.3. However, it can be broken down into two regions A_1 and A_2, a trapezoid and right triangle, respectively. A_2 has

$$base = \frac{x - x_1}{x_2 - x_1} \quad \text{and} \quad height = \frac{x_3 - x}{x_3 - x_1} - \frac{x_2 - x}{x_2 - x_1} = \frac{(x_3 - x_2)(x - x_1)}{(x_3 - x_1)(x_2 - x_1)}.$$

Thus

$$\begin{aligned} area\,(A_2) &= \frac{1}{2} base * height \\ &= \frac{1}{2}\frac{(x - x_1)^2(x_3 - x_2)}{(x_2 - x_1)^2(x_3 - x_1)}. \end{aligned}$$

The dimensions for A_1 are

$$height = \frac{x_2 - x}{x_2 - x_1}, \quad base = \frac{x - x_0}{x_2 - x_0}, \quad top = \frac{x - x_1}{x_2 - x_1},$$

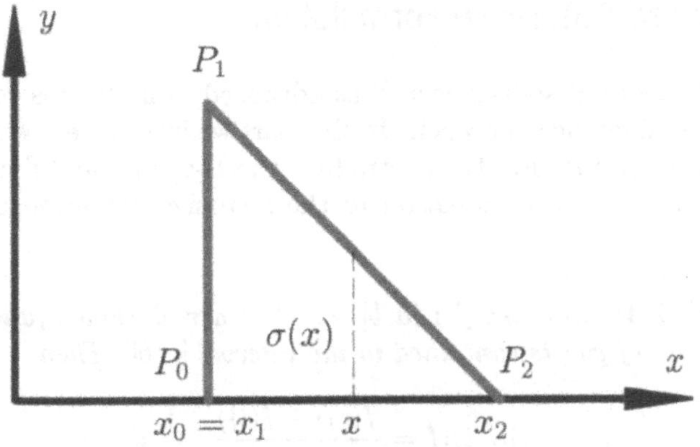

Figure 11.4. Multiple knot in the linear simplex.

where x is the value at which we want to evaluate M. Hence

$$\begin{aligned} area\,(A_1) &= height * \left[\frac{base + top}{2}\right] \\ &= \frac{1}{2}\frac{x_2 - x}{x_2 - x_1}\left[\frac{x - x_0}{x_2 - x_0} + \frac{x - x_1}{x_2 - x_1}\right]. \end{aligned}$$

Thus,

$$M_3(x) = \frac{3(x_2 - x)}{(x_2 - x_1)(x_3 - x_0)}\left[\frac{x - x_0}{x_2 - x_0} + \frac{x - x_1}{x_2 - x_1}\right].$$

It can be shown that $M_3(x; x_0, x_1, x_2, x_3)$ is equal to the same piecewise polynomial. □

It is clear from the above examples that although the choice of simplex does not affect the final answer, it can certainly affect the complexity of the calculations. We end this result with a calculation for a multiple knot.

Example 11.3. Consider $n = 2$, where $x_0 = x_1$. Let $P_0 = (x_0, 0)$, $P_1 = (x_1, 1) = (x_0, 1)$ and $P_2 = (x_2, 0)$.

For $x_0 < x < x_2$, $M_2(x) = \frac{2(x_2 - x)}{(x_2 - x_1)^2}$. The effect of a double knot on the continuity of the function is clear. At $x = x_0 = x_1$, there is no area inside the triangle. We might think to count the boundary, but if the location of the double knot were reversed, say $x_1 = x_2$, then it would lead to the normalized B-splines not summing to 1 wherever a multiple knot occurred. □

11.2 Divided Difference Formulation

Many proofs about B-splines as well as advanced formulations require use of the divided difference approach. In this part we introduce divided differences and their properties, then introduce the B-spline, and finally prove the equivalence of this formulation to the recursive definition previously introduced.

Definition 11.4. *Suppose* $f : [a,b] \to \boldsymbol{R}$ *is a real valued function and* $\{t_i\}_{i=0}^{n}$ *is a set of points contained in the interval* $[a,b]$. *Then*

$$[t_i, t_j]f = \frac{f(t_j) - f(t_i)}{t_j - t_i}$$

is called a first divided difference *of the function f.*

Definition 11.5. *Define*

$$[t_{i_0}, t_{i_1}, \ldots, t_{i_k}]f = \frac{[t_{i_1}, t_{i_2}, \ldots, t_{i_k}]f - [t_{i_0}, t_{i_1}, \ldots, t_{i_{k-1}}]f}{t_{i_k} - t_{i_0}}$$

to be the kth-order divided difference *of* f *on* $\{t_{i_j}\}_{j=0}^{k}$.

The reason for being interested in divided differences can now be stated.

Definition 11.6. *Given a sequence of points* $\{t_j\}$,

$$B_{i,k}(t; t_i, \ldots, t_{i+k}) = (t_{i+k} - t_i)\,[t_i, \ldots, t_{i+k}]_y\,(y - t)_+^{k-1}$$

is called the ith normalized B-spline of order k *defined over the values* $\{t_i\}$. *Using degree notation,* $\mathcal{B}_{i,k-1}(t; t_i, \ldots, t_{i+k}) = B_{i,k}(t; t_i, \ldots, t_{i+k})$.

We shall show that the functions defined here are the same as those of Definition 6.7. But first we review some properties of divided difference which are relevant.

11.2.1 Divided Differences

Lemma 11.7. *The divided difference is symmetric.*

Proof: Since $[t_i, t_j]f = [t_j, t_i]f$, the order of the elements appearing in the brackets is irrelevant. ∎

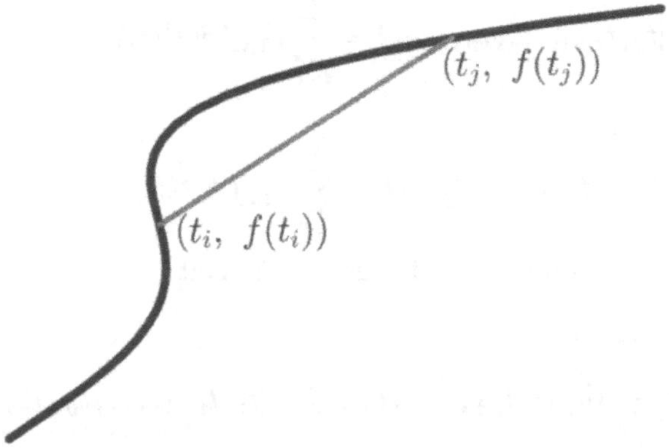

Figure 11.5. Chord connecting the points $(t_i, f(t_i))$ and $(t_j, f(t_j))$.

$[t_i, t_j]f$ is the slope of the chord connecting the points $(t_i, f(t_i))$, and $(t_j, f(t_j))$ for any function f. Thus whenever f is differentiable at x_i, $\lim_{t_j \to t_i}[t_i, t_j]f = f'(t_i)$.

We use this limiting value and adopt the convention that $[t_i, t_i]f = f'(t_i)$. In general, using the limiting values as definitions for divided differences with the same point appearing multiple times yields $[t_i, \dots, t_{i+j}]\, f = \frac{f^{(j)}(t_i)}{j!}$, if $t_i = \cdots = t_{i+j}$.

Lemma 11.8. *The divided difference is a linear functional, i.e.,* $[u, v](af + bg) = a\,([u, v]f) + b\,([u, v]g)$ *and hence the higher order differences are also linear.*

Proof: The proof is straightforward and is left as an exercise. ■

Theorem 11.9. *Let* $\mathbf{t} = \{t_i\}_{i=0}^{k}$. *Define* $q_i = \max\{j : t_{i+j} = t_i\}$. *For any subscripts* p, r, $0 \le p \le r \le k$, *there exists constants,* $\{c_i\}$ *and* $\{q_i\}$, *depending only on* $\mathbf{t}$, p, *and* r, *such that*

$$[t_p, t_{p+1}, \dots, t_r]f = \sum_{i=0}^{k} c_i f^{(q_i)}(t_i).$$

Proof: If $r - p = 0$, then it follows by definition of the 0th divided difference. Also, if $t_p = \cdots = t_r$, it also clearly follows, since $[t_p, t_{p+1}, \dots, t_r]f = \frac{f^{(r-p)}(t_r)}{(r-p)}!$ Now suppose it is true for $r - p < j$, and $t_p < t_r$. Then there exist constant sequences $\{c_{1,i}\}$ and $\{c_{2,i}\}$ so that

$$[t_p, t_{p+1}, \ldots, t_{p+j-1}]f = \sum_{i=0}^{k} c_{1,i} f^{(q_{i,1})}(t_i)$$

and

$$[t_{p+1}, t_{p+2}, \ldots, t_{p+j}]f = \sum_{i=0}^{k} c_{2,i} f^{(q_{i,2})}(t_i),$$

since they are divided differences of order $j-1$. But

$$\begin{aligned}
&[t_p, t_{p+1}, \ldots, t_{p+j}]f \\
&\quad = \frac{1}{t_{p+j} - t_p}\left([t_{p+1}, t_{p+2}, \ldots, t_{p+j}]f - [t_p, t_{p+1}, \ldots, t_{p+j-1}]f\right) \\
&\quad = \frac{1}{t_{p+j} - t_p}\left(\sum_{i=0}^{k} c_{2,i} f^{(q_i)}(t_i) - \sum_{i=0}^{k} c_{1,i} f^{(q_i)}(t_i)\right) \\
&\quad = \frac{1}{t_{p+j} - t_p}\sum_{i=0}^{k}(c_{2,i} - c_{1,i}) f^{(q_i)}(t_i),
\end{aligned}$$

as desired. ∎

If l is a linear function, $l(x) = mx + d$, then $[u, v]l$ is its slope, m, for any choice of values of u and v in $[a, b]$.

$$\begin{aligned}
[t_i, t_{i+1}, t_{i+2}]l &= \frac{[t_{i+1}, t_{i+2}]l - [t_i, t_{i+1}]l}{t_{i+2} - t_i} \\
&= \frac{m - m}{t_{i+2} - t_i} \\
&= 0.
\end{aligned}$$

Thus $[t_i, \ldots, t_{i+k}]\, f = 0$, for $k > 1$ if f is linear. In general, we may say:

Theorem 11.10. *Given a function $f(x)$ defined on $[a, b]$ and abscissas $\{x_i\}$ in $[a, b]$, define the kth degree polynomial p_k so that*

$$p_k(x) = a_{0,k} + a_{1,k}(x - x_0) + \cdots + a_{k,k}(x - x_0)\cdots(x - x_{k-1}),$$

with coefficients chosen so that $p_k(x_i) = f(x_i)$, $i = 0, \ldots, k$. That is, p_k interpolates f at the $k+1$ points x_i, $i = 0, \ldots, k$. Then

$$a_{i,k} = [x_0, \ldots, x_i]\, f.$$

Proof: For $i = 0$, $a_{0,k} = f(x_0) = [x_0]f$ for any $k > 0$. Hence, call $a_0 = f(x_0)$. Now suppose the hypothesis is true for $j < k$, that is, for every polynomial $q(x) = \sum_{i=0}^{k-1} c_i(x - x_{j_0}) \cdots (x - x_{j_{i-1}})$ which interpolates $\{(x_{j_i}, f(x_{j_i}))\}_{i=0}^{k-1}$ then $c_i = [x_{j_0}, x_{j_1}, \ldots, x_{j_i}]f$. This induction hypothesis results in $a_i = a_{i,j} = [x_0, \ldots, x_i]\, f$ for $i = 0, \ldots, j$, and $j < k$.

Letting $p_{k-1}(x)$ be as above, define $r_{k-1}(x)$ to be the polynomial that agrees with f on $x_1, \ldots, x_k$. Thus, the induction hypothesis gives $r_{k-1}(x) = b_0 + b_1(x - x_1) + \cdots + b_{k-1}(x - x_1) \cdots (x - x_{k-1})$ where $b_i = [x_1, \ldots, x_{i+1}]\, f$. The kth degree polynomial $\frac{x - x_0}{x_k - x_0} r_{k-1}(x) + \frac{x_k - x}{x_k - x_0} p_{k-1}(x)$ interpolates f at $x_1, x_2, \ldots, x_{k-1}$ since $p_i = r_i = f(x_i)$, $i = 1, \ldots, k-1$, and it further interpolates f at x_0 and x_k by the design of the blend of p_{k-1} and r_{k-1}. Since a kth degree interpolating polynomial of $k + 1$ points is unique,

$$p_k(x) = \frac{x - x_0}{x_k - x_0} r_{k-1}(x) + \frac{x_k - x}{x_k - x_0} p_{k-1}(x).$$

Now, the coefficient of x^k in $p_k(x)$ is $a_{k,k}$, so

$$\begin{aligned} a_{k,k} &= \frac{b_{k-1} - a_{k-1,k-1}}{x_k - x_0} \\ &= \frac{[x_1, \ldots, x_k]\, f - [x_0, \ldots, x_{k-1}]\, f}{x_k - x_0} \\ &= [x_0, \ldots, x_k]\, f. \end{aligned}$$

Thus, the second subscript can be omitted, and p_n has the form

$$p_n(x) = \sum_{i=0}^{n} [x_0, \ldots, x_i]\, f \prod_{j=0}^{i-1} (x - x_j).$$

■

This is called the *Newton form* of the interpolating polynomial.

Results following from this lemma are:

Corollary 11.11. *If f is a polynomial of degree m, and $n > m$, then*

$$[x_0, \ldots, x_n]\, f = 0. \tag{11.1}$$

Proof: The interpolating polynomial for $(n+1)$ positions of a polynomial of degree m is the original polynomial whenever $n > m$. Thus the coefficients of higher degree terms must be zero, and $a_n = [x_0, \ldots, x_n]\, f = 0$, for $n > m$. ■

Corollary 11.12. *If f is a polynomial of degree m, then $[x_0, \ldots, x_m]\, f = [t_0, \ldots, t_m]\, f$ for any sequences $\{x_i\}$ and $\{t_i\}$.*

Proof: Since f is a polynomial of degree m, then it can be put in the Newton form using the points $x_0, \ldots, x_m$. The coefficient of $(x-x_0)\cdots(x-x_{m-1})$ is $[x_0, \ldots, x_m]\, f$. Since this is the only term with the power x^m, the coefficient of x^m in the power basis is $[x_0, \ldots, x_m]\, f$. Analogously, f can be put in the Newton form using the points $t_0, \ldots, t_m$. From this we get that the coefficient of x^m in the power basis is $[t_0, \ldots, t_m]\, f$. By the uniqueness of the coefficients of a polynomial in a power basis, one obtains the result of the corollary. ■

Theorem 11.13. Leibnitz's Formula. *If $f(x) = g(x)h(x)$ then for distinct values x_i, …, x_{i+n},*

$$[x_i, \ldots, x_{i+n}]\, f = \sum_{j=0}^{n} [x_i, \ldots, x_{i+j}]\, g\, [x_{i+j}, \ldots, x_{i+n}]\, h.$$

Proof: The proof is by induction. For $n = 1$,

$$\begin{aligned}
&[x_i, x_{i+1}]f \\
&= \frac{g(x_{i+1})h(x_{i+1}) - g(x_i)h(x_i)}{x_{i+1} - x_i} \\
&= \frac{g(x_{i+1})h(x_{i+1}) - h(x_{i+1})g(x_i) + h(x_{i+1})g(x_i) - g(x_i)h(x_i)}{x_{i+1} - x_i} \\
&= \frac{(g(x_{i+1}) - g(x_i))\, h(x_{i+1}) + (h(x_{i+1}) - h(x_i))\, g(x_i)}{x_{i+1} - x_i} \\
&= \frac{(g(x_{i+1}) - g(x_i))}{x_{i+1} - x_i} h(x_{i+1}) + \frac{(h(x_{i+1}) - h(x_i))\, g(x_i)}{x_{i+1} - x_i} \\
&= [x_i, x_{i+1}]g[x_{i+1}]h + [x_i]g[x_i, x_{i+1}]h \\
&= \sum_{j=0}^{1} [x_i, x_{i+j}]g[x_{i+j}, x_{i+1}]h,
\end{aligned}$$

and the initial step of the recursion is proved. Now, assume the conclusion is true for divided differences of order $n-1$. We show it true for divided differences of order n.

$$[x_i, \ldots, x_{i+n}]\, f = \frac{1}{x_{i+n} - x_i} \left([x_{i+1}, \ldots, x_{i+n}]\, f - [x_i, \ldots, x_{i+n-1}]\, f \right).$$

Both terms in the difference on the right are $(n-1)^{st}$ order divided differences, so the induction hypothesis may be applied.

$$\begin{aligned}
&(x_{i+n} - x_i)\,[x_i, \ldots, x_{i+n}]\,f \\
&\quad = \sum_{j=0}^{n-1} \left([x_{i+1}, x_{i+1+j}]g[x_{i+1+j}, x_{i+n}]h - [x_i, x_{i+j}]g[x_{i+j}x_{i+n-1}]h \right).
\end{aligned}$$

Each term in the above summation can be manipulated in the same manner as we did for the $n = 1$ case.

$$\begin{aligned}
&[x_{i+1}, x_{i+1+j}]g[x_{i+1+j}, x_{i+n}]h - [x_i, x_{i+j}]g[x_{i+j}, x_{i+n-1}]h \\
&\quad = [x_{i+1}, x_{i+1+j}]g[x_{i+1+j}, x_{i+n}]h - [x_i, x_{i+j}]g[x_{i+1+j}, x_{i+n}]h \\
&\qquad + [x_i, x_{i+j}]g[x_{i+1+j}, x_{i+n}]h - [x_i, x_{i+j}]g[x_{i+j}, x_{i+n-1}]h \\
&\quad = (x_{i+1+j} - x_i)[x_i, x_{i+1+j}]g[x_{i+1+j}, x_{i+n}]h \\
&\qquad + (x_{i+n} - x_{i+j})[x_i, x_{i+j}]g[x_{i+j}, x_{i+n}]h.
\end{aligned}$$

Substituting, we get

$$\begin{aligned}
&(x_{i+n} - x_i)\,[x_i, \ldots, x_{i+n}]\,f \\
&\quad = \sum_{j=0}^{n-1}(x_{i+1+j} - x_i)[x_i, x_{i+1+j}]g[x_{i+1+j}, x_{i+n}]h \\
&\qquad\qquad + \sum_{j=0}^{n-1}(x_{i+n} - x_{i+j})[x_i, x_{i+j}]g[x_{i+j}, x_{i+n}]h \\
&\quad = \sum_{j=1}^{n}(x_{i+j} - x_i)[x_i, x_{i+j}]g[x_{i+j}, x_{i+n}]h \\
&\qquad\qquad + \sum_{j=0}^{n-1}(x_{i+n} - x_{i+j})[x_i, x_{i+j}]g[x_{i+j}, x_{i+n}]h \\
&\quad = (x_{i+n} - x_i)\,[x_{i+n}, x_i]g[x_i, x_{i+n}]h \\
&\qquad + \sum_{j=1}^{n-1}(x_{i+n} - x_i)[x_i, x_{i+j}]g[x_{i+j}, x_{i+n}]h \\
&\qquad + [x_i, x_{i+n}]g[x_{i+n}, x_{i+n}]h
\end{aligned}$$

which proves the result. ■

A final definition and theorm reformulate divided differences as integrals over simplices.

Definition 11.14. *Let* $S^n = \{(t_0, \ldots, t_n) : t_i \geq 0$ *and* $\sum t_i = 1\}$. S^n *is called the* standard n-simplex.

Theorem 11.15. Hermite Gennochi. *For a function* $f \in C^{(n)}$,

$$[x_0, \ldots, x_n]\, f = \int_{S^n} f^{(n)}(x_0 t_0 + \cdots + x_n t_n) dt_1 \ldots dt_n.$$

Proof: The proof will proceed by induction on n. Since $t_0 + t_1 = 1$,

$$\begin{aligned}
\int_{S^1} f'(x_0 t_0 + x_1 t_1) dt_1 &= \int_{S^1} f'\left(x_0(1-t_1) + x_1 t_1\right) dt_1 \\
&= \int_0^1 f'\left(x_0 + (x_1 - x_0)t_1\right) dt_1 \\
&= \frac{1}{x_1 - x_0} f\left(x_0 + (x_1 - x_0)t_1\right)\Big|_{t_1=0}^{1} \\
&= \frac{1}{x_1 - x_0}\left(f(x_1) - f(x_0)\right) \\
&= [x_0, x_1] f.
\end{aligned}$$

The higher order parts of this proof are left as an exercise for the reader. ■

11.2.2 Divided Differences for B-Splines

Theorem 11.16. *The normalized B-spline defined by Definition 11.6 is the same as the recursively defined B-spline.*

Proof: Let $\mathcal{B}_{i,k-1}(t)$ continue to designate the recursively defined normalized B-spline, and let

$$\beta_{i,k}(t; t_i, \ldots, t_{i+k}) = (t_{i+k} - t_i)\, [t_i, \ldots, t_{i+k}]_y\, (y - t)_+^{k-1}.$$

We shall show that $\mathcal{B}_{i,k-1}(t) = \beta_{i,k}(t)$ for all t, k, and i. We shall use the Leibnitz formula for divided differences, Theorem 11.13. To use the formula we set

$$f(y) = (y - t)_+^{k-1}, \quad g(y) = (y - t) \quad \text{and} \quad h(y) = (y - t)_+^{k-2}.$$

Then $f(y) = g(y)h(y)$ and

$$\beta_{i,k}(t) = (t_{i+k} - t_i)\, [t_i, \ldots, t_{i+k}]\, f,$$

and

$$\beta_{i,k-1}(t) = (t_{i+k-1} - t_i)\, [t_i, \ldots, t_{i+k-1}]\, h.$$

By Leibnitz's formula:

$$[t_i,\ldots,t_{i+k}]\, f(y) = \sum_{j=0}^{k} [t_i,\ldots,t_{i+j}]\, g(y)\, [t_{i+j},\ldots,t_{i+k}]\, h(y).$$

Now g is just a simple linear polynomial, and so

$$[t_i,\ldots,t_{i+j}]\, g(y) = [t_i,\ldots,t_{i+j}]\,(y-t) = 0, \qquad \text{for } j > 1.$$

Hence,

$$\begin{aligned}
&[t_i,\ldots,t_{i+k}]\, f(y) \\
&= [t_i]g(y)\,[t_i,\ldots,t_{i+k}]\, h(y) + [t_i,t_{i+1}]g(y)\,[t_{i+1},\ldots,t_{i+k}]\, h(y) \\
&= [t_i]g(y)\left(\frac{[t_{i+1},\ldots,t_{i+1+k-1}]\, h(y) - [t_i,\ldots,t_{i+k-1}]\, h(y)}{t_{i+k}-t_i}\right) \\
&\quad + [t_i,t_{i+1}]g(y)\,[t_{i+1},\ldots,t_{i+k}]\, h(y) \\
&= \left(\frac{[t_i]g(y)}{t_{i+k}-t_i} + [t_i,t_{i+1}]g(y)\right)[t_{i+1},\ldots,t_{i+1+k-1}]\, h(y) \\
&\quad - [t_i]g(y)\frac{[t_i,\ldots,t_{i+k-1}]\, h(y)}{t_{i+k}-t_i} \\
&= \left(\frac{(t_i-t)}{t_{i+k}-t_i}+1\right)[t_{i+1},\ldots,t_{i+k}]\, h(y) \\
&\quad + \frac{(t-t_i)}{t_{i+k}-t_i}\,[t_i,\ldots,t_{i+k-1}]\, h(y) \\
&= \frac{(t_{i+k}-t)}{t_{i+k}-t_i}\frac{\mathcal{B}_{i+1,k-2}(t)}{t_{i+k}-t_{i+1}} + \frac{(t-t_i)}{t_{i+k}-t_i}\frac{\mathcal{B}_{i,k-2}(t)}{t_{i+k-1}-t_i} \\
&= \frac{1}{t_{i+k}-t_i}\left(\frac{(t_{i+k}-t)}{t_{i+k}-t_{i+1}}\beta_{i+1,k-1}(t) + \frac{(t-t_i)}{t_{i+k-1}-t_i}\beta_{i,k-1}(t)\right)
\end{aligned}$$

which is the same recursive relationship as the $\mathcal{B}_{i,k-1}$ have. Noting that $\beta_{i,1}(t)$ is equal to $\mathcal{B}_{i,0}(t)$, the equivalence becomes complete. ∎

Using this definition we observe that by Theorem 11.9, for a given j, $0 \le j \le n$, there exist coefficients c_i, $i = 0, \ldots, k$ depending on $t_j, \ldots, t_{j+k}$ such that

$$\mathcal{B}_{i,k-1}(t) \;=\; (t_{i+k}-t_i)\sum_{j=0}^{k} c_j(k-1)\cdots(k-q_{i,j})(t_{i+j}-t)_+^{k-q_{i,j}-1} \tag{11.2}$$

where $q_{i,j}$ is defined as in Theorem 6.24.

Corollary 11.17. *When $m_i = 1$ for all i,*

$$\mathcal{B}_{i,n-1}(t; t_i, \ldots, t_{i+n}) = (t_{i+n} - t_i) \sum_{j=0}^{n} \frac{(t_{i+j} - t)_+^{n-1}}{\omega'_{i,n}(t_{j+i})}$$

where

$$\omega_{i,n} = \prod_{p=0}^{n} (t - t_{i+p}).$$

Proof: When $n = 1$, $\omega'_{i,n}(t) = (t - t_{i+1}) + (t - t_i)$. The right hand side of the equation is then equal to

$$\begin{aligned}(t_{i+1} - t_i) \sum_{j=0}^{1} \frac{(t_{i+j} - t)_+^0}{\omega'_{i,1}(t_{j+i})} &= (t_{i+1} - t_i) \left(\frac{(t_i - t)_+^0}{t_i - t_{i+1}} + \frac{(t_{i+1} - t)_+^0}{t_{i+1} - t_i} \right) \\ &= (t_{i+1} - t)_+^0 - (t_i - t)_+^0.\end{aligned}$$

But, this function evaluates to 1 if $t \in [t_i, t_{i+1})$ and zero otherwise, which is just $\mathcal{B}_{i,0}(t)$.

Now, let us use induction. It is easy to show that:

$$\omega'_{i,n-1}(t_{i+j}) = \prod_{p=0, p \neq j}^{n-1} (t_{i+j} - t_{i+p}).$$

Using the induction hypothesis

$$\mathcal{B}_{i,n-2}(t) = (t_{i+n-1} - t_i) \sum_{j=0}^{n-1} \frac{(t_{j+i} - t)_+^{n-2}}{\omega'_{i,n-1}(t_{i+j})},$$

we must show the conclusion follows.

Using the recursive definition for $\mathcal{B}_{i,n-1}(t)$:

$$\begin{aligned}&\mathcal{B}_{i,n-1}(t) \\ &= \frac{t - t_i}{t_{i+n-1} - t_i} \mathcal{B}_{i,n-2}(t) + \frac{t_{i+n} - t}{t_{i+n} - t_{i+1}} \mathcal{B}_{i+1,n-2}(t) \\ &= \frac{t - t_i}{t_{i+n-1} - t_i} \left((t_{i+n-1} - t_i) \sum_{j=0}^{n-1} \frac{(t_{j+i} - t)_+^{n-2}}{\omega'_{i,n-1}(t_{i+j})} \right) \\ &\quad + \frac{t_{i+n} - t}{t_{i+n} - t_{i+1}} \left((t_{i+n} - t_{i+1}) \sum_{j=0}^{n-1} \frac{(t_{j+i+1} - t)_+^{n-2}}{\omega'_{i+1,n-1}(t_{i+1+j})} \right)\end{aligned}$$

$$
\begin{aligned}
&= (t-t_i)\sum_{j=0}^{n-1}\frac{(t_{j+i}-t)_+^{n-2}}{\omega'_{i,n-1}(t_{i+j})} + (t_{i+n}-t)\sum_{j=0}^{n-1}\frac{(t_{j+i+1}-t)_+^{n-2}}{\omega'_{i+1,n-1}(t_{i+1+j})} \\
&= (t-t_i)\frac{(t_i-t)_+^{n-2}}{\omega'_{i,n-1}(t_i)} + (t-t_i)\sum_{j=1}^{n-1}\frac{(t_{j+i}-t)_+^{n-2}}{\omega'_{i,n-1}(t_{i+j})} \\
&\quad + (t_{i+n}-t)\sum_{j=0}^{n-2}\frac{(t_{j+i+1}-t)_+^{n-2}}{\omega'_{i+1,n-1}(t_{i+1+j})} + (t_{i+n}-t)\frac{(t_{j+n}-t)_+^{n-2}}{\omega'_{i+1,n-1}(t_{i+n})} \\
&= (-1)\frac{(t_i-t)_+^{n-1}}{\omega'_{i,n-1}(t_i)} \\
&\quad + \sum_{j=1}^{n-1}(t_{j+i}-t)_+^{n-2}\left(\frac{(t-t_i)}{\omega'_{i,n-1}(t_{i+j})} + \frac{(t_{i+n}-t)}{\omega'_{i+1,n-1}(t_{i+j})}\right) \\
&\quad + \frac{(t_{j+n}-t)_+^{n-1}}{\omega'_{i+1,n-1}(t_{i+n})}.
\end{aligned}
$$

One must now determine, for a fixed i and $0 < j < n$,

$$
\begin{aligned}
&\frac{(t-t_i)}{\omega'_{i,n-1}(t_{i+j})} + \frac{(t_{i+n}-t)}{\omega'_{i+1,n-1}(t_{i+j})} \\
&= \frac{(t-t_i)}{\prod_{p=0,p\neq j}^{n-1}(t_{i+j}-t_{i+p})} + \frac{(t_{i+n}-t)}{\prod_{p=1,p\neq j}^{n}(t_{i+j}-t_{i+p})} \\
&= \frac{(t-t_i)(t_{i+j}-t_{i+n})}{\prod_{p=0,p\neq j}^{n}(t_{i+j}-t_{i+p})} + \frac{(t_{i+n}-t)(t_{i+j}-t_i)}{\prod_{p=0,p\neq j}^{n}(t_{i+j}-t_{i+p})} \\
&= \frac{(t-t_i)(t_{i+j}-t_{i+n}) + (t_{i+n}-t)(t_{i+j}-t_i)}{\prod_{p=0,p\neq j}^{n}(t_{i+j}-t_{i+p})} \\
&= \frac{(t_{i+j}-t)(t_{i+n}-t_i)}{\prod_{p=0,p\neq j}^{n}(t_{i+j}-t_{i+p})}.
\end{aligned}
$$

Substituting this simplification into the equation yields:

$$
\begin{aligned}
B_{i,n}(t) &= (-1)\frac{(t_i-t)_+^{n-1}}{\omega'_{i,n-1}(t_i)} + \sum_{j=1}^{n-1}(t_{j+i}-t)_+^{n-2}\left(\frac{(t_{i+j}-t)(t_{i+n}-t_i)}{\prod_{p=0,p\neq j}^{n}(t_{i+j}-t_{i+p})}\right) \\
&\quad + \frac{(t_{j+n}-t)_+^{n-1}}{\omega'_{i+1,n-1}(t_{i+n})}
\end{aligned}
$$

which easily reduces to the conclusion. ■

11.3 Fast Evaluation of Polynomials Using Divided Differences

It is frequently desired to evaluate a polynomial at many points inside a fixed interval. Clearly, when the degree of the polynomial is larger than one, this can lead to many operations. The general method for polynomial evaluation is called *Horner's Rule*, or *synthetic division*. This method's performance can be significantly speeded up in special case where the points are evenly spaced resulting in an algorithm requiring n adds to evaluate a polynomial of degree n, after some initial setup operations.

First, suppose that it is desired to evaluate the nth degree polynomial $p(x)$ at many evenly spaced points separated by a spacing h. Let abscissa values have the form $x_j = x_0 + jh$. Then, $p(x)$ can be represented as

$$p(x) = \sum_{i=0}^{n} a_{k,i} \prod_{j=0}^{i-1} (x - x_{k+j}), \qquad \text{where } a_{k,i} = [x_k, \ldots, x_{k+i}]\, p.$$

We find the algorithm by defining

$$\beta_{k,0} = p(x_k) \qquad \text{and} \qquad \beta_{k,i} = \beta_{k+1,i-1} - \beta_{k,i-1}. \tag{11.3}$$

$\beta_{k,1} = p(x_{k+1}) - p(x_k)$, is called the first *forward difference*, and in general $\beta_{k,i}$ is the ith forward difference at x_k. It is already known that $a_{k,0} = [x_k]p = p(x_k) = \beta_{k,0}$, for all k. Now assume an induction hypothesis, i.e., that $a_{k,i-1} = \frac{\beta_{k,i-1}}{(i-1)!h^{i-1}}$, for all k. Then

$$\begin{aligned}
a_{k,i} &= \frac{[x_{k+1}, \ldots, x_{k+i}]\, p - [x_k, \ldots, x_{k+i-1}]\, p}{x_{k+i} - x_k} \\
&= \frac{[x_{k+1}, \ldots, x_{k+i}]\, p - [x_k, \ldots, x_{k+i-1}]\, p}{ih} \\
&= \frac{a_{k+1,i-1} - a_{k,i-1}}{ih} \\
&= \frac{\frac{\beta_{k+1,i-1}}{(i-1)!h^{i-1}} - \frac{\beta_{k,i-1}}{(i-1)!h^{i-1}}}{ih} \\
&= \frac{\beta_{k+1,i-1} - \beta_{k,i-1}}{i!h^i} \\
&= \frac{\beta_{k,i}}{i!h^i}.
\end{aligned}$$

How do $\beta_{k,i}$, $i = 0, \ldots, n$, relate to $\beta_{k-1,i}$, $i = 0, \ldots, n$? That is, we ask the question, is there a usable relation between the ith forward difference at x_k and the ith forward difference at x_{k-1}. If there is, it will be possible to exploit it to form a computational algorithm.

$\beta_{k,n} = \beta_{k-1,n}$ since $a_{k,n}$ is constant for all k. (It is an nth order divided difference on a polynomial of degree n.) Further, $\beta_{k-1,i} = \beta_{k,i-1} - \beta_{k-1,i-1}$ implies $\beta_{k,i} = \beta_{k-1,i} + \beta_{k-1,i+1}$, $i = 0, \ldots, n-1$, which is the final critical piece to make a fast algorithm [48].

Algorithm 11.18.

Step 1: Find $\beta_{0,i}$ for $i = 0, \ldots, n$.
Step 2: Set $\beta_i = \beta_{0,i}$ for $i = 0, \ldots, n$. Set $k = 0$.
Step 3: β_0 is $p(x_k)$.
Step 4: Set $\beta_i = \beta_i + \beta_{i+1}$ which yields $\beta_{k+1,i}$, $i = 0, \ldots, n-1$.
Set $k = k + 1$.
Step 5: Repeat steps 3 and 4 until the function is evaluated for all abscissa.

Example 11.19. Evaluate $p(x) = 2x^3 + x^2 - 4x + 1$ starting at $x_0 = 0$ with $h = .25$.

$$
\begin{aligned}
p(0) &= 1 & p(.25) &= 3/32 & p(.5) &= -1/2 \\
p(.75) &= -19/32 & p(1) &= 0 & p(1.25) &= 47/32 \\
p(1.5) &= 4
\end{aligned}
$$

and so forth. □

Step 1:

$$
\begin{aligned}
&\beta_{0,0} = 1 \\
&\beta_{0,1} = -\tfrac{29}{32} \quad \beta_{1,1} = -\tfrac{19}{32} \quad \beta_{2,1} = -\tfrac{3}{32} \\
&\beta_{0,2} = \tfrac{10}{32} \quad \beta_{1,2} = \tfrac{16}{32} \\
&\beta_{0,3} = \tfrac{6}{32}
\end{aligned}
$$

2:	$\beta_0 = 1$	$\beta_1 = -\frac{29}{32}$	$\beta_2 = \frac{10}{32}$	$\beta_3 = \frac{6}{32}$	$k = 0$
3:					$p(x_0) = 1$
4:	$\frac{3}{32}$	$-\frac{19}{32}$	$\frac{16}{32}$	1	
3:					$p(x_1) = \frac{3}{32}$
4:	$-\frac{16}{32}$	$-\frac{3}{32}$	$\frac{22}{32}$	2	
3:					$p(x_2) = -\frac{1}{2}$
4:	$-\frac{19}{32}$	$\frac{19}{32}$	$\frac{28}{32}$	3	
3:					$p(x_3) = -\frac{19}{32}$
4:	0	$\frac{47}{32}$	$\frac{34}{32}$	4	
3:					$p(x_4) = 0$
4:	$\frac{47}{32}$	$\frac{81}{32}$	$\frac{40}{32}$	5	
3:					$p(x_5) = \frac{47}{32}$
4:	$\frac{128}{32}$	$\frac{121}{32}$	$\frac{46}{32}$	6	
3:					$p(x_6) = 4$

So far, the method seems to have only desirable characteristics. After the initial set up, one need only perform n adds to obtain the polynomial value. What are the drawbacks of this method? Alas, simply that while this example was done using exact arithmetic, the computer uses floating point arithmetic. Since each new answer depends on the previous answer, the error is cumulative, and it is easy to see that the farther away from x_0 one gets and the greater k is, the greater the error becomes. Carrying more significant digits can push off the day when error becomes intolerable, but come it must. However, when the polynomial is changing slowly between the first and last points, this method yields good results.

11.3.1 Fast Calculations with Bézier Curves

An immediate question to ask is, "Can this method conveniently be used with the Bézier curve method?" We must only add a few preprocessing steps. For a Bézier curve $\gamma(t)$ of degree n, evaluate γ at t_i, $i = 0, \ldots, n$ using the recursive curve method of Algorithm 5.1. Find the $\beta_{0,j}$, $j = 1, \ldots, n$ from these values using Equation 11.3. Then continue as in the algorithm above.

11.3.2 Fast Evaluation of B-Spline Curves

The same algorithm developed in Section 11.3 which can be used in the Bézier method, Section 11.3.1 can be adapted to provide fast evaluation of B-spline curves.

The problem is to evaluate $\gamma(v_m)$, a B-spline curve of degree $k-1$. This can be done using the fast polynomial evaluation algorithm of Section 11.3 with the constraint that the algorithm must be restarted for each distinct polynomial segment, that is, it must be restarted for evaluations of $t \in [t_i, t_{i+1})$, whenever $t_i < t_{i+1}$.

1. For a fixed i, set $h = \frac{t_{i+1}-t_i}{M}$ and where $v_m = t_i + mh$, $0 \le m \le M-1$. Then $t_i \le v_m < t_{i+1}$.

2. Evaluate $\gamma(v_i)$, $i = 0, \ldots, k-1$, using Algorithm 6.4.

3. Then proceed to evaluate $\gamma(v_i)$, $i = k, \ldots, M-1$, using the algorithms developed in Section 11.3 as modified for Bézier curves in Section 11.3.1.

4. Increment i as needed to obtain another distinct interval and return to step 1.

Note that while fast, this algorithm can have numerical instabilities, since the error compounds as the number of points evaluated by it increases. If the degree of the B-spline is low, such as two or three, not much setup overhead is required before evaluation just takes two or three additions. For a quick look at such a curve, the method is very suitable.

11.4 Inverse Fourier Transform Definition of Splines

This view was first introduced by Schoenberg [74]. If one has data $\{y_n\}_{n=-\infty}^{\infty}$ then one can view it as sampled data evenly spaced over time. At the time the data of concern were values of the drag coefficient of a projectile as a function of velocity. The data is empirical, with all inherent noise and inaccuracies, but very smooth approximations to the data, and very accurate first and second derivatives were needed. A formal Lagrange type interpolation formula for infinite, evenly spaced data is

$$F(x) = \sum y_n L(x-n), \qquad \text{where } L(x) = \frac{\sin(\pi x)}{\pi x}.$$

This $L(x)$ has the property that $L(0) = 1$, but $L(i) = 0$ for all other integers. The function $F(x)$ then interpolates all values of y. This is called cardinal series representation. Note the effect of the data is limited to one blending function.

One may want to modify this idea to use it for smoothing the data. Consider a function $L(x)$, similar to the last one, but, such that $L(x) = L(-x)$ and for $|x| > p$, $L(x) = 0$, one can transform the original sequence $\{y_n\}$ into a new function $\{F(x)\}$ which can be written

$$F(x) = \sum_{r=-\infty}^{\infty} y_r L(x-r).$$

The summation is really finite since $L = 0$ outside of an interval centered at $x = 0$. This is an example of a linear transformation. In signal processing it is called *time invariant filtering*, with the L the impulse response function. One then wants to ask the question, *under what conditions is F actually a smoothed version of the set of data $\{y_n\}$?* It is in this context that mathematical splines are introduced. Using properties of Fourier transforms and convolutions, the above formula reads

$$\mathcal{F}[F](\omega) = \mathcal{F}[y](\omega)\mathcal{F}[L](\omega).$$

In this light, the function $\mathcal{F}[L]$ is the frequency response to the system. A smoothing filter is one that is essentially a low pass filter. This means that

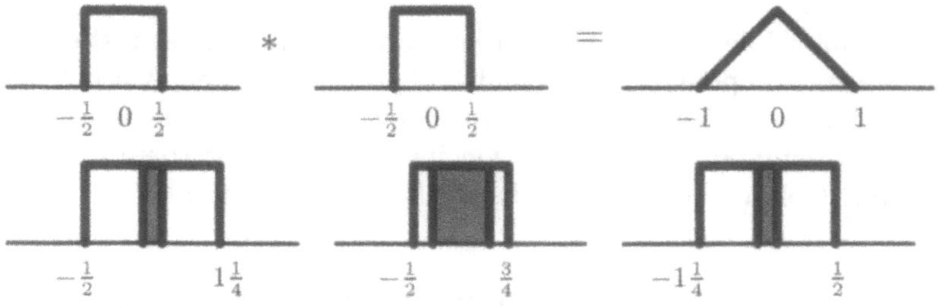

Figure 11.6. Convolution of constant B-splines to get linear B-spline.

$\mathcal{F}[L](\omega)$ should be small or zero for large ω. Define

$$M_k(x) = \frac{1}{2\pi}\int_{-\infty}^{\infty}\left(\frac{\sin(\omega/2)}{\omega/2}\right)^k e^{i\omega x}d\omega$$

for $k = 1, 2, \ldots,$ then

$$\mathcal{F}[M_k](\omega) = \left(\frac{\sin(\omega/2)}{\omega/2}\right)^k.$$

Schoenberg shows that

$$M_k(x) = \frac{1}{(k-1)!}[-k/2, -k/2+1, \ldots, k/2]_y(y-x)_+^{k-1}.$$

Using the M's in the above formula instead of the L's, Schoenberg established the now well-known properties of the functions M and their smoothing and approximation behavior. In this situation, all $(k-1)^{st}$ degree B-splines are translates of this one basic form. It is very interesting to note that because of the properties of convolutions, $\mathcal{F}[M_k] = \mathcal{F}[M_{k-j}]\mathcal{F}[M_j]$ which means

$$M_k(x) = \int M_{k-j}(t)M_j(x-t)dt, \qquad j = 1, \ldots, k-1.$$

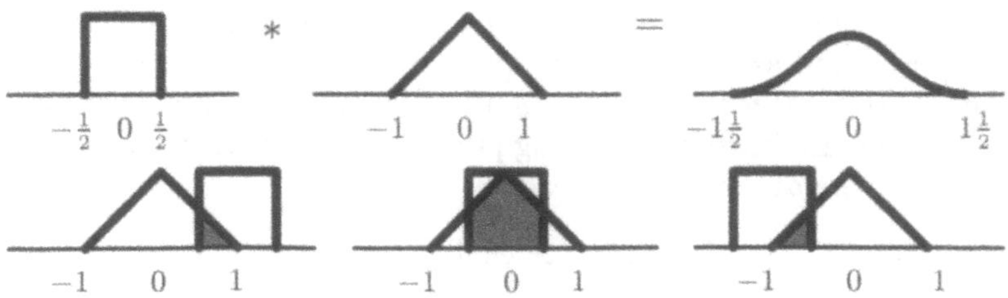

Figure 11.7. Convolution of constant with linear to get quadratic B-spline.

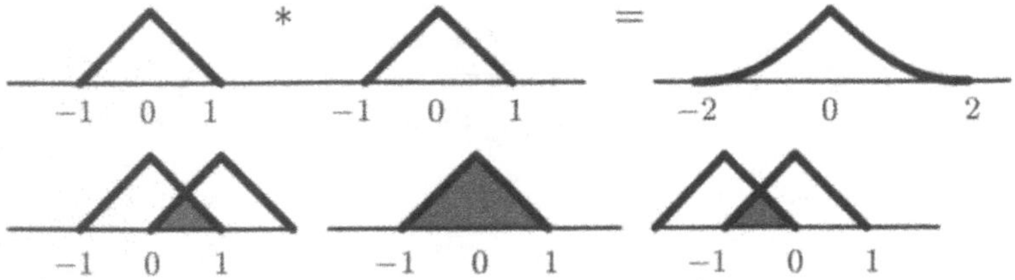

Figure 11.8. Convolution of linear B-splines to get cubic B-spline.

Thus, with knots every unit distance, one can build up all other degrees as a convolutions of lower degrees. We show several sequences.

One can see here that areas under piecewise constants change linearly, areas under piecewise linear change quadratically, etc. As is shown, an order four B-spline (the cubic) with uniform knots can be arrived at as a convolution of a constant B-spline and a parabolic B-spline (convolution of orders one and three) or as a convolution of two linear B-splines (convolution of orders two and two). Thus the piecewise polynomial nature of the B-spline becomes clear. Similarly, the local nature of the support is evident. Using this approach, the relationships to problems and terminalogy in digital signal processing can be seen. Note, however, that this convolutional relationship no longer holds if the knots are not evenly spaced.

Exercises

1. Prove Lemma 11.8. Hint: First show that it is true for first order divided differences and then show for higher orders using induction.

2. Finish the proof of Theorem 11.15, the Hermite Gennochi Theorem.

III

Surfaces

12

Differential Geometry for Surfaces

The purpose of this study is, as with curves, to investigate shape properties of parametric surfaces. From a given point on a curve, it is possible to move in only the positive or negative direction along the curve. However, it is possible to move in any of 2π direction from a given point on a curve. We shall present differential properties of surfaces and some well-known operators on them, and how these properties can be used to gain information about the shape of a surface. As in the case of curves, we shall try to understand more complex surfaces in terms of simpler ones, namely linear surfaces, i.e., planes, and quadric surfaces. We can also try to understand certain properties of surfaces by looking at classes of curves which lie in the surfaces. We address two major conceptual issues. One is to learn as much as possible about a surface by computing its differential characteristics. The other is to determine which of these characteristics is invariant under surface reparameterization. That is, we seek to determine which characteristics are *intrinsic* to the surface rather than being based on a particular parametric representation.

12.1 Regular Surfaces

Definition 12.1. *A* regular parametric representation *of class* $C^{(k)}$, *for* $k \geq 1$, *is a mapping* $\sigma : U \to W$, *where* $U \subset \mathbb{R}^2$ *is an open set and* $W = \sigma(U) \subset \mathbb{R}^3$ *such that* $\sigma \in C^{(k)}$ *and* $\frac{\partial \sigma}{\partial u_1} \times \frac{\partial \sigma}{\partial u_2} \neq 0$ *at all points* $\mu \in U$.

In order that $\frac{\partial \sigma}{\partial u_1} \times \frac{\partial \sigma}{\partial u_2} \neq 0$ for all $\mu \in U$, the partial derivatives must be linearly independent at all $\mu \in U$. Thus, neither one can be the zero vector, nor can one be a scalar multiple of the other.

Example 12.2. For $f : \mathbb{R}^2 \to \mathbb{R}^1$, a scalar-valued bivariate function, let $\xi = (x_1, x_2) \in \mathbb{R}^2$, and $\sigma : \mathbb{R}^2 \to \mathbb{R}^3$ where $\sigma_f(\xi) = (x_1, x_2, f(\xi))$. If f is $C^{(k)}$, so is σ_f. Note that

$$\frac{\partial \sigma_f}{\partial x_1} = \left(1, 0, \frac{\partial f}{\partial x_1}\right)$$

and

$$\frac{\partial \sigma_f}{\partial x_2} = \left(0, 1, \frac{\partial f}{\partial x_2}\right).$$

Hence $\frac{\partial \sigma_f}{\partial x_1} \times \frac{\partial \sigma_f}{\partial x_2} = \left(-\frac{\partial f}{\partial x_1}, -\frac{\partial f}{\partial x_2}, 1\right)$, which is never the zero vector. Thus, an explicit $C^{(1)}$ surface can always be represented as a regular parametric representation. Sometimes an explicit surface is called a *height field.* □

Example 12.3. Let $\mu = (u_1, u_2) \in \mathbb{R}^2$ with $\sigma(\mu) = (u_1, u_2, u_1 u_2)$. Then,

$$\frac{\partial \sigma}{\partial u_1} = (1, 0, u_2)$$

and

$$\frac{\partial \sigma}{\partial u_2} = (0, 1, u_1).$$

Hence $\frac{\partial \sigma}{\partial u_1} \times \frac{\partial \sigma}{\partial u_2} = (-u_2, -u_1, 1)$ which is never the zero vector. □

Example 12.4. Let $U = \{\mu = (u_1, u_2) \in \mathbb{R}^2 : \|\mu\| < 1\}$ with $\sigma(\mu) = \left(u_1, u_2, \sqrt{1 - (u_1)^2 - (u_2)^2}\right)$. From Example 12.2, it is clear that this is a regular parametric representation of an open hemisphere. □

As we have seen, approximation of univariate functions with higher continuity near one value of the domain is facilitated by Taylor's theorem and the polynomial. While the bivariate case is more complex, there is a corresponding bivariate Taylor's theorem which we shall find helpful in determining features of regular surfaces.

Theorem 12.5. Taylor's Theorem. *Let* $f : \mathbb{R}^2 \to \mathbb{R}^1$*. For* $\boldsymbol{a} = (a_1, a_2)$ *in the domain of* f*, and* $r > 0$*, define* $C_{\boldsymbol{a},r} = \{\xi = (x_1, x_2) : \|\xi - \boldsymbol{a}\| < r\}$.

Assume $f \in C^{(n+1)}$ *on* $C_{\boldsymbol{a},r}$. *Then, for* $\xi \in C_{\boldsymbol{a},r}$ *and* $\boldsymbol{\delta_\xi} = (\delta_1, \delta_2) = (x_1 - a_1, x_2 - a_2)$,

$$f(\xi) = \sum_{j=0}^{n} \frac{1}{j!} \left(\delta_1 \frac{\partial}{\partial x_1} + \delta_2 \frac{\partial}{\partial x_2} \right)^j f(\boldsymbol{a}) + R_{\boldsymbol{a},n}$$

where

$$\left(\delta_1 \frac{\partial}{\partial x_1} + \delta_2 \frac{\partial}{\partial x_2} \right)^j = \sum_{i=0}^{j} \binom{j}{i} (\delta_1)^i (\delta_2)^{j-i} \frac{\partial^j}{\partial {x_1}^i \partial {x_2}^{j-i}}$$

and

$$\begin{aligned} R_{\boldsymbol{a},n} &= \int_0^1 \frac{(1-t)^n}{n!} \left(\delta_1 \frac{\partial}{\partial x_1} + \delta_2 \frac{\partial}{\partial x_2} \right)^{n+1} f(a_1 + \delta_1 t, a_2 + \delta_2 t)\, dt \\ &= \frac{1}{(n+1)!} \left(\delta_1 \frac{\partial}{\partial x_1} + \delta_2 \frac{\partial}{\partial x_2} \right)^{n+1} f(a_1 + \delta_1 \theta, a_2 + \delta_2 \theta) \end{aligned}$$

where θ *is a value between zero and one.*

As in the case for curves we would like to determine conditions for which a reparameterization of a regular parametric representation of a surface leads to another regular parametric representation of the same surface.

Suppose that U is an open subset of $\mathbb{R}^2$, and $\sigma : U \to \mathbb{R}^3$ is a regular parametric representation of class $C^{(k)}$. Let $\mu = (u_1, u_2) \in U$ and $\nu = (v_1, v_2) \in V \subset \mathbb{R}^2$, V open. Suppose $\mu(\nu) = (u_1(\nu), u_2(\nu)))$ be a $C^{(k)}$ mapping from $V \to U$. By the chain rule

$$\frac{\partial \sigma}{\partial v_j} = \frac{\partial \sigma}{\partial u_1} \frac{\partial u_1}{\partial v_j} + \frac{\partial \sigma}{\partial u_2} \frac{\partial u_2}{\partial v_j}, \qquad j = 1, 2.$$

So,

$$\begin{aligned} \frac{\partial \sigma}{\partial v_1} \times \frac{\partial \sigma}{\partial v_2} &= \left(\frac{\partial \sigma}{\partial u_1} \frac{\partial u_1}{\partial v_1} + \frac{\partial \sigma}{\partial u_2} \frac{\partial u_2}{\partial v_1} \right) \times \left(\frac{\partial \sigma}{\partial u_1} \frac{\partial u_1}{\partial v_2} + \frac{\partial \sigma}{\partial u_2} \frac{\partial u_2}{\partial v_2} \right) \\ &= \frac{\partial \sigma}{\partial u_1} \times \frac{\partial \sigma}{\partial u_1} \frac{\partial u_1}{\partial v_2} \frac{\partial u_1}{\partial v_1} + \frac{\partial \sigma}{\partial u_1} \times \frac{\partial \sigma}{\partial u_2} \frac{\partial u_2}{\partial v_2} \frac{\partial u_1}{\partial v_1} \\ &\quad + \frac{\partial \sigma}{\partial u_2} \times \frac{\partial \sigma}{\partial u_1} \frac{\partial u_1}{\partial v_2} \frac{\partial u_2}{\partial v_1} + \frac{\partial \sigma}{\partial u_2} \times \frac{\partial \sigma}{\partial u_2} \frac{\partial u_2}{\partial v_2} \frac{\partial u_2}{\partial v_1} \\ &= \frac{\partial \sigma}{\partial u_1} \times \frac{\partial \sigma}{\partial u_2} \frac{\partial u_2}{\partial v_2} \frac{\partial u_1}{\partial v_1} + \frac{\partial \sigma}{\partial u_2} \times \frac{\partial \sigma}{\partial u_1} \frac{\partial u_1}{\partial v_2} \frac{\partial u_2}{\partial v_1} \\ &= \frac{\partial \sigma}{\partial u_1} \times \frac{\partial \sigma}{\partial u_2} \left(\frac{\partial u_2}{\partial v_2} \frac{\partial u_1}{\partial v_1} - \frac{\partial u_1}{\partial v_2} \frac{\partial u_2}{\partial v_1} \right) \\ &= \frac{\partial \sigma}{\partial u_1} \times \frac{\partial \sigma}{\partial u_2} \frac{\partial (u_1, u_2)}{\partial (v_1, v_2)}. \end{aligned}$$

Definition 12.6. *If $\rho = (r_1, \ldots, r_n)$ is a $C^{(k)}$ mapping from $U \subset \mathbb{R}^m$ to $\mathbb{R}^n$, and $\boldsymbol{a} \in U$ is fixed, then the matrix*

$$J_\mu(\rho)_{\boldsymbol{a}} = \left(\frac{\partial r_i}{\partial u_j}\right) = \begin{bmatrix} \left.\dfrac{\partial r_1}{\partial u_1}\right|_{\mu=\boldsymbol{a}} & \left.\dfrac{\partial r_2}{\partial u_1}\right|_{\mu=\boldsymbol{a}} & \cdots & \left.\dfrac{\partial r_n}{\partial u_1}\right|_{\mu=\boldsymbol{a}} \\ \vdots & \vdots & \ddots & \vdots \\ \left.\dfrac{\partial r_1}{\partial u_m}\right|_{\mu=\boldsymbol{a}} & \left.\dfrac{\partial r_2}{\partial u_m}\right|_{\mu=\boldsymbol{a}} & \cdots & \left.\dfrac{\partial r_n}{\partial u_m}\right|_{\mu=\boldsymbol{a}} \end{bmatrix}$$

is called the Jacobian matrix *of the mapping at $\boldsymbol{a}$. If $m = n$, the Jacobian matrix is square. Its determinant,*

$$|J_\mu(\rho)| = \frac{\partial(r_1, r_2, \ldots, r_n)}{\partial(u_1, u_2, \ldots, u_n)}$$

is called the Jacobian determinant.

Thus, we have derived,

Lemma 12.7. *Suppose that U is an open subset of $\mathbb{R}^2$, and $\sigma : U \to \mathbb{R}^3$ is a regular parametric representation of class $C^{(k)}$. Let $\mu = (u_1, u_2) \in U$ and $\nu = (v_1, v_2) \in V$, and suppose $\mu(\nu) = (u_1(\nu), u_2(\nu)))$ is a $C^{(k)}$ mapping from $V \to U$ where V is an open subset of $\mathbb{R}^2$. Then the mapping $\rho(\nu) = \sigma(\mu(\nu))$ from V to $\mathbb{R}^3$ is a regular surface if and only if for all $\nu \in V$, the Jacobian determinant of the mapping from V to U is nonzero, that is,*

$$\frac{\partial(u_1, u_2)}{\partial(v_1, v_2)} \neq 0.$$

Definition 12.8. *A $C^{(k)}$* coordinate transformation *is a $C^{(k)}$ one-to-one, onto function $\psi : V \to U$ of open sets in $\mathbb{R}^2$ whose inverse $i_\psi : U \to V$ is also of class $C^{(k)}$.*

Definition 12.9. *A $C^{(k)}$* simple surface, *also known as a* coordinate patch, *is a regular parametric representation which is a one-to-one function.*

Example 12.10. Suppose one has a planar curve $\gamma(t) = (x_1(t), 0, x_3(t))$, where $x_1(t) > 0$, such that γ is a regular one-to-one curve. Define $\rho(t, \theta) = (x_1(t)\cos\theta, x_1(t)\sin\theta, x_3(t))$, where $(t, \theta) \in R \times R_b$, R is some open interval and $R_b = (-b, b)$, the open interval from $-b$ to b. The domain is an open rectangle, and the surface ρ is called a *surface of revolution* about the z-axis. If $0 < b \leq \pi$, then the surface ρ is one-to-one and in $C^{(1)}$. Analogous definitions exist for surfaces of revolution about an arbitrary axis as well as about the x- and z- axes.

$$\begin{aligned}\frac{\partial \rho}{\partial t} &= \left(\frac{dx_1}{dt}\cos\theta, \frac{dx_1}{dt}\sin\theta, \frac{dx_3}{dt}\right),\\ \frac{\partial \rho}{\partial \theta} &= (-x_1\sin\theta, x_1\cos\theta, 0),\end{aligned}$$

and

$$\begin{aligned}\frac{\partial \rho}{\partial t}\times\frac{\partial \rho}{\partial \theta} &= \left(-x_1\frac{dx_3}{dt}\cos\theta, -x_1\frac{dx_3}{dt}\sin\theta, x_1\frac{dx_1}{dt}\cos^2\theta + x_1\frac{dx_1}{dt}\sin^2\theta\right)\\ &= \left(-x_1\frac{dx_3}{dt}\cos\theta, -x_1\frac{dx_3}{dt}\sin\theta, x_1\frac{dx_1}{dt}\right).\end{aligned}$$

The cross product has magnitude

$$\begin{aligned}&\left\|\frac{\partial \rho}{\partial t}\times\frac{\partial \rho}{\partial \theta}\right\|\\ &= \sqrt{(x_1)^2\left(\frac{dx_3}{dt}\right)^2\cos^2\theta + (x_1)^2\left(\frac{dx_3}{dt}\right)^2\sin^2\theta + (x_1)^2\left(\frac{dx_1}{dt}\right)^2}\\ &= |x_1|\sqrt{\left(\frac{dx_3}{dt}\right)^2 + \left(\frac{dx_1}{dt}\right)^2}.\end{aligned}$$

Since $(x_1(t), 0, x_3(t))$ is a regular space curve and one to one,

$$\left\|\tfrac{d}{dt}(x_1, 0, x_3)\right\| = \sqrt{\left(\frac{dx_3}{dt}\right)^2 + \left(\frac{dx_1}{dt}\right)^2} > 0,$$

so $\frac{\partial\rho}{\partial t}\times\frac{\partial\rho}{\partial\theta} \neq 0$ and ρ is a simple surface.

$\frac{\partial\rho}{\partial\theta}$ has no components in the $\mathbf{e}^3$ direction (the z direction), and furthermore, $\left\langle\frac{\partial\rho}{\partial t}, \frac{\partial\rho}{\partial\theta}\right\rangle = 0$, so the partials are orthogonal. The t curves are called *circles of latitude* and the θ curves are called *meridians*.

If $b > \pi$, then the map ρ is no longer one-to-one, so although it is still a regular parametric representation, it is not a simple surface. □

Definition 12.11. *Suppose σ is a simple surface, $\sigma : U \to \mathbb{R}^3$, and let $\boldsymbol{a} \in U$. Then $\gamma_1(t) = \sigma(a_1 + t, a_2)$ and $\gamma_2(t) = \sigma(a_1, a_2 + t)$ are called* isoparametric curves *in the surface σ through $\sigma(\boldsymbol{a})$.*

$$\gamma_1'(t) = \left.\frac{\partial \sigma}{\partial u_1}\right|_{\mu=\boldsymbol{a}+(t,0)}$$
$$\gamma_2'(t) = \left.\frac{\partial \sigma}{\partial u_2}\right|_{\mu=\boldsymbol{a}+(0,t)}$$

whenever $(a_1+t, a_2) \in U$ *or* $(a_1, a_2+t) \in U$, *respectively.*

Lemma 12.12. *If* σ *is a simple surface,* $\sigma : U \to \boldsymbol{R}^3$, *and* $\psi : V \to U$ *is a coordinate transformation, then* $\phi(\nu) = \sigma(\psi(\nu))$ *is a simple surface from* $V \to \boldsymbol{R}^3$ *and has the same graph as* σ.

Lemma 12.13. $C^{(k)}$ *coordinate transformations form an equivalence relation on the set of coordinate patches.*

The proof of this lemma is left as an exercise for the reader.

Example 12.14. Given a simple surface $\sigma : U \to \boldsymbol{R}^3$, and a constant $\boldsymbol{a}$, define $V = U - \boldsymbol{a} = \{\mu - \boldsymbol{a} : \mu \in U\}$. Then, $\mu = \psi(\nu) = \nu + \boldsymbol{a}$, and since $\psi(\nu)$ is just a translation, it is a coordinate transformation. Hence, $\sigma(\psi(\nu)) = \sigma(\nu + \boldsymbol{a})$ is a simple surface. □

Example 12.15. Suppose $V = (0,1) \times (0,1)$ and $U = (0,2) \times (3,4)$, and $\psi(\nu) = \left(2v_1, (v_2)^2 + 3\right)$. Then ψ is one-to-one, onto, and is $C^{(k)}$ for any positive k, and $i_\psi(\mu) = \left(u_1/2, \sqrt{u_2 - 3}\right)$.

The Jacobian matrix of ψ with respect to ν is

$$J_\nu(\psi) = \begin{bmatrix} 2 & 0 \\ 0 & 2v_2 \end{bmatrix},$$

with Jacobian determinant $4v_2$. □

Example 12.16. Suppose $V = \{\nu : \|\nu\| < 1\} = U$ and for any fixed value of θ, set $\psi_\theta(\nu)) = (v_1 \cos\theta - v_2 \sin\theta, v_2 \cos\theta + v_1 \sin\theta)$. Then

$$J_\nu(\psi_\theta) = \begin{bmatrix} \cos\theta & \sin\theta \\ -\sin\theta & \cos\theta \end{bmatrix}$$

and the Jacobian determinant has value 1. This coordinate transformation is a rotation. □

Example 12.17. Consider a rotation applied to the surface of Example 12.3, $\sigma(\mu) = u_1 u_2$. Then,

$$\begin{aligned}\sigma(\psi(\nu)) &= (v_1\cos\theta - v_2\sin\theta)(v_2\cos\theta + v_1\sin\theta)\\ &= v_1v_2(\cos^2\theta - \sin^2\theta) + (v_1^2 - v_2^2)\cos\theta\sin\theta,\end{aligned}$$

which, for $\theta = \frac{\pi}{4}$ equals

$$\sigma(\psi(\nu)) = \frac{v_1^2 - v_2^2}{2}. \qquad \square$$

Example 12.18. Let $V = \{\nu = (v_1, v_2) : \|\nu\| < 1,\ \ 0 < v_2\}$ and $\psi(\nu) = \left(v_1, \sqrt{1-(v_1)^2-(v_2)^2}\right)$. ψ maps V into V and

$$J_\nu(\psi) = \begin{bmatrix} 1 & \frac{-v_1}{\sqrt{1-(v_1)^2-(v_2)^2}} \\ 0 & \frac{-v_2}{\sqrt{1-(v_1)^2-(v_2)^2}} \end{bmatrix}$$

with determinant $-v_2/\sqrt{1-(v_1)^2-(v_2)^2}$. This determinant is nonzero and well defined whenever $0 \neq v_2$ and $\|\nu\| < 1$. This is just mapping a half solid disk without boundaries onto a half disk without boundaries. $\square$

The classes of regular parametric representations can also form an equivalence relationship under more complex rules which will allow the inverse function to be just locally defined. The term *regular surface* also specifies these types of relations. In this chapter, unless otherwise specified, we shall be using the result of Lemma 12.13 and looking at surfaces locally.

Lemma 12.19. *If σ is a coordinate patch on domain U, then for each $\boldsymbol{a} \in U$,*

$$\left\{ \frac{\partial\sigma}{\partial u_1}\bigg|_{\mu=\boldsymbol{a}}, \frac{\partial\sigma}{\partial u_2}\bigg|_{\mu=\boldsymbol{a}}, \left(\frac{\partial\sigma}{\partial u_1} \times \frac{\partial\sigma}{\partial u_2}\right)\bigg|_{\mu=\boldsymbol{a}} \right\}$$

forms a basis for $\mathbb{R}^3$.

Proof: This result follows immediately from the linear independence of the partials implied since their cross product is nonzero everywhere in U. ■

The moving trihedron (Definition 4.20) imposed intrinsic right-handed orthogonal coordinate systems for $\mathbb{R}^3$ at each point on a regular curve. We are now in a position to define right handed coordinate systems for $\mathbb{R}^3$ at each point of a coordinate patch. Unfortunately, these systems will not necessarily form orthogonal bases for $\mathbb{R}^3$. The parameter domain U

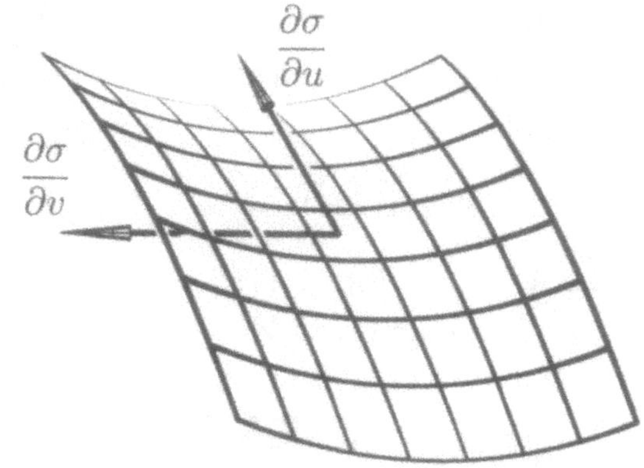

Figure 12.1. Coordinate system on the simple surface σ.

has an order imposed from u_1 to u_2. The ordered triple consisting of $\{\frac{\partial \sigma}{\partial u_1}, \frac{\partial \sigma}{\partial u_2}, \frac{\partial \sigma}{\partial u_1} \times \frac{\partial \sigma}{\partial u_2}\}$ forms a right-handed coordinate system.

Example 12.20. Consider the regular surface of Example 12.4, and the coordinate transformation of Example 12.18. For

$$V = \{ \nu = (v_1, v_2) : \|\nu\| < 1 \text{ and } 0 < v_2 \},$$

create a new surface by setting $\psi(\nu) = \left(v_1, \sqrt{1 - (v_1)^2 - (v_2)^2}\right)$. Then $\sigma(\psi(\nu)) = \left(v_1, \sqrt{1 - (v_1)^2 - (v_2)^2}, v_2\right)$ which is a quarter hemisphere with the same image. □

12.2 Tangents to Surfaces

Suppose that we designate the points in the domain of σ as $\mu = (u_1, u_2) \in U$. Now suppose α maps an interval into U. Designate $\alpha(t)$ as $(u_1(t), u_2(t))$. Define $\gamma(t) = \sigma(\alpha(t)) = \sigma(u_1(t), u_2(t))$ where σ is a regular surface and $\mu(t) = (u_1(t), u_2(t))$ is a regular planar curve with image in the domain of γ. By the chain rule

$$\frac{d\sigma}{dt} = \frac{\partial \sigma}{\partial u_1}\frac{du_1}{dt} + \frac{\partial \sigma}{\partial u_2}\frac{du_2}{dt}.$$

Since σ is a regular surface, $\frac{\partial \sigma}{\partial u_1} \times \frac{\partial \sigma}{\partial u_2} \neq 0$, so $\frac{\partial \sigma}{\partial u_1} \neq 0$ and $\frac{\partial \sigma}{\partial u_2} \neq 0$. Further, since γ is a regular curve, $\gamma'(t) = (\frac{du_1}{dt}, \frac{du_2}{dt}) \neq (0,0)$. Thus, $\frac{d\sigma}{dt} \neq 0$ for all values of t, and $\sigma(\mu(t))$ is a regular curve whenever $\sigma(\mu)$ is a regular surface and $\gamma(t) = (u_1(t), u_2(t))$ is a regular curve.

From the above derivation, the tangent vector to every regular curve in the simple surface σ at the point $\boldsymbol{a} = (a_1, a_2)$ can be written as a linear combination of

$$\frac{\partial\sigma}{\partial u_1}\bigg|_{\mu=\boldsymbol{a}} \quad \text{and} \quad \frac{\partial\sigma}{\partial u_2}\bigg|_{\mu=\boldsymbol{a}},$$

and hence is in the vector space spanned by those two vectors. Since a two-dimensional subspace of $\mathbb{R}^3$ is a plane, we have

Definition 12.21. *The* vector space of the tangent plane, $\boldsymbol{T_{\sigma,a}}$, *to a simple surface* $\sigma : U \to \mathbb{R}^3$ *at a point* $\sigma(\boldsymbol{a})$ *is the plane spanned by*

$$\left\{ \frac{\partial\sigma}{\partial u_1}\bigg|_{\mu=\boldsymbol{a}}, \frac{\partial\sigma}{\partial u_2}\bigg|_{\mu=\boldsymbol{a}} \right\}.$$

Thus, this is the plane through the origin with (nonunit) normal vector

$$\left(\frac{\partial\sigma}{\partial u_1} \times \frac{\partial\sigma}{\partial u_2}\right)\bigg|_{\mu=\boldsymbol{a}}.$$

The tangent plane *of* σ *at* $\sigma(\boldsymbol{a})$ *is the plane through the point* $\sigma(\boldsymbol{a})$ *with normal vector*

$$\left(\frac{\partial\sigma}{\partial u_1} \times \frac{\partial\sigma}{\partial u_2}\right)\bigg|_{\mu=\boldsymbol{a}}.$$

The elements of the tangent plane at $\sigma(\boldsymbol{a})$ are

$$\sigma(\boldsymbol{a}) + t_1 \frac{\partial\sigma}{\partial u_1}\bigg|_{\mu=\boldsymbol{a}} + t_2 \frac{\partial\sigma}{\partial u_2}\bigg|_{\mu=\boldsymbol{a}}.$$

For a given direction vector $\boldsymbol{k} = (k_1, k_2)$, consider $\gamma_{\boldsymbol{k}}(t) = (u_1(t), u_2(t)) = (a_1 + tk_1, a_2 + tk_2)$. We already know that this is a regular curve, and hence for a simple surface σ, $\sigma(\gamma_{\boldsymbol{k}}(t))$ is a regular curve with derivative

$$\frac{d\sigma}{dt} = \frac{\partial\sigma}{\partial u_1}k_1 + \frac{\partial\sigma}{\partial u_2}k_2. \tag{12.1}$$

Lemma 12.22. *The* directional derivative *of a simple surface* σ *in the direction* $\boldsymbol{k}$ *at the point* $\boldsymbol{a} = (a_1, a_2)$ *is*

$$D_{\boldsymbol{k}}\sigma(\boldsymbol{a}) = k_1 \frac{\partial\sigma}{\partial u_1}\bigg|_{\mu=\boldsymbol{a}} + k_2 \frac{\partial\sigma}{\partial u_2}\bigg|_{\mu=\boldsymbol{a}}.$$

Proof: $D_{\boldsymbol{k}}\sigma(\boldsymbol{a}) = \frac{d\sigma}{dt}(\boldsymbol{a})$ from its definition. Hence by Equation 12.1, the result follows. ■

We have shown that all directional derivatives are tangent vectors to some regular curve in the image of σ and are in the tangent plane. We have shown containment of the set of tangent vectors to regular curves in σ at $\boldsymbol{a}$ in that plane.

Corollary 12.23. *The directional derivative of the surface σ in the direction ν is a tangent vector to the surface. All tangent vectors are also directional derivatives.*

Proof: We have shown one direction and need only show the other. It is subsumed in the proof of the lemma below. ∎

To fully describe the tangent plane, we must show that any element in the tangent plane is a tangent vector to a regular curve in σ at $\boldsymbol{a}$.

Lemma 12.24. *A vector $\boldsymbol{t}$ is in the vector space of the tangent plane of the simple surface σ at $\sigma(\boldsymbol{a})$ if and only if it is a tangent vector of some regular curve in the surface σ at the point $\sigma(\boldsymbol{a})$.*

Proof: We have already shown that if $\boldsymbol{t}$ is a tangent vector of a regular curve in the simple surface σ at $\sigma(\boldsymbol{a})$ then it is an element in the vector space of the tangent plane. We must show only the other direction.

Since $\boldsymbol{t}$ is in the vector space of the tangent plane, $\boldsymbol{t} = t_1\frac{\partial\sigma}{\partial u_1} + t_2\frac{\partial\sigma}{\partial u_2}$. Define the planar curve $\gamma(t) = (u_1(t), u_2(t)) = (a_1 + tt_1, a_2 + tt_2) = \boldsymbol{a} + t(t_1, t_2)$. Then $\sigma(\gamma(t))$ is a regular curve in the surface σ through the point $\sigma(\boldsymbol{a})$ with tangent vector $\boldsymbol{t}$. Thus, the result is proved. ∎

The set of vectors in the vector space of the tangent plane at $\sigma(\boldsymbol{a})$ are called *tangent vectors* to the surface σ at $\sigma(\boldsymbol{a})$. The elements of the tangent vector space of σ at $\sigma(\boldsymbol{a})$, $\boldsymbol{t}$, are written

$$\boldsymbol{t} = t_1 \left.\frac{\partial\sigma}{\partial u_1}\right|_{\mu=\boldsymbol{a}} + t_2 \left.\frac{\partial\sigma}{\partial u_2}\right|_{\mu=\boldsymbol{a}}.$$

Let $\psi = (\psi_1, \psi_2) : V \to U$ be a $C^{(k)}$ coordinate transformation and $\psi(\boldsymbol{b}) = \boldsymbol{a}$. Define $\beta(\nu) = \sigma(\psi(\nu))$. Since the first partials of β define its tangent plane at $\nu = \boldsymbol{b}$, it is necessary to find them. The chain rule can be applied giving

$$\left.\frac{\partial\beta}{\partial v_i}\right|_{\nu=\boldsymbol{b}} = \left.\left(\frac{\partial\sigma}{\partial u_1}\frac{\partial\psi_1}{\partial v_i} + \frac{\partial\sigma}{\partial u_2}\frac{\partial\psi_2}{\partial v_i}\right)\right|_{\mu=\boldsymbol{a},\nu=\boldsymbol{b}}, \qquad i = 1, 2. \tag{12.2}$$

It can be shown through direct calculation that

$$\left(\frac{\partial \beta}{\partial v_1} \times \frac{\partial \beta}{\partial v_2}\right)\bigg|_{\nu=b} = \left(\frac{\partial \psi_1}{\partial v_1}\frac{\partial \psi_2}{\partial v_2} - \frac{\partial \psi_2}{\partial v_1}\frac{\partial \psi_1}{\partial v_2}\right)\bigg|_{\nu=b} \left(\frac{\partial \sigma}{\partial u_1} \times \frac{\partial \sigma}{\partial u_2}\right)\bigg|_{\mu=a}. \tag{12.3}$$

If the scalar multiplying the cross product is positive, the two tangent planes have the same normal. If the scalar multiplying the cross product is negative, then the normals point in opposite directions. In both cases, however, the tangent planes must be the same. That scalar is just the Jacobian determinant of ψ, and since ψ is a $C^{(1)}$ coordinate transformation, its value is nonzero.

Definition 12.25. *The* unit normal *to the regular surface* σ *at* $\sigma(\mathbf{a})$ *is*

$$n_{\mathbf{a}} = \frac{\left(\frac{\partial \sigma}{\partial u_1} \times \frac{\partial \sigma}{\partial u_2}\right)\Big|_{\mu=a}}{\left\|\left(\frac{\partial \sigma}{\partial u_1} \times \frac{\partial \sigma}{\partial u_2}\right)\Big|_{\mu=a}\right\|}.$$

Corollary 12.26. *The dimension of the vector space of the tangent plane is two.*

In practice, the term *tangent plane* is used to mean the tangent plane and also the vector space of the tangent plane interchangeably. When confusion might result, it will be clarified which one is intended.

12.3 First Fundamental Form

In this section we continue to study the relationship between regular curves in a surface and the surfaces which contain them.

12.3.1 Arc Length of a Curve on a Simple Surface

Let $\mu(t) = (u_1(t), u_2(t))$ and $\gamma(t) = \sigma(\mu(t))$ be defined so γ is a regular curve in a simple surface σ.

Since the arc length of γ between a fixed point a and t is

$$s = s(t) = \int_a^t \left\|\frac{d\gamma(t)}{dt}\right\| dt,$$

we know that the rate of change of the arc length is $\left\|\frac{d\gamma(t)}{dt}\right\|$. Hence

$$\begin{aligned}
\left(\frac{ds}{dt}\right)^2 &= \left\|\frac{d\gamma(t)}{dt}\right\|^2 = \left\langle \frac{d\gamma(t)}{dt}, \frac{d\gamma(t)}{dt} \right\rangle \\
&= \left\langle \sum_i \frac{\partial\sigma}{\partial u_i}\frac{du_i}{dt}, \sum_j \frac{\partial\sigma}{\partial u_j}\frac{du_j}{dt} \right\rangle \\
&= \sum_i \sum_j \frac{du_i}{dt}\left\langle \frac{\partial\sigma}{\partial u_i}, \frac{\partial\sigma}{\partial u_j} \right\rangle \frac{du_j}{dt} \\
&= \sum_i \sum_j \frac{du_i}{dt} g_{i,j} \frac{du_j}{dt} \\
&= \left[\frac{du_1}{dt} \; \frac{du_2}{dt} \right] \boldsymbol{G} \left[\frac{du_1}{dt} \; \frac{du_2}{dt} \right]^T
\end{aligned}$$

where $g_{i,j} = \left\langle \frac{\partial\sigma}{\partial u_i}, \frac{\partial\sigma}{\partial u_j} \right\rangle$ and $\boldsymbol{G} = (g_{i,j})$. By symmetry properties of inner products, $g_{1,2} = g_{2,1}$. When this quadratic form is expanded, usually, one writes $E = g_{1,1}$, $F = g_{1,2}$ and $G = g_{2,2}$ to get

$$\begin{aligned}
\left(\frac{ds}{dt}\right)^2 &= E\left(\frac{du_1}{dt}\right)^2 + 2F\frac{du_1}{dt}\frac{du_2}{dt} + G\left(\frac{du_2}{dt}\right)^2 \\
&= \frac{1}{E}\left[\left(E\frac{du_1}{dt} + F\frac{du_2}{dt}\right)^2 + (EG - F^2)\left(\frac{du_2}{dt}\right)^2\right].
\end{aligned}$$

A quadratic form is called positive definite if its values are nonnegative and it is zero only at the origin, namely when $\left(\frac{du_1}{dt}, \frac{du_2}{dt}\right) = (0,0)$. We already know that the above quadratic form is always positive since the arc length is a strictly increasing function for all regular curves. Since $E > 0$ and $G > 0$, the only way that it can be ensured that the form is positive for all $\mu(t)$ is if $EG - F^2 > 0$ at all times. But, $\det(\boldsymbol{G}) = EG - F^2$, so we have shown that the determinant of $\boldsymbol{G}$ is always positive.

Lemma 12.27. *The determinant of* $\boldsymbol{G}$ *is always positive.*

Definition 12.28. *The quadratic form*

$$E\left(\frac{du_1}{dt}\right)^2 + 2F\frac{du_1}{dt}\frac{du_2}{dt} + G\left(\frac{du_2}{dt}\right)^2$$

is called the first fundamental form *and is frequently denoted by*

$$\mathrm{I}\left(\frac{du_1}{dt}, \frac{du_2}{dt}\right) = \mathrm{I}\left(\frac{d\mu}{dt}\right).$$

Example 12.29. Using the simple surface of Example 12.4

$$\begin{aligned}\frac{\partial\sigma}{\partial u_1}(\mu) &= \left(1, 0, \frac{-u_1}{\sqrt{1-(u_1)^2-(u_2)^2}}\right),\\ \frac{\partial\sigma}{\partial u_2}(\mu) &= \left(0, 1, \frac{-u_2}{\sqrt{1-(u_1)^2-(u_2)^2}}\right).\end{aligned}$$

Thus,

$$\begin{aligned} g_{1,1} &= 1+\frac{(u_1)^2}{1-(u_1)^2-(u_2)^2}\\ &= \frac{1-(u_2)^2}{1-(u_1)^2-(u_2)^2},\\ g_{1,2} &= g_{2,1}\\ &= \frac{u_1u_2}{1-(u_1)^2-(u_2)^2},\\ g_{2,2} &= 1+\frac{(u_2)^2}{1-(u_1)^2-(u_2)^2},\\ &= \frac{1-(u_1)^2}{1-(u_1)^2-(u_2)^2}\end{aligned}$$

and

$$G = \frac{1}{1-(u_1)^2-(u_2)^2}\begin{bmatrix} 1-(u_2)^2 & u_1u_2 \\ u_1u_2 & 1-(u_1)^2 \end{bmatrix}.$$

□

Example 12.30. For $\sigma(\mu) = (u_1+u_2, u_1u_2, u_1-u_2)$,

$$\frac{\partial\sigma}{\partial u_1}(\mu) = (1, u_2, 1) \quad \text{and} \quad \frac{\partial\sigma}{\partial u_2}(\mu) = (1, u_1, -1).$$

$$\frac{\partial\sigma}{\partial u_1}\times\frac{\partial\sigma}{\partial u_2} = (-u_2-u_1, 2, u_1-u_2)$$

which is never zero, so the domain can be all or any part of $\boldsymbol{R}^2$.

$$\begin{aligned} g_{1,1}(\mu) &= 2+(u_2)^2,\\ g_{1,2}(\mu) &= g_{2,1} = u_1u_2,\\ g_{2,2}(\mu) &= 2+(u_1)^2.\end{aligned}$$

□

The matrix $\boldsymbol{G}$ is not dependent on the particular regular curve that is chosen but only on the point in the surface for which the partials are calculated. The matrix $\boldsymbol{G}$ is then independent of the choice of regular curves and is intrinsic to the surface in this parameterization. We now study when the first fundamental form is invariant under coordinate transformations.

12.3.2 Invariance of the First Fundamental Form

We now suppose $\gamma(t)$ is a regular map into $V \subset \mathbb{R}^2$, that $\mu = \psi(\nu) = (u_1(v_1, v_2), u_2(v_1, v_2))$ is a coordinate transformation and that σ is a simple surface. $\beta(\nu) = \sigma(\psi(\nu))$ is another regular parametric representation of the same surface which is also one-to-one. Since $\psi(\nu)$ is a coordinate transformation, the inverse function, $i_\psi(\mu)$, exists. Further let $\alpha(t) = \beta(\gamma(t))$. Since the first fundamental form is just the evaluation of the magnitude squared of the derivative of the arc length, and arc length is invariant under regular changes of variables, it is clear that the first fundamental form is invariant under coordinate transformations. However, the matrix, $\boldsymbol{G}$, is not invariant, nor are the derivative vectors.

$$
\begin{aligned}
\frac{d\alpha}{dt} &= \sum_{i=1}^{2} \frac{\partial \beta}{\partial v_i}\frac{dv_i}{dt} \\
&= \sum_{i=1}^{2}\sum_{j=1}^{2} \frac{\partial \sigma}{\partial u_j}\frac{\partial u_j}{\partial v_i}\frac{dv_i}{dt}.
\end{aligned}
$$

Let us consider first the elements of $\boldsymbol{G}^r$, the matrix of the first fundamental form for β.

$$
\begin{aligned}
g^r_{i,j} &= \left\langle \frac{\partial \beta}{\partial v_i}, \frac{\partial \beta}{\partial v_j} \right\rangle \\
&= \left\langle \textstyle\sum_{k=1}^{2} \dfrac{\partial \sigma}{\partial u_k}\dfrac{\partial u_k}{\partial v_i}, \textstyle\sum_{r=1}^{2} \dfrac{\partial \sigma}{\partial u_r}\dfrac{\partial u_r}{\partial v_j} \right\rangle \\
&= \sum_{k=1}^{2}\sum_{r=1}^{2} \left\langle \frac{\partial \sigma}{\partial u_k}\frac{\partial u_k}{\partial v_i}, \frac{\partial \sigma}{\partial u_r}\frac{\partial u_r}{\partial v_j} \right\rangle \\
&= \sum_{k=1}^{2}\sum_{r=1}^{2} \left\langle \frac{\partial \sigma}{\partial u_k}, \frac{\partial \sigma}{\partial u_r} \right\rangle \frac{\partial u_k}{\partial v_i}\frac{\partial u_r}{\partial v_j} \\
&= \sum_{k=1}^{2}\sum_{r=1}^{2} g_{k,r}\frac{\partial u_k}{\partial v_i}\frac{\partial u_r}{\partial v_j} \\
&= \begin{bmatrix} \dfrac{\partial u_1}{\partial v_i} & \dfrac{\partial u_2}{\partial v_i} \end{bmatrix} \boldsymbol{G} \begin{bmatrix} \dfrac{\partial u_1}{\partial v_j} & \dfrac{\partial u_2}{\partial v_j} \end{bmatrix}^T.
\end{aligned}
$$

Thus,

$$
\boldsymbol{G}^r = J_\nu(\mu)\boldsymbol{G}\left(J_\nu(\mu)\right)^T.
$$

In general, E, F, and G, the coefficients of the first fundamental form, are not invariant under coordinate transformations. Next, consider $\left(\frac{ds_\alpha}{dt}\right)^2$.

$$\begin{aligned}
\left(\frac{ds_\alpha}{dt}\right)^2 &= \left\langle \frac{d\beta(\nu(t))}{dt}, \frac{d\beta(\nu(t))}{dt} \right\rangle \\
&= \left[\begin{array}{cc} \frac{dv_1}{dt} & \frac{dv_2}{dt} \end{array}\right] G^* \left[\begin{array}{cc} \frac{dv_1}{dt} & \frac{dv_2}{dt} \end{array}\right]^T \\
&= \left[\begin{array}{cc} \frac{dv_1}{dt} & \frac{dv_2}{dt} \end{array}\right] J_\nu(\mu) G \left(J_\nu(\mu)\right)^T \left[\begin{array}{cc} \frac{dv_1}{dt} & \frac{dv_2}{dt} \end{array}\right]^T.
\end{aligned}$$

So

$$\begin{aligned}
\left[\begin{array}{cc} \frac{dv_1}{dt} & \frac{dv_2}{dt} \end{array}\right] J_\nu(\mu) &= \left[\begin{array}{cc} \frac{dv_1}{dt}\frac{\partial u_1}{\partial v_1} + \frac{dv_2}{dt}\frac{\partial u_1}{\partial v_2} & \frac{dv_1}{dt}\frac{\partial u_2}{\partial v_1} + \frac{dv_2}{dt}\frac{\partial u_2}{\partial v_2} \end{array}\right] \\
&= \left[\begin{array}{cc} \frac{du_1}{dt} & \frac{du_2}{dt} \end{array}\right].
\end{aligned}$$

Finally,

$$\left(\frac{ds_\alpha}{dt}\right)^2 = \left[\begin{array}{cc} \frac{du_1}{dt} & \frac{du_2}{dt} \end{array}\right] G \left[\begin{array}{cc} \frac{du_1}{dt} & \frac{du_2}{dt} \end{array}\right]^T.$$

Theorem 12.31. *The first fundamental form is invariant under coordinate transformations.*

12.3.3 Angles between Tangent Vectors

When two regular curves intersect at a point in space, it is sometimes necessary to find the angle between the tangent vectors of the two curves at the point of intersection. Since the curvilinear coordinate system is rarely orthogonal, an analogous process to finding the projections in the x, y, and z directions is inappropriate. $\frac{\partial \sigma}{\partial u_1}$ and $\frac{\partial \sigma}{\partial u_2}$ need not be orthogonal at any point in the domain; in fact, they rarely are. When both curves are regular curves in a surface, one can establish a relationship which facilitates the calculation of this angle measurement.

If $\boldsymbol{t}^1$ and $\boldsymbol{t}^2$ are two tangent vectors in $\boldsymbol{T}_\sigma$ at μ, then

$$\begin{aligned}
\boldsymbol{t}^1 &= t_1^1 \frac{\partial \sigma}{\partial u_1} + t_2^1 \frac{\partial \sigma}{\partial u_2}, \\
\boldsymbol{t}^2 &= t_1^2 \frac{\partial \sigma}{\partial u_1} + t_2^2 \frac{\partial \sigma}{\partial u_2},
\end{aligned}$$

so $\langle \boldsymbol{t^1}, \boldsymbol{t^2} \rangle = \|\boldsymbol{t^1}\|\|\boldsymbol{t^2}\| \cos\theta$, where θ as the angle between $\boldsymbol{t^1}$ and $\boldsymbol{t^2}$. Then

$$\begin{aligned} \langle \boldsymbol{t}^1, \boldsymbol{t}^2 \rangle &= \sum_i \sum_j t_i^1 t_j^2 \left\langle \frac{\partial \sigma}{\partial u_i}, \frac{\partial \sigma}{\partial u_j} \right\rangle \\ &= \sum_i \sum_j t_i^1 t_j^2 g_{i,j} \\ &= \left[\begin{array}{cc} t_1^2 & t_2^2 \end{array}\right] \boldsymbol{G} \left[\begin{array}{cc} t_1^1 & t_2^1 \end{array}\right]^T \end{aligned} \tag{12.4}$$

which is a more general version of the quadratic form defining the first fundamental form. Some define this generalized form *as* the first fundamental form, and use the one we have defined as a restricted case. Hence,

$$\begin{aligned} \cos\theta &= \frac{\langle \boldsymbol{t}^1, \boldsymbol{t}^2 \rangle}{(\|\boldsymbol{t}^1\|\,\|\boldsymbol{t}^2\|)} \\ &= \frac{\sum_i \sum_j t_i^1 t_j^2 g_{i,j}}{\sqrt{\sum_i \sum_j t_i^1 t_j^1 g_{i,j}} \sqrt{\sum_i \sum_j t_i^2 t_j^2 g_{i,j}}}. \end{aligned}$$

Lemma 12.32. *The unnormalized vector* $\frac{\partial \sigma}{\partial u_1} \times \frac{\partial \sigma}{\partial u_2}$ *to the simple surface* σ *at* μ *has magnitude*

$$\sqrt{|\boldsymbol{G}|}.$$

The proof is left as an exercise for the reader.

Lemma 12.33. *The first partials of* σ *at* μ *are orthogonal if and only if* $g_{1,2} = 0$*. The* i^{th} *partial has unit length if and only if* $g_{i,i} = 1$.

Proof:

$$\begin{aligned} \left\langle \frac{\partial \sigma}{\partial u_i}, \frac{\partial \sigma}{\partial u_j} \right\rangle &= g_{i,j} \\ &= \left\|\frac{\partial \sigma}{\partial u_i}\right\| \left\|\frac{\partial \sigma}{\partial u_j}\right\| \cos\theta, \end{aligned}$$

where θ is the angle between the first partials. Setting $i = 1$ and $j = 2$, the result follows. Using the same reasoning, if $i = j = 1$, then $\theta = 0$, and again the result follows. ∎

12.3.4 Surface Area of a Simple Surface

So far, the matrix of the first fundamental form has been used only in relation to the elements of the tangent plane. This section investigates a use of the $\boldsymbol{G}$ matrix to determine information about the surface itself.

To calculate the area of a particular simple surface requires derivation. Using traditional ideas, we first try to approximate the surface by many small elements whose areas are easily computed and for which, in the limit, the sum of the areas converges to the area of the surface.

For a surface σ at $\sigma(\boldsymbol{a})$, let du_1 and du_2 be small positive real numbers. Since $\sigma \in C^{(1)}$, for small enough du_1 and du_2,

$$\left.\frac{\partial\sigma}{\partial u_1}\right|_{\mu=\boldsymbol{a}} du_1 \approx \sigma(a_1 + du_1, a_2) - \sigma(\boldsymbol{a}),$$

where $\approx$ stands for approximately equal to and

$$\left.\frac{\partial\sigma}{\partial u_2}\right|_{\mu=\boldsymbol{a}} du_2 \approx \sigma(a_1, a_2 + du_2) - \sigma(\boldsymbol{a}).$$

We wish to calculate the area of the surface element bounded by the parametric rectangle $[a_1, a_1 + du_1] \times [a_2, a_2 + du_2]$. This is the surface region bounded by the four parametric curves with endpoints $\sigma(\boldsymbol{a})$ to $\sigma(a_1 + du_1, a_2)$; $\sigma(a_1 + du_1, a_2)$ to $\sigma(a_1 + du_1, a_2 + du_2)$; $\sigma(\boldsymbol{a})$ to $\sigma(a_1, a_2 + du_2)$; and $\sigma(a_1, a_2 + du_2)$ to $\sigma(a_1 + du_1, a_2 + du_2)$. For small enough $d\mu = du_1 du_2$, this surface subregion can be approximated by a parallelogram (see Figure 12.2) in the tangent plane bounded by the points

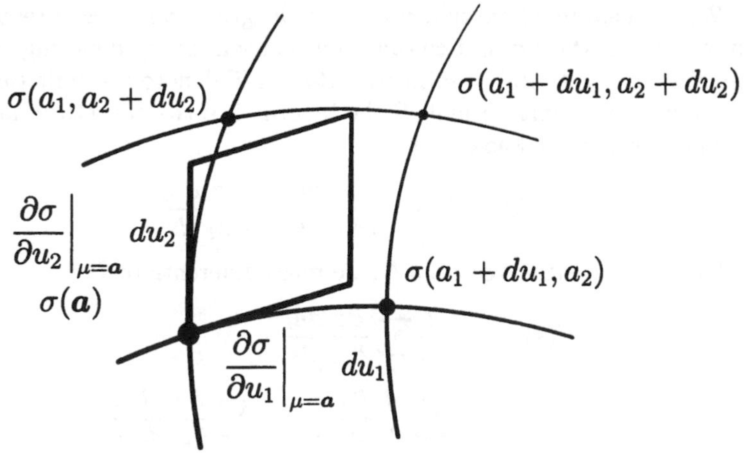

Figure 12.2. Approximate surface element.

$$\sigma(\boldsymbol{a}), \qquad \sigma(\boldsymbol{a}) + \left.\frac{\partial\sigma}{\partial u_1}\right|_{\mu=\boldsymbol{a}} du_1,$$

$$\sigma(\boldsymbol{a}) + \left.\frac{\partial\sigma}{\partial u_2}\right|_{\mu=\boldsymbol{a}} du_2, \qquad \sigma(\boldsymbol{a}) + \left.\frac{\partial\sigma}{\partial u_1}\right|_{\mu=\boldsymbol{a}} du_1 + \left.\frac{\partial\sigma}{\partial u_2}\right|_{\mu=\boldsymbol{a}} du_2.$$

The area of that parallelogram is just

$$\begin{aligned}\left\| \left.\frac{\partial\sigma}{\partial u_1}\right|_{\mu=\boldsymbol{a}} du_1 \times \left.\frac{\partial\sigma}{\partial u_2}\right|_{\mu=\boldsymbol{a}} du_2 \right\| &= \left\| \left.\left(\frac{\partial\sigma}{\partial u_1} \times \frac{\partial\sigma}{\partial u_2}\right)\right|_{\mu=\boldsymbol{a}} \right\| du_1 du_2 \\ &= \sqrt{|\boldsymbol{G}|_{\mu=\boldsymbol{a}}}\, du_1 du_2.\end{aligned}$$

Adding the areas together and taking the limit yields

$$\text{Area} = \iint \sqrt{|\boldsymbol{G}|}\, du_1 du_2.$$

12.4 The Second Fundamental Form and Curves in the Surface

We have seen that the first fundamental form gives information about curves in the surface, and the relationship of intrinsic properties of the curves to an intrinsic property of the surface. It would seem reasonable to suspect that second order geometry of all curves in the surface would relate to other intrinsic properties of the surface.

We shall assume here that γ is an arc length parametrized regular $C^{(k)}$ curve in the surface σ and continue our convention by denoting the parameterization by s, $\gamma(s) = \sigma(\mu(s))$. Also let T denote the unit tangent to γ, N denote the unit normal, B the binormal, κ the curvature, and τ the torsion. We already know that

$$T = \gamma'(s) = \frac{\partial\sigma}{\partial u_1}\frac{du_1}{ds} + \frac{\partial\sigma}{\partial u_2}\frac{du_2}{ds}.$$

To find the curvature vector, κN, we must differentiate T.

$$\begin{aligned}\gamma''(s) &= \frac{d}{ds}\sum_i \frac{\partial\sigma}{\partial u_i}\frac{du_i}{ds} \\ &= \sum_i \frac{d}{ds}\frac{\partial\sigma}{\partial u_i}\frac{du_i}{ds} + \sum_i \frac{\partial\sigma}{\partial u_i}\frac{d}{ds}\frac{du_i}{ds} \\ &= \sum_i \frac{d}{ds}\frac{\partial\sigma}{\partial u_i}\frac{du_i}{ds} + \sum_i \frac{\partial\sigma}{\partial u_i}\frac{d^2u_i}{ds^2}\end{aligned}$$

and

$$\begin{aligned}\frac{d}{ds}\frac{\partial\sigma}{\partial u_i}(u_1(s),u_2(s)) &= \frac{\partial}{\partial u_1}\frac{\partial\sigma}{\partial u_i}\frac{du_1}{ds}+\frac{\partial}{\partial u_2}\frac{\partial\sigma}{\partial u_i}\frac{du_2}{ds}\\ &= \frac{\partial^2\sigma}{\partial u_1\partial u_i}\frac{du_1}{ds}+\frac{\partial^2\sigma}{\partial u_2\partial u_i}\frac{du_2}{ds}\\ &= \sum_j\frac{\partial^2\sigma}{\partial u_j\partial u_i}\frac{du_j}{ds}.\end{aligned}$$

Then,

$$\gamma''(s) = \sum_j\sum_i\frac{\partial^2\sigma}{\partial u_j\partial u_i}\frac{du_j}{ds}\frac{du_i}{ds}+\sum_i\frac{\partial\sigma}{\partial u_i}\frac{d^2u_i}{ds^2}. \qquad (12.5)$$

Since $\gamma''(s)=\kappa N$ is the curvature vector, we would like to decompose this vector into components depending only on this particular curve and the surface.

Definition 12.34. *The* intrinsic normal *to a curve γ in the surface σ is $\boldsymbol{S}=n\times T$, where n is the unit normal to the surface at that point and T is the unit tangent to the curve at the point.*

Note that the intrinsic normal is orthogonal to both the tangent to the curve and the unit normal to the surface. Since the intrinsic normal is orthogonal to the surface normal, it must lie in the tangent plane. The intrinsic normal need not be the curve normal N.

Example 12.35. Consider the hemisphere surface, $\sigma(\mu)$ and curve $\gamma(s)$ on σ, where,

$$\begin{aligned}\sigma(\mu) &= \left(u_1,u_2,\sqrt{1-(u_1)^2-(u_2)^2}\right),\\ \gamma(s) &= \left(\frac{1}{\sqrt{2}}\cos s\sqrt{2},\frac{1}{\sqrt{2}}\sin s\sqrt{2},\frac{1}{\sqrt{2}}\right).\end{aligned}$$

We study the relationship of T, S and n.

$$\begin{aligned}\gamma'(s) &= T\\ &= \left(-\sin s\sqrt{2},\cos s\sqrt{2},0\right);\\ T' &= \left(-\sqrt{2}\cos s\sqrt{2},-\sqrt{2}\sin s\sqrt{2},0\right)\end{aligned}$$

so

$$N=\frac{T'}{\|T'\|}=\left(-\cos s\sqrt{2},-\sin s\sqrt{2},0\right) \qquad \text{and} \qquad \kappa=\sqrt{2}.$$

At $s = \frac{\pi}{2\sqrt{2}}$, $\gamma(s) = \left(0, 1/\sqrt{2}, 1/\sqrt{2}\right)$ and $N = (0, -1, 0)$. Since,

$$\begin{aligned} \frac{\partial \sigma}{\partial u_1} &= \left(1, 0, \frac{-u_1}{\sqrt{1-(u_1)^2-(u_2)^2}}\right), \\ \frac{\partial \sigma}{\partial u_2} &= \left(0, 1, \frac{-u_2}{\sqrt{1-(u_1)^2-(u_2)^2}}\right), \end{aligned}$$

the surface normal is $n = \left(u_1, u_2, \sqrt{1-(u_1)^2-(u_2)^2}\right)$.

At $\mu = \left(0, 1/\sqrt{2}\right)$, $\sigma(\mu) = \left(0, 1/\sqrt{2}, 1/\sqrt{2}\right)$, and $n = \left(0, 1/\sqrt{2}, 1/\sqrt{2}\right)$. Clearly, $N \neq n$ and

$$\begin{aligned} \boldsymbol{S} &= n \times T \\ &= \left(0, 1/\sqrt{2}, 1/\sqrt{2}\right) \times (-1, 0, 0) \\ &= \left(0, -1/\sqrt{2}, 1/\sqrt{2}\right). \end{aligned}$$

□

It is clear that the same curve may lie on many surfaces and that each of those surfaces might have a very different tangent plane at corresponding points along the curve. The intrinsic normals of the curve are different in each of these surfaces. It is a function of both the curve and the surface in which it is embedded.

We now return to the problem of determining the projection of the curve's curvature vector in the tangent plane. Since both T and $\boldsymbol{S}$ are in the tangent plane, and $\langle T, \boldsymbol{S}\rangle = 0$, they form an orthonormal basis for the tangent plane. So the projection of γ'' in the tangent plane can be decomposed as a sum of a vector in the T and a vector in the $\boldsymbol{S}$ direction, with coefficients determined by finding the inner product of $\gamma'' = \kappa N$ with T and $\boldsymbol{S}$, respectively. The three vectors $\{T, \boldsymbol{S}, n\}$ form an orthonormal basis for $\mathbb{R}^3$, at each point on the curve.

$$\gamma''(s) = \langle \gamma'', T\rangle T + \langle \gamma'', \boldsymbol{S}\rangle \boldsymbol{S} + \langle \gamma'', n\rangle n.$$

In the T direction,

$$\begin{aligned} \langle \gamma''(s), T(s)\rangle &= \langle \kappa(s) N(s), T(s)\rangle \\ &= \kappa(s)\langle N(s), T(s)\rangle \\ &= 0. \end{aligned}$$

Hence, the projection of γ'' into the tangent plane must be a scalar multiple of $\boldsymbol{S}$, the intrinsic normal, so

$$\gamma''(s) = \langle \gamma''(s), \boldsymbol{S}\rangle \boldsymbol{S} + \langle \gamma''(s), n\rangle n.$$

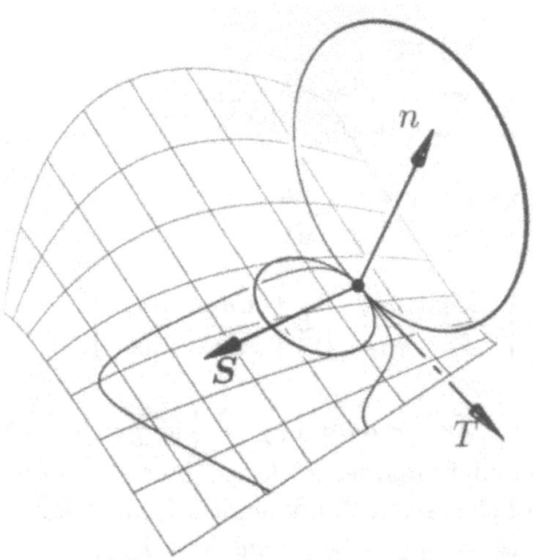

Figure 12.3. The normal and geodesic curvatures.

Definition 12.36. *If $\gamma(s)$ is an arc length parametrized regular curve in the simple surface σ, denote by $\kappa_n(s) = \langle \gamma''(s), n \rangle$, the portion of the curvature vector of the curve in the direction of the surface normal. This is called the* normal curvature *of the curve. Denote by $\kappa_g(s) = \langle \gamma''(s), \boldsymbol{S} \rangle$, the portion of the curvature vector of the curve in the direction of the curve's intrinsic normal in this surface. This is called the* geodesic curvature *of the curve.*

We observe that the normal curvature of the curve in the surface can be negative.

$\gamma''(s) = \kappa(s)N(s) = \kappa_n(s)n + \kappa_g(s)\boldsymbol{S}$, and since n and $\boldsymbol{S}$ are orthogonal, $\kappa^2 = (\kappa_n)^2 + (\kappa_g)^2$.

Definition 12.37. *A* geodesic curve *on the surface σ is a unit speed regular curve in σ with geodesic curvature κ_g equal to zero everywhere along the curve.*

Note that when $\kappa_g = 0$, $N = +n$, that is, the surface normal and the curve normal define the same planes. Thus, a regular curve in a surface is also a geodesic curve in a simple surface if its rectifying planes (see Definition 4.21) are the same as the tangent planes for the corresponding surface points.

Next we look at the projection of the curve's curvature vector in the direction of the surface normal.

$$
\begin{aligned}
\kappa_n(s) &= \langle \gamma''(s), n\rangle \\
&= \sum_j \sum_i \left\langle \frac{\partial^2 \sigma}{\partial u_j \partial u_i}, n \right\rangle \frac{du_j}{ds}\frac{du_i}{ds} + \sum_i \left\langle \frac{\partial \sigma}{\partial u_i}, n \right\rangle \frac{d^2 u_i}{ds^2} \\
&= \sum_j \sum_i L_{j,i} \frac{du_j}{ds}\frac{du_i}{ds} \\
&= \begin{bmatrix} \frac{du_1}{ds} & \frac{du_2}{ds} \end{bmatrix} \boldsymbol{L} \begin{bmatrix} \frac{du_1}{ds} & \frac{du_2}{ds} \end{bmatrix}^T . \qquad (12.6)
\end{aligned}
$$

Definition 12.38. *The scalars* $L_{i,j} = \left\langle \frac{\partial^2 \sigma}{\partial u_j \partial u_i}, n \right\rangle$ *are called the* coefficients of the second fundamental form, *and the matrix,* $\boldsymbol{L} = (L_{i,j})$ *is called the* matrix of the second fundamental form. *The coefficients are also written* $L = L_{1,1}$, $M = L_{1,2} = L_{2,1}$, *and* $N = L_{2,2}$.

For an arbitrary tangent vector $d\mu = \frac{du_1}{dt}\frac{\partial \sigma}{\partial u_1} + \frac{du_2}{dt}\frac{\partial \sigma}{\partial u_2}$, *the form*

$$
\mathrm{II}\left(\frac{du_1}{dt}, \frac{du_2}{dt}\right) = \begin{bmatrix} \frac{du_1}{dt} & \frac{du_2}{dt} \end{bmatrix} \boldsymbol{L} \begin{bmatrix} \frac{du_1}{dt} & \frac{du_2}{dt} \end{bmatrix}^T
$$

is called the fundamental form.

Lemma 12.39. $L_{i,j} = -\left\langle \frac{\partial \sigma}{\partial u_i}, \frac{\partial n}{\partial u_j} \right\rangle$.

Proof: Since $0 \equiv \left\langle \frac{\partial \sigma}{\partial u_i}, n \right\rangle$ for all values of μ,

$$
\begin{aligned}
0 &\equiv \frac{\partial}{\partial u_j}\left\langle \frac{\partial \sigma}{\partial u_i}, n \right\rangle \\
&\equiv \left\langle \frac{\partial^2 \sigma}{\partial u_j \partial u_i}, n \right\rangle + \left\langle \frac{\partial \sigma}{\partial u_i}, \frac{\partial n}{\partial u_j} \right\rangle
\end{aligned}
$$

and the result follows. ■

Example 12.40. We calculate the coefficients of the second fundamental form for the surface of Example 12.30, $\sigma(\mu) = (u_1 + u_2, u_1 u_2, u_1 - u_2)$.

$$
\begin{aligned}
\frac{\partial \sigma}{\partial u_1} &= (1, u_2, 1), \\
\frac{\partial \sigma}{\partial u_2} &= (1, u_1, -1), \quad \text{and} \\
\frac{\partial \sigma}{\partial u_1} \times \frac{\partial \sigma}{\partial u_2} &= (-u_2 - u_1, 2, u_1 - u_2).
\end{aligned}
$$

$$\left\|\frac{\partial\sigma}{\partial u_1}\times\frac{\partial\sigma}{\partial u_2}\right\| = \sqrt{(u_1+u_2)^2+4+(u_1-u_2)^2}, \quad \text{so}$$

$$n = \frac{1}{\sqrt{(u_1+u_2)^2+4+(u_1-u_2)^2}}(-u_2-u_1, 2, u_1-u_2).$$

$$\frac{\partial^2\sigma}{\partial {u_1}^2} = (0,0,0), \qquad \frac{\partial^2\sigma}{\partial u_1\partial u_2} = (0,1,0);$$
$$\frac{\partial^2\sigma}{\partial u_2\partial u_1} = (0,1,0), \qquad \frac{\partial^2\sigma}{\partial {u_2}^2} = (0,0,0),$$

and

$$\boldsymbol{L} = \frac{2}{\sqrt{(u_1+u_2)^2+4+(u_1-u_2)^2}}\begin{bmatrix} 0 & 1 \\ 1 & 0 \end{bmatrix}.$$

□

The second fundamental form is defined for a given parametric representation. As in the case of the first fundamental form it can be shown that

Theorem 12.41. *The second fundamental form is invariant under coordinate transformations which have a positive Jacobian.*

Since the surface normal flips directions when the coordinate transformation has a negative Jacobian, it is clear that the coefficients will change sign. The proof of this theorem is left as an exercise.

Theorem 12.42. *If γ^1 and γ^2 are two arc length parametrized curves in σ with tangent vector T at the same intersection point then they both have the same normal curvature at that point.*

Proof: The normal curvature of the arc length curves depends only on the coefficients of the second fundamental form and the tangent vectors for the curves. The matrix of the second fundamental form is determined by the surface geometry and surface parameterization only. If both curves have the same tangent vector, then the result follows. ■

Theorem 12.43. *For an arc length parametrized curve γ with curve normal N in surface σ, let θ be the angle between N and n. Then*

$$\begin{aligned}\kappa_n(s) &= \langle\gamma''(s), n\rangle \\ &= \kappa(s)\langle N, n\rangle \\ &= \kappa(s)\cos\theta.\end{aligned}$$

Usually, a regular curve in a surface is not specified by its arc length parameterization. Hence we need to determine a formulation for the normal curvature of an arbitrary regular curve at a point in the surface. Now, suppose we want to consider a different regular parameterization of γ, say, $t = t(s)$. Then

$$\left(\frac{du_1}{dt}, \frac{du_2}{dt}\right) = \left(\frac{du_1}{ds}, \frac{du_2}{ds}\right)\frac{ds}{dt},$$

by the chain rule, and rewriting Equation 12.6 gives

$$\begin{aligned}
\kappa(s)\cos\theta &= L_{1,1}\left(\frac{du_1}{ds}\right)^2 + 2L_{1,2}\frac{du_1}{ds}\frac{du_2}{ds} + L_{2,2}\left(\frac{du_2}{ds}\right)^2 \\
&= L\left(\frac{du_1}{ds}\right)^2 + 2M\frac{du_1}{ds}\frac{du_2}{ds} + N\left(\frac{du_2}{ds}\right)^2 \\
&= \left[L\left(\frac{du_1}{dt}\right)^2 + 2M\frac{du_1}{dt}\frac{du_2}{dt} + N\left(\frac{du_2}{dt}\right)^2\right]\left(\frac{dt}{ds}\right)^2 \\
&= \frac{L\left(\frac{du_1}{dt}\right)^2 + 2M\frac{du_1}{dt}\frac{du_2}{dt} + N\left(\frac{du_2}{dt}\right)^2}{E\left(\frac{du_1}{dt}\right)^2 + 2F\frac{du_1}{dt}\frac{du_2}{dt} + G\left(\frac{du_2}{dt}\right)^2} \\
&= \frac{\mathrm{II}\left(\frac{d\mu}{dt}\right)}{\mathrm{I}\left(\frac{d\mu}{dt}\right)}.
\end{aligned}$$

Thus, no matter what the parameterization, the normal curvature of a regular curve embedded in a regular surface can be written as the ratio of the second fundamental form to the first fundamental form.

Corollary 12.44. *If γ^1 and γ^2 are two regular curves in the simple surface σ which intersect at some point, and at that point have the same unit tangent vectors and the same unit normals, then they have the same curvature.*

Consider the set $\mathcal{S}_{\boldsymbol{a}}$ of all $C^{(2)}$ regular curves $\mu(t) = (u_1(t), u_2(t))$ contained in U through the point $\boldsymbol{a}$, such that $\left\|\frac{d\mu}{dt}\right\| = 1$ at $\boldsymbol{a}$. Given a direction $(\cos\alpha, \sin\alpha)$, there is a curve in $\mathcal{S}_{\boldsymbol{a}}$ such that $\frac{d\mu}{dt} = (\cos\alpha, \sin\alpha)$, for example, $\mu(t) = (a_1 + t\cos\alpha, a_2 + t\sin\alpha)$. Thus, $Q = \left\{\frac{d\mu}{dt} : \mu \in \mathcal{S}_{\boldsymbol{a}}\right\} = S^1$, the unit circle in the parametric plane. The normal curvature κ_n is a continuous function of $\frac{d\mu}{dt}$ on the closed and bounded set Q. Hence, κ_n has a maximum and a minimum over this set.

Definition 12.45. *The maximum and minimum values of the normal curvature are called the* principal curvatures. *The unit vectors in the directions for which these values are attained are called the* principal directions *of the surface.*

Since each $\mu(t)$ is in $C^{(2)}$, κ_n is continuously differentiable as a function of $\frac{d\mu}{dt}$. For now, set $\nu = (v_1, v_2) = \frac{d\mu}{dt}$, and write $k(\nu) = \kappa_n(\nu)$. To solve for the directions of maximum and minimum values, we find the two partial derivative equations of k and solve for the values of ν which are roots of both partials simultaneously.

$$\frac{\partial k}{\partial v_1} = \frac{\frac{\partial \mathtt{II}}{\partial v_1}\mathtt{I} - \frac{\partial \mathtt{I}}{\partial v_1}\mathtt{II}}{\mathtt{I}^2},$$
$$\frac{\partial k}{\partial v_2} = \frac{\frac{\partial \mathtt{II}}{\partial v_2}\mathtt{I} - \frac{\partial \mathtt{I}}{\partial v_2}\mathtt{II}}{\mathtt{I}^2}.$$

Since $\mathtt{I}$ is nonzero, this can be done. First set $\frac{\partial k}{\partial v_1} = 0$ and $\frac{\partial k}{\partial v_2} = 0$. We must solve for the values of v_1, v_2, and k for which this can happen. Multiply each equation through by $1/\mathtt{I}$ to get

$$\begin{aligned} 0 &= \frac{\partial \mathtt{II}}{\partial v_1} - \frac{\partial \mathtt{I}}{\partial v_1}\frac{\mathtt{II}}{\mathtt{I}} \\ &= \frac{\partial \mathtt{II}}{\partial v_1} - k\frac{\partial \mathtt{I}}{\partial v_1}. \end{aligned} \tag{12.7}$$

Analogously,

$$0 = \frac{\partial \mathtt{II}}{\partial v_2} - k\frac{\partial \mathtt{I}}{\partial v_2}. \tag{12.8}$$

Now,

$$\frac{\partial \mathtt{II}}{\partial v_1} = 2Lv_1 + 2Mv_2,$$
$$\frac{\partial \mathtt{II}}{\partial v_2} = 2Mv_1 + 2Nv_2,$$

and

$$\frac{\partial \mathtt{I}}{\partial v_1} = 2Ev_1 + 2Fv_2,$$
$$\frac{\partial \mathtt{I}}{\partial v_2} = 2Fv_1 + 2Gv_2.$$

Substituting these values into Equations 12.7 and 12.8 gives

$$\begin{aligned} Lv_1 + Mv_2 - k(Ev_1 + Fv_2) &= 0, \\ Mv_1 + Nv_2 - k(Fv_1 + Gv_2) &= 0. \end{aligned}$$

Regrouping,

$$\begin{aligned} (L - kE)v_1 + (M - kF)v_2 &= 0, \\ (M - kF)v_1 + (N - kG)v_2 &= 0. \end{aligned} \tag{12.9}$$

Since the extremal values of k, the normal curvature, cannot be 0, and the maximum and minimum must occur at $\nu = (v_1, v_2) \in Q$, the determinant of the coefficient matrix to (v_1, v_2) must be zero, i.e.,

$$\begin{vmatrix} L - kE & M - kF \\ M - kF & N - kG \end{vmatrix} = 0.$$

Expanding the determinant gives

$$\left(EG - F^2\right) k^2 - (GL + EN - 2FM)k + \left(LN - M^2\right) = 0. \tag{12.10}$$

Let the solutions to this quadratic equation be κ_1 and κ_2. Since $EG - F^2 > 0$, this is a proper quadratic equation. Further, the roots must be real values since the normal curvature must take on both maximum and minimum values over Q. If the normal curvature is a constant function in all directions, the maximum equals the minimum so $\kappa_1 = \kappa_2$. Otherwise, the values must be distinct. Using properties of quadratic functions,

$$\begin{aligned} \kappa_1\kappa_2 &= \frac{LN - M^2}{EG - F^2}, \\ \kappa_1 + \kappa_2 &= \frac{GL + EN - 2FM}{EG - F^2}. \end{aligned} \tag{12.11}$$

Definition 12.46. *The quantity $\kappa_1\kappa_2$ is called the* Gaussian curvature. *The quantity $(\kappa_1 + \kappa_2)/2$ is called the* mean curvature.

We shall show later that they are invariants of the surface.

Once the principal values are known by solving the quadratic equation, Equation 12.10, one can calculate the principal directions by solving Equation 12.9 and normalizing the result to have magnitude 1. Proceeding, using the first line of the equation, for $i = 1, 2$,

$$v_2{}^i = -\frac{(L - \kappa_i E)v_1{}^i}{M - \kappa_i F}$$

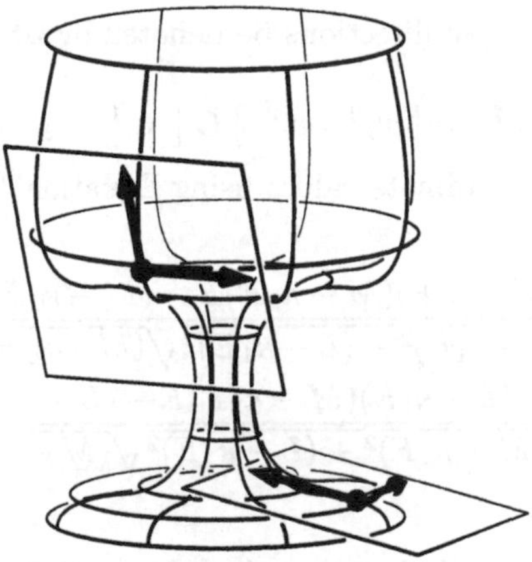

Figure 12.4. Principal directions on a surface.

and

$$\begin{aligned} 1 &= \left\| v^i \right\|^2 \\ &= \left(v_1{}^i\right)^2 + \left(-\frac{(L-\kappa_i E)v_1{}^i}{M-\kappa_i F}\right)^2 \\ &= \left[1+\left(\frac{L-\kappa_i E}{M-\kappa_i F}\right)^2\right]\left(v_1{}^i\right)^2 \\ &= \frac{(M-\kappa_i F)^2+(L-\kappa_i E)^2}{(M-\kappa_i F)^2}\left(v_1{}^i\right)^2. \end{aligned}$$

Thus,

$$\begin{aligned} v_1{}^i &= \frac{M-\kappa_i F}{\sqrt{(M-\kappa_i F)^2+(L-\kappa_i E)^2}} \\ v_2{}^i &= -\frac{L-\kappa_i E}{\sqrt{(M-\kappa_i F)^2+(L-\kappa_i E)^2}}. \end{aligned} \tag{12.12}$$

Note that the second equation of Equation 12.9 can equally well be used to solve for $v_2{}^i$. Also, one could have solved for $v_1{}^i$ instead of $v_2{}^i$. Finally, this form of solution can lead to instabilities, as we shall see in one of the following examples.

Lemma 12.47. *If $\kappa_1 \neq \kappa_2$, the principal directions at a point on a surface are orthogonal.*

Proof: Let the principal directions be denoted by $\boldsymbol{v}^1$ and $\boldsymbol{v}^2$. By Equation 12.4,

$$\langle \boldsymbol{v}^1, \boldsymbol{v}^2 \rangle = \begin{bmatrix} v_1{}^2 & v_2{}^2 \end{bmatrix} \boldsymbol{G} \begin{bmatrix} v_1{}^1 & v_2{}^1 \end{bmatrix}^T.$$

Substituting for the coordinate values using Equation 12.12 and by Equations 12.11

$$\begin{aligned} \langle \boldsymbol{v}^1, \boldsymbol{v}^2 \rangle &= \frac{(M-\kappa_1 F)(M-\kappa_2 F)E-(M-\kappa_1 F)(L-\kappa_2 E)F}{\sqrt{(M-\kappa_1 F)^2+(L-\kappa_1 E)^2}\sqrt{(M-\kappa_2 F)^2+(L-\kappa_2 E)^2}} \\ &\quad - \frac{(L-\kappa_1 E)(M-\kappa_2 F)F-(L-\kappa_1 E)(L-\kappa_2 E)G}{\sqrt{(M-\kappa_1 F)^2+(L-\kappa_1 E)^2}\sqrt{(M-\kappa_2 F)^2+(L-\kappa_2 E)^2}} \\ &= 0. \end{aligned}$$

That means that the angle between the two pricipal direction vectors is ninety degrees, and the principal direction vectors are orthogonal. ■

12.4.1 Examples

Example 12.48. Let $r > 0$ be a fixed real number and

$$\sigma(\mu) = (r\cos u_1 \cos u_2, r\sin u_1 \cos u_2, r\sin u_2).$$

Then, $(r\cos u_1 \cos u_2)^2 + (r\sin u_1 \cos u_2)^2 + (r\sin u_2)^2 = r^2$, and the surface is a sphere.

$$\begin{aligned} \frac{\partial \sigma}{\partial u_1} &= (-r\sin u_1 \cos u_2, r\cos u_1 \cos u_2, 0), \\ \frac{\partial \sigma}{\partial u_2} &= (-r\cos u_1 \sin u_2, -r\sin u_1 \sin u_2, r\cos u_2), \end{aligned}$$

$$g_{1,1} = \left\langle \frac{\partial \sigma}{\partial u_1}, \frac{\partial \sigma}{\partial u_1} \right\rangle = r^2 \cos^2 u_2, \qquad \text{and}$$

$$g_{2,2} = \left\langle \frac{\partial \sigma}{\partial u_2}, \frac{\partial \sigma}{\partial u_2} \right\rangle = r^2.$$

$$\begin{aligned} g_{1,2} &= \left\langle \frac{\partial \sigma}{\partial u_1}, \frac{\partial \sigma}{\partial u_2} \right\rangle \\ &= r^2 (\sin u_1 \cos u_1 \cos u_2 \cos u_2 - \sin u_1 \cos u_1 \cos u_2 \cos u_2) \\ &= 0 = g_{2,1}. \end{aligned}$$

$$\boldsymbol{G} = \begin{bmatrix} r^2 \cos^2 u_2 & 0 \\ 0 & r^2 \end{bmatrix}.$$

For the second fundamental form:

$$\begin{aligned}
\frac{\partial^2 \sigma}{\partial {u_1}^2} &= (-r\cos u_1 \cos u_2, -r\sin u_1 \cos u_2, 0),\\
\frac{\partial^2 \sigma}{\partial u_1 \partial u_2} &= (r\sin u_1 \sin u_2, -r\cos u_1 \sin u_2, 0),\\
\frac{\partial^2 \sigma}{\partial {u_2}^2} &= (-r\cos u_1 \cos u_2, -r\sin u_1 \cos u_2, -r\sin u_2),
\end{aligned}$$

and

$$\frac{\partial \sigma}{\partial u_1} \times \frac{\partial \sigma}{\partial u_2} = r^2 \cos u_2 (\cos u_1 \cos u_2, \sin u_1 \cos u_2, \sin u_2).$$

Then $\boldsymbol{n} = (\cos u_1 \cos u_2, \sin u_1 \cos u_2, \sin u_2)$ as expected, so

$$\boldsymbol{L} = \begin{bmatrix} -r\cos^2 u_2 & 0 \\ 0 & -r \end{bmatrix}.$$

The Gaussian curvature is

$$\begin{aligned}
\kappa_1 \kappa_2 &= \frac{|\boldsymbol{L}|}{|\boldsymbol{G}|}\\
&= \frac{r^2 \cos^2 u_2}{r^4 \cos^2 u_2} = r^{-2}.
\end{aligned}$$

Since

$$\begin{aligned}
\kappa_1 + \kappa_2 &= \frac{GL + EN - 2FM}{|\boldsymbol{G}|}\\
&= -\frac{2}{r},
\end{aligned}$$

the mean curvature is $-1/r$. □

Example 12.49. A saddle surface with parameterization $\sigma(\mu) = (u_1, u_2, u_1 u_2)$.

$$\begin{aligned}
\frac{\partial \sigma}{\partial u_1} &= (1, 0, u_2),\\
\frac{\partial \sigma}{\partial u_2} &= (0, 1, u_1),\\
g_{1,1} &= \left\langle \frac{\partial \sigma}{\partial u_1}, \frac{\partial \sigma}{\partial u_1} \right\rangle = 1 + (u_2)^2,\\
g_{2,2} &= \left\langle \frac{\partial \sigma}{\partial u_2}, \frac{\partial \sigma}{\partial u_2} \right\rangle = 1 + (u_1)^2,
\end{aligned}$$

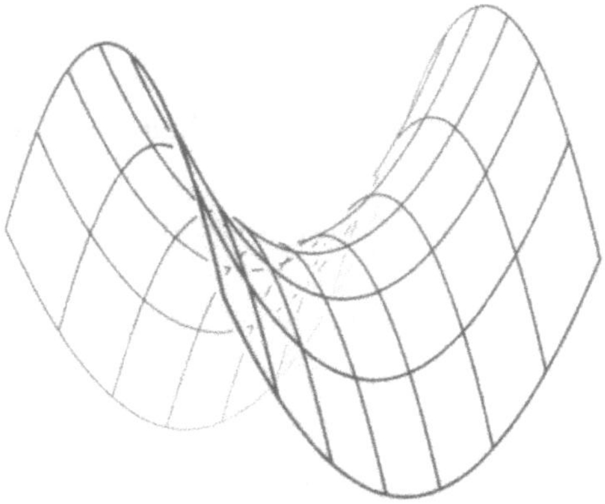

Figure 12.5. A saddle surface.

and $g_{1,2} = g_{2,1} = u_1 u_2$, so

$$G = \begin{bmatrix} 1+(u_2)^2 & u_1u_2 \\ u_1u_2 & 1+(u_1)^2 \end{bmatrix}.$$

For the second fundamental form:

$$\begin{aligned} \frac{\partial^2\sigma}{\partial {u_1}^2} &= (0,0,0), \\ \frac{\partial^2\sigma}{\partial u_1 \partial u_2} &= (0,0,1), \\ \frac{\partial^2\sigma}{\partial {u_2}^2} &= (0,0,0), \qquad \text{and} \\ \frac{\partial\sigma}{\partial u_1} \times \frac{\partial\sigma}{\partial u_2} &= (-u_2,-u_1,1). \end{aligned}$$

Hence,

$$\begin{aligned} \boldsymbol{n} &= \sqrt{\frac{1}{1+(u_1)^2+(u_2)^2}},(-u_2,-u_1,1), \\ \boldsymbol{L} &= \frac{1}{\sqrt{1+(u_1)^2+(u_2)^2}} \begin{bmatrix} 0 & 1 \\ 1 & 0 \end{bmatrix}. \end{aligned}$$

Since $|\boldsymbol{G}| = 1 + (u_1)^2 + (u_2)^2$ and $|\boldsymbol{L}| = -\frac{1}{1+(u_1)^2+(u_2)^2}$, the Gaussian curvature is

$$\kappa_1\kappa_2 = -\left(\frac{1}{1+(u_1)^2+(u_2)^2}\right)^2$$

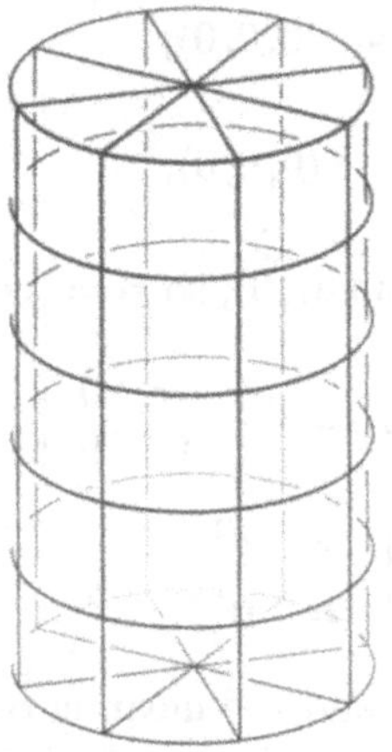

Figure 12.6. A cylinder of radius r.

and the mean curvature is

$$\kappa_1 + \kappa_2 = -\frac{2u_1u_2}{[1 + (u_1)^2 + (u_2)^2]^{3/2}}.$$

While computation of the principal directions is straightforward, the equations get messy in symbolic computation. □

Example 12.50. $\sigma(\mu) = (r\cos u_1, r\sin u_1, u_2)$ represents a cylinder of radius r.

$$\frac{\partial\sigma}{\partial u_1} = (-r\sin u_1, r\cos u_1, 0) \quad \text{and} \quad \frac{\partial\sigma}{\partial u_2} = (0,0,1).$$

$$\begin{aligned} g_{1,1} &= \left\langle \frac{\partial\sigma}{\partial u_1}, \frac{\partial\sigma}{\partial u_1} \right\rangle = r^2, \\ g_{2,2} &= \left\langle \frac{\partial\sigma}{\partial u_2}, \frac{\partial\sigma}{\partial u_2} \right\rangle = 1, \\ g_{1,2} &= g_{2,1} = 0. \end{aligned}$$

$$\boldsymbol{G} = \begin{bmatrix} r^2 & 0 \\ 0 & 1 \end{bmatrix}.$$

For the second fundamental form:

$$\frac{\partial^2\sigma}{\partial {u_1}^2} = (-r\cos u_1, -r\sin u_1, 0),$$

$$\begin{aligned} \frac{\partial^2 \sigma}{\partial u_1 \partial u_2} &= (0,0,0), \\ \frac{\partial^2 \sigma}{\partial {u_2}^2} &= (0,0,0), \end{aligned}$$

and $\frac{\partial \sigma}{\partial u_1} \times \frac{\partial \sigma}{\partial u_2} = (r\cos u_1, r\sin u_1, 0)$, so $\boldsymbol{n} = (\cos u_1, \sin u_1, 0)$.

$$\begin{aligned} \boldsymbol{L} &= \begin{bmatrix} -r & 0 \\ 0 & 0 \end{bmatrix}, \\ |\boldsymbol{G}| &= r^2, \\ |\boldsymbol{L}| &= 0. \end{aligned}$$

The Gaussian curvature is 0, and the mean curvature is $-1/(2r)$.

Since $\kappa_1\kappa_2 = 0$ and $\kappa_1 + \kappa_2 = -1/r$, we find that $\kappa_1 = 0$ and $\kappa_2 = -1/r$. Solving for $\boldsymbol{v}^1$ we find that

$$\begin{bmatrix} -r & 0 \\ 0 & 0 \end{bmatrix} \nu = \boldsymbol{0},$$

so ${v_1}^1 = 0$, and since $\left\|\boldsymbol{v}^1\right\| = 0$, then ${v_2}^1 = 1$. Thus, the first principal direction is $\frac{\partial \sigma}{\partial u_2}$. The second principal direction has $\boldsymbol{v}^2 = (1,0)$ and so is $\frac{\partial \sigma}{\partial u_1}$. While this is not shown directly, it follows from the orthogonality of principal directions, as shown in Theorem 12.47. □

12.4.2 The Osculating Paraboloid

In the study of general regular curves, we determined the Frenet frame and then studied the behavior of the curve by studying the rates of change of the Frenet frame. The rate of change and direction of change of the unit tangent vector has proven very informative about the curve, giving the curvature information. A second approach to deriving local information about the curve looked at the second order Taylor approximation to the regular curve. It was discovered to lie in the osculating plane, and share tangent vector and curvature with the regular curve at the tangent point. Hence, this second order approximation can be used for viewing the behavior of the curve locally. For a surface, the first order approximation is the tangent plane. Its normal and rate of change in different directions gives a great deal of information about the behavior and geometry of the surface, as we shall determine mathematically. If we expand the simple surface into its second order bivariate Taylor approximation, the surface will be a quadric surface. We shall assume that the simple surface is at least $C^{(k)}$ where $k \geq 2$.

By defining the function

$$\begin{aligned}\delta_{\boldsymbol{a},\sigma}(\nu) &= \left(\frac{\partial\sigma}{\partial u_1}v_1 + \frac{\partial\sigma}{\partial u_2}v_2\right)\bigg|_{\mu=\boldsymbol{a}} \\ &\quad + \frac{1}{2}\left(\frac{\partial^2\sigma}{\partial {u_1}^2}(v_1)^2 + 2\frac{\partial^2\sigma}{\partial u_1\partial u_2}v_1v_2 + \frac{\partial^2\sigma}{\partial {u_2}^2}(v_2)^2\right)\bigg|_{\mu=\boldsymbol{a}}\end{aligned}$$

the Taylor approximation to σ at $\boldsymbol{a}$ can be written

$$\sigma(\boldsymbol{a}+\nu) = \sigma(\boldsymbol{a}) + \delta_{\boldsymbol{a},\sigma}(\nu) + R_{\boldsymbol{a},2}(\nu),$$

where $R_{\boldsymbol{a},2}$ is the remainder term. Usually, ν is small, and $\sigma \in C^{(3)}$ for this expansion to be valid. δ is a quadric function in each of its coordinate functions and the domain of δ is an open set around the origin. While this is helpful, we would like to measure several bits of information relating how *explicit* σ is relative to the tangent plane. Remember, the tangent plane at $\boldsymbol{a}$ can be specified as $T_{\boldsymbol{a},\sigma} = \left\{ \boldsymbol{t} \in \mathbb{R}^3 : \langle \boldsymbol{t} - \sigma(\boldsymbol{a}), n\rangle = 0 \right\}$. We shall analyze the behavior of σ, up to second order smoothness, with respect to the tangent plane at $\boldsymbol{a}$ by analyzing the behavior of δ in the n direction. Let $n = n(\boldsymbol{a})$ be the normal of σ at $\mu = \boldsymbol{a}$, and define

$$\rho(\nu) = \langle \delta(\nu), n\rangle.$$

Then,

$$\begin{aligned}\rho(\nu) &= \left(\left\langle \frac{\partial\sigma}{\partial u_1}, n\right\rangle v_1 + \left\langle \frac{\partial\sigma}{\partial u_2}, n\right\rangle v_2\right)\bigg|_{\mu=\boldsymbol{a}} \\ &\quad + \frac{1}{2}\left(\left\langle \frac{\partial^2\sigma}{\partial {u_1}^2}, n\right\rangle (v_1)^2 + 2\left\langle \frac{\partial^2\sigma}{\partial u_1\partial u_2}, n\right\rangle v_1v_2 \right. \\ &\qquad\qquad \left. + \left\langle \frac{\partial^2\sigma}{\partial {u_2}^2}, n\right\rangle (v_2)^2\right)\bigg|_{\mu=\boldsymbol{a}} \\ &= \frac{1}{2}\left(\left\langle \frac{\partial^2\sigma}{\partial {u_1}^2}, n\right\rangle (v_1)^2 + 2\left\langle \frac{\partial^2\sigma}{\partial u_1\partial u_2}, n\right\rangle v_1v_2 \right. \\ &\qquad\qquad \left. + \left\langle \frac{\partial^2\sigma}{\partial {u_2}^2}, n\right\rangle (v_2)^2\right)\bigg|_{\mu=\boldsymbol{a}} \\ &= \frac{1}{2}\left(L_{1,1}(v_1)^2 + 2L_{1,2}v_1v_2 + L_{2,2}(v_2)^2\right) \\ &= \frac{1}{2}\mathrm{II}(v_1, v_2)\end{aligned}$$

since $\left\langle \frac{\partial\sigma}{\partial u_1}\Big|_{\mu=\boldsymbol{a}}, n\right\rangle = 0$ and $\left\langle \frac{\partial\sigma}{\partial u_2}\Big|_{\mu=\boldsymbol{a}}, n\right\rangle = 0$.

Definition 12.51. *The surface, $\rho(\nu)$, as a function of ν measures the approximate distance of σ from the tangent plane and is called the* osculating paraboloid.

To clarify the discussion, we can think of the function $\psi(\nu) = \sigma(\boldsymbol{a}+\nu)$. The domain of ψ, which includes the origin, is a translated version of the domain of σ. Also, since $\delta_{\mathbf{0},\psi} = \delta_{\boldsymbol{a},\sigma}$, the osculating paraboloid of ψ at $\mathbf{0}$ is the same as the osculating paraboloid of σ at $\boldsymbol{a}$. ν can be restricted to values with magnitude small enough so that $R_{\boldsymbol{a},2}(\nu)$ is uniformly small, so that the behavior of σ is well-defined by the behavior of δ. The second order characteristics of σ are given by the second fundamental form. Now, ρ is a measure of how σ behaves with respect to the tangent plane, and in particular, it is a local measure of shape of the surface. If ρ is positive for all permissible values of ν, then locally σ lies on the positive side of the tangent plane; analogous results hold if ρ is negative for all permissible values of ν. When ρ changes sign we must look further. When ρ is considered as the parametric function $(v_1, v_2, \rho(\nu))$, the shape of ρ near $\rho(\mathbf{0})$ gives information about the shape of σ near $\sigma(\boldsymbol{a})$. For these reasons, we shall analyze the shape of ρ.

Factoring the osculating paraboloid gives:

$$\rho(\nu) = \frac{1}{2L}\left[(Lv_1 + Mv_2)^2 + (LN - M^2)(v_2)^2\right].$$

Since the first term is always nonnegative, the information about whether ρ is positive, negative, or both in the neighborhood around $\boldsymbol{a}$ is always determined by the determinant of $\boldsymbol{L}$ which equals $LN - M^2$. When contour curves are formed by setting ρ equal to a constant (making it an implicit function), each equation becomes that of a conic section, called the *dupin indicatrix.* The type of conic is fixed by the value of the discriminant, which remains fixed and the same on all contours. The discriminant, then, equals $(2M)^2 - 4LN = 4(M^2 - LN) = -4\left(|\boldsymbol{L}|\right)$. Hence, the type of all the contours is determined by the determinant of the matrix of the second fundamental form. The sign of this determinant also determines the sign of the Gaussian curvature of the surface. There are four possibilities: 1) $|\boldsymbol{L}| > 0$; 2) $|\boldsymbol{L}| < 0$; 3) $|\boldsymbol{L}| = 0$, but not all of L, M, and N are equal to zero; and 4) $|\boldsymbol{L}| = 0$, and $L = M = N = 0$. We discuss each of these cases separately.

If $|\boldsymbol{L}| > 0$, then the discriminant of the contours is negative, and all the curves are ellipses, and ρ is an *elliptic paraboloid.* This also implies that the second fundamental form is definite, and that the surface σ, in a neighborhood of $\sigma(\boldsymbol{a})$, lies entirely on one side of the tangent plane, touching it only at $\sigma(\boldsymbol{a})$. If the form is positive definite, the surface lies on

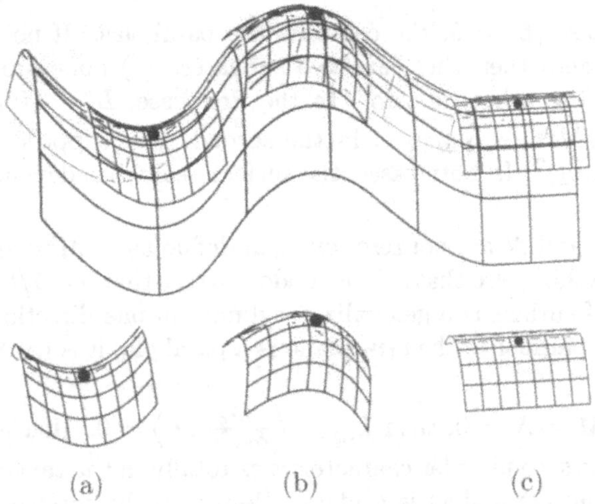

Figure 12.7. (a) Hyperbolic surface; (b) elliptical surface; (c) parabolic surface.

the $+\boldsymbol{n}$ direction of the tangent plane. If it is negative definite, the surface lies on the $-\boldsymbol{n}$ direction of the tangent plane.

If $|\boldsymbol{L}| < 0$, then the discriminant of the contours is positive, and all the curves are hyperbolas. The surface ρ is a *hyperbolic paraboloid*. It intersects the $\rho = 0$ plane (the tangent plane of the surface) in the conic section

$$\begin{aligned} 0 &= \left[(Lv_1 + Mv_2)^2 + \left(LN - M^2\right) {v_2}^2\right] \\ &= \left[\left(L\frac{v_1}{v_2} + M\right)^2 + \left(LN - M^2\right)\right] \\ &= L\left[L\left(\frac{v_1}{v_2}\right)^2 + 2M\frac{v_1}{v_2} + N\right] \\ &= L\left(\frac{v_1}{v_2}\right)^2 + 2M\frac{v_1}{v_2} + N \end{aligned}$$

as long as $L \neq 0$ and $v_2 \neq 0$. But this is just the conic form of a degenerate hyperbola consisting of two lines crossed at the origin. Thus the tangent plane is cut by ρ along these crossed lines and they partition the tangent plane into four regions. The surface ρ, and hence σ then changes from the positive side of the tangent plane to the negative side, and back, alternating at these crossed lines.

In the case $|\boldsymbol{L}| = 0$, the contours are parabolas. If not all of L, M, and N are zero then there are two subcases: (a) none are zero; (b) M and one of the others is zero. In the first case, $L = M^2/N$, and $\rho = 1/(2NM^2)\left(M^2 v_1 + NMv_2\right)^2$; in the second case, suppose L is nonzero. Then $\rho = L(v_1)^2$. In both cases, the surface really has dependence on only one variable.

If L, M, and N are nonzero, one can define $w_1 = M^2 v_1 + NMv_2$, and w_2 to be anything else that is independent to see that $\rho = 1/(2NM^2)(w_1)^2$. This kind of surface is a generalized cylinder in one direction. Since each of the slices parallel to the $v_1 v_2$ plane is a parabola, it is called a parabolic cylinder.

If $L = M = N = 0$, then $L_{i,j} = \left\langle \frac{\partial^2 \sigma}{\partial v_i \partial v_j}, n \right\rangle = 0$. Hence, the original surface σ has second order characteristics totally in the tangent plane, and the osculating paraboloid is a plane. Points on the surface at which this condition is satisfied are called *planar points*. Since very locally the shape of σ is flat, which we already knew, not much additional can be learned. However,

Lemma 12.52. *If all the points of a surface σ are planar points, then the surface is a plane.*

Proof: Since $\left\langle \frac{\partial \sigma}{\partial v_i}, n \right\rangle = 0$, by the definition of n, taking the partial derivative with respect to v_j yields a zero result, and

$$\begin{aligned} 0 &= \left\langle \frac{\partial^2 \sigma}{\partial v_i \partial v_j}, n \right\rangle + \left\langle \frac{\partial \sigma}{\partial v_i}, \frac{\partial n}{\partial v_j} \right\rangle \\ &= \left\langle \frac{\partial \sigma}{\partial v_i}, \frac{\partial n}{\partial v_j} \right\rangle. \end{aligned}$$

For this last equation to hold, the normal to the tangent plane has no rate of change at all points in the surface, and so the tangent plane is constant for all points in the surface. The only way this can hold is if the surface *is* the tangent plane. ■

Example 12.53. The surface of Example 12.48 has $|\boldsymbol{L}| > 0$, and positive Gaussian curvature, so it is elliptic.

The surface of Example 12.49 has $|\boldsymbol{L}| < 0$, so the surface is hyperbolic.

Finally, the surface of Example 12.50 has $|\boldsymbol{L}| = 0$, but not all the elements are 0, so it is a parabolic cylinder. □

12.5 Surfaces

Most of the effort so far has been devoted to describing the properties of simple surfaces or coordinate patches. The theorems can be proved straightforwardly for this type of surface. Unfortunately, many surfaces cannot be described as surface patches. However, they frequently can be described as *surfaces.*

Definition 12.54. *A* $C^{(k)}$ surface S *in* $\mathbb{R}^3$ *is a subset of* $\mathbb{R}^3$ *such that there exists a collection of coordinate patches such that for each point* $P \in S$*, there is a proper* $C^{(k)}$ *coordinate patch containing* P *whose image is completely in the surface. For* $\sigma : U \to S$ *and* $\psi : V \to S$ *are two such coordinate patches; let* $U_\sigma = \sigma(U)$ *and* $V_\psi = \psi(V)$*. Define* σ^{-1} *and* ψ^{-1} *to be their inverse functions. Whenever* $U_\sigma \cap V_\psi = S_{\sigma,\psi}$ *is nonempty, the map* $\psi^{-1}(\sigma) : U^* \to V$ *must be a* $C^{(k)}$ *coordinate transformation, where* $U^* = \sigma^{-1}(\sigma(U) \cap \psi(V))$*. Let* $V^* = \psi^{-1}(\sigma(U) \cap \psi(V))$*, then* $\psi^{-1}[\sigma] : U^* \to V^*$ *must be a* $C^{(k)}$ *coordinate transformation.*

The surface of revolution of Example 12.10 is a *surface* under this definition since one can select two coordinate patches which cover the image. Let σ be defined over $\theta \in (-\pi, \pi)$ and ψ be defined over $\theta \in (0, 2\pi)$.

Definition 12.55. *A surface* S *is* orientable *if there is a continuous function* $\alpha : S \to S^2$*, the unit sphere with* $\alpha(P)$ *orthogonal to* S *at* P *for all points on the surface.*

The normal vector defined over a simple surface or coordinate patch clearly satisfies this requirement. When a surface is not one patch, this further requirement must hold for the surface to be orientable. When surfaces are used in computer aided geometric design to construct three-dimensional objects, they must have an *inside* and an *outside.* This means that they must be orientable. Basically, all *surfaces* are locally orientable since they can be covered with patches. The need for a *global* condition is what differentiates this definition.

Example 12.56. The function

$$\sigma(\mu) = (\cos u_1, \sin u_1, 0) + u_2(\sin(u_1/2)\cos u_1, \sin(u_1/2)\sin u_1, \cos(u_1/2))$$

is called a Möbius band. The Möbius band is not orientable, as we shall show.

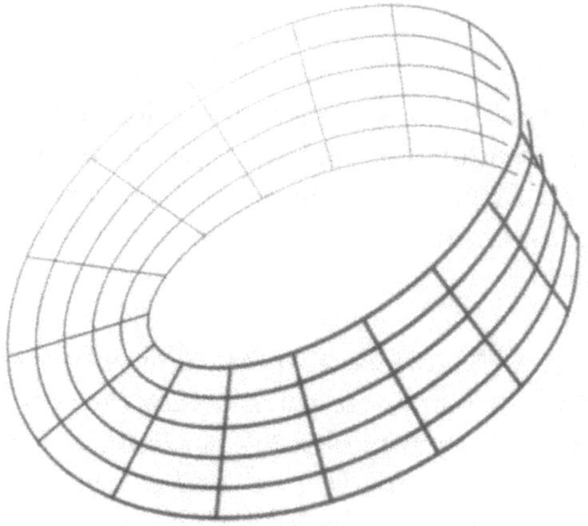

Figure 12.8. A Mobius band.

$$\frac{\partial \sigma}{\partial u_1}(u_1, 0) = (-\sin u_1, \cos u_1, 0),$$
$$\frac{\partial \sigma}{\partial u_2}(u_1, 0) = (\sin(u_1/2)\cos u_1, \sin(u_1/2)\sin u_1, \cos(u_1/2)).$$

Thus, the cross product of the two first partials along $(u_1, 0)$ has magnitude 1 and is a normal vector so

$$n = (\cos u_1 \cos u_1/2, \sin u_1 \cos u_1/2, -\sin u_1/2).$$

Consider the coordinate patch domain as being $(-\pi, \pi) \times (-1, 1)$. Then,

$$\lim_{u_1 \to -\pi} n(u_1, 0) = (0, 0, 1) \quad \text{while} \lim_{u_1 \to \pi} n(u_1, 0) = (0, 0, -1).$$

Further, there is no method of defining normals of the patches in a consistent way which will allow the normal vector function to be continuous everywhere. □

Exercises

1. Prove Lemma 12.13.

2. Consider a curve in the (r, z) plane given by $r = r(u) > 0$, $z = z(u)$. If the curve is rotated about the z-axis, we obtain a surface of revolution that can be parametrized in the following way:

$$f(u, v) = (r(u)\cos v, r(u)\sin v, z(u)) \qquad -\pi < v < \pi$$

- Prove that $\frac{\partial f}{\partial u}$ is perpendicular to $\frac{\partial f}{\partial v}$ everywhere. Such a surface f is called an "orthogonal net".
- Find the matrix of the first fundamental form for f.
- Find the matrix of the second fundamental form for f.
- Find the principal curvatures, Gaussian curvature, and mean curvature for f.
- For $r(u) = 2 + \cos u$ and $z(u) = \sin u$, apply your results above to find the matrices of the first and second fundamental forms, and the various curvatures.

3. Consider the surface patch given by $f(u_1, u_2) = (u_1 \cos u_2, u_1 \sin u_2, 0)$, for $u_1 > 0$, and $-\pi < u_2 \leq \pi$. Let $P = (2, 0, 0)$, which lies in the surface. It is necessary to find the parameter values corresponding to P for each parameterization below.
   (a) Find the matrix of the first fundamental form for this patch. What is the matrix at P?
   (b) Consider the curve lying in the surface described by $\mu(t) = (u_1(t), u_2(t)) = (3, t)$, for $0 \leq t \leq \pi$. Use the first fundamental form to find the length of the curve $f(\mu(t))$.
   (c) Consider the patch $h(\nu) = (v_1, v_2, 0)$, which describes the same surface. Find the matrix of the first fundamental form of this patch. What is the matrix of the first fundamental form at P? Find the coordinate transformation of ν in terms of μ. Find $\nu(t)$ that will describe the same curve as $\mu(t)$. Find the length of the curve using the first fundamental form for this new patch.
   (d) Let $\mathtt{I}_f(\frac{du}{dt})$ denote the first fundamental form induced by f, and $\mathtt{I}_h(\frac{dv}{dt})$ denote the first fundamental form induced by h. Show that $\mathtt{I}_h(\frac{d\nu(\mu)}{dt}) = \mathtt{I}_f(\frac{d\mu}{dt})$.
4. Derive Equation 12.3 from Equation 12.2.
5. Complete the computation of $\left\| \frac{\partial \sigma}{\partial u_1} \times \frac{\partial \sigma}{\partial u_2} \right\|$ which completes the proof of Lemma 12.32.
6. Prove Theorem 12.41. Hint: Use Lemma 12.39.
7. Programming Exercise: Develop a program for viewing surface geometry. Your program should display a parametric surface (by drawing iso-parametric curves), and the surface geometry moving along any such curve picked by the user. The following geometric entities should be displayable:

- The two partial derivatives,
- The unit surface normal,
- The tangent plane at the point,
- principal curvatures and principal directions. You could display them by drawing osculating circles in the plane determined by the principal direction and the surface normal. You might want to animate the motion along a curve, or to allow for a still frame study at a particular location.

13

Surface Representations

Requirements for shape representation by surfaces arise from the same types of circumstances as with curves. Sometimes the designer has the shape of a complex object in mind and must quantify that shape with a mathematical representation. Other times, shapes are required to meet certain constraints, such as to pass through specific points in space or to pass through certain curves already created. Sometimes the points or curves being used as a starting point for surface creation have some regularity or neighborhood associativity, while at other times they are more randomly distributed. In this chapter we introduce representation forms that can be used to solve such problems for subsets of general problems. However, design involves more than just representation of the surface. We must be able to create models and query them for rendering, analysis, and manufacture. Further, for representations to be useful, the designers must find them easy to manipulate and modify. Hence, even though we introduce a variety of representations, we will stress and show how to both interpolate and approximate surfaces using mainly *parametric tensor product* surface representations.

13.1 Surface Representations

Consider a domain $D \subset \boldsymbol{R}^2$, an open region or the closure of an open region, and the collection of bivariate functions

$$\mathcal{H} = \{\, h_j(u,v) : j = 0, \ldots,\; N, h_j \in C^{(0)}, (u,v) \in D \,\} .$$

Then for any natural number ℓ, $span_\ell(\mathcal{H}) = \left\{ \sum_{j=0}^{N} c_j h_j(u,v) : c_j \in \mathbb{R}^\ell \right\}$, is a function space. If $\ell = 1$, the surfaces are scalar surfaces, i.e., *height fields* . If $\ell = 2$ the surfaces are planar surfaces, some of which may be coordinate transformations. If $\ell = 3$, $span_\ell(\mathcal{H})$ consists of parametric three-dimensional surfaces.

In order for the the elements of $span_\ell(\mathcal{H})$ to be useful in graphics, modeling, and manufacturing, users must be able to specify (and later modify) coefficients for the functions in $\mathcal{H}$, and compute differential characteristics (tangents, normals, curvatures). Shape might be specified by requiring interpolation or approximation to point position and normal sets (a traditional approach), by interactively manipulating the coefficients to attain aesthically pleasing shape, by satisfying more general optimization criteria while also satisfying some constraints, or by interactive manipulation of the coefficients until an aesthetically pleasing form is created. One typical characteristic of the functions in $\mathcal{H}$ is their linear independence that allow for unique solutions and consistent treatment.

13.2 Tensor Product Surfaces

Definition 13.1. *Consider $\mathcal{F}$ and $\mathcal{G}$, two sets of univariate functions, with interval domains U and V, respectively,*

$$\mathcal{F} = \{f_i(u)\}_{i=0}^{m}, \qquad \mathcal{G} = \{g_j(v)\}_{j=0}^{n}.$$

A surface formed by $h(u,v) = \sum_{j=0}^{n} \sum_{i=0}^{m} c_{i,j} f_i(u) g_j(v)$ is called a tensor product surface *with domain $U \times V$. If $c_{i,j} \in \mathbb{R}^3$ for all i, j, then h is a parametric surface.*

Example 13.2. Consider the linear blending functions $\mathcal{F} = \{f_0(u) = 1-u, f_1(u) = u\}$, and $\mathcal{G} = \{g_0(v) = 1-v, g_1(v) = v\}$, with domain $\mathcal{D} = [0,1] \times [0,1]$. The tensor product surface

$$h(u,v) = p_{0,0} f_0(u) g_0(v) + p_{0,1} f_0(u) g_1(v) + p_{1,0} f_1(u) g_0(v) + p_{1,1} f_1(u) g_1(v)$$

is a bilinear surface. While it is at most linear in each of the variables, factors of uv appear, so it is a quadratic (bivariate) polynomial. Investigating its properties, we see that

$$h(0,0) = p_{0,0} \quad h(0,1) = p_{0,1},$$
$$h(1,0) = p_{1,0} \quad h(1,1) = p_{1,1}.$$

Hence this function interpolates the values of the four coefficients. Further, for every fixed value of $v = \hat{v}$,

$$\begin{aligned} h(u, \hat{v}) &= (p_{0,0}g_0(\hat{v}) + p_{0,1}g_1(\hat{v}))\, f_0(u) + (p_{1,0}g_0(\hat{v}) + p_{1,1}g_1(\hat{v}))\, f_1(u) \\ &= P_{0,\hat{v}}(1-u) + P_{1,\hat{v}}u, \end{aligned}$$

where

$$\begin{aligned} P_{0,\hat{v}} &= p_{0,0}g_0(\hat{v}) + p_{0,1}g_1(\hat{v}), \\ P_{1,\hat{v}} &= p_{1,0}g_0(\hat{v}) + p_{1,1}g_1(\hat{v}). \end{aligned}$$

Hence, the isoparametric curves in u for fixed v are straight lines. In particular, the boundary curves $h(u,0)$ and $h(u,1)$ are straight lines connecting $p_{0,0}$ to $p_{1,0}$ and $p_{0,1}$ to $p_{1,1}$, respectively. Similarly, it can be shown that the boundary curves in v for fixed u are straight lines, in particular, the boundary curves $h(0,v)$ and $h(1,v)$ connecting $p_{0,0}$ to $p_{0,1}$ and $p_{1,0}$ to $p_{1,1}$, respectively. □

Definition 13.3. *Consider* $\mathcal{P} = \left\{ P_{i,j} \in \mathbb{R}^3 : 0 \le i \le m, \quad 0 \le j \le n \right\}$ *with the collections of functions*

$$\mathcal{F} = \{\theta_{i,m}(a,b;u)\}_{i=0}^{m} \qquad \mathcal{G} = \{\theta_{j,n}(c,d;v)\}_{j=0}^{n}$$

(see Definition 5.7). The parametric surface

$$\sigma_{m,n}(u,v) = \sum_{j=0}^{n} \sum_{i=0}^{m} P_{i,j}\theta_{i,m}(a,b;u)\theta_{j,n}(c,d;v)$$

is called a degree m–by–degree n tensor product Bézier surface *over the rectangular domain* $[a,b] \times [c,d] \subset \mathbb{R}^2$.

Recall that when a domain is not specified, it is assumed to range from zero to one. From properties of univariate Bernstein polynomials,

$$\begin{aligned} \sigma_{m,n}(u,0) &= \sum_{i=0}^{m} \sum_{j=0}^{n} P_{i,j}\theta_{j,n}(0)\theta_{i,m}(u) \\ &= \sum_{i=0}^{m} P_{i,0}\theta_{i,m}(u); \\ \sigma_{m,n}(u,1) &= \sum_{i=0}^{m} \sum_{j=0}^{n} P_{i,j}\theta_{j,n}(1)\theta_{i,m}(u) \\ &= \sum_{i=0}^{m} P_{i,n}\theta_{i,m}(u); \end{aligned}$$

$$\begin{aligned}\sigma_{m,n}(0,v) &= \sum_{j=0}^{n}\sum_{i=0}^{m} P_{i,j}\theta_{i,m}(0)\theta_{j,n}(v)\\ &= \sum_{j=0}^{n} P_{0,j}\theta_{j,n}(v);\\ \sigma_{m,n}(1,v) &= \sum_{j=0}^{n}\sum_{i=0}^{m} P_{i,j}\theta_{i,m}(1)\theta_{j,n}(v)\\ &= \sum_{j=0}^{n} P_{m,j}\theta_{j,n}(v).\end{aligned}$$

Thus, the four boundary curves reduce to Bézier curves over the polygons $\{P_{i,0}\}_{i=0}^{m}$ and $\{P_{i,n}\}_{i=0}^{m}$, $\{P_{0,j}\}_{j=0}^{n}$, and $\{P_{m,j}\}_{j=0}^{n}$, respectively. Furthermore since each Bézier curve interpolates its endpoints, σ interpolates the four corner points. Note that the bilinear surface in Example 13.2 is a special case of a Bézier surface.

Definition 13.4. *Consider collections of functions* $\mathcal{F} = \{\mathcal{B}_{i,\kappa_u,\tau_u}(u)\}_{i=0}^{m}$, *and* $\mathcal{G} = \{\mathcal{B}_{j,\kappa_v,\tau_v}(v)\}_{j=0}^{n}$, *where* τ_u *is a knot vector with* $m + \kappa_u + 2$ *elements, such that* $\tau_{u,\kappa_u} < \tau_{u,\kappa_u+1}$, $\tau_{u,m} < \tau_{u,m+1}$, *and* $\tau_{u,i} < \tau_{u,i+\kappa_u+1}$ *for all* i; τ_v *is a knot vector with* $n + \kappa_v + 2$ *elements, such that* $\tau_{v,\kappa_v} < \tau_{v,\kappa_v+1}$, $\tau_{v,n} < \tau_{v,n+1}$, *and* $\tau_{v,j} < \tau_{v,j+\kappa_v+1}$ *for all* j.

Then if $\mathcal{P} = \{ P_{i,j} \in \mathbb{R}^3 : 0 \le i \le m, \quad 0 \le j \le n \}$,

$$\sigma(u,v) = \sum_{j=0}^{n}\sum_{i=0}^{m} P_{i,j}\mathcal{B}_{i,\kappa_u,\tau_u}(u)\mathcal{B}_{j,\kappa_v,\tau_v}(v)$$

is a degree κ_u–by–degree κ_v tensor product B-spline surface *over the rectangular domain* $[\tau_{u,\kappa_u}, \tau_{u,m+1}) \times [\tau_{v,\kappa_v}, \tau_{v,n+1}) \subset \mathbb{R}^2$.

Note that the degrees used in the tensor product may be different in the u and v directions, and the knot vecotrs may differ as well. For example, one could have open quadratic in u and floating cubic in v. Further, for any fixed value of u, say $u = \hat{u}$, $\sigma(\hat{u}, v)$ is a B-spline curve of degree κ_v with knot vector $\boldsymbol{\tau}_v$. Similarly, for a fixed value of v, say $v = \hat{v}$, $\sigma(u, \hat{v})$ is a B-spline curve of degree κ_u with knot vector $\boldsymbol{\tau}_u$.

Example 13.5. Let $\boldsymbol{\tau}_u = \boldsymbol{\tau}_v = \{0, 0, 0, 1, 2, 2, 2\}$. We draw the corresponding blending functions for $\kappa_u = \kappa_v = 2$ in Figures 13.1-13.5. Because of symmetries between the $\boldsymbol{\tau}_u$ and $\boldsymbol{\tau}_v$ knot vectors, we need only consider $\mathcal{B}_{i,2}(u)\mathcal{B}_{j,2}(v)$ when $i \le j$. □

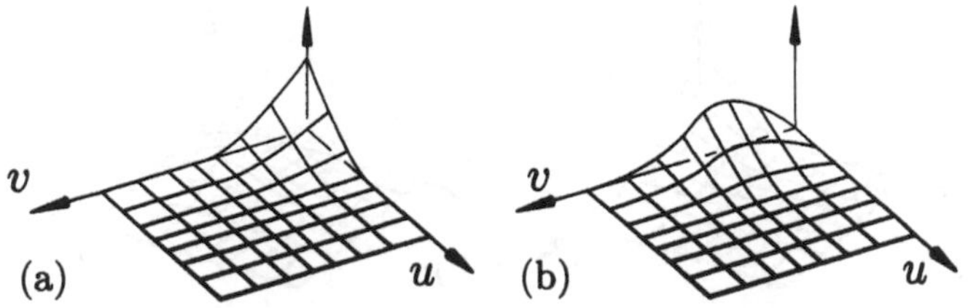

Figure 13.1. (a) $\mathcal{B}_{0,2}(u)\mathcal{B}_{0,2}(v)$; (b) $\mathcal{B}_{0,2}(u)\mathcal{B}_{1,2}(v)$.

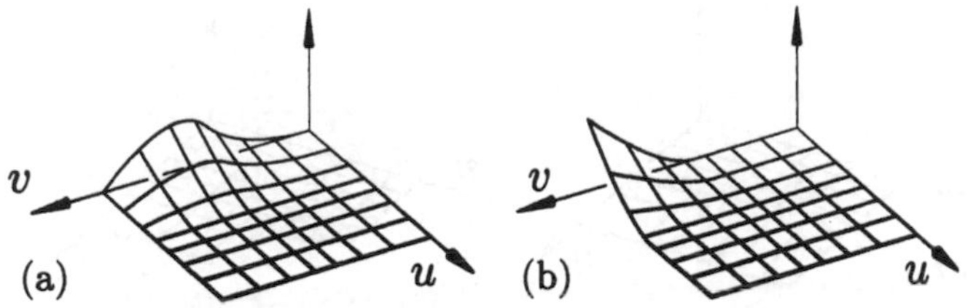

Figure 13.2. (a) $\mathcal{B}_{0,2}(u)\mathcal{B}_{2,2}(v)$; (b) $\mathcal{B}_{0,2}(u)\mathcal{B}_{3,2}(v)$.

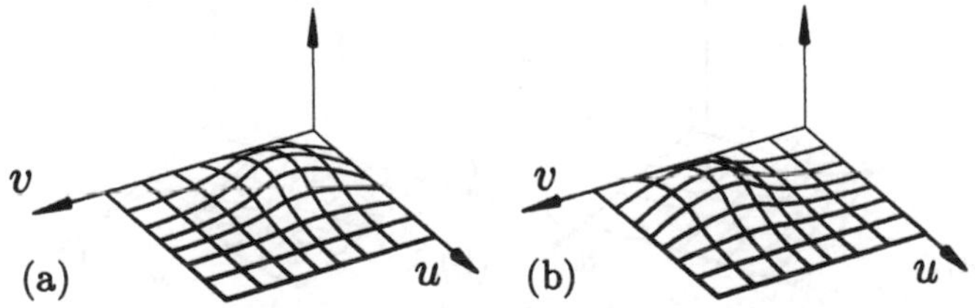

Figure 13.3. (a) $\mathcal{B}_{1,2}(u)\mathcal{B}_{1,2}(v)$; (b) $\mathcal{B}_{1,2}(u)\mathcal{B}_{2,2}(v)$.

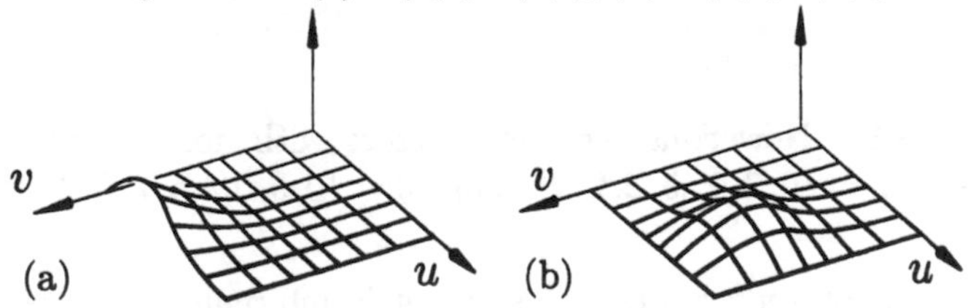

Figure 13.4. (a) $\mathcal{B}_{1,2}(u)\mathcal{B}_{3,2}(v)$; (b) $\mathcal{B}_{2,2}(u)\mathcal{B}_{2,2}(v)$.

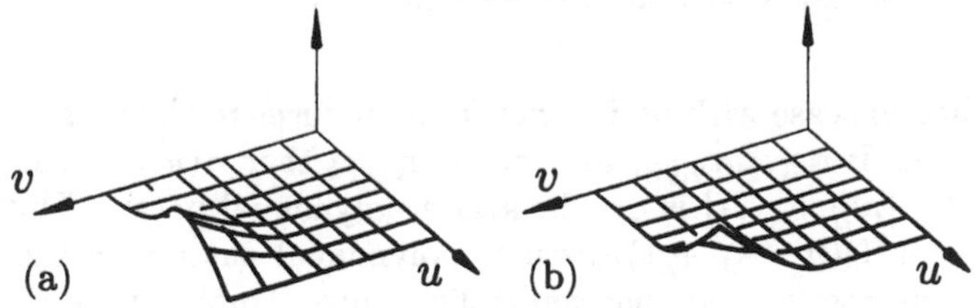

Figure 13.5. (a) $\mathcal{B}_{2,2}(u)\mathcal{B}_{3,2}(v)$; (b) $\mathcal{B}_{3,2}(u)\mathcal{B}_{3,2}(v)$.

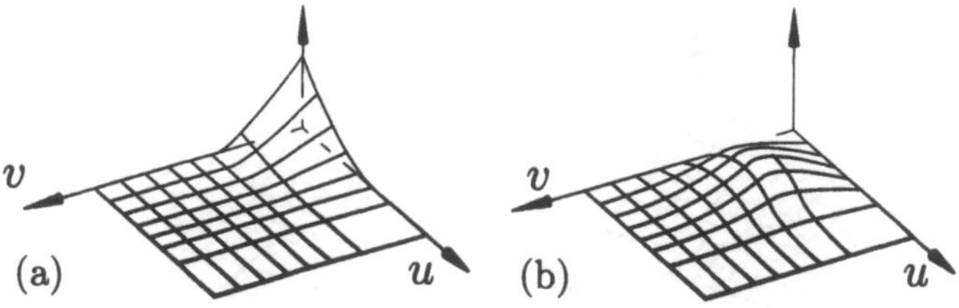

Figure 13.6. (a) $\mathcal{B}_{0,2}(u)\mathcal{N}_{0,2}(v)$; (b) $\mathcal{B}_{1,2}(u)\mathcal{N}_{1,2}(v)$.

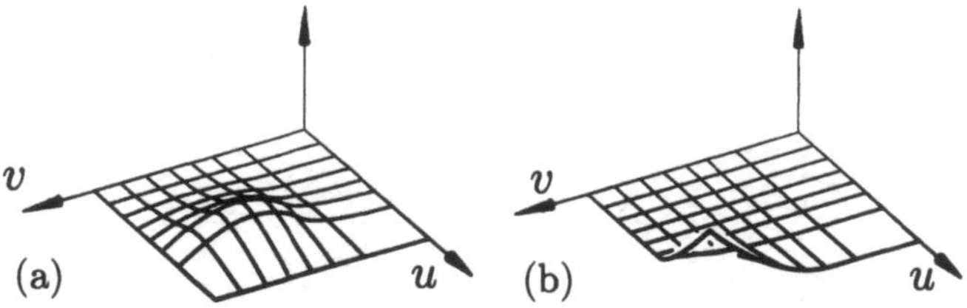

Figure 13.7. (a) $\mathcal{B}_{2,2}(u)\mathcal{N}_{2,2}(v)$; (b) $\mathcal{B}_{3,2}(u)\mathcal{N}_{3,2}(v)$.

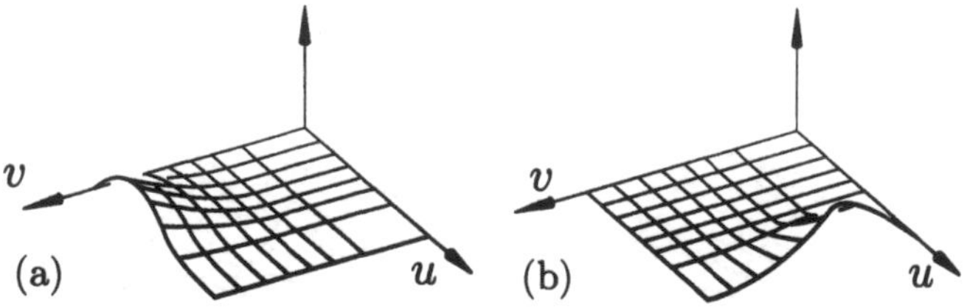

Figure 13.8. (a) $\mathcal{B}_{1,2}(u)\mathcal{N}_{3,2}(v)$; (b) $\mathcal{B}_{3,2}(u)\mathcal{N}_{1,2}(v)$.

Example 13.6. Now consider knot vectors with nonuniformly spaced knot values, $\boldsymbol{\tau}_u = \{0, 0, 0, .5, 2, 2, 2\}$ and $\boldsymbol{\tau}_v = \{0, 0, 0, 1.5, 2, 2, 2\}$ for $\kappa_u = \kappa_v = 2$.

Here, the knot vectors are not symmetric internally and are not the same in both directions. The blending functions are skewed accordingly. Figures 13.6-13.8 show several of these functions. □

Note that the case with no interior knots reduces to the tensor product Bézier surface. If $\tau_{u,i} < \tau_{u,i+1}$ and $\tau_{v,j} < \tau_{v,j+1}$ then $\sigma(u, v)$ with $(u, v) \in [\tau_{u,i}, \tau_{u,i+1}] \times [\tau_{v,j}, \tau_{v,j+1}]$ is one bivariate polynomial patch. Thus, there are $(m - \kappa_u + 1)(n - \kappa_v + 1)$ distinct polynomial patches in the surface $\sigma(u, v)$ if each interior knot has multiplicity one. Further note that there are exactly $(\kappa_u+1)\times(\kappa_v+1)$ coefficients for each patch. The basis functions carry the continuity conditions.

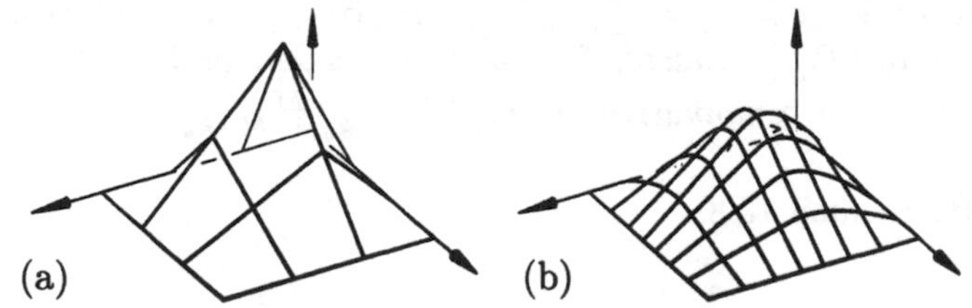

Figure 13.9. (a) A parametric B-spline mesh; (b) a surface using the knot vectors $\boldsymbol{u} = \boldsymbol{v} = \{0, 0, 0, 1, 2, 2, 2\}$.

Definition 13.7. *The collection* $\{P_{i,j}\}$ *is called the* control mesh *for the B-spline surface.*

The definition of the convex hull of a control mesh is similar to the convex hull of the control polygon for a curve. It is simply the smallest convex set in $\mathbb{R}^3$ containing the mesh. Since four points in space can have four different planes through any three of the points, the number of planes which contribute to finding the convex hull of a surface is large. Note however, that the convex hull of any finite set of points is a polyhedron, and thus the convex hull of the control mesh is a polyhedron. The convex hull property then is important but difficult to compute.

Lemma 13.8. *Suppose* $\sigma(u,v) = \sum_{j=0}^{n}\sum_{i=0}^{m} P_{i,j}\mathcal{B}_{i,\kappa_u,\tau_u}(u)\mathcal{B}_{j,\kappa_v,\tau_v}(v)$ *is a tensor product B-spline surface. For* $u \in [\tau_{u,\kappa_u}, \tau_{u,m+1})$, *and* $v \in [\tau_{v,\kappa_v}, \tau_{v,n+1})$,

$$1 = \sum_{j=0}^{n}\sum_{i=0}^{m} \mathcal{B}_{i,\kappa_u,\tau_u}(u)\mathcal{B}_{j,\kappa_v,\tau_v}(v).$$

Proof: For $u \in [\tau_{u,\kappa_u}, \tau_{u,m+1})$ and for $v \in [\tau_{v,\kappa_v}, \tau_{v,n+1})$,

$$\begin{aligned} 1 &= \sum_{i=0}^{m} \mathcal{B}_{i,\kappa_u,\tau_u}(u), \\ 1 &= \sum_{j=0}^{n} \mathcal{B}_{j,\kappa_v,\tau_v}(v). \end{aligned}$$

Thus,

$$\begin{aligned} \sum_{j=0}^{n}\sum_{i=0}^{m} \mathcal{B}_{i,\kappa_u,\tau_u}(u)\mathcal{B}_{j,\kappa_v,\tau_v}(v) &= \sum_{j=0}^{n}\left(\sum_{i=0}^{m} \mathcal{B}_{i,\kappa_u,\tau_u}(u)\right)\mathcal{B}_{j,\kappa_v,\tau_v}(v) \\ &= \sum_{j=0}^{n} \mathcal{B}_{j,\kappa_v,\tau_v}(v) \\ &= 1. \end{aligned}$$

■

Theorem 13.9. *$\sigma(u,v) = \sum_{i=0}^{m}\sum_{j=0}^{n} P_{i,j}\mathcal{B}_{j,k_v}(v)\mathcal{N}_{i,k_u}(u)$ is a convex combination of $\{P_{i,j}\}$. Indeed, for $u_M \le u < u_{M+1}$ and $v_N \le v < v_{N+1}$, $\sigma(u,v)$ is a convex combination of $\{P_{i,j}\}_{j=N-k_v,\, i=M-k_u}^{N,\quad M}$.*

Proof: By Lemma 13.8,

$$\sum_{i=0}^{n}\sum_{j=0}^{m} \mathcal{B}_{j,k_v}(v)\mathcal{N}_{i,k_u}(u) = 1.$$

For arbitrary (u,v) in the domain, find M and N such that $u_M \le u < u_{M+1}$ and $v_N \le v < v_{N+1}$. Then $\mathcal{N}_{i,k_u}(u) = 0$, $i < M - k_u$ and $i > M$, and $\mathcal{B}_{j,k_v}(v) = 0$ for $j < N - k_v$ and $j > N$. So $\mathcal{N}_{i,k_u}(u)\mathcal{B}_{j,k_v}(v) \neq 0$ when $i \in \{M - k_u, M - k_u + 1, \ldots, M\}$ and $j \in \{N - k_v, N - k_v + 1, \ldots, N\}$. ■

Extending the variation diminishing property to surfaces is a difficult notion to comprehend. Unlike general curves, surfaces can intersect a plane in points, curves, and coincident subsurface regions. There may not be a way to count the *number* of intersections.

While a tensor product surface approximation which is derived from curve approximation schemes exhibiting the variation diminishing property is well behaved and also has no extraneous wiggles, a definition for the corresponding phenomena is not well defined. In particular, a polygon, which defines a spline curve, is also a curve, but the control mesh which defines a surface is not a surface. One must first define a surface through the mesh points. Then one must determine a generalized definition which would keep the intent. It has not been done.

Modeling with Bézier and B-spline representations for surfaces are discussed in the chapters that follow.

Example 13.10. Consider the linear blending functions $\mathcal{F} = \{f_0(u) = 1 - u, f_1(u) = u\}$, and the cubic Hermite functions from Equation 10.3, $\mathcal{G} = \{g_0(v) = h_{0,0}(v), g_1(v) = h_{1,0}(v), g_2(v) = h_{0,1}(v), g_3(v) = h_{1,1}(v)\}$, where

$$\begin{array}{ll} h_{0,0}(v) = (2v+1)(v-1)^2 & h_{1,0}(v) = (-2v+3)v^2 \\ h_{0,1}(v) = v(v-1)^2 & h_{1,1}(v) = v^2(v-1). \end{array} \tag{13.1}$$

Then

$$\begin{array}{llll} f_0(0) = 1 & f_0(1) = 0 & f_0'(0) = -1 & f_0'(1) = -1 \\ f_1(0) = 0 & f_1(1) = 1 & f_1'(0) = 1 & f_1'(1) = 1 \end{array}$$

and by the properties of the collection h,

$$\begin{array}{llll} g_0(0)=1 & g_0(1)=0 & g_0'(0)=0 & g_0'(1)=0 \\ g_1(0)=0 & g_1(1)=1 & g_1'(0)=0 & g_1'(1)=0 \\ g_2(0)=0 & g_2(1)=0 & g_2'(0)=1 & g_2'(1)=0 \\ g_3(0)=0 & g_3(1)=0 & g_3'(0)=0 & g_3'(1)=1. \end{array}$$

Consider the surface $\sigma(u,v)=\sum_{i=0}^{1}\sum_{j=0}^{3} f_i(u)a_{i,j}g_j(v)$ where $a_{i,j}\in \mathbb{R}^3$. Then $\frac{\partial\sigma}{\partial u}=(a_{1,0}-a_{0,0})g_0(v)+(a_{1,1}-a_{0,1})g_1(v)+(a_{1,2}-a_{0,2})g_2(v)+(a_{1,3}-a_{0,3})g_3(v)$, and

$$\begin{array}{ll} \sigma(0,0)=a_{0,0} & \sigma(1,0)=a_{1,0} \\ \sigma(0,1)=a_{0,1} & \sigma(1,1)=a_{1,1} \\ \dfrac{\partial\sigma}{\partial u}(0,0)=a_{1,0}-a_{0,0} & \dfrac{\partial\sigma}{\partial u}(1,0)=a_{1,0}-a_{0,0} \\ \dfrac{\partial\sigma}{\partial u}(0,1)=a_{1,1}-a_{0,1} & \dfrac{\partial\sigma}{\partial u}(1,1)=a_{1,1}-a_{0,1} \\ \dfrac{\partial\sigma}{\partial v}(0,0)=a_{0,2} & \dfrac{\partial\sigma}{\partial v}(1,0)=a_{1,2} \\ \dfrac{\partial\sigma}{\partial v}(0,1)=a_{0,3} & \dfrac{\partial\sigma}{\partial v}(1,1)=a_{1,3} \\ \dfrac{\partial^2\sigma}{\partial u\partial v}(0,0)=a_{1,2}-a_{0,2} & \dfrac{\partial^2\sigma}{\partial u\partial v}(1,0)=a_{1,2}-a_{0,2} \\ \dfrac{\partial^2\sigma}{\partial u\partial v}(0,1)=a_{1,3}-a_{0,3} & \dfrac{\partial^2\sigma}{\partial u\partial v}(1,1)=a_{1,3}-a_{0,3} \end{array}$$

Hence the resulting tensor product surface interpolates the points $a_{0,0}$, $a_{1,0}$, $a_{0,1}$, and $a_{1,1}$. The partials with respect to v are given at the four corner points by $a_{0,2}$, $a_{1,2}$, $a_{0,3}$, and $a_{1,3}$. But since the interpolant in the u direction is exact only to 0th order (position), no derivative information is interpolated in that direction. This tensor product surface is a cubic Hermite interpolant (to the end points) in the v direction, and a linear interpolant in the u direction. □

Definition 13.11. *Suppose $\mathcal{F}$ and $\mathcal{G}$ are both cubic Hermite functions. Then $\sigma(u,v)=\sum_{j=0}^{3}\sum_{i=0}^{3} c_{i,j}f_i(u)g_j(v)$ is a bicubic surface in the* tensor product bicubic Hermite representation.

Example 13.12. Suppose $\sigma(u,v)=\sum_{j=0}^{3}\sum_{i=0}^{3} c_{i,j}f_i(u)g_j(v)$ is a bicubic surface in the tensor product Hermite representation.

Then,

$$\begin{array}{llll}
\sigma(0,0) = c_{0,0} & \sigma(1,0) = c_{1,0} & \dfrac{\partial \sigma}{\partial u}(0,0) = c_{2,0} & \dfrac{\partial \sigma}{\partial u}(1,0) = c_{3,0} \\
\sigma(0,1) = c_{0,1} & \sigma(1,1) = c_{1,1} & \dfrac{\partial \sigma}{\partial u}(0,1) = c_{2,1} & \dfrac{\partial \sigma}{\partial u}(1,1) = c_{3,1} \\
\dfrac{\partial \sigma}{\partial v}(0,0) = c_{0,2} & \dfrac{\partial \sigma}{\partial v}(1,0) = c_{1,2} & \dfrac{\partial^2 \sigma}{\partial u \partial v}(0,0) = c_{2,2} & \dfrac{\partial^2 \sigma}{\partial u \partial v}(1,0) = c_{3,2} \\
\dfrac{\partial \sigma}{\partial v}(0,1) = c_{0,3} & \dfrac{\partial \sigma}{\partial v}(1,1) = c_{1,3} & \dfrac{\partial^2 \sigma}{\partial u \partial v}(0,1) = c_{2,3} & \dfrac{\partial^2 \sigma}{\partial u \partial v}(1,1) = c_{3,3}
\end{array}$$

Thus, the tensor product bicubic Hermite representation automatically interpolates position, both first partials, and the cross partial at the four corner points of the domain. Note carefully that the coefficient subscripts are not related to geometric positioning. A mesh formed by such coefficients provides no intuition as to the shape of the surface. □

13.3 Evaluating Surfaces and Partial Derivatives and Rendering Isocurves

There are generic properties of tensor product surfaces which we discuss here. One important part of representing surfaces is to be able to evaluate surfaces and their derivatives. Suppose $h(u, v)$ is a tensor product surface as defined as above, and it is necessary to evaluate the curve at $(\hat{u}, \hat{v})$ in the domain.

$$\begin{aligned}
h(\hat{u}, \hat{v}) &= \sum_{j=0}^{n} \sum_{i=0}^{m} c_{i,j} f_i(\hat{u}) g_j(\hat{v}) \\
&= \sum_{j=0}^{n} \gamma_j(\hat{u}) g_j(\hat{v}) \qquad (13.2)
\end{aligned}$$

where

$$\gamma_j(\hat{u}) = \sum_{i=0}^{m} c_{i,j} f_i(\hat{u}). \qquad (13.3)$$

Analogously,

$$h(\hat{u}, \hat{v}) = \sum_{i=0}^{m} \psi_i(\hat{v}) f_i(\hat{u}), \qquad (13.4)$$

$$\psi_i(\hat{v}) = \sum_{j=0}^{n} c_{i,j} g_j(\hat{v}). \qquad (13.5)$$

Thus, surface evaluation can be performed by performing $n+1$ curve evaluations in u and one curve evaluation in v, or $m+1$ curve evaluations in v and one curve evaluation in u. Thus, if the blending functions f and g are selected to be independent and to have fast evaluation algorithms, the resulting tensor product surface will have fast evaluation algorithms.

Now let let $F(u)$, C, and $G(v)$ be the $m+1$ row vector, $(m+1)\times(n+1)$ matrix and $n+1$ column vector defined as follows:

$$F(u) = \begin{bmatrix} f_0(u) & f_1(u) & \dots & f_{m-1}(u) & f_m(u) \end{bmatrix}; \tag{13.6}$$

$$C = \begin{bmatrix} c_{0,0} & c_{0,1} & \dots & c_{0,n-1} & c_{0,n} \\ c_{1,0} & c_{1,1} & \dots & c_{1,n-1} & c_{1,n} \\ & & \vdots & & \\ c_{m-1,0} & c_{m-1,1} & \dots & c_{m-1,n-1} & c_{m-1,n} \\ c_{m,0} & c_{m,1} & \dots & c_{m,n-1} & c_{m,n} \end{bmatrix}; \tag{13.7}$$

$$G(v) = \begin{bmatrix} g_0(v) & g_1(v) & \dots & g_{n-1}(v) & g_n(v) \end{bmatrix}^t. \tag{13.8}$$

Then, by Equation 13.3

$$\gamma(u) = \begin{bmatrix} \gamma_0(u) & \gamma_1(u) & \dots \gamma_{n-1}(u) & \gamma_n(u) \end{bmatrix} = F(u)C \tag{13.9}$$

and by Equation 13.5,

$$\psi(v) = \begin{bmatrix} \psi_0(v) & \psi_1(v) & \dots \psi_{m-1}(v) & \psi_m(v) \end{bmatrix}^t = CG(v). \tag{13.10}$$

So

$$\sigma(u,v) = \gamma(u)G(v) \tag{13.11}$$

$$= F(u)\psi(v). \tag{13.12}$$

Suppose that the surface is to be drawn with isoparametric curves (Definition 12.11). Suppose $h(\hat{u}, v)$ is the curve to be drawn. Evaluating Equation 13.9 gives $\gamma(\hat{u})$, which is equivalent to evaluating Equation 13.2 for each value of j ($n+1$ times). Then drawing the curve in v is simply evaluating Equation 13.11 as a curve. Analogously, an isoparametric curve in u with fixed $\hat{v}$ can be drawn using Equations 13.10 and 13.12.

Example 13.13. Let $f_0(u) = 1-u$, $f_1(u) = u$, $g_0(v) = (1-v)^2$, $g_1(v) = 2v(1-v)$, $g_2(v) = v^2$, and define

$$\begin{aligned} h(u,v) &= 3f_0(u)g_0(v) - 2f_1(u)g_0(v) + 1f_0(u)g_1(v) \\ &\quad +3f_1(u)g_1(v) + 2f_0(u)g_2(v) + 1f_1(u)g_2(v). \end{aligned}$$

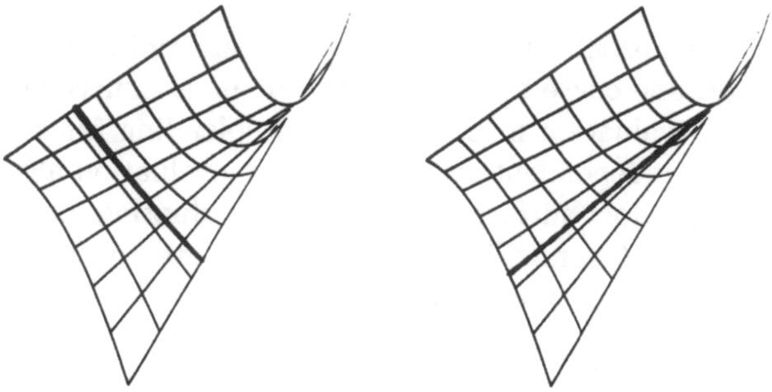

Figure 13.10. Extraction of isoparametric curves in a tensor product. (See Example 13.13.)

To rewrite in matrix form, set

$$\begin{aligned} F(u) &= \begin{bmatrix} 1-u & u \end{bmatrix} \\ G(v) &= \begin{bmatrix} (1-v)^2 & 2v(1-v) & v^2 \end{bmatrix}^t. \end{aligned}$$

Consider the computation necessary to draw the isoparametric curve $u = 1/3$. $F(1/3) = \begin{bmatrix} 2/3 & 1/3 \end{bmatrix}$, so

$$\begin{aligned} \gamma(1/3) &= F(1/3)C \\ &= \begin{bmatrix} 2/3 & 1/3 \end{bmatrix} \begin{bmatrix} 3 & 1 & 2 \\ -2 & 3 & 1 \end{bmatrix} \\ &= \begin{bmatrix} 4/3 & 5/3 & 5/3 \end{bmatrix}. \end{aligned}$$

Thus, the isocurve is

$$\begin{aligned} h(1/3, v) &= \gamma(1/3)G(v) \\ &= \begin{bmatrix} 4/3 & 5/3 & 5/3 \end{bmatrix} G(v) \\ &= (4/3)g_0(v) + (5/3)g_1(v) + (5/3)g_2(v). \end{aligned}$$

The isoparametric curve $u = 1/3$ is drawn in Figure 13.10 in dark gray. Now, consider the isocurve for $v = 1/3$. $G(1/3) = \begin{bmatrix} 4/9 & 4/9 & 1/9 \end{bmatrix}^t$, so

$$\begin{aligned} \psi(1/3) &= CG(1/3) \\ &= \begin{bmatrix} 3 & 1 & 2 \\ -2 & 3 & 1 \end{bmatrix} \begin{bmatrix} 4/9 \\ 4/9 \\ 1/9 \end{bmatrix} \\ &= \begin{bmatrix} 2 \\ 5/9 \end{bmatrix}. \end{aligned}$$

Thus, the isocurve is

$$\begin{aligned} h(u,1/3) &= F(u)\psi(1/3) \\ &= F(u)\begin{bmatrix} 2 \\ 5/9 \end{bmatrix} \\ &= (2)f_0(u) + (5/9)f_1(u). \end{aligned}$$

The isoparametric curve $v = 1/3$ is drawn in light gray. □

Now consider evaluating partial derivatives of a tensor product surface h.

$$\frac{\partial h}{\partial u} = \frac{\partial F}{\partial u}CG(v) \tag{13.13}$$

$$= \sum_{j=0}^{n}\sum_{i=0}^{m} \frac{\partial f_i(u)}{\partial u} c_{i,j} g_j(v); \tag{13.14}$$

$$\frac{\partial h}{\partial v} = F(u)C\frac{\partial G(v)}{\partial v} \tag{13.15}$$

$$= \sum_{j=0}^{n}\sum_{i=0}^{m} f_i(u) c_{i,j} \frac{\partial g_j(v)}{\partial v}. \tag{13.16}$$

Once again, the characteristics of the tensor product surface are used to shorten the computation of the first partial derivatives.

13.3.1 Bézier Surface Evaluation

Let $F(u) = \Theta_m(u) = \begin{bmatrix} \theta_{0,m}(u) & \ldots \theta_{m,m}(u) \end{bmatrix}$ and let $G(v) = \Theta_n^t(v)$. Then, by Equations 13.13–13.16,

$$\sigma(u,v) = \Theta_m(u)\boldsymbol{P}\Theta_n^t(v).$$

Let $\Psi(v) = \boldsymbol{P}\Theta_n^t(v)$. Then

$$\psi_i(v) = \sum_{j=0}^{n} P_{i,j}\theta_{j,n}(v), \qquad \text{for } i = 0, \ldots, m.$$

Each of these is a Bézier curve. Then $\sigma(u,v) = \Theta_m(u)\Psi(v)$. For each fixed v, the right hand side of this equation is a Bézier curve. So, a Bézier surface can be evaluated using the constructive algorithm already developed for the Bézier curve. However the control points are not from the original surface mesh of points. Rather, there is a distinct polygon for each value of v. Isoparametric curves in constant v are just Bézier curves with control polygon $\{\psi_i(v)\}$, none of which, in general, are from the mesh defining the surface.

By Equations 13.13 through 13.16, the partial derivatives are

$$\begin{aligned}
\frac{\partial\sigma}{\partial u}(u,v) &= \sum_{i=0}^{m}\sum_{j=0}^{n}\theta'_{i,m}(u)P_{i,j}\theta_{j,n}(v)\\
&= m\sum_{j=0}^{n}\sum_{i=0}^{m-1}\left(P_{i+1,j}-P_{i,j}\right)\theta_{i,m-1}(u)\theta_{j,n}(v), \qquad \text{and}\\
\frac{\partial\sigma}{\partial v}(u,v) &= \sum_{i=0}^{m}\sum_{j=0}^{n}\theta_{i,m}(u)P_{i,j}\theta'_{j,n}(v)\\
&= n\sum_{i=0}^{m}\sum_{j=0}^{n-1}\left(P_{i,j+1}-P_{i,j}\right)\theta_{j,n-1}(v)\theta_{i,m}(u).
\end{aligned}$$

Let $\mathcal{D}_m$ be the $m\times(m+1)$ matrix with $d_{i,i}=-1$, and $d_{i,i+1}=1$, for $i=0,\ldots,\ m-1$, and zeros in all other places. That is

$$\mathcal{D}_m = \begin{bmatrix} -1 & 1 & 0 & 0 & \ldots & 0 & 0 & 0\\ 0 & -1 & 1 & 0 & \ldots & 0 & 0 & 0\\ & & & & \vdots & & & \\ 0 & 0 & 0 & 0 & \ldots & -1 & 1 & 0\\ 0 & 0 & 0 & 0 & \ldots & 0 & -1 & 1 \end{bmatrix}.$$

Then

$$\frac{\partial\sigma}{\partial u}(u,v) = m\Theta_{m-1}(u)\mathcal{D}_m\boldsymbol{P}\Theta_n^t(v)$$

and

$$\frac{\partial\sigma}{\partial v}(u,v) = n\Theta_m(u)\boldsymbol{P}\mathcal{D}_n^t\Theta_{n-1}^t(v).$$

The Bézier surface and both partial derivatives can be evaluated by the following algorithm, using Algorithm 5.18 as a foundation.

Algorithm 13.14. *To evaluate the $m\times n$ degree Bézier surface σ with mesh points $\{P_{i,j} : j=0,\ \ldots,\ n;\ i=0,\ldots,\ m\}$ and its partial derivatives at the point $(\hat{u},\hat{v})$, apply the evaluation path below or interchange the roles of u and v ($i(m)$ and $j(n)$, respectively).*

Step 1: Evaluate $2*(m+1)$ Bézier curves at $v = \hat{v}$. That is, for $i = 0, \ldots, m$, use Algorithm 5.18 to get

$$\begin{aligned} Q_{i,\hat{v}} &= \sum_{j=0}^{n} P_{i,j}\theta_{j,n}(\hat{v}) \\ Q'_{i,\hat{v}} &= \sum_{j=0}^{n} P_{i,j}\theta'_{j,n}(\hat{v}). \end{aligned}$$

Step 2: Using Algorithm 5.18 at $u = \hat{u}$ for simultaneous evaluation of curve and derivative, evaluate the following three Bézier curves to get a surface value and partial derivative values.

$$\begin{aligned} \sigma(\hat{u},\hat{v}) &= \sum_{i=0}^{m} Q_{i,\hat{v}}\theta_{i,m}(\hat{u}) \\ \tfrac{\partial\sigma}{\partial u}(\hat{u},\hat{v}) &= m\sum_{i=0}^{m} Q_{i,\hat{v}}\theta'_{i,m}(\hat{u}) \\ \tfrac{\partial\sigma}{\partial v}(\hat{u},\hat{v}) &= \sum_{i=0}^{m} Q'_{i,\hat{v}}\theta_{i,m}(\hat{u}). \end{aligned}$$

```
// Evaluates a Bezier surface and partials at (u,v) = (û,v̂).
//procedure Bez_eval(R, n, t, γ,γp) to evaluate
// a Bezier curve and its derivative where:
// γ(t) = Σ_{i=0}^n R_i θ_{i,n}(t).
procedure Bez_Surf-Der_Eval(P, m, n, û, v̂, σ,σu,σv )
    // P[i;] denotes the i^th row of the Bezier mesh.
    // P[;j] denotes the j^th column of the Bezier mesh.
    for i = 0 to m
        Bez_eval(P[i;], n, v̂, Qi,Qpi )
    Bez_eval(Q, m, û, σ,σu )
    Bez_eval(Qp, m − 1, û, σv,σuv )
end
```

Notice that the program example also gives $\frac{\partial^2\sigma}{\partial u \partial v}$ for one extra subtraction. The symmetry of the tensor product allows the roles of u and v (and hence i and j) to be interchanged in this algorithm, resulting in an equivalent algorithm. The choice of which parameter to evaluate first is based on several issues, one of which is the relative degree. Step 1 requires evaluation of $m + 1$ curves, each of degree n, and step 2 requires evaluation of one curve of degree m. It is seemingly irrelevant which one to do first unless a great many points are to be evaluated in some pattern. Such examples include drawing isoparametric curves in the surface and also in creating triangle strips for fast hardware surface drawing.

Example 13.15. Use Algorithm 13.14, to draw isocurves. Suppose the surface σ is $m \times n$ degree in u and v, respectively, with mesh $P_{i,j}$, $i = 0, \ldots, m$, $j = 0, \ldots, n$, and that it has been decided to draw $(m+1)$ isoparametric lines in constant u and $(n+1)$ isoparametric lines in constant v, $\alpha_i = i/m$, $i = 0, \ldots, m$, and $\beta_j = j/n$, $j = 0, \ldots, n$. Finally suppose that a fineness number N has been set so that $\delta_u = 1/(mN)$, and $\delta_v = 1/(nN)$. Thus, $N\delta_u = \alpha_{i+1} - \alpha_i = 1/m$ and $N\delta_v = \beta_{i+1} - \beta_i = 1/n$. □

```
// Evaluates isoparametric lines in constant u with
// (nN + 1) points.
// iso_u( i, k ) are the values for the i^th isoparametric
// curve with fixed u as v varies.
// iso_v( j, k ) are the values for the j^th isoparametric
// curve with fixed v, as u varies.
Procedure Surface_Isolines(P, m, n, N, iso_u, iso_v )
    δ_u = 1/(mN);
    δ_v = 1/(nN);
    for i = 0 to m, α_i = i/m;
    for j = 0 to n, β_j = j/n;
    for i = 0 to m begin // iso lines with constant u.
        for j = 0 to n, Bez_eval( P[;j], m, α_i, Q_j )
        for k = 0 to (Nn), Bez_eval( Q, n, kδ_v, iso_u(i,k) )
    end { i loop }
    for j = 0 to n begin // iso lines with constant v.
        for i = 0 to m, Bez_eval( P[i;], n, β_j, Q_i )
        for k = 0 to (Nm), Bez_eval( Q, m, kδ_u, iso_v(j,k) )
    end { j loop }
end
```

13.3.2 Evaluation of Tensor Product B-Spline Surfaces

We seek to evaluate $\sigma(a,b)$, where $(a,b) \in [u_{\kappa_u}, u_{m+1}] \times [v_{\kappa_v}, u_{n+1}]$. In studying B-spline curves of degree κ, it was determined that for any parameter value, u, there are at most $\kappa + 1$ basis functions which can be nonzero. Thus, the curve was the sum of at most $\kappa + 1$ terms. Analogously, a given control point P_J, occurs as a coefficient over at most $\kappa + 1$ intervals. The surface situation is more complex. Given a point (u, v), there are integers μ and ν such that $u_\mu \le u < u_{\mu+1}$ and $v_\nu \le v < v_{\nu+1}$, meaning that the only basis functions having nonzero image are $\mathcal{B}_{j,\kappa_v}(v)\mathcal{N}_{i,\kappa_u}(u)$, $i = \mu - \kappa_u, \ldots, \mu$, $j = \nu - \kappa_v, \ldots, \nu$. Thus at any point (u, v), $\sigma(u, v) = \sum_{i=\mu-\kappa_u}^{\mu} \sum_{j=\nu-\kappa_v}^{\nu} P_{i,j}\mathcal{B}_{j,\kappa_v}(v)\mathcal{N}_{i,\kappa_u}(u)$. There are $(\kappa_u + 1)(\kappa_v + 1)$ possible terms in the summation. Conversely, each mesh point $P_{i,j}$ can appear in the summation for up to $(\kappa_u + 1)(\kappa_v + 1)$ possible patches.

We now show how the general tensor product evaluation methods are modified to take advantage of the characteristics of B-splines.

$$
\begin{aligned}
\sigma(a, b) &= \sum_{i=0}^{m}\sum_{j=0}^{n} \mathcal{N}_{i,\kappa_u}(a)P_{i,j}\mathcal{B}_{j,\kappa_v}(b) \\
&= \sum_{i=0}^{m} \mathcal{N}_{i,\kappa_u}(a)\left(\sum_{j=0}^{n} P_{i,j}\mathcal{B}_{j,\kappa_v}(b)\right) \\
&= \sum_{i=0}^{m} Q_{i,b}\mathcal{N}_{i,\kappa_u}(a),
\end{aligned}
$$

where $Q_{i,b} = \sum_{j=0}^{n} P_{i,j}\mathcal{B}_{j,\kappa_v}(b)$.

The following algorithm outlines a method to first compute the $(\kappa_u + 1)$ necessary $Q_{i,b}$ using Algorithm 6.4 for the v direction $(\kappa_u + 1)$ times, and then to compute $\sigma(a, b)$, using Algorithm 6.4 for the u direction.

Algorithm 13.16.

Step 1: Find ν such that $v_\nu \le b < v_{\nu+1}$.

Step 2: Find μ such that $u_\mu \le a < u_{\mu+1}$.
Then $\sigma(a, b) = \sum_{i=\mu-\kappa_u}^{\mu} Q_{i,b}\mathcal{N}_{i,\kappa_u}(a)$,
where $Q_{i,b} = \sum_{j=\nu-\kappa_v}^{\nu} P_{i,j}\mathcal{B}_{j,\kappa_v}(b)$.

Step 3: Using Theorem 6.4, evaluate $Q_{i,b}$, $i = \mu - \kappa_u, \ldots, \mu$.

Step 4: Use the same algorithm to evaluate $\sigma(a, b)$ with the $Q_{i,b}$'s which were evaluated in Step 3.

If $\kappa_u = \kappa_v = k$, the algorithm computes Algorithm 6.4 $(k + 2)$ times. Since Algorithm 6.4 has $O(k^2)$ operations, that means this combination is $O(k^3)$.

The algorithm for computing a single point may be used to evaluate an isocurve in the surface by

Algorithm 13.17. *To evaluate $\sigma(u, b)$, use Algorithm 13.16 for each value of u to be found. Clearly Step 1 and Step 3 of the algorithm need only be performed one time since the second variable is fixed. Step 2 must*

be repeated one time for each domain interval that is drawn, while Step 4 must be repeated once for each surface point drawn.

Clearly the roles of u and v in Theorem 13.16 and its corollary can be reversed if it is desired to evaluate a curve in the surface with a constant u parameter value.

Example 13.18. Suppose $\kappa_u = 2$ and $\kappa_v = 3$, $\boldsymbol{u} = \{0,0,0,1,2,3,3,3\}$, and $\boldsymbol{v} = \{1,1,1,1,4,4,4,4\}$. Then σ will be a quadratic-by-cubic open spline surface, and σ will have the form

$$\sigma(u,v) = \sum_{i=0}^{4}\sum_{j=0}^{3}\mathcal{N}_{i,2}(u)P_{i,j}\mathcal{B}_{j,3}(v).$$

To evaluate the surface along the curve $\sigma(u,2)$:

$$Q_{i,2} = \sum_{j=0}^{3} P_{i,j}\mathcal{B}_{j,3}(2),$$

and

$$\begin{aligned} Q_{0,2} &= \textstyle\sum_{j=0}^{3} P_{0,j}\mathcal{B}_{j,3}(2), & Q_{1,2} &= \textstyle\sum_{j=0}^{3} P_{1,j}\mathcal{B}_{j,3}(2),\\ Q_{2,2} &= \textstyle\sum_{j=0}^{3} P_{2,j}\mathcal{B}_{j,3}(2), & Q_{3,2} &= \textstyle\sum_{j=0}^{3} P_{3,j}\mathcal{B}_{j,3}(2),\\ Q_{4,2} &= \textstyle\sum_{j=0}^{3} P_{4,j}\mathcal{B}_{j,3}(2). \end{aligned}$$

Now, the curve evaluation for $\sigma(u,2)$ can proceed as if it were an independent curve. □

Frequently, isoparametric curves in both u and v are graphed to obtain a sense of the behavior of the surface. See Figure 13.11.

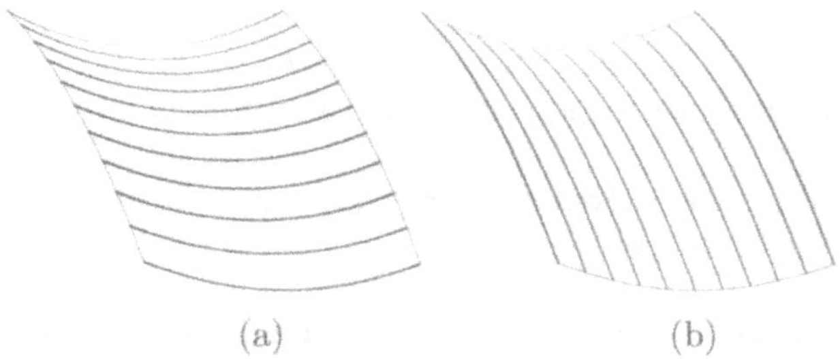

Figure 13.11. Isoparametric curves: (a) constant v; (b) constant u.

13.4 Uniform Refinement for Uniform Floating Spline Surfaces

Many surface modeling queries and modifications require geometric computations that can be supported by algorithms which use B-spline surface refinement as a central tool. Surface-surface intersection, rendering, and multiresolution modeling are several such applications. In Chapter 16 we present specific algorithms to carry out such geometric computations as well as algorithms for arbitrary refinement. In Section 7.2.1 we developed algorithms for subdivision curves; that is, algorithms for uniform refinement of uniform floating spline curves. In this section we generalize those results to regular subdivision surfaces.

13.4.1 Uniform Biquadratic Subdivision

Recall that for curves, quadratic uniform refinement is given by:

$$Q_m = \begin{cases} \frac{1}{4}P_{i-2} + \frac{3}{4}P_{i-1} & m = 2i-1 \\ \frac{3}{4}P_{i-1} + \frac{1}{4}P_i & m = 2i. \end{cases} \tag{13.17}$$

Consider a uniform floating form tensor product surface,

$$\sigma(u,v) = \sum_j \sum_i \mathcal{B}_{i,2}(u) P_{i,j} \mathcal{B}_{j,2}(v).$$

Considered as functions of u, Equation 13.17 can be applied for each fixed j resulting in

$$\sigma(u,v) = \sum_j \sum_m \mathcal{N}_{m,2}(u) \overline{Q}_{m,j} \mathcal{B}_{j,2}(v),$$

where

$$\overline{Q}_{m,j} = \begin{cases} \frac{1}{4}P_{i-2,j} + \frac{3}{4}P_{i-1,j} & m = 2i-1 \\ \frac{3}{4}P_{i-1,j} + \frac{1}{4}P_{i,j} & m = 2i. \end{cases}$$

Now, considering the surface as functions of v, for each fixed m gives

$$\sigma(u,v) = \sum_n \sum_m \mathcal{N}_{m,2}(u) Q_{m,n} \mathcal{N}_{n,2}(v)$$

with

$$Q_{m,n} = \begin{cases} \frac{1}{4}\hat{Q}_{m,j-2} + \frac{3}{4}\hat{Q}_{m,j-1} & n = 2j-1 \\ \frac{3}{4}\hat{Q}_{m,j-1} + \frac{1}{4}\hat{Q}_{m,j} & n = 2j. \end{cases}$$

Thus we have obtained a new control mesh which can be written in terms of the old.

$$Q_{m,n} = \begin{cases} \frac{1}{16}P_{i-2,j-2} + \frac{3}{16}P_{i-1,j-2} \\ \qquad + \frac{3}{16}P_{i-2,j-1} + \frac{9}{16}P_{i-1,j-1} & m = 2i-1,\ n = 2j-1 \\ \frac{3}{16}P_{i-1,j-2} + \frac{1}{16}P_{i,j-2} + \frac{9}{16}P_{i-1,j-1} + \frac{3}{16}P_{i,j-1} & m = 2i,\ n = 2j-1 \\ \frac{3}{16}P_{i-2,j-1} + \frac{9}{16}P_{i-1,j-1} + \frac{1}{16}P_{i-2,j} + \frac{3}{16}P_{i-1,j} & m = 2i-1,\ n = 2j \\ \frac{9}{16}P_{i-1,j-1} + \frac{3}{16}P_{i,j-1} + \frac{3}{16}P_{i-1,j} + \frac{1}{16}P_{i,j} & m = 2i,\ n = 2j \end{cases} \tag{13.18}$$

For each i and j, $\{P_{i-1,j-1}, P_{i,j-1}, P_{i,j}, P_{i-1,j}\}$ forms a quadrilateral, called a *face* in the original control mesh. We define

$$\begin{aligned} F_{i-1,j-1} &= \frac{1}{4}(P_{i-1,j-1} + P_{i,j-1} + P_{i,j} + P_{i-1,j}) \\ V_{i-1,j-1} &= \frac{1}{2}(P_{i-1,j-1} + P_{i,j-1}) \\ H_{i-1,j-1} &= \frac{1}{2}(P_{i-1,j-1} + P_{i-1,j}) \end{aligned} \tag{13.19}$$

as the barycenter, or *face average*, the middle of the vertical edge, and the middle of the horizontal edge, respectively. Then the formula in Equation 13.18 can be rewritten as

$$Q_{m,n} = \begin{cases} \frac{1}{4}(P_{i-1,j-1} + V_{i-2,j-1} + H_{i-1,j-2} + F_{i-2,j-2}) & m = 2i-1,\ n = 2j-1 \\ \frac{1}{4}(P_{i-1,j-1} + H_{i-1,j-2} + V_{i-1,j-1} + F_{i-1,j-2}) & m = 2i,\ n = 2j-1 \\ \frac{1}{4}(P_{i-1,j-1} + V_{i-2,j-1} + H_{i-1,j-1} + F_{i-2,j-1}) & m = 2i-1,\ n = 2j \\ \frac{1}{4}(P_{i-1,j-1} + V_{i-1,j-1} + H_{i-1,j-1} + F_{i-1,j-1}) & m = 2i,\ n = 2j \end{cases} \tag{13.20}$$

Each $Q_{m,n}$ is the barycenter of one of the ephemeral faces defined by

$$\begin{gathered} \{P_{i-1,j-1}, V_{i-2,j-1}, F_{i-2,j-2}, H_{i-1,j-2}\}, \\ \{P_{i-1,j-1}, H_{i-1,j-2}, F_{i-1,j-2}, V_{i-1,j-1}\}, \\ \{P_{i-1,j-1}, H_{i-1,j-1}, F_{i-2,j-1}, V_{i-2,j-1}\}, \\ \{P_{i-1,j-1}, V_{i-1,j-1}, F_{i-1,j-1}, H_{i-1,j-1}\}. \end{gathered}$$

For each m and n, $\mathbf{Q}_{m,n} = \{Q_{m,n}, Q_{m,n+1}, Q_{m+1,n}, Q_{m+1,n+1}\}$ forms a quadrilateral in the new mesh. If each quadrilateral is called a *face*, then Figure 13.12 shows the relationship of new mesh faces to the original mesh faces. The results are shown for m and n both even and m and n both odd in Figure 13.12 (a). Figure 13.12 (b) shows the cases for m even, n

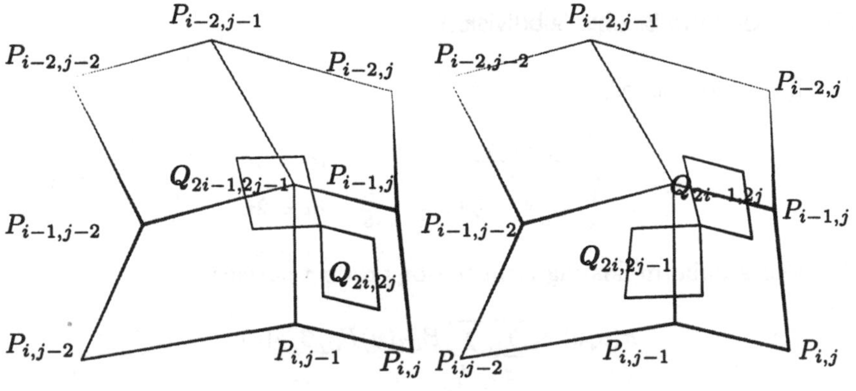

Figure 13.12. Relation of refined quadratic mesh to original.

odd and m odd and n even, respectively. Figure 13.13 shows a refinement sequence.

Observe that just as the quadratic curve refinement is *edge* focused, bi-quadratic surface refinement is *face* focused. That is, each face in the new control mesh is *associated with* either a face, edge, or vertex in the original control mesh. Then,

$$\mathit{face}\boldsymbol{Q}_{m,n} \leftarrow \begin{cases} \mathit{face}\,\{P_{i,j}, P_{i,j+1}, P_{i+1,j}, P_{i+1,j+1}\} & m = 2(i-1), n = 2(j-1) \\ \mathit{edge}\,[P_{i,j}, P_{i,j+1}] & m = 2(i-1)-1, n = 2(j-1) \\ \mathit{edge}\,[P_{i,j}, P_{i+1,j}] & m = 2(i-1), n = 2(j-1)-1 \\ \mathit{vertex}\ P_{i,j} & m = 2(i-1)-1, n = 2(j-1)-1 \end{cases}$$

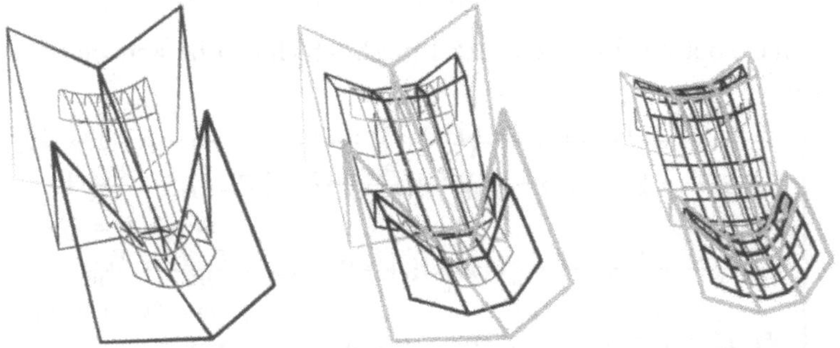

Figure 13.13. Regular refinement for floating uniform biquadratic surface. Previous mesh is shown in light gray.

13.4.2 Uniform Bicubic Subdivision

Recall that for curves,

$$Q_m = \begin{cases} \frac{P_{i-2}+P_{i-1}}{2} & m = 2i-1 \\ \frac{P_{i-2}}{8} + \frac{3}{4}P_{i-1} + \frac{P_i}{8} & m = 2i. \end{cases} \quad (13.21)$$

Consider a uniform floating form tensor product surface,

$$\sigma(u,v) = \sum_j \sum_i \mathcal{B}_{i,3}(u) P_{i,j} \mathcal{B}_{j,3}(v).$$

Considered as functions of u, Equation 13.21 can be applied for each fixed j resulting in

$$\sigma(u,v) = \sum_j \sum_m \mathcal{N}_{m,3}(u) \overline{Q}_{m,j} \mathcal{B}_{j,3}(v),$$

where

$$\overline{Q}_{m,j} = \begin{cases} \frac{P_{i-2,j}+P_{i-1,j}}{2} & m = 2i-1 \\ \frac{P_{i-2,j}}{8} + \frac{3}{4}P_{i-1,j} + \frac{P_{i,j}}{8} & m = 2i. \end{cases}$$

Now, considering the surface as functions of v, for each fixed m gives

$$\sigma(u,v) = \sum_n \sum_m \mathcal{N}_{m,3}(u) Q_{m,n} \mathcal{N}_{n,3}(v),$$

with

$$Q_{m,n} = \begin{cases} \frac{\hat{Q}_{m,j-2}+\hat{Q}_{m,j-1}}{2} & n = 2j-1 \\ \frac{\hat{Q}_{m,j-2}}{8} + \frac{3}{3}\hat{Q}_{m,j-1} + \frac{\hat{Q}_{m,j}}{8} & n = 2j. \end{cases}$$

Thus we have obtained a new control mesh which can be written in terms of the old.

$$Q_{m,n} = \begin{cases} \frac{P_{i-2,j-2}}{4} + \frac{P_{i-1,j-2}}{4} + \frac{P_{i-2,j-1}}{4} + \frac{P_{i-1,j-1}}{4} & m = 2i-1, n = 2j-1 \\ \frac{P_{i-2,j-2}}{16} + \frac{3}{8}P_{i-1,j-2} + \frac{P_{i,j-2}}{16} + \frac{P_{i-2,j-1}}{16} + \frac{3}{8}P_{i-1,j-1} + \frac{P_{i,j-1}}{16} & \\ & m = 2i, n = 2j-1 \\ \frac{P_{i-2,j-2}}{16} + \frac{P_{i-1,j-2}}{16} + \frac{3}{8}P_{i-2,j-1} + \frac{3}{8}P_{i-1,j-1} + \frac{P_{i-2,j}}{16} + \frac{P_{i-1,j}}{16} & \\ & m = 2i-1, n = 2j \\ \frac{P_{i-2,j-2}}{64} + \frac{3}{32}P_{i-1,j-2} + \frac{P_{i,j-2}}{64} + \frac{3}{32}P_{i-2,j-1} + \frac{9}{16}P_{i-1,j-1} & \\ + \frac{3}{32}P_{i,j-1} + \frac{P_{i-2,j}}{64} + \frac{3}{32}P_{i-1,j} + \frac{P_{i,j}}{64} & m = 2i, n = 2j \end{cases}$$

$$(13.22)$$

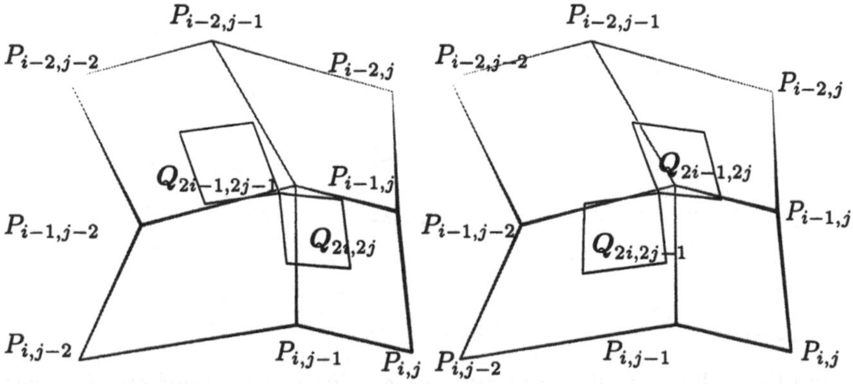

Figure 13.14. Relation of refined cubic mesh to original.

Rewriting Equation 13.22 using the definitions in Equation 13.19,

$$Q_{m,n} = \begin{cases} F_{i-2,j-2} & m = 2i-1, n = 2j-1 \\ \frac{1}{2}H_{i-1,j-2} + \frac{1}{4}F_{i-2,j-2} + \frac{1}{4}F_{i-1,j-2} & m = 2i, n = 2j-1 \\ \frac{1}{2}V_{i-2,j-1} + \frac{1}{4}F_{i-2,j-2} + \frac{1}{4}F_{i-2,j-1} & m = 2i-1, n = 2j \\ \frac{1}{4}P_{i-1,j-1} + \frac{1}{2}\frac{(H_{i-1,j-2}+H_{i-1,j-1}+V_{i-2,j-1}+V_{i-1,j-1})}{4} & \\ \quad + \frac{1}{4}\frac{(F_{i-2,j-2}+F_{i-1,j-2}+F_{i-1,j-1}+F_{i-2,j-1})}{4} & m = 2i, n = 2j. \end{cases} \tag{13.23}$$

Once again, $Q_{m,n} = \{Q_{m.n+1}, Q_{m+1,n}, Q_{m+1,n+1}\}$ forms a face in the new mesh. Figure 13.14 shows the relationship of new mesh vertices and faces to the original mesh. Again, the results are shown for m and n both even and m and n both odd in Figure 13.14 (a). The cases for m even, n odd and m odd, n even, respectively are shown in Figure 13.14 (b). Figure 13.15 shows a refinement sequence.

Observe that just as the cubic curve refinement is *vertex* focused, bicubic surface refinement is *vertex* focused. That is, for each face, edge, and vertex in the original control mesh, there is a vertex in the new control mesh.

$$\textit{vertex}\ Q_{m,n} \leftarrow \begin{cases} \textit{face}\ \{P_{i-2,j-2}, P_{i-1,j-2}, P_{i-1,j-1}, P_{i-2,j-1}\} & \\ & m = 2i-1, n = 2j-1 \\ \textit{edge}\ [P_{i-2,j-1}, P_{i-1,j-1}] & m = 2i-1, n = 2j \\ \textit{edge}\ [P_{i-1,j-2}, P_{i-1,j-1}] & m = 2i, n = 2j-1 \\ \textit{vertex}\ P_{i-2,j-2}, & m = 2i, n = 2j \end{cases}$$

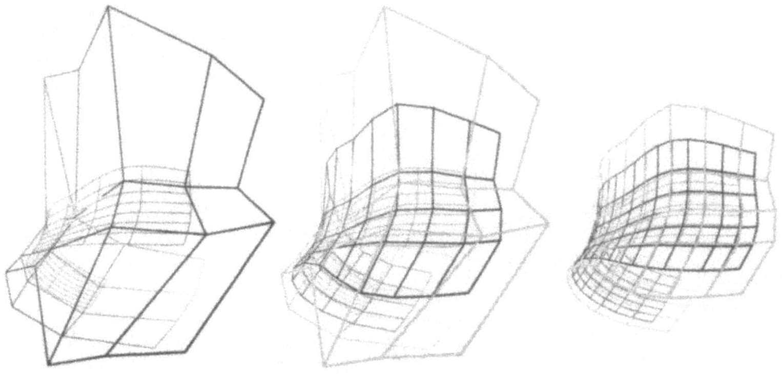

Figure 13.15. Regular refinement for floating uniform bicubic surfaces. Previous mesh is shown in light gray.

13.5 Matrix Expansions

13.5.1 B-Splines and Bézier

Another approach to evaluation uses a modified matrix approach [56]. This approach decomposes the tensor product spline, making note that for a specific point (a, b), and μ and ν as in Algorithm 13.16,

$$\sigma(a,b) = \sum_{i=\mu-\kappa_u}^{\mu} \sum_{j=\nu-\kappa_v}^{\nu} \mathcal{N}_{i,\kappa_u}(a) P_{i,j} \mathcal{B}_{j,\kappa_v}(b).$$

If

$$\begin{aligned} \boldsymbol{\mathcal{N}}_{\mu,\kappa_u}(a) &= \left[\begin{array}{cccc} \mathcal{B}_{\mu-\kappa_u,\kappa_u}(a) & \mathcal{B}_{\mu-\kappa_u+1,\kappa_u}(a) & \cdots & \mathcal{B}_{\mu,\kappa_u}(a) \end{array}\right]^t, \\ \boldsymbol{\mathcal{B}}_{\nu,\kappa_v}(b) &= \left[\begin{array}{cccc} \mathcal{B}_{\nu-\kappa_v,\kappa_v}(b) & \mathcal{B}_{\nu-\kappa_v+1,\kappa_v}(b) & \cdots & \mathcal{B}_{\nu,\kappa_v}(b) \end{array}\right]^t, \end{aligned}$$

then

$$\sigma(a,b) = \boldsymbol{\mathcal{N}}^t_{\mu,\kappa_u}(a) \boldsymbol{P}_{\mu,\nu} \boldsymbol{\mathcal{B}}_{\nu,\kappa_v}(b),$$

where

$$\boldsymbol{P}_{\mu,\nu} = \left[\begin{array}{cccc} P_{\mu-\kappa_u,\nu-\kappa_v} & P_{\mu-\kappa_u,\nu-\kappa_v+1} & \cdots & P_{\mu-\kappa_u,\nu} \\ P_{\mu-\kappa_u+1,\nu-\kappa_v} & P_{\mu-\kappa_u+1,\nu-\kappa_v+1} & \cdots & P_{\mu-\kappa_u+1,\nu} \\ & & \vdots & \\ P_{\mu,\nu-\kappa_v} & P_{\mu,\nu-\kappa_v+1} & \cdots & P_{\mu,\nu} \end{array}\right].$$

Each of the B-splines can be evaluated using Algorithm 11.16. The algorithm requires two application of Algorithm 11.16, which have $O\left(n^2\right)$ operations each, and one vector-matrix multiplication, $O\left(n^2\right)$ operations, and one vector inner product, $O(n)$ operations. Hence, it has $O\left(n^2\right)$ operations. While direct surface evaluation is $O(n^3)$ as the discussion after Algorithm 13.16 shows, we can see that recursive basis evaluation and matrix multiplication have fewer operations as the degree or size increases.

13.5.2 Bézier to Power Basis Evaluation Using Matrices

If $\Theta(t) = \begin{bmatrix} \theta_{0,3}(t) & \theta_{1,3}(t) & \theta_{2,3}(t) & \theta_{3,3}(t) \end{bmatrix}$, then in matrix notation, a bicubic Bézier patch can be written as

$$\sigma_{3,3}(u,v) = \sum_{i=0}^{3}\sum_{j=0}^{3} P_{i,j}\theta_{i,3}(u)\theta_{j,3}(v) = \Theta(u)P\Theta^t(v),$$

where

$$P = \begin{bmatrix} P_{0,0} & P_{0,1} & P_{0,2} & P_{0,3} \\ P_{1,0} & P_{1,1} & P_{1,2} & P_{1,3} \\ P_{2,0} & P_{2,1} & P_{2,2} & P_{2,3} \\ P_{3,0} & P_{3,1} & P_{3,2} & P_{3,3} \end{bmatrix}.$$

The Θ matrix must now be decomposed into the power basis.

$$\begin{aligned} \theta_{i,3}(t) &= \tbinom{3}{i} t^i (1-t)^{3-i} \\ &= \binom{3}{i} t^i \sum_{j=0}^{3-i} \binom{3-i}{j} (-t)^{3-i-j} \\ &= \sum_{j=0}^{3-i} (-1)^{3-i-j} \binom{3-i}{j} \binom{3}{i} t^{3-j}, \end{aligned}$$

so

$$\Theta(t) = \begin{bmatrix} t^3 & t^2 & t & 1 \end{bmatrix} Q,$$

where

$$Q = \begin{bmatrix} (-1)\binom{3}{0}\binom{3}{0} & \binom{3}{1}\binom{3}{0} & (-1)\binom{3}{2}\binom{3}{0} & \binom{3}{3}\binom{3}{0} \\ \binom{2}{0}\binom{3}{1} & (-1)\binom{2}{1}\binom{3}{1} & \binom{2}{2}\binom{3}{1} & 0 \\ (-1)\binom{1}{0}\binom{3}{2} & \binom{1}{1}\binom{3}{2} & 0 & 0 \\ \binom{0}{0}\binom{3}{3} & 0 & 0 & 0 \end{bmatrix}.$$

Putting it all together,

$$\sigma(u,v) = UQPQ^tV^t,$$

where $U = \begin{bmatrix} u^3 & u^2 & u & 1 \end{bmatrix}$, and V is defined analogously. The Q matrix is clearly invertible, so once again it is clear that every bicubic surface over the unit square can be written as a Bézier patch.

13.5.3 Bicubic Hermite Surface Patches

Let

$$\begin{aligned} U &= \begin{bmatrix} u^3 & u^2 & u & 1 \end{bmatrix}, \\ V &= \begin{bmatrix} v^3 & v^2 & v & 1 \end{bmatrix}. \end{aligned}$$

Every bicubic surface can be written using the power basis as UCV^t where C is the matrix of coefficients.

Suppose a bicubic surface, $\sigma(u,v)$, is in the Hermite tensor product representation. The first step to obtaining its form in the power basis is to find its matrix representation.

$$\begin{aligned} \sigma(u,v) &= \sum_{j=0}^{3}\sum_{i=0}^{3} c_{i,j} f_i(u) g_j(v) \\ &= FCG^t \\ &= UMCM^tV^t, \end{aligned}$$

where

$$\begin{aligned} F &= \begin{bmatrix} f_0(u) & f_1(u) & f_2(u) & f_3(u) \end{bmatrix} \\ &= \begin{bmatrix} h_{0,0}(u) & h_{1,0}(u) & h_{0,1}(u) & h_{1,1}(u) \end{bmatrix} \\ &= UM, \\ G &= \begin{bmatrix} g_0(v) & g_1(v) & g_2(v) & g_3(v) \end{bmatrix} \\ &= \begin{bmatrix} h_{0,0}(v) & h_{1,0}(v) & h_{0,1}(v) & h_{1,1}(v) \end{bmatrix}^t \\ &= M^tV^t, \\ C &= (c_{i,j}) \end{aligned}$$

and

$$M^t = \begin{bmatrix} 2 & -2 & 1 & 1 \\ -3 & 3 & -2 & -1 \\ 0 & 0 & 1 & 0 \\ 1 & 0 & 0 & 0 \end{bmatrix}, \tag{13.24}$$

as a direct expansion of the cubic Hermite basis functions. (See Equations 13.1.)

The coefficient matrix for the power basis then is MCM^t. The matrix M (and hence M^t) is nonsingular and thus has an inverse.

Given any bicubic surface, $\sigma(u,v)$ written in the bicubic power basis as $\sigma(u,v) = UGV^t$, we can write it in form of a bicubic Hermite patch by writing

$$C = M^{-1}G(M^t)^{-1},$$

and then

$$\sigma(u,v) = F(u)CG^t(v).$$

Example 13.19. Given the surface $\sigma(u,v) = 3u^3v^2 + 2u^2v^2 + uv + 1$, we shall convert it to the bicubic Hermite representation. Start by writing $\sigma(u,v)$ in matrix form:

$$\sigma(u,v) = \begin{bmatrix} u^3 & u^2 & u & 1 \end{bmatrix} \begin{bmatrix} 0 & 3 & 0 & 0 \\ 0 & 2 & 0 & 0 \\ 0 & 0 & 1 & 0 \\ 0 & 0 & 0 & 1 \end{bmatrix} \begin{bmatrix} v^3 \\ v^2 \\ v \\ 1 \end{bmatrix}.$$

The completion of this is left as an exercise. □

13.5.4 Comparing Coefficients in Bicubic Representations

Since both Bézier surfaces and Hermite patches are tensor product polynomial surfaces, a comparison of their attributes is in order. Just as higher order Hermite interpolation is possible for curves, higher order tensor product Hermite patches are possible. If we were to consider only the information necessary for the higher order tensor product Hermite patch, we would see a pattern evolve. The bilinear case (degree one in each variable) requires one piece of information at each of the four corner points, its position. The bicubic requires four pieces of information at each of the four corner points (position, two first partials, twists), and results from a tensor product application of cubic Hermite interpolation, which needs only position and tangent. The biquintic generalized Hermite patch requires nine pieces of information at each of the four corner points. Interpolating a curve using this representation requires position, first derivative, and second derivative at each of two points, but the tensor product surface form requires position, the two first partials, the three second partials, two of the third partials, and one of the fourth partials. While it is straightforward to work out the details of this method, it becomes terribly tedious. Most applications focus on using the bicubic patch.

To derive the information necessary for the Hermite patch, we analyze the relationship of the Bézier mesh points to the coefficients of the Hermite patch (the derivative information). We showed in Section 13.3.1 that for an arbitrary $m \times n$ Bézier surface $\sigma(u, v)$, (degree m in the u direction and degree n in the v direction) that

$$\begin{aligned}
\sigma(0,0) &= P_{0,0} & \sigma(1,0) &= P_{m,0} \\
\frac{\partial}{\partial u}\sigma(0,0) &= m(P_{1,0} - P_{0,0}) & \frac{\partial}{\partial u}\sigma(1,0) &= m(P_{m,0} - P_{m-1,0}) \\
\frac{\partial}{\partial v}\sigma(0,0) &= n(P_{0,1} - P_{0,0}) & \frac{\partial}{\partial v}\sigma(1,0) &= n(P_{m,1} - P_{m,0}) \\
\sigma(0,1) &= P_{0,n} & \sigma(1,1) &= P_{m,n} \\
\frac{\partial}{\partial u}\sigma(0,1) &= m(P_{1,n} - P_{0,n}) & \frac{\partial}{\partial u}\sigma(1,1) &= m(P_{m,n} - P_{m-1,n}) \\
\frac{\partial}{\partial v}\sigma(0,1) &= n(P_{0,n} - P_{0,n-1}) & \frac{\partial}{\partial v}\sigma(1,1) &= n(P_{m,n} - P_{m,n-1}).
\end{aligned}$$

Thus, given a bicubic Bézier surface, the positions are given by the corner mesh points and the first partial derivatives at each of the four corner points can be evaluated. We also see that given the positions and first partials, the locations of twelve of the sixteen points of the Bézier mesh are determined uniquely. Finally we consider the cross partials of the surface σ for an arbitrary Bézier surface.

$$\begin{aligned}
\frac{\partial^2}{\partial v \partial u}\sigma(0,0) &= \sum_{j=0}^{n}\sum_{i=0}^{m} P_{i,j}\left(\frac{d}{du}\theta_{i,m}(u)\right)\bigg|_{(0,0)}\left(\frac{d}{dv}\theta_{j,n}(v)\right)\bigg|_{(0,0)} \\
&= \sum_{j=0}^{n}\sum_{i=0}^{m} P_{i,j}\theta'_{i,m}(0)\theta'_{j,n}(0) \\
&= \sum_{j=0}^{1}\sum_{i=0}^{1} P_{i,j}\theta'_{i,m}(0)\theta'_{j,n}(0) \\
&= mn\left(P_{1,1} - (P_{1,0} + P_{0,1}) + P_{0,0}\right).
\end{aligned}$$

If the mesh points are known, then it is clear that the cross partial is known. Conversely, if the cross partial is known, and the points $P_{1,0}$, $P_{0,1}$ and $P_{0,0}$ have already been determined from the position and first partial information, it is clear that $P_{1,1}$ can be uniquely determined.

We can show via analogous means that

$$\begin{aligned}
\frac{\partial^2}{\partial v \partial u}\sigma(1,0) &= mn\left[P_{m,1} - (P_{m-1,1} + P_{m,0}) + P_{m-1,0}\right] \\
\frac{\partial^2}{\partial v \partial u}\sigma(0,1) &= mn\left[P_{1,n} - (P_{0,n} + P_{1,n-1}) + P_{0,n-1}\right] \\
\frac{\partial^2}{\partial v \partial u}\sigma(1,1) &= mn\left[P_{m,n} - (P_{m-1,n} + P_{m,n-1}) + P_{m-1,n-1}\right].
\end{aligned}$$

Thus, when $m = n = 3$, if the coefficients are known for either one of the tensor product Hermite surface or the tensor product Bézier surface, the coefficients for the other can be found easily.

Note that a matrix of position values is needed in order to use the Bézier surface method. While the result can be a bicubic, instead of using the derivative information required for the Hermite method, this method uses additional position information. If a degree $n \times m$ Bézier surface is used to approximate a surface $f(u, v)$, then the Bézier surface does not interpolate the derivatives of the function σ at the corners of the square, while the Hermite interpolant patch would. So these are not equivalent approximation methods, in just the same way as Hermite interpolation is not identical to Bernstein approximation.

13.6 Other Polynomial-Based Surface Representations

13.6.1 Simplex Splines

In Section 11.1, B-splines were presented as projections of higher-dimensional simplices projected onto $\mathbb{R}^1$. In this section we extend this idea to surfaces by projecting higher-dimensional simplices onto the plane. We continue with the notation in Section 11.1. Select $n > 0$. Define $\boldsymbol{P} : \mathbb{R}^n \to \mathbb{R}^2$ by $\boldsymbol{P}(t_1, \ldots, t_n) = (t_1, t_2)$, the projection operator on the first two coordinates. Given $\{\boldsymbol{x}_0, \ldots, \boldsymbol{x}_n\} \in \mathbb{R}^2$, not all in a single straight line, choose points P_i in $\mathbb{R}^n$ such that $\boldsymbol{P}[P_i] = \boldsymbol{x}_i$ and the sequence $\{P_0, \ldots, P_n\}$ form the vertices of an arbitrary n-simplex, σ. Note, however that the simplex obtained is only one of the infinite number possible. For the purpose of example, suppose $\boldsymbol{x}_0, \boldsymbol{x}_1, \boldsymbol{x}_2$ are three points in arbitrary position (they form a triangle with positive area). Choose $P_0 = (x_{0,1}, x_{0,2}, 0, \ldots, 0)$, $P_1 = (x_{1,1}, x_{1,2}, 0, \ldots, 0)$, $P_2 = (x_{2,1}, x_{2,2}, 0, \ldots, 0)$. Let $P_i = (x_{i,1}, x_{i,2}, 0, \ldots, \delta_{i,j}, 0, \ldots, 0)$, $i = 3, \ldots, n$, where $\delta_{i,j} = 1$ if $i = j$ and $\delta_{i,j} = 0$ otherwise, i.e., if $n = 3$, $P_0 = (x_{0,1}, x_{0,2}, 0)$, $P_1 = (x_{1,1}, x_{1,2}, 0)$, $P_2 = (x_{2,1}, x_{2,2}, 0)$, $P_3 = (x_{3,1}, x_{3,2}, 1)$. It can be shown that $P_i - P_0$, $i = 1, \ldots, n$, considered as vectors, form a basis for $\mathbb{R}^n$, and thus $\{P_i\}$, $i = 0, \ldots, n$ form the vertices of an n-simplex. For the example where $n = 3$,

$$\begin{aligned}
P_1 - P_0 &= (x_{1,1} - x_{0,1}, x_{1,2} - x_{0,2}, 0) \\
P_2 - P_0 &= (x_{2,1} - x_{0,1}, x_{2,2} - x_{0,2}, 0) \\
P_3 - P_0 &= (x_{3,1} - x_{0,1}, x_{3,2} - x_{0,2}, 1)
\end{aligned}$$

Define $\sigma(u, v) = \{\, P \in \sigma : \boldsymbol{P}(P) = (u, v) \,\}$. Then $\sigma(u, v)$ is an $(n - 2)$-dimensional subset of σ.

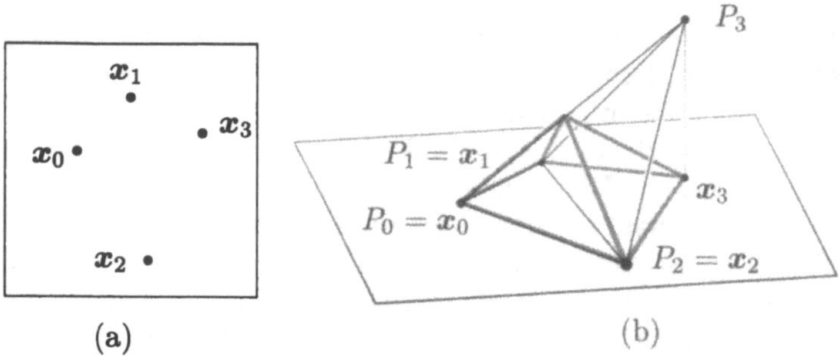

Figure 13.16. (a) x_0, x_1, x_2, x_3 in the plane; (b) their liftings to P_0, P_1, P_2, P_3 (and associated simplex spline in gray).

Set

$$M_{n,2}(u,v) = \frac{vol_{n-2}(\sigma(u,v))}{vol_n(\sigma)}.$$

Clearly $M_{n,2}$ depends on n and $x_0, \ldots, x_n$. (See Figure 13.16.) It is a piecewise polynomial of degree $n-2$. Further, the projections of the edges of the simplex σ into $\mathbb{R}^2$ form the boundaries of each piecewise region. At first glance it appears that it also depends on the particular simplex σ used. It can be shown that this is not the case. That is, $M_{n,2}$ is independent of the choice of σ as long as $\boldsymbol{P}(\{P_0, \ldots, P_n\}) = \{x_0, \ldots, x_n\}$.

Notice that the domain of $M_{n,2}(u,v)$ is the convex hull of the points $\{x_1, x_2, x_2, x_3, \}$, which can look like either of the images in Figure 13.17 (a). Notice that the spline is different on each of its different domains. While it appears that each subdomain forms a triangle, this is true only for $n = 3$. See Figure 13.17 (b) for $n = 5$. While this is an intrinsically multivariate spline, it becomes difficult to determine desirable properties, such as basis independence, and convex hull properties, and is theoretically challenging.

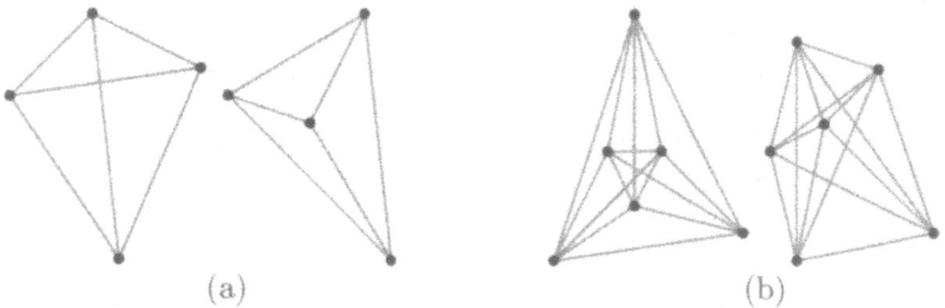

Figure 13.17. (a) Two different domains for $M_{3,2}$; (b) two different domains for $M_{5,2}$.

13.6.2 Box Splines

Instead of lifting the points to form a simplex, another *lifting* can be performed to get a class of intrinsically multivariate, yet easier to analyze surfaces. We present the material for a subclass of the general case and refer readers to [30] for a theoretical and comprehensive treatment. Consider vectors $\boldsymbol{e}_1 = (1,0)$, $\boldsymbol{e}_2 = (0,1)$, and $\boldsymbol{e}_2 = (1,1)$ in the plane. Consider a multiplicity vector $m = (m_1, m_2, m_3)$. Let $n = m_1 + m_2 + m_3$. Let $\mathcal{X} = \{\boldsymbol{x}_i\}_{i=1}^n$ be the collection of vectors that has $\boldsymbol{e}_i$ present m_i times, $i = 0, \ldots, 2$. Let $\mathcal{Y} = \{\boldsymbol{y}_i\}_{i=1}^n \subset \mathbb{R}^n$ be such that $\boldsymbol{P}(\boldsymbol{y}_i) = \boldsymbol{x}_i$, and define

$$\mathcal{B} = \left\{ \sum_{i=1}^{n} \nu_i \boldsymbol{y}_i : 0 \leq \nu_i \leq 1, i = 1, \ldots, n \right\}.$$

$\mathcal{B}$ is just a generalized n-dimensional cube (box). Then

$$M(u, v|\boldsymbol{x}_1, \ldots, \boldsymbol{x}_n) = \frac{vol_{n-2}(\{\boldsymbol{y} \in \mathcal{B} : \boldsymbol{P}(\boldsymbol{y}) = (u,v)\})}{vol_n(\mathcal{B})}$$

is a *box spline.* It is a piecewise degree $(n-2)$ bivariate polynomial with the projections of the edges of the n-dimensional box forming the boundaries of each polynomial piece. Figure 13.18 (a) shows the $\mathbb{R}^3$ parallelopiped and the domain onto which the $m = (1,1,1)$ box spline projects. The box shown in Figure 13.18 (a) can be used to cover $\mathbb{R}^2$, providing a tiling of height one. Thus translations of that single function can be used to provide a collection of blending functions over a locally rectangular grid. The resulting surface will keep the convex hull property since at every (u, v), the blending functions sum to one. Discrete box splines [21] produce algorithms for refining box spline surface meshes and have led to the $C^{(2)}$continuous, $m = (2,2,2)$ box spline being used to support design and rendering of subdivision surfaces, presented in Chapter 20. The $m = (2,2,2)$ box spline is shown in Figure 13.18 (b).

13.6.3 Bézier Triangles

Another type of representation, a generalization of the Bézier representation, is used to represent a surface over a triangular domain. It is defined over the triangular planar domain $\Delta = \{(u,v) : u, v \geq 0,\ u + v \leq 1\}$. Given the three points in the plane, $O = (0,0)$, $P_1 = (1,0)$, and $P_2 = (0,1)$, then $\Delta = \{wO + uP_1 + vP_2 : u, v, w \geq 0,\ u + v + w = 1\}$. Notice that (u, v, w) are the barycentric coordinates (see Section 1.4) of the point in the triangle.

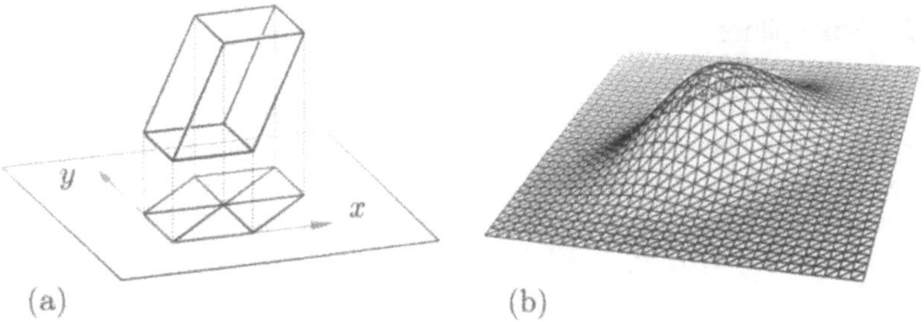

Figure 13.18. (a) $\mathbb{R}^3$ parallelopiped and the domain onto which the $m = (1, 1, 1)$ box spline projects; (b) $m = (2, 2, 2)$ box spline.

Let $n > 0$ be an integer and $\boldsymbol{i} = (i_1, i_2, i_3)$ be a triple of integers, such that $n = |\boldsymbol{i}| = i_1 + i_2 + i_3$. Then for $0 \le u, v, w \le 1$,

$$\begin{aligned}
(u + v + w)^n &= \sum_{i_1=0}^{n} \binom{n}{i_1} u^{i_1} (v + w)^{n-i_1} \\
&= \sum_{i_1=0}^{n} \binom{n}{i_1} u^{i_1} \sum_{i_2=0}^{n-i_1} \binom{n - i_1}{i_2} v^{i_2} w^{n-i_1-i_2} \\
&= \sum_{i_1=0}^{n} \sum_{i_2=0}^{n-i_1} \binom{n}{i_1} \binom{n - i_1}{i_2} u^{i_1} v^{i_2} w^{n-i_1-i_2} \\
&= \sum_{i_1=0}^{n} \sum_{i_2=0}^{n-i_1} \frac{n!}{i_1!(n - i_1)!} \frac{(n - i_1)!}{i_2!(n - i_1 - i_2)!} u^{i_1} v^{i_2} w^{n-i_1-i_2} \\
&= \sum_{|(i_1, i_2, i_3)|=n} \frac{n!}{i_1! i_2! i_3!} u^{i_1} v^{i_2} w^{i_3}.
\end{aligned}$$

Definition 13.20. *Let $n > 0$ be an integer and $\boldsymbol{i} = (i_1, i_2, i_3)$ be a triple of integers, such that $n = |\boldsymbol{i}| = i_1 + i_2 + i_3$. Then for $0 \le u, v, w \le 1$,*

$$\begin{aligned}
\theta_{\boldsymbol{i},n}(u, v, w) &= \binom{n}{\boldsymbol{i}} u^{i_1} v^{i_2} w^{i_3} \\
&= \frac{n!}{i_1! i_2! i_3!} u^{i_1} v^{i_2} w^{i_3},
\end{aligned}$$

where $i_3 = n - i_1 - i_2$, is called a trivariate Bernstein polynomial.

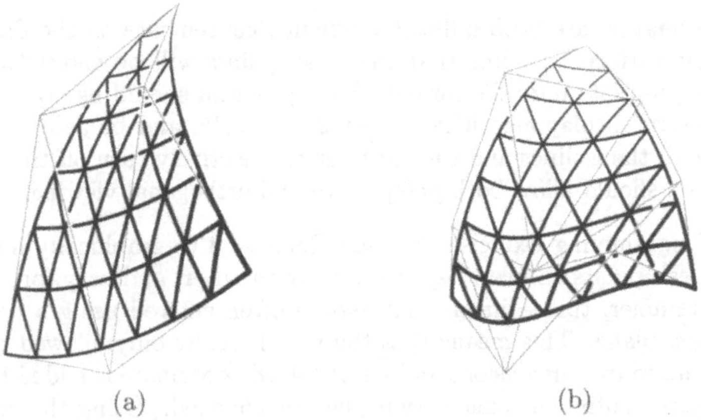

Figure 13.19. (a) Quadratic and (b) cubic triangular patches. The control meshes are shown in thin lines.

Definition 13.21. *For the case,* $0 \le w = 1 - u - v, \;\; u, v \ge 0$, *the domain is restricted to a planar triangle. In that case the surface*

$$\sigma(u,v,w) = \sum_{\boldsymbol{i}=n} P_{\boldsymbol{i}} \theta_{\boldsymbol{i},n}(u,v,w),$$

where $P_{\boldsymbol{i}} \in \mathbb{R}^3$, *is called a* parametric triangular Bézier *surface.*

Subdivision algorithms have been developed to support the use of this representation in geometric modeling similar to those in Algorithm 17.2 for curves and tensor product surfaces. Examples of triangular patches can be seen in Figure 13.19. For a detailed presentation, see [37].

Exercises

1. Find the matrix M^{-1}, where M^t is given in Equation 13.24.
2. Modify Algorithm 13.15 to calculate a grid of function values. Make your algorithm take computational advantage of the tensor product form, as this algorithm does.
3. Convert $\sigma(u,v) = 3u^3v^2 + 2u^2v^2 + uv + 1$ via matrices to both the Bézier and Hermite patch forms.
4. Programming Exercise: Write a program to design tensor product Bézier surfaces. Since entering and displaying three-dimensional in-

formation are both difficult problems, concentrate on the display of the surface. Assume that the mesh points will be entered in some appropriate data file format. Your program should use the isoparametric display algorithms in Example 13.15, and be able to display both the isolines and the mesh, or to selectively turn off the display. You should allow both perspective and orthogonal viewing.

5. Programming Exercise: Expand Exercise 4 to graphically input the mesh points. Since it is intuitive for the user, and easy for the programmer, three-dimensional data is often entered using a construction plane. This means that the user is really only allowed to enter data in two dimensions and a third fixed coordinate is added to every point, until the construction plane is changed. Using this method, one can define a set of mesh rows (or columns) by entering each one as a two-dimensional polygon with fixed z (or x or y) values, in the construction plane. The interaction to edit the mesh is also easier in this direction. One very nice method of interaction with this scheme of input is the two-window display. One window is reserved for the two-dimensional entering or editing a row of the mesh, and the other window provides a three-dimensional view of the mesh, with the standard rotation, translations, scaling, and perspective transformations available to change that view. Your program should

   - choose the degree for each direction of the surface,
   - choose in which direction the mesh will be entered (which direction will initially constrained to be planar).
   - allow for editing of the points.

6. Consider the cubic uniform floating B-spline basis defined with knots at the integers. If $u, v \in [0, 1]$, use a matrix representation to express the surface $\sigma(u, v) = \sum_i \sum_j P_{i,j}\mathcal{B}_{j,3}(v)\mathcal{B}_{i,3}(u)$ in terms of the power basis for u and v. Use the translational property to represent σ for values of u and/or v outside the unit interval.

7. Programming Exercise: Augment the surface geometry program of Chapter 12, Exercise 7 to compute and display geometry of tensor product B-spline surfaces. The splines may be of any order and have arbitrary end conditions. The user should be able to display geometric objects separately or together as he chooses, and should even have the option of toggling the display of the surface and/or control mesh.

14

Fitting Surfaces

As with curves, the surface modeling problem may be considered first as an approximation problem, that is, to fit a surface to predefined data. Within this framework one finds the concepts of interpolation and approximation, the same as with curves. The other approach is to view a surface as an *ab initio* design problem, that is one wants to fit to an idea of a shape. This second approach requires that the specification method be as friendly and intuitive as possible while allowing a wide variety of shape possibilities. The methods might include construction operators for simple shapes, surfaces of revolution, and simple swept surfaces. The types of surfaces one might want to include are, of course, sculptured surfaces, as well as the more traditional boundaries of spheres, ellipsoids, cones, cylinders, pie wedges, etc. We shall approach the problem in an analogous fashion to the curve problem by first studying the problem from the first viewpoint and then adapting those methods to the second viewpoint.

14.1 Interpolating Tensor Product Surfaces

Consider the surface interpolation problem for a tensor product surface h, defined as in Definition 13.1, for which one wants $h(\xi_k) = \phi_k$, $k = 0, \ldots, q$, where $\xi_k = (u_k, v_k)$. What are the restrictions on the collections of functions $\mathcal{F} = \{f_i\}$ and $\mathcal{G} = \{g_j\}$, the parameter values, ξ_k, and the number of data points, $q + 1$, to form a unique interpolant. In this most general form, the linear system $HC = \Phi$ is created where,

$$\begin{aligned} H &= (h_{k,p}); \quad h_{k,(n+1)*i+j} = f_i(u_k)g_j(v_k); \\ C &= \left[\begin{array}{cccccccc} c_{0,0} & \ldots & c_{0,n} & c_{1,0} & \ldots & c_{1,n} & \ldots & c_{m,n} \end{array}\right]^T; \\ \Phi &= \left[\begin{array}{cccc} \phi_0 & \phi_1 & \ldots & \phi_q \end{array}\right]^T. \end{aligned}$$

To be solvable, H must be a square matrix, so $q+1 = (m+1)(n+1)$, and it must have a nonzero determinant, i.e., $|H| \neq 0$. For a given collection of blending functions, there are many possible data sets for which this problem is not uniquely solvable. Furthermore, in this general case, the matrix can be quite large. To interpolate 100 unstructured points, when possible, requires a matrix with dimension 100×100. By imposing a grid structure on the abscissa values that are interpolated, it is possible to take advantage of tensor product surface characteristics.

14.2 Interpolation to a Grid of Data

Suppose $D = U \times V$, where $U, V \subset \mathbf{R}^1$ are intervals, so D is a rectangular domain. Let $u_i \in U$ such that $u_0 < u_1 < \ldots < u_m$, and $v_j \in V$ such that $v_0 < v_1 < \ldots < v_n$. Consider solving the interpolation problem for data values arranged on the grid $\xi_{(n+1)*i+j} = (u_i, v_j)$. In this case we can write

$$\sum_{j=0}^{n}\sum_{i=0}^{m} f_i(u_p)c_{i,j}g_j(v_q) = h_{p,q}$$

for unknowns $c_{i,j}$, for all p, q. Rewriting using matrix notation gives

$$\boldsymbol{H} = \boldsymbol{F}\boldsymbol{C}\boldsymbol{G}^t \tag{14.1}$$

where,

$$\boldsymbol{H} = \begin{bmatrix} h_{0,0} & h_{0,1} & \ldots & h_{0,n-1} & h_{0,n} \\ h_{1,0} & h_{1,1} & \ldots & h_{1,n-1} & h_{1,n} \\ & & \vdots & & \\ h_{m-1,0} & h_{m-1,1} & \ldots & h_{m-1,n-1} & h_{m-1,n} \\ h_{m,0} & h_{m,1} & \ldots & h_{m,n-1} & h_{m,n} \end{bmatrix};$$

$$\boldsymbol{F} = \begin{bmatrix} f_0(u_0) & f_1(u_0) & \ldots & f_{m-1}(u_0) & f_m(u_0) \\ f_0(u_1) & f_1(u_1) & \ldots & f_{m-1}(u_1) & f_m(u_1) \\ & & \vdots & & \\ f_0(u_{m-1}) & f_1(u_{m-1}) & \ldots & f_{m-1}(u_{m-1}) & f_m(u_{m-1}) \\ f_0(u_m) & f_1(u_m) & \ldots & f_{m-1}(u_m) & f_m(u_m) \end{bmatrix};$$

$$G = \begin{bmatrix} g_0(v_0) & g_1(v_0) & \dots & g_{n-1}(v_0) & g_n(v_0) \\ g_0(v_1) & g_1(v_1) & \dots & g_{n-1}(v_1) & g_n(v_1) \\ & & \vdots & & \\ g_0(v_{n-1}) & g_1(v_{n-1}) & \dots & g_{n-1}(v_{n-1}) & g_n(v_{n-1}) \\ g_0(v_n) & g_1(v_n) & \dots & g_{n-1}(v_n) & g_n(v_n) \end{bmatrix}.$$

Solving the interpolation problem can be decomposed into solving two sequential interpolation problems, each of which is much smaller and simpler. Now, even though $\boldsymbol{C}$ is unknown, we can rewrite the equations and set $r_{p,j} = \sum_{i=0}^{m} f_i(u_p)c_{i,j}$ to get the matrix equations $\boldsymbol{R} = \boldsymbol{FC}$, and $\boldsymbol{H} = \boldsymbol{RG}^t$. So, since $\boldsymbol{H}$ and $\boldsymbol{G}$ are known, if $\boldsymbol{G}$ is invertible, $\boldsymbol{R} = \boldsymbol{H}(\boldsymbol{G}^t)^{-1}$. This is equivalent to solving the interpolation problem for $n+1$ different curves. Then, $\boldsymbol{R} = \boldsymbol{FC} = \boldsymbol{H}(\boldsymbol{G}^t)^{-1}$. Thus, it is possible to solve for $\boldsymbol{C}$ as long as $\boldsymbol{F}$ is invertible:

$$\boldsymbol{C} = \boldsymbol{F}^{-1}\boldsymbol{H}(\boldsymbol{G}^t)^{-1}. \tag{14.2}$$

14.2.1 Nodal Interpolation

Suppose $\mathcal{F} = \{\mathcal{B}_{i,\kappa_u}(u)\}_{i=0}^{m}$ where the $\mathcal{B}$ are defined over a prespecified knot vector $\boldsymbol{u} = \{u_i\}_{i=0}^{m+\kappa_u+1}$ and $\mathcal{G} = \{\mathcal{N}_{j,\kappa_v}(u)\}_{i=0}^{n}$ where the $\mathcal{N}$ are defined over a prespecified knot vector $\boldsymbol{v} = \{v_j\}_{j=0}^{n+\kappa_v+1}$. Let $\boldsymbol{u}^*$ and $\boldsymbol{v}^*$ be the $\kappa_u{}^{th}$ and $\kappa_v{}^{th}$ node vectors with respect to $\boldsymbol{u}$ and $\boldsymbol{v}$, respectively, defined by Definition 9.3). Suppose the data to be interpolated is given as $\{(u_i^*, v_j^*, h_{i,j})\}$. By Equation 9.5,

$$\boldsymbol{F} = \boldsymbol{B}_{\kappa_u,\boldsymbol{u}^*} = \begin{bmatrix} \mathcal{B}_{0,\kappa_u}(u_0^*) & \dots & \mathcal{B}_{m,\kappa_u}(u_0^*) \\ & \vdots & \\ \mathcal{B}_{0,\kappa_u}(u_m^*) & \dots & \mathcal{B}_{m,\kappa_u}(u_m^*) \end{bmatrix}; \tag{14.3}$$

$$\boldsymbol{G} = \boldsymbol{N}_{\kappa_v,\boldsymbol{v}^*} = \begin{bmatrix} \mathcal{N}_{0,\kappa_v}(v_0^*) & \dots & \mathcal{N}_{n,\kappa_v}(v_0^*) \\ & \vdots & \\ \mathcal{N}_{0,\kappa_v}(v_n^*) & \dots & \mathcal{N}_{n,\kappa_v}(v_n^*) \end{bmatrix}. \tag{14.4}$$

From the Schoenberg-Whitney theorem (see Theorem 9.2), both $\boldsymbol{F}$ and $\boldsymbol{G}$ are invertible. This type of interpolation is *nodal* surface interpolation.

Example 14.1. Interpolate data $\{(i/m, j/n, h_{i,j})\}_{i=0,j=0}^{m,n}$ with a polynomial surface in the Bernstein-Bézier representation. Since a Bézier surface is a special case of a tensor product B-spline surface, this problem is the surface equivalent of Example 9.4.

$$F = \Theta_{m,*} = \begin{bmatrix} \theta_{0,m}(0) & \cdots & \theta_{m,m}(0) \\ \theta_{0,m}(1/m) & \cdots & \theta_{m,m}(1/m) \\ & \vdots & \\ \theta_{0,m}((m-1)/m) & \cdots & \theta_{m,m}((m-1)/m) \\ \theta_{0,m}(1) & \cdots & \theta_{m,m}(1) \end{bmatrix}$$

and $G = \Theta_{n,*}$ For the case $m = n = 3$, we get

$$F = \begin{bmatrix} 1 & 0 & 0 & 0 \\ 8/27 & 4/9 & 2/9 & 1/27 \\ 1/27 & 2/9 & 4/9 & 8/27 \\ 0 & 0 & 0 & 1 \end{bmatrix} = G.$$

□

14.2.2 Complete Cubic Spline Interpolation

Since the data is again on a grid we seek to use the method of Section 9.1.1 and decompose the problem into solving sequential one-dimensional interpolation problems. For this type of interpolation, knot vectors are not given. The knot vectors are determined from the parameter values of the data. Hence, we assume parametric values $\{u_i\}_{i=0}^m$ in u and $\{v_j\}_{j=0}^n$ in v such that the position data is of the form $\{(u_i, v_j, F_{i,j})\}_{i=0,j=0}^{m,\ n}$ Suppose the interpolant is called $S(u, v)$, with isocurves $\gamma_j(u) = S(u, v_j), j = 0, \ldots, n$ and $\beta_i(v) = S(u_i, v), i = 0, \ldots, m$. As a complete spline interpolant $S(u, v)$ also interpolates tangents. We investigate the isocurves to understand the behavior of S.

γ_j is the interpolant to data $(u_i, F_{i,j})$ and as a complete spline interpolant, it also must interpolate a tangent value at u_0 and u_m. Let $s_{0,j}^u$ and $s_{m,j}^u$ be the tangent values to γ_j at the beginning and end values of u as required for complete spline interpolation, for each j. That is,

$$\begin{aligned} \gamma_j'(u_0) &= s_{0,j}^u & j = 0, \ldots, n, \\ \gamma_j'(u_m) &= s_{m,j}^u & j = 0, \ldots, n. \end{aligned}$$

Similarly, β_i is the interpolant to data $(v_j, F_{i,j})$ with interpolated tangents $s_{i,0}^v$ and $s_{i,n}^v$, so

$$\begin{aligned} \beta_i'(v_0) &= s_{i,0}^v & i = 0, \ldots, m, \\ \beta_i'(v_n) &= s_{i,n}^v & i = 0, \ldots, m. \end{aligned}$$

Now, at parameter values (u_0, v_0),

$$\begin{aligned} S(u_0, v_0) &= \gamma_0(u_0) \\ &= \beta_0(v_0); \\ \tfrac{\partial S}{\partial u}(u_0, v_0) &= \gamma_0'(u_0); \\ \tfrac{\partial S}{\partial v}(u_0, v_0) &= \beta_0'(v_0). \end{aligned}$$

At the other three *corner* points of the parametric interpolation grid, the position and two first partials must also be known.

We consider the grid for which $m = n = 1$. According to Example 9.1, the two knot vectors are $\boldsymbol{\mu} = \{u_0, u_0, u_0, u_0, u_1, u_1, u_1, b_u\}$, where $b_u > u_1$, and $\boldsymbol{\nu} = \{v_0, v_0, v_0, v_0, v_1, v_1, v_1, b_v\}$, where $b_v > v_1$. The two point basis function matrix for complete cubic spline interpolation is, by Example 9.1

$$\boldsymbol{C}_u = \begin{bmatrix} 1 & 0 & 0 & 0 \\ \mathcal{B}'_{0,3,\boldsymbol{\mu}}(u_0) & \mathcal{B}'_{1,3,\boldsymbol{\mu}}(u_0) & 0 & 0 \\ 0 & 0 & \mathcal{B}'_{2,3,\boldsymbol{\mu}}(u_1) & \mathcal{B}'_{1,3,\boldsymbol{\mu}}(u_1) \\ 0 & 0 & 0 & 1 \end{bmatrix}.$$

Call $\boldsymbol{C}_v$ the corresponding matrix for the $\boldsymbol{\nu}$ knot vector. Write a tensor product cubic spline surface in this space,

$$\sigma(u, v) = \sum_{i=0}^{3} \sum_{j=0}^{3} P_{i,j} \mathcal{B}_{i,3,\boldsymbol{\mu}}(u) \mathcal{B}_{j,3,\boldsymbol{\nu}}(v).$$

Now consider the matrix D, where

$$\begin{aligned} D &= \boldsymbol{C}_u \boldsymbol{P} \boldsymbol{C}_v^t \\ &= \begin{bmatrix} A_{1,1} & A_{1,2} \\ A_{2,1} & A_{2,2} \end{bmatrix}, \end{aligned} \tag{14.5}$$

and

$$\begin{aligned} A_{1,1} &= \begin{bmatrix} P_{0,0} & \sum_j P_{0,j} \mathcal{B}'_{j,3,\boldsymbol{\nu}}(v_0) \\ \sum_i P_{i,0} \mathcal{B}'_{i,3,\boldsymbol{\mu}}(u_0) & \sum_{i,j} P_{i,j} \mathcal{B}'_{i,3,\boldsymbol{\mu}}(u_0) \mathcal{B}'_{j,3,\boldsymbol{\nu}}(v_0) \end{bmatrix} \\ A_{1,2} &= \begin{bmatrix} \sum_j P_{0,j} \mathcal{B}'_{j,3,\boldsymbol{\nu}}(v_1) & P_{0,3} \\ \sum_{i,j} P_{i,j} \mathcal{B}'_{i,3,\boldsymbol{\mu}}(u_0) \mathcal{B}'_{j,3,\boldsymbol{\nu}}(v_1) & \sum_i P_{i,3} \mathcal{B}'_{i,3,\boldsymbol{\mu}}(u_0) \end{bmatrix} \\ A_{2,1} &= \begin{bmatrix} \sum_i P_{i,0} \mathcal{B}'_{i,3,\boldsymbol{\mu}}(u_1) & \sum_{i,j} P_{i,j} \mathcal{B}'_{i,3,\boldsymbol{\mu}}(u_1) \mathcal{B}'_{j,3,\boldsymbol{\nu}}(v_0) \\ P_{3,0} & \sum_j P_{3,j} \mathcal{B}'_{j,3,\boldsymbol{\nu}}(v_0) \end{bmatrix} \\ A_{2,2} &= \begin{bmatrix} \sum_{i,j} P_{i,j} \mathcal{B}'_{i,3,\boldsymbol{\mu}}(u_1) \mathcal{B}'_{j,3,\boldsymbol{\nu}}(v_1) & \sum_i P_{i,3} \mathcal{B}'_{i,3,\boldsymbol{\mu}}(u_1) \\ \sum_j P_{3,j} \mathcal{B}'_{j,3,\boldsymbol{\nu}}(v_1) & P_{3,3} \end{bmatrix} \end{aligned}$$

Then,

$$D = \begin{bmatrix} \sigma(u_0, v_0) & \dfrac{\partial \sigma(u_0, v_0)}{\partial v} & \dfrac{\partial \sigma(u_0, v_1)}{\partial v} & \sigma(u_0, v_1) \\ \dfrac{\partial \sigma(u_0, v_0)}{\partial u} & \dfrac{\partial^2 \sigma(u_0, v_0)}{\partial u \partial v} & \dfrac{\partial^2 \sigma(u_0, v_1)}{\partial u \partial v} & \dfrac{\partial \sigma(u_0, v_1)}{\partial u} \\ \dfrac{\partial \sigma(u_1, v_0)}{\partial u} & \dfrac{\partial^2 \sigma(u_1, v_0)}{\partial u \partial v} & \dfrac{\partial^2 \sigma(u_1, v_1)}{\partial u \partial v} & \dfrac{\partial \sigma(u_1, v_1)}{\partial u} \\ \sigma(u_1, v_0) & \dfrac{\partial \sigma(u_1, v_0)}{\partial v} & \dfrac{\partial \sigma(u_1, v_1)}{\partial v} & \sigma(u_1, v_1) \end{bmatrix}$$

Hence, if $\boldsymbol{C}_u$ and $\boldsymbol{C}_v$ are invertible, the coefficient mesh arrived at by solving the above equation gives

$$\boldsymbol{P} = \boldsymbol{C}_u^{-1} D (\boldsymbol{C}_v^t)^{-1}. \tag{14.6}$$

The corresponding surface $\sigma(u, v)$ interpolates four corner cross partials, as well as the four corner position values and the values of the two first partials at each of the corners. Thus, complete cubic spline interpolation for surfaces requires that each corner point have as data:

$$\begin{matrix} F_{i,j} & s^v_{i,j} \\ s^u_{i,j} & s^{u,v}_{i,j} \end{matrix} \qquad i = 0, 1, \text{ and } j = 0, 1.$$

Equation 14.5 is the matrix form of the four corner point complete bi-cubic spline interpolant if $\mathcal{F} = \boldsymbol{D}$ is defined as

$$\mathcal{F} = \begin{bmatrix} F_{0,0} & s^v_{0,0} & s^v_{0,1} & F_{0,1} \\ s^u_{0,0} & s^{uv}_{0,0} & s^{uv}_{0,1} & s^u_{0,1} \\ s^u_{1,0} & s^{uv}_{1,0} & s^{uv}_{1,1} & s^u_{1,1} \\ F_{1,0} & s^v_{1,0} & s^v_{1,1} & F_{1,1} \end{bmatrix}. \tag{14.7}$$

A generalization of Equation 14.5 appropriate for $m, n > 1$ is obtained using the $\boldsymbol{B}$ in Equation 9.4. That is, let $B_i(u_j) = \mathcal{B}_{i,3,\boldsymbol{\mu}}(u_j)$. $\boldsymbol{B}_u = (b_{i,j})$ is an $(m+3) \times (m+3)$ matrix whose entries are all zeros (0) except as follows:

$b_{0,0} = 1$

$b_{1,0} = \mathcal{B}'_0(u_0) \qquad b_{1,1} = \mathcal{B}'_1(u_0)$

$b_{i,i-1} = \mathcal{B}_{i-1}(u_{i-1}) \qquad b_{i,i} = \mathcal{B}_i(u_{i-1}) \qquad b_{i,i+1} = \mathcal{B}_{i+1}(u_{i-1})$ for $i = 2, \ldots, m$

$b_{m+1,m+1} = \mathcal{B}'_{m+1}(u_m) \qquad b_{m+1,m+2} = \mathcal{B}'_{m+2}(u_m)$

$b_{m+2,m+2} = 1$

Let $\boldsymbol{B}_v$ be defined analogously over v. Then,

$$\mathcal{F} = \boldsymbol{B}_u \boldsymbol{C} \boldsymbol{B}_v^t. \tag{14.8}$$

Finally, $\mathcal{F}$ can be written in terms of nine submatrices,

$$\mathcal{F} = \begin{bmatrix} Q_{0,0} & Q_{0,1} & Q_{0,2} \\ Q_{1,0} & Q_{1,1} & Q_{1,2} \\ Q_{2,0} & Q_{2,1} & Q_{2,2} \end{bmatrix}, \tag{14.9}$$

where

$$\begin{aligned} Q_{0,0} &= \begin{bmatrix} F_{0,0} & s^v_{0,0} \\ s^u_{0,0} & s^{uv}_{0,0} \end{bmatrix}, & Q_{0,2} &= \begin{bmatrix} s^v_{0,n} & F_{0,n} \\ s^{uv}_{0,n} & s^u_{0,n} \end{bmatrix}, \\ Q_{2,0} &= \begin{bmatrix} s^u_{m,0} & s^{uv}_{m,0} \\ F_{m,0} & s^v_{m,0} \end{bmatrix}, & Q_{2,2} &= \begin{bmatrix} s^{uv}_{m,n} & s^u_{m,n} \\ s^v_{m,n} & F_{m,n} \end{bmatrix}. \end{aligned} \tag{14.10}$$

The four *edge* submatrices are

$$Q_{0,1} = \begin{bmatrix} F_{0,1} & F_{0,2} & \cdots & F_{0,n-2} & F_{0,n-1} \\ s^u_{0,1} & s^u_{0,2} & \cdots & s^u_{0,n-2} & s^u_{0,n-1} \end{bmatrix},$$

$$Q_{1,0} = \begin{bmatrix} F_{1,0} & s^v_{1,0} \\ F_{2,0} & s^v_{2,0} \\ \vdots & \vdots \\ F_{m-2,0} & s^v_{m-2,0} \\ F_{m-1,0} & s^v_{m-1,0} \end{bmatrix}, \quad Q_{1,2} = \begin{bmatrix} s^v_{1,n} & F_{1,n} \\ s^v_{2,n} & F_{2,n} \\ \vdots & \vdots \\ s^v_{m-2,n} & F_{m-2,n} \\ s^v_{m-1,n} & F_{m-1,n} \end{bmatrix},$$

$$Q_{2,1} = \begin{bmatrix} s^u_{m,1} & s^u_{m,2} & \cdots & s^u_{m,m-2} & s^u_{m,n-1} \\ F_{m,1} & F_{m,2} & \cdots & F_{m,n-2} & F_{m,n-1} \end{bmatrix},$$

and finally,

$$Q_{1,1} = \begin{bmatrix} F_{1,1} & F_{1,2} & \cdots & F_{1,n-2} & F_{1,n-1} \\ F_{2,1} & F_{2,2} & \cdots & F_{2,n-1} & F_{2,n-1} \\ & & \vdots & & \\ F_{m-2,1} & F_{m-2,2} & \cdots & F_{m-2,n-2} & F_{m-2,n-1} \\ F_{m-1,1} & F_{m-1,2} & \cdots & F_{m-1,n-2} & F_{m-1,n-1} \end{bmatrix}.$$

To solve for the coefficient matrix $\boldsymbol{C}$, the control mesh for the interpolating surface, it is necessary only to invert $\boldsymbol{B}_u$ and $\boldsymbol{B}_v^t$ and then $\boldsymbol{C} = \boldsymbol{B}_u^{-1}\boldsymbol{\mathcal{F}}(\boldsymbol{B}_v^t)^{-1}$. It is frequently the case that edge cross-tangents are unknown, and the cross partials, or *twists*, as they are called, are also unknown. However, they must be estimated to use the surface version of complete spline interpolation. There is continuing work in devising good estimation techniques for partials. [79, 8, 9].

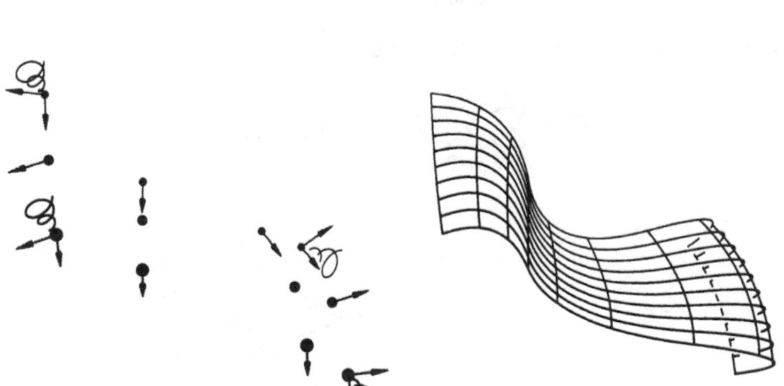

(a) (b)

Figure 14.1. Interpolation. (a) data for complete interpolation; (b) interpolant.

14.3 Surface Approximation

14.3.1 Schoenberg Variation Diminishing Approximation

The Schoenberg variation diminishing spline approximation, presented in Section 9.4 can be generalized to a tensor product approximation form. That is, for f a bivariate function, let $f_{i,j} = f(u^*_{i,\kappa_u}, v^*_{j,\kappa_v})$. The variation diminishing tensor product spline approximation to f is

$$V[f](u,v) = \sum_i \sum_j f_{i,j} \mathcal{B}_{i,\kappa_u}(u) \mathcal{N}_{j,\kappa_v}(v). \tag{14.11}$$

Example 14.2. Bézier Surfaces as Approximations.
Thus,

$$V[F] = \sum_i \sum_j F(i/m, j/n) \theta_{i,m}(u) \theta_{j,n}(v).$$ □

14.3.2 Least Squares Surface Data Fitting

In this section we consider the problem of fitting data of the form

$$\{(u_i, v_j, f_{i,j})\}_{i=0,j=0}^{M,N}$$

with a tensor product B-spline surface of degrees κ_u and κ_v with knot vectors $\boldsymbol{\mu}$ and $\boldsymbol{\nu}$, in u and v respectively. Let the B-splines over $\boldsymbol{\mu}$ be denoted as $\mathcal{B}_k(u) = \mathcal{B}_{k,\kappa_u,\boldsymbol{\mu}}(u)$ and over $\boldsymbol{\nu}$ be denoted as $\mathcal{N}_q(v) = \mathcal{B}_{q,\kappa_v,\boldsymbol{\nu}}$. The the best fit will have the form

$$\sigma(u,v) = \sum_{k=0}^{m} \sum_{q=0}^{n} P_{k,q} \mathcal{B}_k(u) \mathcal{N}_q(v).$$

The error to be minimized is defined as

$$E = \sum_{i=0}^{M} \sum_{j=0}^{N} \left(f_{i,j} - \sigma(u_i, v_j)\right)^2,$$

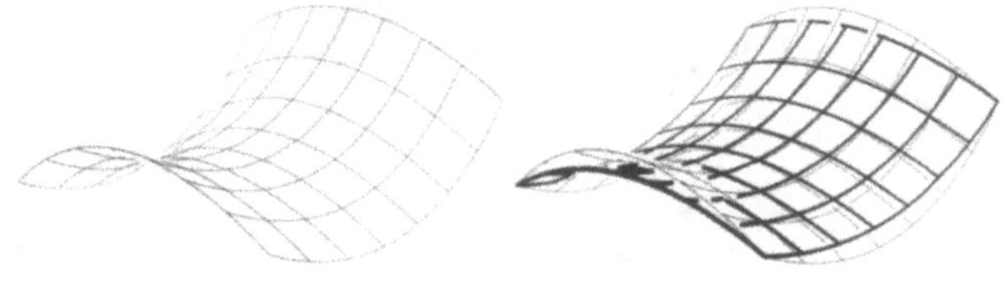

Figure 14.2. A surface and its tensor product Bernstein approximation.

where the $\{P_{k,q}\}$ are unknown. The derivation follows that for curves given in Section 9.3.

For a fixed but arbitrary pair (I, J), differentiating with respect to $P_{I,J}$ yields

$$\begin{aligned}
\frac{\partial E}{\partial P_{I,J}} &= 2\sum_{i=0}^{M}\sum_{j=0}^{N}(f_{i,j} - \sigma(u_i, v_j))\left(-\frac{\partial\sigma(u_i,v_j)}{\partial P_{I,J}}\right) \\
&= 2\sum_{i=0}^{M}\sum_{j=0}^{N}(f_{i,j} - \sigma(u_i, v_j))\,(-\mathcal{B}_I(u_i)\mathcal{N}_J(v_j)) \\
&= -2\sum_{i=0}^{M}\sum_{j=0}^{N} f_{i,j}\mathcal{B}_I(u_i)\mathcal{N}_J(v_j) + 2\sum_{i=0}^{M}\sum_{j=0}^{N}\sigma(u_i, v_j))\mathcal{B}_I(u_i)\mathcal{N}_J(v_j).
\end{aligned}$$

Setting the partials to zero gives

$$\begin{aligned}
&\sum_{i=0}^{M}\sum_{j=0}^{N} f_{i,j}\mathcal{B}_I(u_i)\mathcal{N}_J(v_j) \\
&\quad = \sum_{i=0}^{M}\sum_{j=0}^{N}\sigma(u_i, v_j))\mathcal{B}_I(u_i)\mathcal{N}_J(v_j) \\
&\quad = \sum_{k}\sum_{q}\left(\sum_{i=0}^{M}\mathcal{B}_k(u_i)\mathcal{B}_I(u_i)\right) P_{k,q}\left(\sum_{j=0}^{N}\mathcal{N}_q(v_j)\mathcal{N}_J(v_j)\right).
\end{aligned}$$

Define $\mathcal{L}_u = (\ell_{u,i,j})$ as $\ell_{u,I,k} = \sum_{i=0}^{M}\mathcal{B}_k(u_i)\mathcal{B}_I(u_i)$. Analogously, define $\mathcal{L}_v = (\ell_{v,i,j})$ as $\ell_{u,J,q} = \sum_{j=0}^{N}\mathcal{N}_q(v_j)\mathcal{N}_J(v_j)$. Finally, $\mathcal{F} = (\mathcal{F}_{I,J})$ where $\mathcal{F}_{I,J} = \sum_{i=0}^{M}\sum_{j=0}^{N} f_{i,j}\mathcal{B}_I(u_i)\mathcal{N}_J(v_j)$. Now the least squares system can be written in matrix notation as

$$\mathcal{F} = \mathcal{L}_u P \mathcal{L}_v^t$$

If $\mathcal{L}_u$ and $\mathcal{L}_v^t$ are invertible, the system is uniquely solvable. Note that this requires $M \geq m$ and $N \geq n$. That condition is insufficient, for if there are no data points over the support of any one of the blending functions, the system will have a row and column of zeros in one of the $\mathcal{L}$ matrices, which means that the matrix is not invertible.

This tensor product representation for least squares is valid only for data arranged in a grid.

14.4 Interpolating Arbitrary Boundary Curves: A Linear Operator Approach

We next develop surface interpolation operators that can fill the gaps between curves and give a full surface representation.

Suppose a function $F(u,v)$ is given, and it is desired to interpolate it along the boundaries of the unit square by a simple surface. That is, it is desired that this interpolant reproduce the function $F(u,v)$ along the curves $F(u,0)$, $F(u,1)$, $F(0,v)$, and $F(1,v)$. This problem may be termed *transfinite* interpolation, since it is desired to reproduce not zero-dimensional data (points) but one-dimensional data (curves). As the simplest case to address, remember linear interpolation of two points. That is given points z_0 and z_1, the interpolant would be $(1-u)z_0 + uz_1$. We may view this as a linear operator on any curve $f(u)$ that has $f(0) = z_0$ and $f(1) = z_1$. Using operator notation, define

$$\boldsymbol{P}[f](u) = (1-u)f(0) + uf(1).$$

We can extend this notation to bivariate surfaces by applying $\boldsymbol{P}$ to the function as a curve in u and separately as a curve in v by defining

$$\begin{aligned} \boldsymbol{P_u}[F](u,v) &= (1-u)F(0,v) + uF(1,v), \\ \boldsymbol{P_v}[F](u,v) &= (1-v)F(u,0) + vF(u,1). \end{aligned} \qquad (14.12)$$

What properties does the surface $\boldsymbol{P_u}[F]$ have? $\boldsymbol{P_u}[F](0,v) = F(0,v)$ so $\boldsymbol{P_u}[F]$ reproduces F along $u = 0$. $\boldsymbol{P_u}[F](1,v) = F(1,v)$ and $\boldsymbol{P_u}[F]$ reproduces F along $u = 1$.

Further,

$$\begin{aligned} \boldsymbol{P_u}\big[\boldsymbol{P_u}[F]\big](u,v) &= \boldsymbol{P_u}\big[(1-u)F(0,v) + uF(1,v)\big] \\ &= (1-u)F(0,v) + uF(1,v) \\ &= \boldsymbol{P_u}[F](u,v). \end{aligned}$$

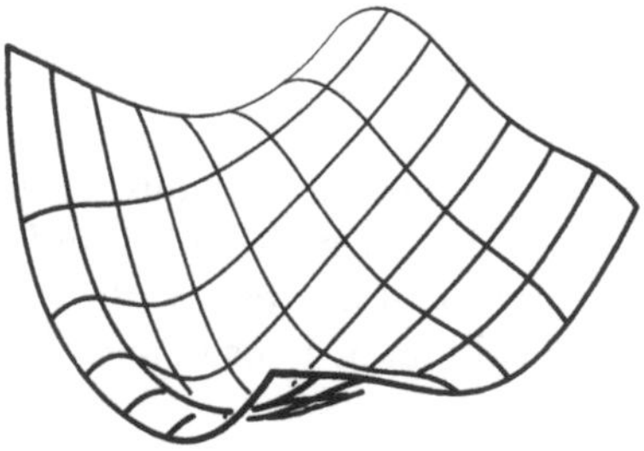

Figure 14.3. An original surface function.

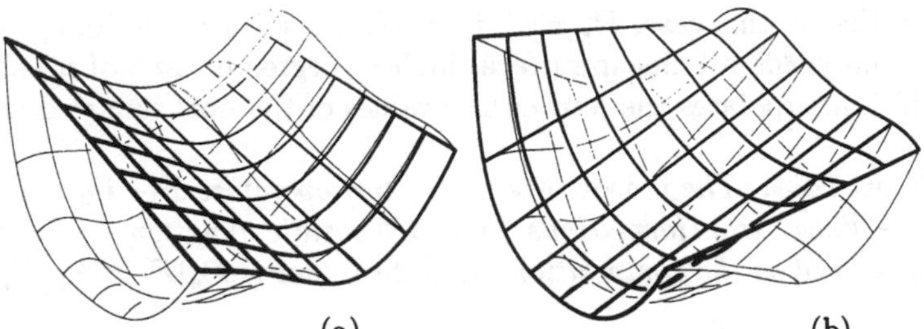

Figure 14.4. (a) $\boldsymbol{P_u}[F]$; (b) $\boldsymbol{P_y}[F]$, for surface of Figure 14.3 (in thin lines).

In general, we may say

Definition 14.3. *An operator* $\boldsymbol{P}$ *is called* idempotent *if* $\boldsymbol{PP}[F] = \boldsymbol{P}[F]$. *That is, it acts as the identity on all functions in its image subset.*

We have shown that linear blending of two functions in the manner of $\boldsymbol{P_u}$ is idempotent. Also, for any fixed value of v, $\boldsymbol{P_u}[F]$ is a straight line connecting $F(0,v)$ with $F(1,v)$. Unfortunately,

$$\boldsymbol{P_u}[F](u,0) = (1-u)F(0,0) + uF(1,0),$$

and

$$\boldsymbol{P_u}[F](u,1) = (1-u)F(0,1) + uF(1,1).$$

Thus, it is clear that $\boldsymbol{P_u}[F]$ does not reproduce $F(u,0)$ nor $F(u,1)$. Analogously, $\boldsymbol{P_v}[F]$ reproduces F along $v = 0$ and $v = 1$, and is idempotent, but does not reproduce the function along $u = 0$ or $u = 1$.

There are now two operators which reproduce distinct parts of the boundary, although neither reproduces all of the boundary. Since one wants to interpolate the whole boundary, it is natural to ask if there is a way to combine these operators in order to effect the desired result.

The composition of two operators P_1 and P_2 is defined as

$$\boldsymbol{P_\bullet}[F] = \boldsymbol{P_1}\big(\boldsymbol{P_2}[F]\big),$$

where $\boldsymbol{P_2}$ is applied first and $\boldsymbol{P_1}$ is applied to its result. When $\boldsymbol{P_1}$ and $\boldsymbol{P_2}$ commute, the order is irrelevant. In this section, we use $\boldsymbol{P_u}$ and $\boldsymbol{P_v}$ as the two operators. Then,

$$\begin{aligned}
\boldsymbol{P_\bullet}[F](u,v) &= \boldsymbol{P_u}\big[(1-v)F(u,0) + vF(u,1)\big] \\
&= (1-u)\big[(1-v)F(0,0) + vF(0,1)\big] \\
&\quad + u\big[(1-v)F(1,0) + vF(1,1)\big].
\end{aligned}$$

One sees that in this case, $\boldsymbol{P}_\bullet[F](i,j) = F(i,j)$, when $i, j \in \{0,1\}$. Thus, composition yields a bilinear surface (highest degree in each of u and v is one) which interpolates the four corner values of the unit square domain.

Definition 14.4. *The* tensor product *of two operators* $\boldsymbol{P}_1$, $\boldsymbol{P}_2 : \boldsymbol{S} \to \boldsymbol{S}$, *which are linear transformations from a bivariate function space,* $\boldsymbol{S}$, *to itself, is defined as the composition of the two, that is* $\boldsymbol{P}_\bullet[F] = \boldsymbol{P}_1\big[\boldsymbol{P}_2[F]\big]$.

This definition bears a formal resemblance to the logical and operation. We shall see that the properties $\boldsymbol{P}_\bullet[F]$ has are the intersection of the properties that $\boldsymbol{P}_1$ and $\boldsymbol{P}_2$ have separately.

Definition 14.5. *The* Boolean sum *of two operators* $\boldsymbol{P}_1$, $\boldsymbol{P}_2 : \boldsymbol{S} \to \boldsymbol{S}$, *which are linear transformations from a function space, S, to itself, is defined as*

$$\boldsymbol{P}_1 \oplus \boldsymbol{P}_2 = \boldsymbol{P}_1 + \boldsymbol{P}_2 - \boldsymbol{P}_1\boldsymbol{P}_2.$$

The similarity between the definition of Boolean sum and the logical or *operation clarifies the choice of name. We shall see that the properties which the Boolean sum has are in some sense the union of the properties of* $\boldsymbol{P}_1$ *and* $\boldsymbol{P}_2$ *separately.*

Forming the Boolean sum of $\boldsymbol{P}_u$ and $\boldsymbol{P}_v$ yields

$$\begin{aligned} \boldsymbol{P}_\oplus[F] &= \boldsymbol{P}_u \oplus \boldsymbol{P}_v[F] \\ &= (1-u)F(0,v) + uF(1,v) + (1-v)F(u,0) + vF(u,1) \\ &\quad - (1-u)\big[(1-v)F(0,0) + vF(0,1)\big] \\ &\quad - u\big[(1-v)F(1,0) + vF(1,1)\big]. \end{aligned}$$

It is straightforward to show that

$$\begin{aligned} \boldsymbol{P}_\oplus[F](u,0) &= F(u,0) \\ \boldsymbol{P}_\oplus[F](u,1) &= F(u,1) \\ \boldsymbol{P}_\oplus[F](0,v) &= F(0,v), \quad \text{and} \\ \boldsymbol{P}_\oplus[F](1,v) &= F(1,v). \end{aligned}$$

Thus the Boolean sum has eight terms and interpolates all four sides of the patch. It is *bilinearly blended* and is called the *bilinearly blended Coons patch.*

When using linear operators, tensor products and Boolean sums for approximation and for new operator building, the next two theorems are helpful.

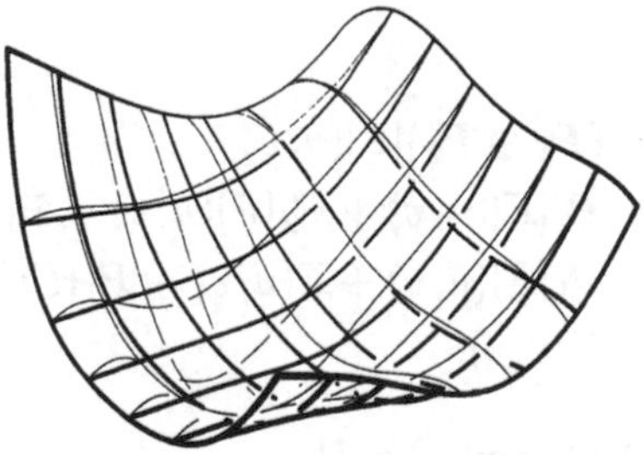

Figure 14.5. $(\boldsymbol{P_u} \oplus \boldsymbol{P_v})[F]$ for surface of Figure 14.3 (in thin lines).

Theorem 14.6. The Remainder Theorem. *Suppose $\boldsymbol{P_1}$ and $\boldsymbol{P_2}$ are linear operators. Then,*

- *The error of the Boolean sum is the tensor product of the error operator.*
- *The error of the tensor product is the Boolean sum of the error operator.*

Proof: Consider the function F on which $\boldsymbol{P_1}$ and $\boldsymbol{P_2}$ operate. Now, $\boldsymbol{I}$, the identity operator is an operator also, which means that $\boldsymbol{R_i}[F] = \mathtt{I}[F] - \boldsymbol{P_i}[F] = F - \boldsymbol{P_i}[F]$, $i = 1, 2$, and $\boldsymbol{R_i}[F]$ are the errors between F and its approximation by the respective operator.

$$\begin{aligned} \boldsymbol{R}_{12} &= \mathtt{I} - (\boldsymbol{P_1} \oplus \boldsymbol{P_2}) \\ &= (\mathtt{I} - \boldsymbol{P_1} - \boldsymbol{P_2} + \boldsymbol{P_1}\boldsymbol{P_2}) \\ &= (\mathtt{I} - \boldsymbol{P_1})(\mathtt{I} - \boldsymbol{P_2}) \\ &= \boldsymbol{R_1}\boldsymbol{R_2}. \end{aligned}$$

The second part of the proof follows from analogous arguments and is left for the reader. ■

The next theorem is due to Barnhill and Gregory [4].

Theorem 14.7. *Suppose $\boldsymbol{P_1}$ and $\boldsymbol{P_2}$ are linear operators, and D is the domain of $\boldsymbol{P_1} \oplus \boldsymbol{P_2}$. Let*

$$\begin{aligned} Q_1 &= \{\, F : \boldsymbol{P_2}[F] \equiv F \,\} \quad \textit{and} \\ Q_2 &= \{\, (u,v) \in S \text{ (unit square)} : \boldsymbol{P_1}[G](u,v) = G(u,v) \text{ for all } G \in D \,\}. \end{aligned}$$

If $\boldsymbol{P} = \boldsymbol{P_1} \oplus \boldsymbol{P_2}$, $\boldsymbol{P}[F] \equiv F$, for all $F \in Q_1$, and $\boldsymbol{P}[G](u,v) = G(u,v)$ for all $(u,v) \in Q_2$, and all $G \in D$.

Proof: For $F \in Q_1$,

$$\begin{aligned} \boldsymbol{P}[F](u,v) &= (\boldsymbol{P_1} \oplus \boldsymbol{P_2})[F](u,v) \\ &= \boldsymbol{P_1}[F](u,v) + \boldsymbol{P_2}[F](u,v) - \boldsymbol{P_1}\left[\boldsymbol{P_2}[F]\right](u,v) \\ &= \boldsymbol{P_1}[F](u,v) + F(u,v) - \boldsymbol{P_1}[F](u,v) \\ &= F(u,v), \end{aligned}$$

as desired. For $(u,v) \in Q_2$ and $G \in D$,

$$\begin{aligned} \boldsymbol{P}[G](u,v) &= (\boldsymbol{P_1} \oplus \boldsymbol{P_2})[G](u,v) \\ &= \boldsymbol{P_1}[G](u,v) + \boldsymbol{P_2}[G](u,v) - \boldsymbol{P_1}\left[\boldsymbol{P_2}[G]\right](u,v) \\ &= G(u,v) + \boldsymbol{P_2}[G](u,v) - \left[\boldsymbol{P_2}[G]\right](u,v) \\ &= G(u,v). \end{aligned}$$

■

These conditions are not the same. The first condition provides a result about operators reproducing *functions*. In fact it helps us identify which functions can be reproduced exactly by the Boolean sum of two simpler operators. The second condition tells us that the operator will reproduce all functions in its domain at specific subsets of points, but may not reproduce any single function everywhere. When $\boldsymbol{P_1}$ and $\boldsymbol{P_2}$ commute under tensor product, and hence under Boolean sum, the resulting Boolean sum reproduces the set of functions which is the union of what each operator can reproduce separately. Similarly, the set of points at which all functions will be reproduced is the union of the separate sets of points for each operator. This result clearly shows that the theorem merely provides inclusion; it does not specify the whole extent.

Example 14.8. Set $\boldsymbol{P_1} = \boldsymbol{P_u}$ and $\boldsymbol{P_2} = \boldsymbol{P_v}$, where $\boldsymbol{P_u}$ and $\boldsymbol{P_v}$ are defined as in Equation 14.12. Then

$$\begin{aligned} Q_1 &= \{\, f(u,v) : f(u,v) = au + bv + cuv + d\,\}; \\ Q_2 &= \{(0,0),(0,1),(1,0),(1,1)\}. \end{aligned}$$

□

14.4.1 Coons Patches

The next derivation uses the cubic Hermite blending functions derived in Section 10.1.1. Define

$$\begin{aligned} \boldsymbol{H_u}[F] &= h_{0,0}(u)F(0,v) + h_{0,1}(u)\tfrac{\partial F}{\partial u}(0,v) + h_{1,0}(u)F(1,v) + h_{1,1}(u)\tfrac{\partial F}{\partial u}(1,v); \\ \boldsymbol{H_v}[F] &= h_{0,0}(v)F(u,0) + h_{0,1}(v)\tfrac{\partial F}{\partial v}(u,0) + h_{1,0}(v)F(u,1) + h_{1,1}(v)\tfrac{\partial F}{\partial v}(u,1). \end{aligned}$$

The expression $\frac{\partial^{i+j}}{\partial v^i \partial u^j} F(a,b)$, means to take the jth derivative of $F(u,v)$ with respect to u, then take the ith derivative of $F_{j,0}(a,v)$ with respect to v and then to evaluate at $(u,v) = (a,b)$.

Now,

$$\begin{aligned} \boldsymbol{H_u}[F](0,v) &= h_{0,0}(0)F(0,v) + h_{0,1}(0)F_u(0,v) \\ &\quad + h_{1,0}(0)F(1,v) + h_{1,1}(0)F_u(1,v) \\ &= F(0,v) \end{aligned}$$

and

$$\begin{aligned} \frac{\partial}{\partial u}\boldsymbol{H_u}[F](0,v) &= h'_{0,0}(0)F(0,v) + h'_{0,1}(0)F_u(0,v) \\ &\quad + h'_{1,0}(0)F(1,v) + h'_{1,1}(0)F_u(1,v) \\ &= F_u(0,v). \end{aligned}$$

These two calcuations show that $\boldsymbol{H_u}$ interpolates the surface F and its partial in the u direction along the boundary $(0,v)$. Analogous calculations will show that $\boldsymbol{H_u}$ also interpolates the surface F and its partial in the u direction along the boundary $(1,v)$ and that $\boldsymbol{H_v}$ interpolates the surface and its partial in the v direction along the boundaries $(u,0)$ and $(u,1)$.

Along the constant variable line, $v = 0$,

$$\boldsymbol{H_u}[F](u,0) = h_{0,0}(u)F(0,0) + h_{0,1}(u)F_u(0,0) + h_{1,0}(u)F(1,0) + h_{1,1}(u)F_u(1,0)$$

and $\boldsymbol{H_u}$ is a cubic function in u which interpolates position and u partial at both $F(0,0)$ and $F(1,0)$. If F is a cubic or lower degree polynomial along $(u,0)$, that property means that $\boldsymbol{H_u}$ reproduces F along that curve line. Otherwise, F is not reproduced. Analogously, $\boldsymbol{H_u}(u,1)$ is a cubic function in u which interpolates the u partial at both $F(0,1)$ and $F(1,1)$, but generally does not reproduce F along the rest of the $(u,1)$ boundary. Thus, $\boldsymbol{H_u}$, in general, interpolates the surface F along two boundaries which are opposite to each other.

It can also be shown in an analogous manner that $\boldsymbol{H_v}[F](0,v)$ and $\boldsymbol{H_u}[F](1,v)$ are cubic functions in v, and so cannot interpolate F unless F is also cubic along the appropriate boundaries. The result is that

Theorem 14.9. *For functions $F \in C^{(1)}[S]$, where S is the unit square:*

- $\boldsymbol{H_u}$ *interpolates the boundary curves and normal derivatives over $(0,v)$ and $(1,v)$. Additionally, if $F(u,0)$ and $F(u,1)$ are cubic functions,* $\boldsymbol{H_u}$ *interpolates those boundary curves as well.*
- $\boldsymbol{H_v}$ *interpolates the boundary curves and normal derivatives over $(u,0)$ and $(u,1)$. Additionally, if $F(0,v)$ and $F(1,v)$ are cubic functions,* $\boldsymbol{H_v}$ *interpolates those boundary curves as well.*

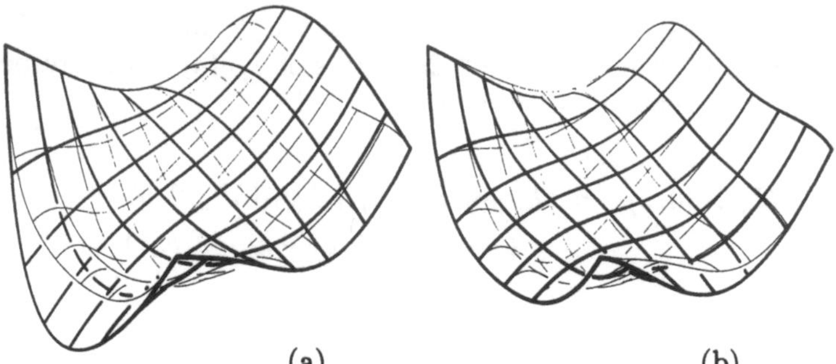

Figure 14.6. (a) $\boldsymbol{H_u}[F]$; (b) $\boldsymbol{H_v}[F]$ for the surface in Figure 14.3 (in thin lines).

- *If in addition, F is a cubic along the four boundaries of the unit square, both* $\boldsymbol{H_u}[F]$ *and* $\boldsymbol{H_v}[F]$ *reproduce F along those four boundary curves.*

Here we derive the tensor product operator and find its attributes. Henceforth the assumption is made that $\frac{\partial F}{\partial u}$ and $\frac{\partial F}{\partial v}$ exist at all the appropriate points. It is further assumed that the cross derivatives, $\frac{\partial^2 F}{\partial u \partial v}$ and $\frac{\partial^2 F}{\partial v \partial u}$ also exist.

$$\begin{aligned}\boldsymbol{H_u H_v}[F] &= \sum_{i=0}^{1} h_{i,0}(u)\left(\sum_{j=0}^{1} h_{j,0}(v)F(i,j) + \sum_{j=0}^{1} h_{j,1}(v)\tfrac{\partial F}{\partial v}(i,j)\right)\\ &+ \sum_{i=0}^{1} h_{i,1}(u)\left(\sum_{j=0}^{1} h_{j,0}(v)\tfrac{\partial F}{\partial u}(i,j) + \sum_{j=0}^{1} h_{j,1}(v)\frac{\partial^2 F}{\partial u \partial v}(i,j)\right)\end{aligned}$$

$$\begin{aligned}\boldsymbol{H_u H_v}[F] &= \sum_{i=0}^{1}\sum_{j=0}^{1} h_{i,0}(u)h_{j,0}(v)F(i,j) + \sum_{i=0}^{1}\sum_{j=0}^{1} h_{i,0}(u)h_{j,1}(v)\tfrac{\partial F}{\partial v}(i,j)\\ &+ \sum_{i=0}^{1}\sum_{j=0}^{1} h_{i,1}(u)h_{j,0}(v)\tfrac{\partial F}{\partial u}(i,j) + \sum_{i=0}^{1}\sum_{j=0}^{1} h_{i,1}(u)h_{j,1}(v)\frac{\partial^2 F}{\partial u \partial v}(i,j).\end{aligned} \tag{14.13}$$

If the assumption is made that $\frac{\partial^2 F}{\partial u \partial v} = \frac{\partial^2 F}{\partial v \partial u}$, then $\boldsymbol{H_u H_v}[F] = \boldsymbol{H_v H_u}[F]$. Equation 14.13 clearly shows that $\boldsymbol{H_u H_v}$ is a cubic in each coordinate direction for all fixed values of the other variable.

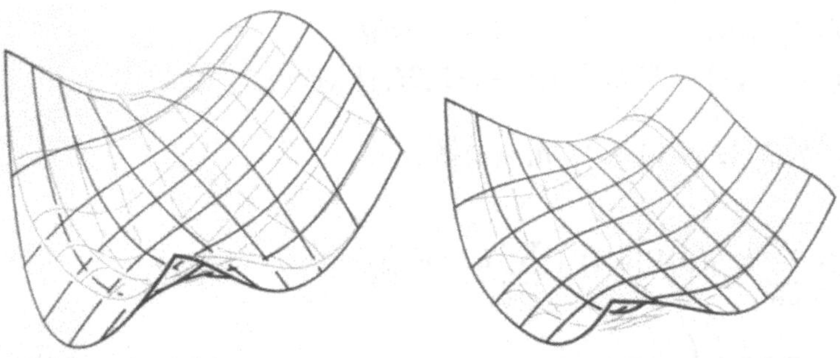

Figure 14.7. (a) $\boldsymbol{H_u H_v}$; (b) $\boldsymbol{H_u} \oplus \boldsymbol{H_v}$, both for the surface in Figure 14.3 (in thin lines).

Setting $\beta(u,v) = \boldsymbol{H_u H_v}[F](u,v)$ and substituting in the definitions, it can be shown that

$$\begin{aligned}
\beta(0,0) &= F(0,0), & \beta(1,0) &= F(1,0),\\
\beta(0,1) &= F(0,1), & \beta(1,1) &= F(1,1)\\
\tfrac{\partial \beta}{\partial u}(0,0) &= \tfrac{\partial F}{\partial u}(0,0), & \tfrac{\partial \beta}{\partial u}(1,0) &= \tfrac{\partial F}{\partial u}(1,0),\\
\tfrac{\partial \beta}{\partial u}(0,1) &= \tfrac{\partial F}{\partial u}(0,1), & \tfrac{\partial \beta}{\partial u}(1,1) &= \tfrac{\partial F}{\partial u}(1,1)\\
\tfrac{\partial \beta}{\partial v}(0,0) &= \tfrac{\partial F}{\partial v}(0,0), & \tfrac{\partial \beta}{\partial v}(1,0) &= \tfrac{\partial F}{\partial v}(1,0),\\
\tfrac{\partial \beta}{\partial v}(0,1) &= \tfrac{\partial F}{\partial v}(0,1), & \tfrac{\partial \beta}{\partial v}(1,1) &= \tfrac{\partial F}{\partial v}(1,1)\\
\tfrac{\partial^2 \beta}{\partial u \partial v}(0,0) &= \tfrac{\partial^2 F}{\partial u \partial v}(0,0), & \tfrac{\partial^2 \beta}{\partial u \partial v}(1,0) &= \tfrac{\partial^2 F}{\partial u \partial v}(1,0),\\
\tfrac{\partial^2 \beta}{\partial u \partial v}(0,1) &= \tfrac{\partial^2 F}{\partial u \partial v}(0,1), & \tfrac{\partial^2 \beta}{\partial u \partial v}(1,1) &= \tfrac{\partial^2 F}{\partial u \partial v}(1,1).
\end{aligned}$$

Thus, sixteen independent values from the surface and its derivative surfaces determine the tensor product surface, which interpolates them all. It is a bicubic surface (cubic in each direction as a function of the appropriate variable). The tensor product, however, does not reproduce the function F unless F is a bicubic surface. Since a bicubic has precisely sixteen degrees of freedom, it is reproduced exactly by β.

Let us form the Boolean sum of $\boldsymbol{H_u}$ and $\boldsymbol{H_v}$ to see if it interpolates any surface at its boundaries.

$$
\begin{aligned}
&\boldsymbol{H_u} \oplus \boldsymbol{H_v}[F](u,v) \\
&\quad = \boldsymbol{H_u}[F](u,v) + \boldsymbol{H_v}[F](u,v) - \boldsymbol{H_u}\boldsymbol{H_v}[F](u,v) \\
&\quad = \sum_{i=0}^{1} h_{i,0}(u)F(i,v) + \sum_{i=0}^{1} h_{i,1}(u)\tfrac{\partial F}{\partial u}(i,v) \\
&\qquad + \sum_{i=0}^{1} h_{i,0}(v)F(u,i) + \sum_{i=0}^{1} h_{i,1}(v)\tfrac{\partial F}{\partial v}(u,i) \\
&\qquad - \sum_{i=0}^{1}\sum_{j=0}^{1}\Bigg(h_{i,0}(u)h_{j,0}(v)F(i,j) + h_{i,0}(u)h_{j,1}(v)\tfrac{\partial F}{\partial v}(i,j) \\
&\qquad\qquad + h_{i,1}(u)h_{j,0}(v)\tfrac{\partial F}{\partial u}(i,j) + h_{i,1}(u)h_{j,1}(v)\frac{\partial^2 F}{\partial u \partial v}(i,j)\Bigg).
\end{aligned}
$$

Once again, if $\frac{\partial^2 F}{\partial u \partial v}(i,j) = \frac{\partial^2 F}{\partial v \partial u}(i,j)$, for $i,j = 0,1$, then $\boldsymbol{H_u} \oplus \boldsymbol{H_v} = \boldsymbol{H_v} \oplus \boldsymbol{H_u}$, otherwise the Boolean sum is not commutative.

Theorem 14.10. *For surfaces F defined on the unit square such that $\frac{\partial F}{\partial u}$ and $\frac{\partial F}{\partial v}$ exist along the boundaries and $\frac{\partial^2 F}{\partial u \partial v}$ and $\frac{\partial^2 F}{\partial v \partial u}$ exist and are equal at the four corner points, $\boldsymbol{H_u} \oplus \boldsymbol{H_v}$ interpolates all four boundary curves and their cross boundary derivatives.*

Proof: Consider $\boldsymbol{H_u} \oplus \boldsymbol{H_v}[F]$. Since $\boldsymbol{H_u}$ interpolates position and partial in the u direction along $(0,v)$ and $(1,v)$, by Theorem 14.7 $\boldsymbol{H_u} \oplus \boldsymbol{H_v}[F]$ also interpolates position and partial in the u direction along $(0,v)$ and $(1,v)$. Analogously, for $\boldsymbol{H_v} \oplus \boldsymbol{H_u}[F]$. Since $\boldsymbol{H_v}$ interpolates position and partial in the v direction along $(u,0)$ and $(u,1)$, by Theorem 14.7 $\boldsymbol{H_v} \oplus \boldsymbol{H_u}[F]$ also interpolates position and partial in the v direction along $(u,0)$ and $(u,1)$. $\boldsymbol{H_v}$ and $\boldsymbol{H_u}$ commute for this F under composition since the mixed partials are equal, which gives $\boldsymbol{H_u} \oplus \boldsymbol{H_v}[F] = \boldsymbol{H_v} \oplus \boldsymbol{H_u}[F]$, and the interpolation properties are then the union of the individual interpolation properties. ■

14.4.2 Using the Tensor Product Patch

To use the Coons patch in its full generality requires the specification of eight curves and four additional values. The eight curves are $\sigma(u,0)$, $\sigma(u,1)$, $\sigma(0,v)$, $\sigma(1,v)$, $\sigma_v(u,0)$, $\sigma_v(u,1)$, $\sigma_u(0,v)$, and $\sigma_u(1,v)$. The four additional points are $\sigma_{uv}(0,0)$, $\sigma_{uv}(0,1)$, $\sigma_{uv}(1,0)$ and $\sigma_{uv}(1,1)$. The four cross-partials are called the *twist vectors*. Intuitively one can discuss this value as a measure of how the boundary changes its normal directions at

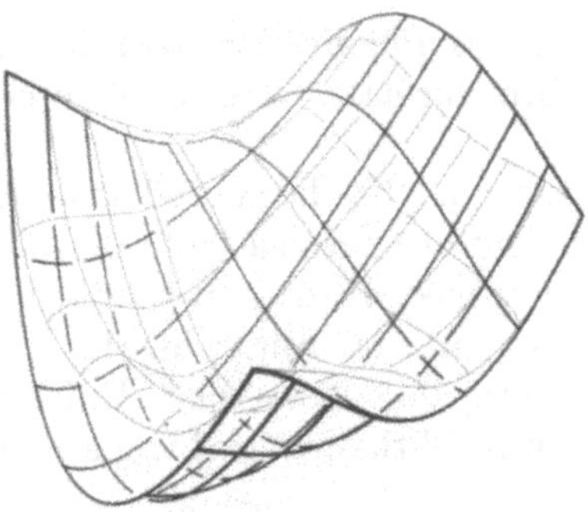

Figure 14.8. Hermite patch with twists specified as zero for the surface in Figure 14.3 (in thin lines).

the corner points. The four position curves are easy to understand, and the four normal curves (perpendicular in parametric space to the direction of motion) specify the tangency conditions to the boundary that the interpolating surface will have.

In practice, often the simplified tensor product scheme is used. We have seen that it is identical to the tensor product Hermite patch. This method requires four vector quantities at each of the four corner points of the unit square; in all, sixteen values are needed. Thus, full curve information is not needed, and the method is intrinsically simpler. When the B-spline basis is used for this surface, it is the complete interpolant over the four corner points. Further, since the resulting surface is a bicubic surface, it has all the properties of polynomials. The Boolean sum surface is bicubically blended, but is not a bicubic surface unless S is.

Theorem 14.11. *If the function S is a bicubic function, then* $\boldsymbol{H_u} \oplus \boldsymbol{H_v}[S] = \boldsymbol{H_u}\boldsymbol{H_v}[S] = S$.

Since the role of the twist vectors and reasonable values for them are not intuitive, one of the earliest procedures was to set all four of them equal to zero. The effect of this was to produce local flats in the surface around the four corner points, which was usually totally undesirable. Later work has focused on developing estimating programs to first estimate values for the twists, and to apply then either the Boolean sum Coons method or the tensor product Coons method.

Finally, the partial derivatives specified are all with respect to the parameters u and v. Thus, the four corner positions can be anywhere in space. When absolute distance is small, the same partials give a much different surface than when the absolute distance is large.

The Coons patch was the pioneering work in the area of parametric representation of freeform surfaces. The tensor product scheme achieved wide attention.

The tensor product scheme has been used as a design scheme. The designer must specify the four values at each of the four parametric locations. The bicubic surface is returned. If the user wants to change a point's position, frequently the other three quantities for that parametric value might be changed as well. Figure 14.8 shows why. Unfortunately, many people have found the specification of the required values unintuitive and difficult.

Exercises

1. Complete the proof of Theorem 14.6.

2. Show that the operator $\boldsymbol{H_u}$ interpolates the surface F and its partial in the u direction along the boundary $(1, v)$. Show that the operator $\boldsymbol{H_v}$ interpolates the surface F and its partial in the v direction along the boundaries $(u, 0)$ and $(u, 1)$.

3. If the position, two partial derivatives, and twist vector are specified at each of the four corner points of a Bézier surface, $\sigma(u, v)$, find the sixteen corner control points as functions of them.

15

Modeling with B-Spline Surfaces

As seen in the last chapter, the tensor product spline surface formalism is no more complicated than tensor product Bézier surfaces. Just as the Bézier curve is a special case of the B-spline curve, the tensor product Bézier surface is a special case of the tensor product B-spline surface. In this chapter, we investigate modeling with tensor product B-spline surfaces, always noting the case for Bézier where applicable. We show both B-spline and Bézier surfaces over the same surface mesh as one means of comparison.

The tensor product parametrizes a surface in an inherently *rectangular* way, which is not intrinsic to some surfaces, such as a hemisphere. We show typical ways to *bend* the parametric formulation to match the desired shape, and also study the effects of these parametrizations on the geometry.

15.1 Generalized Cylinder and Extrusion

We shall first determine a mesh for a straightforward case. We shall use the generalized cylinder. The surface will only be piecewise parabolic, since we will not determine the rational coefficients. However, the style would be the same for that case.

A cylinder is a circle "crossed" with a straight line, so we use a 3×1 B-spline surface which is periodic in the cubic direction and open in the linear direction. Hence $\kappa_u = 3$ and $\kappa_v = 1$. In particular, let the u-knot vector be $\boldsymbol{u} = \{0, 1, 2, 3, 4, 5, 6\}$ and the v knot vector be $\boldsymbol{v} = \{0, 0, 1, 1\}$. Remember that for the periodic case this means that there are six piecewise cubic segments and that "modulo" knot arithmetic identifies the knot values u_6

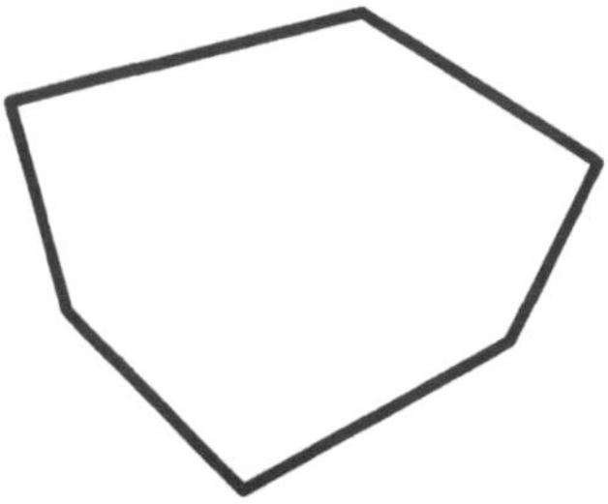

Figure 15.1. The S polygon.

and u_0. Without loss of generality, consider the points $S_i = (x_i, y_i)$, $i = 0, \ldots, 5$, which lie in the $x - y$ plane and when connected in order, with S_5 connected to S_0, form a closed polygon. The cross of a curve with a line is also known as an *extrusion*. (See Figure 15.1.)

If it is necessary to have this figure extending from $z = 0$ to $z = 5$, then form the $\boldsymbol{P}$ control mesh such that

$$\begin{aligned} P_{i,0} &= (x_i, y_i, 0), \qquad i = 0, \ldots, 5, \qquad \text{and} \\ P_{i,1} &= (x_i, y_i, 5), \qquad i = 0, \ldots, 5. \end{aligned}$$

The resulting surface is shown in Figure 15.2.

It is possible to obtain other shapes from this basic shape since this surface has only one linear piece in v. For example, using the results in Theorem 5.31, it is possible to raise the degree of the surface in the v direction (see Section 5.6) and manipulate vertices of this new surface to obtain other shapes. One can keep the beginning and ending curves the same by changing just the interior control vertices.

We have already seen that a curve with an open knot vector "crossed" or *tensored* with a curve having a periodic knot vector creates a surface shaped as a generalized cylinder, since each cross section is topologically equivalent to a circle and each longitude is topologically equivalent to a line.

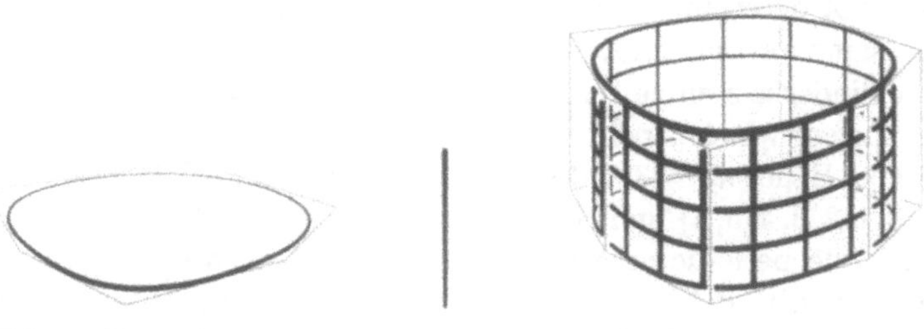

Figure 15.2. Generalized cylindrical surface with the cross section from Figure 15.1.

By analogy, the possible surfaces which result from the tensor product of two curves with periodic knot vectors are all topologically equivalent to a torus. Remember that since circles, ellipses, and hyperbolas can all be represented exactly as rational quadratics, it is possible to obtain an exact representation for cylinders and generalized cylinders with elliptical cross sections. One must determine only the rational component to solve this problem correctly.

15.2 Ruled Surfaces

Suppose $\gamma_0(u) = \sum P_{i,0}\mathcal{B}_{i,\kappa_u}(u)$ and $\gamma_1(u) = \sum P_{i,1}\mathcal{B}_{i,\kappa_u}(u)$ are two parametric spline curves of the same degree κ_u over the same knot vector $\boldsymbol{u}$. For each u in the shared domain of γ_0 and γ_1, consider the line segment from $\gamma_0(u)$ to $\gamma_1(u)$,

$$\sigma(u,v) = (1-v)\gamma_0(u) + v\gamma_1(u). \tag{15.1}$$

For $\boldsymbol{v} = \{0,0,1,1\}$ the first degree B-splines are $\mathcal{N}_{0,1}(v) = (1-v)$ and $\mathcal{N}_{1,1}(v) = v$. Then Equation 15.1 can be rewritten as

$$\begin{aligned} \sigma^u(u,v) &= \mathcal{N}_{0,1}(v)\sum_i P_{i,0}\mathcal{B}_{i,k}(u) + \mathcal{N}_{1,1}(v)\sum_i P_{i,1}\mathcal{B}_{i,k}(u) \\ &= \sum_{j=0}^{1}\sum_i P_{i,j}\mathcal{B}_{i,k}(u)\mathcal{N}_{j,1}(v). \end{aligned} \tag{15.2}$$

Analogously, given two spline curves $\beta_0(v) = \sum_j Q_{0,j}\mathcal{B}_{j,\kappa_v}(v)$ and $\beta_1(v) = \sum_j Q_{1,j}\mathcal{B}_{j,\kappa_v}(v)$, of the same degree κ_v over the same knot vector $\boldsymbol{v}$, let $\boldsymbol{u} = \{0,0,1,1\}$ define B-splines $\mathcal{N}_{0,1}(u) = (1-u)$ and $\mathcal{N}_{1,1}(u) = u$. Then

$$\sigma^v(u,v) = \sum_j \sum_{i=0}^{1} Q_{i,j}\mathcal{N}_{i,1}(u)\mathcal{B}_{j,\kappa_v}(v).$$

This type of surface is call a *ruled surface.* (See Figure 15.3.)

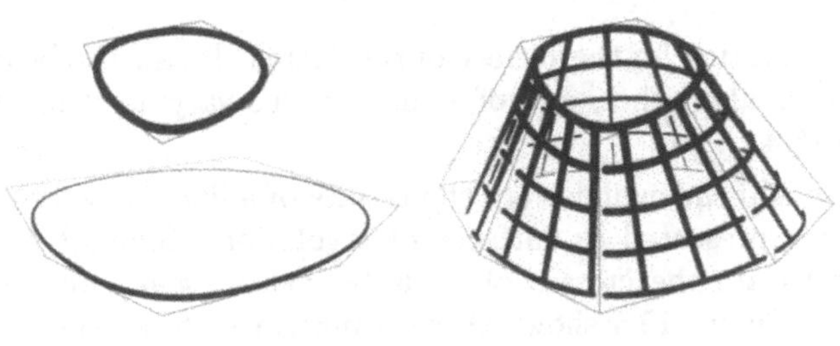

Figure 15.3. A ruled surface.

15.3 Surface of Revolution

A torus is a circle "crossed" with another circle. We can construct a *surface of revolution* from some given cross section, $\gamma(t)$, in the x-z plane, by crossing $\gamma(t)$ with a circle, $c(u)$. Without loss of generality, assume that $c(u)$ is a unit circle in the x-y plane that is constructed as the compound B-spline curve having four 90-degree arcs, $c(u) = \sum_{i=0}^{9} A_i \mathcal{B}_{i,2,\boldsymbol{\mu}}(u)$, with knot vector

$$\boldsymbol{\mu} = \{0,0,0,1,1,2,2,3,3,4,4,4\},$$

and

$$\begin{aligned} A_0 &= (1,0,0,1) \\ A_1 &= (\sqrt{2},\sqrt{2},0,\sqrt{2}) \\ A_2 &= (0,1,0,1) \\ A_3 &= (-\sqrt{2},\sqrt{2},0,\sqrt{2}) \\ A_4 &= (-1,0,0,1) \\ A_5 &= (-\sqrt{2},-\sqrt{2},0,\sqrt{2}) \\ A_6 &= (0,-1,0,1) \\ A_7 &= (\sqrt{2},-\sqrt{2},0,\sqrt{2}) \\ A_8 &= (1,0,0,1). \end{aligned}$$

Then, each control point $Q_j = (x_j, 0, z_j, 1)$ with cross section, $\gamma(t) = \sum_j Q_j \mathcal{B}_{j,\kappa,\tau}(t)$ is crossed with the B-spline curve representing the circle to get

$$\sigma(u,t) = \sum_{i,j} (x_j A_i + (0,0,z_j,0))\, \mathcal{B}_{i,2,\boldsymbol{u}}(u)\mathcal{B}_{j,\kappa,\boldsymbol{\tau}}(t), \qquad (15.3)$$

as the representation for the surface of revolution. It can also be considered as an extrusion along the z-axis of a curve with a varying cross-section[12]. See Figure 15.4.

The cone and the cylinder are the results of a line crossed by the circle that can be represented as surfaces of revolution. Similarly, the sphere and the torus are the crosses of a circle with a (semi)circle and circle, respectively. Figure 15.5 shows these primitive surfaces along with their constructed control meshes, as surfaces of revolution.

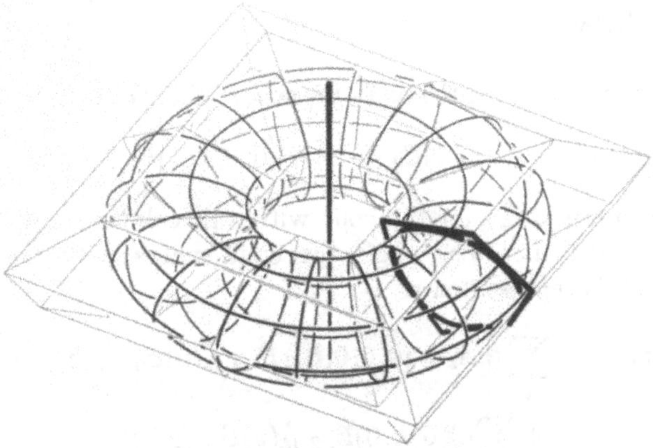

Figure 15.4. A surface of revolution.

15.3.1 The Cylinder

Suppose a circle is represented as in Equation 15.3. Consider a rational cross section curve with homogeneous coordinate representation

$$Q(t) = (r, 0, 0, 1)\mathcal{B}_{0,1,\boldsymbol{\tau}}(t) + (r, 0, h, 1)\mathcal{B}_{1,1,\boldsymbol{\tau}}(t).$$

According to Equation 15.3 the equation for the cylinder of radius r and height h, with axis along the $z-$axis is

$$\begin{aligned}\sigma(u,t) &= \sum_i (rA_i + (0,0,0,0))\,\mathcal{B}_{i,2,\boldsymbol{\mu}}(u)\mathcal{B}_{0,1,\boldsymbol{\tau}}(t) \\ &\quad + \sum_i (rA_i + (0,0,h,0))\,\mathcal{B}_{i,2,\boldsymbol{\mu}}(u)\mathcal{B}_{1,1,\boldsymbol{\tau}}(t).\end{aligned}$$

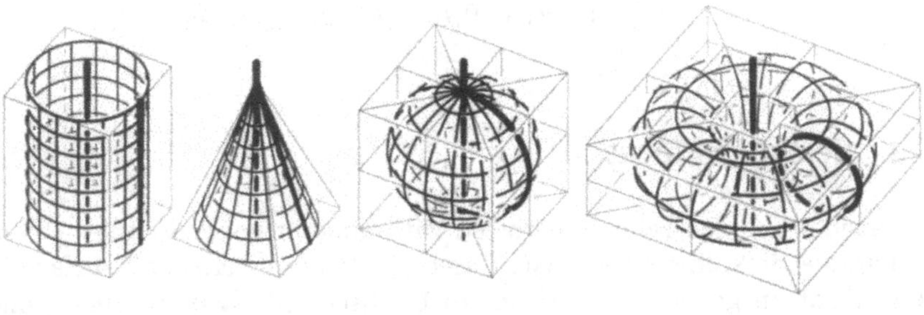

Figure 15.5. The cylinder, cone, sphere and torus represented as surfaces of revolution.

15.3.2 The Cone

Again the circle is represented as in Equation 15.3. This time,

$$Q(t) = (r, 0, 0, 1)\mathcal{B}_{0,1,\boldsymbol{\tau}}(t) + (0, 0, h, 1)\mathcal{B}_{1,1,\boldsymbol{\tau}}(t).$$

We seek a representation for the cone with axis coincident to the $z-$axis with apex at $(0, 0, h)$. The angle the cone makes with the $z-$axis is $\arctan(r/h)$. It is represented as

$$\begin{aligned}\sigma(u,t) &= \sum_i (rA_i + (0,0,0,0))\, \mathcal{B}_{i,2,\boldsymbol{\mu}}(u)\mathcal{B}_{0,1,\boldsymbol{\tau}}(t) \\ &\quad + \sum_i (0,0,h,0)\mathcal{B}_{i,2,\boldsymbol{\mu}}(u)\mathcal{B}_{1,1,\boldsymbol{\tau}}(t) \\ &= \sum_i rA_i\mathcal{B}_{i,2,\boldsymbol{\mu}}(u)\mathcal{B}_{0,1,\boldsymbol{\tau}}(t) + (0,0,h,0)\mathcal{B}_{1,1,\boldsymbol{\tau}}(t).\end{aligned}$$

15.3.3 The Truncated Cone

Again the circle is represented as in Equation 15.3. This time,

$$Q(t) = (r_0, 0, 0, 1)\mathcal{B}_{0,1,\boldsymbol{\tau}}(t) + (r_1, 0, z_1, 1)\mathcal{B}_{1,1,\boldsymbol{\tau}}(t).$$

It is represented as

$$\begin{aligned}\sigma(u,t) &= \sum_i (r_0A_i + (0,0,0,0))\, \mathcal{B}_{i,2,\boldsymbol{\mu}}(u)\mathcal{B}_{0,1,\boldsymbol{\tau}}(t) \\ &\quad + \sum_i (r_1A_i + (0,0,z_1,0))\, \mathcal{B}_{i,2,\boldsymbol{\mu}}(u)\mathcal{B}_{1,1,\boldsymbol{\tau}}(t) \\ &= \sum_i r_0A_i\mathcal{B}_{i,2,\boldsymbol{\mu}}(u)\mathcal{B}_{0,1,\boldsymbol{\tau}}(t) \\ &\quad + \sum_i (r_1A_i + (0,0,z_1,0))\, \mathcal{B}_{i,2,\boldsymbol{\mu}}(u)\mathcal{B}_{1,1,\boldsymbol{\tau}}(t).\end{aligned}$$

15.4 Sweep Surfaces

We consider one general curve crossed with another general curve. Given two parametric spline curves $\gamma_0(u)$ and $\gamma_1(v)$, their "crossed" surface is not rational, in general, for reasons to be discussed shortly. Hence, one must resort to approximation methods much like those for computation of offsets. In fact, there is a significant relationship to offsets.

Let

$$T_1(v) = \frac{\gamma_1'(v)}{\|\gamma_1'(v)\|}, \quad B_1(v) = \frac{\gamma_1'(v) \times \gamma_1''(v)}{\|\gamma_1'(v) \times \gamma_1''(v)\|}$$

and let $N_1(v) = B_1(v) \times T_1(v)$ (recall Section 4.3) be the unit tangent, binormal, and normal vector fields of $\gamma_1(v)$. One way to orient $\gamma_0(u)$ is to use the *moving trihedron* at v, $(T_1(v),\ N_1(v),\ B_1(v))$. Let M_v designate the matrix for the rotation that takes the canonical x-y-z coordinate frame to the $\{TNB\}$ coordinate frame; that is, M_v maps the column vector

$$\gamma_0(u) = \begin{bmatrix} x(u) & y(u) & z(u) \end{bmatrix}^t$$

to a column vector in the TNB coordinate system. Then,

$$\sigma(u, v) = M_v \gamma_0(u) + \gamma_1(v)$$

is the *sweep* of $\gamma_0(u)$ in space along $\gamma_1(v)$. $\gamma_1(v)$ is called the *axis curve* and $\gamma_0(u)$ is called the *cross section curve.* Unfortunately, in general, $(T_1(v),\ N_1(v),\ B_1(v))$ are not rational functions of v, so M_v and consequently $\sigma(u, v)$ are not rational as functions of v.

This general cross surface, known as a *sweep surface*, is quite general. When $B_1(v)$ is constant, $\gamma_1(v)$ is a planar curve, and the problem is reduced to approximating the unit normal curve $N_1(v)$ in the plane. $N_1(v)$ is the curve needed to form the offset of γ_1. When $N_1(v)$ is constant, the problem of creating the sweep surface is reduced to constructing a linear extrusion. When the curvature vector ($\kappa(v)N_1(v)$) is constant, $\gamma_1(v)$ is an arc so the problem is reduced to constructing a surface of revolution.

Approximation methods for sweep surfaces are many [11, 47, 14, 80, 24]. One can *position* $\gamma_0(u)$ at key locations along $\gamma_1(u)$ with *proper* orientation and position, and treat each such position as one row in the control mesh of the constructed surface. The difficulties with such positioning schemes stem from the need to detect and identify key locations and orientations, as well as the need to determine the error in this rational approximation.

The Frenet frame method of orienting the sweep is not suitable for all curves. The Frenet frame is unstable near axis curve inflection points, and is undefined on embedded linear segments. Since a sweep is a generalization of an extrusion, one can choose to form an extrusion over the linear segment. In that case it is necessary to formulate an appropriate orientation which is consistent with the sweep surface at the endpoints. This is not possible except in certain circumstances. Much research has been devoted to seeking better orienatation schemes [11, 80, 47]. Finally, more degrees of freedom in sweep surface construction can be gained by allowing the cross section curve, $\gamma_0(u)$, to be scaled, shaped, or modified as it is swept along $\gamma_1(v)$. Figure 15.6 shows some examples for general sweeps [12].

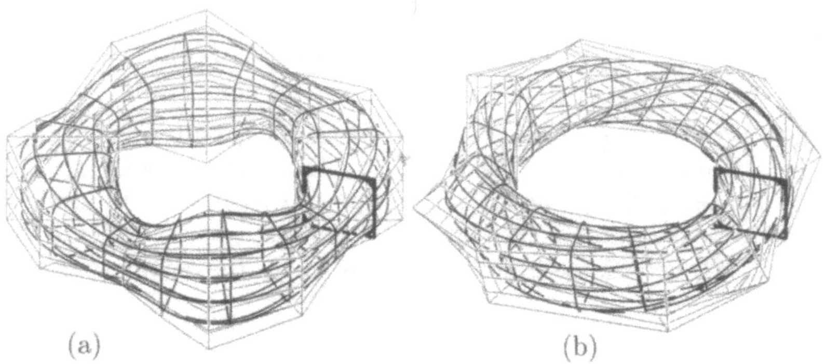

Figure 15.6. A sweep of a rounded square around a circular path. In (a), the cross section is scaled while in (b) the orientation frame twists.

15.5 Normals and Offset Surfaces

Many applications require computing tangent planes to the surface. Examples range from computer graphics lighting and shading models to mathematical analysis and cutter path geometry for numerical control. The Gaussian curvature and mean curvature give important clues to light reflection behavior and, in conjunction with geodesic curvature, can be useful in computing aerodynamic properties, such as wind drag and turbulence.

For a tensor product spline surface, $\sigma(u, v)$ the unit normal surface,

$$n(u, v) = \frac{\frac{\partial \sigma(u,v)}{\partial u} \times \frac{\partial \sigma(u,v)}{\partial v}}{\left\| \frac{\partial \sigma(u,v)}{\partial u} \times \frac{\partial \sigma(u,v)}{\partial v} \right\|}$$

cannot be written as a rational piecewise polynomial because it has a square root (the magnitude of the cross product) in its denominator. Hence, derived surfaces which depend on the unit normal, as well as the unit normal surface must be approximated.

The true offset surface, say offset by a distance r, is one such derived surface. Its definition is $\sigma_r(u, v) = \sigma(u, v) + rn(u, v)$. In practice, approximation techniques are used for all but the simplest surfaces.

Approximation methods for offsets are many. One simple yet stable approach was proposed by [18]. Consider a planar B-spline curve $c(t) = \sum_i P_i \mathcal{B}_{i,\kappa,\boldsymbol{t}}(t)$ and consider the nodes vector $\boldsymbol{t}^*$ (Definition 9.3). The planar offset curve, c_r, by amount r, is approximated, according to [18], as

$$c_r(t) = \sum_i (P_i + rN(t_i^*))\, \mathcal{B}_{i,\kappa,\boldsymbol{t}}(t),$$

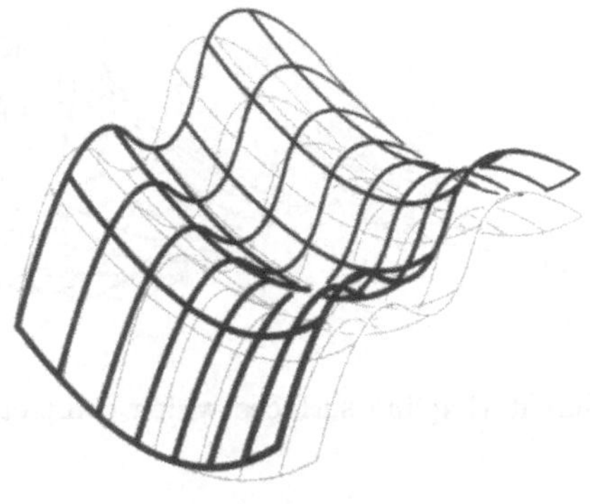

Figure 15.7. A surface (thick lines) with its offset surface.

where $t^*_{i,\kappa} \in \boldsymbol{t}^*_\kappa$ and $N(t)$ is the unit normal field of $c(t)$. Evaluating c at each i gives $\{c(t^*_i)\}_i$, an easily determined set of points on the curve that are known to be close to the control points for each i, respectively, even though each $c(t^*_i)$ is not necessarily the closest point on the curve to P_i. This approach can easily be extended to surfaces.

Given surface $\sigma(u,v) = \sum_i \sum_j P_{i,j}\mathcal{B}_{i,\kappa_u,\boldsymbol{u}}(u)\mathcal{N}_{j,\kappa_v,\boldsymbol{v}}(v)$, the planar offset surface, $\sigma_r(u,v)$, by amount r, is approximated as

$$\sigma_r(u,v) = \sum_i \sum_j \left(P_{i,j} + rn(u^*_i, v^*_j)\right) \mathcal{B}_{i,\kappa_u,\boldsymbol{u}}(u)\mathcal{N}_{j,\kappa_v,\boldsymbol{v}}(v).$$

See Figure 15.7 for an example.

15.5.1 Given Four Sides

Another frequent example of surface construction is one with a closed, smooth boundary such as the one in Figure 15.8. When the boundary is planar and convex such as this one, with certain attributes, it is very straightforward to create a surface with these boundaries. However, non-convex boundaries can yield self-intersecting surfaces. The flat, self-intersecting, surface in Figure 15.10 (b) is the result of a Boolean sum (see Section 14.4) of the flat boundary curves provided in Figure 15.10 (a).

Figure 15.8. Rounded convex and planar boundaries.

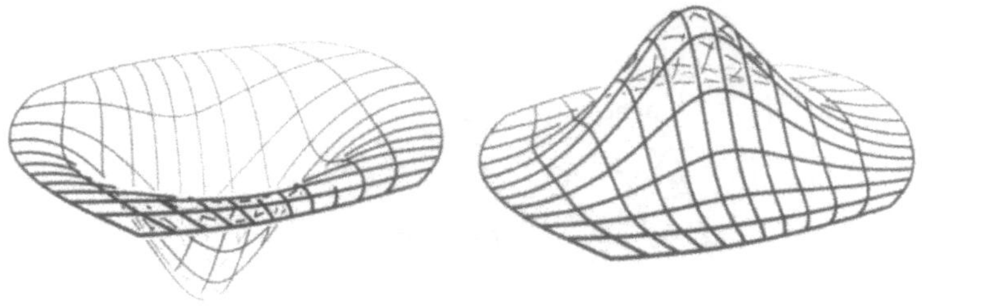

Figure 15.9. Two different B-spline surfaces which complete the boundary of Figure 15.8.

Suppose the boundary is composed of four open spline curves, $\gamma_0(u)$, $\gamma_1(u)$, $\beta_0(v)$, and $\beta_1(v)$, with control polygons $\{P_{i,0}\}_{i=0}^m$, $\{P_{i,1}\}_{i=0}^m$, and $\{Q_{0,j}\}_{j=0}^n$, and $\{Q_{1,j}\}_{j=0}^n$, respectively. Suppose, in addition, that γ_0 and γ_1 have the same knot vector $\boldsymbol{u}$ and degree κ_u. Also, let β_0 and β_1 have the same knot vector $\boldsymbol{v}$ and degree κ_v. Finally we assume that

$$\begin{aligned} P_{0,0} &= Q_{0,0} \\ P_{m,0} &= Q_{1,0} \\ P_{0,1} &= Q_{0,n} \\ P_{m,1} &= Q_{1,n}. \end{aligned}$$

We can define the surface $\sigma(u,v)$, with degree κ_u in the u direction and degree κ_v in the v direction, using $\boldsymbol{u}$ as the knot vector in the u direction, and $\boldsymbol{v}$ as the knot vector in the v direction. The surface $\sigma(u,v)$ will have a mesh that is constrained around the boundaries to be:

$$\begin{aligned} M_{i,0} &= P_{i,0}, & i &= 0,\ldots,\ m \\ M_{i,n} &= P_{i,1}, & i &= 0,\ldots,\ m \\ M_{0,j} &= Q_{0,j}, & j &= 0,\ldots,\ n \\ M_{m,j} &= Q_{1,j}, & j &= 0,\ldots,\ n. \end{aligned}$$

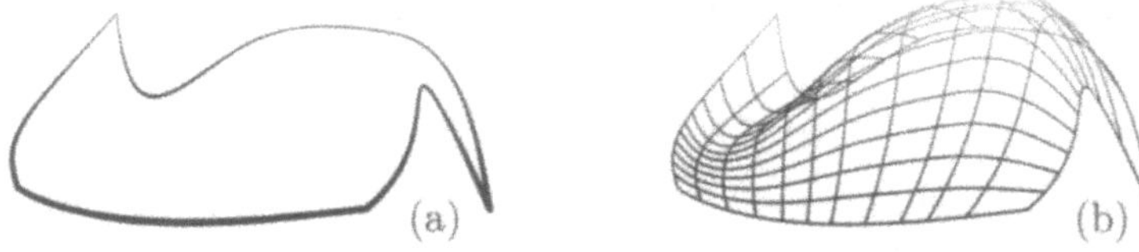

Figure 15.10. (a) A non-convex planar boundary is more likely to yield a (b) self-intersecting flat surface!

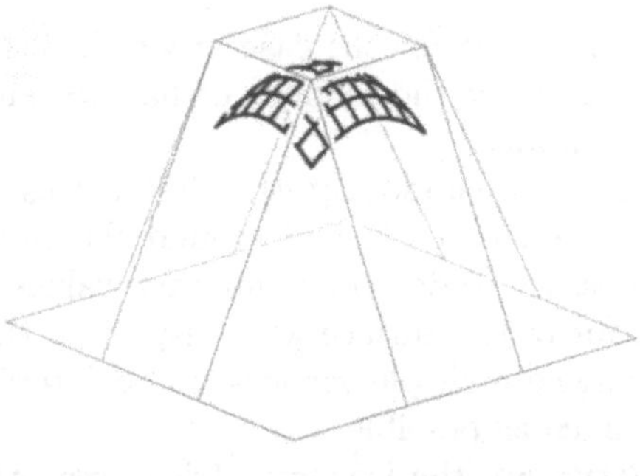

Figure 15.11. A floating spline surface and mesh.

The rest of the mesh values can be chosen to suit esthetic or design requirements, subject of course to their projections lying inside the convex hull of the boundary polygons. This choice of interior mesh points can be made by the user or as the solution to mathematically-posed optimization constraints.

15.5.2 Floating Surface

A surface with floating knot vectors in both u and v, called a *floating surface* has been used historically. In Chapter 13, we derived a straightforward matrix evaluation scheme for the case of the uniform floating spline surface.

Given integers $\kappa_u, \kappa_v \geq 0$, and knot vectors $\boldsymbol{u}$ and $\boldsymbol{v}$ with $m + \kappa_u + 2$ and $n + \kappa_v + 2$ elements, respectively, the domain for the tensor product B-spline space is $[u_{\kappa_u}, u_{m+1}] \times [v_{\kappa_v}, v_{n+1}]$. We study the case $\kappa_u = \kappa_v = 3$, a bicubic spline, and assume that $u_i < u_{i+1}$ for all i, and $v_j < v_{j+1}$ for all j.

A floating surface, of course, inherits its characteristics from the floating curve formulation on which it is based. However, it is more complicated. Since open end conditions do not apply, the surface does not interpolate the four "corner" points of the mesh, and the boundary rows (columns) of the mesh are not the control points for the isoparametric spline curve on the boundaries. Instead those boundary curves require meshes themselves, just as the interior isoparametric curves do.

The size of the mesh necessary to define a single floating bicubic polynomial patch is exactly the same as to define a single open bicubic, a four-by-four mesh with 16 elements.

Just as in the univariate case, multiple vertices may be used to imitate the effects of multiple knots. However, the complexity is greater. We search for the conditions under which multiple vertices effect some sort of "open" boundary type conditions, and, indeed, even what that means in this context. We shall search for the case in which the final mesh has 16 distinct elements, and study the number of patches and their geometric distribution over the surface.

At present, the smallest subdomain that fulfills this requirement is unknown, so we study the surface in and around the values $[u_M, u_{M+R}] \times [v_N, v_{N+T}]$. It is necessary to determine minimal values for M, R, N, and T so that the behavior of the surface with respect to the mesh having 16 distinct points (and some multiple vertices, to be determined), will act as an open surface as much as possible.

The four boundaries are the isoparametric curves $\sigma(u_M, v)$, $\sigma(u, v_N)$, $\sigma(u_{M+R}, v)$, and $\sigma(u, v_{N+T})$. We first analyze $\sigma(u, v_N)$.

$$\begin{aligned}\sigma(u, v_N) &= \sum_i \sum_j P_{i,j} \mathcal{B}_{j,3}(v_N) \mathcal{N}_{i,3}(u) \\ &= \sum_i \sum_{j=N-3}^{N-1} P_{i,j} \mathcal{B}_{j,3}(v_N) \mathcal{N}_{i,3}(u).\end{aligned}$$

Since the tensor product B-spline functions are continuous, $\mathcal{B}_{N,3}(v_N) = 0$, and further since the curve lies in the convex hull of those points, the curve lies in a strip of the mesh, three columns wide. Hence, suppose those three points coalesce. That is,

$$P_{i,j} = P_{i,N-1}, \qquad j = N-3, N-2.$$

In that case,

$$\begin{aligned}\sigma(u, v_N) &= \sum_i P_{i,N-1} \sum_{j=N-3}^{N-1} \mathcal{B}_{j,3}(v_N) \mathcal{N}_{i,3}(u) \\ &= \sum_i P_{i,N-1} \mathcal{N}_{i,3}(u),\end{aligned}$$

and the boundary curve is determined by the triple mesh points along the "bottom". Hence, we see that $N \geq 3$, if the smallest subscript on the knot vector is 0, and $P_{i,0} = P_{i,1} = P_{i,2}$, for all i.

Analogously, if

$$P_{i,j} = P_{M-1,j}, \qquad i = M-3, M-2,$$

then,

$$\begin{aligned}\sigma(u_M, v) &= \sum_j P_{M-1,j} \sum_{i=M-3}^{M-1} \mathcal{N}_{i,3}(u_M) \mathcal{B}_{j,3}(v) \\ &= \sum_j P_{M-1,j} \mathcal{B}_{j,3}(v).\end{aligned}$$

Again, $M \geq 3$, and $P_{0,j} = P_{1,j} = P_{2,j}$, for all j. Thus far we see that two boundaries of the mesh must have each of three points coalescing in order to have the boundary curve be a B-spline curve depending only on the "boundary mesh". Each of those B-spline curves is then also a floating curve with multiple vertices at the ends. The "lower" corner, then, has $P_{2,2} = P_{i,j}$, $0 \leq i \leq 2$, and $0 \leq j \leq 2$.

Now consider the upper end.

$$\begin{aligned}\sigma(u, v_{N+T}) &= \sum_i \sum_j P_{i,j}\mathcal{B}_{j,3}(v_{N+T})\mathcal{N}_{i,3}(u)\\ &= \sum_i \sum_{j=N+T-3}^{N+T-1} P_{i,j}\mathcal{B}_{j,3}(v_{N+T})\mathcal{N}_{i,3}(u).\end{aligned}$$

Again, coalesce the vertices to get

$$P_{i,j} = P_{i,N+T-1}, \qquad j = N+T-3, N+T-2.$$

So,

$$\begin{aligned}\sigma(u, v_{N+T}) &= \sum_i P_{i,N+T-3} \sum_{j=N+T-3}^{N+T-1} \mathcal{B}_{j,3}(v_{N+T})\mathcal{N}_{i,3}(u)\\ &= \sum_i P_{i,N+T-3}\mathcal{N}_{i,3}(u).\end{aligned}$$

Now, to have any distinct mesh points, and hence nondegenerate patches, $N+T-3 > N-1$, and so $T > 2$ is the minimal condition and $P_{i,N+T-3} = P_{i,N+T-2} = P_{i,N+T-1}$, for all i. In particular, the previous boundary conditions give $P_{2,N+T-3} = P_{i,j}$, $0 \leq i \leq 2$, and $N+T-3 \leq j \leq N+T-1$, another vertex which will have multiplicity nine. Analogously,

$$\begin{aligned}\sigma(u_{M+R}, v) &= \sum_j \sum_i P_{i,j}\mathcal{N}_{i,3}(u_{M+R})\mathcal{B}_{j,3}(v)\\ &= \sum_j \sum_{i=M+R-3}^{M+R-1} P_{i,j}\mathcal{N}_{i,3}(u_{M+R})\mathcal{B}_{j,3}(v).\end{aligned}$$

Again, coalesce the vertices to get

$$P_{i,j} = P_{M+R-1,j}, \qquad i = M+R-3, M+R-2.$$

Then,

$$\begin{aligned}\sigma(u_{M+R}, v) &= \sum_j P_{M+R-3,j} \sum_{i=M+R-3}^{M+R-1} \mathcal{N}_{i,3}(u_{M+R})\mathcal{B}_{j,3}(v)\\ &= \sum_j P_{M+R-3,j}\mathcal{B}_{j,3}(v).\end{aligned}$$

Again, we require the surface to be nondegenerate, so $M + R - 3 > M - 1$, and $R > 2$. This last condition combined with the others gives $P_{M+R-3,2} = P_{i,j}$, $0 \le j \le 2$, and $M + R - 3 \le i \le M + R - 1$, and also that $P_{M+R-3,N+T-3} = P_{i,j}$, $N + T - 3 \le j \le N + T - 1$, and $M + R - 3 \le i \le M + R - 1$. Each one has a multiplicity of nine. The configuration, then, is,

$$\sigma(u, v) = \sum_{i=M-3}^{M+R-1} \sum_{j=N+T-3}^{N+T-1} P_{i,j}\mathcal{B}_{j,3}(v)\mathcal{N}_{i,3}(u).$$

Or, minimally

$$\begin{aligned}
\sigma(u, v) &= \sum_{i=0}^{5}\sum_{j=0}^{5} P_{i,j}\mathcal{B}_{j,3}(v)\mathcal{N}_{i,3}(u) \\
&= \sum_{i=0}^{2}\sum_{j=0}^{2} P_{i,j}\mathcal{B}_{j,3}(v)\mathcal{N}_{i,3}(u) + \sum_{i=0}^{2}\sum_{j=3}^{5} P_{i,j}\mathcal{B}_{j,3}(v)\mathcal{N}_{i,3}(u) \\
&\quad + \sum_{i=3}^{5}\sum_{j=0}^{2} P_{i,j}\mathcal{B}_{j,3}(v)\mathcal{N}_{i,3}(u) + \sum_{i=3}^{5}\sum_{j=3}^{5} P_{i,j}\mathcal{B}_{j,3}(v)\mathcal{N}_{i,3}(u) \\
&= P_{2,2}\sum_{i=0}^{2}\sum_{j=0}^{2} \mathcal{B}_{j,3}(v)\mathcal{N}_{i,3}(u) + P_{2,3}\sum_{i=0}^{2}\sum_{j=3}^{5} \mathcal{B}_{j,3}(v)\mathcal{N}_{i,3}(u) \\
&\quad + P_{3,2}\sum_{i=3}^{5}\sum_{j=0}^{2} \mathcal{B}_{j,3}(v)\mathcal{N}_{i,3}(u) + P_{3,3}\sum_{i=3}^{5}\sum_{j=3}^{5} \mathcal{B}_{j,3}(v)\mathcal{N}_{i,3}(u).
\end{aligned}$$

This minimal mesh has four distinct points, each of which has a multiplicity of nine. This is shown in Figure 15.12.

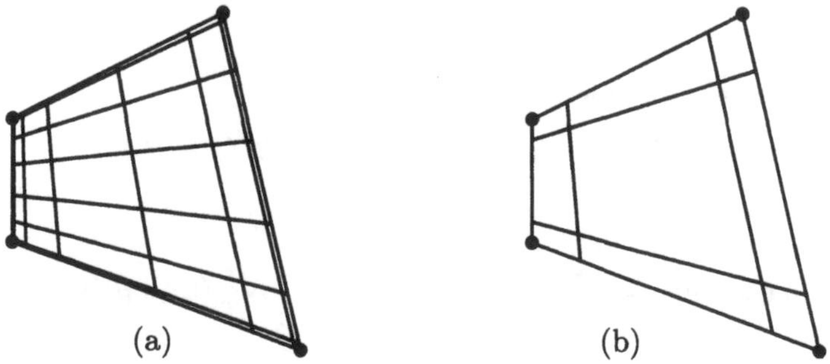

(a) (b)

Figure 15.12. (a) Minimal nondegenerate mesh and surface; (b) knot lines shown.

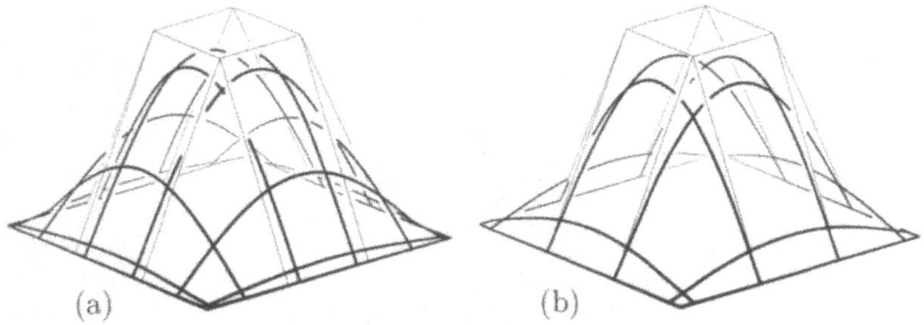

Figure 15.13. (a) A "four-by-four" floating mesh simulating open conditions; (b) Same surface as (a) with knot lines. (See also Figure 15.11.)

We can determine, empricially from the figure but also from the mathematics, that this formulation with just four distinct points has exactly nine different bicubic patch pieces. If we allow a nonminimal mesh, then let $T = R = 5$. The parametric domain is $[u_3, u_8] \times [v_3, v_8]$, and the mesh has 64 points, of which exactly 16 points are distinct. The four corner points each still have multiplicity nine, and, in addition, the two edge vertices along each of the four boundaries have multiplicity three. Only the four internal points have multiplicity one. This surface has exactly 25 bicubic patches! (See Figure 15.13.)

Consider any of the multiple vertex surfaces over $[u_3, u_4] \times [v_3, v_4]$. Remembering that over that interval, $\sum_{j=0}^{3} \mathcal{B}_{j,3}(v) = 1$ and $\sum_{i=0}^{3} \mathcal{N}_{i,3}(u) = 1$,

$$\begin{aligned}
\sigma(u,v) &= \sum_{i=0}^{3}\sum_{j=0}^{3} P_{i,j}\mathcal{B}_{j,3}(v)\mathcal{N}_{i,3}(u) \\
&= \sum_{i=0}^{2}\sum_{j=0}^{2} P_{i,j}\mathcal{B}_{j,3}(v)\mathcal{N}_{i,3}(u) + \sum_{i=0}^{2} P_{i,3}\mathcal{B}_{3,3}(v)\mathcal{N}_{i,3}(u) \\
&\quad + \sum_{j=0}^{2} P_{3,j}\mathcal{B}_{j,3}(v)\mathcal{N}_{3,3}(u) + P_{3,3}\mathcal{B}_{3,3}(v)\mathcal{N}_{3,3}(u) \\
&= P_{2,2}\sum_{i=0}^{2}\sum_{j=0}^{2} \mathcal{B}_{j,3}(v)\mathcal{N}_{i,3}(u) + P_{2,3}\sum_{i=0}^{2} \mathcal{B}_{3,3}(v)\mathcal{N}_{i,3}(u) \\
&\quad + P_{3,2}\sum_{j=0}^{2} \mathcal{B}_{j,3}(v)\mathcal{N}_{3,3}(u) + P_{3,3}\mathcal{B}_{3,3}(v)\mathcal{N}_{3,3}(u) \\
&= P_{2,2}\left(1-\mathcal{N}_{3,3}(u)\right)\left(1-\mathcal{B}_{3,3}(v)\right) + P_{2,3}\mathcal{B}_{3,3}(v)\left(1-\mathcal{N}_{3,3}(u)\right) \\
&\quad + P_{3,2}\left(1-\mathcal{B}_{3,3}(v)\right)\mathcal{N}_{3,3}(u) + P_{3,3}\mathcal{B}_{3,3}(v)\mathcal{N}_{3,3}(u),
\end{aligned}$$

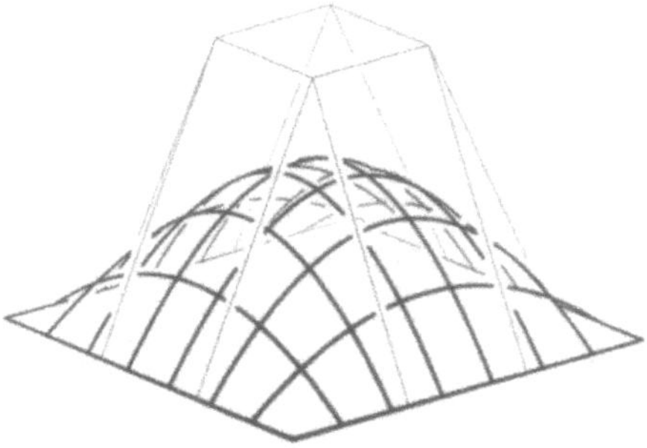

Figure 15.14. Corresponding B-spline surface with open ends and same mesh as in Figure 15.11, for comparison.

which is clearly a bilinear surface. In general, for a floating bicubic B-spline surface which has multiple vertices at the boundaries, the corner "patches" are bilinear surfaces and all the "side" "patches" are either linear-by-cubic or cubic-by-linear patches. They meet the interior "patches" in the surface with $C^{(2)}$ continuity. Formerly, this form of B-spline surface was used in order to take advantage of the properties of the uniform floating B-spline functions, namely that they were all translates. The goal was to evaluate the surface as a moving window of matrix multiplications. Since current practice leans towards using subdivision for rendering, multiple vertex representations are seen less frequently.

15.6 Piecing together the Surfaces

Suppose one has four tensor product surfaces meeting at a single point as in Figure 15.15. We consider the simplest cases, bicubic surfaces with no internal knots.

The four points Q, R, S, and T are common to two of the four patches, with W being common to all four. The goal is to sew the patches together to obtain the desired continuity. Immediately it is evident that different conditions apply depending on whether one must sew together explicit surface patches or parametric patches, as was also the case for sewing together piecewise explicit cubic Hermite curves or piecewise parametric cubic Hermite (or Bézier) curves.

We shall elaborate on the most common patch configuration, the piecewise bicubic tensor product, in either the Bézier surface formulation or the Hermite patch formulation. We suppose that each of the individual patches is parametrized over the unit square $[0, 1] \times [0, 1]$, and that

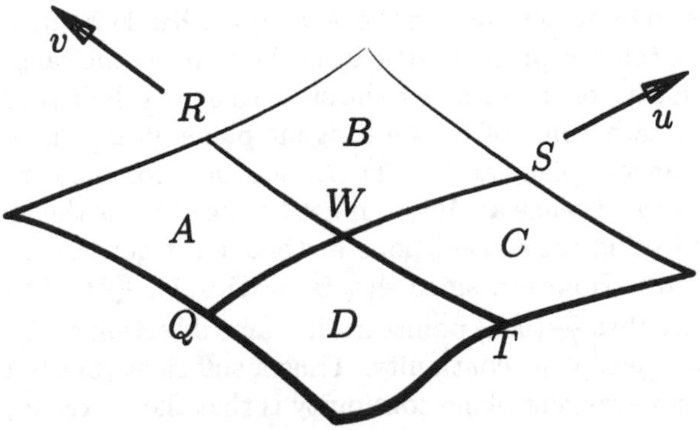

Figure 15.15. Four patches meeting at a single point.

$$
\begin{aligned}
Q &= A(0,0) = D(0,1)\\
R &= A(1,1) = B(0,1)\\
S &= B(1,0) = C(1,1)\\
T &= C(0,0) = D(1,0)
\end{aligned}
$$

and $W = A(1,0) = B(0,0) = C(0,1) = D(1,1)$.

To provide continuity between the patch boundaries, we must have

$$
\begin{aligned}
A(1,v) = B(0,v) \quad &\text{and} \quad D(1,v) = C(0,v),\\
D(u,1) = A(u,0) \quad &\text{and} \quad C(u,1) = B(u,0).
\end{aligned}
$$

In Bézier formulation, for u direction continuity:

$$
\begin{aligned}
Q &= P_{A,0,0} = P_{D,0,3} & P_{A,1,0} &= P_{D,1,3}\\
W &= P_{A,3,0} = P_{D,3,3} & P_{A,2,0} &= P_{D,2,3}\\
W &= P_{B,0,0} = P_{C,0,3} & P_{B,1,0} &= P_{C,1,3}\\
S &= P_{B,3,0} = P_{C,3,3} & P_{B,2,0} &= P_{C,2,3}
\end{aligned}
$$

and for v direction continuity:

$$
\begin{aligned}
T &= P_{D,3,0} = P_{C,0,0} & P_{D,3,1} &= P_{C,0,1}\\
W &= P_{D,3,3} = P_{C,0,3} & P_{D,3,2} &= P_{C,0,2}\\
W &= P_{A,3,0} = P_{B,0,0} & P_{A,3,1} &= P_{B,0,1}\\
R &= P_{A,3,3} = P_{B,0,3} & P_{A,3,2} &= P_{B,0,2}.
\end{aligned}
$$

Thus, matching those mesh points will lead to a continuous piecewise bicubic surface. However, the surface will not have any higher order continuity. If a smoother surface is needed, more must be done. In particular, if the surface is to be smooth across the $A(u,0) = D(u,1)$ boundary, for each value of u, the tangent plane at $A(u,0)$ is the same as the tangent plane at $D(u,1)$. Similarly, for the surface to be smooth across the $B(u,0) = C(u,1)$ boundary, for each value of u, the tangent plane at $B(u,0)$ must be the same as the tangent plane at $C(u,1)$. In addition, for their meeting to be smooth, the tangent plane at $A(1,0)$ must be the same as the tangent plane at $B(0,0)$. Meeting those conditions in their most general form is rather complicated [10]. However, since $A(u,0) = D(u,1)$, $\frac{\partial A}{\partial u}(u,0) = \frac{\partial D}{\partial u}(u,1)$, so we need only that $\frac{\partial A}{\partial v}(u,0)$ points in the same direction as $\frac{\partial D}{\partial v}(u,1)$ in order to have tangent plane continuity. Thus a sufficient (but not necessary) condition to have tangent plane continuity is that there exist $a_{AD}, a_{BC} > 0$ such that

$$\tfrac{\partial A}{\partial v}(u,0) = a_{AD}\tfrac{\partial D}{\partial v}(u,1) \quad\text{and}\quad \tfrac{\partial B}{\partial v}(u,0) = a_{BC}\tfrac{\partial C}{\partial v}(u,1).$$

Since the parametric derivatives are vectors, the directions need only be the same for the appropriate boundary partners, and a_{AD} and a_{BC} could be any positive scalar, except now we must consider the point W. Since $\frac{\partial}{\partial v}A(1,0) = \frac{\partial}{\partial v}B(0,0)$, the two scalars must be equal, that is,

$$\tfrac{\partial A}{\partial v}(u,0) = a_v\tfrac{\partial D}{\partial v}(u,1) \qquad\text{and}\qquad \tfrac{\partial B}{\partial v}(u,0) = a_v\tfrac{\partial C}{\partial v}(u,1). \tag{15.4}$$

Analogously,

$$\tfrac{\partial A}{\partial u}(1,v) = a_u\tfrac{\partial B}{\partial u}(0,v) \qquad\text{and}\qquad \tfrac{\partial D}{\partial u}(1,v) = a_u\tfrac{\partial C}{\partial u}(0,v). \tag{15.5}$$

Rewriting using Bézier formulation gives

$$\begin{aligned}\tfrac{\partial A}{\partial v}(u,0) &= 3\sum_{i=0}^{3}[P_{A,i,1} - P_{A,i,0}]\theta_i(u),\\ \tfrac{\partial D}{\partial v}(u,1) &= 3\sum_{i=0}^{3}[P_{D,i,3} - P_{D,i,2}]\theta_i(u),\end{aligned}$$

and by Equation 15.4,

$$\sum_i [P_{A,i,1} - P_{A,i,0}]\theta_i(u) = a_v \sum_i [P_{D,i,3} - P_{D,i,2}]\theta_i(u).$$

Since the Bézier blending functions form a basis for cubics, this means that

$$P_{A,i,1} - P_{A,i,0} = a_v\{P_{D,i,3} - P_{D,i,2}\}, \qquad i = 0,1,2,3. \tag{15.6}$$

Analogously,

$$P_{B,i,1} - P_{B,i,0} = a_v\{P_{C,i,3} - P_{C,i,2}\}, \qquad i = 0,1,2,3. \tag{15.7}$$

And for u,

$$\begin{aligned} P_{A,3,j} - P_{A,2,j} &= a_u\{P_{B,1,j} - P_{B,0,j}\}, \qquad j = 0,1,2,3, \quad (15.8) \\ P_{D,3,j} - P_{D,2,j} &= a_u\{P_{C,1,j} - P_{C,0,j}\}, \qquad j = 0,1,2,3. \quad (15.9) \end{aligned}$$

Equations 15.6 and 15.7 require that the points $P_{D,i,2}$, $P_{D,i,3} = P_{A,i,0}$, and $P_{A,i,1}$ all lie along a straight line, for each i and that $P_{C,i,2}$, $P_{C,i,3} = P_{B,i,0}$, and $P_{B,i,1}$ also each lie along a straight line for each i. While the seven lines for these coefficients may vary, the relative placements must be dictated by the value a_v, so they are not totally independent. The analogous results follow from Equations 15.8 and 15.9. However, further important conditions must be noted.

The points

$$\begin{matrix} P_{A,2,0}, & P_{A,3,0}, & P_{A,2,1} \text{ and } & P_{A,3,1} \\ P_{B,0,0}, & P_{B,1,0}, & P_{B,0,1} \text{ and } & P_{B,1,1} \\ P_{C,0,3}, & P_{C,0,2}, & P_{C,1,3} \text{ and } & P_{C,1,2} \\ P_{D,2,3}, & P_{D,3,3}, & P_{D,2,2} \text{ and } & P_{D,3,2} \end{matrix}$$

are used in equations for the partials in both directions. Note that there are really only nine distinct points, but the six lines through the various points must all intersect each other at appropriate locations, or the surface will not be "smooth".

We now study the number of degrees of freedom in such a configuration and also learn more about the values a_u and a_v. Suppose $D(u, v)$ is given and we want to add $C(u, v)$. Further assume that a_u and a_v are arbitrary nonnegative real numbers.

Since continuity is necessary,

$$P_{C,0,j} = P_{D,3,j}, \qquad j = 0,1,2,3;$$

and since $\frac{\partial}{\partial u}D(1, v) = a_u \frac{\partial}{\partial u}C(0, v)$,

$$P_{C,1,j} = P_{C,0,j} + (1/a_u)\left(P_{D,3,j} - P_{D,2,j}\right), \qquad j = 0,1,2,3.$$

Thus, half of the mesh points of C are determined by its continuity requirements with D and the scalar value of a_u.

Analogously, since continuity is necessary, to add $A(u, v)$,

$$P_{A,i,0} = P_{D,i,3}, \qquad i = 0,1,2,3;$$

and since $\frac{\partial}{\partial v}A(u, 0) = a_v \frac{\partial}{\partial v}D(u, 1)$,

$$P_{A,i,1} = P_{A,i,0} + a_v\left(P_{D,i,3} - P_{D,i,2}\right), \qquad j = 0,1,2,3.$$

Again, half of the mesh points of A are determined by its continuity requirements with D and the scalar value a_v.

Now, suppose that the other half of the meshes for A and C have been determined in some manner. So far there are $8+8$ total degrees of freedom, plus the scalar freedoms of a_u and a_v. Consider the flexibility in the selection of B. Continuity requirements with C uniquely determine $P_{B,i,0}$, $i = 0, 1, 2, 3$; and derivative continuity requirements with C along with a_v determine $P_{B,i,1}$, $i = 0, 1\ 2, 3$, in terms of the coefficients of C. Note that means that $P_{B,i,1}$ and $P_{B,i,0}$, $i = 0, 1$, are really completely determined by D and a_v. Analogously, continuity requirements with A uniquely determine $P_{B,0,j}$, $j = 0, 1, 2, 3$, and derivative continuity requirements with A, along with a_u determine $P_{B,1,j}$, $j = 0, 1, 2, 3$, in terms of the coefficients of C. Again note that $P_{B,1,j}$ and $P_{B,0,j}$, $j = 0, 1$, are uniquely determined by the coefficients of D and a_u.

Again, $P_{B,1,1}$ shows up as uniquely determined in two distinct ways.

$$\begin{aligned} P_{B,1,1} &= P_{B,1,0} + a_v\left(P_{C,1,3} - P_{C,1,2}\right) \\ &= P_{B,0,1} + (1/a_u)\left(P_{A,3,1} - P_{A,2,1}\right). \end{aligned}$$

We shall trace one of these definitions to explicitly determine $P_{B,1,1}$ in terms of P_D. Since

$$\begin{aligned} P_{B,1,0} &= P_{C,1,3}, \\ P_{B,1,1} &= P_{C,1,3} + a_v(P_{C,1,3} - P_{C,1,2}). \end{aligned}$$

Further

$$\begin{aligned} P_{C,1,3} &= P_{D,3,3} + (1/a_u)(P_{D,3,3} - P_{D,2,3}), \\ P_{C,1,2} &= P_{D,3,2} + (1/a_u)(P_{D,3,2} - P_{D,2,2}), \end{aligned}$$

and

$$\begin{aligned} P_{B,1,1} &= P_{D,3,3} + (1/a_u)(P_{D,3,3} - P_{D,2,3}) + a_v\left(P_{D,3,3} - P_{D,3,2}\right) \\ &\quad + \frac{a_v}{a_u}\left(P_{D,3,3} - P_{D,2,3} - P_{D,3,2} + P_{D,2,2}\right). \end{aligned}$$

Thus, given a_u and a_v, we see that four of the coefficients of B are determined uniquely by D, four of the coefficients of B are determined uniquely by the coefficients of C which are independent of D, and four of the coefficients of B are determined uniquely by the coefficients of A which are independent of D. Thus, B has exactly four degrees of freedom left! In both the Bézier formulation which has been used for this derivation, and the Coons formulation, these continuity constraints must be carried along "by hand". In the next chapter we shall once again see how B-splines free the systems designer and user from worrying about these difficulties.

Finally, note that in the special case that $a_u = a_v = 1$, the surface is $C^{(1)}$ continuous at the point W. Otherwise, the surfaces are clearly visually $C^{(1)}$. The surfaces are also $C^{(1)}$ in some regular parametrization, however, there is no easily defined "arc length" parametrization for surfaces in which this is true.

Exercises

1. Constants a_u and a_v defined in Equations 15.4 and 15.5 were used to insure geometric continuity in the appropriate parametric directions. Is it necessary that these values be constant? If they were made functions, what conditions would they have to meet?

2. Suppose that it is desired to piece four piecewise bicubic surfaces together with *tangent plane continuity.* What conditions would it be necessary to enforce on the partial derivatives along the four seam lines? *Tangent plane continuity* means that a point on the boundary has the same tangent plane no matter which side of the boundary it is calculated on.

IV

Advanced Techniques

16

Subdivision and Refinement for Splines

The design and modelling of three-dimensional objects clearly has many parts. Much effort has been spent discussing representational methods for curves and tensor product surfaces as well as their various attributes which might affect user interaction. An extremely important assumption has been implicit throughout. Design of three-dimensional objects sometimes requires the object to satisfy constraints of a hard kind, such as aerodynamic or hydrodynamic analyses, thermo or stress analyses, or variational constraints. Sometimes there are only aesthetic constraints; they are the most difficult since they are the least analytic. However, all design spirals have a human designer involved. Further, the designer's needs to have a visualization capability must be satisfied. This visualization capability may be needed to see if the computer-generated model really *looks* the way his internal vision said it should, that is for the geometrical shape, or it may be necessary for the visualization of nongeometric information such as the results of stress analyses, or it may be necessary to see the interaction of the various parts of the model. For all of these reasons visualization methods that allow all levels from the coarse rough look to a high management presentation quality are necessary.

All algorithms for visualization of B-splines have thus far concerned techniques for evaluation of the splines. The question of how many points are needed and what their distribution should be to obtain best available quality has not been asked, much less answered. B-splines have been used to represent surfaces and model objects whose sole purpose is to present a visual illusion; the medium *is* the message. Most computer graphics methods for drawing shaded, high quality, surfaces on raster devices require more information than just a few evaluation points. Today, graphical models are highly sophisticated and include information as to a lighting model and a model of surface characteristics. At the least, one must have a surface normal as well as surface evaluation. Techniques for rendering surfaces on graphical devices have fallen into two main categories: those that operate directly on the surface and those that require the surfaces to be divided by some means into a faceted (polyhedral) model. Since the polyhedral model is really supposed to represent smoothly-surfaced objects, many very tiny facets are needed to represent the object if the viewer is close. However, if the viewer is far away, the brute force method requires performance of much more computation than is justifiable. Thus, it would be desirable to have a database of polyhedra which changes to suit the particular image being created.

Additionally, the free-form surface design process is usually top down. That is, a rough idea of the shape is known. After that shape is attained, the designer generally would like to refine his ideas and modify the curve in only certain parts. Being local, B-spline representations allow for this type of modification. However, it is rare that, at the beginning of the design process, the designer will know the number and location of the degrees of freedom (knots) required to attain the final design. Over-specification of knots and parameters at the initial stages will require the designer to manipulate parameters, which will not be needed until later, in a very cumbersome and laborious way. It may, in fact, inhibit the attainment of the final design by causing the designer to stumble over his own feet, so to speak. Under-specification of parameters will prevent attainment of the final design. The designer will have to settle for something less than he really would like. Since splines are piecewise polynomials, one should be able to insert knots at will at arbitrary locations to attain immediate flexibility. In 1980 Lane and Riesenfeld [49] developed algorithms for the subdivision of Bézier curves. The computation of the new B-spline coefficients, however, required cumbersome matrix solutions to linear systems of equations. In 1980, Cohen, Lyche, and Riesenfeld [20] used the theory of discrete B-splines [28, 76] to provide a basis for the theoretical development of several general recursive computational algorithms. The algorithms presented in both papers used geometric properties of the Bézier curves and B-spline curves to allow dynamically adaptive methods, useful for may purposes, including curve and surface rendering and finding intersections. We discuss the algorithms and the strategy of refinement and *divide and conquer* embedded within them.

Suppose $\boldsymbol{\tau}$ and $\boldsymbol{t}$ are knot vectors, with lengths $n+\kappa+2$ and $m+\kappa+2$, respectively, $m > n \geq \kappa$. If $\boldsymbol{t}$ contains the elements of $\boldsymbol{\tau}$ with the same or greater multiplicity, then $\mathcal{S}_{\kappa,\boldsymbol{\tau}} \subset \mathcal{S}_{\kappa,\boldsymbol{t}}$, by Theorem 7.7. $\boldsymbol{t}$ is called a

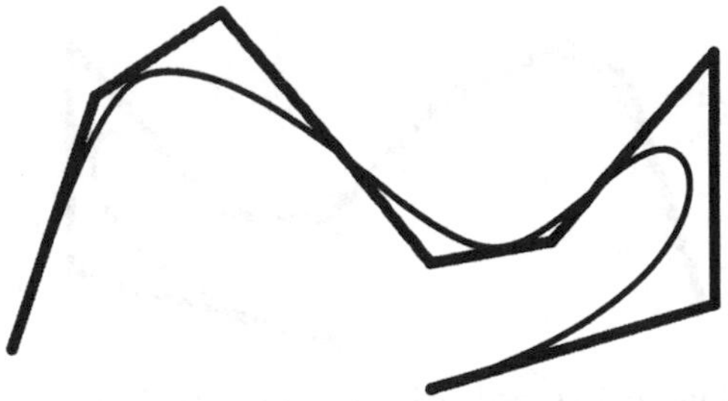

Figure 16.1. Curve and original polygon; $\boldsymbol{\tau} = \{0,0,0,0,1,4,5,7,10,10,10,10\}$.

refinement of $\boldsymbol{\tau}$. Thus, if

$$\gamma(t) = \sum_{i=0}^{n} P_i \mathcal{B}_{i,\kappa}(t) \in \mathcal{S}_{\kappa,\boldsymbol{\tau}},$$

there exist D_j's such that

$$\gamma(t) = \sum_{j=0}^{m} D_j \mathcal{N}_{j,\kappa}(t) \in \mathcal{S}_{\kappa,\boldsymbol{\tau}},$$

where $\{\mathcal{B}_{i,\kappa}\}_i$ and $\{\mathcal{N}_{j,\kappa}\}_j$ are the B-spline functions of degree κ with respect to $\boldsymbol{\tau}$ and $\boldsymbol{t}$, respectively. Corollary 7.9 developed an algorithm for an arbitrary $\boldsymbol{\tau}$ knot vector with the $\boldsymbol{t}$ knot vector having a single additional knot value. We used that approach to prove Theorem 7.7 by repeated applications of that constructive algorithm.

For floating uniform quadratic and cubic curves and floating uniform bi-quadratic and bi-cubic surfaces, we have already developed algorithms for finding those coefficients. (See Sections 7.2.1 and 13.4.)

In this chapter we seek to determine constructive algorithms that can be applied to an arbitrary refinement knot vector $\boldsymbol{t}$ of an arbitrary knot vector $\boldsymbol{\tau}$. Figure 16.1 shows a polygon and cubic spline curve with knot vector $\tau = \{0, 0, 0, 0, 1, 4, 5, 7, 10, 10, 10, 10\}$ which has no apparent symmetries. Figure 16.2 shows the polygon for the same curve represented with an additional knot at $t = 2$, that is, $\boldsymbol{t} = \{0, 0, 0, 0, 1, 2, 4, 5, 7, 10, 10, 10, 10\}$. Figure 16.3 shows the polygon for the refined curve where $\boldsymbol{t} = \{0, 0, 0, 0, .5, 1, 2.5, 4, 4.5, 5, 6, 7, 8.5, 10, 10, 10, 10\}$. Figure 16.4 has $\boldsymbol{t} = \{0, 0, 0, 0, .25, .75, 1, 2, 3, 4, 4.4, 4.7, 5, 6, 6, 7, 8, 9, 10, 10, 10, 10\}$. Figure 16.5 studies a slightly different phenomenon. Here knots of multiplicity four (the order of the curve) are inserted at several locations, which will effect spline subdivision (this will be discussed later). It has knot vector $\boldsymbol{t} = \{0, 0, 0, 0, 1, 2, 2, 2, 2, 4, 5, 7, 7, 7, 7, 10, 10, 10, 10\}$. Figure 16.6 shows the *original* polygon and knot vector as the result

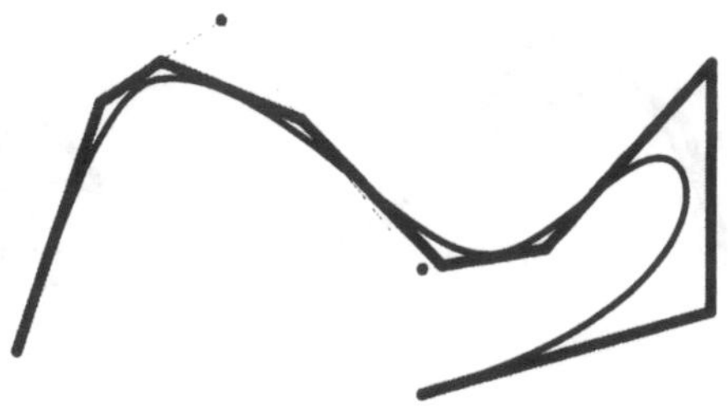

Figure 16.2. Original and refined polygons for insertion of a single new knot; $t = \{0, 0, 0, 0, 1, 2, 4, 5, 7, 10, 10, 10, 10\}$.

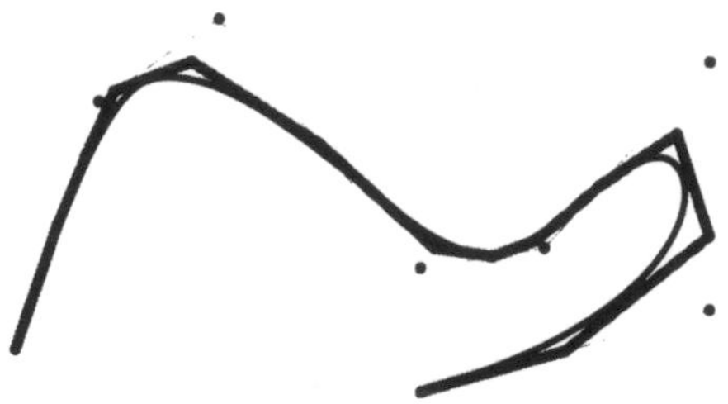

Figure 16.3. Original and refined, new knot halfway between two distinct τ's, $t = \{0, 0, 0, 0, .5, 1, 2.5, 4, 4.5, 5, 6, 7, 8.5, 10, 10, 10, 10\}$.

Figure 16.4. Original and refined, two new knots between two distinct τ's; $t = \{0, 0, 0, 0, .25, .75, 1, 2, 3, 4, 4.4, 4.7, 5, 6, 6, 7, 8, 9, 10, 10, 10, 10\}$.

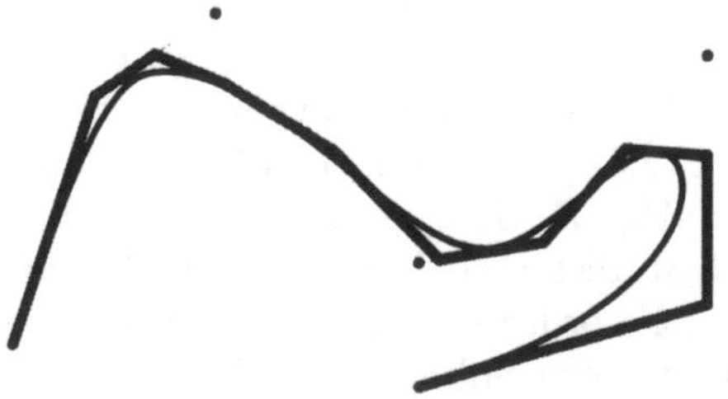

Figure 16.5. Original and refined polygons, multiple knot insertion at a single position; $t = \{0, 0, 0, 0, 1, 2, 2, 2, 2, 4, 5, 7, 7, 7, 7, 10, 10, 10, 10\}$.

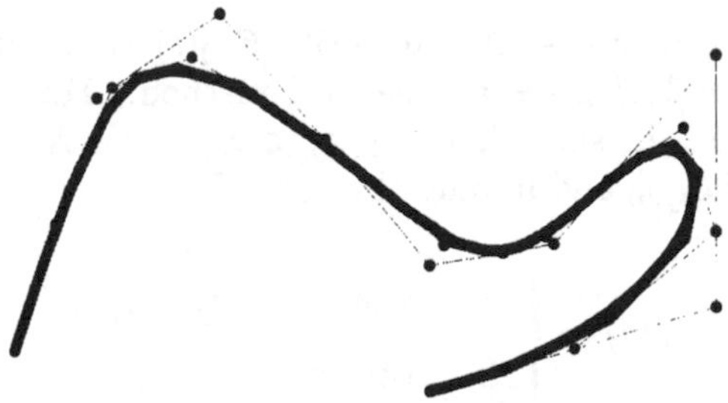

Figure 16.6. Refining the refined curve of Figure 16.3 in the same way; $\boldsymbol{t} = \{0, 0, 0, 0, .25, .5, .75, 1, 1.75, 2.5, 3.25, 4, 4.25, 4.5, 4.75, 5, 5.5, 6, 6.5, 7, 7.75, 8.5, 9.25, 10, 10, 10, 10\}$.

of Figure 16.3, that is $\boldsymbol{\tau} = \{0, 0, 0, 0, .5, 1, 2.5, 4, 4.5, 5, 6, 7, 8.5, 10, 10, 10, 10\}$. It creates a new $\boldsymbol{t} = \{0, 0, 0, 0, .25, .5, .75, 1, 1.75, 2.5, 3.25, 4, 4.25, 4.5, 4.75, 5, 5.5, 6, 6.5, 7, 7.75, 8.5, 9.25, 10, 10, 10, 10\}$. It is particularly interesting to note the shape and length of the sides of these successive polygons. Their convergence characteristics to the curve, as well as their rate of convergence, has recently been shown, and will be discussed in Section 16.2.

For an arbitrary knot vector $\boldsymbol{\tau}$ and an arbitrary refinement $\boldsymbol{t}$, the case in which, for a value of I, $0 \le I \le n$, $P_q = \delta_{I,q}$ corresponds to writing each $\mathcal{B}_{I,\kappa}$ as a sum of the $\mathcal{N}$ functions. We give the coefficients of the $\mathcal{N}$'s special names for that case. First, however, we develop some basic intuition by looking at the arbitrary constant and linear cases.

For $\kappa = 0$,

$$\mathcal{B}_{i,0}(t) = \begin{cases} 1 & \text{for } \tau_i \le t < \tau_{i+1} \\ 0 & \text{otherwise,} \end{cases}$$

and

$$\mathcal{N}_{j,0}(t) = \begin{cases} 1 & \text{for } t_j \le t < t_{j+1}, \\ 0 & \text{otherwise.} \end{cases}$$

Fix i, and assume $\tau_i < \tau_{i+1}$. Let $j_{i,1} + 1$ be the smallest subscript j such that $\tau_i < t_j$, and let $j_{i,2}$ be the largest subscript j such that $t_j < \tau_{i+1}$. Then,

$$\tau_i = t_{j_{i,1}} < t_{j_{i,1}+1} \le \cdots \le t_{j_{i,2}} < t_{j_{i,2}+1} = \tau_{i+1},$$

so

$$\mathcal{B}_{i,0}(t) = \sum_{q=j_{i,1}}^{j_{i,2}} \mathcal{N}_{q,0}(t). \tag{16.1}$$

But now, suppose it is desirable to write $\mathcal{B}_{i,0}(t)$ as a curve in terms of all $\mathcal{N}$'s. Equation 16.1 gives a a general method. For a given j, there exists exactly one i, i_j, such that $\tau_i \leq t_j < \tau_{i+1}$, and $\mathcal{N}_{j,0}$ appears in the decomposition of $\mathcal{B}_{i_j,0}$, and in only that one. Let

$$\alpha_{i,0}(j) = \begin{cases} 1, & \text{when } \tau_i \leq t_j < \tau_{i+1} \\ 0 & \text{otherwise.} \end{cases} \tag{16.2}$$

Then Equation 16.1 can be rewritten,

$$\mathcal{B}_{i,0}(t) = \sum_j \alpha_{i,0}(j)\mathcal{N}_{j,0}(t). \tag{16.3}$$

Observe that this equation is true even when $\tau_i = \tau_{i+1}$. Since that is true, then there can be no j for which $\tau_i \leq t_j < \tau_{i+1}$, so $\alpha_{i,0}(j) = 0$ for all j.

Also, for all piecewise constant functions in $\mathcal{S}_{0,\boldsymbol{\tau}}$,

$$\begin{aligned} \gamma_0(t) &= \sum_i P_i \mathcal{B}_{i,0}(t) \\ &= \sum_i P_i \left(\sum_j \alpha_{i,0}(j)\mathcal{N}_{j,0}(t) \right) \\ &= \sum_j \left(\sum_i P_i \alpha_{i,0}(j) \right) \mathcal{N}_{j,0}(t). \end{aligned}$$

By first developing formulations to write each $\mathcal{B}_{i,1}$ as a summation over all $\mathcal{N}_{j,1}$, it is possible to give a constructive algorithm for writing functions in $\mathcal{S}_{1,\boldsymbol{\tau}}$ as functions in $\mathcal{S}_{1,\boldsymbol{t}}$. We search for coefficients $\alpha_{i,1}(j)$ such that for all i, $\mathcal{B}_{i,1}(t) = \sum_j \alpha_{i,1}(j)\mathcal{N}_{j,1}(t)$.

By Equation 9.15 in Example 9.17,

$$\mathcal{B}_{i,1}(t) = \sum_j \mathcal{B}_{i,1}(t_{j+1})\mathcal{N}_{j,1}(t).$$

From their definitions,

$$\begin{aligned} \mathcal{B}_{i,0}(t_{j+1}) &= \begin{cases} 1 & \tau_i \leq t_{j+1} < \tau_{i+1} \\ 0 & \text{otherwise} \end{cases} \\ &= \alpha_{i,0}(j+1) \end{aligned}$$

and

$$\begin{aligned} \mathcal{B}_{i+1,0}(t_{j+1}) &= \begin{cases} 1 & \tau_{i+1} \le t_{j+1} < \tau_{i+2} \\ 0 & \text{otherwise} \end{cases} \\ &= \alpha_{i+1,0}(j+1); \end{aligned}$$

$$\begin{aligned} \alpha_{i,1}(j) &= \mathcal{B}_{i,1}(t_{j+1}) \\ &= \frac{t_{j+1} - \tau_i}{\tau_{i+1} - \tau_i}\mathcal{B}_{i,0}(t_{j+1}) + \frac{\tau_{i+2} - t_{j+1}}{\tau_{i+2} - \tau_{i+1}}\mathcal{B}_{i+1,0}(t_{j+1}) \\ &= \frac{t_{j+1} - \tau_i}{\tau_{i+1} - \tau_i}\alpha_{i,0}(j+1) + \frac{\tau_{i+2} - t_{j+1}}{\tau_{i+2} - \tau_{i+1}}\alpha_{i+1,0}(j+1). \quad (16.4) \end{aligned}$$

Thus, we have one step towards defining the general case. However now, consider the following five cases.

1. $t_j \le t_{j+1} < \tau_i$ or $\tau_{i+2} \le t_j \le t_{j+1}$;
2. $t_j < t_{j+1} = \tau_i$;
3. $\tau_i \le t_j \le t_{j+1} < \tau_{i+1}$ or $\tau_{i+1} \le t_j \le t_{j+1} < \tau_{i+2}$;
4. $\tau_i \le t_j < t_{j+1} = \tau_{i+1}$;
5. $\tau_{i+1} \le t_j < t_{j+1} = \tau_{i+2}$.

In Case 1, $\mathcal{B}_{i,0}(t_{j+1}) = 0 = \alpha_{i,0}(j)$ and $\mathcal{B}_{i+1,0}(t_{j+1}) = 0 = \alpha_{i+1,0}(j)$, so

$$\begin{aligned} \alpha_{i,1}(j) &= \mathcal{B}_{i,1}(t_{j+1}) \\ &= \frac{t_{j+1} - \tau_i}{\tau_{i+1} - \tau_i}\mathcal{B}_{i,0}(t_{j+1}) + \frac{\tau_{i+2} - t_{j+1}}{\tau_{i+2} - \tau_{i+1}}\mathcal{B}_{i+1,0}(t_{j+1}) \\ &= \frac{t_{j+1} - \tau_i}{\tau_{i+1} - \tau_i}\alpha_{i,0}(j) + \frac{\tau_{i+2} - t_{j+1}}{\tau_{i+2} - \tau_{i+1}}\alpha_{i+1,0}(j). \quad (16.5) \end{aligned}$$

In Case 2, $\mathcal{B}_{i+1,0}(t_{j+1}) = 0 = \alpha_{i+1,0}(j)$ and

$$\frac{t_{j+1} - \tau_i}{\tau_{i+1} - \tau_i}\mathcal{B}_{i,0}(t_{j+1}) = \frac{t_{j+1} - \tau_i}{\tau_{i+1} - \tau_i}\alpha_{i,0}(j),$$

so Equation 16.5 is valid for this case.

Case 3 occurs when both t_j and t_{j+1} are in the same span of $\boldsymbol{\tau}$. In that case, $\mathcal{B}_{i,0}(t_{j+1}) = \alpha_{i,0}(j+1) = \alpha_{i,0}(j)$ and $\mathcal{B}_{i+1,0}(t_{j+1}) = \alpha_{i+1,0}(j+1) = \alpha_{i+1,0}(j)$, so once again Equation 16.5 is valid.

In Case 4, $\alpha_{i,1}(j) = \mathcal{B}_{i,1}(t_{j+1}) = \mathcal{B}_{i,0}(\tau_{i+1}) + \mathcal{B}_{i+1,0}(\tau_{i+1}) = 1$ since only one of $\mathcal{B}_{i,0}(t_{j+1})$ and $\mathcal{B}_{i+1,0}(\tau_{i+1})$ can be nonzero. Further, in Case 4,

$$1 = \frac{t_{j+1} - \tau_i}{\tau_{i+1} - \tau_i}\alpha_{i,0}(j) + \frac{\tau_{i+2} - t_{j+1}}{\tau_{i+2} - \tau_{i+1}}\alpha_{i+1,0}(j),$$

so once again Equation 16.5 is true. Finally, in Case 5, $\alpha_{i,0}(\jmath) = 0$, $\alpha_{i+1,0}(\jmath) = 1$, $\alpha_{i,0}(\jmath) = 0$, $\alpha_{i+1,0}(\jmath) = 1$, and $\frac{\tau_{i+2}-t_{j+1}}{\tau_{i+2}-\tau_{i+1}} = 0$. Since $\mathcal{B}_{i,1}(t_{j+1}) = \mathcal{B}_{i,1}(\tau_{i+2}) = 0$, Equation 16.5 is valid. Thus we have shown that

$$\begin{aligned}
\alpha_{i,1}(\jmath) &= \mathcal{B}_{i,1}(t_{j+1}) \\
&= \frac{t_{j+1}-\tau_i}{\tau_{i+1}-\tau_i}\alpha_{i,0}(\jmath+1) + \frac{\tau_{i+2}-t_{j+1}}{\tau_{i+2}-\tau_{i+1}}\alpha_{i+1,0}(\jmath+1) \quad (16.6)\\
&= \frac{t_{j+1}-\tau_i}{\tau_{i+1}-\tau_i}\alpha_{i,0}(\jmath) + \frac{\tau_{i+2}-t_{j+1}}{\tau_{i+2}-\tau_{i+1}}\alpha_{i+1,0}(\jmath). \quad (16.7)
\end{aligned}$$

In moving to arbitrary κ, the problem gets more complex. From Example 9.16,

$$\mathcal{B}_{i,\kappa}(t) = \sum_j \lambda_{j,\kappa,\boldsymbol{t}_{\kappa}^*}(\mathcal{B}_{i,\kappa})\mathcal{N}_{j,\kappa}(t)$$

where $\lambda_{j,\kappa,\boldsymbol{t}_{\kappa}^*}(\mathcal{B}_{i,\kappa})$ is defined in Definition 9.12. Hence,

$$\begin{aligned}
\alpha_{i,\kappa}(\jmath) &= \lambda_{j,\kappa,\boldsymbol{t}_{\kappa}^*}(\mathcal{B}_{i,\kappa}) \\
&= \sum_{r=0}^{\kappa}(-1)^r\psi_{j,\kappa}^{(\kappa-r)}(t_{\kappa,j}^*)\mathcal{B}_{i,\kappa}^{(r)}(t_{\kappa,j}^*),
\end{aligned}$$

where

$$\psi_{j,\kappa}(y) = \begin{cases} 1, & \text{for } \kappa = 0,\\ \frac{\prod_{r=1}^{\kappa}(y-t_{j+r})}{\kappa!}, & \text{for } \kappa > 0 \end{cases}$$

and

$$t_{\kappa,j}^* = \frac{1}{\kappa}\sum_{r=1}^{\kappa} t_{j+r}.$$

This representation is not easily computed. We seek a more straightforward approach, and look to generalizing the forms in Equations 16.6 and 16.7. We get two distinct possibilities.

Define the two generalizations as

$$\beta_{i,\kappa,\boldsymbol{\tau},\boldsymbol{t}}(\jmath) = \begin{cases} 1 & \text{if } \tau_i \le t_j < \tau_{i+1} \text{ and } \kappa = 0\\ 0 & \text{otherwise, if } \kappa = 0\\ \frac{t_{j+1}-\tau_i}{\tau_{i+\kappa}-\tau_i}\beta_{i,\kappa-1,\boldsymbol{\tau},\boldsymbol{t}}(\jmath+1) & \\ \quad + \frac{\tau_{i+\kappa+1}-t_{j+1}}{\tau_{i+\kappa+1}-\tau_{i+1}}\beta_{i+1,\kappa-1,\boldsymbol{\tau},\boldsymbol{t}}(\jmath+1) & \kappa > 0, \end{cases} \quad (16.8)$$

and

$$\eta_{i,\kappa,\boldsymbol{\tau},\boldsymbol{t}}(j) = \begin{cases} 1 & \text{if } \tau_i \leq t_j < \tau_{i+1} \text{ and } \kappa = 0 \\ 0 & \text{otherwise, if } \kappa = 0 \\ \frac{t_{j+\kappa}-\tau_i}{\tau_{i+\kappa}-\tau_i}\eta_{i,\kappa-1,\boldsymbol{\tau},\boldsymbol{t}}(j) & \\ \quad + \frac{\tau_{i+\kappa+1}-t_{j+\kappa}}{\tau_{i+\kappa+1}-\tau_{i+1}}\eta_{i+1,\kappa-1,\boldsymbol{\tau},\boldsymbol{t}}(j) & \kappa > 0. \end{cases} \tag{16.9}$$

From Equations 16.2, 16.6, and 16.7, for all i and j, and for $r = 0, 1$,

$$\alpha_{i,r}(j) = \beta_{i,r}(j) = \eta_{i,r}(j)$$

In this chapter, we reserve the coefficient name "α" for the scalars such that

$$\mathcal{B}_{i,\kappa}(t) = \sum_j \alpha_{i,\kappa}(j)\mathcal{N}_{j,\kappa}(t).$$

In what follows, we determine if either of the generalizations can be shown to be constructive definitions of the α's that we seek. First, however, we show that the two generalizations give the same values.

Lemma 16.1. *Given* $\boldsymbol{t}$ *a refinement of* $\boldsymbol{\tau}$, *for* $\{\eta_{i,\kappa}(j)\}$ *defined in Equation 16.9, and* $\{\beta_{i,\kappa}(j)\}$ *defined in Equation 16.8, then for all* i, j, *and* $\kappa > 0$,

$$\eta_{i,\kappa}(j) = \beta_{i,\kappa}(j).$$

Proof: The case for $\kappa = 1$, is shown above.

Now, let $\kappa > 1$, and suppose it is true that for $r < \kappa$,

$$\eta_{i,r}(j) = \beta_{i,r}(j)$$

then,

$$\begin{aligned} \eta_{i,\kappa}(j) &= \frac{t_{j+\kappa}-\tau_i}{\tau_{i+\kappa}-\tau_i}\eta_{i,\kappa-1}(j) + \frac{\tau_{i+\kappa+1}-t_{j+\kappa}}{\tau_{i+\kappa+1}-\tau_{i+1}}\eta_{i+1,\kappa-1}(j) \\ &= \frac{t_{j+\kappa}-\tau_i}{\tau_{i+\kappa}-\tau_i}\beta_{i,\kappa-1}(j) + \frac{\tau_{i+\kappa+1}-t_{j+\kappa}}{\tau_{i+\kappa+1}-\tau_{i+1}}\beta_{i+1,\kappa-1}(j) \\ &= \frac{t_{j+\kappa}-\tau_i}{\tau_{i+\kappa}-\tau_i}\left(\frac{t_{j+1}-\tau_i}{\tau_{i+\kappa-1}-\tau_i}\beta_{i,\kappa-2}(j+1)\right. \\ &\qquad \left. + \frac{\tau_{i+\kappa}-t_{j+1}}{\tau_{i+\kappa}-\tau_{i+1}}\beta_{i+1,\kappa-2}(j+1)\right) \end{aligned}$$

$$
\begin{aligned}
& + \frac{\tau_{i+\kappa+1} - t_{j+\kappa}}{\tau_{i+\kappa+1} - \tau_{i+1}} \left(\frac{t_{j+1} - \tau_{i+1}}{\tau_{i+\kappa} - \tau_{i+1}} \beta_{i+1,\kappa-2}(j+1) \right. \\
& \qquad \left. + \frac{\tau_{i+\kappa+1} - t_{j+1}}{\tau_{i+\kappa+1} - \tau_{i+2}} \beta_{i+2,\kappa-2}(j+1) \right) \\
= & \ \frac{t_{j+1} - \tau_i}{\tau_{i+\kappa} - \tau_i} \left(\frac{t_{j+\kappa} - \tau_i}{\tau_{i+\kappa-1} - \tau_i} \beta_{i,\kappa-2}(j+1) \right. \\
& \qquad \left. + \frac{\tau_{i+\kappa} - t_{j+\kappa}}{\tau_{i+\kappa} - \tau_{i+1}} \beta_{i+1,\kappa-2}(j+1) \right) \\
& + \frac{t_{j+\kappa} - t_{j+1}}{\tau_{i+\kappa} - \tau_{i+1}} \beta_{i+1,\kappa-2}(j+1) - \frac{t_{j+\kappa} - t_{j+1}}{\tau_{i+\kappa} - \tau_{i+1}} \beta_{i+1,\kappa-2}(j+1) \\
& + \frac{\tau_{i+\kappa+1} - t_{j+1}}{\tau_{i+\kappa+1} - \tau_{i+1}} \left(\frac{t_{j+\kappa} - \tau_{i+1}}{\tau_{i+\kappa} - \tau_{i+1}} \beta_{i+1,\kappa-2}(j+1) \right. \\
& \qquad \left. + \frac{\tau_{i+\kappa+1} - t_{j+\kappa}}{\tau_{i+\kappa+1} - \tau_{i+2}} \beta_{i+2,\kappa-2}(j+1) \right) \\
= & \ \frac{t_{j+1} - \tau_i}{\tau_{i+\kappa} - \tau_i} \left(\frac{t_{j+\kappa} - \tau_i}{\tau_{i+\kappa-1} - \tau_i} \eta_{i,\kappa-2}(j+1) \right. \\
& \qquad \left. + \frac{\tau_{i+\kappa} - t_{j+\kappa}}{\tau_{i+\kappa} - \tau_{i+1}} \eta_{i+1,\kappa-2}(j+1) \right) \\
& + \frac{\tau_{i+\kappa+1} - t_{j+1}}{\tau_{i+\kappa+1} - \tau_{i+1}} \left(\frac{t_{j+\kappa} - \tau_{i+1}}{\tau_{i+\kappa} - \tau_{i+1}} \eta_{i+1,\kappa-2}(j+1) \right. \\
& \qquad \left. + \frac{\tau_{i+\kappa+1} - t_{j+\kappa}}{\tau_{i+\kappa+1} - \tau_{i+2}} \eta_{i+2,\kappa-2}(j+1) \right) \\
= & \ \frac{t_{j+1} - \tau_i}{\tau_{i+\kappa} - \tau_i} \eta_{i,\kappa-1}(j+1) + \frac{\tau_{i+\kappa+1} - t_{j+1}}{\tau_{i+\kappa+1} - \tau_{i+1}} \eta_{i+1,\kappa-1}(j+1) \\
= & \ \frac{t_{j+1} - \tau_i}{\tau_{i+\kappa} - \tau_i} \beta_{i,\kappa-1}(j+1) + \frac{\tau_{i+\kappa+1} - t_{j+1}}{\tau_{i+\kappa+1} - \tau_{i+1}} \beta_{i+1,\kappa-1}(j+1) \\
= & \ \beta_{i,\kappa}(j)
\end{aligned}
$$

and the lemma is proved. ■

16.1 Refinement Theorems and Algorithms

When there is no confusion over the choice of $\boldsymbol{\tau}$ and $\boldsymbol{t}$ those subscripts will be omitted from use, just as they are for continuous B-splines.

Since the generalizations of the degree 1 coefficients are equivalent, we immediately investigate whether these constructively defined sequences of coefficients are the α's we seek.

For $r > 1$, define $R_{i,r}(t) = \sum_j \eta_{i,r}(j)\mathcal{N}_{j,r}(t)$. Now let $\kappa > 1$ be fixed, and suppose the induction hypothesis is true for $r < \kappa$, that is $\mathcal{B}_{i,r}(t) = R_{i,r}(t)$. Then, for a given i,

$$\begin{aligned} R_{i,\kappa}(t) &= \sum_j \eta_{i,\kappa}(j)\mathcal{N}_{j,\kappa}(t) \\ &= \sum_j \eta_{i,\kappa}(j)\left(\frac{t-t_j}{t_{j+\kappa}-t_j}\mathcal{N}_{j,\kappa-1}(t) + \frac{t_{j+\kappa+1}-t}{t_{j+\kappa+1}-t_{j+1}}\mathcal{N}_{j+1,\kappa-1}(t)\right) \\ & \qquad\qquad\qquad\qquad\qquad\qquad\qquad\qquad\qquad\qquad\qquad\qquad\qquad\qquad (16.10) \\ &= \sum_j \frac{t-t_j}{t_{j+\kappa}-t_j}\eta_{i,\kappa}(j)\mathcal{N}_{j,\kappa-1}(t) \\ &\quad + \sum_j \frac{t_{j+\kappa+1}-t}{t_{j+\kappa+1}-t_{j+1}}\eta_{i,\kappa}(j)\mathcal{N}_{j+1,\kappa-1}(t) \qquad (16.11) \\ &= \mathcal{I}_1 + \mathcal{I}_2 \qquad (16.12) \end{aligned}$$

We consider each of the summations in Equation 16.11 separately.

$$\begin{aligned} \mathcal{I}_1 &= \sum_j \frac{t-t_j}{t_{j+\kappa}-t_j}\eta_{i,\kappa}(j)\mathcal{N}_{j,\kappa-1}(t) \qquad (16.13) \\ &= \sum_j \frac{t-t_j}{t_{j+\kappa}-t_j}\left(\frac{t_{j+k}-\tau_i}{\tau_{i+k}-\tau_i}\eta_{i,\kappa-1}(j)\right. \\ &\qquad \left. + \frac{\tau_{i+k+1}-t_{j+k}}{\tau_{i+k+1}-\tau_{i+1}}\eta_{i+1,\kappa-1}(j)\right)\mathcal{N}_{j,\kappa-1}(t) \qquad (16.14) \\ &= \sum_j \frac{t-t_j}{t_{j+\kappa}-t_j}\frac{t_{j+k}-\tau_i}{\tau_{i+k}-\tau_i}\eta_{i,\kappa-1}(j)\mathcal{N}_{j,\kappa-1}(t) \\ &\quad + \sum_j \frac{t-t_j}{t_{j+\kappa}-t_j}\frac{\tau_{i+k+1}-t_{j+k}}{\tau_{i+k+1}-\tau_{i+1}}\eta_{i+1,\kappa-1}(j)\mathcal{N}_{j,\kappa-1}(t) \qquad (16.15) \\ &= I_{1,1} + I_{1,2} \end{aligned}$$

$$\begin{aligned} \mathcal{I}_2 &= \sum_j \frac{t_{j+\kappa+1}-t}{t_{j+\kappa+1}-t_{j+1}}\eta_{i,\kappa}(j)\mathcal{N}_{j+1,\kappa-1}(t) \qquad (16.16) \\ &= \sum_j \frac{t_{j+\kappa+1}-t}{t_{j+\kappa+1}-t_{j+1}}\left(\frac{t_{j+1}-\tau_i}{\tau_{i+\kappa}-\tau_i}\eta_{i,\kappa-1}(j+1)\right. \\ &\qquad \left. + \frac{\tau_{i+\kappa+1}-t_{j+1}}{\tau_{i+\kappa+1}-\tau_{i+1}}\eta_{i+1,\kappa-1}(j+1)\right)\mathcal{N}_{j+1,\kappa-1}(t) \qquad (16.17) \end{aligned}$$

$$
\begin{aligned}
&= \sum_j \frac{t_{j+\kappa}-t}{t_{j+\kappa}-t_j}\left(\frac{t_j-\tau_i}{\tau_{i+\kappa}-\tau_i}\eta_{i,\kappa-1}(j)\right. \\
&\qquad \left. + \frac{\tau_{i+\kappa+1}-t_j}{\tau_{i+\kappa+1}-\tau_{i+1}}\eta_{i+1,\kappa-1}(j)\right)\mathcal{N}_{j,\kappa-1}(t) \qquad (16.18) \\
&= \sum_j \frac{t_{j+\kappa}-t}{t_{j+\kappa}-t_j}\frac{t_j-\tau_i}{\tau_{i+\kappa}-\tau_i}\eta_{i,\kappa-1}(j)\mathcal{N}_{j,\kappa-1}(t) \\
&\qquad + \sum_j \frac{t_{j+\kappa}-t}{t_{j+\kappa}-t_j}\frac{\tau_{i+\kappa+1}-t_j}{\tau_{i+\kappa+1}-\tau_{i+1}}\eta_{i+1,\kappa-1}(j)\mathcal{N}_{j,\kappa-1}(t) \qquad (16.19) \\
&= I_{2,1}+I_{2,2}
\end{aligned}
$$

Now since $R_{i,\kappa}(t) = I_{1,1} + I_{1,2} + I_{2,1} + I_{2,2}$, we rearrange the order of summation and first compute $I_{1,1} + I_{2,1}$ and $I_{1,2} + I_{2,2}$.

$$
\begin{aligned}
I_{1,1}+I_{2,1} &= \sum_j \frac{t-t_j}{t_{j+\kappa}-t_j}\frac{t_{j+k}-\tau_i}{\tau_{i+k}-\tau_i}\eta_{i,\kappa-1}(j)\mathcal{N}_{j,\kappa-1}(t) \\
&\qquad + \sum_j \frac{t_{j+\kappa}-t}{t_{j+\kappa}-t_j}\frac{t_j-\tau_i}{\tau_{i+\kappa}-\tau_i}\eta_{i,\kappa-1}(j)\mathcal{N}_{j,\kappa-1}(t) \\
&= \frac{t-\tau_i}{\tau_{i+\kappa}-\tau_i}\sum_j \eta_{i,\kappa-1}(j)\mathcal{N}_{j,\kappa-1}(t) \\
&= \frac{t-\tau_i}{\tau_{i+\kappa}-\tau_i}\mathcal{B}_{i,\kappa-1}(t), \qquad (16.20)
\end{aligned}
$$

and

$$
\begin{aligned}
I_{1,2}+I_{2,2} &= \sum_j \frac{t-t_j}{t_{j+\kappa}-t_j}\frac{\tau_{i+k+1}-t_{j+k}}{\tau_{i+k+1}-\tau_{i+1}}\eta_{i+1,\kappa-1}(j)\mathcal{N}_{j,\kappa-1}(t) \\
&\qquad + \sum_j \frac{t_{j+\kappa}-t}{t_{j+\kappa}-t_j}\frac{\tau_{i+\kappa+1}-t_j}{\tau_{i+\kappa+1}-\tau_{i+1}}\eta_{i+1,\kappa-1}(j)\mathcal{N}_{j,\kappa-1}(t) \\
&= \frac{\tau_{i+\kappa+1}-t}{\tau_{i+\kappa+1}-\tau_{i+1}}\sum_j \eta_{i+1,\kappa-1}(j)\mathcal{N}_{j,\kappa-1}(t) \\
&= \frac{\tau_{i+\kappa+1}-t}{\tau_{i+\kappa+1}-\tau_{i+1}}\mathcal{B}_{i+1,\kappa-1}(t). \qquad (16.21)
\end{aligned}
$$

So we see that,

$$
\begin{aligned}
R_{i,\kappa}(t) &= (I_{1,1}+I_{2,1})+(I_{1,2}+I_{2,2}) \\
&= \frac{t-\tau_i}{\tau_{i+\kappa}-\tau_i}\mathcal{B}_{i,\kappa-1}(t) + \frac{\tau_{i+\kappa+1}-t}{\tau_{i+\kappa+1}-\tau_{i+1}}\mathcal{B}_{i+1,\kappa-1}(t) \qquad (16.22) \\
&= \mathcal{B}_{i,\kappa}(t). \qquad (16.23)
\end{aligned}
$$

Equation 16.22 is a result of substitution from Equations 16.20 and 16.21, while Equation 16.23 follows from the recursive definition. We have shown that the recursively defined η's are the coefficients we seek.

Definition 16.2. *Define*

$$\alpha_{i,\kappa,\boldsymbol{\tau},\boldsymbol{t}}(j) = \begin{cases} 1 & \text{if } \tau_i \le t_j < \tau_{i+1} \text{ and } \kappa = 0 \\ 0 & \text{otherwise, if } \kappa = 0 \\ \frac{t_{j+\kappa}-\tau_i}{\tau_{i+\kappa}-\tau_i}\alpha_{i,\kappa-1,\boldsymbol{\tau},\boldsymbol{t}}(j) & \\ \quad + \frac{\tau_{i+\kappa+1}-t_{j+\kappa}}{\tau_{i+\kappa+1}-\tau_{i+1}}\alpha_{i+1,\kappa-1,\boldsymbol{\tau},\boldsymbol{t}}(j) & \kappa > 0. \end{cases}$$

The $\alpha_{i,\kappa}(j) = \alpha_{i,\kappa,\boldsymbol{\tau},\boldsymbol{t}}(j)$ *are called* discrete B-splines of degree κ on $\boldsymbol{t}$ with knots $\boldsymbol{\tau}$.

The basis refinement theorem follows from what has been shown above.

Theorem 16.3. *Let* $\boldsymbol{t}$ *be a knot vector refinement of* $\boldsymbol{\tau}$, *and let* $\mathcal{B}$ *be the B-splines over* $\boldsymbol{\tau}$, *while* $\mathcal{N}$ *are the B-splines over* $\boldsymbol{t}$. *Then, for* $\kappa > 0$ *an integer,*

$$\mathcal{B}_{i,\kappa} = \sum_j \alpha_{i,\kappa}(j)\mathcal{N}_{j,\kappa}(t).$$

Lemma 16.4. *Given* $\boldsymbol{t}$ *a refinement of* $\boldsymbol{\tau}$, *for* $\{\alpha_{i,\kappa}(j)\}$ *defined in Definition 16.2 then for all* i, j, *and* $\kappa > 0$,

$$\alpha_{i,\kappa}(j) = \frac{t_{j+1}-\tau_i}{\tau_{i+\kappa}-\tau_i}\alpha_{i,\kappa-1,\boldsymbol{\tau},\boldsymbol{t}}(j+1) + \frac{\tau_{i+\kappa+1}-t_{j+1}}{\tau_{i+\kappa+1}-\tau_{i+1}}\alpha_{i+1,\kappa-1,\boldsymbol{\tau},\boldsymbol{t}}(j+1).$$

Proof: The proof follows from Lemma 16.1. ∎

The following properties immediately follow from the definition of $\alpha_{i,\kappa}(j)$.

Lemma 16.5.

1. *For* $0 \le j \le m$, *let* J *be such that* $\tau_J \le t_j < \tau_{J+1}$. *Then* $\alpha_{i,\kappa}(j) = 0$ *for* $i \notin \{J-\kappa, \ldots, J\}$. *That is, for any* j *there are at most* $\kappa+1$ *discrete B-splines which can be nonzero.*

2. $\alpha_{i,\kappa}(j) \ge 0$.

3. $\sum_{i=0}^{n} \alpha_{i,\kappa}(j) = 1, \qquad \tau_\kappa \le t_j < \tau_{n+1}.$

Proof: The proof is left as an exercise. ∎

Corollary 16.6. *If $\gamma(t) = \sum_i P_i \mathcal{B}_{i,\kappa,\boldsymbol{\tau}}(t)$, and $\mathbf{t}$ is a refinement of $\boldsymbol{\tau}$, then, if $t_j < t_{j+\kappa+1}$ for all j,*

$$\gamma(t) = \sum_j D_j \mathcal{N}_{j,\kappa,\mathbf{t}}(t)$$

where

$$D_j = \sum_{i=J-\kappa}^{J} P_i \alpha_{i,\kappa}(j).$$

Proof:

$$\begin{aligned}\gamma(t) &= \sum_i P_i \mathcal{B}_{i,\kappa}(t) \\ &= \sum_i P_i \left(\sum_j \alpha_{i,\kappa}(j) \mathcal{N}_{j,\kappa}(t) \right) \\ &= \sum_j \left(\sum_i P_i \alpha_{i,\kappa}(j) \right) \mathcal{N}_{j,\kappa}(t).\end{aligned}$$

It remains to determine, for a given j, the set of i at which $\alpha_{i,\kappa}(j) \neq 0$. This is done by induction. Let j be fixed, but arbitrary. There exists exactly one subscript J of $\boldsymbol{\tau}$ for which $\tau_J \le t_j < \tau_{J+1}$. Hence, $\{ i : \alpha_{i,0}(j) > 0 \} = \{J\}$.

Suppose that for $k < \kappa$, $\{ i : \alpha_{i,k}(j) > 0 \} \subseteq \{J - k, J - k + 1, \ldots, J\}$. Then, since

$$\alpha_{i,\kappa}(j) = \frac{t_{j+\kappa} - \tau_i}{\tau_{i+\kappa} - \tau_i} \alpha_{i,\kappa-1,\boldsymbol{\tau},\mathbf{t}}(j) + \frac{\tau_{i+\kappa+1} - t_{j+\kappa}}{\tau_{i+\kappa+1} - \tau_{i+1}} \alpha_{i+1,\kappa-1,\boldsymbol{\tau},\mathbf{t}}(j),$$

$\alpha_{i,\kappa}(j)$ might be nonzero only if

$$\max\left(\alpha_{i,\kappa-1,\boldsymbol{\tau},\mathbf{t}}(j), \alpha_{i+1,\kappa-1,\boldsymbol{\tau},\mathbf{t}}(j) \right) > 0,$$

so, $\{ i : \alpha_{i,\kappa}(j) > 0 \}) \subseteq \{ i : \alpha_{i,\kappa-1}(j) > 0 \} \cup \{ i : \alpha_{i+1,\kappa-1}(j) > 0 \}$.

By the induction hypothesis,

$$\begin{aligned}\{ i : \alpha_{i,\kappa-1}(j) > 0 \}) &\subseteq \{J - \kappa + 1, J - \kappa + 2, \ldots, J\} \\ \{ i : \alpha_{i+1,\kappa-1}(j) > 0 \}) &\subseteq \{J - \kappa, J - \kappa + 1, \ldots, J - 1\}.\end{aligned}$$

Hence $\{ i : \alpha_{i,\kappa}(j) > 0 \}) \subseteq \{J - \kappa, J - \kappa + 1, \ldots, J - 1, J\}$. ∎

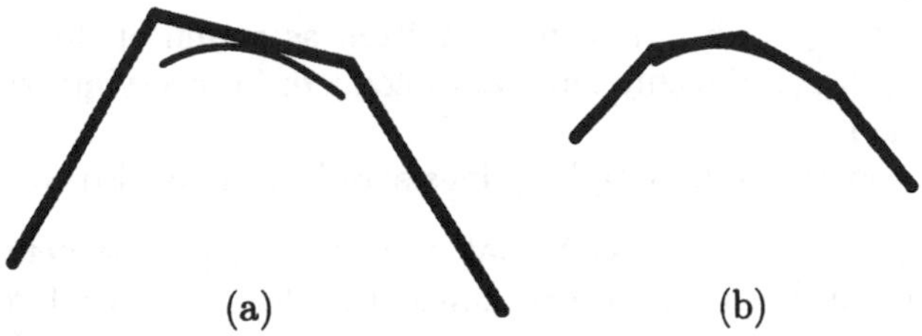

Figure 16.7. (a) Floating polygon and curve, $\boldsymbol{\tau} = \{0, 1, 2, 3, 4, 5, 6, 7\}$; (b) refined polygon and curve, $\boldsymbol{t} = \{1.5, 2, 2.5, 3, 3.5, 4, 4.5, 5, 5.5\}$.

In its complete generality, the results of Corollary 16.6 provide a framework for arbitrary refinement and can be used to effect subdivision of B-spline curves, all in one step. Figures 16.1 through 16.6 illustrate computational use of Corollary 16.6 in refinement.

In all the definitions and proofs, we have assumed that $t_j < t_{j+\kappa+1}$. We must also consider the case in which that is not so. This is really two special subcases: 1) $\tau_i < \tau_{i+\kappa+1}$, and $t_j \in [\tau_i, \tau_{i+\kappa+1})$, $t_j = t_{j+\kappa+1}$, and 2) $\tau_i = \tau_{i+\kappa+1}$, $t_j = t_{j+\kappa+1}$. In the first case, $\mathcal{B}_{i,\kappa}(t)$ has a non-empty interval of support, but $\mathcal{N}_{j,\kappa}(t) \equiv 0$. In the second case, $\mathcal{B}_{i,\kappa}(t) = \mathcal{N}_{j,\kappa}(t) \equiv 0$. While it would seem that in these instances the values $\alpha_{i,\kappa}(j)$ are not relevant, or should, at best be zero, one can define these values appropriately to be used in proofs of certain theoretical results concerning discrete splines, refinement, and algorithmic optimization of these results [57]. Research [56] indicates that these results and methods are optimal for certain types of calculations.

Example 16.7. Refinement for Floating Splines. Recall that a spline curve $\gamma(t) = \sum_{i=0}^{n} P_i \mathcal{B}_{i,\kappa}(t)$ for knot vector $\{\tau_i\}_{i=0}^{n+\kappa+1}$ is called a floating spline when the first $\kappa + 1$ values of the knot vector are not equal and the last $\kappa + 1$ values are not equal. As usual the domain of the function is $[\tau_\kappa, \tau_{n+1})$. For this curve $\mathcal{B}_{0,\kappa}$ influences the curve if and only if $\tau_\kappa \neq \tau_{\kappa+1}$, since that is the only span of its support which is over the domain. Analogously, $\mathcal{B}_{n,\kappa}$ influences the curve if and only if $\tau_n \neq \tau_{n+1}$. We wish to refine the curve γ and represent it over the knot vector $\{t_j\}_{j=0}^{m+\kappa+1}$, where $m > n$, and $\boldsymbol{t}$ is a refinement of $\boldsymbol{\tau}$.

This seemingly straightforward application of Corollary 16.6 can be doomed to failure. The reason is that the corollary uses J_j such that $\tau_{J_j} \le t_j < \tau_{J_j+1}$, and then computes $\alpha_{i,\kappa}(j)$ for $i = J_j - \kappa, \ldots, J_j$. This requires the existence of $\tau_{J_i+1-\kappa+r}$, $r = 0, \ldots, 2\kappa - 1$. If $j = 0$ then $\tau_0 = t_0$, but there do not exist $\mathcal{B}_{-\kappa,\kappa}, \ldots, \mathcal{B}_{-1,\kappa}$ for which one needs to find coefficients.

Now, since $t_\kappa \le \tau_\kappa$, $\mathcal{B}_{0,\kappa}(t)$ can be written as a sum of the B-splines $\mathcal{N}_{j,\kappa}(t)$ over $[\tau_\kappa, \tau_{\kappa+1})$, the only interval of concern for every function with domain $[\tau_\kappa, \tau_{n+1})$.

The first property of discrete B-splines should be amended to include:

> for $0 \le j \le m$, let J be such that $\tau_J \le t_j < \tau_{J+1}$, and define $M_J = \max(0, J-\kappa)$ and $U_J = \min(J, n)$. Then $\alpha_{i,\kappa}(j) = 0$ for $i \notin \{M_J, \ldots, U_J\}$.

Analogous results hold for the upper end of the domain.

Theorem 16.3, Corollary 16.6, and the results of Example 16.7 combine to yield

Algorithm 16.8. Oslo Algorithm 1. *For integers $\kappa \ge 1$ and j and J, with $M_J = \max\{0, J-\kappa\}$, and $U_J = \min\{J, n\}$, let $\tau_{M_J}, \ldots, \tau_{U_J+\kappa}$ and $t_{j+1}, \ldots, t_{j+\kappa}$ be given such that $\tau_J \le t_j < \tau_{J+1}$. Then,*

$$\alpha_{i,0}(j) = \begin{cases} 1, & \tau_i \le t_j < t_{j+1} \le \tau_{i+1}, \\ 0, & \textit{otherwise}. \end{cases}$$

Moreover, for $\kappa \ge 1$ and all i, j,

$$\alpha_{i,\kappa}(j) = \begin{cases} \frac{t_{j+\kappa}-\tau_i}{\tau_{i+\kappa}-\tau_i}\alpha_{i,\kappa-1}(j) + \frac{\tau_{i+\kappa+1}-t_{j+\kappa}}{\tau_{i+\kappa+1}-\tau_{i+1}}\alpha_{i+1,\kappa-1}(j), & i \in \{M_J, \ldots, U_J\} \\ 0, & \textit{otherwise}. \end{cases}$$

Algorithm 16.9. Oslo Algorithm 2. *Suppose τ, t, $\mathcal{B}$, $\mathcal{N}$, κ, P_i, γ, and D_j are as in Corollary 16.6. Then*

$$D_j = \sum_{i=J-\kappa+w}^{J} P_{i,j}^{[w]} \alpha_{i,\kappa-w}(j),$$

where

$$P_{i,j}^{[w]} = \begin{cases} P_{i,j}^{[w-1]} \frac{t_{j+\kappa-w+1}-\tau_i}{\tau_{i+\kappa-w+1}-\tau_i} + P_{i-1,j}^{[w-1]} \frac{\tau_{i+\kappa-w+1}-t_{j+\kappa-w+1}}{\tau_{i+\kappa-w+1}-\tau_i}, & \\ & 0 < w \le \kappa, J-\kappa+w \le i \le J \\ P_i, & w = 0. \end{cases}$$

Finally,

$$D_j = P_{J,j}^{[\kappa]}.$$

16.1.1 Special Case Algorithm

Theorem 7.8 and Corollary 7.9 developed efficient methods to determine the new coefficients of the curve when the refined knot vector has exactly one added element. In this section we develop an algorithm for the coefficients when the refined knot vector has a single element added multiple times.

It is rare that only a single knot is added. A frequent application however is to subdivide the spline curve. That is, to break a B-spline curve of degree κ on $[a,b]$, $\gamma(t)$, into nonoverlapping spline curves $\gamma_1(t)$ over $[a,c]$ and $\gamma_2(t)$ over $[c,b]$ whose unioned image curve is equal to $\gamma(t)$. This is done by inserting a knot with multiplicity $\kappa+1$, since at that value the refined curve has parametric continuity $C^{(-1)}$. This means that subdivision can be effected by selecting either a totally new value or a previous knot value and supplementing its multiplicity to obtain a multiplicity of $\kappa+1$. If t_p is such a knot, $t_{p-1} < t_p = \cdots = t_{p+\kappa} < t_{p+\kappa+1}$. D_i, $i \le p-1$ are the control points of one of the subdivided curves and D_i, $i \ge p$ are the control points of the other subdivided curve. This subdivision can be effected by applying Corollary 7.9 serially, or by calculating discrete spline coefficients.

Let $\{\tau_i\}$ be the original knot sequence, and suppose it is desired to add a single knot of multiplicity κ at a value t^*, to subdivide the curve into two distinct B-spline curves. Suppose $\tau_q \le t^* < \tau_{q+1}$. If $t^* = \tau_q$, suppose that τ_q has multiplicity r, that is, where $\tau_{q-r+1} = \cdots = \tau_q = t^*$. If $\tau_q < t^*$, set $r = 0$. Then the new sequence $\boldsymbol{t}$ has

$$t_j = \begin{cases} \tau_j, & j \le q \\ t^* & q+1 \le j \le q+\kappa-r+1 \\ \tau_{j-\kappa+r+1} & q+\kappa-r+2 \le j. \end{cases}$$

For $j \le q-\kappa$, there are no new knots, in exactly the same way as for single knot insertion, and hence only one discrete B-spline is nonzero, $\alpha_{j,\kappa}(j) = 1$, and $D_j = P_j$. For $j \ge q+\kappa+1-2r$, there are also only old knots, again in an analogous proof to that for single knot insertion, and $\alpha_{j-\kappa+r-1,\kappa}(j) = 1$, so $D_j = P_{j-\kappa+r-1}$.

For $q-\kappa+1 \le j \le q+\kappa-2r$, set $j = q-\kappa+p$, $p = 1, \ldots, 2(\kappa-r)$. We see that for $p = 1$, there is one new knot (and so two nonzero discrete B-splines), for $p = 2$, there are two new knots (and so three nonzero discrete B-splines), $\ldots$, for $p = \kappa-r$ there are $\kappa-r$ new knots (and so $\kappa-r+1$ nonzero discrete B-splines). For $p = (\kappa-r)+1$ there are $(\kappa-r)$ new knots (and so $\kappa-r+1$ nonzero discrete B-splines). Continuing in this way, for $p = 2(\kappa-r)$ there is one new knot (and so two nonzero discrete B-splines) to use to compute the new polygon points. Hence, if $r = 0$, $\kappa + 1$ knots are inserted, and there are $2(\kappa)$ polygon points which must be computed.

This information is automatically computed as part of the efficient Oslo algorithms, and no computation of zeros is done.

16.1.2 Subdivision of Bézier Curves

We begin by deriving the new coefficients which result from splitting a cubic Bézier curve at an arbitrary point.

Example 16.10. Subdivision for Cubic Bézier Curves. Let $a < b$, and let the parametric domain of the curve be the interval from a to b, and $\{P_i\}_{i=0}^{3}$ be its control polygon. The cubic B-spline knot vector to represent the Bézier curve is $\boldsymbol{\tau} = \{\tau_i\}$, where

$$\tau_i = \begin{cases} a & i = 0, \ldots, 3, \\ b & i = 4, \ldots, 7. \end{cases}$$

Let c be any value, $a < c < b$. Form the knot vector $\boldsymbol{t} = \{t_j\}$ with

$$t_j = \begin{cases} a & j = 0, \ldots, 3, \\ c & j = 4, \ldots, 6, \\ b & j = 7, \ldots, 10. \end{cases}$$

Observe that we have added a knot of multiplicity three at c. We use Algorithm 16.9 to determine the seven new control points. In this example $\kappa = 3$, and $J = 3$ for $j = 0, \ldots, 6$, that is, J is a constant fixed value for every value of j. Consider $j = 0$.

$$\begin{aligned} P_{i,0}^{[1]} &= P_{i,0}^{[0]} \frac{t_{0+3-1+1} - \tau_i}{\tau_{i+3-1+1} - \tau_i} + P_{i-1,0}^{[0]} \frac{\tau_{i+3-1+1} - t_{0+3-1+1}}{\tau_{i+3-1+1} - \tau_i} \\ &= P_{i,0}^{[0]} \frac{a - a}{b - a} + P_{i-1,0}^{[0]} \frac{b - a}{b - a} \\ &= P_{i-1,0}^{[0]} \\ &= P_{i-1} \qquad\qquad i = 1, \ldots, 3, \\ P_{i,0}^{[2]} &= P_{i,0}^{[1]} \frac{t_{0+3-2+1} - \tau_i}{\tau_{i+3-2+1} - \tau_i} + P_{i-1,0}^{[1]} \frac{\tau_{i+3-2+1} - t_{0+3-2+1}}{\tau_{i+3-2+1} - \tau_i} \\ &= P_{i-1,0}^{[1]} \\ &= P_{i-2} \qquad\qquad i = 2, \ldots, 3, \end{aligned}$$

$$\begin{aligned} P^{[3]}_{3,0} &= P^{[2]}_{3,0}\frac{t_{0+3-3+1}-\tau_3}{\tau_{3+3-3+1}-\tau_3} + P^{[2]}_{3-1,0}\frac{\tau_{3+3-3+1}-t_{0+3-3+1}}{\tau_{3+3-3+1}-\tau_3} \\ &= P^{[2]}_{3-1,0} \\ &= P_0. \end{aligned}$$

So $D_0 = P_0$.

Now, for $j = 1, 2, 3$

$$\begin{aligned} P^{[1]}_{i,j} &= P_i\frac{t_{j+3-1+1}-\tau_i}{\tau_{i+3-1+1}-\tau_i} + P_{i-1}\frac{\tau_{i+3-1+1}-t_{j+3-1+1}}{\tau_{i+3-1+1}-\tau_i} \\ &= P_i\frac{c-a}{b-a} + P_{i-1}\frac{b-c}{b-a} && (16.24) \\ &= P^{[1]}_i \qquad\qquad i = 1, \ldots, 3, && (16.25) \\ P^{[2]}_{i,j} &= P^{[1]}_i\frac{t_{j+3-2+1}-\tau_i}{\tau_{i+3-2+1}-\tau_i} + P^{[1]}_{i-1}\frac{\tau_{i+3-2+1}-t_{j+3-2+1}}{\tau_{i+3-2+1}-\tau_i} \\ & \qquad\qquad i = 2, 3. && (16.26) \end{aligned}$$

Now Equation 16.25 follows from Equation 16.24 because there is no discernible dependence on j. The same substitutions cannot be used in Equation 16.26. For $i = 2, 3$,

$$\begin{aligned} P^{[2]}_{i,j} &= P^{[1]}_i\frac{t_{j+3-2+1}-\tau_i}{\tau_{i+3-2+1}-\tau_i} + P^{[1]}_{i-1}\frac{\tau_{i+3-2+1}-t_{j+3-2+1}}{\tau_{i+3-2+1}-\tau_i} \\ &= \begin{cases} P^{[1]}_i\frac{a-a}{b-a} + P^{[1]}_{i-1}\frac{b-a}{b-a} & j = 1 \\ P^{[1]}_i\frac{c-a}{b-a} + P^{[1]}_{i-1}\frac{b-c}{b-a} & j = 2, 3. \end{cases} \\ &= \begin{cases} P^{[1]}_{i-1} & j = 1 \\ P^{[1]}_i\frac{c-a}{b-a} + P^{[1]}_{i-1}\frac{b-c}{b-a} & j = 2, 3. \end{cases} \end{aligned}$$

Then,

$$\begin{aligned} D_j = P^{[3]}_{3,j} &= P^{[2]}_{3,j}\frac{t_{j+1}-\tau_3}{\tau_4-\tau_3} + P^{[2]}_{2,j}\frac{\tau_4-t_{j+1}}{\tau_4-\tau_3} \\ &= P^{[2]}_{3,j}\frac{t_{j+1}-a}{b-a} + P^{[2]}_{2,j}\frac{b-t_{j+1}}{b-a} \end{aligned}$$

so,

$$D_j = P^{[3]}_{3,j} = \begin{cases} P^{[2]}_{2,1} = P^{[1]}_1 & j = 1 \\ P^{[2]}_{2,2} & j = 2 \\ P^{[2]}_{3,3}\frac{c-a}{b-a} + P^{[2]}_{2,3}\frac{b-c}{b-a} & j = 3. \end{cases}$$

Now for $j = 4,5,6$, $t_{j+3-1+1} = t_{j+3} = b$, so,

$$\begin{aligned} P_{i,j}^{[1]} &= P_{i,j}^{[0]} \frac{t_{j+3-1+1} - \tau_i}{\tau_{i+3-1+1} - \tau_i} + P_{i-1,j}^{[0]} \frac{\tau_{i+3-1+1} - t_{j+3-1+1}}{\tau_{i+3-1+1} - \tau_i} \\ &= P_{i,j}^{[0]} \frac{b-a}{b-a} + P_{i-1,j}^{[0]} \frac{b-b}{b-a} \\ &= P_{i,j}^{[0]} \\ &= P_i. \end{aligned}$$

Again, the cases differ at the second level for $i = 2,3$,

$$\begin{aligned} P_{i,j}^{[2]} &= P_{i,j}^{[1]} \frac{t_{j+3-2+1} - \tau_i}{\tau_{i+3-2+1} - \tau_i} + P_{i-1,j}^{[1]} \frac{\tau_{i+3-2+1} - t_{j+3-2+1}}{\tau_{i+3-2+1} - \tau_i} \\ &= \begin{cases} P_{i,4}^{[1]} \frac{c-a}{b-a} + P_{i-1,j}^{[1]} \frac{b-c}{b-a}, & j = 4 \\ P_{i,j}^{[1]} \frac{b-a}{b-a} + P_{i-1,j}^{[1]} \frac{b-b}{b-a} & j = 5,6 \end{cases} \\ &= \begin{cases} P_i \frac{c-a}{b-a} + P_{i-1} \frac{b-c}{b-a}, & j = 4 \\ P_i & j = 5,6. \end{cases} \end{aligned}$$

Finally,

$$\begin{aligned} D_j = P_{3,j}^{[3]} &= P_{3,j}^{[2]} \frac{t_{j+1} - a}{b-a} + P_{2,j}^{[2]} \frac{b - t_{j+1}}{b-a}, \\ D_4 = P_{3,4}^{[3]} &= P_{3,4}^{[2]} \frac{c-a}{b-a} + P_{2,4}^{[2]} \frac{b-c}{b-a}, \\ D_5 = P_{3,5}^{[3]} &= P_{3,5}^{[2]} \frac{c-a}{b-a} + P_{2,5}^{[2]} \frac{b-c}{b-a} \\ &= P_3 \frac{c-a}{b-a} + P_2 \frac{b-c}{b-a}, \\ D_6 = P_{3,6}^{[3]} &= P_3. \end{aligned}$$

Then $\{D_i\}_{i=0}^{3}$ is the polygon for a Bézier curve on the interval $[a,c]$ and $\{D_i\}_{i=3}^{6}$ is the polygon for a Bézier curve on the interval $[c,b]$. This example can be generalized to Theorem 16.12. □

Using Definition 5.7, we defined the Bézier curve over an arbitrary interval.

Definition 16.11. *We call $\theta[P_0, \ldots, P_n; a, b](t) = \sum_{i=0}^{n} P_i \theta_{i,n}(a,b;t)$, the Bézier curve with control points $\{P_i\}$ on the interval $[a,b]$.*

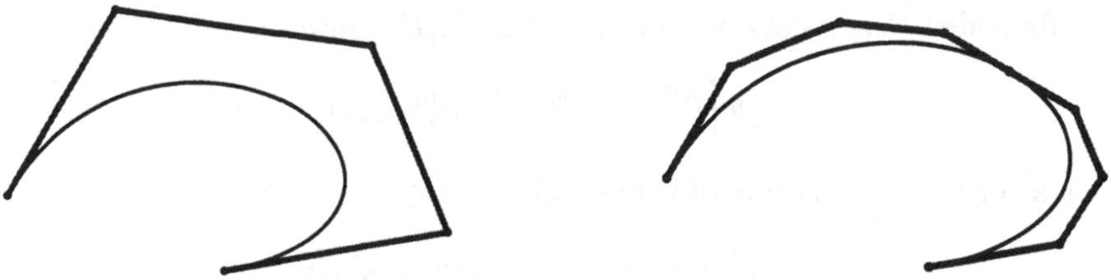

(a) (b)

Figure 16.8. (a) A Bézier curve; (b) the same Bézier curve subdivided.

Theorem 16.12. The Bézier Curve Subdivision Theorem.
For $\theta[\boldsymbol{P}; a, b]$ defined as above then

$$\theta[\boldsymbol{P}; a, b](t) = \begin{cases} \theta\left[P_0^{[0]}, P_1^{[1]}, \ldots, P_n^{[n]}; a, c\right](t) \\ \theta\left[P_n^{[n]}, P_n^{[n-1]}, \ldots, P_n^{[0]}; c, b\right](t) \end{cases}$$

where

$$P_i^{[k]} = \begin{cases} \frac{\left[(b-c)P_{i-1}^{[k-1]} + (c-a)P_i^{[k-1]}\right]}{(b-a)} & k > 0,\ k \le i \le n \\ P_i & k = 0. \end{cases}$$

This theorem follows directly from refinement properties of this special case B-spline curve. However, a separate proof using only properties of Bézier curves appears at the end of this chapter in Section 16.4.

If a Bézier curve of degree n is written in terms of the Bernstein polynomials, it is defined everywhere by its values on $(n+1)$ points, so the two

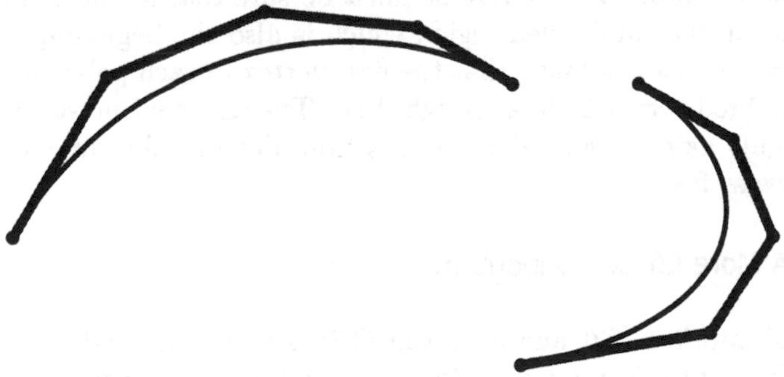

Figure 16.9. The individual Bézier polygons with their curves—separated for clarity.

subdivided Bézier curves are identical on $[a, b]$. However

$$\theta\left[P_0^{[0]}, \ldots, P_{n-1}^{[n-1]}, P_n^{[n]}; a, c\right]$$

is a convex combination of values only for $t \in (a, c)$ and

$$\theta\left[P_n^{[n]}, P_n^{[n-1]}, \ldots, P_n^{[0]}; c, b\right](t)$$

is a convex combination of values only for $t \in (c, b)$.

After the application of this process, the curve is defined as the union of the two separate Bézier curves, even though it is really just one. Now instead of $n + 1$ polygon points, there are $2n + 2$ points, $n + 1$ for each part. However, two of those points, the last of the first polygon and the first of the second polygon, are really the same point. The curve in $[a, c]$ is contained in the convex hull of

$$\left\{P_0^{[0]}, \ldots, P_{n-1}^{[n-1]}, P_n^{[n]}\right\}$$

while the curve in $[c, b]$ is contained in the convex hull of

$$\left\{P_n^{[n]}, \ldots, P_n^{[1]}, P_n^{[0]}\right\}.$$

In the context of Bézier curves, then, subdivided curves will allow for a refinement approach to design. The approach would be to design as close to the ideal curve as possible with a single Bézier curve. Then subdivide that curve into regions needing further specification and those that are completely represented. The newly created Bézier curves each have their own $n+1$ coefficients which will allow for manipulation. Note however that the designer is responsible for keeping all continuity requirements. That is, if the designer wants a $C^{(1)}$curve he must be sure that at the end point of a polygon for one subdivided region which is also the beginning point for the next region, care is taken that the first vertex in each polygon together with the shared vertex lie in a straight line. The care that must be taken is in fact highly akin to that which results from defining the curve originally in a piecewise fashion.

16.1.3 A More Efficient Algorithm

Lyche and Morken [57] analyzed the Oslo Algorithms and developed a systematic way to decide which discrete B-splines are nonzero for a given j, and then calculate them optimally. We skip the development of these results, and just state and use the results. The interested reader is referred to [57] and [55]. The results are stated in a form which is suitable for both open and floating end conditions.

Given a knot sequence $\boldsymbol{t}$ and a subsequence $\boldsymbol{\tau}$ of $\boldsymbol{t}$, for each fixed j, consider $\{t_{j+1}, \ldots, t_{j+\kappa}\}$ and let $z_{j,q}$, $q = 1, \ldots, h_j$ be the distinct values in that set, $z_{j,q} < z_{j,q+1}$ for each q, each occurring with multiplicity $r_{j,q}$, $q = 1, \ldots, h_j$. If $z_{j,q}$ occurs in the whole τ sequence $s_{j,q}$ times, then sequences $\omega_j = \{\omega_{j,1}, \ldots, \omega_{j,\kappa-\nu}\}$ and $\xi_j = \{\xi_{j,1}, \ldots, \xi_{j,\nu}\}$ can be defined so that ω_j, called the *old knot sequence*, has $\sum_{q=1}^{h_j} \min(s_{j,q}, r_{j,q})$ elements in it which are the values $z_{j,q}$ each repeated $\min(s_{j,q}, r_{j,q})$ times in increasing order. Let $\nu_{j,q} = \max(r_{j,q} - s_{j,q}, 0)$, then ξ_j has $\nu = \sum_{i=1}^{h_j} \nu_{j,i}$ elements, which are the values $z_{j,i}$ each repeated $\nu_{j,i}$ times in increasing order. ξ_j is called the *new knot sequence*. For each j, let $\boldsymbol{\tau}'_j$ and $\boldsymbol{t}'_j$ denote the sequences obtained by removing the old knot sequence, ω_j from $\boldsymbol{\tau}$ and $\boldsymbol{t}$, respectively.

Theorem 16.13. *Define J'_j as the unique integer such that $\tau_J = \tau'_{J'} \le t_j = t' < \tau'_{J'+1}$; let ω_j, ξ_j, z, ν_j, $\boldsymbol{\tau}'_j$, and $\boldsymbol{t}'_j$ be the same as above. Then*

$$\alpha_{i,\kappa,\boldsymbol{\tau},\boldsymbol{t}}(j) = \alpha_{i,\nu,\boldsymbol{\tau}'_j,\boldsymbol{t}'_j}(j).$$

Moreover, $\alpha_{i,\kappa,\boldsymbol{\tau},\boldsymbol{t}}(j)$ is nonzero if and only if $i \in \{J'_j - \nu_j, J'_j - \nu_j + 1, \ldots, J'_j\}$.

Example 16.14. For $\kappa = 3$ and $\boldsymbol{\tau} = \{0,0,0,0,2,4,4,4,4\}$ and $\boldsymbol{t} = \{0,0,0,0,1,2,3,4,4,4,4\}$, find $\alpha_{i,\kappa}(j)$ for all i and j.

Set $j = 0$ and consider $\{t_1, t_2, t_3\} = \{0,0,0\}$. $z_{0,1} = 0$ and $h_0 = 1$, $r_{0,1} = 3$. But $s_{0,1} = 3$ also, so $\omega_0 = \{0,0,0\}$ and there are no new knots; thus ξ_j has no elements in it, and $\nu = 0$. $\boldsymbol{\tau}' = \{0,2,4,4,4,4,\}$ and $\boldsymbol{t}' = \{0,1,2,3,4,4,4,4,\}$ and $J'_j = 0$. Then $\alpha_{i,\kappa,\boldsymbol{\tau},\boldsymbol{t}}(0) \neq 0$ if and only if $i = 0$. Since for a given j, $\sum_i \alpha_{i,\kappa}(j) = 1$, $\alpha_{0,\kappa}(0) = 1$.

Now let $j = 1$. $\{t_2, t_3, t_4\} = \{0,0,1\}$, so $\omega_1 = \{0,0\}$, $\xi_1 = \{1\}$, and $\nu = 1$. $\boldsymbol{\tau}' = \{0,0,2,4,4,4,4,\}$ and $\boldsymbol{t}' = \{0,0,1,2,3,4,4,4,4,\}$. $J' = 1$, so

$$\begin{aligned}
\alpha_{1,3,\boldsymbol{\tau},\boldsymbol{t}}(1) &= \alpha_{1,1,\boldsymbol{\tau}',\boldsymbol{t}'}(1) \\
&= \frac{t'_{1+1} - \tau'_1}{\tau'_{1+1} - \tau'_1}\alpha_{1,0,\boldsymbol{\tau}',\boldsymbol{t}'}(1) + \frac{\tau'_{1+2} - t'_{1+1}}{\tau'_{1+2} - \tau'_{1+1}}\alpha_{2,0,\boldsymbol{\tau}',\tau}(1) \\
&= \frac{t'_2 - \tau'_1}{\tau'_2 - \tau'_1} \\
&= \frac{1-0}{2-0} \\
&= 1/2.
\end{aligned}$$

Since $\sum_i \alpha_{i,3}(1) = 1$, $\alpha_{0,3}(1) = 1/2$.

For,

- $j = 2$, $\{t_3, t_4, t_5\} = \{0, 1, 2\}$, so $\omega_2 = \{0, 2\}$, $\xi_2 = \{1\}$, and $\nu = 1$. $\boldsymbol{\tau}' = \{0, 0, 0, 4, 4, 4, 4\}$ and $\boldsymbol{t}' = \{0, 0, 0, 1, 3, 4, 4, 4, 4\}$. $J' = 2$, so $\alpha_{i,3}(2) > 0$ for $i = 1, 2$.
- $j = 3$, $\{t_4, t_5, t_6\} = \{1, 2, 3\}$, so $\omega_3 = \{2\}$, $\xi_3 = \{1, 3\}$, and $\nu = 2$. $\boldsymbol{\tau}' = \{0, 0, 0, 0, 4, 4, 4, 4\}$ and $\boldsymbol{t}' = \{0, 0, 0, 0, 1, 3, 4, 4, 4, 4\}$. $J' = 3$, so $\alpha_{i,3}(3) > 0$ for $i = 1, 2, 3$.
- $j = 4$, $\{t_5, t_6, t_7\} = \{2, 3, 4\}$, so $\omega_4 = \{2, 4\}$, $\xi_4 = \{3\}$, and $\nu = 1$. $\boldsymbol{\tau}' = \{0, 0, 0, 0, 4, 4, 4\}$ and $\boldsymbol{t}' = \{0, 0, 0, 0, 1, 3, 4, 4, 4\}$. $J' = 3$, so $\alpha_{i,3}(3) > 0$ for $i = 2, 3$.
- $j = 5$, $\{t_6, t_7, t_8\} = \{3, 4, 4\}$, so $\omega_5 = \{4, 4\}$, $\xi_5 = \{3\}$, and $\nu = 1$. $\boldsymbol{\tau}' = \{0, 0, 0, 0, 2, 4, 4\}$ and $\boldsymbol{t}' = \{0, 0, 0, 0, 1, 2, 3, 4, 4\}$. $J' = 4$, so $\alpha_{i,3}(5) > 0$ for $i = 3, 4$.
- $j = 6$, $\{t_7, t_8, t_9\} = \{4, 4, 4\}$, so $\omega_6 = \{4, 4, 4\}$, ξ_6 is empty and $\nu = 0$. $\boldsymbol{\tau}' = \{0, 0, 0, 0, 2, 4\}$ and $\boldsymbol{t}' = \{0, 0, 0, 0, 1, 2, 3, 4\}$. $J' = 4$, so $\alpha_{i,3}(6) > 0$ for $i = 4$, which means that $\alpha_{4,3}(6) = 1$. □

The algorithms derived are very efficient in finding $\alpha_{i,\kappa}(j)$, see [57]. In [57] efficient algorithms for finding D_j's, the new control points, are also presented.

Algorithm 16.15. Efficient Oslo Algorithm I. *Let* $j \in \{m_1, \ldots, m_2\}$.

Step 1. Find J' *such that* $\tau'_{J'} \leq t_j < \tau'_{J'+1}$.

Step 2. Figure out the old and new knots. Place the new knots in ξ.

Step 3. Figure out the reduced alpha's for $i \in \{J' - \nu, \ldots, J'\}$.
For all other values of i, $\alpha_{i,\kappa}(j) = 0$.

A pseudo-code version of this algorithm follows. This is set up for clarity, not utmost efficiency.

```
Procedure Refine(τ, t, κ, alpha_mx)
    /* κ is the degree
    /* τ is the original knot vector
    /* t is the refined knot vector
```

(*continued on next page*)

```
    alpha_mx ← 0
    For each j (to size of t)
        J' ← FindJprime(τ, t, κ, j)
        FindNewKnots (τ, t, κ, j, J', ξ, ν)
        Compute_R_alphas(τ, t, κ, j,ξ, ν, J', alpha_mx)
end (alpha_mx is returned)

Procedure FindJprime(τ, t, κ, j)
    Find J such that τ_J ≤ t_j < τ_{J+1}
    If the value τ_J occurs
        in τ p times and
        in the set {t_{j+1},...,t_{j+κ}} r times,
    then J' ← J − min(p,r)
return J'
end

Procedure FindNewKnots (τ, t, κ, j, J', ξ, ν)
    /* Put values of the ''new'' knots in ξ.*/
    ih ← J' + 1
    ν ← 0
    for p = 1 to κ
        if t_{j+p} = τ_{ih}
            then ih ← ih + 1
        else
            ν ← ν + 1;
            ξ_ν ← t_{j+p}
    end {p loop}
    return ξ, ν end

Procedure Compute_R_alphas(τ, t, κ, j,ξ, ν, J', alpha_mx)
    /* Figure out reduced alpha's*/
    ah_{κ,1} ← 1
    for p = 1 to ν
        β1 ← 0
        if p > J' then
            β1 ← (ξ_p − τ_0) * ah_{κ−J',p}/(τ_{p+κ−ν} − τ_0);
        i_l ← max(1, J' − p + 1);
        i_u ← min(J', n + ν − p);
        for i = i_l to i_u
            d1 ← ξ_p − τ_i;  d2 ← τ_{i+p+κ−ν} − ξ_p;
            β ← ah_{i+κ−J',p}/(d1 + d2);
            ah_{i+κ−J'−1,p+1} ← d2 * β + β1;
            β1 ← d1 * β
        end {i loop}
```

(continued on next page)

```
        ah_{i_u+κ-J',p+1} ← β1;
        if i_u < J' then
            ah_{i_u+κ-J',p+1} ← β1 + (τ_{n+κ+1} - ξ_p)
            *ah_{i_u+κ-J'+1,p}/(τ_{n+κ+1} - τ_{i_u+1})
    end {p loop}
    for i = i_l to i_u
        alpha_mx(i, j) = ah_{i+κ-J',ν+1}
end
```

16.2 Convergence of Refinement Strategies

The empirical evidence of Figures 16.3 through 16.6 is that the sequence piecewise linear curves formed by the control polysons of the successively refined knot sectors converges to the original B-spline curve. It is implied by [29] and explicitly shown in [19], using a general method for showing convergence over certain types of spline sequences, that the convergence properties of a sequence of polygons arising from successive refinement of a B-spline curve (including Bézier curves and subdivision) depends on the successive sequence of refined knot vectors in the following way.

Theorem 16.16. *Let $\gamma(t) = \sum P_i \mathcal{B}_{i,\kappa}(t)$ be a B-spline curve with knot vector $\mathbf{t}_0$. Suppose that $\mathbf{t}_j = \{t_{j,p}\}_p$ is a refinement of $\mathbf{t}_{j-1} = \{t_{j-1,q}\}_q$, that is, the knot vector $\mathbf{t}_{j-1}$ is a subset of the knot vector $\mathbf{t}_j$, including multiplicities. Let $\Delta_j = \max_q |t_{j,q} - t_{j,q-1}|$. Further, let $\boldsymbol{D}_j(t)$ be the piecewise linear curve representing a polygon of the jth refinement vector. Then*

$$\left\|\gamma(t) - \boldsymbol{D}_j\right\| \leq C\left(\|\Delta_j\|\right)^2 \|\gamma'\|$$

where C is a constant depending on neither the knot vectors nor the function γ. Further if γ'' is not continuous everywhere, it is known to be bounded, and the piecewise norm is still defined.

Thus, for example, form $\boldsymbol{t}_j$ from $\boldsymbol{t}_{j-1}$ by the rule that $\boldsymbol{t}_j$ is all the values of $\boldsymbol{t}_{j-1}$ with the following new ones added in the proper order: If $t_{j-1,q} \neq t_{j-1,q+1}$, insert a new knot half way between them. Then $\Delta_j = (1/2)^j \Delta_0$, and the refinement converges as $(1/2)^{2j}$, a rather fast refinement.

Further note that the Bézier subdivision at the midpoint of each successive interval is a slight variation of this algorithm where one inserts a knot of multiplicity $\kappa+1$ instead of multiplicity one. The rate of convergence is unchanged. We derive a purely geometric approach to show the convergence of Bézier subdivision at the interval midpoints.

Given a particular Bézier polygon, $\boldsymbol{P}$, defined on the interval $[a, b]$, let $\psi^0[\boldsymbol{P}]$ be the piecewise linear function given by the original polygon. Suppose that the subdivision on the curve over the interval $[a, b]$ is performed at the midpoint. Let $\psi^1[\boldsymbol{P}]$ be the piecewise linear function formed with vertices defined by concatenating together the control polygons for the two subdivided curves

$$\left\{P_0^{[0]}, \ldots, P_n^{[n]}\right\} \text{ and } \left\{P_n^{[n]}, \ldots, P_n^{[0]}\right\}.$$

It has $2n+1$ distinct points. Continue in this way, defining $\psi^k[P]$ to be a piecewise linear function defined by the ordered vertices of the control polygons of the 2^k Bézier curves whose composite is the original curve. That is, the ith subdivided Bézier curve at level k is over the interval $\left[a+i(b-a)/2^k, a+(i+1)(b-a)/2^k\right]$ and has vertices $\rho_{k,in+j}$, $j=0, \ldots, n$, for $i=0, \ldots, 2^k-1$. So, ψ^k is the piecewise linear function defined over the ρ and we shall write $\psi^k = \left\{\rho_{k,in+j} : j=0, \ldots, n;\ i=0, \ldots, 2^k-1\right\}$. We see that ψ^k has $2^k n+1$ distinct points which define it.

Theorem 16.17. $\lim_{k\to\infty} \psi^k[P] = \theta[P]$. *That is, the polyline consisting of the union of all the subpolygons converges to the Bézier curve.*

We first prove several lemmas which will make the proof of the theorem easier.

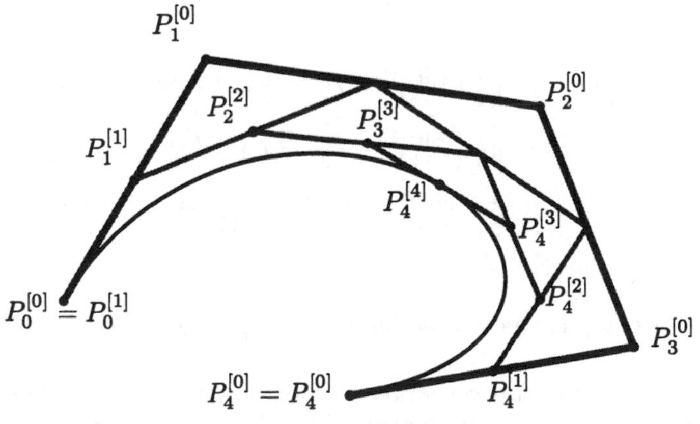

Figure 16.10. ψ^0 and ψ^1 and Bézier curve formed by $\boldsymbol{P}$.

Figure 16.11. (a) ψ^2 and Bézier curve formed by $\boldsymbol{P}$; (b) ψ^2 and Bézier curve formed by $\boldsymbol{P}$.

Lemma 16.18. *If $\theta[P_0, \ldots, P_n](a, b; t)$ is a Bézier curve, define $C = \max\{\, \|P_{j+1} - P_j\| : j = 0, \ldots, n-1\}$. If $P_j^{[i]}$ are defined by the rule in Theorem 16.12, then $\left\|P_{j+1}^{[i]} - P_j^{[i]}\right\| \le C$, for $j = i, \ldots, n-1$, $i = 0, \ldots, n-1$.*

Proof: By induction on the superscript, for $i = 1$,

$$\begin{aligned}
\left\|P_{j+1}^{[1]} - P_j^{[1]}\right\| &= \left\|\tfrac{1}{2}\left(P_{j+1}^{[0]} + P_j^{[0]}\right) - \tfrac{1}{2}\left(P_j^{[0]} + P_{j-1}^{[0]}\right)\right\| \\
&= \left\|\tfrac{1}{2}\left(P_{j+1}^{[0]} - P_j^{[0]}\right) + \tfrac{1}{2}\left(P_j^{[0]} - P_{j-1}^{[0]}\right)\right\| \\
&\le \frac{1}{2}\left\|P_{j+1}^{[0]} - P_j^{[0]}\right\| + \frac{1}{2}\left\|P_j^{[0]} - P_{j-1}^{[0]}\right\| \\
&\le C.
\end{aligned}$$

Now, suppose that the conclusion has been shown for superscripts up to $i-1$. Then,

$$\begin{aligned}
\left\|P_{j+1}^{[i]} - P_j^{[i]}\right\| &= \left\|\tfrac{1}{2}\left(P_{j+1}^{[i-1]} + P_j^{[i-1]}\right) - \tfrac{1}{2}\left(P_j^{[i-1]} + P_{j-1}^{[i-1]}\right)\right\| \\
&= \left\|\tfrac{1}{2}\left(P_{j+1}^{[i-1]} - P_j^{[i-1]}\right) + \tfrac{1}{2}\left(P_j^{[i-1]} - P_{j-1}^{[i-1]}\right)\right\| \\
&\le \frac{1}{2}\left\|P_{j+1}^{[i-1]} - P_j^{[i-1]}\right\| + \frac{1}{2}\left\|P_j^{[i-1]} - P_{j-1}^{[i-1]}\right\| \\
&\le C.
\end{aligned}$$

where the last step is true by the induction hypothesis. ■

Lemma 16.19. *Any two consecutive vertices of ψ^k are no farther apart than $2^{-k}C$, where C is independent of k. That is, if A and B are two consecutive vertices of ψ^k then $\|A - B\| \le 2^{-k}C$.*

Proof: Since ψ^k comes from the polygons of the subdivided curves, we know that consecutive vertices *always* come from the same subdivided curve.

We shall prove the result by induction on k, the level of subdivision.

Let $\left\|\psi^k\right\| = \max\left\{ \|\rho_{k,m+1} - \rho_{k,m}\| : m = 0, \ldots, 2^k n - 1 \right\}$.

First, consider ψ^1. $\rho_{1,i} = P_i^{[i]}$, $i = 0, \ldots, n$, and $\rho_{1,i} = P_n^{[2n-i]}$, $i = n, \ldots, 2n$. Let $m < n$.

$$\begin{aligned}
\|\rho_{1,m+1} - \rho_{1,m}\| &= \left\| P_{m+1}^{[m+1]} - P_m^{[m]} \right\| \\
&= \left\| \tfrac{1}{2} P_{m+1}^{[m]} + \tfrac{1}{2} P_m^{[m]} - P_m^{[m]} \right\| \\
&= \frac{1}{2} \left\| P_{m+1}^{[m]} - P_m^{[m]} \right\| \\
&\leq \frac{1}{2} C,
\end{aligned}$$

by Lemma 16.18, when

$$C = \max\{ \|P_{i+1} - P_i\| : i = 0, \ldots, n-1 \}.$$

Now, suppose $m \geq n$.

$$\begin{aligned}
\|\rho_{1,m+1} - \rho_{1,m}\| &= \left\| P_n^{[2n-m-1]} - P_n^{[2n-m]} \right\| \\
&= \left\| P_n^{[2n-m-1]} - \tfrac{1}{2} P_n^{[2n-m-1]} - \tfrac{1}{2} P_{n-1}^{[2n-m-1]} \right\| \\
&= \frac{1}{2} \left\| P_n^{[2n-m-1]} - P_{n-1}^{[2n-m-1]} \right\| \qquad (16.27) \\
&\leq \frac{1}{2} C,
\end{aligned}$$

again using Lemma 16.18. Hence, $\left\|\psi^1\right\| \leq C/2$. Now, assume the conclusion is true for $r < k$, that is, $\left\|\psi^r\right\| \leq 2^{-r} C$. Now we show it true for $r = k$.

The vertices in ψ^k are defined by subdividing the Bézier polygons in ψ^{k-1}. We see that $\rho_{k,2in+j}$, $j = 0, \ldots, 2n$, are formed by subdividing the Bézier curve with control polygon $\rho_{k-1,in+j}$, $j = 0, \ldots, n$, where $i = 0, \ldots, 2^{k-1} - 1$, respectively.

We shall prove the result for $\|\rho_{k,2in+j+1} - \rho_{k,2in+j}\|$, $j = 0, \ldots, 2n-1$ for each fixed i. Let us then fix i, and call $Q_j = \rho_{k-1,in+j}$, $j = 0, \ldots, n$. Then by the Subdivision Theorem, Theorem 16.12, $\rho_{k,2in+j} = Q_j^{[j]}$, $j = 0, \ldots, n$, and $\rho_{k,(2i+1)n+j} = Q_n^{[n-j]}$, $j = 0, \ldots, n$.

By Equation 16.27,

$$\begin{aligned}
&\max_{j=0,\ldots,2n}\left\{\left\|\rho_{k,2in+j+1}-\rho_{k,2in+j}\right\|\right\}\\
&=\max_{j=0,\ldots,n-1}\left\{\left\|Q_{j+1}^{[j+1]}-Q_j^{[j]}\right\|,\left\|Q_n^{[n-j-1]}-Q_n^{[n-j]}\right\|\right\}\\
&\leq\frac{1}{2}\left\|\psi[Q_0,\ldots,Q_n]\right\|
\end{aligned}$$

and using the induction hypothesis,

$$\begin{aligned}
&\leq \left(C/2^{k-1}\right)/2,\\
&\leq C/2^k.
\end{aligned}$$

Since this is proved for all i, the conclusion of the lemma holds for all k. ■

Proof: The subdivision theorem showed that over each subinterval $\left[a+(b-a)i/2^k, a+(b-a)(i+1)/2^k\right]$, the Bézier curve resulting from the appropriate subcollection of $\psi^k[P]$ is identical to the original, $\theta[P]$. We denote this by $\theta\left(\psi^k[P];a,b\right)=\theta(P;a,b)$.

An arbitrary value u in the original interval is then contained in an infinite sequence of intervals, $\left[a+i_{u,k}\frac{b-a}{2^k}, a+(i_{u,k}+1)\frac{b-a}{2^k}\right]$, for which

$$\lim_{k\to\infty}\left[a+i_{u,k}\frac{b-a}{2^k}, a+(i_{u,k}+1)\frac{b-a}{2^k}\right]=\{u\}.$$

Hence, the curve value, $\theta_n[P;a,b](u)$ lies within the convex hull of the $n+1$ vertices of ψ^k which correspond to the Bézier polygon over

$$\left[a+i_{u,k}\frac{b-a}{2^k}, a+(i_{u,k}+1)\frac{b-a}{2^k}\right], \text{ for each } k.$$

Since $\left\|\psi^k\right\| = O\left(\frac{1}{2^k}\right)$, the spacial extent of the convex hull of each Bézier polygon over $\left[a+i\frac{b-a}{2^k}, a+(i+1)\frac{b-a}{2^k}\right]$, all i and k, gets smaller and converges to zero.

Consider the subsequence of polygons corresponding to the intervals containing u. $\theta(P;a,b)(u)$ is contained in all of them, for all k. Further, if any other curve point were contained in all of them, say $\theta_n(P;a,b)(u^*)$, then u^* would be in

$$\lim_{k\to\infty}\left[a+i_{u,k}\frac{b-a}{2^k}, a+(i_{u,k}+1)\frac{b-a}{2^k}\right].$$

Since u is the only point in that intersection, $\theta(P;a,b)(u)$ is the only point in the intersection of the convex hulls of the Bézier polygons of these selected subintervals. Hence, the polygonal approximations converge. Thus, $\lim_{k\to\infty}\psi^k[P]=\theta_n[P]$. ■

16.3 Degree Raising for Splines

We have seen that the refinement process allows the introduction of new local degrees of freedom into a spline curve. The price paid for this is that once these parameters are used, the continuity of the curve decreases at the knot values, the amount depending on the multiplicity of the new knot. This is clearly a generalization of the subdivision property defined for Bézier curves, where all knowledge carried by the basis functions is lost, except for continuity. We showed that for Bézier curves there is another method of increasing the number of parameters—raising the degree of the representation of the curve. For each degree that the representation is raised, one new parameter is added.

We ask if there is a simple method for obtaining such a capability with splines, and if so, how many degrees of freedom (i.e., parameters) are added? Straightforward possibilities include:

1. Find the appropriate knot vector, and by evaluating the B-spline curve and enough of its derivatives at appropriate locations, interpolate this information to fit a B-spline curve of one degree higher on the appropriate knot vector.

2. Find the proper knot vector and solve using quasi-interpolants.

3. Break the B-spline curve into its separate Bézier curve components and degree-raise each of them.

Method 1 requires solving the interpolation problem and does not provide a nice closed form solution in terms of the knot vector and the original control points. Method 2 is closed form, but the quasi-interpolants are rather difficult to use computationally. While methods 1 and 2 each theoretically give the correct minimal degree-raised B-spline curve, method 3 does not. One is reduced to just the piecewise Bézier knowledge. Again, any higher-order continuity must be carried explicitly.

We find the knot vector of the degree-raised B-spline curve. That information will tell the additional number of degrees of freedom. Suppose $\gamma(t)$ is a B-spline curve defined over knot vector $\boldsymbol{\tau}$ of degree κ, where $\boldsymbol{\tau}$ has $\{u_i\}_{i=0}^{p}$ as distinct values and each u_i has multiplicity m_i. So, $\boldsymbol{\tau}$ has $\sum_{i=0}^{p} m_i$ elements, and γ has $n+1 = \sum_{i=0}^{p} m_i - (\kappa+1)$ distinct terms in its sum. If one represents γ as a B-spline curve of degree $\kappa+1$, one must insure that the continuity, as guaranteed by the knot vector, is the same. The basis functions can be guaranteed to be of continuity at most $C^{(\kappa-m_i)}$ at u_i, and hence the same is true for general γ. Suppose u_i has multiplicity q_i in the new knot vector, $\boldsymbol{t}$. Then $(\kappa+2)-1-q_i = \kappa - m_i$, and so $q_i = m_i + 1$, and all curves γ in the original representation have

$$\begin{aligned}\sum_{i=0}^{p} q_i - (\kappa+2) &= \sum_{i=0}^{p}(m_i+1) - (\kappa+2)\\ &= \sum_{i=0}^{p} m_i - \kappa + p - 1\\ &= n+1+p\end{aligned}$$

parameters. Thus the number of new parameters must be one less than the total number of distinct knots. The following theorem has been shown in [23, 22]:

Theorem 16.20. *If $\gamma(t) = \sum_{i=0}^{n} P_i \mathcal{B}_{i,\kappa}(t)$ with knot vector $\boldsymbol{\tau}$ defining the B-splines, and $\mathbf{t}$ is defined as above, with B-splines $\{\mathcal{N}_{j,\kappa+1}(t)\}_{j=0}^{n+p}$. Then*

$$\gamma(t) = \sum_{j=0}^{n+p} R_j \mathcal{N}_{j,\kappa+1}(t)$$

where

$$R_j = \frac{1}{\kappa+1}\sum_{i=0}^{n} P_i \Lambda_{i,\kappa}(j)$$

and

$$\Lambda_{i,\kappa}(j) = \begin{cases} \alpha_{i,0}(j) & \text{for } \kappa = 0\\ \frac{t_{j+\kappa+1}-\tau_i}{\tau_{i+\kappa}-\tau_i}\Lambda_{i,\kappa-1}(j) + \frac{\tau_{i+\kappa+1}-t_{j+\kappa+1}}{\tau_{i+\kappa+1}-\tau_{i}+1}\Lambda_{i+,\kappa-1}(j) + \alpha_{i,\kappa}(j) & \\ & \text{for } \kappa > 0\end{cases}$$

The Λ are strongly dependent on the properties of discrete splines, and hence it can be shown that $\Lambda_{i,\kappa}(j) \geq 0$, and $\sum_i \Lambda_{i,\kappa}(j) \equiv \kappa + 1$, for all κ, and j. Further the Λ are only nonzero for certain local values of i.

Thus, one now has a stable, constructive, closed form method for computing the control vertices of the new B-spline curve in terms of the old knot vector, the new knot vector, and the original control polygon.

The sequence in Figures 16.13 and 16.14 shows a degree-raising sequence for the cubic B-spline curve in Figure 16.12

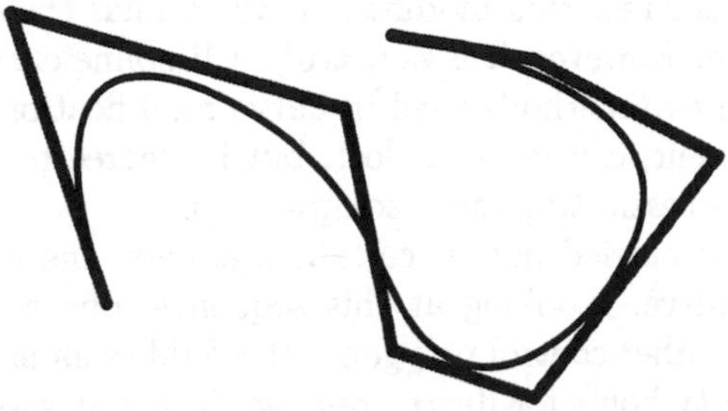

Figure 16.12. Original cubic curve and polygon.

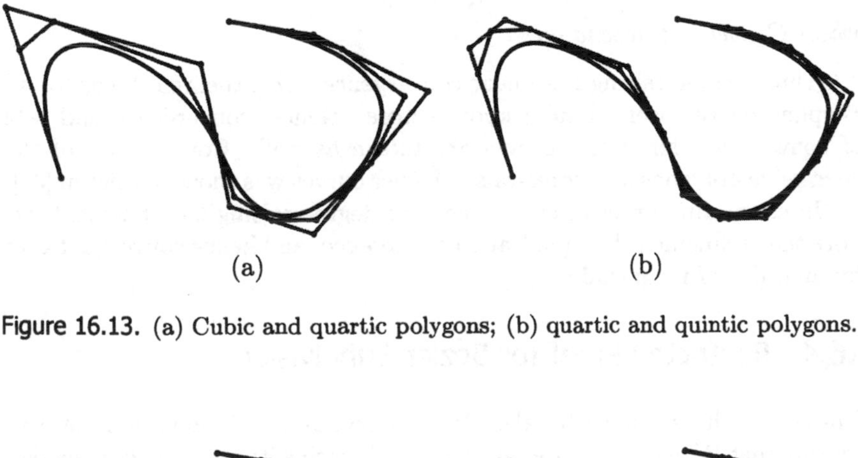

(a) (b)

Figure 16.13. (a) Cubic and quartic polygons; (b) quartic and quintic polygons.

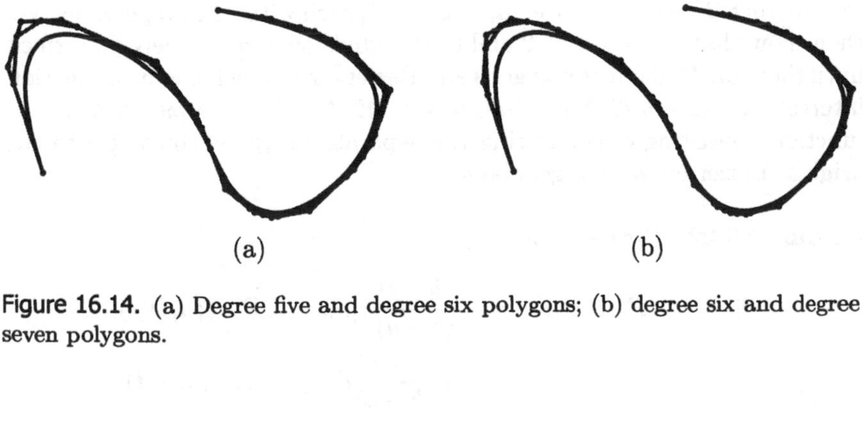

(a) (b)

Figure 16.14. (a) Degree five and degree six polygons; (b) degree six and degree seven polygons.

The degree-raised curves have exactly the same continuity properties as the original curve. Hence, if one of the new parameters is used to modify the curve locally, the resulting modified curve *still* has the same continuity as the initial curve. However, it is now truly a B-spline curve of one degree higher. Thus, the two methods used in curve modification have the trade off that in refinement, continuity is lost, but in degree raising, the degree becomes higher, necessitating more computation.

If refinement is carried out in certain manners, the control polygons converge to the curve. Looking at this sequence, one might ascertain a trend towards smoother control polygons. It would seem natural to wonder if the same property holds for degree raising. In [19] it was shown:

Theorem 16.21. *If γ is a B-spline curve of degree κ that has been degree raised p times and $\boldsymbol{D}_{n+p}$ denotes the control polygon of the degree-raised curve, then*

$$\|\gamma - \boldsymbol{D}_{n+p}\| \leq C(1/p)\|\gamma\|.$$

where C does not depend on γ.

Thus, degree raising has linear convergence. This theorem holds for all B-spline curves over all valid knot vectors. Hence, convergence and rate of convergence for Bézier curves are shown as well. Convergence of the successive polygons for degree-raised Bézier curves was shown earlier in [38].

In comparing order of convergence of degree raising to refinement, we note that refinement has quadratic convergence, and hence converges faster by an order of magnitude.

16.4 Restricted Proof for Bézier Subdivision

The goal of this section is to take a Bézier curve over $[a, b]$ and represent it as two distinct Bézier curves over $[a, c]$ and $[c, b]$ each with its own polygon. We then show that if $c = (a+b)/2$ and the subdivision is recursively performed until there are 2^k distinct polygons and Bézier curves each having a function interval $\left[a + (a+b)i2^{-k}, a + (a+b)(i+1)2^{-k}\right]$, then the piecewise linear function consisting of connecting the separate polygons converges to the original Bézier curve as k gets large.

Lemma 16.22. *For $t \in [a, b]$,*

$$\begin{aligned}\theta[P_0, \ldots, P_n; a, b](t) &= \frac{(b-t)}{(b-a)}\theta[P_0, \ldots, P_{n-1}; a, b](t) \\ &\quad + \frac{(t-a)}{(b-a)}\theta[P_1, \ldots, P_n; a, b](t).\end{aligned}$$

Proof: Following the proof of Theorem 5.4, it is easy to show that

$$\theta_{k,n}(a,b;t) = \begin{cases} 1 & \text{for } n=0 \\ \frac{t-a}{b-a}\theta_{k-1,n-1}(a,b;t) + \frac{b-t}{b-a}\theta_{k,n-1}(a,b;t) & \\ & \text{for } n>0,\ 0<k<n \\ \frac{b-t}{b-a}\theta_{0,n-1}(a,b;t) & \text{for } n>0,\ k=0 \\ \frac{t-a}{b-a}\theta_{n-1,n-1}(a,b;t) & \text{for } k=n>0. \end{cases} \tag{16.28}$$

Now

$$\begin{aligned} \sum_{i=0}^{n} P_i\theta_{i,n}(a,b;t) &= \sum_{i=0}^{n} P_i \left[\frac{t-a}{b-a}\theta_{i-1,n-1}(a,b;t) + \frac{b-t}{b-a}\theta_{i,n-1}(a,b;t)\right] \\ &= \frac{t-a}{b-a}\sum_{i=0}^{n} P_i\theta_{i-1,n-1}(a,b;t) + \frac{b-t}{b-a}\sum_{i=0}^{n} P_i\theta_{i,n-1}(a,b;t) \\ &= \frac{t-a}{b-a}\sum_{i=0}^{n-1} P_{i+1}\theta_{i,n-1}(a,b;t) + \frac{b-t}{b-a}\sum_{i=0}^{n-1} P_i\theta_{i,n-1}(a,b;t). \end{aligned}$$

Since the left side is $\theta[P_0,\ldots,P_n;a,b]$, and the right side is the conclusion of the lemma, the lemma is proved. ∎

We follow this up restating Theorem 16.12. *For $\theta[\mathbf{P};a,b]$ defined as above then*

$$\theta[\mathbf{P};a,b](t) = \begin{cases} \theta\left[P_0^{[0]}, P_1^{[1]}, \ldots, P_n^{[n]}; a, c\right](t) \\ \theta\left[P_n^{[n]}, P_n^{[n-1]}, \ldots, P_n^{[0]}; c, b\right](t) \end{cases}$$

where

$$P_i^{[k]} = \begin{cases} \frac{\left[(b-c)P_{i-1}^{[k-1]} + (c-a)P_i^{[k-1]}\right]}{(b-a)} & k>0,\ k\le i\le n \\ P_i & k=0. \end{cases}$$

Proof: We first show the theorem is true for $t \in [a,c]$, by induction on n, the degree of the curve and for arbitrary c, $a<c<b$.

If $n = 1$,

$$\begin{aligned} \theta[P_0, P_1; a, b](t) &= \frac{[(b-t)P_0 + (t-a)P_1]}{(b-a)} \quad \text{(by definition)} \\ &= \frac{(c-t)P_0 + (t-a)\frac{(b-c)P_0+(c-a)P_1}{b-a}}{c-a} \quad (16.29) \\ &= \theta\left[P_0, \frac{(b-c)P_0 + (c-a)P_1}{b-a}; a, c\right](t) \quad (16.30) \\ &= \theta[P_0^{[0]}, P_1^{[1]}; a, c](t). \end{aligned}$$

Now, assume the theorem holds for all $k < n$. Equation 16.31, follows from Lemma 16.22. Equation 16.32 follows from the induction hypothesis.

$$\begin{aligned} &\theta[P_0, \ldots, P_n; a, b](t) \\ &= \frac{(b-t)\theta[P_0, \ldots, P_{n-1}; a, b](t) + (t-a)\theta[P_1, \ldots, P_n; a, b](t)}{b-a} \quad (16.31) \\ &= \frac{b-t}{b-a}\theta\left[P_0^{[0]}, \ldots, P_{n-1}^{[n-1]}; a, c\right](t) + \frac{t-a}{b-a}\theta\left[P_{1+0}^{[0]}, \ldots, P_{1+(n-1)}^{[n-1]}; a, c\right](t) \\ & \quad (16.32) \\ &= \frac{b-t}{b-a}\sum_{i=0}^{n-1} P_i^{[i]}\theta_{i,n-1}(a, c; t) + \frac{t-a}{b-a}\sum_{i=0}^{n-1} P_{1+i}^{[i]}\theta_{i,n-1}(a, c; t) \\ &= \sum_{i=0}^{n-1}\left(\frac{b-t}{b-a}P_i^{[i]} + \frac{t-a}{b-a}P_{1+i}^{[i]}\right)\theta_{i,n-1}(a, c; t). \quad (16.33) \end{aligned}$$

Now using the results from $n = 1$, Equations 16.29 and 16.30,

$$\begin{aligned} &= \sum_{i=0}^{n-1}\left[\frac{c-t}{c-a}P_i^{[i]} + \frac{t-a}{c-a}\left(\frac{(b-c)P_i^{[i]} + (c-a)P_{i+1}^{[i]}}{b-a}\right)\right]\theta_{i,n-1}(a, c; t) \\ &= \sum_{i=0}^{n-1}\left(\frac{c-t}{c-a}P_i^{[i]} + \frac{t-a}{c-a}P_{i+1}^{[i+1]}\right)\theta_{i,n-1}(a, c; t) \\ &= \frac{c-t}{c-a}\sum_{i=0}^{n-1} P_i^{[i]}\theta_{i,n-1}(a, c; t) + \frac{t-a}{c-a}\sum_{i=0}^{n-1} P_{i+1}^{[i+1]}\theta_{i,n-1}(a, c; t) \\ &= \frac{c-t}{c-a}\theta\left[P_0^{[0]}, \ldots, P_{n-1}^{[n-1]}; a, c\right](t) + \frac{t-a}{c-a}\theta\left[P_1^{[1]}, \ldots, P_n^{[n]}; a, c\right](t) \\ &= \theta\left[P_0^{[0]}, \ldots, P_n^{[n]}; a, c\right](t), \qquad \text{by Lemma 16.22.} \end{aligned}$$

We now show the theorem for the other part of the interval.

Once again, set $n = 1$ and

$$\begin{aligned}\theta[P_0, P_1; a, b](t) &= \frac{(b-t)P_0 + (t-a)P_1}{(b-a)} \quad \text{(by definition)}\\ &= \frac{(b-t)\frac{(b-c)P_0+(c-a)P_1}{b-a} + (t-c)P_1}{b-c} \qquad (16.34)\\ &= \theta[P_1^{[1]}, P_1^{[0]}; c, b](t). \qquad (16.35)\end{aligned}$$

The argument for the right end of the interval proceeds the same way as above for the left side.

$$\begin{aligned}&\theta[P_0, \ldots, P_n; a, b](t)\\ &= \frac{(b-t)\theta[P_0, \ldots, P_{n-1}; a, b](t) + (t-a)\theta[P_1, \ldots, P_n; a, b](t)}{b-a} \qquad (16.36)\\ &= \frac{b-t}{b-a}\theta\left[P_{n-1}^{[n-1]}, \ldots, P_{n-1}^{[0]}; c, b\right](t)\\ &\quad + \frac{t-a}{b-a}\theta\left[P_{1+(n-1)}^{[n-1]}, \ldots, P_{1+(n-1)}^{[0]}; c, b\right](t) \qquad (16.37)\\ &= \frac{b-t}{b-a}\sum_{i=0}^{n-1} P_{n-1}^{[n-1-i]}\theta_{i,n-1}(c,b;t) + \frac{t-a}{b-a}\sum_{i=0}^{n-1} P_n^{[n-1-i]}\theta_{i,n-1}(c,b;t)\\ &= \sum_{i=0}^{n-1}\left(\frac{b-t}{b-a}P_{n-1}^{[n-1-i]} + \frac{t-a}{b-a}P_n^{[n-1-i]}\right)\theta_{i,n-1}(c,b;t)\\ &= \sum_{i=0}^{n-1}\left(\frac{b-t}{b-c}P_n^{[n-i]} + \frac{t-c}{b-c}P_n^{[n-1-i]}\right)\theta_{i,n-1}(c,b;t)\\ &= \sum_{i=0}^{n} P_n^{n-i}\theta_{i,n}(c,b;t). \qquad (16.38)\end{aligned}$$

The conclusion follows since

$$\begin{aligned}&\frac{b-t}{b-a}P_{n-1}^{[n-1-i]} + \frac{t-a}{b-a}P_n^{[n-1-i]}\\ &= \frac{(b-t)\frac{(b-c)P_{n-1}^{[n-1-i]}+(c-a)P_n^{[n-1-i]}}{b-a} + (t-c)P_n^{[n-1-i]}}{b-c} \qquad (16.39)\\ &= \frac{b-t}{b-c}P_n^{[n-i]} + \frac{t-c}{b-c}P_n^{[n-1-i]}, \qquad (16.40)\end{aligned}$$

where Equations 16.39 and 16.40 follow from applications of Equations 16.34 and 16.35. ■

Exercises

1. Prove Lemma 16.5.

2. Given a B-spline surface $\sigma(u,v)$ with degree κ_u and κ_v, figure out an algorithm to extract the knot vector and polygon for the isoparametric curve $\sigma(u^*,v)$, where u^* is an arbitrary fixed value in the u-domain.

3. Suppose $\gamma_1(t)$ and $\gamma_2(t)$ are B-spline curves over knot vectors $\boldsymbol{t}_1$ and $\boldsymbol{t}_2$, respectively, of degree k and m, respectively, such that their domains are the same, but $k < m$. Create a surface between them. (Hint: first degree raise γ_1 until it has degree m, then refine them both to have a shared knot vector. Then use a ruled surface algorithm.)

4. Let $\boldsymbol{\tau}$ be an arbitrary knot vector and $\boldsymbol{t}$ an arbitrary refinement.

   (a) find $\alpha_{i,0}(j)$ for all i, j.

   (b) If $\gamma(t) = \sum_i P_i \mathcal{B}_{i,0,\boldsymbol{\tau}}(t)$, find its new coefficients in the refined basis, that is, find D_j in terms of P_i, $\boldsymbol{\tau}$, and $\boldsymbol{t}$ where $\gamma(t) = \sum_j D_j \mathcal{N}_{j,0,\boldsymbol{t}}(t)$.

5. One may be tempted to conclude from the linear case that evaluation of the discrete spline at j gives the same value as evaluating the continuous spline at t_j. That is not true as this problem shows. Let $\boldsymbol{\tau} = \{0,0,0,1,2,2,2\}$ and $\boldsymbol{t} = \{0,0,0,1,1,2,2,2\}$. Evaluate the quadratic discrete splines $\alpha_{3,2}(3)$ and $\alpha_{3,2}(4)$.

6. Consider a quadratic spline curve with control points P_i, $i = 0, \ldots, 3$, and knot vector $\boldsymbol{\tau} = \{0,0,0,1,3,3,3\}$. Use the Oslo Algorithm to find D_j, for $j = 0, \ldots, 4$, when $\boldsymbol{t} = \{0,0,0,1,2,3,3,3\}$.

7. Programming Exercise: Augment the Bézier program written for Chapter 5 Exercise 8 to use Algorithm 17.2. The subdivision algorithm will be used in two different ways: 1) to adaptively subdivide for rendering the curves, and 2) to allow the user to split a curve (anywhere he chooses) and manipulate either (or both) of the new curves. Therefore, the program must handle more than one curve. For comparison purposes, allow the user to choose how a curve is to be rendered, i.e., by subdivision, by point evaluation, or by both (using different colors, if available).

17

Subdivision and Refinement in Modeling

It is often reiterated that the local properties of B-splines make them very useful for representing geometry. The ability to support general, stable, rapid refinement algorithms help make them supportive of hierarchical design. For instance, in Figure 17.1 (a), the gross shape of the curve with the shown knot vector is desirable; however, some fine detail is lacking. In Figure 17.1(b), the knot vector has been supplemented generating a new control polygon for the same curve. Finally in Figure 17.2 the curve is truly modified. The flexibility introduced by the pseudoknots has been used, and those knots are real. Note that the modified curve is identical to the original in its gross shape, but has just the desired local perturbations. We present a general algorithm which implements Algorithm 16.9.

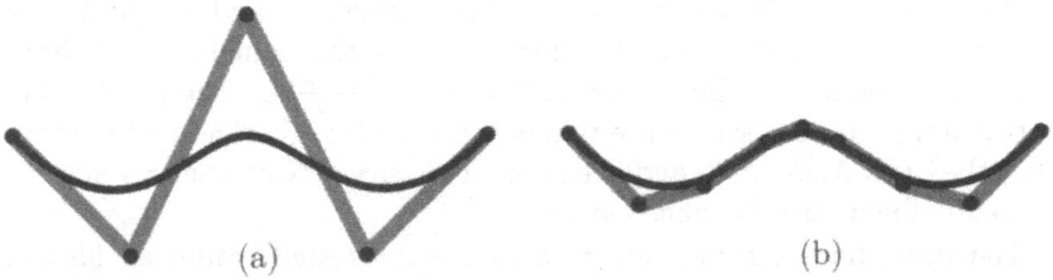

Figure 17.1. (a) Initial curve and control polygon, $\boldsymbol{\tau} = \{0, 0, 0, 0, 1, 2, 2, 2, 2\}$; (b) refined control polygon, $\boldsymbol{t} = \{0, 0, 0, 0, 0.6, 0.8, 1, 1.2, 1.4, 2, 2, 2, 2\}$.

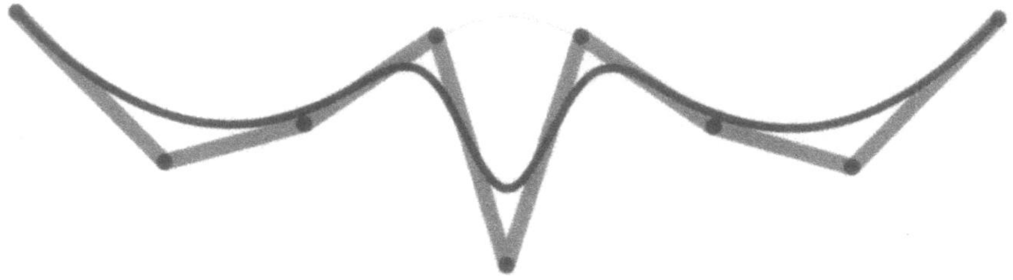

Figure 17.2. Modified control polygon and resulting curve.

While this is efficient for curves, when we present surface subdivision, we implement an algorithm with the approach of Algorithm 16.8 in its more efficient form (Algorithm 16.13). We present special algorithms optimized for Bézier curves and surfaces as well.

17.1 Subdivision of B-Spline Curves

The algorithm for subdivision can be an application of the rule for insertion of a single multiple knot (which effects splitting the curve into two spline subcurves) or it can occur with multiple instances of multiple knots (splitting the curve into several spline subcurves).

17.1.1 General Algorithm

Suppose γ, an open B-spline curve of degree κ, is defined by knot vector $\boldsymbol{\tau}$ and control polygon $\boldsymbol{P} = \{P_0, \ldots, P_n\}$. Further suppose that one wants to subdivide the curve into two open B-spline curves, γ_Q and γ_R (of degree κ) at the middle of its parametric range. Since the domain of γ is from τ_κ to τ_{n+1}, the subdividing point will be $t^* = \frac{\tau_\kappa + \tau_{n+1}}{2}$. The subdivision process must first determine the refined knot vector, then refine the curve using the Oslo Algorithm, and then split the curve. Note that γ_Q and γ_R are well-defined open B-spline curves.

The procedures outlined are to give the interested reader an idea of implementation, but are by no means highly efficient. The implementation of the refinement is left to the designer. This is because there may be some applications for which the special purpose algorithm is more useful to implement, and others in which it is more desirable to apply the general algorithm to this specific instance.

Algorithm 17.1.

```
Procedure SplineCurvesplit(P, n, τ, κ, t*, Q, μ, q, R, ρ, r)

/* P is the polygon of the original curve γ with
/* n+1 vertices;
/* τ is the knot vector of the original curve γ;
/* κ is the degree of the spline curves;
/* Calculated:
/* Q is the polygon of the 'left' B-spline subcurve with
/* q+1 vertices;
/* R is the polygon of the 'right' B-spline subcurve with
/* r+1 vertices;
/* μ is the knot vector of the new 'left' curve;
/* ρ is the knot vector of the new 'right' curve;

    Refineknotvector( τ, n, κ, t*, t, p, N )
    Refinepolygon( κ, P, n, τ, N, t, D )
    SplitSpline( p, κ, D, N, t, Q, μ, q, R, ρ, r )

end

Procedure Refineknotvector(τ, n, κ, t*, t, p, N)

    Find J such that τ_J ≤ t* < τ_{J+1}.
    Find p−1 such that τ_{p−1} < t* but τ_p ≥ t*.
    For i = 0 to J
    t_i = τ_i
    addnum = κ + 1 − J + (p − 1)
/* tells how many new copies of t* to add */
    N = n + addnum
    For i = J+1 to J + addnum
        t_i = t*
    For i = J+1+addnum to N + κ + 1
        t_i = τ_{i−addnum}

end
```

(continued on next page)

```
Procedure SplitSpline(p, κ, D, N, t, Q, μ, q, R, ρ, r)

    /* Left Curve */
    q = p − 1
    For i = 0 to q
        Q_i = D_i
    For i = 0 to q + κ + 1
        μ_i = t_i

    /* RightCurve */
    r = N − p
    For i = 0 to r
        R_i = D_{i+p}
    For i = 0 to r + κ + 1
        ρ_i = t_{i+p}

end
```

Algoritm 17.1 (continued)

The algorithm for Bézier curves first appeared in [49], and was based on Theorem 16.12.

Algorithm 17.2. Bézier Curve Splitting. *Given polynomial coefficients $\boldsymbol{P} = [P_0, \ldots, P_n]$ for a Bézier curve over $[a, b]$, split the curve at c, and find the polygons $\boldsymbol{Q} = [Q_0, \ldots, Q_n]$ and $\boldsymbol{R} = [R_0, \ldots, R_n]$ such that*

$$\theta_n[\boldsymbol{P}; a, b](t) = \theta_n\left[\boldsymbol{Q}; a, c\right](t) = \theta_n\left[\boldsymbol{R}; c, b\right](t).$$

```
Procedure BezCurvesplit(P, n, a, b, c, Q, R)

    /* Step 1:  (Initialization):  */
    Q ← P;  R_n ← P_n.
          tu = (c−a)/(b−a),  tl = 1 − tu
    /* Step 2:  Compute the coefficients in a double loop */
          for j = 1 to n begin
              for k = n to j begin
                  Q_k ← tl * Q_{k−1} + tu * Q_k
              end
              R_{n−j} ← Q_n
          end
```

In the following applications, we refer to both of these algorithms with the generic title *Curvesplit*, and assume that the information necessary for computation is included with the curve information.

17.2 Curve Generation with Geometric Subdivision

In this section we show how using the B-spline curve leads to both ease of manipulation, and with the subdivision algorithms, yields a rendering algorithm which is guaranteed to always carry the shape information in either image space or object space.

In Section 16.2 it was shown that repeated subdivision uniformly over the domain of the original curve causes the subdivided polygons to converge to the curve. However, not all parts of the curve are equally far from the polygon and certainly not from the kth level subdivided polygons. A very flat section of curve has subdivided polygons with flat convex hulls. Since the curve must lie in the convex hull of the polygons, and since B-spline curves are variation diminishing, checking the flatness of the convex hull gives guaranteed information about the flatness of the curve.

Algorithm 17.3. Curve Generation. *For B-spline curve γ, in either B-spline or Bézier form, set up a stack called A.*

Step 1: This initializes the algorithm by setting stack A to empty.

Step 2: Test for convergence If $crit(\gamma) < tol$, go to Step 6.

Step 3: Subdivide γ using the `Curvesplit` *procedure into $\boldsymbol{Q}$ and $\boldsymbol{R}$.*

Step 4: Push $\boldsymbol{R}$ onto the stack A.

Step 5: Set $\boldsymbol{P} \leftarrow \boldsymbol{Q}$ and go to Step 2.

Step 6: Draw the appropriate desired lines. Either the piecewise linear segments of the polygon or just the one line segment from the first to last control polygon point are often used.

Step 7: Pop a curve off the stack A to remove the last previously computed $\boldsymbol{R}$ and set it equal to γ by $\gamma \leftarrow A$. Return to Step 2. If the stack is empty, the algorithm terminates since the curve is drawn.

Frequently used convergence criteria include

- the length of the longest side of the polygon is smaller than a tolerance,
- the total length of the sides of the polygon is smaller than a tolerance, and
- the maximum distance of the intermediate vertices from the line segment joining the first and last vertices is smaller than a tolerance.

Writing this sequence of operations recursively,

```
Procedure Curve_Gen ( γ )
    Are the points of γ within tolerance?
    if yes,
        Draw_Crv( γ )
    else begin
        Curvesplit( γ, Q, R )
        Curve_Gen( Q )
        Curve_Gen( R )
    end {else}
end
```

Figures 17.3 through 17.5 illustrate an application of this algorithm.

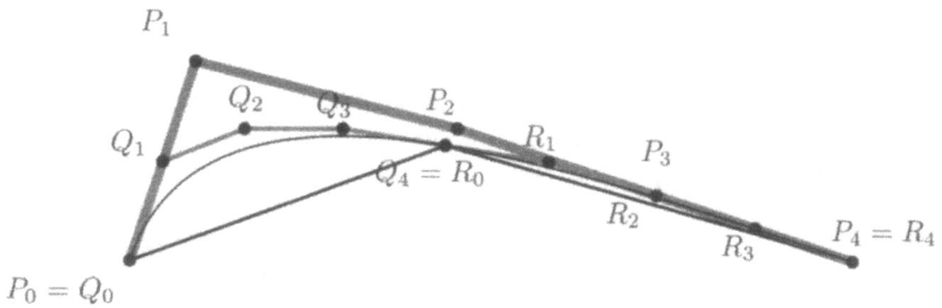

Figure 17.3. Original $\boldsymbol{P}$; finding Q and R.

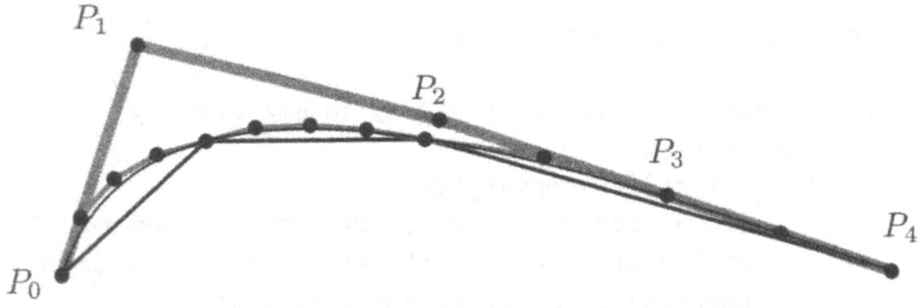

Figure 17.4. New $\boldsymbol{P}$; finding Q and R.

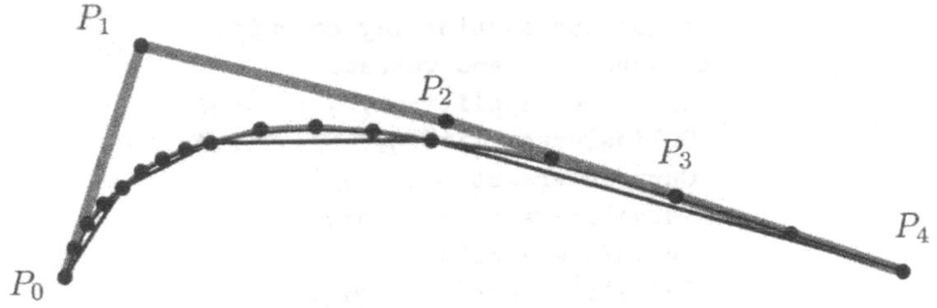

Figure 17.5. New $\boldsymbol{P}$; drawing an approximation.

17.3 Curve-Curve Intersection

An important problem is to find the intersections of two curves. Even for explicit polynomial curves, this cannot be done analytically for curves of degree higher than four. That is, for arbitrary polynomials of degree five and higher, no general formula can be given expressing the roots in terms of combinations of radicals of rational functions of the coefficients. Particular functions may be amenable to such a decomposition. For parametrically defined curves, one wants to find the actual coordinates of the intersection and also find the distinct parametric values associated with that point on each curve.

Given γ_1 and γ_2, B-spline curves, assume their data structures carry degree, knotvector, and polygon information. The following algorithm finds all points of intersection of the two curves.

The implementation of this algorithm clearly requires analytic methods for determining whether convex hulls overlap and whether the convex hulls can be approximated to necessary tolerance by the appropriate combination of desired straight lines. Further a stable method for computing the intersection of two line segments is required.

```
Procedure CurveIntersect(γ1, γ2)

    If the convex hulls of γ1 and γ2 do not overlap
        Then return
        Else, test for convergence:
            If the convex hulls can be approximated to
            within tolerance by the straight line segment
            connecting the first and last points,
            call them L1 and L2,
                output the intersection of L1 and L2.
                This approximates the intersection of the
                curves to within tolerance.
                This intersection may be empty.
            Else, subdivide and retest:
                SplineCurvesplit( γ1, γ1,1, γ1,2)
                SplineCurvesplit( γ2, γ2,1, γ2,2)
                CurveIntersect(γ1,1, γ2,1)
                CurveIntersect(γ1,1, γ2,2)
                CurveIntersect(γ1,2, γ2,1)
                CurveIntersect(γ1,2, γ2,2)

end
```

Figures 17.6 through 17.8 demonstrate this process. First the convex hulls of the original curves are compared. Then the convex hulls of the subdivided curves. Finally the remaining pieces are treated separately, and the process is recursively solved.

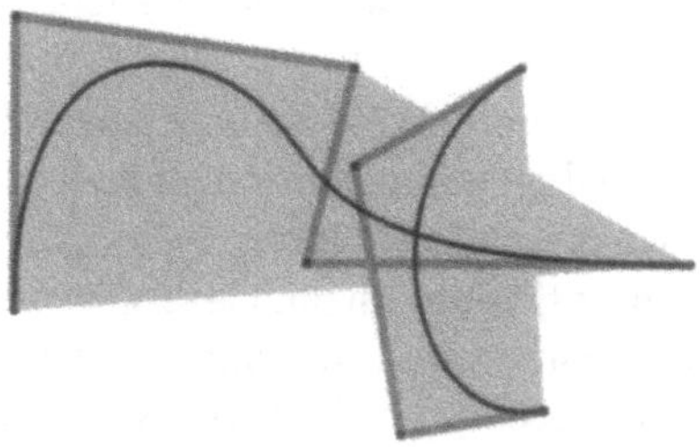

Figure 17.6. Curves γ_1 and γ_1—convex hulls intersect.

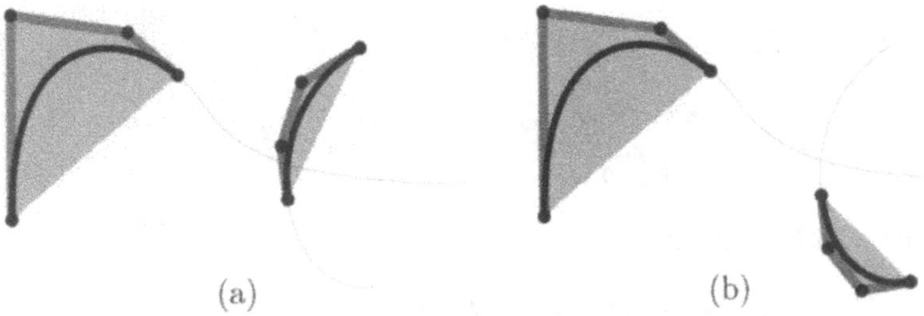

Figure 17.7. (a) Comparing $\gamma_{1,1}$ and $\gamma_{2,1}$; (b) comparing $\gamma_{1,1}$ and $\gamma_{2,2}$.

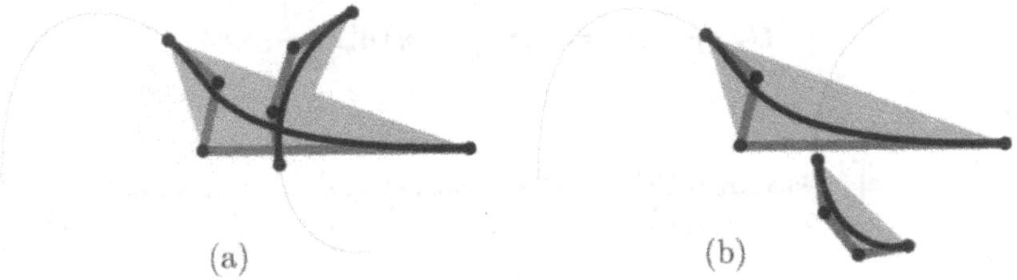

Figure 17.8. (a) Comparing $\gamma_{1,2}$ and $\gamma_{2,1}$; (b) comparing $\gamma_{1,2}$ and $\gamma_{2,2}$.

17.4 Surface Refinement and Subdivision

Let $\sigma(u,v) = \sum_{j=0}^{n_v} \sum_{i=0}^{n_u} P_{i,j} \mathcal{B}_{i,\kappa_u,\boldsymbol{\tau}_u}(u) \mathcal{B}_{j,\kappa_v,\boldsymbol{\tau}_v}(v)$. If one wanted to refine at several values of u, that could be accomplished by merely forming the correct $\boldsymbol{t_u}$ knot vector; thus, if one wishes to subdivide simultaneously at r different values, $\boldsymbol{\tau}_u \cup \{c_{u_1,j}, \ldots, c_{u_r,j}\}_{j=0}^{n_u}$. Then one would consider the *curves* $\sum P_{i,j} \mathcal{B}_{i,\kappa_u}(u)$ and subdivide by refining into $\sum_{q=0}^{m_u} D_{q,j} \mathcal{N}_{q,\kappa_u,\boldsymbol{t}_u}(u)$, where $\{\mathcal{N}_{q,\kappa_u,\boldsymbol{t}_u}(u)\}$ are the B-splines over the refined knot vector, yielding $\sigma(u,v) = \sum\sum D_{q,j} \mathcal{N}_{q,\kappa_u,\boldsymbol{t}_u}(u) \mathcal{B}_{j,\kappa_v,\boldsymbol{\tau}_v}(v)$.

17.4.1 A Matrix Approach

As was shown in [56], a modified matrix approach is more efficient for computation. Let $r \in \{u, v\}$, and let

$$\begin{aligned} \boldsymbol{B}_{\kappa_r,\boldsymbol{\tau}_r}(r) &= \left[\; \mathcal{B}_{0,\kappa_r,\boldsymbol{\tau}_r}(r) \;\; \mathcal{B}_{1,\kappa_r,\boldsymbol{\tau}_r}(r) \;\; \ldots \;\; \mathcal{B}_{n_r,\kappa_r,\boldsymbol{\tau}_r}(r) \; \right], \\ \boldsymbol{N}_{\kappa_r,\boldsymbol{t}_r}(r) &= \left[\; \mathcal{N}_{0,\kappa_r,\boldsymbol{t}_r}(r) \;\; \mathcal{N}_{1,\kappa_r,\boldsymbol{t}_r}(r) \;\; \ldots \;\; \mathcal{N}_{m_r,\kappa_r,\boldsymbol{t}_r}(r) \; \right]. \end{aligned}$$

We have shown that

$$\sigma(u,v) = \sum_{i=0}^{n_u}\sum_{j=0}^{n_v} P_{i,j}\mathcal{B}_{j,\kappa_v,\tau_v}(v)\mathcal{B}_{i,\kappa_u,\tau_u}(u),$$

which in turn can be written:

$$\sigma(u,v) = \boldsymbol{B}_{\kappa_u,\boldsymbol{\tau}_u}(u)\boldsymbol{P}\boldsymbol{B}^t_{\kappa_v,\boldsymbol{\tau}_v}(v).$$

But,

$$\boldsymbol{B}_{\kappa_u,\boldsymbol{\tau}_u}(u) \quad = \quad \boldsymbol{N}_{\kappa_u,\boldsymbol{t}_u}(u)A^t_{\boldsymbol{\tau}_u,\boldsymbol{t}_u}$$

where

$$A^t_{\boldsymbol{\tau}_u,\boldsymbol{t}_u} \quad = \quad \begin{bmatrix} \alpha_{0,\kappa_u,\tau_u,t_u}(0) & \alpha_{1,\kappa_u,\tau_u,t_u}(0) & \cdots & \alpha_{n_u,\kappa_u,\tau_u,t_u}(0) \\ \alpha_{0,\kappa_u,\tau_u,t_u}(1) & \alpha_{1,\kappa_u,\tau_u,t_u}(1) & \cdots & \alpha_{n_u,\kappa_u,\tau_u,t_u}(1) \\ & & \vdots & \\ \alpha_{0,\kappa_u,\tau_u,t_u}(m_u) & \alpha_{1,\kappa_u,\tau_u,t_u}(m_u) & \cdots & \alpha_{n_u,\kappa_u,\tau_u,t_u}(m_u) \end{bmatrix}$$

Also,

$$\boldsymbol{B}^t_{\kappa_v,\boldsymbol{\tau}_v} \quad = \quad A_{\boldsymbol{\tau}_v,\boldsymbol{t}_v}\boldsymbol{N}^t_{\kappa_v,\boldsymbol{t}_v}$$

where

$$A_{\boldsymbol{\tau}_v,\boldsymbol{t}_v} \quad = \quad \begin{bmatrix} \alpha_{0,\kappa_v,\tau_v,t_v}(0) & \alpha_{0,\kappa_v,\tau_v,t_v}(1) & \cdots & \alpha_{0,\kappa_v,\tau_v,t_v}(m_v) \\ \alpha_{1,\kappa_v,\tau_v,t_v}(0) & \alpha_{1,\kappa_v,\tau_v,t_v}(1) & \cdots & \alpha_{1,\kappa_v,\tau_v,t_v}(m_v) \\ & & \vdots & \\ \alpha_{n_v,k_v,\tau_v,t_v}(0) & \alpha_{n_v,k_v,\tau_v,t_v}(1) & \cdots & \alpha_{n_v,k_v,\tau_v,t_v}(m_v) \end{bmatrix}.$$

If

$$\boldsymbol{P} = \begin{bmatrix} P_{\mu-\kappa_u,\nu-\kappa_v} & P_{\mu-\kappa_u,\nu-\kappa_v+1} & \cdots & P_{\mu-\kappa_u,\nu} \\ P_{\mu-\kappa_u+1,\nu-\kappa_v} & P_{\mu-\kappa_u+1,\nu-\kappa_v+1} & \cdots & P_{\mu-\kappa_u+1,\nu} \\ & & \vdots & \\ P_{\mu,\nu-\kappa_v} & P_{\mu,\nu-\kappa_v+1} & \cdots & P_{\mu,\nu} \end{bmatrix},$$

then

$$\sigma(u,v) = \boldsymbol{N}_{\kappa_u,\boldsymbol{t_u}}A^t_{\boldsymbol{\tau}_u,\boldsymbol{t_u}}\boldsymbol{P}A_{\boldsymbol{\tau}_v,\boldsymbol{t_v}}\boldsymbol{N}^t_{\kappa_v,\boldsymbol{t_v}}. \tag{17.1}$$

Equation 17.1 creates a matrix form and shows that the new mesh, represented by the $\boldsymbol{D}$ matrix, is written

$$\boldsymbol{D} = A^t_{\boldsymbol{\tau}_u,\boldsymbol{t_u}}\boldsymbol{P}A_{\boldsymbol{\tau}_v,\boldsymbol{t_v}}. \tag{17.2}$$

The more general capability would be to introduce more flexibility into the surface representation by refining the knot vectors in certain areas. The procedure outlined above for calculating the alpha matrices and then performing matrix multiplication to evaluate the new mesh points is equally applicable for straightforward refinement.

If instead, it is necessary to subdivide a surface into four subsurfaces, the strategy developed below incorporates the matrix approach.

We give an algorithm for subdividing an open B-spline surface about given values u and v in the u and v directions, respectively. Here, the surface σ has control mesh $\boldsymbol{P}$, knot vectors $\boldsymbol{\mu}_u$, $\boldsymbol{\nu}_v$, and orders m and n, in the u and v directions, respectively. A decision is made to split σ in both parametric directions. The split will be at the computed midpoint of each knot vector into surfaces σ_i, $i = 1, \ldots, 4$. To facilitate the algorithm, we assume all the information for each surface is passed with the pointer to its name. The pseudo-code implementations in this chapter assume that

```
Procedure SplineSurfsplit(σ, σ1,1, σ1,2, σ2,1, σ2,2 )
```

is an implementation of Algorithm 17.4.

Algorithm 17.4.

Step 1. Create the refining knot vectors $\boldsymbol{u}$ *and* $\boldsymbol{v}$*, using procedure* `Refineknotvector`*, as defined in Algorithm 17.1.1.*

Step 2. Find the refining alpha matrices $A^t_{\boldsymbol{\mu},\boldsymbol{u}}$ *and* $A_{\boldsymbol{\nu},\boldsymbol{v}}$.

Step 3. Perform the multiplication, $\boldsymbol{D} = A^t_{\boldsymbol{\mu},\boldsymbol{u}}\boldsymbol{P}A_{\boldsymbol{\nu},\boldsymbol{v}}$.

Step 4. Now, split the refined surfaces into its four component surfaces: First the resulting mesh is broken into two meshes by treating the mesh $\boldsymbol{D}$ *as "curve polygons" in one direction using the method of procedure* `SplitSpline`*. Then the same process is applied to both intermediate surfaces in the other direction. The knot vectors are treated only once per direction. This leaves four surfaces.*

The method of calculation then for rendering, refinement in support of hierarchical design and in support of intersection operators, is to calculate the discrete splines for each of the tensor product directions. The refinement vectors in each direction are decided by the type of problem at hand. Then perform the matrix multiplication indicated above to get the $\boldsymbol{D}$ matrix, the mesh coefficients in the new tensor product basis. Note that if no refinement is desired in a particular direction, then it is unnecessary to apply the operator, and only one matrix multiplication is performed.

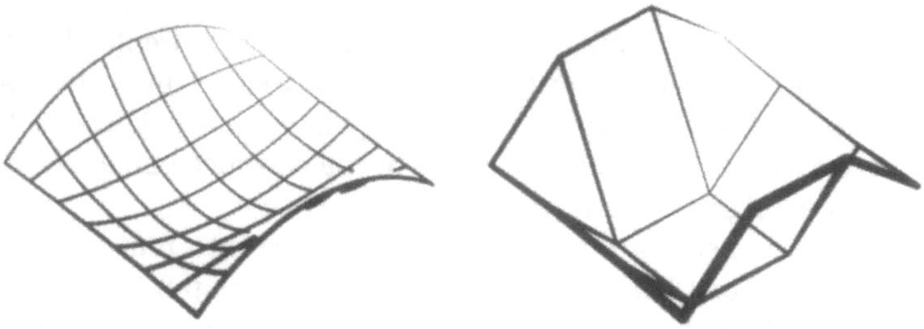

(a) (b)

Figure 17.9. A bicubic tensor product spline surface.

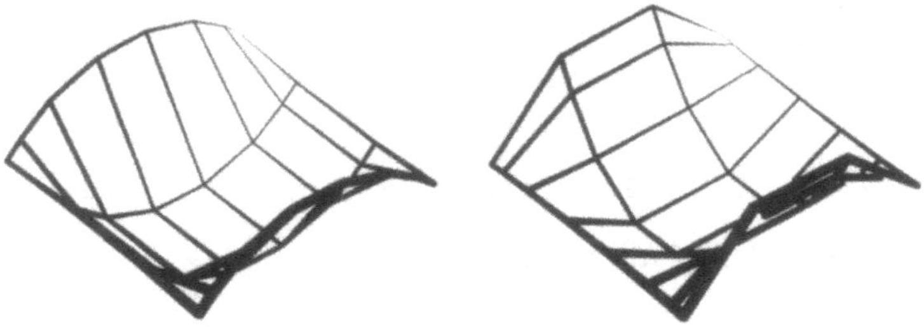

(a) (b)

Figure 17.10. (a) Refinement in the u direction; (b) refinement in v direction.

In Figures 17.9–17.11, only the control meshes are shown. Figures 17.12–17.13 show surface subdivison. Note that since the surface is a tensor product surface, the control mesh must be rectangular, and hence for each knot addition, there is one additional *row* or *column* of control mesh, not just one new point.

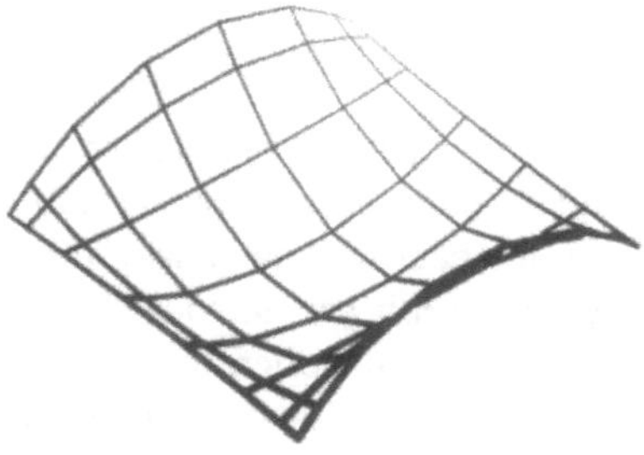

Figure 17.11. Simultaneous refinements in both directions.

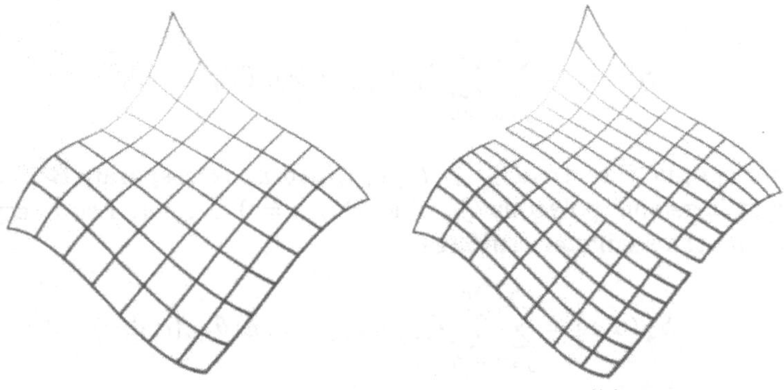

(a) (b)

Figure 17.12. (a) Original; (b) subdivision into Q and R.

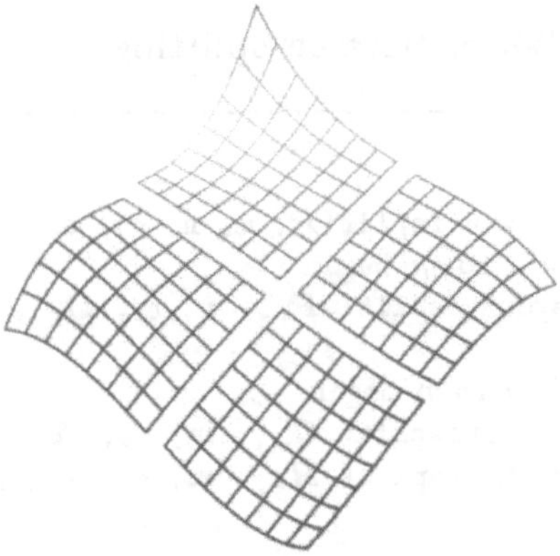

Figure 17.13. Subdivision into $\sigma_{1,1}$, $\sigma_{1,2}$, $\sigma_{2,1}$, $\sigma_{2,2}$.

Example 17.5. Curve Subdivision Approach to Bézier Surfaces. The Bézier surface is a single bi-polynomial patch. The subdivision process can find two distinct partitions of this patch into two subpatches, or can find four patches whose whole is the original patch. The Subdivision Theorem and the consequent algorithmn for curve splitting can be extended to the splitting of tensor product Bézier surfaces in a straightforward way.

Given a mesh $\boldsymbol{P} = \{P_{i,j}\}$ which corresponds to a Bézier surface over $[0, 1] \times [0, 1]$ of degree m by n, the strategy is to first split the surface in the u direction getting meshes $Q = \{Q_{i,j}\}$ and $R = \{R_{i,j}\}$ and then split the surfaces represented by mesh Q and then mesh R in the v direction resulting in surfaces with meshes A and B, and C and D, respectively.

$$S(u,v) = \sum_{j=0}^{n} \sum_{i=0}^{m} P_{i,j}\theta_{i,m}(a,b;u)\theta_{j,n}(c,d;v).$$

For each specific j, we treat $\sum_{i=0}^{m} P_{i,j}\theta_{i,m}(a,b;u)$ as a separate Bézier curve and split it getting polygons $Q_{i,j}$ and $R_{i,j}$, $i = 0, \ldots, m$. Let j vary from 0 to n and the results are surfaces.

$$S_Q(u,v) = \sum_{j=0}^{n} \sum_{i=0}^{m} Q_{i,j}\theta_{i,m}(a, c_u; u)\,\theta_{j,n}(c,d;v)$$

and

$$S_R(u,v) = \sum_{j=0}^{n} \sum_{i=0}^{m} R_{i,j}\theta_{i,m}(c_u, b; u)\,\theta_{j,n}(c,d;v).$$

Algorithm 17.6. Bézier Surface Splitting.

```
Procedure BezSurfsplit(P, m, n, c_u, c_v, A, B, C, D)
    For i = 0 to m begin
        BezCurvesplit( P_{i,*}, n, c_v, Q_{i,*}, R_{i,*} )
    end
    For j = 0 to n begin
        BezCurvesplit( Q_{*,j}, m, c_u, A_{*,j}, C_{*,j} )
        BezCurvesplit( R_{*,j}, m, c_u, B_{*,j}, D_{*,j} )
    end
end
```

The notation $\boldsymbol{P}_{i,*}$ denotes the vector of elements in the ith row of the mesh $\boldsymbol{P}$, and $\boldsymbol{P}_{*,j}$ denotes the vector of elements in the jth column of the mesh $\boldsymbol{P}$.

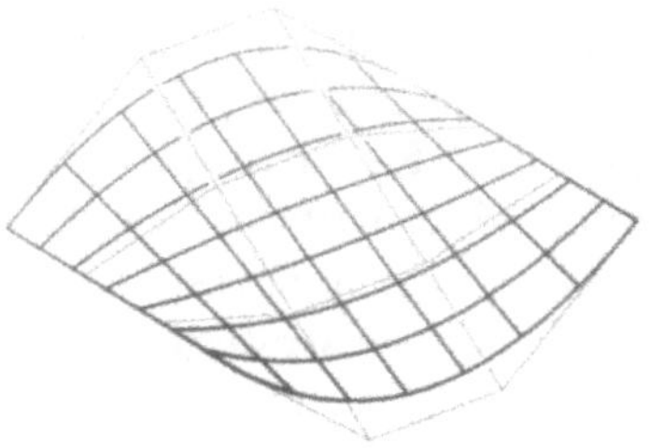

Figure 17.14. A bicubic Bézier surface with control mesh.

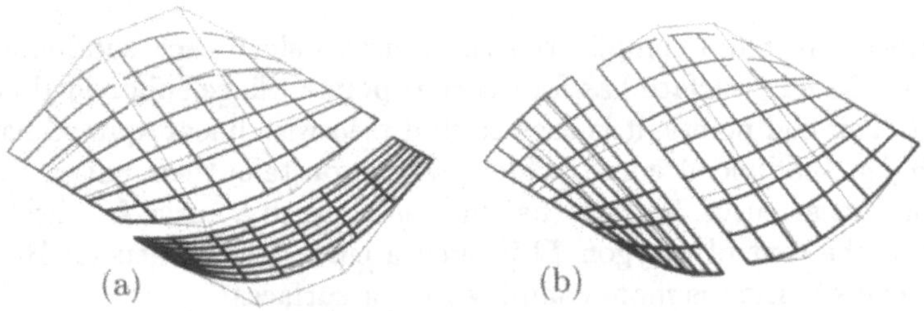

Figure 17.15. (a) Subdivision at $u = 1/4$; (b) Subdivision at $v = 1/3$.

The above algorithm can be generalized to allow for surface splitting around any arbitrary values. We illustrate this capability using a single bicubic Bézier patch illustrated in Figure 17.14 by its control mesh. Figure 17.15 shows various subdivision patterns.

The order of the operation is irrelevant since S is a tensor product surface and these operations commute.

17.5 Rendering Open Surfaces

17.5.1 Raster Images

The rendering of B-spline surfaces, as well as curves, is important. Algorithm 17.7 outlines such a procedure. Again the surface σ carries all the information of mesh, degrees, and knot vectors.

Algorithm 17.7.

```
Procedure SplineSurfgen(σ)
    Are the points of the mesh of σ within tolerance?
    if yes,
        Draw_mesh( σ )
    else begin
        SplineSurfsplit( σ, σ1, σ2, σ3, σ4 )
        SplineSurfgen( σ1 )
        SplineSurfgen( σ2 )
        SplineSurfgen( σ3 )
        SplineSurfgen( σ4 )
    end {else}
end
```

Surfaces are more complicated than curves since each subdivided B-spline (or Bézier) surface has four corner points. Since subdivided curves meet at their end points, it is easy to find piecewise linear approximations to each *flat* subdivided curve, which, when you take their union, form a piecewise linear curve. In fact the curve ψ^k is such a curve for the Bézier case, and the control polygon $\boldsymbol{D}$ is such a curve for the general B-spline curve. The situation is more complicated for surfaces.

The decision of whether the mesh for a subdivided surface is flat enough is more complicated than the decision for a curve, since each quadrilateral region of the mesh, $P_{i,j}$, $P_{i+1,j}$, $P_{i,j+1}$, and $P_{i+1,j+1}$, corresponding to the linear segment in the curve case, is *not* planar (linear)! One strategy is to use the convex hull test. That is, the test requires that the convex hull must be flat enough by some measure. Finding the convex hull for a parametric surface is a far more complicated and time consuming task than for curves. An alternative is to find a bounding box for the convex hull, which, of course, also includes the surface, and test the bounding box. Since bounding boxes are quite easy to find, this test is not onerous. Unfortunately, the bounding box can be substantially larger than the convex hull, so the algorithm might subdivide the surface more than is necessary.

However, suppose it has been determined that the subdivided piece of surface is flat enough. Then one still must determine further attributes in order to make it feasible to render this surface on a raster display. In particular, most graphical rendering algorithms [62] require that the database be polygons. Even though one might want to imitate the curve case and use the B-spline mesh, that is impossible. While the whole mesh may be flat enough to approximate the surface, each quadrilateral is not planar. System developers must determine the polygons to be used. Even if only the four corner points of the submesh are used, it is still impossible to fit just one plane through them. Also, the user must somehow determine a strategy which will be consistent from one subdivided piece to its neighbor, so a least-square fit is not suitable. Sometimes people just treat the

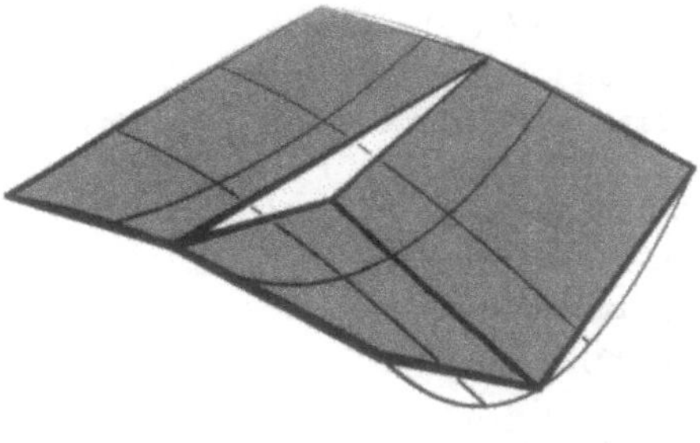

Figure 17.16. Adjacent subdivided subsurfaces with their corner points.

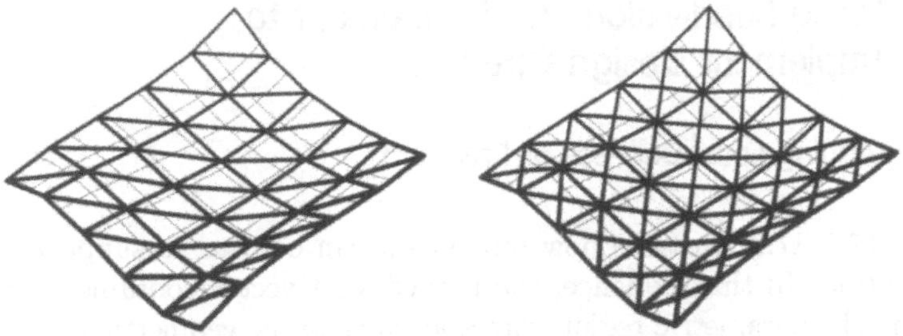

Figure 17.17. A planar fitting based on two different tilings.

quadrilaterals as if they were planar polygons. Sometimes this strategy is successful; other times the method fails. Certainly this is not a general strategy.

When two sides along a subboundary are subdivided to a different level, the black hole or cracking problem can result. Along one side of the boundary, a single line segment represents the edge, but along the other side, two or more line segments might represent the edge. This is called the *black hole problem*; it is illustrated in Figure 17.16.

One way of determining a consistent planar-fitting capability based on the data is to determine two triangles, each of which are defined by the diagonal vertices and one of the remaining corner points. It does not matter which of the diagonals is chosen. There are many which work very well. The patterns in Figure 17.17 are just two of them.

17.5.2 Line Drawings

Frequently, just the control mesh of a B-spline surface is drawn as a line-drawing. However, that sometimes does not give high enough resolution. For example, the control mesh of a sphere can be a box!

A common method is to draw isoparametric curves in the surface using subdivision to extract the curve representations from the surface mesh, and then to draw each curve. This can be performed in both directions, so the line drawings can be done.

The choices of which isoparametric curves to draw have varied solutions. Enough curves must be drawn to convey the surface shape, but drawing too many overfills the screen. One possibility, with justification in theory, is to draw the curves at the nodes in the appropriate directions. Suppose that $\boldsymbol{u}$ is the knot vector in the u-direction, and $\{u^*_{i,\kappa}\}_i$ are the nodes in the u direction. One frequent solution renders the surface with isoparametric curves at the nodes, $\sigma(u^*_i, v)$, for all i.

17.6 Using Subdivision and Refinement to Implement Design Operations

17.6.1 Adding Local Degrees of Freedom

Figures 17.1–17.2 illustrate how refinement can be used to support design modification. In this sequence, the refined knot vector contains new knot values in the parametric region corresponding to the geometric area which requires change. Once the new representation has been computed, the user can modify the curve to the desired shape, either by manually moving control points or by using higher-order design operators which can move groups of points according to goal function requirements.

There are several distinct problems with the Bézier curve form when used in hierarchical design that the use of subdivision can help to overcome. First, a single Bézier curve of degree n is a single polynomial and so it has exactly $n+1$ degrees of freedom. Extra degrees of freedom cannot be added to regions of interest. For example, suppose the shape of Figure 17.18 (a) has been designed using a Bézier curve, but it is the shape in Figure 17.18 (b) that the user wants. Note that the second curve has identically the same shape as the first, except for one small region. A single Bézier curve cannot be used to generate that shape since the first shape has no degrees of freedom left, and further, changing a single control vertex results in a *global* change to the curve.

Now consider Figure 17.19. The Bézier curve was subdivided at two different places into three new Bézier curves. The internal vertices of the middle control polygon can be moved at the will of the designer, leaving the two exterior Bézier curves unchanged. However, when *any* vertices are moved, the interior curve will no longer have the same higher-degree

Figure 17.18. (a) Initial shape design; (b) desired final shape.

Figure 17.19. (a) Subdivided curve; (b) modified middle control polygon.

continuity with the exterior segments. However if the last and first control vertices are left unchanged, the curves will have $C^{(1)}$continuity with each other. If the original curve is a cubic, the interior curve would then have no freedom left to move control points. A way around this is to design the initial gross shape with whatever curve develops. Then the user can degree raise that curve using Theorem 5.31, and apply subdivision and hierarchical design. Otherwise, the user can apply subdivision first and determine that there is not enough freedom to move vertices and keep continuity. He might then apply degree raising to individual segments.

Exercises

1. In Section 17.5.2 it was suggested to draw isoparametric curves at the nodes. Suppose they are extracted using the results of Exercies 2 in Chapter 16. Should each curve be rendered separately using subdivision? Can you suggest a modified subdivision method which can be used on all the curves simultaneously?

18

Set Operations to Effect Modeling

Sometimes there are shapes a designer wants to attain with surfaces which cannot be modeled easily as a single tensor product B-spline, nor in fact a *single* surface patch in any form, be it explicit, implicit, parametric, or over rectangular or triangular domain. In these situations it is sometimes possible for the designer to model the shape as a compound shape, modeling each part as a B-spline (or more restrictive Bézier) surface. Sometimes only stylized versions of those shapes can be modeled in that way, but real parts designed to specification cannot be. Sometimes, not even stylized parts can be modeled that way. In these situations, Boolean operations are frequently useful.

We define a *complete model* as one which is completely defined by its boundaries and is realizable. A sculptured surface, a plane, and a line, while the building blocks of models, are not complete models. They are mathematical descriptions. A surface has no thickness and is a boundary. An infinite plane divides space into two parts, *above* and *below*, the same way an infinite line divides the plane into two parts.

A model is a *solid model* if it trichotomizes space. That is, given a point in space, we must be able to decide, if the point is *inside* the model, *outside* the model, or *on the boundary* of the model. In the plane, a simple closed curve is one that is defined to not cross itself. The conclusion of the Jordan Curve Theorem is that the plane is divided by a simple closed curve into two distinct regions: inside the curve and outside the curve. Mathematically, those points on the boundary are not considered interesting, but they are to the designer since most objects are defined by their boundary surfaces. If a curve is not closed, this partition does not occur. If a closed curve is

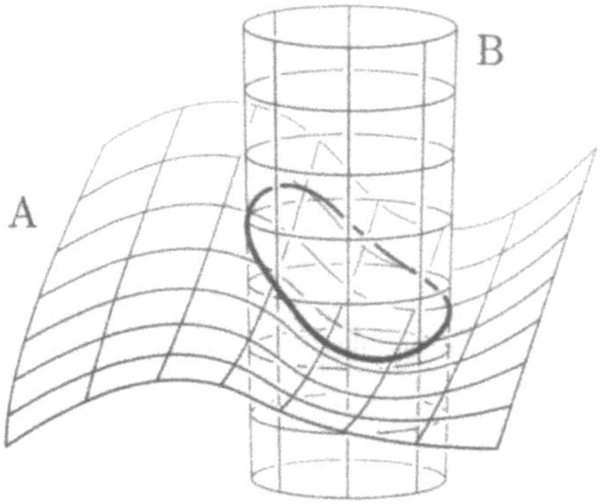

Figure 18.1. Two surfaces intersecting each other.

defined as a piecewise curve, in terms of sections from various previously modeled curves, until this piecing together is done, there is no inside. If a model is incompletely specified, as in the process of design, one cannot trichotomize space. However, answering inside or outside is very important in the process of building the boundary of the part.

For example, consider the surfaces of Figure 18.1. As a resulting shape, one could easily want the part of surface B which is above surface A. This is frequently called a *trimmed surface.* Namely, surface B has been trimmed by surface A. One could also, just as easily want the part of B which is below surface A. Since the roles of A and B can be interchanged, we see there are four basic classifications of A and B with respect to each other. Although it is quite easy to determine which of the parts we want visually, it is not so simple computationally. The terms above and below are difficult to define. If a part is complete and realizable with the boundaries defined by sculptured surfaces, then one might use the terms inside and outside which can be defined analytically for orientable surfaces.

Sometimes the intersection of two finite extent surfaces may not divide each other into distinct subpieces (see Figure 18.2). A piece of paper with a square cut out, while well defined, went through several stages during which each edge of the square was cut out. During those stages it was not well defined. A part with sculptured boundaries may be modeled in much the same way. It may not be completely defined and *realizable* until all the final boundary surfaces have been defined and the final intersections determined. Until all the intersections are known, the original surfaces may not be decomposed into the distinct trimmed subpieces that together form the model boundary. Thomas [82] developed a *cut operator* for use in these situations, and decomposed all Boolean operations in terms of this new operator.

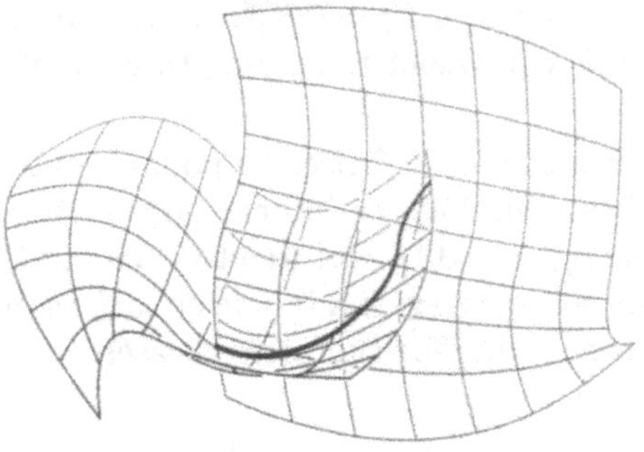

Figure 18.2. The juxtaposition of the two surfaces and the intersection curve.

However, as a first step one must compute the intersection curves. Then one must have a clearly defined method of always deciding which portion(s) of the surface are to be used as the model boundary. Finally, there must be a coherent way of grouping the boundaries so that the system can use this knowlege in other system operations, such as more modeling or analysis or graphics or simulation. Most of these issues are still hot areas of research so in the sections and chapters that follow, we will mostly allude to the problems and present approaches, if only on simpler examples.

18.1 Intersections as Root Finding

Although the discussion centers around methods for finding the intersections of parametric curves and surfaces, methods used for finding intersections of explicit and implicit curves and surfaces can sometimes be adapted or serve as a guide.

Finding the intersections of two explicit curves $f(x)$ and $g(x)$ is equivalent to finding the roots of a third explicit function $h(x) = f(x) - g(x)$. Similarly, finding the intersections of two explicit surfaces $f(x, y)$ and $g(x, y)$ is equivalent to finding the roots of a third surface $h(x, y) = f(x, y) - g(x, y)$. In this case, the roots are not discrete values, generally, but form the equation of an implicit curve in the x-y plane, $h(x, y) = 0$. Hence, finding the intersections between two explicit surfaces is equivalent to graphing an implicit curve.

Finding the intersections of two implicit curves, $f(x, y) = 0$ and $g(x, y) = 0$ is equivalent to finding the points (x, y) which simultaneously solve that system of two nonlinear equations in two unknowns. This approach can be extended to considering the simultaneous intersections of three surfaces given in implicit form as $f(x, y, z) = 0$, $g(x, y, z) = 0$, and $h(x, y, z) = 0$. Finding this solution is equivalent to solving three nonlinear equations in three unknowns.

Surfaces can intersect in sets of discrete points, continuous curves, and shared surface patches. Finding the intersections of two implicit surfaces $f(x, y, z) = 0$ and $g(x, y, z) = 0$ is equivalent to finding the points (x, y, z) which simultaneously solve the system of two equations in three unknowns. We discuss approaches solving for the various cases.

18.2 Lines

First consider the intersection of two planar lines. The implicit form for each line is $a_1x + b_1y + c_1 = 0$ and $a_2x + b_2y + c_2 = 0$. If a point (x, y) is on both, then it satisfies both equations.

Using the homogeneous notation presented in Chapter 1, represent the first line as $\{ (x, y) : (x, y, 1) \cdot (a_1, b_1, c_1)^T = 0 \}$ and the second line as $\{ (x, y) : (x, y, 1) \cdot (a_2, b_2, c_2)^T = 0 \}$. This can be interpreted within $\mathbb{R}^3$. The intersection must be *orthogonal* to both (a_1, b_1, c_1) and (a_2, b_2, c_2), and have a z-coordinate equal to one. Now, $(x_f, y_f, z_f) = (a_1, b_1, c_1) \times (a_2, b_2, c_2)$ is orthogonal to both (a_1, b_1, c_1) and (a_2, b_2, c_2). If $z_f \neq 0$, then $(x_f/z_f, y_f/z_f)$ is on both lines. But, if $z_f = 0$, the lines are parallel and there is no intersection.

18.3 Curve-Curve Intersections by Bisection

Finding the intersection of two explicit curves $f(x)$ and $g(x)$ is equivalent to finding the roots of a third explicit function $h(x) = f(x) - g(x)$. Assume $h \in C^{(0)}$ on $[c, d]$.

The bisection method is a search method which requires two starting values, x_l, $x_u \in [c, d]$, $x_l < x_u$, such that $h(x_l)h(x_u) < 0$. That is, the function h must change sign over the interval (x_l, x_u). Using those values, h is evaluated at the midpoint $m = (x_l + x_u)/2$. If $h(m) = 0$, the root is found. Otherwise, either $h(x_l)h(m) < 0$ or $h(m)h(x_u) < 0$, since $h(m)$ can have the same sign as only one of the two endpoints. The appropriate interval is selected and the test is repeated on that new interval. The termination conditions usually are based on either the absolute or relative size of the interval and the size of the function.

Algorithm 18.1. Root Finding by Bisection.

```
Procedure Root_by_Bisection(h, x_l, x_u, m)
    /* x_l is the lower given endpoint
    /* x_u is the upper given endpoint
    /* h is given function form, used in this procedure.
    /* m is the returned root value

    m ← (x_l + x_u)/2;

    /* Termination criteria
    while ( h(m) ≠ 0 or x_u − x_l > δ ) begin
        if h(x_l)h(m) < 0
            then x_u ← m;
            else x_l ← m;
        m ← (x_l + x_u) /2;
    end {while}
    return ( m )
end {procedure}
```

This method is guaranteed to converge to a result, but the convergence is slow since this is a binary search. With each iteration, the size of the interval is halved, and so unless one chances upon a root, this method will take N iterations where N is the unique smallest integer such that $\delta < (x_u - x_l) \,/\, 2^{N-1}$, but $(x_u - x_l) \,/\, 2^N \leq \delta$. The size of the initial interval determines the number of iterations.

This method requires knowledge about the particular function. If the user only knows the interval $[c, d]$ and $h(c)h(d) > 0$, then the user must find appropriate x_l and x_u, if they exist. If the user does not want to evaluate points in the interval randomly, he must use function-specific information to find the values x_l and x_u, and even so, he may not be able to bracket all the roots that way. The curve $h(u) = u^2$ will never satisfy the conditions, since it is nonnegative everywhere in its domain, but it does have a double root at $u = 0$.

In summary, the positive side:

- The method always converges when the initial conditions are satisfied.
- It requires just one function evaluation per iteration.

On the negative side:

- The method requires the user to bracket roots within specialized types of intervals.
- It is a slow method.
- It may not find all the roots.

18.4 Intersections with Newton-Raphson

18.4.1 Curves

18.4.1.1 Explicit. Again, let $h(x) = f(x) - g(x) \in C^{(2)}$. This method requires an initial guess for the root, x_0. The method approximates the root of h by finding the root of the tangent line at the current guess. That root becomes the new current guess.

The tangent line at $x = x_i$ has equation

$$\ell_i(x) = h(x_i) + h'(x_i)(x - x_i).$$

Then if $h'(x_i) \neq 0$, since x_{i+1} is the root of ℓ_i,

$$0 = h(x_i) + h'(x_i)(x_{i+1} - x_i),$$

and

$$x_{i+1} = x_i - \frac{h(x_i)}{h'(x_i)}. \tag{18.1}$$

This method results in a sequence of values x_i. We analyze when the sequence converges to the root.

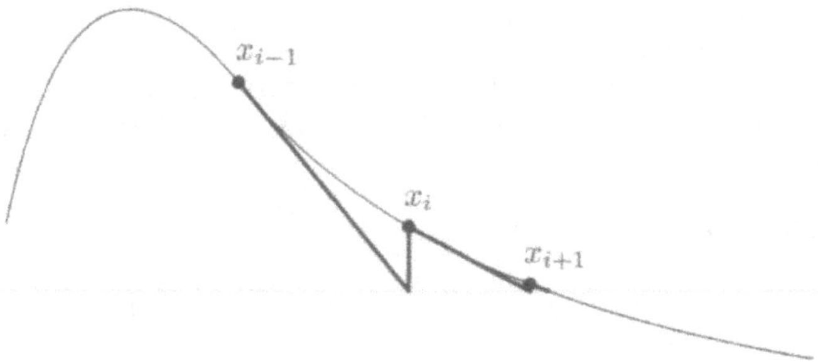

Figure 18.3. Finding x_{i+1} using Newton-Raphson.

Let α be the root that is sought. Call $H(x) = x - h(x)/h'(x)$ the *iteration function* for Newton's method. Then $H(x_i) = x_{i+1}$. If $f \in C^{(2)}$, H is differentiable since $H'(x) = h(x)h''(x) \,/\, (h'(x))^2$. If α is not a multiple root, then $h'(\alpha) \neq 0$, $H'(\alpha) = 0$, and $H(\alpha) = \alpha$. That is, the root is a *fixed point* of H. Repeated application of the mean value theorem gives existence of ζ_j between x_j and α, $j = 0, \ldots, i$, such that

$$\begin{aligned}
x_{i+1} - \alpha &= H(x_i) - H(\alpha) \\
&= (x_i - \alpha)H'(\zeta_i) \\
&= ((x_{i-1} - \alpha)H'(\zeta_{i-1}))\, H'(\zeta_i) \\
&\vdots \\
&= (x_0 - \alpha)\prod_{j=0}^{i} H'(\zeta_j).
\end{aligned}$$

So,

$$|x_{i+1} - \alpha| = |x_0 - \alpha|\,\left|H'(\zeta_i)H'(\zeta_{i-1})\ldots H'(\zeta_1)H'(\zeta_0)\right|.$$

Suppose there exists a real number C such that $\left|H'(\zeta_i)\right| < C$, for all i. Then, $|x_{i+1} - \alpha| < C^{i+1}(x_0 - \alpha)$. Now, if $C < 1$, then $|x_{i+1} - \alpha| < |x_i - \alpha|$. In general, $|x_{i+1} - \alpha| \leq C^{i+1}|x_0 - \alpha|$. The following theorem results.

Theorem 18.2. *Let $|H'(x)| < C$ in the interval containing x_0, x_1, ..., and the root, α. If $C < 1$, then the sequence obtained from the Newton-Raphson method, Equation 18.1, converges to the root α.*

Further, if $h \in C^{(3)}$, and h' is nonzero over the correct interval, then $H \in C^{(2)}$ and, expanding H around the root α of h,

$$\begin{aligned}
H(x) &= H(\alpha) + H'(\alpha)(x - \alpha) + \frac{H''(\zeta)(x - \alpha)^2}{2}, \\
&= \alpha + \frac{H''(\zeta)(x - \alpha)^2}{2},
\end{aligned}$$

where ζ is some value in the interval x and α. Now letting $x = x_i$,

$$x_{i+1} - \alpha = \frac{(x_i - \alpha)^2 H''(\zeta_i)}{2}.$$

Hence, when the function h is three times continuously differentiable and Newton-Raphson converges, we say that it converges quadratically.

In summary, Newton Raphson has positive features:

- The method is relatively fast.
- The method always converges when the conditions on the magnitude of the derivative ratios are satisfied.

On the negative side:

- The method requires the user to have an initial guess with which to start the method. The robustness of the method depends on the guess.
- It may not find all the roots, nor even the one you expect.
- It is necessary that the function be C^2 continuously differentiable in order to apply it.
- The root cannot be a multiple root.
- It may not converge.
- It requires two function evaluations for each level of iteration.

18.4.1.2 m implicit equations in m unknowns. Consider a vector function $\boldsymbol{f} = (f_1, \ldots f_m)^t$, where each coordinate function f_i is a function of variables $x_1, x_2, \ldots x_m$. The problem is to find a single vector $\overline{\boldsymbol{x}} = (\overline{x}_1, \overline{x}_2, \ldots, \overline{x}_m)^t$ so that $f_i(\overline{\boldsymbol{x}}^t) = 0$, $i = 1, \ldots, m$, simultaneously.

Suppose $\boldsymbol{x}^0 = (x_1^0, \ldots, x_m^0)^t$, the initial guess, is given. This method adapts the one-dimensional approach. For each i, the tangent-hyperplane to f_i is determined at $\boldsymbol{x}^0$:

$$\mathcal{L}_i(\boldsymbol{x}) = f_i(\boldsymbol{x}) + \sum_{j=1}^{m} \frac{\partial f_i}{\partial x_j}\bigg|_{\boldsymbol{x}=\boldsymbol{x}^0} \left(x_j - x_j^0\right).$$

Written in matrix form,

$$\begin{bmatrix} \mathcal{L}_1(\boldsymbol{x}) \\ \mathcal{L}_2(\boldsymbol{x}) \\ \vdots \\ \mathcal{L}_m(\boldsymbol{x}) \end{bmatrix} = \boldsymbol{f}|_{\boldsymbol{x}=\boldsymbol{x}^0} + \boldsymbol{J}_{\boldsymbol{f},\boldsymbol{x}^0} \left(\boldsymbol{x} - \boldsymbol{x}^0\right)$$

where $\boldsymbol{J}_{\boldsymbol{f},\boldsymbol{x}^0}$ is the Jacobian of $\boldsymbol{f}$ at $\boldsymbol{x}^0$, that is,

$$\boldsymbol{J}_{\boldsymbol{f},\boldsymbol{x}^0} = \begin{bmatrix} \left.\frac{\partial f_1}{\partial x_1}\right|_{\boldsymbol{x}=\boldsymbol{x}^0} & \cdots & \left.\frac{\partial f_1}{\partial x_m}\right|_{\boldsymbol{x}=\boldsymbol{x}^0} \\ & \vdots & \\ \left.\frac{\partial f_m}{\partial x_1}\right|_{\boldsymbol{x}=\boldsymbol{x}^0} & \cdots & \left.\frac{\partial f_m}{\partial x_m}\right|_{\boldsymbol{x}=\boldsymbol{x}^0} \end{bmatrix}.$$

The simultaneous solution is approximated by the vector $\boldsymbol{x}^1$ which is the simultaneous root to all the tangent-hyperplanes. That is,

$$\boldsymbol{0} = \boldsymbol{f}|_{\boldsymbol{x}=\boldsymbol{x}^0} + \boldsymbol{J}_{\boldsymbol{f},0}\left(\boldsymbol{x}^1 - \boldsymbol{x}^0\right).$$

There exists a unique solution for $\boldsymbol{x}^1$ if and only if the determinant of the Jacobian is nonzero, i.e., $\left|\boldsymbol{J}_{\boldsymbol{f},0}\right| \neq 0$. Let $\mathcal{J} = \mathcal{J}_{\boldsymbol{f},\boldsymbol{x}} = \boldsymbol{J}_{\boldsymbol{x},\boldsymbol{f}}$ be the inverse of $\boldsymbol{J}_{\boldsymbol{f},\boldsymbol{x}^0}$. Then,

$$\boldsymbol{x}^1 = \boldsymbol{x}^0 - \mathcal{J}\, \boldsymbol{f}|_{\boldsymbol{x}=\boldsymbol{x}^0}\,. \tag{18.2}$$

This method has been shown to converge under a variety of constraints. We state several without proof and refer the interested reader to [1].

Theorem 18.3. *Let $\boldsymbol{f}$, $\boldsymbol{J}_{\boldsymbol{f},\boldsymbol{x}}$, and $\mathcal{J}$ be defined as above. Let $\boldsymbol{x}^0 = (x_1^0, \ldots, x_m^0)^t$ be given . Then the sequence $\{\boldsymbol{x}^q\}$ converges to a unique root of $\boldsymbol{f}(\boldsymbol{x}) = \boldsymbol{0}$, where*

$$\boldsymbol{x}^{q+1} = \boldsymbol{x}^q - \mathcal{J}_{\boldsymbol{f},\boldsymbol{x}^q}\, \boldsymbol{f}|_{\boldsymbol{x}=\boldsymbol{x}^q}$$

when one of the following sets of conditions hold.

Set 1:

(a) Suppose there exists an $r > 0$ and

$$B_{r,0} = \left\{\, \boldsymbol{x} = (x_1, \ldots, x_m) \in \mathbb{R}^m : \left|x_i - x_i^0\right| < r, i = 1, \ldots,\ m \right\}$$

such that for all $\boldsymbol{x} \in B_{r,0}$,

$$\left|\frac{\partial^2 f_i}{\partial x_j \partial x_k}\right| \leq K \qquad 1 \leq i, j, k \leq m.$$

(b) $\left|\boldsymbol{J}_{\boldsymbol{f},\boldsymbol{x}^0}\right| \neq 0$.

(c) $\left|f_i(\boldsymbol{x}^0)\right| \leq \eta_0, for 1 \leq i \leq m$.

(d) *Let the elements of* $\mathcal{J}_{\boldsymbol{f},\boldsymbol{x}^0}$ *be denoted by* $a_{i,j}$ *(that is,* $a_{i,j} = \left.\frac{\partial x_i}{\partial f_j}\right|_{\boldsymbol{x}=\boldsymbol{x}^0}$*), and let*

$$\max_i \sum_j |a_{i,j}| \leq \beta_0.$$

(e) $h_0 \leq 1/2$ *and* $\left(1-\sqrt{1-2h_0}\right)\beta_0\eta_0 < rh_0$*, where* h_0 *defined by* $h_0 = \beta_0^2\eta_0 K m^2$.

Set 2:

(a) *Suppose there exists an* $r > 0$ *and* $S_{r,0} = \left\{\boldsymbol{x} \in \boldsymbol{R}^m : \left\|\boldsymbol{x} - \boldsymbol{x}^0\right\| < r\right\}$ *such that for all* $\boldsymbol{x} \in S_{r,0}$,

$$\sum_{i,j,k} \left(\frac{\partial^2 f_i}{\partial x_j \partial x_k}\right)^2 \leq K^2.$$

(b) $\left|J_{\boldsymbol{f},\boldsymbol{x}^0}\right| \neq 0$.

(c) $\left\|\mathcal{J}_{\boldsymbol{f},\boldsymbol{x}^0}\, \boldsymbol{f}|_{\boldsymbol{x}=\boldsymbol{x}^0}\right\| \leq \eta_0$.

(d) *Let the elements of* $\mathcal{J}_{\boldsymbol{f},\boldsymbol{x}^0}$ *be denoted by* $a_{i,j}$ *(that is* $a_{i,j} = \left.\frac{\partial x_i}{\partial f_j}\right|_{\boldsymbol{x}=\boldsymbol{x}^0}$*), and let*

$$\sum_{i,j} (a_{i,j})^2 \leq \beta_0^2.$$

(e) $h_0 \leq 1/2$ *and* $\left(1-\sqrt{1-2h_0}\right)\eta_0 < rh_0$*, where* h_0 *is defined by* $h_0 = \beta_0\eta_0 K$.

18.4.2 Surfaces

The problem of surface-surface intersections is far more complex than that for curves. Whereas curves intersect at discrete points, the intersection of two surfaces is most commonly a collection of possibly disconnected curves.

Consider the issues in solving for the curve of intersection. That is, the roots of $h(x, y) = f(x, y) - g(x, y)$, where $f(x, y)$ and $g(x, y)$ are explicit surfaces, give the domain values at which to evaluate f or g to get the shared intersection curve. h may have no roots, which means that the surfaces do not intersect, or h may be identically zero, in which case the surfaces are identical. Most of the cases lie between these two extremes.

We see that graphing implicit functions, that is, finding the (x, y) that satisfy $h(x, y) = 0$, is equivalent to finding the intersection curve between f and g. But graphing implicit functions requires numerical solutions, so except for the simplest of cases, a closed form representation cannot be found for intersection curves.

Let us consider applying Newton's method directly. The surface analogue would be to find the equation for the tangent plane to h at a given $\boldsymbol{x}^0$,

$$\mathcal{P}_0(\boldsymbol{x}) = h(\boldsymbol{x}^0) + \nabla h(\boldsymbol{x}^0) \begin{bmatrix} x_1 - x_1^0 \\ x_2 - x_2^0 \end{bmatrix}$$

where

$$\nabla h(\boldsymbol{x}^0) = \begin{bmatrix} \left.\frac{\partial h}{\partial x_1}\right|_{\boldsymbol{x}=\boldsymbol{x}^0} & \left.\frac{\partial h}{\partial x_2}\right|_{\boldsymbol{x}=\boldsymbol{x}^0} \end{bmatrix}.$$

There is not a unique solution for a root. Since the intersection of two planes (the tangent plane and the $z = 0$ plane) is a straight line, we cannot determine a unique $\boldsymbol{x}^1$. The extra degree of freedom must be specified in an alternative way.

Example 18.4. One way to constrain the problem is to constrain one of the variables. One might fix one variable in each iteration and alternate which variable may be moved. For example, if i is even, set $x_1^{i+1} = x_1^i$ and solve the above equation for x_2^{i+1}; else if i is odd, set $x_2^{i+1} = x_2^i$ and solve the above equation for x_1^{i+1}. The equations then simplify to

$$\boldsymbol{x}^{i+1} = \boldsymbol{x}^i - \begin{cases} \begin{bmatrix} 0 \\ h(\boldsymbol{x}^i) \Big/ \left(\left.\frac{\partial h}{\partial x_2}\right|_{\boldsymbol{x}=\boldsymbol{x}^i} \right) \end{bmatrix} & \text{when } i \text{ is even,} \\ \begin{bmatrix} h(\boldsymbol{x}^i) \Big/ \left(\left.\frac{\partial h}{\partial x_1}\right|_{\boldsymbol{x}=\boldsymbol{x}^i} \right) \\ 0 \end{bmatrix} & \text{when } i \text{ is odd.} \end{cases}$$

This strategy effectively reduces the two-dimensional problem into two interleaved one-dimensional problems. □

Example 18.5. In this method, the point on the intersection line of the tangent plane and the $z = 0$ plane that is closest (using Euclidean distance) to $(x_1^0, x_2^0, h(\boldsymbol{x}^0))$ is selected as the next iteration point. Under those constraints,

$$\boldsymbol{x}^{i+1} = \boldsymbol{x}^i - h(\boldsymbol{x}^i) \frac{\nabla h(\boldsymbol{x}^i)}{\|\nabla h(\boldsymbol{x}^i)\|}.$$

□

All algorithms for graphing implicit functions are immediately usable.

18.5 Intersections of Parametric Curves and Surfaces

We start with a discussion about the intersections of the boundaries of the boundaries. The boundaries of models are sculptured surfaces and the boundaries of sculptured surfaces are curves. The magnitude of the problems for that case is large enough for an initial impression of the difficulty of the full problem.

18.5.1 Parametric Curves

Given two curves $\gamma_1(u)$ and $\gamma_2(v)$, one wants to find:

1. all the intersection points;
2. the parameter values of u for γ_1 at the intersection points; and
3. the parameter values of v for γ_2 at the intersection points.

The intersection problem has become substantially complicated by using parametric curves. The explicit curve methods discussed above are no longer directly applicable. Finding their intersections amounts to solving $\gamma_1(u) - \gamma_2(v) = (x_1(u) - x_2(v), y_1(u) - y_2(v)), z_1(u) - z_2(u)) = (0, 0, 0)$. This can be rewritten to be

$$
\begin{array}{rclcl}
0 & = & x_3(u,v) & = & x_1(u) - x_2(v) \\
0 & = & y_3(u,v) & = & y_1(u) - y_2(v) \\
0 & = & z_3(u,v) & = & z_1(u) - z_2(v).
\end{array}
$$

Now if both curves are planar and $0 \equiv z_3(u, v)$, finding the intersections of two planar parametric curves is equivalent to finding the intersection points of two implicit curves, and so with an initial guess available, the strategy of Section 18.4.1.2 can be applied.

However, if the curves are space curves, there are three equations in two unknowns. The problem is overconstrained. Sometimes a strategy is adopted that selects any two coordinates and solves that system.

This problem can also be approached in a totally different way that was not available in the explicit case.

18.5.2 Spline Intersections: Divide and Conquer

This method is applicable only to curves which can be represented as B-splines. In Section 17.3, an algorithm for spline intersection based on the divide and conquer approach inherent in the subdivision process is presented. The basic idea of the algorithm is that two B-spline curves can intersect only if the convex hulls of their control polygons intersect. Now, the curves may not intersect even if the control polygons intersect; however, if the convex hulls do not intersect the curves *can not* intersect. The given algorithm keeps subdividing both curves until sections of curve can be eliminated from the possibility of intersection, or until the intersecting convex hulls which might contain an intersection point are determined to be *small enough* to make a last approximation. The *small enough* provision then must allow the curves to be approximated by either their polygons or a single straight line. If the lines intersect, an intersection is claimed, else none is recorded. When the curve is an explicit curve, the values returned for the abscissa are a close approximation, as is the graph point for the intersection. When the curve is parametric, however, that procedure is sufficient for finding the location of the intersection, but not for finding the corresponding parameter values in each curve of that intersection point. If the approximating straight lines are parameterized linearly over the parameter range for those small sections of curve, then the corresponding parameter approximations are easy to determine. Of course, the values given are not any good if the real curve is not close to linear, even if it is flat. For example, if two straight lines intersect at their midpoints, we might guess that the parameter value for each is 1/2. but if each is parameterized as a quadratic B-spline, then the midpoint of the line may not have that parameter value.

For example, let $P_1 = P_0/8 + 7P_2/8$. Then

$$\begin{aligned}
\gamma(1/2) &= P_0\theta_{0,2}(1/2) + P_1\theta_{1,2}(1/2) + P_2\theta_{2,2}(1/2) \\
&= P_0\left(\theta_{0,2}(1/2) + \frac{\theta_{1,2}(1/2)}{8}\right) + P_2\left(\frac{7\theta_{1,2}(1/2)}{8} + \theta_{2,2}(1/2)\right) \\
&= P_0\left((1/2)^2 + \frac{(1/2)^2}{4}\right) + P_2\left(\frac{7(1/2)^2}{4} + (1/2)^2\right) \\
&= \frac{5}{16}P_0 + \frac{11}{16}P_2
\end{aligned}$$

which is not at the midpoint. The geometric location of the intersection point is within whatever tolerance is specified, but there is no check to verify that the parametric value is within a specified tolerance.

18.5.3 Curve-Surface Intersections

Sometimes it becomes necessary to find the intersection of a parametric curve with a parametric surface. We then are searching for the zeros of the parametric function

$$f(t,u,v) = \gamma(t) - \sigma(u,v) = (x(t,u,v), y(t,u,v), z(t,u,v)).$$

This results in finding the roots of a system of three nonlinear equations (the coordinate functions in x, y, and z) in three unknowns (t, u, and v). Since

$$f(t,u,v) = \left[\begin{array}{ccc} x(t,u,v) & y(t,u,v) & z(t,u,v) \end{array}\right]^T,$$

we can use the results of Section 18.4.1.2 for $m = 3$. Let $P = (t,u,v)$ and $P^i = (t_i,u_i,v_i)$. By Equation 18.2, if the Jacobian of f has a nonzero determinant at P^i,

$$P^{i+1} = P^i - \mathcal{J}f(P^i), \tag{18.3}$$

where $\mathcal{J}$ is the inverse of $\boldsymbol{J}_{f,P^i}$, so the next step of the iteration is determined. Conditions for guaranteeing convergence are given in Theorem 18.3. A good initial approximation, as well as well-behaved second partials at each step of the iteration are required for this technique to converge.

18.5.3.1 Subdivision approach. Another approach is to combine subdivision for curves and subdivision for surfaces to get a subdivision approach.

Given a B-spline curve γ with degree m and polygon $\boldsymbol{P}$ and knot vector $\boldsymbol{\tau}$ and a B-spline surface σ of degrees $n \times p$ with control mesh $\boldsymbol{H}$ and a knot vector pairs $(\boldsymbol{\mu},\boldsymbol{\nu})$, the following algorithm finds their intersections. We shall assume that the total information about the geometry is carried by the pointer to the name.

```
Procedure Spline_CrvSrfIntersect(γ, σ)
If the convex hulls of P and H do not overlap return
Else, test for convergence:
    /* If the convex hull of σ is 'flat enough' either
    /* approximate by an approved type of surface,
    /* (such as planes or bilinear), or use the original
    /* surface to intersect directly.
    /* Call the final 'surface' L_H.
    /* If the convex hull of γ is 'flat enough' either
    /* approximate by an approved type of curve, (such as
    /* linear or piecewise linear.  Call the final curve L_P.
    output the intersection of L_P and L_H.
    /* This approximates the intersections to
    /* within tolerance.  This intersection may be empty,
    /* be a single point, multiple isolated points or
    /*a space curve.
```

(*continued on next page*)

```
          /* If L_P is a line and L(H) is a plane, the
          /* intersection is either a point or a segment of
          /* L(P) which is in L(H).
     Else, subdivide and retest:
          /* for lack of other info, subdivide at midpoints of
          /* knotvectors.
          Findmidpoint( τ, t* )
          Findmidpoints( ν, n*, μ, m* )
          SplineCrvsplit( γ, t*, γ1, γ2 )
          SplineSurfsplit( σ, m*, n*, σ1, σ2, σ3, σ4 )
          /* Check for intersections between the children
     For i = 1 to 2
          For j = 1 to 4
               Spline CrvSrfIntersect( γi, σj )
     end {Spline_SurfIntersect}
```

18.5.4 Ray-Surface Intersections

This is a special case of the above problem, when the curve is a straight line. In rendering, ray tracing requires solution of this problem many times just to render a single image. Ray tracing, however, is not concerned with the value of t, but rather *whether* it exists and the surface parameter values (u and v) at the intersection.

Hence, a modified approach can be used which reduces the problem to one of solving two equations in two unknowns.

A straight line in space can be represented parametrically, but it can also be represented as the intersection of two orthogonal planes. There are an infinite number of pairs of planes for which this can be done. For a given line, select a single pair; let the implicit representations be $h_1(x,y,z) = a_1x + b_1y + c_1z + d_1 = 0$, $h_2(x,y,z) = a_2x + b_2y + c_2z + d_2 = 0$. Then the given line is the solution to the set of (x,y,z) which are simultaneously on both implicit surfaces. Suppose the surface is represented as $\sigma(u,v) = (x(u,v), y(u,v), z(u,v))$. Then the simultaneous solutions to

$$\begin{aligned} 0 &= a_1x(u,v) + b_1y(u,v) + c_1z(u,v) + d_1 \\ 0 &= a_2x(u,v) + b_2y(u,v) + c_2z(u,v) + d_2 \end{aligned}$$

give the intersections of the ray with the surface. The approach and methods of Section 18.4.1.2 can be used in its solution, by setting $m = 2$.

18.5.5 Surface-Surface Intersection

In modeling, it is frequently necessary to find the intersection of two parametric surfaces:

$$\begin{aligned}\sigma_1(u,v) &= (x_1(u,v), y_1(u,v), z_1(u,v)),\\ \sigma_2(s,t) &= (x_2(s,t), y_2(s,t), z_2(s,t)).\end{aligned}$$

Equating the coordinate functions gives rise to a system of three nonlinear equations (the coordinate differences of σ_1 and σ_2) in four unknowns (u, v, s, t), which cannot be solved directly. The solution set may be empty, be curves, or be subsurfaces.

18.5.5.1 A numerical approach. Newton's method cannot be applied directly since that would require three equations in three unknowns.

Following previous strategies, one might hold one of the variables u, v, s, or t constant at some initial value and then use the Newton's method equations derived for the parametric curve-surface intersections above. Supposing that $t = t_0$ is held constant, the coordinate functions are

$$\begin{aligned}&(\overline{x}(u,v,s), \overline{y}(u,v,s), \overline{z}(u,v,s))\\ &= (x_1(u,v) - x_2(s,t_0), y_1(u,v) - y_2(s,t_0), z_1(u,v) - z_2(s,t_0)).\end{aligned}$$

Once again, Equation 18.3 can be solved to attain a next guess for u, v, and s. However, this strategy does not give a scientific way of deciding which of the four variables to hold constant. Hence, we are not sure of the quality of the iterated answers. One might be required to use those answers to hold another of the parametric values constant and iterate again.

Another strategy is to approximate the intersection by the intersection of the respective tangent planes to σ_1 and σ_2 at the current estimated value. That intersection is a straight line, so an additional constraint is needed. One typical constraint sets $P_{i+1} = (A_1 + A_2)/2$, where A_1 is the point on the line closest to $\sigma_1(u_i, v_i)$, and A_2 is the point on the line closest to $\sigma_2(s_i, t_i)$. But then since $P_{i+1} = (A_1 + A_2)/2$. But then since P_{i+1} is in both tangent planes, there exist values $(u_{i+1}, v_{i+1}, s_{i+1}, t_{i+1})$ which uniquely solve the equations:

$$\begin{aligned}P_{i+1} &= \sigma_1(u_i, v_i) + \boldsymbol{J}_{\sigma_1,(u_i,v_i)} \begin{bmatrix} u - u_i & v - v_i \end{bmatrix}^t\\ P_{i+1} &= \sigma_2(s_i, t_i) + \boldsymbol{J}_{\sigma_2,(s_i,t_i)} \begin{bmatrix} s - s_i & t - t_i \end{bmatrix}^t.\end{aligned}$$

Those values are used for the next iteration.

18.5.5.2 Subdivision approaches to B-Spline surface intersection. This method parallels the subdivision approach to curve intersection. Given surface α of degree $m \times k$ with mesh $\boldsymbol{P}$ and knot vector pair $(\boldsymbol{\tau}, \boldsymbol{v})$ and surface σ, of degree $n \times p$, with mesh $\boldsymbol{H}$, and knot vector pair $(\boldsymbol{\nu}, \boldsymbol{\mu})$, the following algorithm finds all the intersections of the two surfaces. However, it does not find a linked intersection curve. It assumes that the surfaces intersect in curves and are not the same over any parts of the domains. Assume that the total surface information is carried by the pointer to the name.

```
Procedure Spline_SurfIntersect ( α, σ )
    If the convex hulls of P and H do not overlap return
    Else, test for convergence:
        /* If the convex hulls are 'flat enough' either
        /* approximate by approved types of surfaces,
        /* (such as planes or bilinear), or use the
        /* original surfaces to intersect directly.
        /* Call the final 'surfaces' L_P and L_H.
        output the intersection of L_P and L_H.

        /* This approximates the intersection of the surfaces to
        /* within tolerance.  This intersection may be empty, be
        /* a single point, or be a space curve.  If L_P and
        /* L_H are planes, the intersection is a line segment.
    Else, subdivide and retest:
        /* for lack of other info, subdivide at midpoints of
        /* knotvectors.
        Findmidpoints( τ, t*, υ, u* )
        Findmidpoints( ν, n*, μ, m* )
        SplineSurfsplit( α, t*, u*, α1, α2, α3, α4 )
        SplineSurfsplit( σ, n*, m*, σ1, σ2, σ3, σ4 )
    /* Check for intersections between the 4 × 4 children
    /* surfaces.
    For i = 1 to 4
        For j = 1 to 4
            Spline SurfIntersect( αi, σj )
end {Spline_SurfIntersect}
```

Since the convex hull of a mesh of points is a polyhedron, the determination of intersection is a time-consuming process since the computation of each convex hull is lengthy. Usually a crude approximation to the convex hull is used, such as a *bounding box.* While such an approximation will of necessity include more possible intersections and hence cause the subdivision and intersection testing process to progress deeper down the tree, the

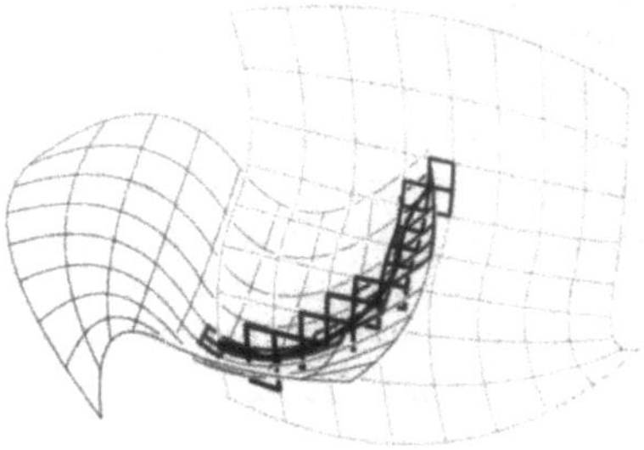

Figure 18.4. Finding the intersection of two spline surfaces with subdivision. (See also Figure 18.2.)

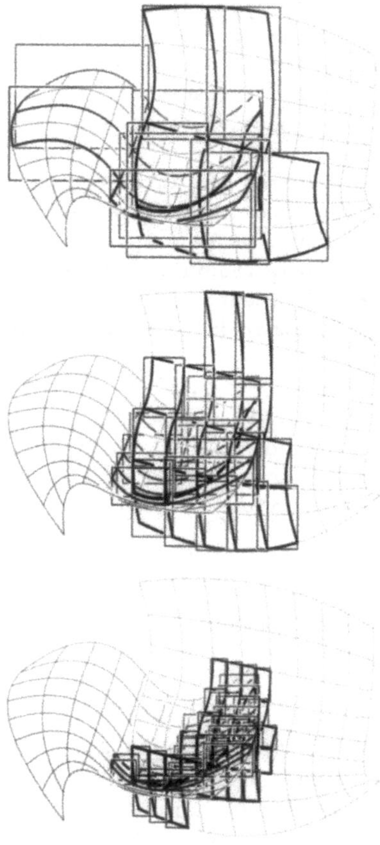

Figure 18.5. Comparing the bounding boxes of children surfaces, at different recursion levels of the subdivision process. (See also Figures 18.2 and 18.4.)

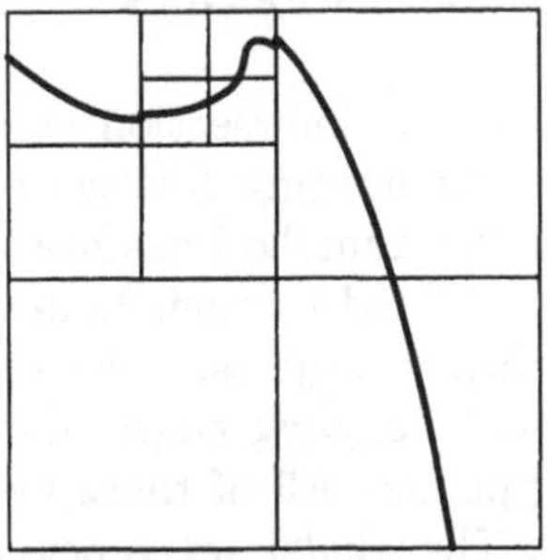

Figure 18.6. Two intersection curve pieces meeting.

rationale is that this additional time is more than compensated for by the time saved in not computing the actual convex hulls.

The output of the above algorithm is an unordered list of line segments (or other curve segments). The next problem for the software developer is to chain these segments together into a piecewise curve which properly approximates the intersection curve. The difficulty of this procedure is caused in part because of floating point arithmetic and the subdivision process.

Figure 18.6 shows how the same point can have two different parametric values caused in part by the approximation process. The curve segment on the left was obtained through a subdivision process that terminated one level higher than the curve segment on the right. The program must decide that these two parametric values represent the same point, and not just two intersection branches which come close to each other.

Another problem occurs when one surface passes through just a vertex of the other. Then just a single point might appear on the list, as shown in Figure 18.7. The difficulty is in deciding whether to make note of this intersection as a zero length curve or to ignore it. Some of the instabilities will be discussed further in the simplified polyhedral/polygon example.

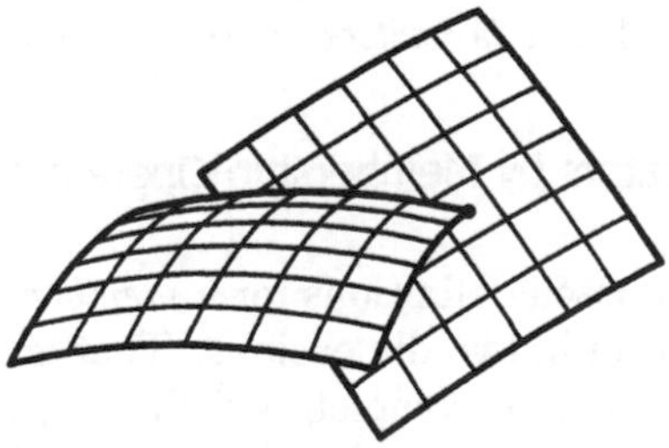

Figure 18.7. Degenerate intersection.

18.6 Boolean Operations on Models

Finding an approximation to the intersection curve is only the first step of many modeling sessions. Computing *Boolean operations* is a necessary part of modeling. Suppose that a method has been defined for determining set membership for objects defined by their boundary descriptions. Since it is desired to model realizable objects, neither models nor their complements are allowed to have dangling edges, dangling surfaces, isolated points, or isolated missing points. All of those items are theoretical entities but cannot be realized. The *regular set* comes closest to attaining these requirements and is used in constructive solid geometry.

Definition 18.6. *A* regular set *is defined as the closure of its interior, written* $\overline{int(A)}$.

That means *regularized Boolean operators* are defined as as $A \cup_R B = \overline{int(A \cup B)}$, $A \cap_R B = \overline{int(A \cap B)}$, and $A -_R B = \overline{int(A - B)}$. The other Boolean difference operations follow.

Definition 18.7. *A point p is an* interior point *to the set P if* $p \in P$, *and there exists a real number* $\epsilon > 0$ *such that if* $\|x - p\| < \epsilon$ *then* $x \in P$.

The set int(P) is the set of all interior points of P.

Definition 18.8. *A point p is called a* limit point *of a set A if for each* $\epsilon > 0$ *there exists a point* $a \in A$ *such that* $\|p - a\| < \epsilon$. *The set of limit points for a set A is called the* closure of A *and is denoted* $\overline{A}$.

It is clear that for every set A, $A \subset \overline{A}$.

Definition 18.9. *A point b is called a* boundary point *of a regular set A if for each* $\epsilon > 0$, *there exists points* $a \in int(A)$ *and* $x \notin A$ *such that* $\|x - b\| < \epsilon$ *and* $\|a - b\| < \epsilon$. *The set of all boundary points of a set A is called the* boundary of A *and is written as* ∂A, *or bdy(A), or* $fr(A)$.

18.6.1 Boolean Operators by Membership Operators

We consider several proposed definitions for a *membership operator*. Let us first consider this problem in two dimensions. That is, let us consider two-dimensional solids which can be completely defined by their boundaries and which have no dangling edges, no isolated points, and no missing isolated points. *Given a set A,* $M(x, A)$ *is a* membership operator *defined by the*

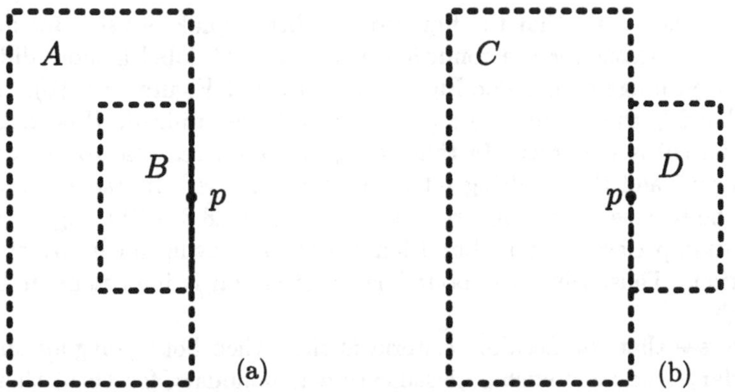

Figure 18.8. Point p on boundary.

rule: $M(x, A) = true,\ (=1)$ *if* $x \in A$, $M(x, A) = false\ (=-1)$ *if* $x \notin A$. Consider A and B to be two polygons as in Figure 18.8 (a), and C and D to be two polygons as in Figure 18.8 (b). Point p is on the boundary of A, B, C, and D. We see that using the obvious set operations, $M(p, A \cap B) =$ true and $M(p, C \cap D) =$ true, but $M(p, A \cap_R B) =$ true and $M(p, C \cap_R D)$ = false. We see that the operators do not give the same results. Since the model resulting from a Boolean operator should be a regular set, we must use the regularized set operators, and for them p is *in* the intersection of A and B but it is *not in* the intersection of C and D. The simple membership operator proposed above cannot be used to define set operations on regular sets, since p is exactly the same in each operand, but not in the result.

Boundary points have special meaning in regular sets, and the membership operator must take that into account. Hence, we redefine it.

Definition 18.10. *Given a regular set A,* $M(x, A)$ *is a* membership operator *defined by the rule:*

$$M(x, A) = \begin{cases} in\,(=1), & if\ x \in int(A), \\ bdy(=0), & if\ x\ is\ a\ boundary\ point\ of\ A \\ out(=-1), & if\ x \notin A. \end{cases}$$

Let us consider a possible redefinition of the intersection operator using this potential membership operator. *For given sets A and B,*

$$M(x, A \cap B) = \begin{cases} in, & if M(x, A) = in\ and\ M(x, B) = in, \\ bdy, & if M(x, A) = bdy\ and\ M(x, B) = in, \\ & or M(x, A) = in\ and\ M(x, B) = bdy, \\ ?? & if M(x, A) = bdy\ and\ M(x, B) = bdy, \\ out & otherwise. \end{cases}$$

It is easy to see that the first two conditions make sense. But finding the correct status for the condition with the "??" label is more difficult. Let us once again consider Figure 18.8 (a) and Figure 18.8 (b). If we set ?? = *bdy* then Figure 18.8 (a) is correctly determined. However, Figure 18.8 (b) is incorrect. In that case p is considered also to be "on the boundary" and the resulting set is not a regular set. If instead ?? = *out* then the reverse conditions are true with Figure 18.8 (b) having a correct membership classification, but Figure 18.8 (a) having an incorrect classification. Thus, while the boundary information is important, it is not enough.

We see that the deciding criteria is that when both polygons share a boundary (the membership classification is boundary for both) then the deciding criteria is whether the polygons are both on the same side of the boundary or whether they are on different sides. If "in" is classified as different sides of the boundary for each of the polygons, then each point on that boundary should be classified "out" of the regular intersection. If "in" is on the same side for both polygons, then each point on the boundary should be classified as "bdy".

Thus the classification function for a regular set must return extra information for boundary points. For each boundary point it must give an indication of where the interior of the set is relative to the boundary point. For a polygon, a single normal pointing to the inside of the polygon might be sufficient for each point along a side. For a shape with curved boundaries, that is not enough. In the curved boundary case, one might think of a neighborhood being returned, that is the set of all points in an arbitrarily small open ball which are also "in" the regular set. Call this set of points $O(p, A)$.

Finally, we define the classification functions as

Definition 18.11. *For given sets A and B,*

$$M(x, A \cap B) = \begin{cases} in, & if\{M(x,A) = in \text{ and } M(x,B) = in\}, \\ bdy, & if\{M(x,A) = bdy \text{ and } M(x,B) = in\}, \\ & \quad or\ \{M(x,A) = in \text{ and } M(x,B) = bdy\}, \\ bdy & if\{M(x,A) = bdy \text{ and } M(x,B) = bdy \\ & \quad \text{and } O(p,A) \cap O(p,B) \neq \emptyset\}, \\ & \\ out & otherwise. \end{cases} \tag{18.4}$$

$$O(x, A \cap B) = O(x, A) \cap O(x, B). \tag{18.5}$$

$$M(x, A \cup B) = \begin{cases} in, & if\{M(x,A) = in \ or \ M(x,B) = in\}, \\ in, & if\{M(x,A) = bdy \ and \ M(x,B) = in\}, \\ & \quad or \ \{M(x,A) = in \ and \ M(x,B) = bdy\}, \\ bdy & if\{M(x,A) = bdy \ and \ M(x,B) = bdy \\ & \quad and \ O(p,A) \cap O(p,B) \neq \emptyset\}, \\ out & otherwise. \end{cases} \tag{18.6}$$

$$O(x, A \cup B) = O(x,A) \cup O(x,B). \tag{18.7}$$

$$M(x, A - B) = \begin{cases} in, & if\{M(x,A) = in \ and \ M(x,B) = out\}, \\ bdy, & if\{M(x,A) = bdy \ and \ M(x,B) = out\}, \\ & \quad or\{M(x,A) = in \ and \ M(x,B) = bdy\}, \\ bdy & if\{M(x,A) = bdy \ and \ M(x,B) = bdy \\ & \quad and \ O(p,A) - O(p,B) \neq \emptyset\}, \\ out & otherwise. \end{cases} \tag{18.8}$$

$$O(x, A - B) = O(x,A) - O(x,B). \tag{18.9}$$

Only the extended membership classification function, including the neighborhood information, is necessary to determine new solids under Boolean operations. However, querying such a function is not an efficient method of computation. Furthermore, if a model is stored in this *unevaluated form*, computation of volume and moments is tedious, and computation of derivatives for boundary properties is almost impossible. However, one can represent a regular set by its boundaries. The boundaries of a regular two-dimensional set is a collection of curves, and the boundaries of a regular three-dimensional set is a collection of surfaces. We shall investigate Boolean operations on regular sets defined by their boundaries.

18.6.2 Evaluating Set Membership Functions

The basis of querying a membership function is the *Jordan Separation Theorem.*

Theorem 18.12. Jordan Separation Theorem. *Every homeomorphic image, $\mathcal{F}^n$ of S^n in $\mathbb{R}^{n+1}$ separates $\mathbb{R}^{n+1}$. That is, $\mathbb{R}^{n+1} - \mathcal{F}^n$ has exactly two components, each of which has $\mathcal{F}^n$ as its complete boundary.*

The gist is that any closed curve in the plane which does not cross itself separates the plane into two regions one bounded and of finite extent and the other unbounded, each of which has as its boundary the closed curve. In three dimensions, that means that any distortion of the sphere separates $\mathbf{R}^3$ into two distinct regions, each of which has the distortion as its complete boundary.

The basic idea used to develop a computation for *member of* uses this property. We shall demonstrate it for the case of a convex shape.

Algorithm 18.13. *Given a point p which is not on the boundary of a shape A,*

Step 1: Select a ray, R, starting at p and extending to infinity.

Step 2: Find all intersections of R and the boundary of A. Since A has finite extent, this will be a finite number unless a portion of the boundary of A and R coincide. If so, select another ray for which this does not happen and repeat.

Step 3: Count the intersections. If an odd number $M(p, A) = in$*, if an even number* $M(p, A) = out$*.*

If p is on the boundary of A, then the intersection of R and ∂A *takes place at p, and that can be determined.*

While this algorithm always works, it is slow and tedious, and every point in space must be tested. Furthermore, one must use knowledge of the extent of A in order to be sure that one has found all intersections of R and ∂A. Such extra knowledge might be found by using a bounding box; the user is then guaranteed that all intersections must take place within the bounding box.

18.6.3 Boolean Operators by Boundary Classification

In this section we assume that regular sets are specified by their boundaries and at each point on the boundary there is a neighborhood function which is the intersection of the interior of a small ball and the set. In this representation one must have some way of calculating the boundaries and neighborhoods of the regular sets $A \cup B$, $A \cap B$, and $A - B$ in terms of the boundaries and neighborhoods of A and B. Consider the membership operator. The boundaries of the intersection are also parts of boundaries of the original sets. One must figure out which parts of boundaries should be used.

The boundaries of $A \cap B$ in Figure 18.8 (a) consist of the boundaries of A which are interior to B and common boundaries of A and B when they are both on the same side of those boundaries. A neighborhood is either basically unchanged or an intersection; however, the orientation is unchanged. In Figure 18.8 (b) the boundaries of the intersection are similarly derived to be the boundaries of D which are interior to C and the common boundaries of C and D when they are both on the same side of those boundaries. According to the membership operator, the boundary of $C \cup D$ is the set of all points on the boundary of C which are outside D, plus all points on the boundary of D which are outside C, plus points on common boundaries when both C and D are on the same side of the boundary. Again, neighborhoods have orientations that are basically unchanged. Finally, by the membership operator, the boundary of $C - D$ is the set of boundaries of C outside of D plus the boundaries of D inside C, and the shared boundaries when C and D are on opposite sides of those boundaries. In this case the orientations of the neighborhoods of the boundary points of D inside C will be reversed, that is the complement part of the small ball defining the neighborhood in D becomes the neighborhood in C.

Definition 18.14. *We define the* boundary classification *function of a set A with respect to a set B as a collection of four sets, that is,*

$$C(A, B) = \{in(\partial A, B), out(\partial A, B), sh(\partial A, \partial B), antis(\partial A, \partial B)\},$$

where

$$\begin{aligned} in(\partial A, B) &= \{p \in \partial A : M(p, B) = in\}_R \\ out(\partial A, B) &= \{p \in \partial A : M(p, B) = out\}_R \\ sh(\partial A, \partial B) &= \{p \in \partial A : M(p, B) = bdy \text{ and } O(p, A) \cap O(p, B) \neq \emptyset\}_R \\ antis(\partial A, \partial B) &= \{p \in \partial A : M(p, B) = bdy \text{ and } O(p, A) \cap B = \emptyset\}_R . \end{aligned}$$

Note that each point in each of the boundary classifications retains its original neighborhood, except for those points in $antis(\partial A, \partial B)$, and they have the intersections of the neighborhoods used in their original definitions as new neighborhoods.

Using Definition 18.14 we can define Boolean operations by defining the resulting boundaries and neighborhoods for each point on the boundary. To facilitate this we define a complementary neighborhood.

Definition 18.15. *Suppose A and B are regular sets and $p \in A \cap \partial B$, that is, p is on the boundary of B and $M(p, A) \in \{in, bdy\}$. Since $O(p, B) = \{q \in S : M(q, B) = in\}$, where S is some appropriately small ball centered at p, we can define*

$$O_c(p, B, A) = \{q \in S : M(q, A) = in \text{ and } M(q, B) = out\}$$

as the complement of $O(p, B)$ in A.

We shall define each of the Boolean operations on boundary-defined sets by defining the new boundaries and the new neighborhoods that go with each.

Definition 18.16.

$$\begin{aligned}\partial(A\cup B) &= \{\, p\in out(\partial A,B) : O(p,A\cup B)=O(p,A)\,\}\\ &\cup \{\, p\in out(\partial B,A) : O(p,A\cup B)=O(p,B)\,\}\\ &\cup \{\, p\in sh(\partial A,\partial B) : O(p,A\cup B)=O(p,B)\cup O(p,A)\,\}\end{aligned}$$

$$\begin{aligned}\partial(A\cap B) &= \{\, p\in in(\partial A,B) : O(p,A\cap B)=O(p,A)\cap B\,\}\\ &\cup \{\, p\in in(\partial B,A) : O(p,A\cap B)=O(p,B)\cap A\,\}\\ &\cup \{\, p\in sh(\partial A,\partial B) : O(p,A\cap B)=O(p,B)\cap O(p,A)\,\}\end{aligned}$$

$$\begin{aligned}\partial(A-B) &= \{\, p\in out(\partial A,B) : O(p,A-B)=O(p,A)\cap B^c\,\}\\ &\cup \{\, p\in in(\partial B,A) : O(p,A-B)=O_c(p,B,A)\,\}\\ &\cup \{\, p\in antis(\partial A,\partial B) : O(p,A-B)=O(p,A)\cap B^c\,\}\end{aligned}$$

The set B^c is the set of all points not *in B.*

Figures 18.9–18.11 portray the classification operations and show the resulting set operations for regular polygons.

Example 18.17. Suppose two solid disks D_1 and D_2 are defined by their boundaries,

$$\begin{aligned}\partial D_1 &= C_1=\{\,(x,y)\in\mathbb{R}^2 : 1-x^2-y^2=0\,\}\ \text{and}\\ \partial D_2 &= C_2=\{\,(x,y)\in\mathbb{R}^2 : 1-(x-1)^2-y^2=0\,\},\end{aligned}$$

respectively, and have neighborhoods

$$\begin{aligned}O((x,y),C_1) &= \{\,(u,v) : \epsilon_1^2>(u-x)^2+(v-y)^2\ \text{and}\ u^2+v^2<1\,\}\\ O((x,y),C_2) &= \{\,(u,v) : \epsilon_2^2>(u-x)^2+(v-y)^2\ \text{and}\ (u-1)^2+v^2<1\,\}.\end{aligned}$$

Let us consider the classification of the boundaries of D_1 with respect to D_2:

$$in(C_1,D_2)=\{\,(x,y)\in C_1 : (x-1)^2+y^2<1\,\}.$$

Now, $(x,y)\in C_1$, means that $y^2=1-x^2$ so we are looking for (x,y) such that

$$\begin{aligned}1 &> (x-1)^2-x^2+1\\ 1 &> -2x+2\\ 1/2 &< x.\end{aligned}$$

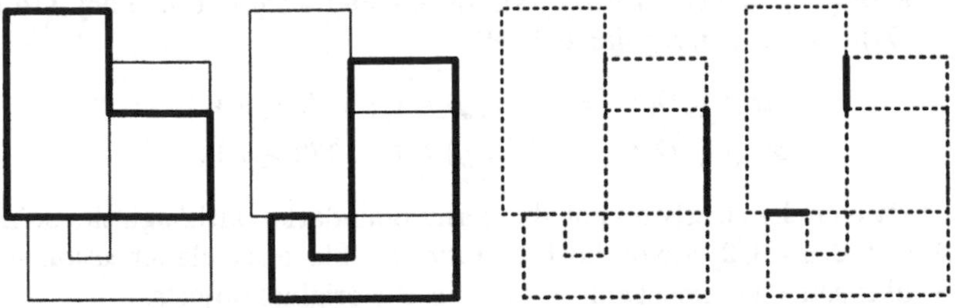

(a) (b) (c) (d)

Figure 18.9. (a) A; (b) B; (c) $sh(\partial A, \partial B)$; (d) $antis(\partial A, \partial B)$.

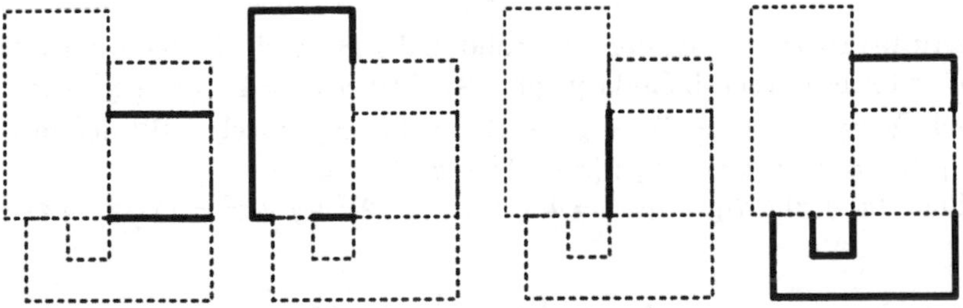

(a) (b) (c) (d)

Figure 18.10. (a) $in(\partial A, B)$; (b) $out(\partial A, B)$; (c) $in(\partial B, A)$; (d) $out(\partial B, A)$.

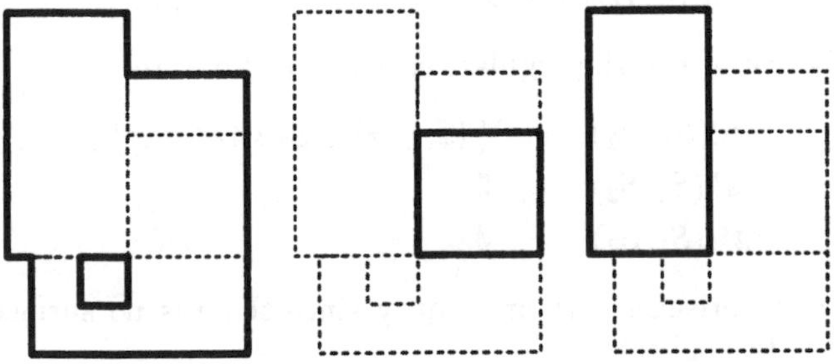

(a) (b) (c)

Figure 18.11. (a) $A \cup B$; (b) $A \cap B$; (c) $A - B$.

The points $(x, y) = (1/2, \pm\sqrt{3}/2)$ are on C_1 and also on C_2. They are in $in(C_1, D_2)$ since it is a regularized set.

$$\begin{aligned} in(C_1, D_2) &= \{ (x, y) \in C_1 : 1/2 \le x \} \\ out(C_1, D_2) &= \{ (x, y) \in C_1 : 1/2 \ge x \} . \end{aligned}$$

The two circles clearly do not share any boundaries, although the points $(x, y) = (1/2, \pm\sqrt{3}/2)$ are on both boundaries. Since the classification sets are regular sets one dimension lower than the original objects,

$$\begin{aligned} sh(C_1, C_2) &= closure\left(int\left((1/2, \sqrt{3}/2), (1/2, -\sqrt{3}/2)\right)\right) \\ &= closure\,(\emptyset) \\ &= \emptyset. \end{aligned}$$

Analogously, $antis(C_1, C_2) = \emptyset$. □

Example 18.18. Consider two solid balls B_1 and B_2 defined by the implicit forms of their boundary spheres, $S_1(x, y, z) = 4 - \left(x^2 + y^2 + z^2\right) = 0$ and $S_2 = 1 - \left((x-2)^2 + y^2 + z^2\right) = 0$, respectively. By definition, $in(S_1, B_2) = \left\{ (x, y, z) \in S_1 : (x-2)^2 + y^2 + z^2 < 1 \right\}$.

Since $(x, y, z) \in S_1$, $y^2+z^2 = 4-x^2$, so $(x-2)^2+y^2+z^2 = (x-2)^2+4-x^2$ and

$$\begin{aligned} 1 &> (x-2)^2 + 4 - x^2 \\ 1 &> -4x + 8 \\ -7 &> -4x \\ 7/4 &< x. \end{aligned}$$

Regularizing the set,

$$in(S_1, B_2) = \{ (x, y, z) \in S_1 : 4/7 \le x \} .$$

Analogously, the rest of the regularized classifications are

$$out(S_1, B_2) = \{ (x, y, z) \in S_1 : 4/7 \ge x \}; \tag{18.10}$$
$$sh(S_1, S_2) = \emptyset; \tag{18.11}$$
$$antis(S_1, S_2) = \emptyset. \tag{18.12}$$

The share and antishare sets are empty since there is no surface that is shared by the two spheres. □

Finding the classifications between two spheres, two cylinders, two boxes, or any two objects that are nameable and have simple definitions can be done analytically, as we did for the two spheres. Finding the classification is a much less well-defined and a much more difficult problem when the two objects are bounded by parametric sculptured surfaces (or parametric free-form curves when the object is a shape in $\mathbb{R}^2$). It entails finding all the intersection curves of a defining boundary surface of one, say S with all the boundary surfaces of the second one, $\{\text{surfaces } T : T \in \partial B\}$, and trimming S into appropriate subpieces bounded by the intersection curves and original boundary curves, and finally classifying each piece with respect to *in*, *out*, *shared*, or *antishared*.

For parametric surfaces, this can only be done with numerical procedures, and then one immediately confronts issues of floating point accuracy, computational techniques, and speed, i.e., robustness. The *shared* and *antishared* determination on surfaces requires deciding if two surfaces in totally different parametric representations are the same. We saw that this is theoretically possible, but at present, computationally infeasible for lack of fast, efficient, robust algorithms.

Example 18.19. Consider two discs bounded by circles in rational quadratic parametric representation. Suppose each has four boundary curves with one fourth the circumference. C_1 has $P_0 = (0,1)$, $P_1 = (1,1)$, and $P_2 = (1,0)$, with $w_0 = w_2 = 1$, and $w_1 = \sec(\pi/4)$. The other three boundaries are defined by the same w values, and values of P_0, P_1, and P_2 rotated by $\pi/2$, π, and $3\pi/2$, respectively. The parameter range on each of the arcs should lie in $t \in [0,1]$. For C_2, the three points defining the first 90 degree arc are $Q_0 = (1/\sqrt{2}, 1/\sqrt{2})$, $Q_1 = (\sqrt{2}, 0)$, and $Q_2 = (1/\sqrt{2}, -1/\sqrt{2})$. Let the w's be the same as for C_1, and consider the boundaries rotated by $\pi/2$, π, and $3\pi/2$. Unless it was known ahead of time that the objects were circles, it would be very difficult to test these parametric curves. Subdivision to approximate the boundaries would give new curve polygons that were close to each, but it would be very difficult to tell if they were identical. Clearly, the closed form gives the easiest results. It would be even more difficult if C_2 were not a complete circle, but just a single arc of a circle with other curves making up its boundaries. □

After seeing the difficulties with parametric boundaries, it is obvious to ask, "Why bother?" Unfortunately, there are many shapes that it is necessary and desirable to model which cannot be modeled as straightforward combinations of named shapes. Examples abound in aerospace, automobiles, and ships, to name big areas, and also in smaller everyday items.

The boundaries of regular three-dimensional sets are regular two-dimensional sets, surfaces; the boundaries of regular two-dimensional sets are regular one-dimensional sets, curves. So there is a dimensional recursion here. To classify the boundaries of three-dimensional sets, one needs to classify the boundaries of two-dimensional sets. To do this, one needs to classify its boundaries, namely the boundaries of one-dimensional sets (points).

18.6.4 Computational Considerations in 2D Classification

Let us first consider the two-dimensional polygonal case with both polygons in the x-y plane. Suppose the edges of polygons A and B are given in a clockwise direction by enumeration of the polygon vertices. Then as one walks the boundary, the inside of the polygon is always on the right side. Since the boundaries always have straight sides, we can designate the neighborhood, $O(p, A)$, of each point on the boundary of A by simply giving a normal vector to the whole edge pointing to the inside of the polygon.

For example, $A = \{P_0, P_1, \ldots, P_m\}$, and edge i is designated $e_i(t) = P_i + t(P_{i+1} - P_i)$, $t \in [0, 1]$, for $i = 0, \ldots,\ m - 1$; $e_m(t) = P_m + t(P_0 - P_m)$, for $t \in [0, 1]$, where $P_i = (x_i, y_i)$. A normal pointing to the inside of the polygon for the whole edge e_i is $n_i = (y_{i+1} - y_i, x_i - x_{i+1})$.

The classification of a boundary (edge) of A with respect to B then amounts to finding the intersections of each boundary of A with all the boundaries of B, and then ascertaining the appropriate classifications for these segments of the edge. The intersections do not necessarily occur in the sequential order of the edges of B.

18.6.4.1 Computing intersections. Figure 18.12 shows some of the typical conditions which can occur during classification. The major effort here is to determine where e_i intersects the edges of B. This can be done for planar line segments using the homogeneous inner product formulation.

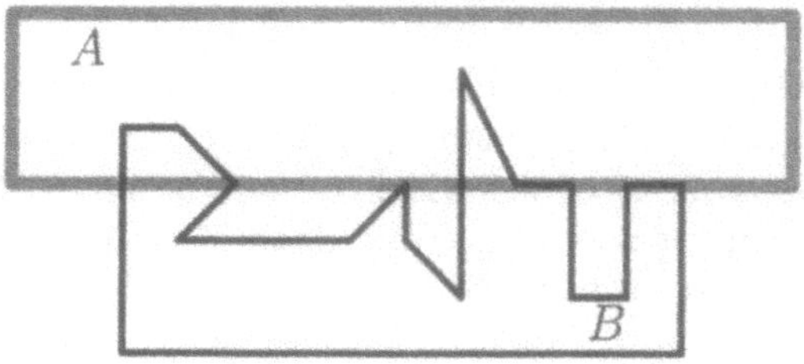

Figure 18.12. Classifying an edge of A with respect to B.

Lemma 18.20. *Suppose a directed line segment in the x-y plane is defined by an edge* $e_i = P_i + t(P_{i+1} - P_i)$. *Let* n_i *represent the "inward" pointing normal to that edge, that is "inside" is on the right while moving from* P_i *to* P_{i+1}. *Define*

$$D_{n,i} = \begin{vmatrix} y_{i+1} & -x_{i+1} \\ y_i & -x_i \end{vmatrix},$$

and let $\hat{n}_i = (n_i, D_{n,i})$, *and* $\hat{P} = (x, y, 1)$ *whenever* $P = (x, y)$. *Then,*

$$P \text{ is on the } \begin{cases} \text{inside of } e_i & \text{if } \left\langle \hat{P}, \hat{n}_i \right\rangle > 0 \\ \text{on } e_i & \text{if } \left\langle \hat{P}, \hat{n}_i \right\rangle = 0 \\ \text{outside of } e_i & \text{if } \left\langle \hat{P}, \hat{n}_i \right\rangle < 0. \end{cases}$$

Lemma 18.21. *Let* e_i, n_i, $\hat{n}_i$, *and* $D_{n,i}$ *be defined as in Lemma 18.20, and define an edge of a polygon B as* $f_j = Q_j + s(Q_{j+1} - Q_j)$. *Define* $\overline{e}_i$ *and* $\overline{f}_j$ *as the infinite length lines containing* e_i *and* f_j, *respectively, as subsegments.*

$$\left\langle \hat{Q}_{j+1}, \hat{n}_i \right\rangle \left\langle \hat{Q}_j, \hat{n}_i \right\rangle \begin{cases} > 0, & \text{then } f_j \text{ does not intersect with } e_i. \\ = 0, & \text{at least one of } Q_j \text{ or } Q_{j+1} \text{ is on } \overline{e}_i. \\ < 0, & \text{then } f_j \text{ intersects } \overline{e}_i. \end{cases}$$

If there is an intersection with $\overline{e}_i$ (or $\overline{f}_j$, respectively) it still remains to be determined if that intersection occurs within the edge itself.

Theorem 18.22. *Let A, B,* e_i f_j, $\overline{e}_i$, $\overline{f}_j$, n_i *be defined as above in Lemmas 18.20 and 18.21, and let* m_j *represent the "inward" pointing normal to edge* f_j, *that is "inside" is on the right while moving from* Q_i *to* Q_{i+1}. *We use the notation* $\overline{e}_i - e_i$ *to mean the complement of* e_i *in* $\overline{e}_i$.

If	*then*	
$\left\langle \hat{Q}_{j+1}, \hat{n}_i \right\rangle \left\langle \hat{Q}_j, \hat{n}_i \right\rangle < 0$ *and* $\left\langle \hat{P}_{i+1}, \hat{m}_j \right\rangle \left\langle \hat{P}_i, \hat{m}_j \right\rangle < 0,$	$e_i \cap f_j$.	(18.13)
$\left\langle \hat{Q}_{j+1}, \hat{n}_i \right\rangle \left\langle \hat{Q}_j, \hat{n}_i \right\rangle < 0$ *and* $\left\langle \hat{P}_{i+1}, \hat{m}_j \right\rangle \left\langle \hat{P}_i, \hat{m}_j \right\rangle > 0,$	$(\overline{e}_i - e_i) \cap f_j$.	(18.14)
$\left\langle \hat{Q}_{j+1}, \hat{n}_i \right\rangle \left\langle \hat{Q}_j, \hat{n}_i \right\rangle = 0$ *and* $\left\langle \hat{P}_{i+1}, \hat{m}_j \right\rangle \left\langle \hat{P}_i, \hat{m}_j \right\rangle < 0,$	Q_j *or* Q_{j+1} *is* on e_i.	(18.15)
$\left\langle \hat{Q}_{j+1}, \hat{n}_i \right\rangle \left\langle \hat{Q}_j, \hat{n}_i \right\rangle = 0$ *and* $\left\langle \hat{P}_{i+1}, \hat{m}_j \right\rangle \left\langle \hat{P}_i, \hat{m}_j \right\rangle > 0,$	Q_j *or* Q_{j+1} *is* on $\overline{e}_i - e_i$.	(18.16)

The parametric values at the intersections are given by

$$t = -\frac{\mid (P_i - Q_j) \quad (Q_{j+1} - Q_j) \mid}{\mid n_i \quad m_j \mid} \tag{18.17}$$

$$s = \frac{\mid (P_{i+1} - P_i) \quad (P_i - Q_j) \mid}{\mid n_i \quad m_j \mid} \tag{18.18}$$

where the notation $\mid \alpha \quad \beta \mid$ *means that the vector* α *is to be used as the first column in the determinant and the vector* β *is to be used as the second.*

If

$$\left\langle \hat{Q}_{j+1}, \hat{n}_i \right\rangle \left\langle \hat{Q}_j, \hat{n}_i \right\rangle = 0 \text{ and } \left\langle \hat{P}_{i+1}, \hat{m}_j \right\rangle \left\langle \hat{P}_i, \hat{m}_j \right\rangle = 0, \tag{18.19}$$

then, we must check if the edges both lie along the same line, or if instead they just share an endpoint.

If Equation 18.19 is true, then one of two conditions holds.

1. Either $|\langle n_i, m_j \rangle| = \|n_i\| \|m_j\|$, in which case the points Q_j, Q_{j+1}, P_i, and P_{i+1} all lie along the same line, but the edges f_j and e_i do not necessarily intersect;

2. or, the inner product condition on the normals does not hold, in which case the edges have a vertex in common.

We start with a discussion of Condition 2 first. In that case, the normals show that the edges are not parallel, but by Equation 18.19, the products of the inner products with both lines equal zero. This means that one inner product with each line must be zero. Suppose

$$\left\langle \hat{Q}_{j+1}, \hat{n}_i \right\rangle = 0 \quad \text{and} \quad \left\langle \hat{P}_i, \hat{m}_j \right\rangle = 0.$$

These two constraints show that Q_{j+1} is somewhere on $\overline{e}_i$, and that also that P_i is on $\overline{f}_j$. If $P_i = Q_{j+1}$ that condition is satisfied. If on the other hand we suppose $P_i \neq Q_{j+1}$, then Q_{j+1} and P_i both are on both $\overline{e}_i$ (with P_{i+1}) and $\overline{f}_j$ (with Q_j), which requires that those lines are identical. That is, the hypothesis of Condition 1 is satisfied.

In general, if the hypothesis of Condition 1 is satisfied, f_j and e_i are parallel. But since Equation 18.19 holds, $\overline{f}_j$ and $\overline{e}_i$ must be the same. Thus, the two edges must overlap.

We find relevant regions of overlap on both lines by writing one in terms of the other, that is, there exist real numbers t_j and t_{j+1} such that

$$Q_j = P_i + t_j(P_{i+1} - P_i) \qquad \text{and} \qquad Q_{j+1} = P_i + t_{j+1}(P_{i+1} - P_i).$$

t_j, t_{j+1}	relationship of edges
$t_j < 0$, $t_{j+1} > 1$	e_i is shared
$t_j > 1$, $t_{j+1} < 0$	e_i is anti-shared, correct the normal
$0 < t_j < t_{j+1} < 1$	f_j is shared
$0 < t_{j+1} < t_j < 1$	f_j is anti-shared, correct the normal
$0 < t_j < 1 < t_{j+1}$	shared boundary goes from Q_j to P_{i+1}
$0 < t_{j+1} < 1 < t_j$	anti- shared boundary goes from Q_j to P_{i+1}
$t_j < 0 < t_{j+1} < 1$	shared boundary goes from P_i to Q_{j+1}
$t_{j+1} < 0 < t_j < 1$	anti-shared boundary goes from P_i to Q_{j+1}

Table 18.1. Possible cases for Theorem 18.22.

One can solve for t_j by solving one linear equation in one unknown by picking either the x-coordinate equation or the y-coordinate equation for Q_j and solving it for t_j. Analogously, one can solve for t_{j+1}. Table 18.1 itemizes the varous cases.

Equations 18.15, 18.16, or 18.19 can lead to a situation which is difficult to resolve. In Figure 18.13 (a), f_j is on one side of e_i, but f_{j+1} is on the opposite side. The vertex Q_{j+1} lies on e_i. In either case, there will be a transition in the classification at that value of t. Figure 18.13 (b) shows the opposite case. That is, f_j and f_{j+1} are both on the same side of e_i and the intersection should be ignored because the classification along e_i does not change. If either of these problems occurs at the edge endpoints, then it is the result of Condition 2. A method to determine which case occurs must be added to the classification: One of Q_j and Q_{j+1} is not on the line represented by e_i, we save the sign, that is,

$$sign(j) = \begin{cases} 1 & if(\hat{Q}_{j+1}, \hat{n}_i) + (\hat{Q}_j, \hat{n}_i)) > 0 \\ -1 & if(\hat{Q}_{j+1}, \hat{n}_i) + (\hat{Q}_j, \hat{n}_i)) < 0 \end{cases} . \tag{18.20}$$

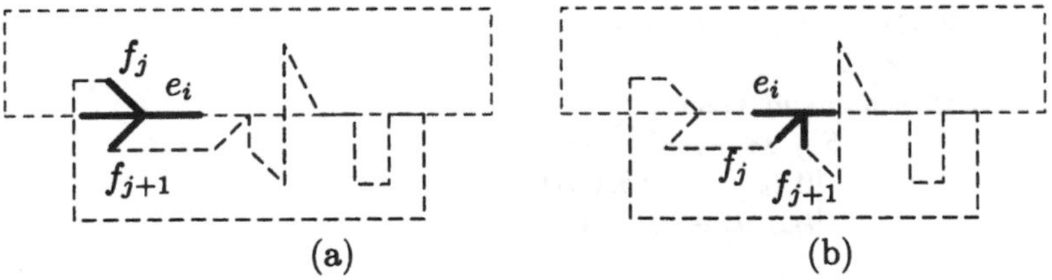

Figure 18.13. (a) Edges on opposite sides; (b) edges on same side, see also Figure 18.12.

18.6.4.2 Orienting the intersections. To classify an edge, e_i, of a polygon A with respect to another polygon B, one can create a list of intersection points and shared or antishared annotations. The following algorithm classifies an edge of a polygon when none of the difficult cases, i.e., shared, antishared, or vertex intersections, is present.

Algorithm 18.23. *To classify edge e_i with normal n_i with respect to polygon B with edges f_j and normals m_j.*

```
Procedure Classify(e_i, n_i, B, in(∂A,B), out(∂A,B),
                          sh(∂A,∂B), antis(∂A,∂B))
/* This algorithm assumes that either
/* Equation 18.13 or Equation 18.14 applies, but
/* that Equations 18.15, 18.16, and 18.19 never apply
Set up an empty list, L.
Find n̂_i.
For each boundary f of B
    Find m̂.
    Solve for t and add it to L.
If size(L) = 0, then
    e_i ∈ out(∂A,B).
    return
Sort the values of t in L from smallest to largest.
If t_size(L) < 0 or t_1 > 1, then
    e_i ∈ out(∂A,B).
    return
else if size(L) = 2 and t_1 < 0 and t_2 > 1, then
    e_i ∈ in(∂A,B).
    return
else
    Set k ← 1
    While t_k < 0, k ← k + 1
    beg ← k
    Set k ← size(L)
    While t_k > 1, k ← k − 1
    end ← k
    if beg = end then
        if beg is even, then
            [0, t_beg] ∈ in(∂A,B)
            [t_beg, 1] ∈ out(∂A,B)
```

(continued on next page)

```
        else
            [0, t_beg] ∈ out(∂A, B)
            [t_beg, 1] ∈ in(∂A, B)
    else if end < beg
        if beg is even, then e_i ∈ in(∂A, B).
        else e_i ∈ out(∂A, B).
    else /* more than one transition /*
        p ← 0,
        if beg = odd then
            class ← "out"
        else /* beg = even/*
            class ← "in"
        For k = beg to end
            Between [p, t_k] edge e_i is in class(∂A, B)
            p ← t_k
            Reverse the value of class
        Between [p, 1] edge e_i is in class(∂A, B)
```

This algorithm is not complete since the shared/antishared case is omitted. Also, the difficult and important cases of a vertex of a side of B lying on e_i as well as a vertex of e_i lying on an edge of B are omitted. The final procedure, classifying two polygons is quite straightforward.

```
Procedure Classify(∂A, B, in(∂A, B), out(∂A, B),
                   sh(∂A, ∂B), antis(∂A, ∂B))
    For each e_i ∈ ∂A
        Classify(e_i, n_i, B, in(∂A, B), out(∂A, bB),
                   sh(∂A, ∂B), antis(∂A, ∂B))
end
```

18.6.4.3 Computational difficulties. Since the straight edges making up polygon boundaries either intersect at a single point, don't intersect at all, or are colinear, the above algorithms enumerate all possibilities. Unfortunately, they have not taken into account inaccuracies which result from using floating point arithmetic.

First, if it turns out that a vertex of one polygon seems to go through an edge of the other polygon, then even a small floating point error can place the vertex to one side or the other of the edge. This results in a problem since the vertex is on two edges, and its relationship to the other polygon is computed once for each edge. Floating point arithmetic could deem the same vertex *on* in one situation and *in* or *out* in the other. It would be difficult to show that all of those classifications are really for the same vertex when it came time to orient the edge classifications to form a new polygon. The situation is even worse when two polygons have vertices which seem to be the same point. The computation involved in determination and then finding parameter values on each of the edges can lead to new errors.

Another problem is the determination of shared or antishared. If two edges have almost the same slope, and, up to numerical accuracy, a vertex of one is on the other, then ill conditioning of the equations makes it extremely difficult to determine if the line equations are the same, or if one should compute an intersection. Usually one resolves this by computing to within some error margin. However, whichever determination is selected, one must make sure that parameter values and vertex classifications are consistent.

Implementation of these algorithms, or any algorithms to effect classification will have to solve all these computational and robustness difficulties.

18.6.5 3D Classification

We consider classifying polyhedrons P_1 and P_2 with respect to each other. It is not possible to order polygonal faces on a polyhedron the way edges can be ordered on a polygon. It is necessary to traverse a data structure in which the faces are stored, and also ensure that faces are tagged after they have been used. Since each face is planar, the plane normal to the face can always be specified to point to the inside of the polyhedron.

Consider face A from polyhedron P_1 with normal n_a. In an analogous way to the polygonal case, one must completely classify A with respect to P_2. This means that it is classified with respect to all the faces of P_2. Suppose the faces of P_2 are ordered by some method and called F_i, each with normal n_i.

If $n_a \times n_i \neq 0$, the planes containing the faces have a well-defined intersection line segments, as in Figure 18.14. The intersection line segments and the orientations for the intersections of all F_i with the plane of A must be calculated. After sorting the line segments, a second complex polygon, B, in the same plane as the polygon A is formed.

Now the two-dimensional classification algorithm can be used to classify A with respect to B. Then, A *in* $P_2 = A \cap B$ and A *out* $P_2 = A - B$. This must be done for each of the faces of P_1. Then the roles must be reversed

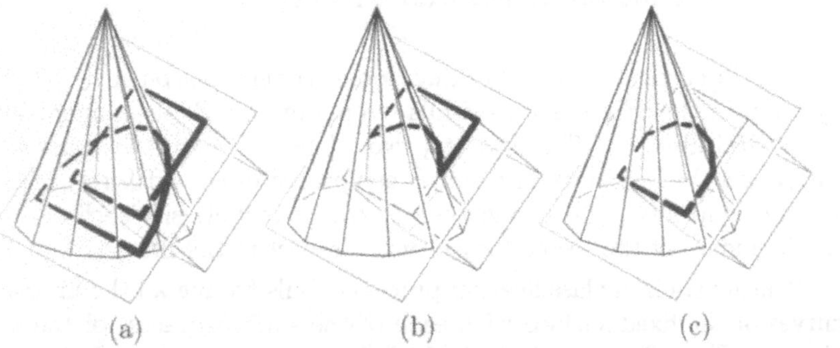

Figure 18.14. The intersection of face A of a cube with a cone P_2 is shown in (a); in (b) A *out* P_2 is shown while (c) presents A *in* P_2.

and each face of P_2 must be completely classified with respect to P_1. This is an n^2 problem since each boundary of P_i must be completely classified with respect to P_j, $i \neq j$.

If $n_a \times n_i = 0$, then face A is parallel to face F_i. One can quickly calculate (up to numerical instabilities) if they are in different planes, and so do not intersect, or if they are in the same plane and require further testing. If they are in the same plane, one must then check normal directions to see if they are in the same or opposite directions as a way of determining if the edges might be partially shared or antishared. Exactly which parts of the faces are shared, antishared, or distinct are determined by using the planar polygonal classification procedure on those two faces. For example, we could transform both polygons into the x-y plane and then find $A \cap F_i$, the shared boundary, and then $A - F_i$ and $F_i - A$ to give *outside* and *inside*, respectively, as computed with respect to just P_2.

We see that robustness of the three-dimensional algorithm is dependent on the robustness of the two-dimensional algorithms, and the stability of the plane-plane intersection algorithm. There are more possible numerically difficult subcases in three-dimensional classification than in two-dimensional classification. A face of a polyhedron can pierce another polyhedron through a seam between the faces as well as through a vertex of a face. A vertex of one polyhedron can lie on a face, edge, or vertex of the other polyhedron. Deciding whether the classification should change (i.e., to count the intersection) has inherent numerical difficulties. Thus, while in principle the problem is really just recursively two-dimensional, in reality all the computational difficulties encountered in the two-dimensional case are present, plus problems unique to three-dimensional computation also occur.

18.7 Booleans on Sculptured Objects

In many important ways performing Boolean operations on objects bounded by sculptured surfaces is very similar to performing Boolean operations on polyhedral objects. The first step would be to completely classify each surface on the boundary of the first sculptured object with respect to the second sculptured object, and then to completely classify each surface on the boundary of the second object with respect to the first.

The generalized classification process entails finding all the intersection curves of the fixed surface with each of the surfaces (faces) of the second object. Then the boundaries, which have also been classified, and the intersection curves are linked together to form boundaries of surfaces which are completely classified in one classification set with respect to the second object. This is done for each surface of the first object. Then the process is repeated classifying the surfaces of the second object with respect to the first. Boolean sculptured models are constructed from these classifications.

In the case of polyhedra with planar faces, knowing the boundaries and a single orientation is sufficient to uniquely define each of the classification parts of the fixed face. When the faces are sculptured, that is no longer true. The surface representation, its boundary curves, and the orientation at each point on the surface is necessary for a unique classification. We have seen that the intersection curves which form parts of the boundaries of these trimmed surfaces usually do not have, even theoretically, a computable parametric closed form solution. However, they form the boundaries of the surfaces in the new model being constructed, and they are not isoparametric curves. These new surfaces cannot be represented as a tensor product B-spline, even if the original surfaces were. Thus, B-spline models can *not* be closed under general Boolean operations. Accurately finding the intersection curves and constructing a representation of these new complex surfaces (which may have many edges and also have holes through them) is a current research problem called the *trimmed surface problem.* Constructing the Boolean models is also an area of active current research in the field.

Exercises

1. Enhance Algorithm 18.23 to correctly classify edges when Equations 18.15, 18.16, or 18.19 are true. Conditions 1 and 2 will be helpful as will Equation 18.20.

2. Programming Exercise: Implement Boolean operations on polygons in the x-y plane. That is, given two polygons A and B, create a new polygon C which can be specified to be one of the following:

$A + B$	(union)
$A * B$	(intersection)
$A - B$	(difference)
$B - A$	(difference)
$(A - B) + (B - A)$	(symmetric difference)

You should allow the user to interactively construct polygons for input. The result should be a polygon, available as input to a further operation. The user should have the option of cleaning up by deleting individual polygons or clearing the whole collection as he chooses. Further, your program should be able to read in data files which will not clear the already defined polygons. Inputs for the Boolean operations should be specifiable interactively. (Graphical picking may not work if one polygon to be used is defined as the intersection or union of two others. How would you decide which to pick?)

19

Model Data Structures

A fundamental concern for handling and processing geometric objects is the proper data structure to use in representing a geometric model. Much like a basic data structure question, one needs to derive a representation for a geometric model that will simplify the common geometric operations.

Traditionally, geometry has been polygonal. Hence, we will start our discussion with an introduction of the *winged-edge* data structure that is a prime example of a representation of a polygonal geometric model. Then, we will continue to explore a similar representation for free-form models.

Considering the types of operations that one is required to apply to a model, set operations (recall Chapter 18) are commonly seen as the more complex and intensive ones. Consider the union operation of $A+B$ and let P_1, $P_2 \in A$ be two adjacent polygons of A sharing an edge e such that P_1 intersects B whereas $P_2 \bigcap B = e \bigcap B = \emptyset$. If the portion of P_1 that contains e is *out* B, P_2 is *out* as well. Hence, the decision of P_2 *out* B could be made if adjacency or neighborhood information is readily available.

A crucial characteristic on which we rely as we examine geometric models is that three-dimensional geometry is typically a 2-manifold. (See Figure 19.1.)

Definition 19.1. *Object $\mathcal{O}$ is an m-manifold if every point $p \in \mathcal{O}$ has an epsilon neighborhood that is homeomorphic to an open disk in $\mathbb{R}^m$.*

Hence, a geometric model $\mathcal{M}$ in $\mathbb{R}^3$ is a 2-manifold if for every point, p, on the surface of $\mathcal{M}$, $\mathcal{M} \bigcap S_\epsilon(p)$ is homeomorphic to an open disk in $\mathbb{R}^2$, where $S_\epsilon(p)$ is a sphere of arbitrarily small radius ϵ centered at p.

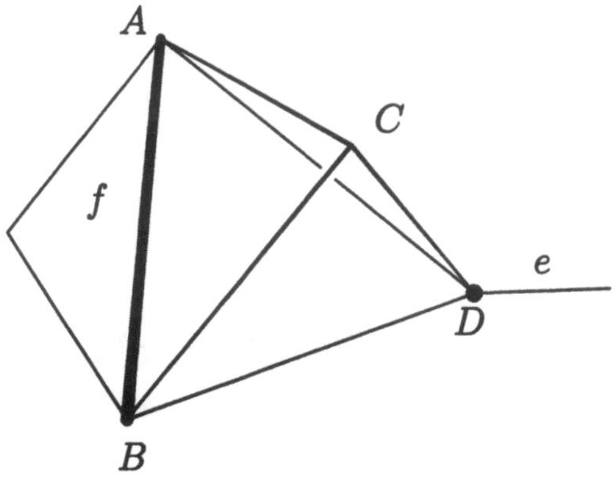

Figure 19.1. The tetrahedral object $ABCD$ has an additional edge e and an extra face f connected to it along the gray areas. Both e and f violate the requirement of a 2-manifold object in Definition 19.1, along the gray areas.

Geometrically, $\mathcal{M} \bigcap S_\epsilon(p)$ can be mapped to an open disk in $\mathbb{R}^2$ by bending and stretching transformations.

19.1 The Winged-Edge Data Structure

An immediate consequence of a polygonal object being a 2-manifold is,

Corollary 19.2. *Given a 2-manifold polygonal solid model, $\mathcal{M}$, an edge e of $\mathcal{M}$ must be shared by exactly two polygons.*

The winged-edge data structure was introduced in the early seventies by Baumgart [7], who was interested in a data structure that could easily provide adjacent and/or neighborhood information for 2-manifold geometry. A solid model in this data structure consists of a list of vertices v_i, a list of edges, e_j, and a list of faces f_k. Further, keeping consistent orientation of edge e_j in a closed loop of edges in a formation of a face, the two different faces that employ edge e_j do so in reverse directions, denoted by the positive and the negative direction of edge e_j.

The data associated with an edge e in the winged edge data structure is seen in Figure 19.2. An edge contains eight references. Two of the references address the two end vertices of e. The edge orientation is given by the order of the vertices. Two more references relate to the two faces sharing e, $PFace(e)$ and $NFace(e)$. Finally, the last four references point to the next edges in the clockwise and counterclockwise directions in both the positive face of e, as $pcw(e)$ and $pccw(e)$, as well as the negative face of e, as $ncw(e)$ and $nccw(e)$. Loops of edges are used to specify the faces, so a single edge is shared by the two faces it bounds, $PFace(e)$ and $NFace(e)$.

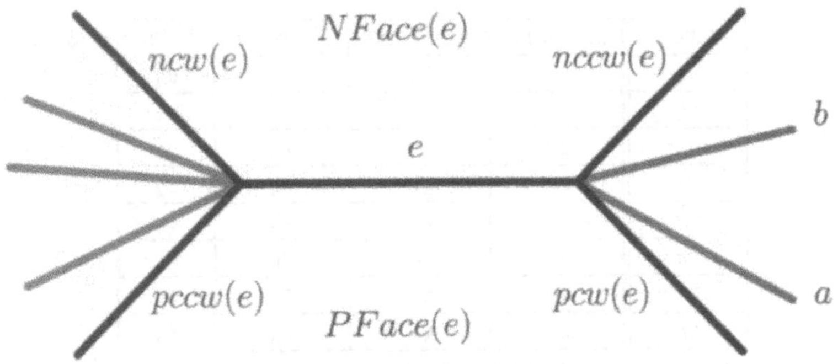

Figure 19.2. The winged-edge data structure—one edge seen from the exterior of the solid model.

While the two end vertices of e, v_1 and v_2 can be shared by more than three edges, the data structure of edge e does not reference these additional edges of v_1 and v_2. See the gray edges in Figure 19.2. Now *edge* $a = ncw(pcw(e))$ (since a is connected to the beginning of edge $pcw(e)$) in $NFace(pcw(e))$, so $pcw(e) = pccw(a)$. Continuing, *edge* $b = pcw(a)$. By iterating on this process, it is possible to reach all the edges and all the faces around the vertex at that end of e. Thus, by specifying a single adjacent edge in each node of the vertex table, it is possible to traverse all edges adjacent to that vertex. Similarly, by specifying a single edge of the loop defining a face, it is possible to traverse through all the edges defining a face.

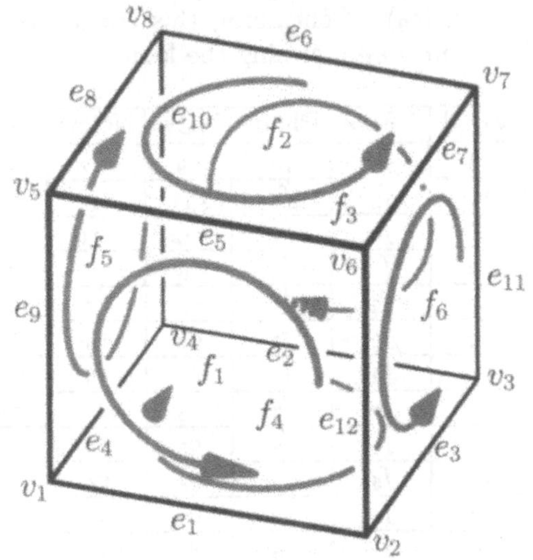

Figure 19.3. A cube represented using the winged-edge data structure.

Vertex	X coord	Y coord	Z coord	Edge
v_1	x_1	y_1	z_1	e_4
v_2	x_2	y_2	z_2	e_1
v_3	x_3	y_3	z_3	e_3
v_4	x_4	y_4	z_4	e_2
v_5	x_5	y_5	z_5	e_8
v_6	x_6	y_6	z_6	e_5
v_7	x_7	y_7	z_7	e_7
v_8	x_8	y_8	z_8	e_6

Table 19.1. The vertex table.

We derive the winged-edge data structure via an example of a model of a cube, seen in Figure 19.3.

Table 19.1 is the vertex table; it contains the Euclidean location of each vertex. The edge table, Table 19.2, provides references to the two vertices of the edge, supplies pointers to the two shared faces, and also points to the four next edges in the clockwise and counterclockwise directions on the two sharing faces.

Finally, we have the face table, Table 19.3. It is sufficient to hold a single edge of the face to recover the entire face as one can trace the entire loop of edges of the face from the references of the next edges of the current edge in the edge table. For example, face f_1 points at edge e_1.

In the edge table, one finds that $f_1 = NFace(e_1)$ and going clockwise $ncw(e_1)$ equals e_9. Similarly, $f_1 = NFace(e_9)$ and hence the next edge in this loop is $e_5 = ncw(e_9)$. Continuing this tracing process, we find $e_{12} = pcw(e_5)$, and $e_1 = ncw(e_{12})$, closing the loop.

Edge	VStart	VEnd	NFace	PFace	NCw	NCcw	PCw	PCcw
e_1	v_2	v_1	f_1	f_4	e_9	e_{12}	e_3	e_4
e_2	v_4	v_3	f_3	f_4	e_{11}	e_{10}	e_4	e_3
e_3	v_3	v_2	f_6	f_4	e_{12}	e_{11}	e_2	e_1
e_4	v_1	v_4	f_5	f_4	e_{10}	e_9	e_1	e_2
e_5	v_6	v_5	f_2	f_1	e_8	e_7	e_{12}	e_9
e_6	v_8	v_7	f_2	f_3	e_7	e_8	e_{10}	e_{11}
e_7	v_7	v_6	f_2	f_6	e_5	e_6	e_{11}	e_{12}
e_8	v_5	v_8	f_2	f_5	e_6	e_5	e_9	e_{10}
e_9	v_1	v_5	f_1	f_5	e_5	e_1	e_4	e_8
e_{10}	v_4	v_8	f_5	f_3	e_8	e_4	e_2	e_6
e_{11}	v_7	v_3	f_6	f_3	e_3	e_7	e_6	e_2
e_{12}	v_6	v_2	f_1	f_6	e_1	e_5	e_7	e_3

Table 19.2. The edge table.

Face	Edge
f_1	e_1
f_2	e_6
f_3	e_2
f_4	e_3
f_5	e_9
f_6	e_7

Table 19.3. The face table.

19.2 Trimmed Surfaces

A polygon can be oriented using its normal. Typically, polyhedral models are defined so that all the normals of all the polygons point into the model or, alternatively, all the normals point outside. For surfaces, the orientation problem is far more difficult. In fact, a single free-form surface can be unorientable. Recall Figure 12.8. For reasons now obvious, the surfaces forming the boundary of a model of a solid must be oriented. Hence, we will require the normals of the surfaces of our model to point into the model, and we will concern ourselves with orientable surfaces only.

It has been recognized that it is difficult, at best, to model certain common shapes using only complete tensor product spline surfaces. For example, it can be quite difficult to find a complete tensor product surface representation of a square planar plate with several holes in it. The planar surface must be subdivided into topologically rectangular domains and each such domain must be fitted with a single tensor product surface, as illustrated in Figure 19.4 (a). Eight such rectangular domains are required in Figure 19.4 (a) in order to construct the plate with the two holes. Unfortunately, there are no algorithms that can automatically perform such a composition when given only the boundary countours.

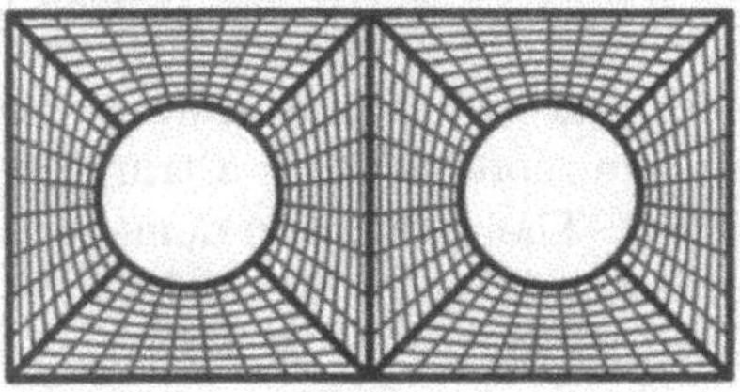

(a)

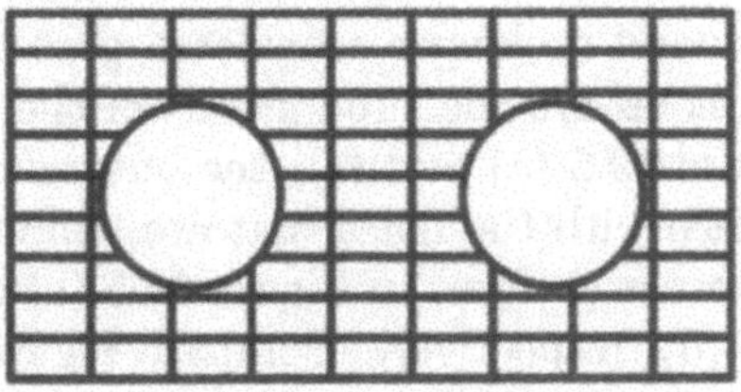

(b)

Figure 19.4. A plate with two holes. Image (a) requires eight regular tensor product surfaces; (b) requires one trimmed tensor product surface.

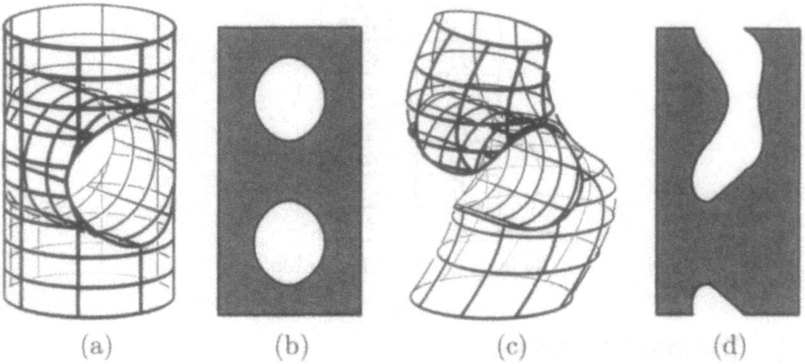

Figure 19.5. (a) A cylinder with a hole through it; (b) the trimmed parametric domain of the outer tensor product surface of the cylinder; (c) a general free-form surface with a hole through it; (d) the trimmed parametric domain of the outer surface; the valid domains are shown in gray.

Clearly, the result of an intersection of two free-form models can yield similar, yet arbitrarily more complex, examples for difficulties in the construction of models. Consider the two examples in Figure 19.5. In Figure 19.5 (a), the topology of the parametric domain of the larger cylinder is similar to that of Figure 19.4 (a), having two holes in its domain. Hence, eight tensor product surfaces would be required to represent the outer cylinder surface, while also required to fit a portion of a cylinder. This fitting problem becomes more complex when the surface is a general free-form surface as can be seen in Figure 19.5 (c). Here, not only the fitting problem of the sub-surfaces is more difficult, but the topology is different.

To overcome these difficulties, an alternative representation—called *a trimmed surface*—was proposed. In a complete tensor product surface, the entire rectangular parametric domain contributes to the actual surface in space. In constrast, in a trimmed surface, only a prescribed sub-domain of the surface is defined to be part of the actual model. Hence, a trimmed surface is a tensor product surface with portions of its domain trimmed away and no longer considered part of the model. In Figure 19.4 (b), the two holes are the subdomains of the surface that were trimmed away. In Figure 19.5 (a) and (c), the outer surfaces are represented as a trimmed surface with the holes that are trimmed away. The parametric domain of the outer surfaces in Figure 19.5 (a) and (c) are shown in Figure 19.5 (b) and (d), respectively. Compare Figure 19.4 (b) with Figure 19.5 (b).

A trimmed surface is therefore a regular tensor product surface with some specification of the domain trimmed away. A common trimmed-surface representation employs a set of two-dimensional simple closed and nested curves in the parametric domain of the surface, $\mathcal{C} = \{C_i(t)\}$, $i =$

$0, \cdots, n-1$. These n curves are also known as the *trimming curves* of the trimmed surface. Then, a point $S(u_0, v_0)$ is considered part of trimmed surface $S(u, v)$ if the ray $(u_0 + t, v_0)$, $t > 0$ intersects $\mathcal{C}$ an odd number of times. (See the Jordan Separation Theorem in Section 18.12.)

Some implementations of trimmed surfaces also require the trimming curves to be oriented so that while moving on $C_i(t)$ in the increased parameter direction t, the trimmed domain is to the left (right). We will not deal with oriented-trimming curves.

Nevertheless, the use of trimmed surfaces introduce constraints on a whole variety of surface-processing procedures. Surface subdivision is now required to split the trimming curves in the parametric domain into the different domains of the sub-trimmed surfaces. A polygonal approximation procedure of the trimmed surface has to trim the polygonal representation to the proper valid domain. Isoparametric curves, that are used to display free-form surfaces as wireframe (see Figure 19.5), must be clipped against the set of trimming curves, $\mathcal{C}$.

Let $S(u_0, v)$ be an isoparametric curve of a trimmed surface and let $\{I_i\}_{i=0}^{n-1}$ be the n intersections of the infinite line (u_0, v) with $\mathcal{C}$, $I_i = (u_0, v_i)$. Assuming that a tangency intersection is considered twice, n must be even. Then, recalling the Jordan Separation Theorem, (Theorem 18.12), $S(u_0, v)$ must be displayed only at subdomains (u_0, v_i) to (u_0, v_{i+1}) for which i is even.

19.3 A Model

We now have a basic idea how polyhedral models can be represented using the winged-edge data structure. Further, we were introduced in Section 19.2 to a representation that can serve as a basic building block replacing the polygon in the free-form model, namely the trimmed surface.

A free-form model now consists of a set of trimmed surfaces. The topology, much like in the polygonal case, must be recoverable. The boundaries of a trimmed surface consist of trimming curves. Even when the trimmed surface contains the entire domain of the tensor product surface, four trimming curves identical to the boundaries of the original domain of the tensor product surface are defined. For example, in Figure 19.6, edge e_1 is a trimming curve along the boundary of the tensor product surface of the body of the Utah teapot, whereas edge e_2 is an edge that is the result of a set-operation computation between the body and the spout of the teapot.

We need the ability to find the surface adjacent to a given surface, along some prescribed trimming boundary. Being a 2-manifold, every boundary

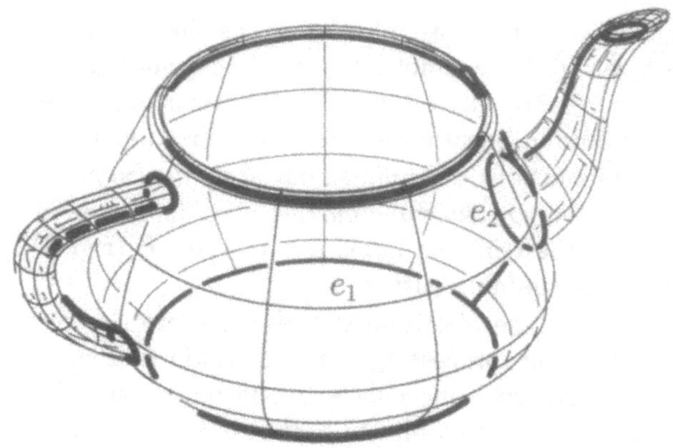

Figure 19.6. The Utah teapot as a trimmed model.

or trimming curve must be shared by exactly two surfaces. Good set operations require an efficient evaluation of the adjacent surface. To trace an intersection curve to the boundary of a trimmed surface, the other trimmed surface adjacent to the boundary must be found in order to continue the computation of the intersection curve. Moreover, the transformation of a free-form model into a piecewise linear approximation could yield inconsistent output along the trimming edges if the approximation is not matched. The gaps that result in a mismatched approximation are also known as black holes. (See Figure 19.7 for examples.)

In order to allow access to the topological information in a model, a data structure must be derived that is similar in nature to the winged-edge data

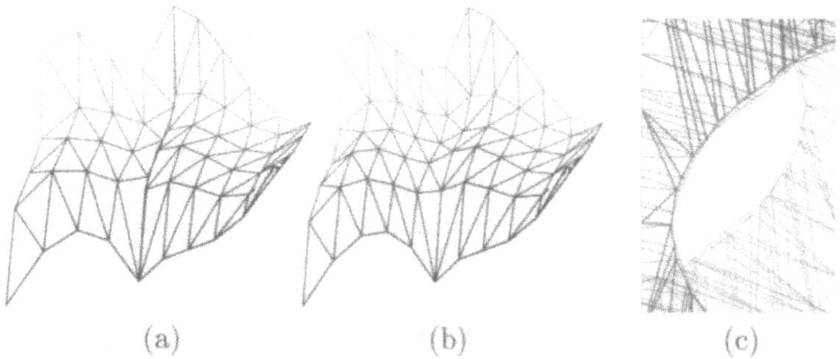

Figure 19.7. Two surfaces sharing an edge: (a) mismatching along a shared boundary edge results in *black holes*; (b) a seamless connection is formed, via a proper matching; (c) a zoomed view over the central area (the hole) of a matchless polygonal approximation of Figure 19.5 (a); the outer cylinder is shown in gray whereas the inner hole is shown in black.

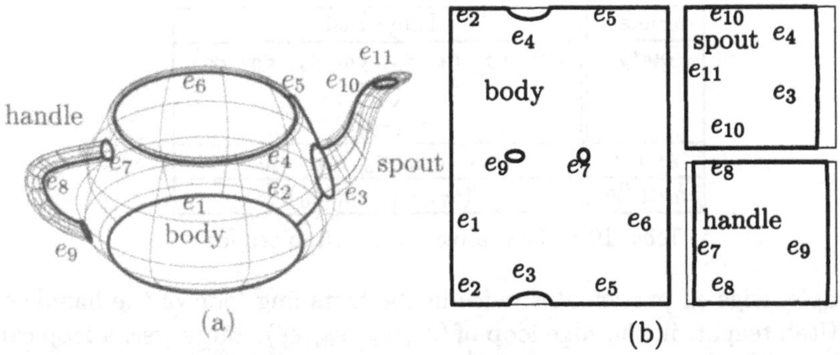

Figure 19.8. (a) The Utah teapot: (a) all trimming edges in the Euclidean space; (b) all trimming edges in the parametric space.

structure. Every edge (trimming curve) references the two faces adjacent to it, PFace and NFace. Further, every edge also lists the previous and the next edges in the two loops of edges in PFace and in NFace. Finally, every face holds a reference to the loops in the face. Unlike the polygonal case, a free-form face could reference more than a single loop of edges.

Consider again the example of the Utah teapot in Figure 19.8. All the edges and faces of this model are listed in both the Euclidean space in Figure 19.8 (a) and in the parametric space in Figure 19.8 (b).

The table of trimming edges of the Utah teapot in Figure 19.8 is given in Table 19.4. Here the edge table is different than the winged-edge data structure for a polygonal object in several respects. Studying our simple example of the Utah teapot, a single edge might be shared twice in the same loop. In other words, NFace and PFace can be the same face. For

Edge	NFace	PFace	NCw	NCcw	PCw	PCcw
e_1	*body*		e_2	e_2		
e_2	*body*	*body*	e_1	e_3	e_4	e_1
e_3	*body*	*spout*	e_2	e_5	e_{10}	e_4
e_4	*body*	*spout*	e_3	e_{10}	e_2	e_5
e_5	*body*	*body*	e_3	e_6	e_6	e_4
e_6	*body*		e_5	e_5		
e_7	*body*	*handle*	e_7	e_7	e_8	e_8
e_8	*handle*	*handle*	e_9	e_{10}	e_6	e_5
e_9	*body*	*handle*	e_8	e_8	e_9	e_9
e_{10}	*spout*	*spout*	e_4	e_{11}	e_{11}	e_3
e_{11}	*spout*		e_{10}	e_{10}		

Table 19.4. Trimming edges of the Utah teapot.

Face	Edge List
body	$(e_1,\ e_2,\ e_4,\ e_5,\ e_6,\ e_5,\ e_3,\ e_2)$ (e_7) (e_9)
spout	$(e_{11},\ e_{10},\ e_3,\ e_4,\ e_{10})$
handle	$(e_7,\ e_8,\ e_9,\ e_8)$

Table 19.5. Face table for the Utah teapot.

example, edge e_8 participates twice in the trimming loop of the handle of the Utah teapot, in the edge loop of $(e_7,\ e_8,\ e_9,\ e_7)$. Moreover, a loop can be formed from only one edge. In the teapot, e_7 and e_9 form two loops of a single edge in the trimming loops of the body of the Utah teapot.

Implicitly, we also allow edges to be shared by only one surface, so an open object can result. This generalization is mostly due to the fact that in the intermediate stages, a model is not necessarily closed. It is possible to construct a single open free-form surface to start with that is clearly not a model. By supporting a representation that permits this intermediate shape, the modeling process can become more intuitive. However, this generalization can result in a non-2-manifold geometry. In the edge table of the Utah teapot, edges e_1, e_6 and e_{11} are all shared by only one surface, having an open teapot at the base of the body (e_1), the cover of the body (e_6) and the opening of the spout (e_{11}).

To complete the representation, we introduce the face table, which lists all the loops of trimming edges of each face. In this representation, a face can have more than one trimming loop. For example, the body of the Utah teapot consists of three loops. In a similar way to the winged-edge data structure, one can list only one edge per loop and trace the entire loop from the edge table. However, for clarity, we list all the edges of every loop in every face, in Table 19.5.

Having studied the representation of a model, it is crucial to distinguish the topology information of the model from its geometric portion. Figure 19.9 portrays this relation. Finally, one should recall that a model in a broad sense could include auxiliary information as well, from the identity of the authoring personnel, through graphics attributes, to manufacturing instructions such as tolerance.

19.4 Non-Manifold Geometry

A boundary representation is capable of representing solids. The boundary prescribes the dichotomy between the inside and the outside of the object. Nonetheless, one can foresee examples where this representation is

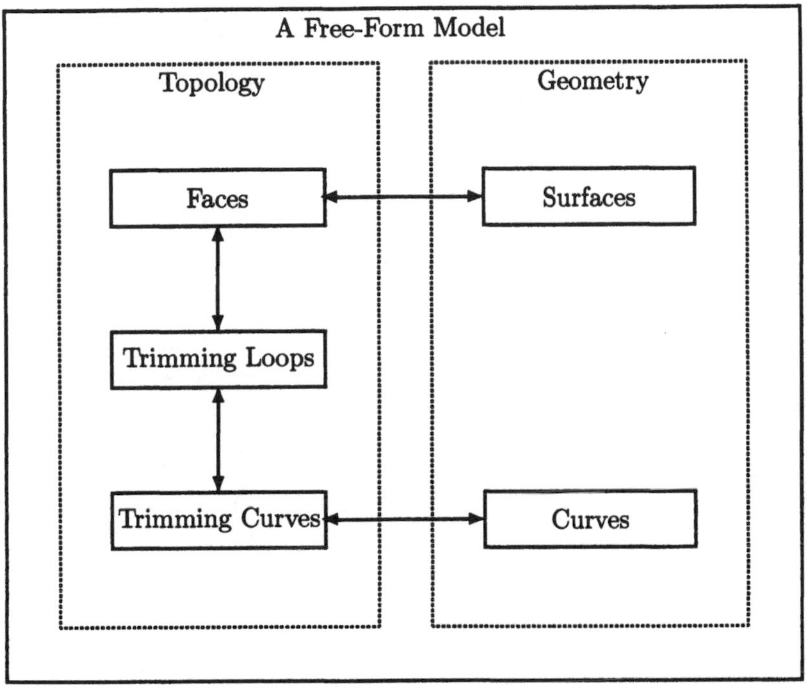

Figure 19.9. The relation between geometry and topology in a model.

too restrictive, as, for example, in Figure 19.1. One is typically interested in handling and manipulating free-form surfaces as open objects and not as solids. Another example is the application of sheet metal. Here, the problem is of developable surfaces that, potentially, have holes in $\mathbb{R}^3$. Yet another example could be a wireframe representation of some geometry. All these examples necessitate the generalization of the winged-edge or the free-form data structures already introduced. Such objects are no longer 2-manifold and a whole variety of operations on these objects become significantly more difficult.

For example, consider the intersection of two arbitrary objects in $\mathbb{R}^3$. If the objects are 2-manifolds, the result of the regularized intersection is the boundary of all points that are inside both objects. In contrast, if the objects are not 2-manifolds, the result of the intersection can be a collection containing both disjoint volumes and faces, and dangling edges! Set operations of non-2-manifold geometry is considered a difficult task and is an active area of contemporary research.

20

Subdivision Surfaces

Arbitrary wireframe configurations arise in many different application areas. The process of laying a surface over the wireframe models—interpolating the original data—is frequently called *fleshing out* wireframe models. It it also useful to use the wireframe as a control mesh for a B-spline surface. In the latter case, the need for a smooth surface is maintained, but the requirement for interpolation is removed, replaced with a requirement for shape control with the convex-hull property and variation-diminishing property.

If each face of the wireframe has exactly four edges uniquely defining it, then it can be fitted with a tensor product surface, such as a B-spline or Coons surface. Smoothness conditions across patches reduce the number of degrees of freedom, but some freedom remains which must be determined by the user. If all faces have four edges and all internal vertices have four edges emanating from them, the grid is said to be a *regular four-grid*, and can be fitted with a B-spline surface. There has been extensive research devoted to finding a good fit when this regularity is not present. Examples for the case where each face has four sides, but the number of edges incident on some vertices is other than four, can be found in [10, 52, 73].

The corner-cutting paradigm [16], extended to surfaces, allows a constructive method of obtaining surfaces from arbitrary networks of lines and vertices. In Section 7.2.1 we showed that a particular form of refinement, when applied to uniform floating spline curves, leads to quadratic and cubic subdivision curves. In Section 13.4 we extended this type of refinement to floating uniform spline curves to obtain regular bi-quadratic and regular bi-cubic subdivision surfaces. The subdivision curves and surfaces obtained were just a different view of spline refinement and curve-surface evaluation. Thus, they do not interpolate specific data. In this chapter we extend those methods to create subdivision surfaces over arbitrary grids leading to Doo-Sabin and Catmull-Clark surfaces. We present generalizations of those methods to support edge curves and allow embedding of creases and darts.

First considered as a two-dimensional extension of Chaikin in [15] and analyzed for continuity in [34], methods to create subdivision surfaces over arbitrary grids have more recently been analyzed in [3] and further extended in the generalized quadratic case in [61]. A subdivision method based on box splines [53] is presented for regular triangular grids, where each vertex of a regular triangular grid has six edges emanating from it. We introduce some of the algorithms and methods.

Research in subdivision surfaces is not isolated to representation schemes for smooth surfaces, creases, fillets, and other geometric attributes (for example, [42, 33]). There is significant research on convergence and smoothness behavior (for example, [70, 77, 85]) and on meeting boundary conditions (for example, [61, 86]). Also, there is research on exact evaluation methods (for example, [81]), fast evaluation methods, adaptive subdivision methods for modeling (for example,[85, 88]), rendering, and texturing surfaces.

As a first step we derive the formulation for refining quadratic and cubic B-spline surfaces using a local matrix approach. Since corner cutting is vertex based, we consider *new* vertices in terms of a single *old* vertex and all the edges incident upon it.

20.1 Catmull-Clark Subdivision Surfaces

20.1.1 Subdivision Curves

Recall (Equation 13.21) that for curves,

$$Q_m = \begin{cases} \frac{P_{i-2}+P_{i-1}}{2} & m = 2i-1 \\ \frac{P_{i-2}}{8} + \frac{3}{4}P_{i-1} + \frac{P_i}{8} & m = 2i. \end{cases}$$

The vertices of the new control polygon can be thought of as *children* of the current control polygon in the following way. The new vertices with even subscripts are children of current vertices, while new vertices with odd subscripts are children of current edges. While it is easy to assign subscripts to curve vertices in an ordered way, that is not true for arbitrary meshes. In order to develop rules for subdivision surfaces which are applicable to

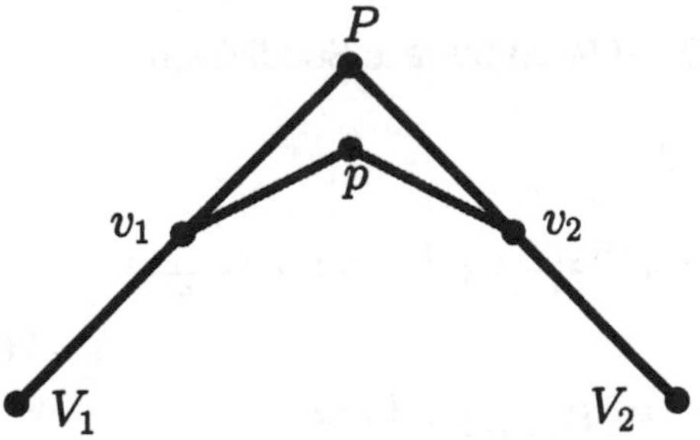

Figure 20.1. The original local polygon is in black. The local polygon resulting from subdivision is in gray.

general meshes, the subdivision algorithms are restated as local *templates*. Thus, a vertex that is the child of a vertex is called a *vertex vertex*, while a vertex that descends from an edge is called an *edge vertex*. An edge vertex has no vertex predecessor. (See Figure 20.1.) Capital letters are used to designate vertices at the current level. Lower case letters are used to designate vertices of the next level. Hence, for the current set of three sequential vertices in the polygon (as in Figure 20.1),

$$p = \frac{V_1}{8} + \frac{3}{4}P + \frac{V_2}{8} \tag{20.1}$$

$$v_1 = \frac{V_1 + P}{2} \tag{20.2}$$

$$v_2 = \frac{P + V_2}{2} \tag{20.3}$$

where p is the child of P and v_i is the child of the edge formed by V_i and P. This can be written in matrix form as

$$\begin{bmatrix} p \\ v_1 \\ v_2 \end{bmatrix} = \begin{bmatrix} 3/4 & 1/8 & 1/8 \\ 1/2 & 1/2 & 0 \\ 1/2 & 0 & 1/2 \end{bmatrix} \begin{bmatrix} P \\ V_1 \\ V_2 \end{bmatrix}. \tag{20.4}$$

If the matrix form is applied serially then a point that is a V_2 becomes the next P, and the current P becomes the next V_1. Hence, each edge vertex is computed as v_2 and then as v_1, as the local focus is shifted. It is important to include it in the new polygon just once.

20.1.2 Regular Quad Mesh Surface Subdivision

From Equation 13.22

$$Q_{m,n} = \begin{cases} \frac{P_{i-2,j-2}}{4} + \frac{P_{i-1,j-2}}{4} + \frac{P_{i-2,j-1}}{4} + \frac{P_{i-1,j-1}}{4}, & \\ & m = 2i-1, n = 2j-1 \\ \frac{P_{i-2,j-2}}{16} + \frac{3}{8}P_{i-1,j-2} + \frac{P_{i,j-2}}{16} & \\ \quad + \frac{P_{i-2,j-1}}{16} + \frac{3}{8}P_{i-1,j-1} + \frac{P_{i,j-1}}{16}, & m = 2i, n = 2j-1 \\ \frac{P_{i-2,j-2}}{16} + \frac{P_{i-1,j-2}}{16} + \frac{3}{8}P_{i-2,j-1} & \\ \quad + \frac{3}{8}P_{i-1,j-1} + \frac{P_{i-2,j}}{16} + \frac{P_{i-1,j}}{16}, & m = 2i-1, n = 2j \\ \frac{P_{i-2,j-2}}{64} + \frac{3}{32}P_{i-1,j-2} + \frac{P_{i,j-2}}{64} + \frac{3}{32}P_{i-2,j-1} & \\ \quad + \frac{9}{16}P_{i-1,j-1} + \frac{3}{32}P_{i,j-1} + \frac{P_{i-2,j}}{64} + \frac{3}{32}P_{i-1,j} + \frac{P_{i,j}}{64} & \\ & n = 2j, m = 2i. \end{cases}$$

In this case, the child vertices with two even subscripts are children of current vertices, and the child vertices with one even subscript and one odd subscript are children of edge vertices. There is a new type of child vertex for a subdivision surface; the child vertices with two odd subscripts are children of current faces, called *face vertices*, and, as shown in Equation 13.23 are located at the face barycenter. Once again a formulation considering only local characteristics is derived. The labeling in Figure 20.2 (a) is used.

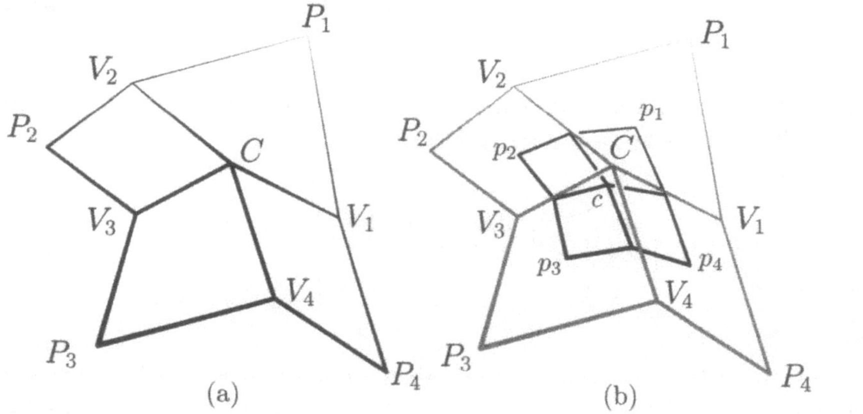

Figure 20.2. (a) Local labeling for a regular mesh; (b) refined mesh and labeling consistent with Equation 20.5.

The current mesh and next mesh around the point P is considered:

$$
\begin{aligned}
\text{face:} \quad p_i &= \tfrac{C+P_i+V_i+V_{i+1}}{4}; \\
\text{edge:} \quad v_i &= \tfrac{3}{8}C + \tfrac{3}{8}V_i + \tfrac{1}{16}V_{i-1} + \tfrac{1}{16}P_{i-1} + \tfrac{1}{16}P_i + \tfrac{1}{16}V_{i+1} \\
&= \tfrac{1}{4}C + \tfrac{1}{4}V_i + \tfrac{1}{4}p_{i-1} + \tfrac{1}{4}p_i; \\
\text{vertex:} \quad c &= \tfrac{9}{16}C + \tfrac{3}{32}\left(V_1+V_2+V_3+V_4\right)\tfrac{1}{64}\left(P_1+P_2+P_3+P_4\right) \\
&= \tfrac{9}{16}C + \tfrac{3}{8}\tfrac{\sum_{i=1}^{4}V_i}{4} + \tfrac{1}{16}\tfrac{\sum_{i=1}^{4}P_i}{4}.
\end{aligned}
\tag{20.5}
$$

In Equations 20.5 the subscripts are assigned modulo 4, i.e., the number of edges emanating from C. p_i is the barycenter of the face defining it. v_i is the barycenter of the ephemeral four-sided face defined by $\{C, p_{i-1}, V_i, p_i\}$. Finally c is a weighted sum of the ephemeral three-sided face defined by C, the barycenter of the edges that emanate from C, and the barycenter of the vertices not connected to C by an edge but defining the faces neighboring C.

Define $\mathcal{V}$ as the barycenter of the vertices emanating from C and $\mathcal{P}$ as the barycenter of the P_i's, that is, the barycenter of the vertices not connected to C by an edge but defining the faces neighboring C. Let $\boldsymbol{p}$ be the barycenter of the newly computed p_i's,

$$\boldsymbol{p} = \frac{\sum_{i=1}^{4} p_i}{4}.$$

Then, the vertex equation of Equation 20.5 can be written in the following two ways.

$$c = \frac{9}{16}C + \frac{3}{8}\mathcal{V} + \frac{1}{16}\mathcal{P} \tag{20.6}$$

$$ = \frac{2}{4}C + \frac{1}{4}\mathcal{V} + \frac{1}{4}\boldsymbol{p}. \tag{20.7}$$

This local formulation can be written in matrix notation. Here, let

$$\boldsymbol{A} = \begin{bmatrix} C & V_1 & V_2 & V_3 & V_4 & P_1 & P_2 & P_3 & P_4 \end{bmatrix}^t \tag{20.8}$$

$$\boldsymbol{a} = \begin{bmatrix} c & v_1 & v_2 & v_3 & v_4 & p_1 & p_2 & p_3 & p_4 \end{bmatrix}^t. \tag{20.9}$$

Then,

$$\boldsymbol{a} = \mathcal{C}\boldsymbol{A}, \tag{20.10}$$

where

$$\mathcal{C} = \begin{bmatrix} \frac{9}{16} & \frac{3}{32} & \frac{3}{32} & \frac{3}{32} & \frac{3}{32} & \frac{1}{64} & \frac{1}{64} & \frac{1}{64} & \frac{1}{64} \\ \frac{3}{8} & \frac{3}{8} & \frac{1}{16} & 0 & \frac{1}{16} & \frac{1}{16} & 0 & 0 & \frac{1}{16} \\ \frac{3}{8} & \frac{1}{16} & \frac{3}{8} & \frac{1}{16} & 0 & \frac{1}{16} & \frac{1}{16} & 0 & 0 \\ \frac{3}{8} & 0 & \frac{1}{16} & \frac{3}{8} & \frac{1}{16} & 0 & \frac{1}{16} & \frac{1}{16} & 0 \\ \frac{3}{8} & \frac{1}{16} & 0 & \frac{1}{16} & \frac{3}{8} & 0 & 0 & \frac{1}{16} & \frac{1}{16} \\ \frac{1}{4} & \frac{1}{4} & \frac{1}{4} & 0 & 0 & \frac{1}{4} & 0 & 0 & 0 \\ \frac{1}{4} & 0 & \frac{1}{4} & \frac{1}{4} & 0 & 0 & \frac{1}{4} & 0 & 0 \\ \frac{1}{4} & 0 & 0 & \frac{1}{4} & \frac{1}{4} & 0 & 0 & \frac{1}{4} & 0 \\ \frac{1}{4} & \frac{1}{4} & 0 & 0 & \frac{1}{4} & 0 & 0 & 0 & \frac{1}{4} \end{bmatrix} \qquad (20.11)$$

Notice that all the edge and face vertices are computed multiple times as the central vertex C moves around the current mesh. For example, a face vertex, p_i is computed at each of its current polygon vertices. It has a different subscript within each of the local templates.

The subdivision process requires specification of which new vertices are connected by edges and which of the new edges are connected to form a face. For the regular quad grid surface, it is straightforward and can be derived from the global information. It is more complex when the mesh has vertices with other than four edges emanating from it, or polygons with other than four edges defining it.

20.1.3 Arbitrary Grid Surface Subdivison

In this section we consider general meshes. They differ from regular meshes in one of two ways. Either a vertex has other than four edges emanating from it or a face is defined by other than four edges. We present generalizations for the formulation in Section 20.1.2. We consider first the situation in which all the faces have four edges, but some vertices have other than four edges emanating from them. Figure 20.3 (a) illustrates that case. Figure 20.3 (b) shows labeling for the new vertices that are defined by the generalized algorithms below.

Equations 20.5 defining p_i and v_i remain unchanged. They depend only on each face having four vertices and each edge defining two faces. The formulation for c has many different generalizations. Consider Equation 20.6 and Equation 20.7 for the regular case.

$$\begin{aligned} c &= \frac{9}{16}C + \frac{3}{8}\mathcal{V} + \frac{1}{16}\mathcal{P} \\ &= \frac{2}{4}C + \frac{1}{4}\mathcal{V} + \frac{1}{4}\mathbf{p}. \end{aligned}$$

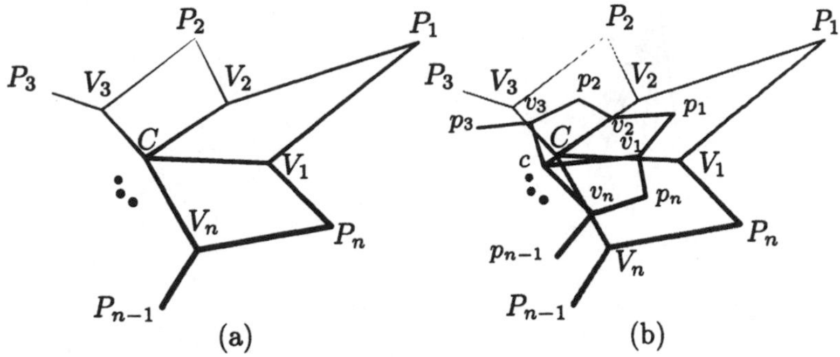

Figure 20.3. (a) A locally irregular mesh with four-sided faces; (b)the refined mesh resulting from application of Equation 20.12.

The number of edges emanating from C appears nowhere in either equation. Rather, each represents c as different weighted sums over three different vertices. $\mathcal{V}$ is just the barycenter of the vertices on edges emanating from C, and $\boldsymbol{p}$ is just the barycenter of the barycenters of the faces surrounding C. Catmull and Clark observed this originally and tried to keep this formulation as the generalization to n vertices emanating from C. p_i is defined identically the same as in Equation 20.5, but the definition of $\boldsymbol{p}$ is modified from its original definition to an exact analogy.

$$\boldsymbol{p} = \frac{\sum_{i=1}^{n} p_i}{n}$$

Then Equation 20.7 is used with the same weights [15]. The result was considered to look too pointy. A second generalization was tried, where p_i and v_i are defined as in Equations 20.5, but c is modified.

$$\begin{aligned} p_i &= \frac{C + P_i + V_i + V_{i+1}}{4}, \\ v_i &= \frac{1}{4}C + \frac{1}{4}V_i + \frac{1}{4}p_{i-1} + \frac{1}{4}p_i, \\ \mathcal{V} &= \frac{\sum_{i=1}^{n} V_i}{n}, \\ \boldsymbol{p} &= \frac{\sum_{i=1}^{n} p_i}{n}, \\ c &= \frac{n-2}{n}C + \frac{1}{n}\mathcal{V} + \frac{1}{n}\boldsymbol{p}. \end{aligned} \tag{20.12}$$

This method (Equations 20.12) has become the widespread subdivision approach of choice. The limit meshes developed from Equations 20.12 have become known as *Catmull-Clark subdivision surfaces.* The size of the

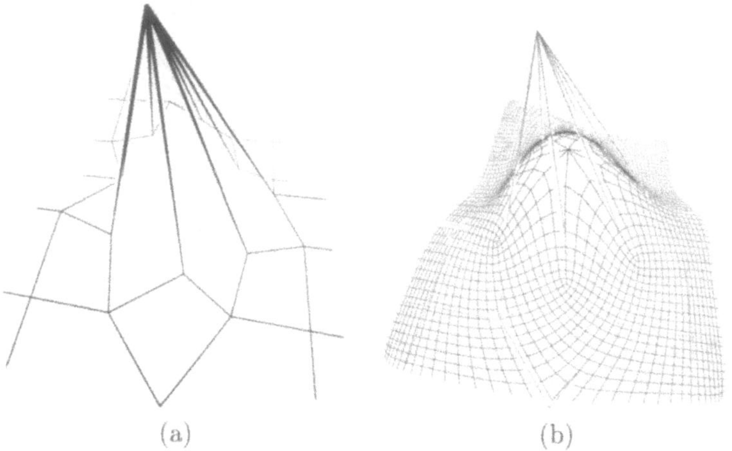

Figure 20.4. (a) A generalized mesh with four-sided faces; (b) surface resulting from repeated application of the second generalization (Equation 20.12.)

quadrilaterals in the new mesh are approximately one-fourth the size in the current mesh with each application of the subdivision process.

Figure 20.4 (a) shows a general mesh. Figure 20.4 (b) shows the surface that results from applying Equation 20.12 to the same mesh.

Now consider the generalization in which each vertex of a face has four edges emanating from it, but the face has n sides where n can be totally general. An example is illustrated in Figure 20.5 (a). Figure 20.5 (b) shows the subdivided mesh quadrilaterals computed with respect to C_{i-2},

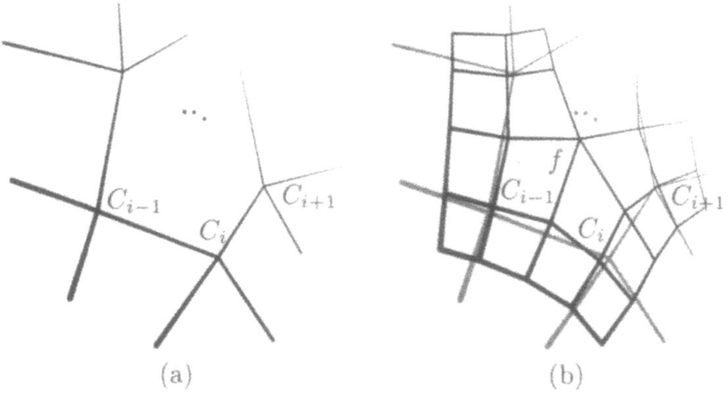

Figure 20.5. (a) A generalized mesh in which each vertex has four edges emanating, but each face can be defined by an arbitrary number of edges; (b) the refined mesh.

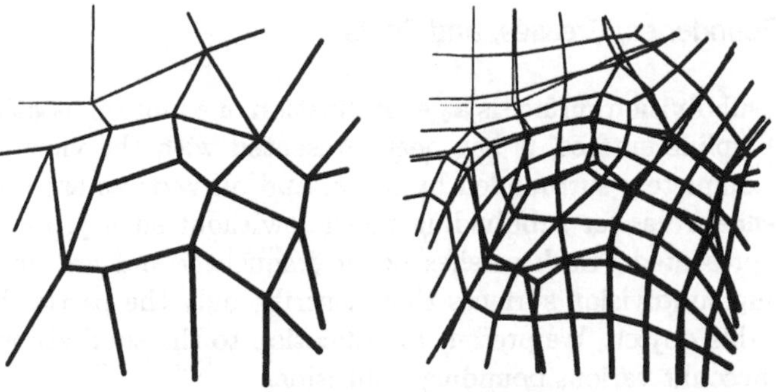

Figure 20.6. (a) Generalized mesh with both types of irregularities; (b) subdivided mesh.

C_{i-1}, C_i, and C_{i+1}. At each vertex the computation is standard, except the barycenter is computed with respect to all the vertices of that face. Equation 20.7 is applied. Notice that the face we consider has m edges (and m vertices) defining it. The face splits at the next level into m regular (quadrilateral) faces surrounding a face vertex with m edges emanating from it.

Now consider a mesh containing faces with other than four vertices and vertices with other than four edges emanating from them, as illustrated in Figure 20.6 (a). Applying Equation 20.7, with face barycenter computed over all the defining vertices results in a new mesh, as in Figure 20.6 (b). The number of vertices having other than four edges emanating is equal to the number of such vertices in the original mesh plus the number of faces in the original mesh having other than four defining edges. All the faces in the resulting mesh are quadrilaterals defined by four vertices.

Definition 20.1. *A vertex in a Catmull-Clark subdivision mesh is called* an extraordinary vertex *if it has other than four edges emanating from it.*

Figure 20.6 shows a subdivision sequence for a Catmull-Clark subdivison surface generated from a general mesh.

Lemma 20.2. *After one application of subdivision, a Catmull-Clark subdivision mesh has all quadrilateral faces.*

Lemma 20.3. *The total number of extraordinary points in a Catmull-Clark subdivision mesh is fixed after one application of subdivision. It is the total of the number present in the original mesh and the number of faces in the original mesh defined by other than four edges.*

20.1.4 Boundaries, Creases, and Darts

Since this subdivision surface is a generalization of a uniform floating tensor product B-spline surface, it has been presented with the view that each vertex is completely surrounded by faces, and all vertices are treated the same. Hence, rules for subdividing meshes without such properties have not been presented. Such meshes occur frequently and are prevalent in representing subdivision surfaces that describe only the partial boundary of a realizable object. We present modification to the subdivision rules to take into account various boundary conditions.

The only irregularities that have existed in the meshes have been caused by initial extraordinary faces or vertices. In this section we present rules to deal with other irregularities that might be desirable in the final rendered subdivision surface, such as interpolation points and curves, creases, and darts. Such irregularities can be handled in a number of different ways.

20.1.4.1 Creases, darts, and interpolation. Edges which are to be creases in the final surface are tagged and the subdivision rules around such edges are modified as presented below. The rules, given with respect to the labels in Figure 20.7, are given in [33]. The rules are:

- Face vertices remain positioned at face centroids, as in Equation 20.12.
- An edge vertex is defined by a different rule if that edge is tagged as a crease. In that case, the edge vertex is defined as the edge midpoint. For example, if edge $\{C_{i-1}, C_i\}$ is tagged a crease, then the new edge point is defined as

$$\frac{C_{i-1} + C_i}{2}.$$

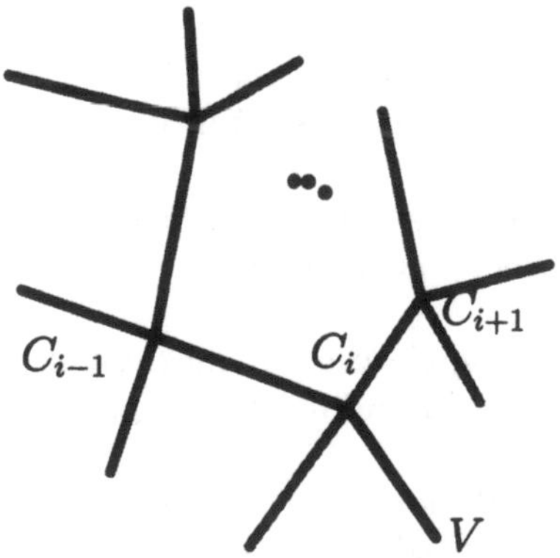

Figure 20.7. Vertex labeling to apply crease rules. The heavier line indicates the crease curve around C_i.

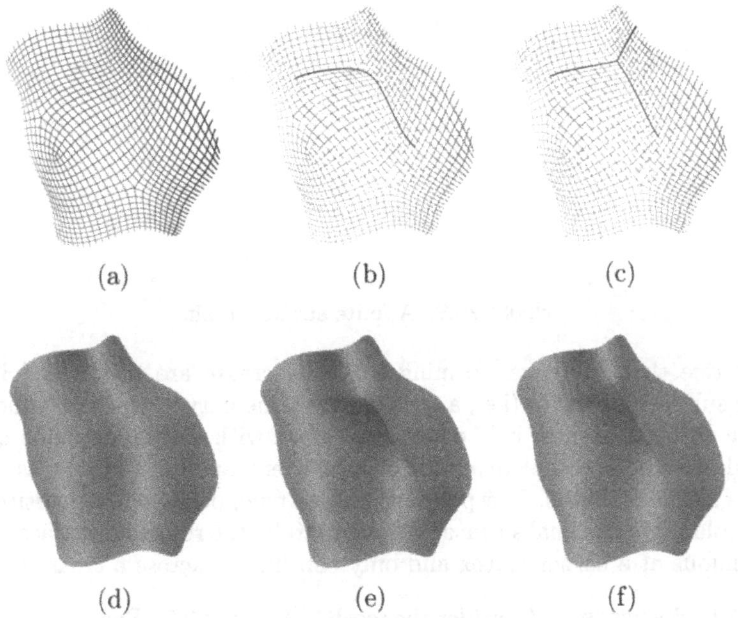

Figure 20.8. Surface showing (a) a dart; (b) a crease; (c) a corner point—all interior to the surface. Parts (d), (e), and (f) are the respective shaded surfaces.

- The new rule for vertex points depends on the number of tagged edges incident on it.

  1. If there are no tagged crease edges incident, then Equation 20.12 is used to determine the child vertex.
  2. If there is one tagged edge incident on it, the vertex is called a *dart* and its child vertex is determined again by Equation 20.12.
  3. If there are two tagged crease edges incident on a vertex, it is called a *crease vertex*. The subdivision curve rule for a vertex-vertex (Equation 20.1) is used to determine the child vertex. Assume in Figure 20.7, that C_i is a crease vertex, along edge $\{C_{i-1}, C_i\}$ and edge$\{C_i, V\}$. The child vertex c_i is defined as

  $$c_i = \frac{1}{8}C_{i-1} + \frac{3}{4}C_i + \frac{1}{8}V.$$

  4. If there are three or more tagged crease edges incident on a vertex, it is called a *corner vertex* and is not moved: $c_i = C_i$.

In Figure 20.8 all three of these irregularities are demonstrated.

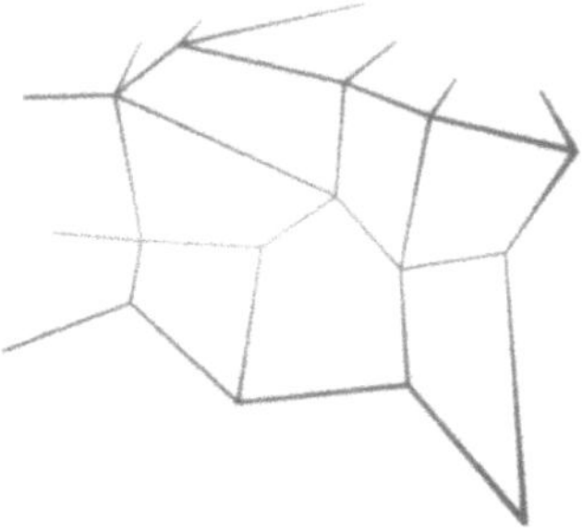

Figure 20.9. A finite surface mesh.

Notice that the edges forming a simple crease are subdivided into a cubic subdivision curve (i.e., a uniform B-spline curve) which is embedded in the surface. Hence, it is possible to start with several B-spline curves that do not cross, complete a mesh around them, and embed them as curves in the surface (that is, interpolate them). Analogously, corner vertices are interpolated in the final surface. Unfortunately, the resulting surface is only continuous at a corner vertex and only continuous across a crease curve.

20.1.4.2 Boundaries. Consider the mesh in Figure 20.9. The vertices along its boundary are not surrounded by faces. Each vertex has at least two edges emanating from it. Some may have four edges, but each such is still an irregular vertex since it is not surrounded by faces. That is, some of the edges emanating from the vertex are used to define only one face. Figure 20.10 (a) shows the result of creating vertices, faces, and edges by applying the basic rules. Notice that a boundary quadrilateral face surrounded on three sides by other faces has two child faces in the subdivided mesh. A boundary quadrilateral face surrounded on two sides by other faces has one child face in the subdivided mesh. Catmull-Clark subdivision

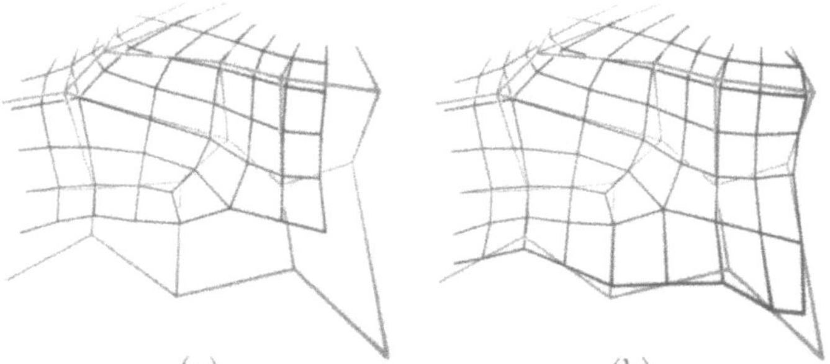

(a) (b)

Figure 20.10. (a) The subdivided mesh obtained from basic rules; (b) the subdivided mesh which marks each boundary edge as a crease.

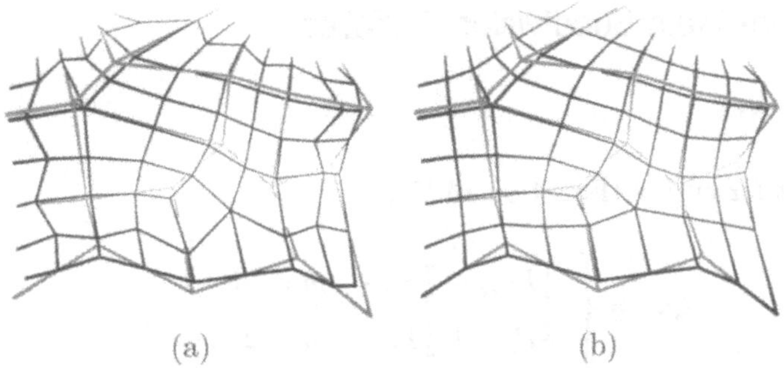

Figure 20.11. (a) Subdivided mesh using boundary markings and modified child edge vertex definition; (b) subdivided mesh using boundary markings and corner vertices.

rules do not extend to a boundary quadrilateral which is adjacent to only one other face.

These rules are modified to treat the boundary as a single uniform floating B-spline curve, as shown in Figure 20.10 (b), by marking each boundary edge as a crease edge. The nearby edge vertices and face vertices can be created in the standard way. An alternative formulation [86] for the edge vertex adjacent to a boundary vertex is for the case when C is a boundary vertex and V_i is an interior vertex sharing an edge with C. If there are just two regular faces adjacent to C, with p_{i-1} and p_i their barycenters, then

$$v_i = \frac{1}{2}C + \frac{1}{4}p_{i-1} + \frac{1}{4}p_{i+1}.$$

This example is shown in Figure 20.11 (a). Yet another variation, which can be applied simultaneously with those above, tags boundary vertices with only two edges emanating from them (i.e., a vertex of just one quadrilateral) as a corner vertex. That way the vertex remains fixed in the mesh. The effect of this is shown in Figure 20.11 (b).

20.1.5 Convergence

The Catmull-Clark subdivision method is a generalization of cubic uniform floating B-spline refinement, so in the regular regions of the mesh the surface has the characteristics of a tensor product bi-cubic uniform floating B-spline surface. It is at extraordinary points that the behavior has been more difficult to predict. Tangent plane continuity for general subdivision algorithms was analyzed by Reif in [70], for Catmull-Clark subdivision it was analyzed by Ball and Storry [3], and for $C^{(1)}$continuity by Peters and Reif [67]. The presented weighting is not the only one that can lead to $C^{(1)}$surfaces, but it is the most widely used.

20.2 Doo-Sabin Subdivision Surfaces

20.2.1 Curves

Recall that for curves (Equation 13.17),

$$Q_m = \begin{cases} \frac{1}{4}P_{i-2} + \frac{3}{4}P_{i-1} & m = 2i-1 \\ \frac{3}{4}P_{i-1} + \frac{1}{4}P_i & m = 2i. \end{cases}$$

Since the ith vertex splits into an edge in the refined mesh, sometimes this type of subdivision algorithm is called a *vertex splitting* algorithm [87]. The ordering of the new vertices to make a new control polygon replaces V_i by the ordered pair defining its child edge. Also, the child of an edge in the control polygon is an edge. Hence in this type of algorithm we consider the behavior on edges, and rewrite Equation 13.17:

$$Q_m = \begin{cases} \frac{3}{4}P_{i-1} + \frac{1}{4}P_i & m = 2i \\ \frac{1}{4}P_{i-1} + \frac{3}{4}P_i & m = 2i+1. \end{cases} \tag{20.13}$$

Another method renames the new vertices in terms of the edges it replaces. In this case, if we call the edge from V_i to V_{i+1} the ith edge, then Figure 20.12 shows

$$\begin{aligned} v_{i,0} &= \frac{3}{4}V_i + \frac{1}{4}V_{i+1} \\ v_{i,1} &= \frac{1}{4}V_i + \frac{3}{4}V_{i+1}. \end{aligned}$$

Hence, the ith edge has $v_{i,0}$ and $v_{i,1}$ as children that form a new edge.

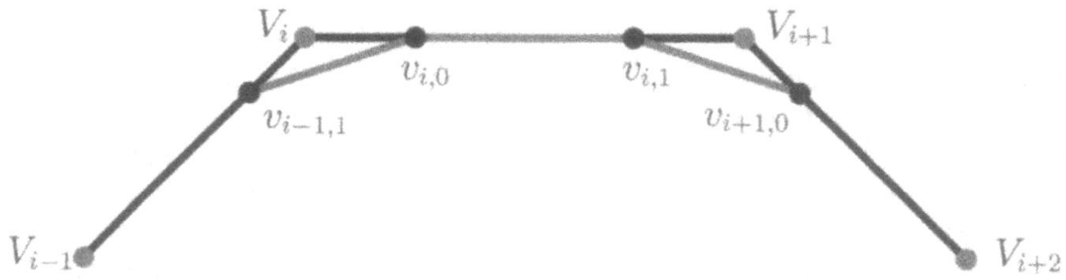

Figure 20.12. Local labeling for original polygon and subdivided polygon.

20.2.2 Regular Quad Mesh Surface Subdivision

In this section, we review bi-quadratic refinement for a regular mesh. Consider the child of the face defined by the ordered vertices $\{C_1, C_2, C_3, C_4\}$, as shown in Figure 20.13.

From Equation 13.18,

$$c_i = \frac{3}{16}C_{i-1} + \frac{9}{16}C_i + \frac{3}{16}C_{i+1} + \frac{1}{16}C_{i+2} \qquad (20.14)$$

where the subscripts are taken modulo 4 to have values between 1 and 4, and $i = 1, \ldots, 4$. In matrix formulation,

$$\mathbf{c} = \mathcal{Q}\mathbf{C}$$

where

$$\mathbf{C} = \begin{bmatrix} C_1 & C_1 & C_3 & C_4 \end{bmatrix}^t, \qquad (20.15)$$

$$\mathbf{c} = \begin{bmatrix} c_1 & c_1 & c_3 & c_4 \end{bmatrix}^t, \qquad (20.16)$$

$$\mathcal{Q} = \begin{bmatrix} 9/16 & 3/16 & 1/16 & 3/16 \\ 3/16 & 9/16 & 3/16 & 1/16 \\ 1/16 & 3/16 & 9/16 & 3/16 \\ 3/16 & 1/16 & 3/16 & 9/16 \end{bmatrix}. \qquad (20.17)$$

Equation 20.14 can be rewritten in terms of the barycenter of the face. Let $\mathcal{C} = \frac{\sum_{i=1}^{4} C_i}{4}$ be the barycenter, and let $\mathcal{A}_i = \frac{C_i + C_{i+1}}{2}$, be the midpoint of the edge between C_i and C_{i+1}.

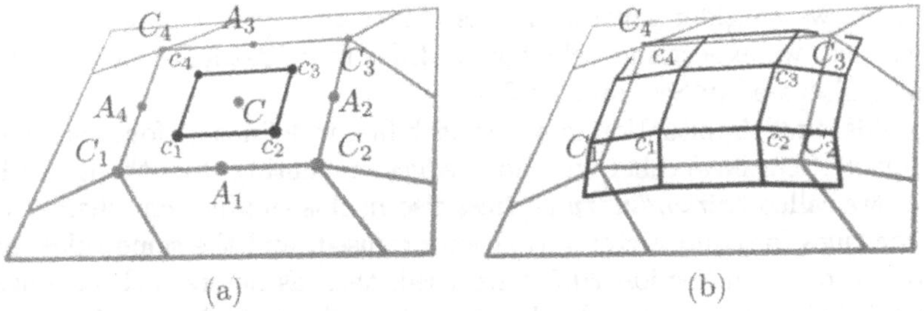

Figure 20.13. (a) Local labeling for a regular mesh and its refined mesh, consistent with Equation 20.14; (b) mesh formation across regular faces.

Then

$$
\begin{aligned}
c_i &= \frac{1}{2}C_i + \frac{1}{8}C_{i-1} + \frac{1}{8}C_{i+1} + \frac{1}{4}\mathcal{C} && (20.18)\\
&= \frac{1}{4}C_i + \frac{1}{4}A_{i-1} + \frac{1}{4}A_i + \frac{1}{4}\mathcal{C} && (20.19)\\
&= \frac{1}{4}C_i + \frac{1}{2}\frac{A_{i-1}+A_i}{2} + \frac{1}{4}\mathcal{C} && (20.20)\\
&= \sum_{j=1}^{4} w_{i,j}C_j \text{ where } w_{i,j} = \begin{cases} 9/16 & |i-j|=0 \\ 3/16 & |i-j|=1 \\ 1/16 & |i-j|>1. \end{cases} && (20.21)
\end{aligned}
$$

Equation 20.19 shows that each new vertex is the barycenter of the ephemeral face formed by C_i, A_{i-1}, A_i, and $\mathcal{C}$.

A new face is formed in three ways. First, it is the child face of a current face, as the equations above show. Notice that for each vertex of the original mesh, there is one new child vertex in each of the current faces adjacent to the current vertex. The new face has exactly the same barycenter as the parent face. In the second method of face formation, those vertices are connected to form a face, ordered by the the way the faces of the current mesh are ordered around the vertex. Finally, each edge has two vertices that determine it, and each edge is adjacent to two faces. Hence, the third method of face formation uses the child vertices in each current adjacent face from each of the current edge-defining vertices to form a new face. Again, the new vertices are ordered by the topology of the current mesh faces and the two vertices defining the edge. New mesh formation is shown in Figure 20.13 (b).

20.2.3 Arbitrary Grid Surface Subdivison

Once again we consider general meshes. Since this type of subdivision is face oriented, we expect the behavior to differ from Catmull-Clark subdivision, as discussed in Section 20.1.3.

Consider first the situation in which each face is defined by four vertices, but some vertices have other than four edges emanating from them. Such vertices are called *extraordinary vertices* also in this subdivision scheme. If the same rules for child vertex formation are used, and the same rules for face formation as in Section 20.2.2 are used, then as before each face has a four-sided child face and each edge has a four-sided child face. However, the child face of each extraordinary vertex has become an *extraordinary face*, i.e., it is defined by other than four vertices.

The other irregularity, and the only one remaining after the first subdivision process, is the case in which each vertex has exactly four edges emanating from it, but some faces are defined by other than four vertices. Suppose it is defined by n vertices. Equations 20.18 and 20.19 can be used directly, as long as the proper barycenter is computed, as the generalized subdivision algorithm for vertex creation with the same rules for face and edge formation. In that case, there is an n-sided child face from each n-sided current face. However, the faces that are the children from vertices and edges are all defined by four child vertices.

Catmull and Clark [15] suggested a generalization based on n, the number of vertices defining the face. Using the same definitions of A_i and C as in Equation 20.20,

$$c_i = \frac{n-3}{n}C_i + \frac{2}{n}\frac{A_{i-1}+A_i}{2} + \frac{1}{n}C \tag{20.22}$$

$$= \sum_{j=1}^{n} w_{i,j}C_j \text{ where } w_{i,j} = \begin{cases} \frac{4n+2}{8n} & |i-j| = 0 \\ \frac{n+2}{8n} & |i-j| = 1 \\ \frac{2}{8n} & |i-j| > 1. \end{cases} \tag{20.23}$$

However, Doo and Sabin [34] showed that the following choice of weights for the new vertices gives a $C^{(1)}$surface. Hence,

Definition 20.4. *The subdivision surface created by defining child vertices*

$$c_i = \sum_{j=1}^{n} w_{i,j}C_i, \text{ where } w_{i,j} = \begin{cases} \frac{n+5}{4n} & i = j \\ \frac{3+2\cos\left(\frac{2\pi(i-j)}{n}\right)}{4n} & i \neq j \end{cases}$$

and forming faces as described in Section 20.2.2 is called a Doo-Sabin subdivision surface. Figure 20.14 shows a subdivison sequence for such a surface.

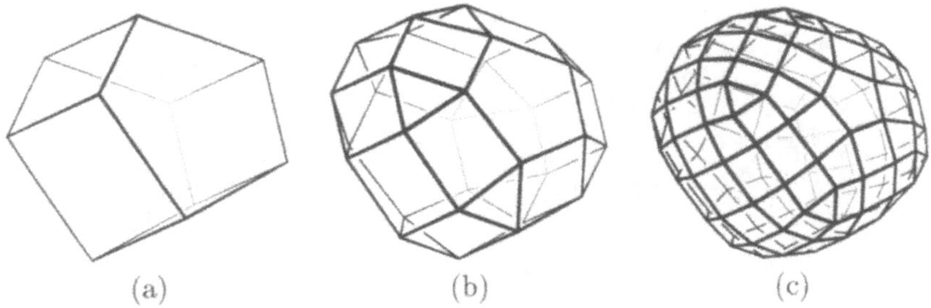

Figure 20.14. (a) Original mesh with irregular faces; (b) one level of subdivision using Definition 20.4; (c) a second level of subdivision.

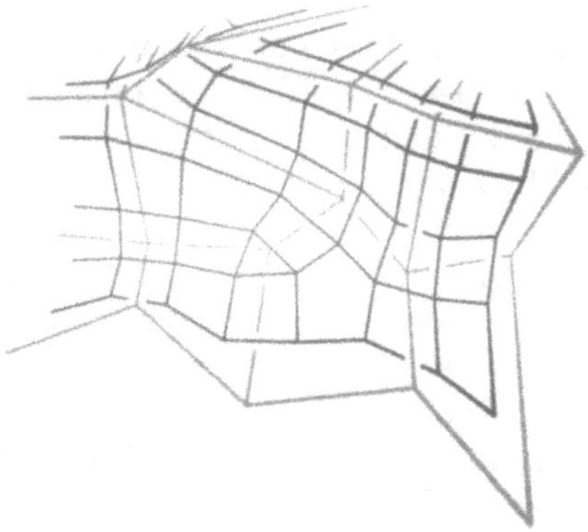

Figure 20.15. A subdivided mesh obtained from basic Doo-Sabin rules.

20.2.4 Boundaries

The same types of special rules created for Catmull-Clark subdivision can be created for Doo-Sabin subdivision in order to embed darts and creases, and to make boundaries behave like the more standard floating uniform bi-quadratic B-spline surface. Again consider the mesh in Figure 20.9. Directly applying the Doo-Sabin rules to the current mesh to create new vertices and connections results in Figure 20.15. As an alternative, the faces with one or more boundary edge can be treated differently.

We use the labeling in Figure 20.16. The child vertices around interior vertices are created using Equation 20.23. Children of edge vertices are created using the uniform quadratic subdivision rules. They are applied to form $c_{2,0}$, $c_{2,1}$, $c_{3,0}$, $c_{3,1}$, $c_{4,0}$, $c_{5,1}$. On the other hand, C_5 is a corner vertex so it stays fixed. (It would seem that $c_{4,1}$ and $c_{5,0}$ are collapsed to C_5.)

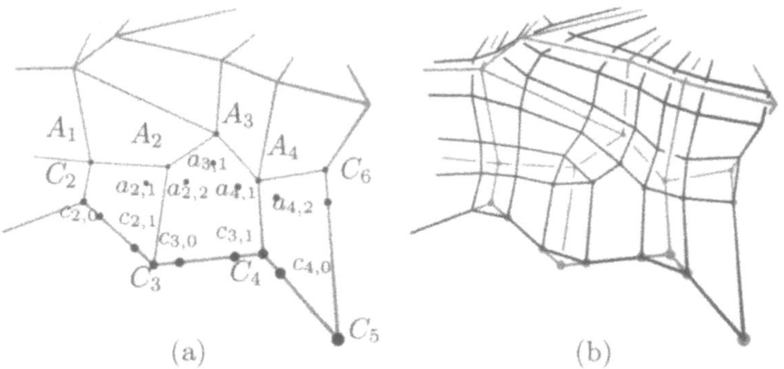

Figure 20.16. (a) Labeling for quadratic boundary conditions; (b) mesh resulting from application of Equation 20.23.

20.2.5 Convergence

Doo-Sabin subdivision is a generalization of uniform quadratic floating B-spline refinement, so in regular regions of the mesh, the surface is $C^{(1)}$. Doo and Sabin [34] showed that their generalized methods are $C^{(1)}$at limit points of extraordinary faces.

20.3 Triangular Subdivision Surfaces

Box splines are an intrinsically multivariate B-spline, so they cannot, in general, be represented as a tensor product of lower-dimensional splines. However, with the advent of discrete box splines and refinement algorithms [21], the regularity of this representation allowed its more general use. A three-direction box spline mesh is composed of triangular faces, each of whose vertices is adjacent to a total of six faces.

Loop [53] generalized the refinement methods, or subdivision surface method, to cases in which the vertices have other than six emanating edges, and proposed the use of this triangle-based subdivision technique. Later creases and corners were introduced by modifying the subdivision rules in the related neighborhood [42, 77]. This section presents rules for arbitrary triangular mesh Loop subdivision, including the modifications for creases and corners.

Like Catmull-Clark and Doo-Sabin schemes, this method approximates the initial mesh. On a regular grid, the subdivision process is equivalent to box spline refinement on the particular three-direction basis function chosen, a $C^{(2)}$quartic, so the surface is $C^{(2)}$. This scheme creates only edge vertices and vertex vertices as children. Consider Figure 20.17. Since an edge is shared by two triangles, the child edge vertex is a symmetric weighed sum of vertices over both faces. Designate by e the child vertex of the edge $\{V_6, C\}$, and c the child vertex of the vertex C. Then

$$\begin{aligned} e &= \frac{3}{8}V_6 + \frac{3}{8}C + \frac{1}{8}V_5 + \frac{1}{8}V_1, \\ c &= \frac{10}{16}C + \frac{6}{16}\frac{\sum_{i=1}^{6} V_i}{6}. \end{aligned} \tag{20.24}$$

Notice that the neighbor vertices are treated symmetrically. The generalization proposed by Loop [53] for an extraordinary vertex with n edges emanating from it has been shown to result in a $C^{(1)}$surface.

Let w be the weight given to the n neighboring vertices where

$$c = (1 - nw)\,C + nw\frac{\sum_{i=1}^{n} V_i}{n}. \tag{20.25}$$

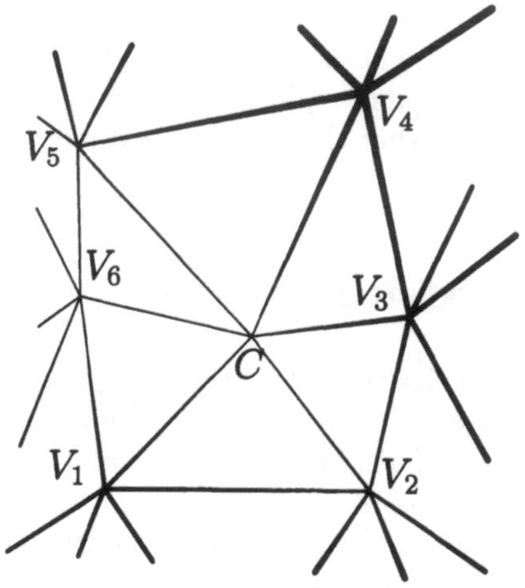

Figure 20.17. Topology of regular triangular mesh.

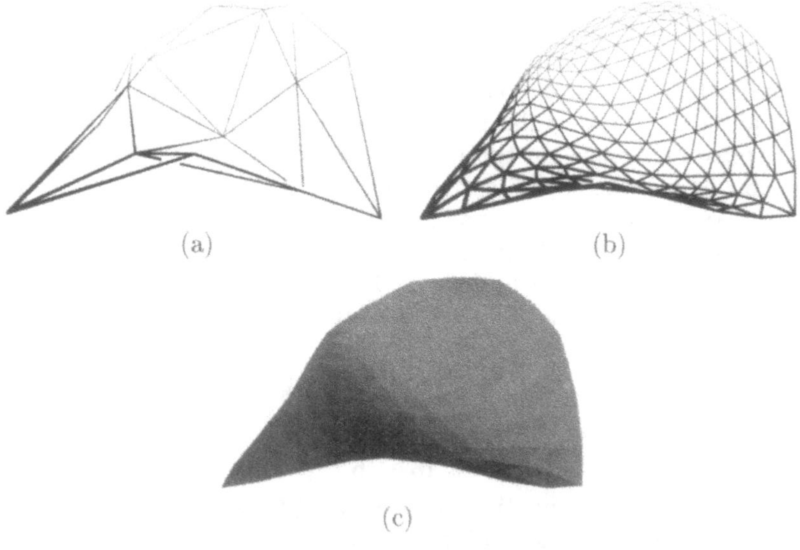

Figure 20.18. (a) A basic mesh; (b) mesh after several subdivisions; (c) part (b) mesh shaded.

The Loop generalization sets

$$w = \frac{\frac{5}{8} - \left(\frac{3}{8} + \frac{1}{4}\cos\left(\frac{2\pi}{n}\right)\right)^2}{n}. \tag{20.26}$$

Figure 20.18 shows a mesh (a), a mesh after several subdivisions (b), and a shaded rendering of (b).

The rules for creating darts, creases, and boundaries for Catmull-Clark subdivision surfaces were modeled on the generalizations for Loop surfaces. The rules in Section 20.1.4.1 are used with no modifications. Thus, these special embedded curves are uniform floating cubic B-splines.

20.4 Other Types of Subdivision Surfaces

With the expanding interest in subdivision surfaces, there has been a significant increase in activity in developing new schemes which are constrained to have prespecified characteristics. Some are intrinsically interpolating schemes; some are created to solve variational constraints. Still others are developed for regular quadrilateral meshes while some are created for regular triangular meshes. Ongoing research has as its goal generalizing these methods to be smooth on arbitrary meshes.

Exercises

1. Programming Exercise: Write a program to permit arbitrary polygons as input and output the refined polygons which are refined according to several of the subdivision methods.

2. Can the winged-edge data structure be used to represent the arbitrary topology necessary for storing these generalized constructive surfaces? Show why or why not. If so, develop a method for automatically generating the data structure for the refined surface from the original grid, either while refinement is being done or before. Note that different methods may be more desirable for quadratic style refinement than for cubic style refinement.

21

Higher-Dimensional Tensor Product B-Splines

The tensor product scheme served us well when we defined surfaces, and it can be further extended to define hypersurfaces in higher dimensions. Here, we will try to explore some of the possibilities of these extended representations.

A curve is a univariate function whereas a surface is a bivariate function. Elevating this dimension to three by generalizing the tensor product surface formulation gives a more general form.

Definition 21.1. *The tensor product B-spline function in three variables is called a* trivariate B-spline function *and has the form*

$$T(u_1, u_2, u_3) = \sum_{i_1} \sum_{i_2} \sum_{i_3} P_{i_1,i_2,i_3} \mathcal{B}_{i_1,\kappa_1,\boldsymbol{\tau}^1}(u_1) \mathcal{B}_{i_2,\kappa_2,\boldsymbol{\tau}^2}(u_2) \mathcal{B}_{i_3,\kappa_3,\boldsymbol{\tau}^3}(u_3).$$

It has variable u_i, degree κ_i, and knot vector $\boldsymbol{\tau}^i$ in the ith dimension.

This approach can be generalized to arbitrary dimension q by determining the vector of polynomial degree in each of the q dimensions, $\boldsymbol{n}$, forming q knot vectors, $\boldsymbol{\tau}^i$, $i = 1, \ldots, q$. Let $\mu = (u_1, u_2, \ldots, u_q) \in \boldsymbol{R}^q$, and let $\boldsymbol{i} = (i_1, i_2, \ldots, i_q)$, where each i_j, $j = 1, \ldots, q$ is an integer. Then we end up with the q-variate tensor product function whose definition is

$$T(\mu) = \sum_{i_1} \sum_{i_2} \cdots \sum_{i_q} P_{\boldsymbol{i}} \mathcal{B}_{i_1,\kappa_1,\boldsymbol{\tau}^1}(u_1) \mathcal{B}_{i_2,\kappa_2,\boldsymbol{\tau}^2}(u_2) \ldots \mathcal{B}_{i_q,\kappa_q,\boldsymbol{\tau}^q}(u_q)$$

of degrees $\kappa_1, \kappa_2, \ldots, \kappa_q$ in each variable (of total degree $\sum_{j=1}^{q} \kappa_j$), and knot sequences $\boldsymbol{\tau}^1, \boldsymbol{\tau}^2, \ldots, \boldsymbol{\tau}^q$, respectively. $T(\mu)$ is a multivariate function from $\mathbb{R}^q$ to $\mathbb{R}^d$, provided that $P_{\boldsymbol{i}} \in \mathbb{R}^d$. If $d > 1$, then T is a vector (parametric) function. Hereafter, we omit the knot sequences and/or the degrees for clarity whenever possible.

The issues and possibilities of computing with general higher-dimensional multivariate functions can be demonstrated for the case $q = 3$, so in the sections that follow, we restrict the presentation to that case.

21.1 Evaluation and Rendering of Trivariates

Suppose it is necessary to evaluate T at $(\hat{u}, \hat{v}, \hat{w})$. First consider evaluating the isoparametric surface with a fixed value of w. For $i = 0, \ldots, n_u$, $j = 0, \ldots, n_v$, let

$$\gamma_{i,j} = \sum_{k=0}^{n_w} P_{i,j,k} B_{k,\kappa_w,\boldsymbol{\tau}^w}(\hat{w}).$$

Then

$$T(u, v, \hat{w}) = \sum_{i=0}^{n_u} \sum_{j=0}^{n_v} \gamma_{i,j} B_{j,\kappa_v,\boldsymbol{\tau}^v}(v) B_{i,\kappa_u,\boldsymbol{\tau}^u}(u)$$

is an isoparametric surface of T, which is just a bivariate tensor product B-spline surface. This can be evaluated with standard surface techniques, either using subdivision and refinement techniques or using the method of evaluating a surface as a series of curves.

Now, consider evaluating an isocurve with fixed values of w and v. Then, for $i = 0, \ldots, n_u$,

$$\begin{aligned} \sigma_i &= \sum_{j=0}^{n_v} \gamma_{i,j} B_{j,\kappa_v,\boldsymbol{\tau}^v}(\hat{v}) \\ &= \sum_{j=0}^{n_v} \sum_{k=0}^{n_w} P_{i,j,k} B_{k,\kappa_w,\boldsymbol{\tau}^w}(\hat{w}) B_{j,\kappa_v,\boldsymbol{\tau}^v}(\hat{v}) \end{aligned}$$

and

$$T(u, \hat{v}, \hat{w}) = \sum_{i=0}^{n_u} \sigma_i B_{i,\kappa_u,\boldsymbol{\tau}^u}(u).$$

Thus, the coefficients σ_i of the isocurve in u are actually points on a B-spline surface (in v and w).

Finally,

$$\begin{aligned}
T(\hat{u}, \hat{v}, \hat{w}) &= \sum_{i=0}^{n_u} \sum_{j=0}^{n_v} \sum_{k=0}^{n_w} P_{i,j,k} \mathcal{B}_{k,\kappa_w,\mathcal{T}^w}(\hat{w}) \mathcal{B}_{j,\kappa_v,\mathcal{T}^v}(\hat{v}) \mathcal{B}_{i,\kappa_u,\mathcal{T}^u}(\hat{u}) \\
&= \sum_{i=0}^{n_u} \sum_{j=0}^{n_v} \gamma_{i,j} \mathcal{B}_{j,\kappa_v,\mathcal{T}^v}(\hat{v}) \mathcal{B}_{i,\kappa_u,\mathcal{T}^u}(\hat{u}) \\
&= \sum_{i=0}^{n_u} \sigma_i \mathcal{B}_{i,\kappa_u,\mathcal{T}^u}(\hat{u}).
\end{aligned}$$

The roles of u, v, and w in the formulation above are interchangeable. More generally, define

$$\gamma_{i,j}(w) = \sum_{k=0}^{n_w} P_{i,j,k} \mathcal{B}_{k,\kappa_w,\mathcal{T}^w}(w),$$

and

$$\begin{aligned}
\sigma_i(v, w) &= \sum_{j=0}^{n_v} \gamma_{i,j}(w) \mathcal{B}_{j,\kappa_v,\mathcal{T}^v}(v) \\
&= \sum_{j=0}^{n_v} \sum_{k=0}^{n_w} P_{i,j,k} \mathcal{B}_{k,\kappa_w,\mathcal{T}^w}(w) \mathcal{B}_{j,\kappa_v,\mathcal{T}^v}(v).
\end{aligned}$$

There are several ways to use the tensor product nature of $T(u, v, w)$ to evaluate it at $(\hat{u}, \hat{v}, \hat{w})$.

- The $n_u + 1$ isoparametric surfaces $\sigma_i(v, w)$ can be evaluated at $(\hat{v}, \hat{w})$, after which the resulting curve in u can be evaluated at $\hat{u}$. The $\sigma_i(\hat{v}, \hat{w})$ are the curve coefficients.
- Alternatively, the $(n_u + 1)(n_v + 1)$ isoparametric curves, $\gamma_{i,j}(w)$ can be evaluated at $w = \hat{w}$. This yields an isoparametric surface which can be evaluated at $(u, v) = (\hat{u}, \hat{v})$.

See Figure 21.1 (a) for an example.

While this evaluation scheme seems complex at first glance, it is useful. For example, it might be necessary to create an isoparametric line drawing of a trivariate function, $T : \mathcal{D} \rightarrow \mathbb{R}^3$, $\mathcal{D} = [0, 1] \times [0, 1] \times [0, 1]$. Using the second strategy proposed above, it is possible to extract the six isoparametric surface boundaries of T as $T(u, v, 0)$, $T(u, v, 1)$, $T(u, 0, w)$, $T(u, 1, w)$, $T(0, v, w)$, $T(1, v, w)$. Then, once obtained, they can be viewed

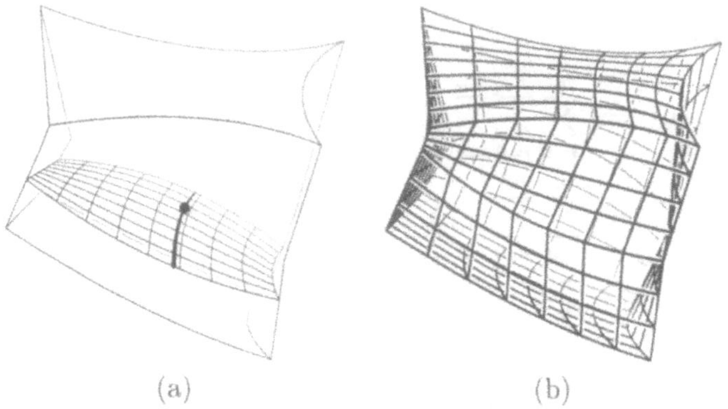

Figure 21.1. (a) A trivariate evaluation as a point on an isoparametric curve of an isoparametric surface of the trivariate; (b) a display that is based on isoparametric curves of the six boundary surfaces of the trivariate.

as B-spline surfaces from which isoparametric curves can be extracted. See Figure 21.1 (b). A shaded display of this trivariate can be derived by extracting a polygonal representation out of these same six boundary surfaces.

Direct rendering of trivariate functions is also possible but is more difficult. For theses cases, the value of the trivariate map usually represents a quantity, such as density or translucency. A ray casting through the volume can integrate the total density and/or opacity that this ray accumulates as it traverses the volume. This computation reduces to a line integral over piecewise polynomial or rational functions for the case of trivariate B-spline functions. See [17, 69].

Another approach to trivariate spline evaluation is to use the subdivision and refinement methods of Chapters 16 and 17.

21.2 Advanced Operations on Trivariates

Because this is a tensor product representation, operations such as degree raising, subdivision, and refinement, can be applied in each of the coordinate direction parameters, in a similar fashion to the way it is done for tensor product surfaces. A trivariate function that is subdivided in u, v, and w would yield eight new trivariate patches. See Figure 21.2. Degree raising, refinement, as well as subdivision can take place in u, v and/or w, independently.

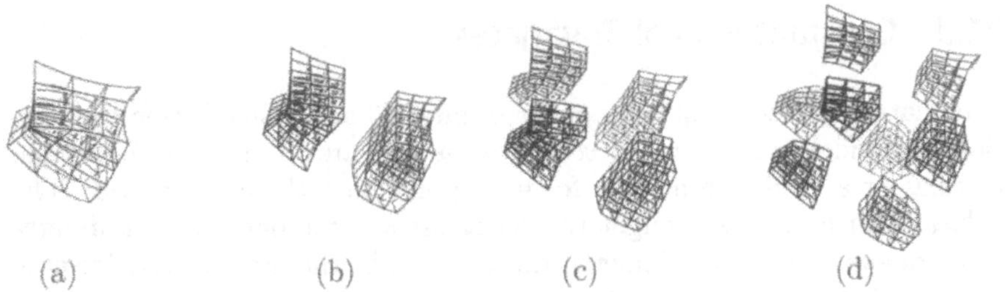

Figure 21.2. (a) A trivariate; (b) subdivided into two subpatches; (c) four subpatches; (d) eight subpatches.

Letting one parameter be constant in surface $\sigma(u, v)$ yields an isoparametric curve. As we already have seen in the evaluation of trivariates, letting one parameter be constant in trivariate $T(u, v, w)$ results in an isoparametric surface as is also shown in Figure 21.3.

Partial derivatives of tensor product trivariates are similarly computed. Let T be $T(u, v, w) : \mathbb{R}^3 \to \mathbb{R}^d$. If $d = 1$, the gradient vector, ∇T is assembled from the partial derivatives. Otherwise, for $d > 1$, ∇T is the Jacobian matrix of size $d \times 3$,

$$\begin{aligned} \nabla T &= \left(\frac{\partial T}{\partial u}, \frac{\partial T}{\partial v}, \frac{\partial T}{\partial w} \right) \\ &= \begin{bmatrix} \sum_i \sum_j \sum_k P_{ijk} \mathcal{B}'_i(u) \mathcal{B}_j(v) \mathcal{B}_k(w), \\ \sum_i \sum_j \sum_k P_{ijk} \mathcal{B}_i(u) \mathcal{B}'_j(v) \mathcal{B}_k(w), \\ \sum_i \sum_j \sum_k P_{ijk} \mathcal{B}_i(u) \mathcal{B}_j(v) \mathcal{B}'_k(w) \end{bmatrix}^T . \end{aligned} \tag{21.1}$$

This matrix plays a major role in many applications that employ trivariates, as will be demonstrated shortly. In Equation 21.1, the $P_{i,j,k}$ are d-dimensional column vectors.

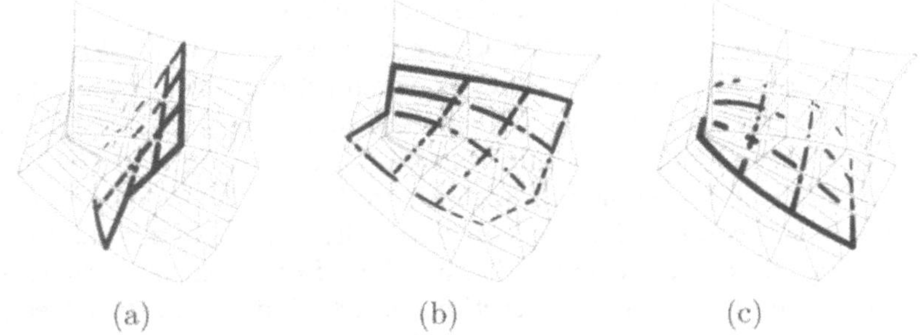

Figure 21.3. A trivariate $T(u, v, w)$ with isoparametric surfaces (thick lines): (a) $\sigma(u, v) = T(u, v, w_0)$; (b) $\sigma(u, w) = T(u, v_0, w)$; (c) $\Sigma(v, w) = T(u_0, v, w)$.

21.3 Constant Sets of Trivariates

Trivariate data sets frequently occur in medical and visualization applications. Volumetric data sets, or *voxel data* as they are often denoted, provide a scalar or a vector data value for each point on a three-space grid. The values of such data sets might represent tissue type, density, or temperature, for example. The volumetric data set can be thought of as a *piecewise constant trivariate spline*, and written

$$T(u,v,w) = \sum_i \sum_j \sum_k P_{i,j,k}\mathcal{B}_{i,0}(u)\mathcal{B}_{j,0}(v)\mathcal{B}_{k,0}(w),$$

where for each i,j,k, $\mathcal{B}_{i,0}(u)\mathcal{B}_{j,0}(v)\mathcal{B}_{k,0}(w)$ is a *piecewise constant* trivariate B-spline basis function, and $\{P_{i,j,k}\}$ is the volumetric data set.

Inherently, the trivariate function carries volumetric information that is difficult to display on two-dimensional screens. Hence, it is typically the case that isosurfaces of the trivariate are displayed instead. A *constant set, or isosurface*, $T(u,v,w) = T_0$, is an implicit surface. Recall from Chapter 2 that implicit curves are difficult to display, so there are even more computational difficulties in querying and displaying implicit surfaces.

Marching cubes [54] is one way to determine isosurfaces for volumetric data sets of constants. The constant sets are extracted by examining a local neighborhood of the eight corners of a single cube in the three-space grid. Note that we used the similar term *isoparametric surface* to denote a different type of surface on a trivariate that is computed by fixing one of the parameters of the trivariate function. In the marching cubes algorithm, each of the twelve edges of the cube is first examined for an intersection with the level of the constant set, T_0, and the proper intersection location along the edge, if any, is linearly interpolated from the two corners of the edge. Then, the faces of the iso-surface inside the cube are approximated from these intersection locations on the twelve edges.

In practice, the marching cubes approach works quite well. While the fact that the piecewise constant data is examined only in a local neighborhood allows rapid computation, it can lead to an ambiguous evaluation of the iso-surface. Consider a cube located at (i,j,k), $i,j,k = 0,1$ and let $T(i,j,k) = (i+j+k) \bmod 2$. (See Figure 21.4.) Then, every face of the cube has two corners along one diagonal with a value of zero and two corners along the other diagonal with a value of one. An iso-surface at $T_0 = \frac{1}{2}$ would result in twelve intersections in the middle of all the edges of the cube. Lacking any knowledge about the content of the interior of the cube, the process of forming faces out of this configuration is, unfortunately, ambiguous.

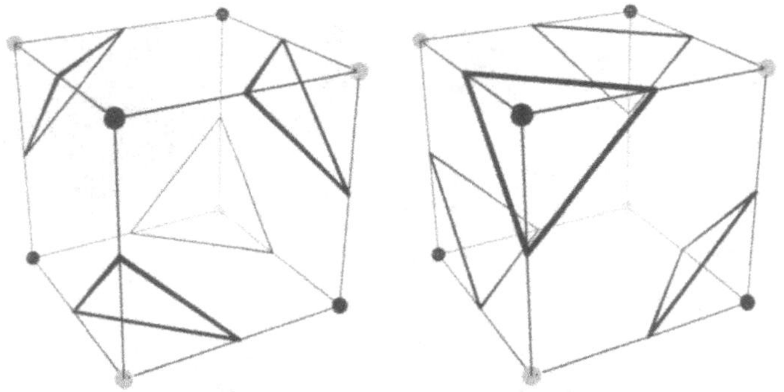

Figure 21.4. Iso-surface extraction from a single cube at level $\frac{1}{2}$. The light corners are at level zero and the black corners at level one. Two possible and equally valid solutions to the iso-surface at level $\frac{1}{2}$ exists and are presented.

21.4 Construction of Trivariates

Clearly, one can examine the same medical data sets introduced in Section 21.3 using higher order B-spline basis functions. The continuous trilinear function,

$$T(u, v, w) = \sum_i \sum_j \sum_k P_{ijk} \mathcal{B}_{i,1}(u) \mathcal{B}_{j,1}(v) \mathcal{B}_{k,1}(w),$$

represents the volumetric data set in a more consistent way. A trilinear representation can greatly alleviate the problem of disambiguating the data set that occurs with the marching cubes method. By having a *continuous* function over (a subset of) $\mathbb{R}^3$, the values that one can extract from the volume are continuous at all locations. Consider again the single cube located at (i, j, k), $i, j, k = 0, 1$ from Section 21.3 having $T(i, j, k) = (i + j + k) \bmod 2$. This time, we employ a trilinear interpolation of the data. Then, computing an iso-surface of the cube can be made more robust since T allows continuously sampling at every single point in the interior of the cube. (See Figure 21.5.) Note that at level 0.5, in Figure 21.5 (c), the resulting surface is no longer a two-manifold and hence an approximation must be employed. Even so, this approximation could be made arbitrary precise with finer sampling.

$T(u, v, w)$ need not be restricted to the linear case. By employing higher-order trivariate functions, it is possible to achieve a smoother, yet non-interpolatory, representation. As an approximation scheme, as the degree gets higher, the fit becomes less accurate—a potential problem, especially in medical imaging. Alternatively, one can employ higher-order trivariate functions to approximate only the surface normals of an iso-

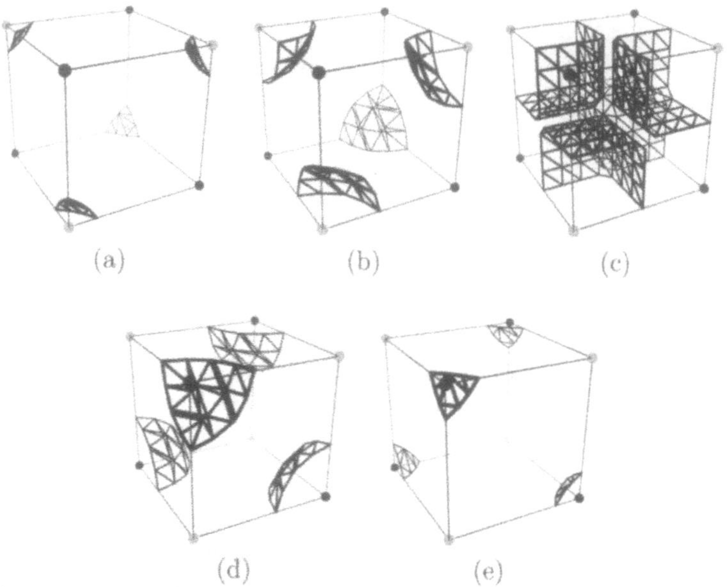

Figure 21.5. Iso-surface extraction from a single cube using a trilinear interpolation at various levels: (a) 0.2; (b) 0.4; (c) 0.5; (d) 0.6; (e) 0.8. The light corners are at level zero and the black corners at level one. Compare with Figure 21.4.

surface $T(u, v, w) = T_0$ without affecting the position of the iso-surface. The normal of an iso-surface can be derived from the gradient of the trivariate at that location (Equation 21.1). Figure 21.6 shows several iso-surfaces, with normals computed using trivariates of different orders.

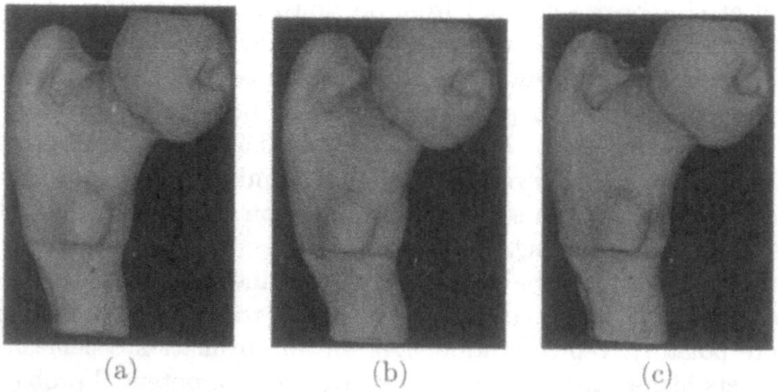

Figure 21.6. An iso-surface rendered using normal computed as gradients of different degrees: (a) flat shading; (b) tricubic; (c) degree 7 trivariate.

21.4.1 Interpolation and Least Squares Fitting

There are many ways to construct and model with trivariates. (See [66, 58].) For example, it is possible to extend the interpolation and approximation methods in Chapter 9 to trivariate functions. Consider the tensor product trivariate T, for which one wants $T(\mu_m) = \phi_m$, $m = 0, \cdots, q$, where $\mu_m = (u_m, v_m, w_m)$.

The $q+1$ interpolation constraints over the trivariate function,

$$T(u_m, v_m, w_m) = T(\mu_m) = \phi_m,$$

immediately reduce to a set of $q+1$ linear constraints over the coefficients of the trivariate, P_{ijk},

$$\sum_{i=0}^{p_u}\sum_{j=0}^{p_v}\sum_{k=0}^{p_w} P_{i,j,k}\mathcal{B}_{i,\kappa_u,\tau^u}(u_m)\mathcal{B}_{j,\kappa_v,\tau^v}(v_m)\mathcal{B}_{k,\kappa_w,\tau^w}(w_m) = \phi_m,$$
$$m = 0, \cdots, q.$$

As in the bivariate surface case, it is a necessary condition that the number of interpolation constraints be equal to the number of coefficients for the system to be solvable. Suppose $q+1 = (p_u+1)(p_v+1)(p_w+1)$, and suppose the system has a nonzero determinant (i.e., the system is solvable). In general, the satisfaction of these constraints is not trivial. However, if the μ_m points are placed on a three-dimensional grid, and following the Schoenberg-Whitney theorem (see Theorem 9.2), the nodal points can be employed in order to guarantee a solution.

If $q+1 > (p_u+1)(p_v+1)(p_w+1)$, a trivariate least squares fit function can fit the data set. Volumetric data is, in many cases, redundant, so it may be possible to represent the same data more compactly. The result of least squares fitting is an L_2 approximation that is global in nature, so it does not have the ability to bound the local error in the L_∞ sense. While this limitation might be prohibitive for medical applications, other areas could benefit from such a compressed trivariate fit. Figure 21.7 presents several examples of a least squares fitting of a volumetric data set of size 64×64×64, using a triquadratic B-spline function of size $8 \times 8 \times 8$, $16 \times 16 \times 16$, and $32 \times 32 \times 32$.

The above methods work equally well when the data $\{\phi_k\}$ is scalar (creating a scalar field) or vector (creating a vector field).

The interpolation of trivariates can be made much more efficient, if the input samples are on a uniform grid. Much like fitting a surface through a set of curves, a trivariate function can be fitted through a set of n surfaces, $\sigma_i(u, v)$, which in turn are fitted through a set of curves.

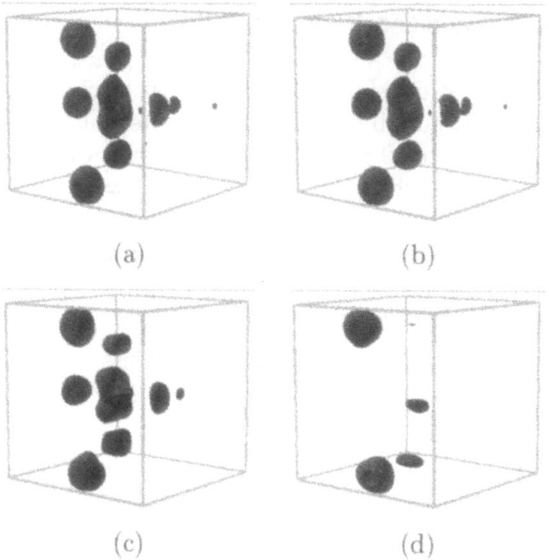

Figure 21.7. A volumetric data set 64 × 64 × 64 least squares fitted by (a) a 32×32×32 triquadratic; (b) a 16×16×16 triquadratic; (c) a 8×8×8 triquadratic; (d) the triquadratic functions are all sampled back into a grid of size 64 × 64 × 64 and then an iso-surface at the same level is extracted and displayed. (Data set courtesy of University of North Carolina, Chapel Hill.)

Suppose the data is of the form $\{(u_m, v_n, w_q, \phi_{m,n,q})\}_{m,n,q}$, $m = 0, \ldots, p_u$, $n = 0, \ldots, p_v$, $q = 0, \ldots, p_w$. Then, for each valid triple (m, n, q),

$$\begin{aligned}
\phi_{m,n,q} &= \sum_{i=0}^{p_u}\sum_{j=0}^{p_v}\sum_{k=0}^{p_w} P_{i,j,k}\mathcal{B}_{k,\kappa_w,\boldsymbol{\tau}^w}(w_q)\mathcal{B}_{j,\kappa_v,\boldsymbol{\tau}^v}(v_n)\mathcal{B}_{i,\kappa_u,\boldsymbol{\tau}^u}(u_m)\\
&= \sum_{i=0}^{p_u}\left(\sum_{j=0}^{p_v}\left(\sum_{k=0}^{p_w} P_{ijk}\mathcal{B}_{k,\kappa_w,\boldsymbol{\tau}^w}(w_q)\right)\mathcal{B}_{j,\kappa_v,\boldsymbol{\tau}^v}(v_n)\right)\mathcal{B}_{i,\kappa_u,\boldsymbol{\tau}^u}(u_m)\\
&= \sum_{i=0}^{p_u}\left(\sum_{j=0}^{p_v} R_{i,j,q}\mathcal{B}_{j,\kappa_v,\boldsymbol{\tau}_v}(v_n)\right)\mathcal{B}_{i,\kappa_u,\boldsymbol{\tau}_u}(u_m)\\
&= \sum_{i=0}^{p_u} Q_{i,n,q}\mathcal{B}_{i,\kappa_u,\boldsymbol{\tau}_u}(u_m),
\end{aligned}$$

where $R_{i,j,q} = \sum_{k=0}^{p_w} P_{i,j,k}\mathcal{B}_{k,\kappa_w,\boldsymbol{\tau}^w}(w_q)$ and $Q_{i,n,q} = \sum_{j=0}^{p_v} R_{i,j,q}\mathcal{B}_{j,\kappa_v,\boldsymbol{\tau}^v}(v_n)$.

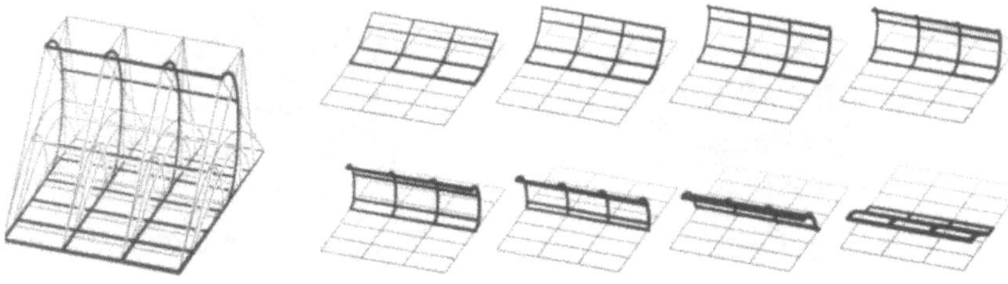

Figure 21.8. (a) An animation of flipping a page created by using a trivariate approximating three key frames of the page state; (b) several snapshots extracted as isoparametric surfaces from the constructed trivariate, at different times.

We start by solving for the $(p_u + 1)(p_v + 1)(p_w + 1)$ coefficients Q_i in

$$\phi_{m,n,q} = \sum_{i=0}^{p_u} Q_{i,n,q} \mathcal{B}_{i,\kappa_u,\mathcal{T}^u}(u_m),$$

forming $(p_v + 1)(p_w + 1)$ curves. We continue by interpolating the $(p_v + 1)(p_w + 1)$ curves into $(p_w + 1)$ surfaces, solving for $R_{i,j,q}$, as in Chapter 14.1.

$$Q_{i,n,q} = \sum_{j=0}^{p_v} R_{i,j,q} \mathcal{B}_{j,\kappa_v,\mathcal{T}^v}(v_n),$$

and we complete the solution by solving for $P_{i,j,k}$,

$$R_{i,j,q} = \sum_{k=0}^{p_w} P_{i,j,k} \mathcal{B}_{k,\kappa_w,\mathcal{T}^w}(w_q).$$

Given q snapshots in time of an elastic surface S, $\{\sigma_i\}_{i=0}^{q-1}$, one could fit a trivariate $T(u, v, t)$ to σ_i so as to construct a continuous transition in time of the elastic motion. An example of this metamorphosis application is presented in Figure 21.8. Several snapshots of flipping a page in a book are provided and a trivariate is fitted to this data set, forming a continuous function in time of a flipped page. The animated sequence is computed by extracting different isoparametric surfaces, $T(u, v, t_k)$ from the trivariate.

21.4.2 Traditional Constructors

Traditional surface constructors such as extrusion (see Section 15.1 for a similar surface construction), ruled surfaces (Section 15.2), surfaces of revolution (Section 15.3), and/or sweep surfaces (Section 15.4) can be extended

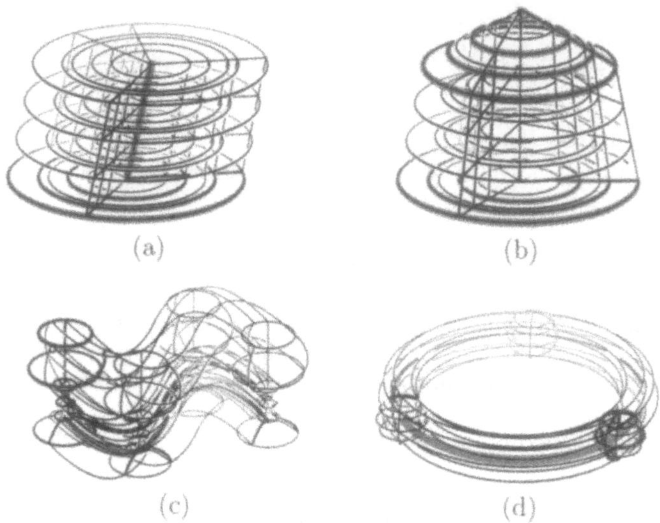

Figure 21.9. (a) Construction of a trivariate extruded volume; (b) ruled volume; (c) translational swept volume; (d) volume of revolution.

to construct trivariate functions. An extruded volume is a surface crossed with a line. Let $\sigma(u, v)$ and V be a parametric spline surface and a unit vector, respectively. Then

$$T(u, v, w) = \sigma(u, v) + wV,$$

represent the volume extruded by surface $\sigma(u, v)$ as the surface is moved in direction V. The *extrusion* of $T(u, v, w)$ is clearly linear in w. Figure 21.9 (a) shows an example of an extruded volume.

Let $\sigma_1(u, v)$ and $\sigma_2(u, v)$ be two parametric spline surfaces in the same space, that is, the same order and knot sequences. Then, the trivariate

$$T(u, v, w) = (1 - w)\sigma_1(u, v) + w\sigma_2(u, v),$$

constructs a *ruled volume* between σ_1 and σ_2. If $\sigma_1(u, v)$ and $\sigma_2(u, v)$ do not share the same function subspace, they both could be elevated into one by raising the degrees of the lower degree trivariate (see Section 16.3) as well as by refining (see Chapters 16 and 17) the coarse trivariate. Figure 21.9 (a) shows an example of a ruled volume.

A swept volume is a surface crossed by a curve. Let $\sigma(u, v)$ and $\gamma(w)$ be a parametric spline surface and a parametric C^2 continuous spline curve. Let

$$T(w) = \frac{\gamma'(w)}{\|\gamma'(w)\|}, \quad B(w) = \frac{\gamma'(w) \times \gamma''(w)}{\|\gamma'(w) \times \gamma''(w)\|},$$

and $N(w) = B(w) \times T(w)$ (recall Section 4.3) be the unit tangent, binormal, and normal vector fields, respectively, of $\gamma(w)$. The *moving trihedron* at w, $(T(w), N(w), B(w))$, can be used to orient $\sigma(u,v)$. Let $M(w)$ be the rotation that takes XYZ to TNB. Then,

$$S(u,v,w) = M(w)[\sigma(u,v)] + \gamma(w),$$

is the sweep of $\sigma(u,v)$ in space along w.

Unfortunately, since $(T(w), N(w), B(w))$ are not rational, in general, neither are $M(v)$ and $\sigma(u,v)$, either. One can keep the orientation frame constant throughout the sweeping process, having $T(u,v,w) = \sigma(u,v) + \gamma(w)$ to yield a rational form. This form of restricted sweep is also known as *translational sweep.* Figure 21.9 (c) presents an example of a translational swept trivariate. If $\gamma(w)$ is a line, the sweep operation is reduced to an extrusion. If $\gamma(w)$ is a circle and $\sigma(u,v)$ follows the moving trihedron of the circle, a volume of revolution will result. (See Figure 21.9 (d).)

21.5 Warping Using Trivariates

A vector trivariate function can serve as a mapping from (a subset of) $\mathbb{R}^3$ to (a subset of) $\mathbb{R}^3$. This observation leads to the suggestion of the use of trivariates to warp (a subset of) $\mathbb{R}^3$ with all the objects in that domain. (For Bézier , see [78]; for splines, see [66].) Let $\sigma(u,v) = (x_\sigma(u,v), y_\sigma(u,v), z_\sigma(u,v))$ and $\gamma(t) = (x_\gamma(t), y_\gamma(t), z_\gamma(t))$ be a surface and a curve in $\mathbb{R}^3$. Then, the composition of

$$T(\sigma(u,v)) = T(x_\sigma(u,v), y_\sigma(u,v), z_\sigma(u,v))$$

or of

$$T(\gamma(t)) = T(x_\gamma(t), y_\gamma(t), z_\gamma(t))$$

creates a warped surface and a warped curve according to the mapping function $T(u,v,w)$. This simple and powerful modeling approach is also quite intuitive since the user can affect the mapping function, $T(u,v,w)$, both directly by editing the control points of $T(u,v,w)$, and indirectly by using a multiresolution decomposition of $T(u,v,w)$ or by imposing constaints on $T(u,v,w)$.

Unfortunately, it is difficult to compute the resulting shape because this composition operation is a polynomial of a multiplicative degree. If the curve is of degree n and the surface is of degree $n \times n$, a trivariate of degree $m \times m \times m$ would, in general, create a warped curve or a warped surface of degree $3nm$ or $3nm \times 3nm$, respectively.

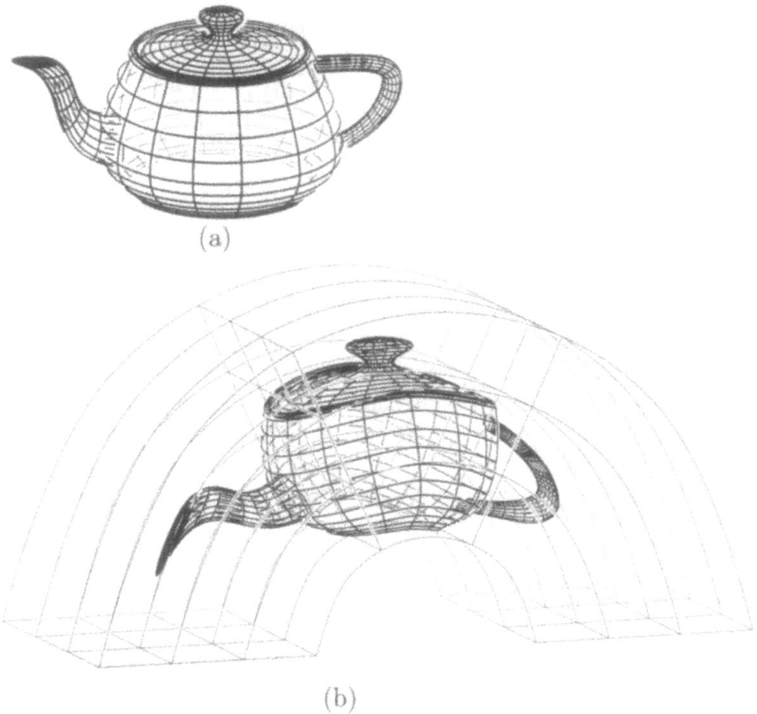

Figure 21.10. (a) The Utah teapot; (b) warped inside a bent B-spline trivariate.

In many instances, such a modeling approach is free style and need not be exact. Then, the above composition can be approximated by applying the mapping function, $T(u, v, w)$, to the control points of the respective curve or surface, or even to the vertices of some polygonal artifact. Clearly, this compromise is inexact but it does converge to the exact curve and/or surface with the aid of refinement. The simple computational properties of this compromise approach make it attractive. Figure 21.10 shows one such example.

21.6 Varying Time

The trivariate function can also be used in representing changes in surfaces. Having a third dimension, one can now represent the motion of a rigid surface in space or even a change or a modification of a free-form surface. Consider two surfaces, $\sigma_1(u, v)$ and $\sigma_2(u, v)$ and let

$$T(u, v, t) = (1 - t)\sigma_1(u, v) + t\sigma_2(u, v). \tag{21.2}$$

Figure 21.11. A continuous metamorphosis between a disk and a wine glasses, using a ruled volume. The trivariate is shown at the two end locations with the respective boundary (input) isoparametric surfaces in bold.

The trivariate in Equation 21.2 is a *ruled-volume* between $\sigma_1(u,v)$ and $\sigma_2(u,v)$. Assume that $\sigma_1(u,v) = \sum_i \sum_j P_{ij}\mathcal{B}_i(u)\mathcal{B}_j(v)$ and $\sigma_2(u,v) = \sum_i \sum_j Q_{ij}\mathcal{B}_i(u)\mathcal{B}_j(v)$ are in the same function space. Then,

$$\begin{aligned} T(u,v,t) &= \sum_i \sum_j \left((1-t)P_{i,j} + tQ_{i,j}\right)\mathcal{B}_i(u)\mathcal{B}_j(v) \\ &= \sum_i \sum_j \sum_{k=0}^{1} R_{i,j,k}\mathcal{B}_i(u)\mathcal{B}_j(v)\mathcal{B}_{k,1,\boldsymbol{t}_w}(t), \end{aligned}$$

where $R_{i,j,0} = P_{i,j}$, $R_{i,j,1} = Q_{i,j}$, and $\boldsymbol{t}_w = (0,0,1,1)$.

The continuity of t suggests that this third axis can also be employed toward a continuous metamorphosis of surface $\sigma_1(u,v)$ into surface $\sigma_2(u,v)$. Figure 21.11 shows an example of a metamorphosis between a disk and a wine glass that was computed using a ruled volume.

Having a representation for the motion of the rigid surface in space, one can try and extract the envelope [39] that is created during such motion. For $t = t_i$, we seek the points on surface $\sigma_i(u,v) = T(u,v,t_i)$ that are tangent to the envelope of the motion. Then, for $t = t_i$ and at the tangency points, $\frac{\partial T}{\partial t}$ is contained in the tangent plane that is spanned by $\frac{\partial T}{\partial u}$ and $\frac{\partial T}{\partial v}$, provided surface $T(u,v,t_i)$ is regular. Hence, this constraint for the three partial derivatives of T to be coplanar is equivalent for the constraint of the Jacobian of $T = (x_T, y_T, z_T)$ to vanish.

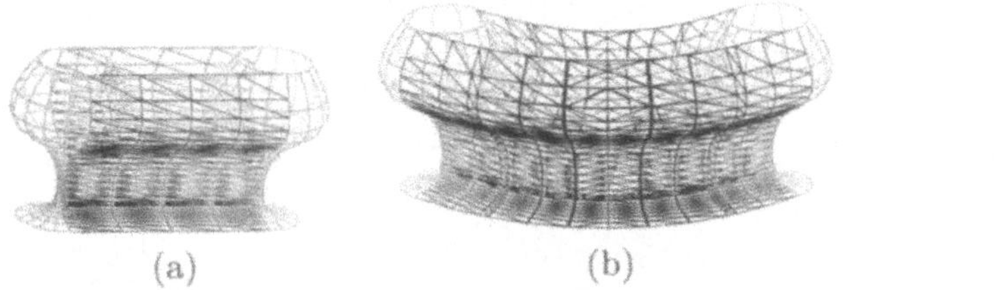

Figure 21.12. The envelope of a (a) translational and (b) rotational motion of a wine glass, computed using the zeros of the Jacobian of a trivariate function.

That is,

$$0 = \begin{vmatrix} \frac{\partial x_T}{\partial u} & \frac{\partial x_T}{\partial v} & \frac{\partial x_T}{\partial t} \\ \frac{\partial y_T}{\partial u} & \frac{\partial y_T}{\partial v} & \frac{\partial y_T}{\partial t} \\ \frac{\partial z_T}{\partial u} & \frac{\partial z_T}{\partial v} & \frac{\partial z_T}{\partial t} \end{vmatrix}.$$

Figure 21.12 shows the envelope surface of both translational and rotational motion of a wine glass. They are computed using the zero set computation of the Jacobian of the trivariate functions.

Bibliography

[1] H. Antosiewicz and W. Rheinboldt. "Numerical analysis and functional analysis." In *Survey of Numerical Analysis*, edited by J. Todd, New York: McGraw-Hill Book Company, 1962.

[2] T. M. Apostol. *Calculus*. New York: John Wiley, 1969.

[3] A. A. Ball and D. J. T. Storry. "Conditions for tangent plane continuity over recursively generated B-spline surfaces." *ACM Transactions on Graphics* 7(2): 83–102 (Apr. 1988).

[4] R. E. Barnhill. "Smooth interpolation over triangles." In *Computer Aided Geometric Design*, edited by R. E. Barnhill and R. F. Riesenfeld, New York: Academic Press, 1974.

[5] P. J. Barry and R. N. Goldman. "Recursive proof of Boehm's knot insertion technique." *Computer Aided Design* 20(4): 181–182 (1988).

[6] R. H. Bartels and J. C. Beatty. "A technique for the direct manipulation of spline curves." In *Proceedings of Graphics Interface*, pp. 33–39, Toronto, Canada: Canadian Human-Computer Communications Society, 1989.

[7] B. G. Baumgart. *Geometric modelling for computer vision*. Report AIM-249, STAN-CS-74-463, Stanford Artificial Intelligence Laboratory, Stanford University, October 1974.

[8] G. H. Behforooz and N. Papamichael. "End conditions for cubic spline interpolation." *Journal Inst. Maths. Applic.* 23: 355–366 (1979).

[9] G. H. Behforooz and N. Papamichael. "End conditions for interpolatory cubic splines with unequally spaced knots." *Journal of Computational and Applied Mathematics* 6(1): 59–65 (1980).

[10] Pierre Bézier . *The Mathematical Basis of the UNISURF CAD System*. Butterworths, 1986.

[11] M. Bloomenthal. *Approximation of sweep surfaces by tensor product B-splines*. Tech Reports UUCS-88-008, University of Utah, 1988.

[12] M. Bloomenthal and R. F. Riesenfeld. "Approximation of sweep surfaces by tensor product NURBS." In *Proceedings of the SPIE Conference on Curves and Surfaces in Computer Vision and Graphics II*, edited by M.J. Silbermann and H.D. Tagare, Bellingham, WA: SPIE Press, 1992.

[13] W. Boehm. "Knot insertion." *Journal of Computational and Applied Mathematics* 6(1): 59–65 (1980).

[14] W. F. Bronsvoort, P. R. van Nieuwenhuizen, and F. H. Post. "Display of profiled sweep objects." *The Visual Computer* 5(3): 147–157 (June 1989).

[15] E. E. Catmull and J. H. Clark. "Recursively generated B-spline surfaces on arbitrary topological meshes." *Computer Aided Design* 10(6): 350–355 (November 1978).

[16] G. M. Chaikin. "An algorithm for high-speed curve generation." *Computer Graphics and Image Processing* 3: 346–349 (1974).

[17] Y.-K. Chang, A. P. Rockwood, and Q. He. "Direct rendering of freeform volumes." *Computer Aided Design* 27(7): 553–558 (1995).

[18] E. Cobb. *Design of Sculptured Surfaces Using the B-Spline Representation.* PhD thesis, University of Utah, 1984.

[19] E. Cohen and Schumaker L. "Rates of convergence of control polygons." *Computer Aided Geometric Design* 2(1-3): 229–235 (1985).

[20] E. Cohen, T. Lyche, and R. F. Riesenfeld. "Discrete B-splines and subdivision techniques in computer-aided geometric design and computer graphics." *Computer Graphics and Image Processing* 14(2): 87–111 (October 1980).

[21] E. Cohen, T. Lyche, and R. F. Riesenfeld. "Discrete box splines and refinement algorithms." *Computer Aided Geometric Design* 1(2): 131–148 (1984).

[22] E. Cohen, T. Lyche, and L. Schumaker. "Algorithms for degree raising of splines." *ACM Transactions on Graphics* 4: 171–181 (1985).

[23] E. Cohen, T. Lyche, and L. Schumaker. "Degree raising for splines." *Journal of Approximation Theory* 46: 170–181 (1986).

[24] S. Coquillart. "A control-point based sweeping technique." *IEEE Computer Graphics and Applications* 7(11): 36–44 (1987).

[25] H. B. Curry and I. J. Schoenberg. "On Polya frequency functions IV: The fundamental spline functions and their limits." *Journal d'Analyse Mathematique* 17: 71–107 (1966).

[26] M. Daehlen and T. Lyche. Decomposition of splines. In *Mathematical Methods in Computer Aided Geometric Design II*, edited by T. Lyche and L. Schumaker, pp. 135–160, Boston: Academic Press, 1992.

[27] M. de Berg, M. van Kreveld, M. Overmars, and O. Schwarzkopf. *Computational Geometry: Algorithms and Applications*. New York: Springer-Verlag, 1997.

[28] C. de Boor. "Splines as linear combinations of B-splines. A survey." In *Approximation Theory II*, edited by G. G. Lorentz, C. K. Chui, and L. L. Schumaker, pp. 1–47, New York: Academic Press, 1976.

[29] C. de Boor. *A Practical Guide to Splines*, volume 27 of *Applied Mathematical Sciences*. New York: Springer-Verlag, 1978.

[30] C. de Boor, K. Hollig, and S Riemenschneider. *Box Splines*, volume 98 of *Applied Mathematical Sciences*. New York: Springer-Verlag, 1993.

[31] C. de Boor and G. Fix. "Spline approximation by quasi-interpolants." *Journal of Approximation Theory* 8: 19–45 (1973).

[32] J. W. Demmel. *Applied Numerical Linear Algebra*. Philadelphia: SIAM, 1997.

[33] T. DeRose, M. Kass, and T. Truong. "Subdivision surfaces in character animation." In *Proc. SIGGRAPH 98, Computer Graphics Proceedings, Annual Conference Series*, edited by Michael Cohen, pp. 85–94, Reading, MA: Addison-Wesley, 1998.

[34] D. Doo and Sabin M. "Behaviour of recursive division surfaces near extraordinary points." *Computer Aided Design* 10(6): 356–360 (November 1978).

[35] G. Elber. *Free Form Surface Analysis using a Hybrid of Symbolic and Numeric Computation*. PhD thesis, University of Utah, 1992.

[36] G. Elber. "Multiresolution curve editing with linear constraints." In *Solid Modeling 2001*, New York: ACM Press, June 2001.

[37] G Farin. *Curves and Surfaces for Computer Aided Geometric Design*. Boston: Academic Press, 1996.

[38] G. E. Farin. *Konstruktion und eigenschaften von Bézier-kurven und Bézier-flaechen*. Technical Report, T.U. Braunschweig, 1977.

[39] I. D. Faux and M. J. Pratt. *Computational Geometry for Design and Manufacture*. Chichester: Ellis Horwood Ltd., 1979.

[40] G. H. Golub and C. F. Van Loan. *Matrix Computation*. Baltimore, MD: The John Hopkins University Press, 1996.

[41] W. J. Gordon and R. F. Riesenfeld. "B-spline curves and surfaces." In *Computer Aided Geometric Design*, edited by Robert E. Barnhill and Richard F. Riesenfeld, pp. 95–126, New York: Academic Press, 1974.

[42] H. Hoppe, T. DeRose, T. Duchamp, M. Halstead, H. Jin, J. McDonald, J. Schweitzer, and W. Stuetzle. "Piecewise smooth surface reconstruction." *Computer Graphics* 28(3): 295–302 (1994).

[43] B. W. Jordan, W. J. Lennon, and B. C. Holm. "An improved algorithm for the generation of nonparametric curves." *IEEE Transactions on Computers* C-22(12): 1052–1060 (December 1973).

[44] S. Karlin. *Total Positivity*, Volume 1. Stanford, CA: Stanford University Press, 1968.

[45] S. Karlin and Z. Ziegler. "Tschebysheffian spline functions." *SIAM J. Numer. Anal.* Series B 3: 514–543 (1966).

[46] R. Kazinnik and G. Elber. "Orthogonal decomposition of non-uniform B-splines using wavelets." *Computer Graphics Forum* 16(3): 27–38 (1997).

[47] F. Klok. "Two moving coordinate frames for sweeping along a 3D trajectory." *Computer Aided Geometric Design* 3(3): 217–219 (1986).

[48] D. Knuth. *The Art of Computer Programming*, Vol 1 & 2. Reading, MA: Addison-Wesley, 1975.

[49] J. M. Lane and R. F. Riesenfeld. "A theoretical development for the computer generation of piecewise polynomial surfaces." *IEEE Transactions on Pattern Analysis and Machine Intelligence* 2(1): 35–46 (January 1980).

[50] J. M. Lane and R. F. Riesenfeld. "A geometric proof of the variation diminishing property of B-spline approximation." *Journal of Approximation Theory* 37(1): 1–4 (1983).

[51] E. Lee and M. Lucian. "Moebius reparametrizations of rational B-splines." *Computer Aided Geometric Design* 8(3): 213–216 (1991).

[52] B. Livingston. *Intersurface Continuity of Solid Models.* Master's thesis, University of Utah, 1990.

[53] C. T. Loop. *Smooth Subdivision Surfaces Based on Triangles.* Master's thesis, University of Utah, 1987.

[54] W. E. Lorensen and H. E. Cline. "Marching cubes: A high resolution 3D surface construction algorithm." *Computer Graphics (Proc. SIGGRAPH '87)* 21(4): 163–169 (July 1987).

[55] T. Lyche. "A note on the Oslo algorithm." *Computer Aided Design* 20(6): 353–355 (1988).

[56] T. Lyche, E. Cohen, and K. Morken. "Knot line refinement algorithms for tensor product B-spline surfaces." *Computer Aided Geometric Design* 2(1-3): 133 –139 (1985).

[57] T. Lyche and K. Morken. "Making the Oslo algorithm more efficient." *SIAM J. Numer. Anal.* 23: 663–675 (1986).

[58] W. Martin and E. Cohen. "Representation and Extraction of Volumetric Attributes using Trivariate Splines: A Mathematical Framework." In *Solid Modeling.*, 2001.

[59] M. J. Marsden. "An identity for spline functions with approximations to variation-diminishing spline approximation." *Journal of Approximation Theory* 3: 7–49 (1970).

[60] K. Morken. "Some identities for products and degree raising of splines." *Constructive Approximation* 7: 195–208 (1991).

[61] Ahmad H. Nasri. "Polyhedral subdivision methods for free-form surfaces." *ACM Transactions on Graphics* 6(1): 29–73 (January 1987).

[62] W. M. Newman and R. F. Sproull. *Principles of Interactive Computer Graphics*. Reading, MA: Addison-Wesley, 1979.

[63] N. Okino, Y. Kakazu and H. Kubo. In *TIPS-1: Technical information processing system for Computer Aided Design and Computer Aided Manufacturing*. In *Computer Languages for Numerical Control*, Amsterdam: North-Holland, 1973.

[64] N. Okino, H. Kubo, and Y. Kakazu. "An integrated CAD/CAM system: TIPS-2." In *Proceedings of the 1976 International Conference on Programming Languages for Numerically Controlled Machine Tools*, Stirling, England, 1976.

[65] N. Okino, Y. Kakazu, H. Kubo, N. Hashimoto, and Y. Shiroma. "Advanced 3D shape description methods in TIPS-1." *Computers in Industry* 3(1,2): 93–104 (March–June 1982).

[66] K. Paik. *Trivariate B-splines*. Master's thesis, University of Utah, 1992.

[67] J. Peters and U. Reif. "Analysis of generalized B-spline subdivision algorithms." *ACM Transactions on Graphics* 16 (1997).

[68] F. Preparata and M. Shamos. *Computational Geometry*. New York: Springer-Verlag, 1985.

[69] A. Raviv and G. Elber. "Interactive direct rendering of trivariate B-spline scalar functions." *IEEE Transactions on Visualization and Computer Graphics*, to appear.

[70] U. Reif. "A unified approach to subdivision algorithms near extraordinary vertices." *Computer Aided Geometric Design* 12(2): 153–174 (1995).

[71] R. F. Riesenfeld. *Applications of B-spline Approximation to Geometric Problems of Computer-Aided Design*. PhD thesis, Syracuse University, Syracuse, N.Y., May 1973. Available as Tech. Report No. UTEC-CSc-73-126, UUCS.

[72] R. F. Riesenfeld and E. Cohen. "General matrix representations for Bézier and B-spline curves." *Computers in Industry* 3(1,2): 9–16 (March 1982).

[73] R. Sarraga. "Computer modeling of surfaces with arbitrary shapes." *IEEE Computer Graphics and Applications* 10: 67–77 (1990).

[74] I. J. Schoenberg. "Contributions to the problem of approximation of equidistant data by analytic functions." *Quarterly Applied Math.* 4(1): 45–99 and 112–141 (1946).

[75] I. J. Schoenberg and A. Whitney. "On Polya frequency functions, III: The positivity of translation determinants with application to the interpolation problem by spline curves." *Trans. Amer. Math. Soc.* 74: 246–259 (1953).

[76] L. L. Schumaker. *Spline Functions: Basic Theory*. New York: John Wiley & Sons, 1981.

[77] J. E. Schweitzer. *Analysis and Application of Subdivision Surfaces*. PhD thesis, University of Washington, 1996.

[78] T. W. Sederberg and S. R. Parry. "Free-form deformation of solid geometric models." *Computer Graphics (Proc. SIGGRAPH '86)* 20(4): 151-160 (August 1986).

[79] Lee Seitelman. "A new user-transparent end condition for cubic spline data fitting." In *SIAM 1977 Fall Meeting*, Philadelphia, PA: SIAM, 1977.

[80] P. Siltanen and C. Woodward. "Normal orientation methods for 3D offset curves, sweep surfaces and skinning." *Computer Graphics Forum* 11: 449–457 (1992).

[81] J. Stam. "Exact evaluation of Catmull-Clark subdivision surfaces at arbitrary parameter values.' In *Proc. Siggraph '98, Computer Graphics Proceedings, Annual Conference Series*, edited by Michael Cohen, pp. 395–404, Reading, MA: Addison-Wesley, 1998.

[82] S. Thomas. *Modelling Volumes Bounded by B-Spline Surfaces*. PhD thesis, University of Utah, 1984.

[83] K. J. Versprille. *Computer-Aided Design Applications of the Rational B-spline Approximation Form*. PhD thesis, Syracuse University, Syracuse, N.Y., February 1975.

[84] F. Yamaguchi. "A new curve fitting method using a CRT computer display." *Computer Graphics and Image Processing* 7(3): 425–437 (June 1978).

[85] D. Zorin. *Stationary Subdivision and Multiresolution Surface Representations*. PhD thesis, Caltech, 1997.

[86] D. Zorin. "Smoothness of subdivision on irregular meshes." *Constructive Approximation* 16(3): 359–397 (2000).

[87] D. Zorin and P. Schroder. Subdivision for modeling and animation. In *Siggraph 2000 Course Notes*. New York: ACM, 2000.

[88] D. Zorin, P. Schroder, and W. Sweldens. "Interactive multiresolution mesh editing." In *Proc. Siggraph '97, Computer Graphics Proceedings, Annual Conference Series*, edited by Turner Whitted, pp. 259–268, Reading, MA: Addison-Wesley, 1997.

Index

Milton Keynes UK
Ingram Content Group UK Ltd.
UKHW020004071024
449327UK00031B/2645

9 780367 447243